AF777098

Flowering Plants of Wales

Flowering Plants of Wales

by

R. G. Ellis

Amgueddfa Genedlaethol Cymru
National Museum of Wales
Caerdydd/Cardiff 1983

First published December, 1983

Typeface: Garamond Roman
Paper: Superior matt art 135 gsm
Printing: South Western Printers, Caerphilly

ISBN 0 7200 0271 0

CONTENTS

To *Maria*
A. E. Wade
H. A. Hyde

LIST OF ILLUSTRATIONS

PLATES
(Colour)

PLATES
(Black and White)

TABLES

FIGURES

MAPS

The grid on these maps is the National Grid taken from Ordnance Survey maps with the permission of the Controller of Her Majesty's Stationery Office.

Foreword

H. A. Hyde and A. E. Wade, respectively the Keeper and Assistant Keeper of Botany of the time, compiled three monographs on the botany of Wales during the nineteen-thirties. The first, issued in 1931, was on *Welsh Timber Trees* (by Hyde alone); the second, in 1934, on *Welsh Flowering Plants*; the third, in 1940, on *Welsh Ferns.*

The three works together provided one of the major botanical contributions from Wales to a decade characterised by much activity in natural history throughout Britain and by the publication of a number of syntheses and summaries.

The book *Welsh Flowering Plants,* although nominally a handbook to the collections in the Welsh National Herbarium, was much more. It provided, surprisingly perhaps, the first complete published list of the flowering plants of the Principality as well as bringing together, again for the first time, analyses of the distribution of the various species in Wales.

The work was revised and largely rewritten by the authors in 1957 and issued as a second edition. And the two editions of the book have since prompted the compilation of a new work on the subject by R. Gwynn Ellis.

The present study differs from the earlier works in the following ways:-

First, it includes all taxa, down to the rank of subspecies, which have been recorded apparently growing wild in Wales, including all casual species, introduced forest trees, hybrids and the microspecies of *Rubus, Taraxacum* and *Hieracium.* This has more than doubled the number of species dealt with from 1343 in the 1957 edition to 2784.

Secondly, it gives the 10 km square as well as the vice-county distribution for each species. This may, depending on the number of squares involved, take the form of maps or lists. The details of locality, recorder, etc., provided for those species which have been recorded from fewer than 10 squares, are taken from a variety of sources and not just from specimens in the Welsh National Herbarium.

Thirdly, as the treatment of plant distribution is much more detailed than in earlier works, it was thought desirable to include chapters on the physical and climatic background in order to help explain some of the patterns of plant distribution.

Fourthly, the Welsh names of plants have been included for the first time. This has been made possible by the publication of definitive lists by the University of Wales Press.

Fifthly, the arrangement of the catalogue now follows that of *Flora Europaea.*

The work clearly reflects the co-operative nature of one aspect at least of the work by the National Museum and is, accordingly, a most appropriate volume to issue during the 75th anniversary of the granting of the Royal Charter to the Museum and when one of the major themes of the celebrations is co-operation.

DOUGLAS A. BASSETT
Director

Acknowledgements

Many people have helped in the preparation of this book and to them all I owe a great debt of thanks : to Dr. H. A. Hyde and Mr. A. E. Wade, authors of *Welsh Flowering Plants* on which this book is based, for their help and encouragement; to R. M. Owens and R. E. Bevins, Department of Geology, National Museum of Wales and J. P. Savidge, Department of Botany, University College of Wales, Aberystwyth, for the chapters on Geology and Climate respectively; to the many specialists who from time to time have examined parts of the collections and have helped with the text of critical groups, especially A. Newton (*Rubus*), Mrs. I. M. Vaughan (*Rosa*), P. F. Yeo (*Euphrasia*), A. J. Richards (*Taraxacum*), P. D. Sell (*Hieracium*), and A. O. Chater, R. W. David and A. C. Jermy (*Carex*); to the many field workers who have contributed records and specimens; to the B.S.B.I. Welsh vice-county recorders, past and present, and those other botanists, who gave so freely of their time and knowledge in checking the text and distribution maps for their particular vice-county or area (especially P. M. Benoit, J. M. Brummitt, A. O. Chater, Miss A. P. Conolly, the late T. A. W. Davis, T. Edmondson, S. B. Evans, T. G. Evans, Q. O. N. Kay, M. Morris, M. Porter, Miss A. C. Powell, R. D. Pryce, Miss D. Pugh, R. H. Roberts, Mrs. I. M. Vaughan, R. G. Woods, G. Wynne); to A. O. Chater who, beside helping in the above ways, also gave willingly advice and constructive criticism on many matters including nomenclature; to the staff of the Biological Records Centre, Monkswood and especially to its former Director, F. H. Perring, who answered innumerable questions, supplied the base maps on which the distribution of individual species is shown, provided a computer print-out of records that had accrued since the publication of the *Atlas of the British Flora*, and prepared maps for the more common species on which the maps used here are based; to the Keeper of Botany, British Museum (Natural History), J. P. M. Cannon and the Keeper of the British Herbarium, R. J. Pankhurst for permission to go through the herbarium extracting Welsh records; to P. D. Sell, Curator of the Herbarium, Botany School, Cambridge, for permission to extract records from the herbarium and his personal field notebooks; and finally to my colleagues in the Department of Botany, National Museum of Wales who helped in many ways. Despite all the help that has been received, all errors in the text or maps remain my sole responsibility.

Further thanks are due to the Director of the Soil Survey of England and Wales, Rothampstead, and the Director General of the Ordnance Survey, for permission to reproduce the Bioclimatic Map (Frontispiece) and to the Director General of the Ordnance Survey for permission to base other maps in this book on Ordnance Survey maps.

Many of the portraits of botanists on Plates 1 and 2 are reproduced from negatives held in the National Museum of Wales. Credit for the remainder is due as follows: 1, Cambridge University Press; 2, Ashmolean Museum, Oxford; 3, Sherardian Professor of Botany, University of Oxford; 4, Mrs C. Grundy; 5, The Royal Society; 8, Journal of Botany, 1911; 16 & 18, The Botanical Society of the British Isles; 21 & 24, Mr W. M. Condry; 22, Miss S. H. Carter.

Figures 1, 2 & 3 were drawn by Miss M. H. Bigwood, Department of Botany, University College of Wales, Aberystwyth. Map 1 was drawn by Mrs L. Norton; maps 2 & 3 by H. M. Reynolds and the dust jacket designed and painted by Miss D. E. Evans, all of the National Museum of Wales.

Finally my thanks to Mrs O. M. Evans, Botany Department Secretary at the National Museum of Wales who typed the whole book at least twice and some parts several times.

A work such as this is somewhat out-of-date even before it is published. The author would welcome additions and/or corrections for incorporation into supplements to be issued every few years. Anyone wanting information on these should contact the Department of Botany, National Museum of Wales, Cardiff.

The Herbarium of the National Museum of Wales

The Department of Botany of the National Museum of Wales first took shape in 1919 but the history of its herbarium may be traced from very small beginnings some ninety years earlier. Charles Conway of Pont-rydyryn formed, during the eighteen-thirties, a small collection of dried plants which after his death was acquired by the then Cardiff Museum. The Herbarium thus begun was added to by John Storrie, who in 1888 gave his own specimens to the Museum. Around the turn of the century important additions were received from two well-known botanists of the period, Mr. Arthur Bennett and the Rev. H. J. Riddelsdell. In 1912, when the entire contents of the Cardiff Museum were taken over by the newly constituted National Museum of Wales the Herbarium was estimated to contain some 3,500 sheets plus an unknown number of unmounted specimens. So far the collection had not advanced beyond the requirement of a local museum, but with the establishment of the Department of which it forms an essential part, the Herbarium grew very rapidly and by 1982 it was estimated to contain about 257,000 specimens. This expansion has been guided by the stated aim of the Department as a part of the Museum, namely the 'complete illustration' of the flora of Wales. This does not imply that the collections shall be exclusively Welsh but that the specimens they contain from outside Wales shall in the main play a secondary or supporting role. In order to facilitate the comparison of Welsh specimens with corresponding material from the rest of the British Isles and Europe, the Welsh collections have been treated as an integral part of the European collections. Specimens of each species are arranged geographically, those from Wales being placed first followed by those from the rest of Britain and Europe.

The collection of British flowering plants totals some 117,000 specimens; these have been acquired from a variety of sources as follows:

1. Gifts. The chief items under this heading are the collections associated with the names of Dr. W. A. Shoolbred, Dr. J. H. Salter, Dr. C. T. and Miss E. Vachell, Mrs M. Richards, Mr R. F. May, Dr. M. S. Percival, née Martin and Mr N. Woodhead. Here also should be mentioned the many donations large and small of specimens collected in Wales specially for the Department. During the years 1919-81 about 45,000 specimens were contributed by 950 donors, many of whom are referred to in the list of recorders in Appendix I on pages 281-284.
2. Many plants have been received through exchange clubs and others have been derived from barter with individual collectors.
3. Departmental collecting has resulted in the addition of c. 13,000 sheets.
4. The following collections have been purchased: J. A. Wheldon (11,278); A. R. Horwood (4,778); J. E. Griffith (3,250).

A fuller account is given by the Keeper of Botany, Mr. S. G. Harrison, in *Nature in Wales,* New series, **1**(2), (for 1982), *39-55,* 1983.

A Short History of Botanical Recording in Wales

Based on articles which first appeared in *Amgueddfa, Bulletin of The National Museum of Wales* (1972-74), and issued separately in 1974 as *Plant Hunting in Wales.*

The first localized records of plants in Wales date from the latter half of the 16th century and are found in a manuscript Herbal attributed to William Salusbury (1520-1584) who lived at Lleweny Hall near Denbigh.

In this Herbal Salusbury gives localities for 10 native species, the first one mentioned is *Althaea officinalis* (Marsh-mallow) which he records from Llansannan near Denbigh. The original manuscript is lost but there is in the National Library of Wales a manuscript copied in 1763. An edited version was published in 1916 called *Llysieulyfr Meddyginiaethol.*

The Salusbury family was evidently very interested in botany as a cousin of William, Sir John Salusbury (1567-1612), owned a copy of *The Herball or General History of Plants* by J. Gerarde (1597) and wrote in its margin details of his own botanical discoveries. Welsh localities for some 20 plants are recorded in this way. Sir John's copy of this *Herball* survives in the Library of Christ Church, Oxford.

The first published records of Welsh plants appeared in the 2nd edition of Gerarde's *Herball*, revised by Thomas Johnson and published in 1633. Some of these records were contributed by an Englishman, George Bowles (c.1604-1671), and the first plant to be mentioned in print as having been found in Wales was *Cicuta virosa* (Cowbane), which Bowles found in "divers ponds in Flintshire". Other records were contributed by Thomas Glynne, a Welshman, who lived at Glynllifon near Caernarvon.

Thomas Johnson (c.1600-1644) was the first botanist of note to visit Wales. He travelled widely in North Wales in 1639 and kept detailed notes on the places he visited and the plants he found. Among these were *Oxyria digyna* (Mountain Sorrel), *Silene acaulis* (Moss Campion) and *Saxifraga nivalis* (Alpine Saxifrage) all recorded for the first time from Britain. Johnson published a record of his Welsh journey in *Mercurii Botanici pars altera* (1641).

John Ray (1627-1705) was possibly the first British botanist who was interested in plants for their own sake. Johnson was an Apothecary and was interested in plants primarily for use in medicines whereas Ray was a University lecturer and Preacher who made it his life's work to discover and describe as many as he could of the unknown plants that abounded in Britain and Europe. He paid two visits to Wales, in 1658 and 1662, and like Johnson kept detailed notes on the plants and places he saw. His plant records were published in three works : *Catalogus Plantarum Angliae, et Insularum Adjacentium*, 1670 (*Catalogue of English Plants*); *Fasciculus Stirpium Britannicarum*, 1688 (*Fasciculus of British Plants*) and *Synopsis Methodica Stirpium Britannicarum*, 1690 (*Synopsis of British Plants*).

Many records of Welsh mountain plants were given to Ray by a Welshman, Edward Llwyd (1660-1709). Llwyd is the most famous of the early Welsh botanists. He explored the mountains of Snowdonia while still in his early twenties and sent long lists to Ray. Due to his efforts nearly 50 plants, mostly upland species, were first recorded in Wales. His best known find was a small white-flowered member of the Lily family which was named *Lloydia serotina* (Snowdon Lily).

The next well documented plant hunting expedition to Wales was by two famous botanists, J. J. Dillenius (1684-1747) and Samuel Brewer (1670-1742). Dillenius was a German who was brought to Britain by W. Sherard in 1721 and became Professor of Botany at Oxford; Brewer had been a cloth manufacturer but became a professional plant hunter. They travelled together to Wales in 1726 and kept diaries of their journeys. Mid-eighteenth century copies are in the Library of the National Museum of Wales. They also collected many plants, some of which are preserved at Oxford and others at the British Museum (Natural History). Brewer remained in North Wales for the next 12 months exploring the mountains of Snowdonia and the fens of Anglesey, searching always for new plants. Details of his finds were sent back to Dillenius who had returned to Oxford. One of his many discoveries was a rock-rose considered then to be new to science; it was named *Helianthemum breweri* (Brewer's Rock-rose) but it is now included in *Tuberaria guttata* and is thought at most to be a variety of this.

The introduction, in the mid 18th century, of the Linnean system of classifying and naming plants, made the cataloguing of the plants of a country a comparatively easy matter and it was not long before the majority of species in the Welsh and British flora had been discovered and listed. So many botanists, well-known national figures as well as obscure local people, had joined in the hunt for plants that it is impossible to name them all.

The last of the historic plant hunting expeditions into Wales took place in the midst of this activity, in 1773. It was made by two famous botanists, Sir Joseph Banks (1743-1820), botanist aboard the Endeavour with Capt. Cook and founder of the colony of New South Wales, and the Rev. John Lightfoot (1735-1788) author of *Flora Scotica*. They journeyed through Wales from south to north and collected specimens along the way many of which survive in the British Museum (Natural History). Lightfoot kept a diary of the journey which was published in the *Journal of Botany* in 1905. They refound many of the plants that had been recorded about a hundred years earlier by Ray and Llwyd, although they failed to find *Lloydia*, probably because they arrived in Snowdonia after it had finished flowering.

By the beginning of the 19th century the exploration of the whole of Britain for new plants had more or less come to an end. It was at this time that a new impetus was given to the study of plants in the field; the rise of the study of plant geography. Botanists came to realize that it was not sufficient merely to list all the plant species in creation, but that they should also try to work out the distribution of each.

One of the first books to deal solely with the distribution of British plants, *The Botanist's Guide through England and Wales*, was published in 1805. In it are listed the uncommon or rare plants of each of the English and Welsh counties. The authors, D. Turner (1775-1858) and L. W. Dillwyn (1778-1855) compiled the lists from previously collected records, some published, others not, and they also added their own records. The number of species recorded from the Welsh counties varied tremendously, from 159 in Caernarvonshire to three lichens in Radnorshire. Thirty years later, in 1835, H. C. Watson published a *New Botanist's Guide.* Watson, an amateur botanist, was the pioneer of the new study of plant geography in Britain. After another 40 years work he introduced the vice-county system of recording plant distribution in his major and last work, *Topographical Botany*, the 2nd edition of which was published in 1883. In this book Watson listed every vascular plant found either as a native or as a well established alien in Britain and provided for each a list of the vice-counties in which it had been recorded (for definition of vice-county (v.c.) see page 35).

The progress of plant recording in Wales will be continued in the following section, *Botany of the Welsh Vice-counties.* Much more detailed information on the history of botanical recording in each of the Welsh counties may be found in an excellent series of articles by the late P. W. Carter (See Appendix II).

Botany of the Welsh Vice-counties

Prior to the 1950's the Welsh vice-counties, with a few notable exceptions, had not received as much attention at the hands of field botanists as had most of the counties of England. Few complete floras or even county lists had been published. Then, in 1954, the B.S.B.I. maps scheme was launched (see page 38). This provided the impetus which was needed for the thorough botanical exploration of all the Welsh vice-counties. One new Flora and several comprehensive vice-county lists have since been published and several new vice-county Floras are in preparation. Reference to all these is made below. A more comprehensive list of relevant books, papers, journals and manuscripts can be found in Appendix II on pages 285-293.

In the sketches of each vice-county that follow, rare plants claim special attention, not as curiosities, but as examples of unusual distribution patterns. They raise innumerable problems but discussion of these is outside the scope of this book.

V.c.35, Monmouthshire (Mons.): 818 native, denizen or colonist; 273 naturalized or established introductions; 165 casual = 1256 species; also 122 microspecies (*Rubus, Taraxacum, Hieracium*) and 93 hybrids. **Mons.** falls naturally into two divisions corresponding with its geology : the western one on the coal measures, and the eastern mainly on Old Red Sandstone. It presents a wide diversity of habitat, ranging from the salt-marshes of the Severn estuary to the moorland of the upper slopes of the Black Mountains. The richest areas in a botanical sense are the lower Wye Valley, the district around Abergavenny (general flora), and the Wentloog and Caldicot Levels (especially aquatic plants). *Euphorbia serrulata* (Upright Spurge), which grows in the lower Wye Valley on both sides of the river, is unknown elsewhere in the British Isles, while *Sorbus eminens* (a Whitebeam) is endemic to woods on limestone in the Wye Valley and the Avon Gorge. *Rumex palustris* (Marsh Dock) is confined in Wales, as a native plant, to the Caldicot Levels. *Potamogeton trichoides* (Hairlike Pondweed) grows in the same area, its only other Welsh localities are in Glamorgan and Anglesey. A small and admittedly incomplete Flora by S. Hamilton was published in 1909. *The Flora of Chepstow* by W. A. Shoolbred was published in 1920 and covers the flora within about a 10 mile radius of Chepstow. The first comprehensive Flora, *The Flora of Monmouthshire* by A. E. Wade, was published in 1970. Recent work by T. G. Evans, the B.S.B.I. vice-county recorder, has greatly increased the number of casuals known.

V.c. 41, Glamorgan (Glam.): 926 + 364 + 550 = 1840 species; 140 microspecies; 137 hybrids. **Glam.** includes a large part of the South Wales coalfield, the flora of which on the whole is rather poor but the southern part (the 'Vale of Glamorgan') is very varied both topographically and geologically. Sand-dunes and salt-marshes are exceptionally well-developed and maritime plants very abundant, particularly in Gower. *Draba aizoides* (Yellow Whitlowgrass) locally abundant on the cliffs of Gower is found nowhere else in the British Isles. *Cirsium tuberosum* (Tuberous Thistle) occurs in cliff pastures in the

1. J. Ray (1627-1705)

2. Edward Lhwyd (1660-1709)

3. J. J. Dillenius (1684-1747)

4. H. Davies (1739-1821)

5. Joseph Banks (1743-1820)

6. L. W. Dillwyn (1778-1855)

7. C. Conway (1797-1860)

8. A. Ley (1842-1911)

9. J. E. Griffith (1843-1933)

10. J. Storrie (1843-1901)

11. C. T. Vachell (1848-1914)

12. W. A. Shoolbred (1852-1928)

Plate 1. Botanists in Wales I

13. R. W. Phillips (1845-1926)

14. D. A. Jones (1861-1936)

15. J. H. Salter (1862-1942)

16. J. A. Wheldon (1862-1924)

17. A. H. Trow (1863-1939)

18. H. J. Riddelsdell (1866-1941)

19. E. Vachell (1879-1948)

20. A. A. Dallman (1883-1963)

21. M.E.A.Richards(1885-1977)

22. P. W. Carter (1898-1971)

23 T. A. W. Davis (1899-1980)

24 V. J. Macnair (1902-1975)

Plate 2. Botanists in Wales II

extreme south of the 'Vale', elsewhere in Britain it is found only in a few sites in southern England. *Suaeda vera* (Shrubby Sea-blite), *Campanula glomerata* (Clustered Bellflower), *Scilla autumnalis* (Autumn Squill), *Poa bulbosa* (Bulbous Meadow-grass), *Agrostis curtisii* (Bristle Bent), *Wolffia arrhiza* (Rootless Duckweed) and *Herminium monorchis* (Musk Orchid) among others, here find (or found in the past) their only stations in Wales. *Aconitum napellus* (Monk's-hood) is abundant beside the River Ely, but rare in the rest of Wales. *Anaphalis margaritacea* (Pearly Everlasting) is one of the many introduced species which South Wales has long been known to possess. First recorded by Edward Llwyd in the 17th century from the Rhymney Valley near Llanbradach, this American species is now well established on waste ground in many parts of Wales.

The very large number of casual species recorded from Glam., some 84% of the Welsh total, is due largely to the work of three men, R. Melville, R. L. Smith and A. E. Wade, who in the early 1920's carried out detailed surveys on the alien flora of the ports of Cardiff and Barry. Their results were published in the *Botanical Society and Exchange Club, Reports* in 1925, 1926 and 1939.

John Storrie's *Flora of Cardiff* (1886), which listed the native and some of the introduced plants in the eastern half of Glamorgan, is still of much historic interest though it has long been superseded by two Floras published almost simultaneously in 1907: *A Flora of Glamorganshire* by H. J. Riddelsdell, published as a supplement to the *Journal of Botany*, 1907; and *The Flora of Glamorgan* edited by A. H. Trow, published in parts in the *Transactions* of the Cardiff Naturalists' Society between 1906 and 1910 and reissued as a book in 1911.

An account of the flora with lists for many localities and a complete list of species by Miss E. Vachell are contained in *The Glamorgan County History* Vol. 1. Dr. Mary E. Gillham has published two volumes in a series of Natural History Studies of various parts of Glamorgan; *The Natural History of Gower* (1977) and *Swansea Bay's Green Mantle* (1982).

A new *Flora of Glamorgan* which will map plant distribution by 5 km squares is (in 1983) in an advanced stage of preparation.

V.c. 42, Breconshire (Brecs.): 755 + 189 + 27 = 971 species; 160 microspecies; 82 hybrids.
Brecs. contains the highest mountains in South Wales (Brecon Beacons with Pen-y-fan 2,906 ft) on which occur a number of 'highland species' such as *Saxifraga oppositifolia* (Purple Saxifrage) and *Festuca vivipara* (Viviparous Fescue) which have their main areas of distribution in Wales in the mountainous regions of the north.

A notable feature of the limestone outcrops in the east are the rare species of Whitebeam, including two, *Sorbus minima* (Lesser Whitebeam) and *Sorbus leyana* (Ley's Whitebeam) which are found nowhere else in the world. Both were first discovered by the Rev. A. Ley. He was vicar of Sellack in Herefordshire from 1878-1908, travelled widely in Wales and contributed much to our knowledge of the flora of this and the neighbouring vice-counties of Mons. and Rads.

A lengthy list of plants recorded from Brecs. by W. B. Barrett was published in the *Journal of Botany* in 1885 and another by R. W. Phillips in 1926 in *The Breconshire Border*. No complete Flora has been published but parts of a manuscript Flora prepared by J. A. Webb have recently been found in University College, Swansea. M. Porter, the B.S.B.I. vice-county recorder, is working on a new Flora and is mapping the distribution of flowering plants and ferns using 5 km squares.

V.c. 43, Radnorshire (Rads.): 669 + 177 + 18 = 864 species; 78 microspecies; 46 hybrids.
Rads. contains the largest proportion of upland of any Welsh county, but its most elevated area, Radnor Forest (the highest point of which stands 2,166 ft. above sea-level), is moorland rather than mountain. There being no large masses of limestone and very little aquatic vegetation, the flora is correspondingly restricted, with fewer species than in any other Welsh vice-county. Stanner Rocks, on the borders of England, is the home of several rare species, including *Gagea bohemica*, found nowhere else in Britain, *Scleranthus perennis* (Perennial Knawel) which elsewhere in Britain occurs only in East Anglia and *Lychnis viscaria* (Sticky Catchfly) found elsewhere in Wales only in Montgomery. *Arabis scabra* (Bristol Rockcress) recorded from this vice-county, is otherwise confined in Britain to limestone rocks near Bristol, and *Potentilla rupestris* (Rock Cinquefoil) which grows in one locality in Rads. and also in Monts., occurs in only

one other site in Britain. No Flora has been published for the county but J. A. Webb prepared in the 1950's a manuscript Flora which is kept in the Library, National Museum of Wales. Miss A. C. Powell, the B.S.B.I. vice-county recorder, maintains a card-index of the flora based on 10 km squares.

V.c. 44, Carmarthenshire (Carms.): 845 + 261 + 88 = 1194 species; 91 microspecies; 63 hybrids. **Carms.** offers a range of plant habitats somewhat similar to that of Glam. but including also an area of mountain, the Black Mountain, with the Carmarthen Van (Bannau Sir Gaer, 2,460 ft.) as its highest point. The general thinning-out of the flora as one goes westward is reflected in the smaller number of species. *Galium parisiense* (Wall Bedstraw) and *Geranium purpureum* (Little-Robin) have both been recorded from Carms. but not elsewhere in Wales. *Lathyrus palustris* (Marsh Pea) is thriving in a marsh near Pembrey but is extinct in its other Welsh localities. *Liparis loeselii* (Fen Orchid) still survives in dune slacks at Laugharne Burrows but has not been seen recently at its other site, Pembrey. It is a feature of dune slacks in Glam., but is otherwise confined in Britain to East Anglia. A list of flowering plants is contained in T. W. Barker's *Natural History of Carmarthenshire* (1905), and in 1969 R. F. May's *A List of the Flowering Plants and Ferns of Carmarthenshire* was published. Mrs. I. M. Vaughan carried out much valuable recording work throughout the county and her successor as B.S.B.I. vice-county recorder, R. D. Pryce, initiated, in 1981, a long term flora project.

V.c. 45, Pembrokeshire (Pembs.): 823 + 227 + 44 = 1094 species; 61 microspecies; 77 hybrids. **Pembs.** includes the most westerly part of Wales. It has a very long and varied coastline, but the area of hill country is small, there are few bodies of fresh water, and the St David's Peninsula in particular is wind-swept and almost treeless; all factors which tend to induce poverty of the flora.

Aster linosyris (Goldilocks Aster) has recently been discovered in South Pembs., elsewhere in Wales it is confined to Gower and Gt Ormes Head. The following plants occur in Pembs. but not as natives elsewhere in Wales : *Lotus subbiflorus* (Hairy Bird's-foot-trefoil), *Limonium transwallianum* and *Limonium paradoxum* (Sea-lavenders), *Centaurium scilloides* (Perennial Centaury), *Pinguicula lusitanica* (Pale Butter-wort) and *Carex appropinquata* (Fibrous Tussock-sedge). No Flora of the county has been published but several valuable lists have appeared, the latest being T. A. W. Davis's *Plants of Pembrokeshire* in 1970. Prior to his death in 1980 Mr. Davis was working on a *Flora of Pembrokeshire* and this work is being continued by his successor as B.S.B.I. vice-county recorder, S. B. Evans.

V.c. 46, Cardiganshire (Cards.): 759 + 262 + 49 = 1070 species; 119 microspecies; 75 hybrids. **Cards.** possesses a rocky coastline with one system of salt-marshes (Dyfi estuary) and few sand dunes. Inland it is diversified topographically but not geologically, and the mountain mass of Pumlumon boasts very few 'highland' species. One of these, *Salix herbacea* (Dwarf Willow), was recently rediscovered there. Bog plants are well developed at Tregaron and Borth and the Rheidol valley contains interesting native woods. *Euphorbia peplis* (Purple Spurge), though now believed to be extinct, has been recorded from Cards. but from nowhere else in Wales. *Hypericum undulatum* (Wavy St. John's-wort) is confined in Wales to this and the adjoining counties of Pembs. and Mer.

A Flora, *The Flowering Plants and Ferns of Cardiganshire,* by J. H. Salter, was published in 1935, and a supplement edited by A. E. Wade appeared in 1952. Much valuable recording work has been done in recent years by A. O. Chater, the B.S.B.I. vice-county recorder.

V.c. 47, Montgomeryshire (Monts.): 748 + 220 + 33 = 1001 species; 57 microspecies; 43 hybrids. **Monts.** possesses no coastline, though it approaches the sea at the head of the Dyfi estuary. It is built up almost entirely of Lower Palaeozoic rocks which rise in the Berwyn Mountains in the extreme north-east to over 2,700 ft. In this area (which includes part of the neighbouring vice-counties of Mer. and Denbs.) *Rubus chamaemorus* (Cloudberry) is found in its only Welsh localities. Breiddin Hill, a very striking mass in the north of the county, has long been known as the home of *Potentilla rupestris* (Rock Cinquefoil), and other rare plants including *Lychnis viscaria* (Sticky Catchfly), *Potentilla crantzii* (Alpine Cinquefoil) and *Veronica spicata* (Spiked Speedwell).

There are no large bodies of freshwater except Lake Vyrnwy, now a reservoir. The Shropshire Union Canal which runs close to the border with England is the only Welsh locality of several rare aquatics including *Potamogeton friesii* (Flat-stalked Pondweed) and *Potamogeton compressus* (Grass-wrack Pondweed).

A manuscript Flora prepared by J. A. Webb in the 1950's is in the National Museum of Wales and a list, *Plants of Montgomeryshire — The Field Records of Janet Macnair*, edited by Mary Hignett and W. S. Lacey, was published in 1977. Miss D. Pugh, B.S.B.I. vice-county recorder, is planning (in 1982) to initiate a new recording scheme which it is hoped will lead eventually to the preparation of a new Flora.

V.c. 48, Merionethshire (Mer.): 769 + 248 + 47 = 1064 species; 93 microspecies; 108 hybrids.
Mer. possesses a considerable length of coastline including dune systems and salt-marshes and, with the possible exception of Caerns., is arguably the most mountainous vice-county in Wales, with the Cadair group (including Cadair Idris, 2,927 ft., and Aran Fawddwy, 2,970 ft.), the Berwyn chain, the Arenigs (Arenig Fawr, 2,800 ft.), and the mountains of Ardudwy. The volcanic ash of Cadair Idris and the greenstone of Rhinog Fawr and the Moelwyns are especially rich in mountain plants with many rarities such as *Oxyria digyna* (Mountain Sorrel); *Thalictrum alpinum* (Alpine Meadow-rue); *Cardaminopsis petraea* (Northern Rock-cress); *Saussurea alpina* (Alpine Saw-wort) and *Poa alpina* (Alpine Meadow-grass), all of which also grow in the mountains of Caerns. but nowhere else in Wales.

A manuscript Flora of Merionethshire for which D. A. Jones was awarded a prize at the National Eisteddfod held at Blaenau Ffestiniog in 1898, is in the Library of the National Museum of Wales. A comprehensive list, *A contribution to a Flora of Merioneth* by P. M. Benoit and Mrs. M. Richards, was published in parts in *Nature in Wales* and a 2nd edition published as a separate booklet in 1963. P. M. Benoit, the B.S.B.I. vice-county recorder, has compiled detailed records of the flora based on 10 km squares.

V.c. 49, Caernarfonshire (Caerns.): 909 + 281 + 88 = 1278 species; 117 microspecies; 102 hybrids.
Caerns. possesses a long and rugged coastline. Its mountain area, Snowdonia, which culminates in the highest peak south of the Tweed (Yr Wyddfa, 3,560 ft.), is an important stronghold of the alpine flora in Britain : out of some 35 'highland species' found in Wales all but 2 occur in Caerns., and 17 in Caerns. alone. As in Mer. this distinctive mountain flora is found almost exclusively on base rich rock composed of volcanic ash. The Gt Ormes Head is the home of a number of rare plants : three of these, *Cotoneaster integerrimus* (Wild Cotoneaster), *Hypochoeris maculata* (Spotted Cat's-ear) and *Orchis ustulata* (Burnt Orchid), are unknown elsewhere in Wales and the first-named is confined in the British Isles to this one area. *Lloydia serotina* (Snowdon Lily) grows on Snowdon and the Glyders, its nearest other locality being in the Swiss Alps. Two species of *Cerastium* (Mouse-ear), four of *Saxifraga* (Saxifrage), two of *Juncus* (Rush), two of *Poa* (Meadow-grass) and three of *Carex* (Sedge) as well as several other species are all confined in Wales to the mountains of Snowdonia.

In 1895 J. E. Griffith published *The Flora of Anglesey and Caernarvonshire.* Since then several lists of parts of the vice-county have been compiled and Miss A. P. Conolly is preparing a Flora of the Lleyn Peninsula which will include distribution maps based on 1 km squares. M. Morris, the B.S.B.I. vice-county recorder, maintains a card index of the flora and has done much valuable work in North Caerns.

V.c. 50, Denbighshire (Denbs.): 859 + 248 + 37 = 1144 species; 158 microspecies; 54 hybrids.
Denbs. has a comparatively short coastline and no high mountains except in the extreme south, but it possesses a wide range of plant habitats, and its coastal belt, standing in the rain shadow of Snowdonia, provides conditions suitable for a great number of species. In the marshy areas on the English border (which belong in the old administrative County of Flintshire but the vice-county of Denbigh) are found several species which do not grow elsewhere in Wales including *Myosurus minimus* (Mousetail), *Oenanthe fluviatilis* (River Water-dropwort) and *Stratiotes aloides* (Water-soldier).

Other species confined in Wales to this vice-county include *Barbarea stricta* (Small-flowered Winter-cress), *Stachys alpina* (Limestone Woundwort), *Gagea lutea* (Yellow Star-of-Bethlehem), *Carex elongata* (Tufted-sedge) and *Ophrys sphegodes* (Early Spider-orchid).

The flowering plants of this and the following vice-county form the subject of a series of important papers by A. A. Dallman published in the *Journal of Botany* between 1906 and 1913. A complete Flora of the two vice-counties by the same author was, at one time, thought to be imminent but unfortunately it never materialized and not even a manuscript is known to exist. J. M. Brummitt, the B.S.B.I. vice-county recorder, is compiling detailed records on a 10 km sq. basis.

V.c. 51, Flintshire (Flints.): 820 + 206 + 48 = 1074 species; 107 microspecies; 45 hybrids.
Flints., the smallest of the Welsh vice-counties, possesses no mountains, but presents considerable diversity in its geology and topography. In particular the Carboniferous limestone is exposed here over a wide area. It should be remembered that the one large and several smaller areas that are detached from the county are included in Denbs. not Flints.

The number of rare species is small compared to the neighbouring counties but *Euphrasia pseudokerneri* (an Eyebright) has not been recorded from elsewhere in Wales and *Senecio cambrensis* (Welsh Groundsel), which was described as a new species in 1955, is recorded from roadsides in this county, Denbs. and Shropshire but nowhere else in the world. Apart from Dallman's writings mentioned under the preceding county, little has been published on the flora.

G. Wynne, the B.S.B.I. vice-county recorder, is now working on a new Flora and is producing distribution maps based on the 'tetrad' or 2 × 2 km square.

V.c. 52, Anglesey (Angl.): 808 + 193 + 36 = 1037 species; 51 microspecies; 91 hybrids.
Angl. is remarkable for its many small lakes and swamps and its lack of woods. One species and three hybrids of *Potamogeton* (Pondweeds) are confined in Wales to this island, as are *Senecio integrifolius* (Field Fleawort), *Taraxacum palustre* s.s. (Narrow-leaved Marsh-dandelion), *Carex depauperata* (Starved Wood-sedge) and *Epipactis dunensis* (Dune Helleborine). *Otanthus maritimus* (Cottonweed) which formerly occurred in Caerns. was seen in Angl. up to 1895; it is now apparently extinct in Britain but still survives in one or two localities in Ireland.

Hugh Davies's *Welsh Botanology* (1813) contains a catalogue of the flowering plants native to the Isle of Anglesey, which is the earliest local flora of any part of Wales. *The Flora of Anglesey and Caernarvonshire*

Table 1. Statistical summary of the flora of the Welsh vice-counties

	Native Denizen Colonist		Naturalized, Established, Introduced, aliens		Casuals		Total Species		Micro-species		Hybrids	
Wales	1078		443		653		2174		325		282	
Glam.	924	(86)*	363	(82)	549	(84)	1836	(84)	140	(43)	135	(48)
Caerns.	909	(84)	280	(63)	88	(13)	1277	(59)	117	(36)	102	(36)
Mons.	818	(76)	273	(62)	164	(25)	1255	(58)	122	(38)	93	(33)
Carms.	845	(78)	261	(59)	87	(13)	1193	(55)	91	(28)	63	(22)
Denbs.	859	(80)	248	(56)	37	(6)	1144	(53)	158	(49)	54	(19)
Pembs.	823	(76)	227	(51)	43	(7)	1093	(50)	61	(19)	77	(27)
Flints.	820	(76)	206	(47)	48	(7)	1074	(49)	107	(33)	45	(16)
Cards.	759	(70)	262	(59)	49	(8)	1070	(49)	119	(37)	74	(26)
Mer.	769	(71)	248	(56)	47	(7)	1064	(49)	93	(29)	108	(38)
Angl.	808	(75)	193	(44)	36	(6)	1037	(48)	51	(16)	95	(34)
Monts.	748	(69)	220	(48)	33	(5)	1001	(46)	57	(18)	43	(15)
Brecs.	755	(70)	189	(43)	27	(4)	971	(45)	160	(49)	82	(29)
Rads.	669	(62)	177	(40)	18	(3)	864	(40)	78	(24)	46	(16)

*Figures in Brackets represent the percentage of the Welsh total

by J. E. Griffith was published in 1895, and in 1982 R. H. Robert's *The Flowering Plants and Ferns of Anglesey* was published by the National Museum of Wales.

The figures which are given at the heads of the preceding paragraphs are summarized in Table 1. It will be seen that Glam. and Caerns. are far richer in native species (including denizens and colonists) than any other vice-county. Rads. (with about 72% of the Glam. total) and the other purely inland vice-counties come together at the foot of the list. Naturalized species and casuals are most numerous in the industrial south and least so in Angl. and the inland vice-counties. Glam. has a longer list of native, naturalized and casual species than has any other vice-county in Wales.

Composition of the Welsh Flora

An analysis of the Welsh flora, based on the status of its constituent species, is shown in Table 2. This table also contains, for comparison, the corresponding figures for the British Isles as a whole, taken from J. E. Dandy's *List of British Vascular Plants,* 1958.

Table 2 The Welsh and general British floras compared.

	WALES		BRITISH ISLES	
	Species	Microspecies	Species	Microspecies
Native	1011	287	1511	688
Aliens *	510	37	626	17
	1521	325	2137	685

*Includes denizens, colonists, naturalized aliens, established aliens and introduced species, but excludes casuals.

These Welsh figures are only roughly comparable with those quoted from Dandy : Many species which in Britain as a whole are regarded as native are believed to be of a lower status in Wales. In fact some species which in the 2nd edition of this work were classed as natives have, in the light of more recent knowledge, been placed in a different category.

The figures show that 67.6% of the species which form the native British flora are found in Wales; if one adds denizens and colonists to the native total the percentage increases to 71%.

Plate 3. Dr. Harold Augustus Hyde (1892-1973)

Plate 4. Arthur Edward Wade (1895-)

Geological Background

by R. M. Owens and R. E. Bevins

Wales has an extremely varied and complex geology, with rocks of many different types, origins and ages exposed at the surface (see map 1). It is this diverse geology which gives the Principality its rich variety of scenery and landforms, and consequently the numerous habitats which ultimately affect the distribution of its flora. Concealed within the rocks themselves there is evidence of over 700,000,000 years of Earth History, showing that the landscape we know today is the product of an immensely long sequence of changes which have given rise to a succession of different environments, ranging from deep oceans to mountain ranges, from arid deserts to arctic wastes. Thus our present day scenery and climate are but an ephemeral phase in the ever-changing, ever-evolving history of the Earth. In addition, for the botanist, the rocks of Wales have a particular intrinsic interest, for they have provided critical pieces of fossil evidence in our understanding of the evolutionary history of plants, as outlined below. In order to describe the different rock types which crop out in Wales, and their effect upon its scenery, it is best to consider them in age sequence. Table 3 shows a simplified geological column, together with important events in the geological evolution of Wales.

The oldest rocks in Wales are of late Precambrian age, and crop out in a number of scattered areas in north Pembrokeshire; near Carmarthen; at Old Radnor; in southwest Lleyn; near Llanberis and most extensively on Anglesey. The strata generally give rise to flat-lying, rather monotonous terrain, such as on Anglesey, or in the St David's area of Pembrokeshire. The sequence includes rocks of both sedimentary and igneous origin, represented now by metamorphosed quartzites, sandstones, mudstones, limestones, volcanic lavas and tuffs, and granites. It has been suggested that the sequences of Anglesey accumulated in and adjacent to a subduction zone, which occurs where one of the Earth's major crustal plates dips down beneath another. The sequences from other parts of Wales suggest that they were formed in island arcs, which are chains of volcanic islands associated with subduction zones.

At the end of Precambrian times, around 570 million years ago, these rocks were uplifted and eroded, and the succeeding sediments were then laid down following a renewed incursion of the sea over an irregular landscape. The sediments which accumulated comprise the rocks of the Cambrian, Ordovician and Silurian systems (collectively called the Lower Palaeozoic) and are exposed over much of North, mid and southwest Wales, with small isolated outcrops near Cardiff, and at Usk. The terms Cambrian, Ordovician, and Silurian, in world-wide usage today, were first coined in Wales, and were derived from the Roman name for Wales (Cambria) and from two ancient Welsh tribes (the Ordovices and the Silures). The sediments were predominantly muds, sands and gravels, with rarer carbonate deposits, giving rise now to a sequence of mudstones, siltstones, sandstones, conglomerates and limestones. Most of these sediments accumulated in a marine trough, the 'Welsh Basin', which was situated along the southern margin of a major ocean, called the Iapetus Ocean.

Volcanicity was extensive in this area during Lower Palaeozoic times, and it is the resistant nature of these ancient volcanic rocks which gives rise to the spectacular mountain scenery of Snowdonia, Cadair Idris and the Aran Mountains, as well as to some of the less spectacular, but nevertheless rugged, hilly regions such as the Carneddau, to the north of Builth Wells, and parts of the Prescelly Hills, in northern Pembrokeshire. In contrast, the sediments are less resistant to erosion than the volcanic rocks, and consequently produce areas of gentler, rounded scenery, as, for example, the great open tracts of country in central Wales and the Denbigh Moors. However, where harder sandstones are thickly developed, such as in the Rhinog Mountains area of the Harlech Dome, more rugged scenery is produced.

Limestones are only rarely developed in the Lower Palaeozoic strata, and at most are thin and of local extent. However, some of the volcanic horizons are carbonate-rich, and it is well established that the arctic-alpine flora of Cwm Idwal in Snowdonia, flourishes on the calcareous rocks of the Middle Basic Volcanic Formation of the Snowdon Volcanic Group.

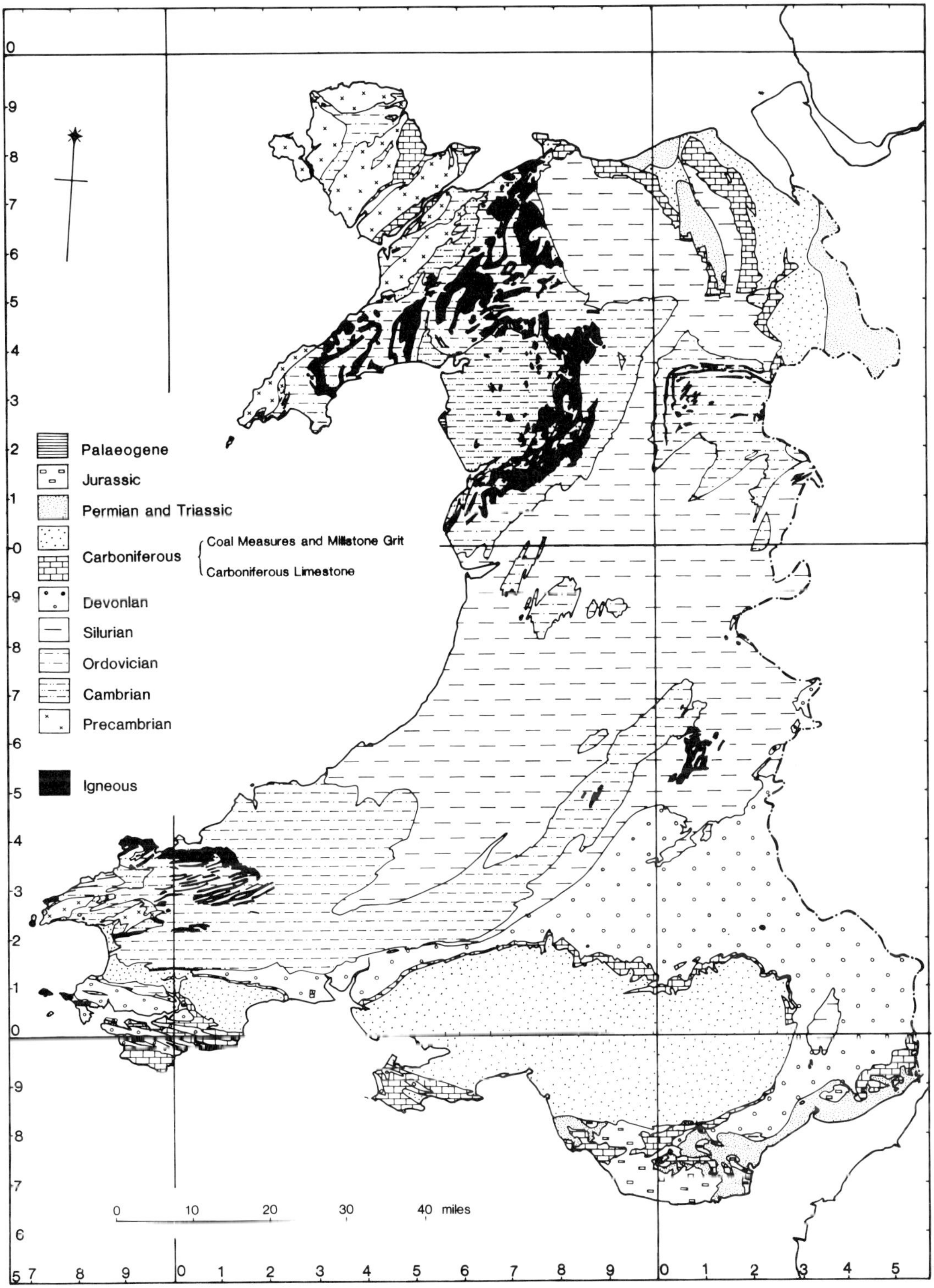

Map 1. Simplified Geological Map of Wales.

Table 3. Geological Time Scale.

ERA		PERIOD	AGE (Millions of years ago)	Summary of geological history of Wales
CENOZOIC	QUATERNARY	Recent or Holocene	0.01 – present	Coastal areas drowned by rising sea level caused by melting ice at the end of the Ice Age. Deposits of alluvium and peat, with further development of present drainage patterns and modern flora and fauna.
		Pleistocene	1.6 – 0.01	The 'Ice Age', with repeated glaciations and milder interglacial periods. The most recent major glaciation reached a maximum 18,000 years ago, and the last local ice left North Wales by 10,000 years ago. Modification of land forms by ice scouring and deposition of glacial drift. First evidence of Man in Wales from Pontnewydd cave, North Wales, about 200,000 years old.
	TERTIARY	Neogene	26 – 1.6	Prolonged, pulsatory uplift and erosion. Late Palaeogene and early Neogene terrestrial sediments known only from Mochras and locally in Gwynedd and south-west Dyfed. Early Palaeogene intrusive ingeous rocks in north-west Wales. Basic landforms and drainage patterns established.
		Palaeogene	65 – 26	
MESOZOIC		Cretaceous	140 – 65	No rocks of this age known in Wales. Different opinions suggest either a persistence of terrestrial conditions, or that the Chalk sea covered much of the country.
		Jurassic	195 – 140	Lower Jurassic marine rocks known only from south-east Glamorgan and Mochras borehole, Harlech, with thick, younger sediments in Bristol Channel. Warm, shallow sea may have transgressed over Wales through the period, but direct evidence is lacking.
		Triassic	230 – 195	Arid and semi-arid terrestrial conditions, with evidence of periodic flash floods. Dinosaur footprints and early mammal remains known from Glamorgan. Marine transgressions in south-east Wales at the very end of the period.
PALAEOZOIC	UPPER	Permian	280 – 230	Uplift and mountain building (Hercynian Orogeny). Erosion across most of Wales.
		Carboniferous	345 – 280	Marine transgression early in period, leading to spread of warm, subtropical seas with rich coral/brachiopod faunas; extensive deposition of carbonate sediments that now form the Carboniferous Limestone. Regression in middle of period, with widespread deltaic deposits. Rich vegetation on coastal plains and deltas in late Carboniferous times; peat accumulated to form coal seams of the Coal Measures.
		Devonian	395 – 345	Uplift and mountain building (Caledonian Orogeny) continued from Silurian times, resulting in the deposition of the terrestrial Old Red Sandstone across most of Wales. Rapid diversificaton of land floras and non-marine fish faunas.
	LOWER	Silurian	435 – 395	Marine muds, silts and sands, with local carbonate sediments. Land with deltas across South Wales. Shallowing and retreat of the sea late in the period, with widespread onset of terrestrial conditions. Earliest fishes in Wales, and first vascular land plants appeared.
		Ordovician	505 – 435	Marine muds, sands, grits and local carbonate sediments. Evidence of extensive vulcanicity in north, mid and south-west Wales. Fossil faunas increasingly diverse, including first corals.
		Cambrian	570 – 505	Transgression of sea into Wales, with deposition of grits, sands and muds. First abundant fossils, including earliest trilobites and brachiopods.
PRECAMBRIAN			4600 – 570	Oldest rocks in Wales dated at about 600-700 million years old, but are possibly considerably older. Evidence of intermittent marine conditions and vulcanicity, with periods of folding, uplift and erosion. Earliest fossils from Wales are 'jellyfish' from Carmarthen of late Precambrian age.

Closure of the Iapetus Ocean to the north of the Welsh Basin during Ordovician and Silurian times resulted from plate convergence, and caused uplift of the Lower Palaeozoic sediments to form the Caledonian mountain chain. At the same time, the rocks were folded and fractured, and in areas of greatest deformation the mudstones were transformed into slates. (These slates have been extensively quarried and mined in areas such as Nantlle, Dinorwic, Bethesda, and Blaenau Ffestiniog in North Wales, Corris in central Wales, and the Prescelly Hills in South Wales.)

Erosion of the Caledonian mountain chain followed uplift, and the lowermost sediments of the succeeding Upper Palaeozoic strata (the Old Red Sandstone) are predominantly marls and siltstones with subordinate coarse-grained sandstones and conglomerates of fluviatile, floodplain or brackish water origin. These are exposed over a wide area of South Wales, including the Carmarthenshire Fans, the Brecon Beacons and the Black Mountains, producing the distinctive red-coloured scarps. By the end of Devonian (Old Red Sandstone) times, much of the topography had been worn down, and marine incursions into the area occurred, witnessed by the presence of marine fossils in the uppermost beds of the Old Red Sandstone. In South Wales these rocks pass upwards without a break into the overlying Carboniferous rocks.

The lowermost deposits of the Carboniferous are limestones which accumulated as lime muds in a shallow, warm, tropical sea. Outcrops of Carboniferous Limestone are spectacular, and form both the great cliffed coastlines of the Gt Ormes Head in North Wales, south Pembrokeshire and the Gower peninsula in South Wales, and the barren karst topography which rims the South Wales Coalfield, and which occurs on either side of the Vale of Clwyd, passing southwards to the Eglwyseg Rocks above Llangollen. Extensive cave systems are well developed in these thickly bedded limestones, particularly along the northern side of the South Wales Coalfield, as for example, the Dan yr Ogof cave system; smaller caves, for example Minchin Hole, on the Gower peninsula, and Pontnewydd Cave, near Denbigh, are well-known sites for early human habitation.

The succeeding Carboniferous strata (the Millstone Grit and the Coal Measures) are a sequence of mudstones, siltstones and sandstones with coals. They accumulated in a deltaic environment, from a system of rivers draining from the north and northeast towards an open sea which existed to the south (over present-day Devon). The coal seams represent decayed vegetable matter derived from the lush tropical forests which existed at this time, and it is these seams which have made Wales one of the most famous coal-mining areas in the world, and which were largely responsible for the great industrial development. The South and North Wales Coalfields are the best-known coal outcrops of Wales, but similar coal-bearing strata also occur in the westward extension of the South Wales Coalfield, around Haverfordwest, as well as on Anglesey.

Rocks of Mesozoic age are of limited distribution in Wales. In the north, sandstones of Triassic age (and possibly also of Permian age) crop out in the Vale of Clwyd, and in a narrow band in the eastern parts of Denbighshire and Flintshire adjacent to the English border. In South Wales, Triassic rocks (once again predominantly sandstones) occur along the Severn Estuary in Monmouthshire, and in southeast Glamorgan, with scattered outcrops in the Vale of Glamorgan. Virtually all of these Triassic rocks were deposited in an arid desert environment, although the very youngest Triassic rocks (of Rhaetian age) which crop out in the Vale of Glamorgan are of brackish or marine origin and are the precursor of a marine episode represented by the Lower Lias, of Jurassic age, which is made up of an alternating sequence of limestones and shales. Following the Lias, there is a gap in the geological record on mainland Wales until the Quaternary. Some authorities suggest that Wales was covered by sediments (including chalk) of late Jurassic and Cretaceous age, which were subsequently eroded away during a period of upwarping and faulting in the early Tertiary. Isolated remnants of Tertiary (early Neogene) sediments have been found in south Pembrokeshire and also near Harlech, where they were encountered in the Mochras Borehole, overlying a great thickness of Lower Lias strata. More recently, a thin sequence of sediments of possible Tertiary age has been identified as a result of shallow drilling near St David's, in Pembrokeshire.

The present day landscape evolved over the last 20 million years through the action of rivers, sea and ice, a time which saw repeated periods of uplift and erosion leading to the development of three distinct plateaux (high: 1700 — 1900 feet, middle: 1200 — 1600 feet and low: 700 — 1000 feet) as well as a series of coastal platforms (at 200, 300, 400 and 600 feet). The latter are generally taken to be benches of marine planation, whilst the origin of the former is still a subject of contention. Some workers propose that they are the product of river erosion, others favour marine planation. Much of the present day landscape of upland Wales is dominated by these plateaux.

Large areas of Wales are covered by a veneer of Quaternary sediments which are of glacial, interglacial or postglacial origin, and deposited within the last half-million years. Some of the oldest of these, the so-called 'Older Drift', date from before the last major interglacial period (approximately 115—128,000 years ago, and known as the Ipswichian). At times during this period the sea level was up to 15 metres above the present, which gave rise to the raised beach deposits such as those of Red Wharf Bay, Anglesey, and the widespread '*Patella* beach' in South Wales. Cave deposits about 200,000 years old at Pontnewydd in North Wales have recently provided the earliest evidence of Man in Wales.

The last glaciation, the Devensian, which reached its maximum some 18,000 years ago, covered much of Wales (with the exception of southern Glamorgan, eastern Monmouthshire and southern Pembrokshire) and was responsible for most of the important glacial landforms preserved today. These include features of deposition such as moraines, eskers, drumlins and in particular the widely distributed boulder clay, as well as erosional landforms such as cirques, arrêtes and U-shaped and hanging valleys. In many cases preexisting drainage patterns were modified during the Quaternary; lakes formed in over-deepened valleys, in valleys dammed by moraines, or in cirque basins. There are many excellent examples of the latter in Snowdonia, of which Cwm Idwal is probably the most renowned. In some cases the impeding of drainage has caused the creation of extensive bogs, such as Cors Tregaron.

In late Devensian times, when periglacial conditions prevailed, protalus drifts and solifluction deposits such as 'head' were formed, and this was also the main period of scree development. In the postglacial period, from 10,000 years B.P.* onwards, the present day floras and faunas became established. Certain dune, alluvium, peat deposits, submerged forests and raised beaches date from this period, and reflect both changing climate and sea levels, gradually arriving at the present disposition of land and sea and today's climate.

Present-day Wales is essentially an upland area, a major component of 'highland Britain' with 36% lying above the 800 foot contour. Apart from the major elements of relief, attention should be drawn to the numerous features of micro-relief, especially in the upland areas, which are so important for local habitats. These reflect the structure, composition and hardness of underlying strata resulting in strike ridges, small scale scarps, depressions and notching and benching derived from strong cross-jointing. Micro-relief also has an influence on soil development. This however is a complex process, in which local bedrock, the nature and origin of underlying drift and aspects of the drainage pattern all play a vital part. Commonly the type of soil which has developed is influenced not by bedrock but by the veneer of drift deposits covering it.

In addition to the diversity of rock types outlined above, a long history of mining has left its mark on the landscape and the flora developed upon it. The waste tips, the remains of winding and pumping gear, of crushing mills and settling tanks, and of the smelters are all testimony to this important era in the industrial history of Wales. The metal-rich, toxic soils around the mines and developed above concealed mineral veins commonly possess anomalous plant communities. Possibly one of the best examples of such a community occurs at Coed y Brenin, north of Dolgellau, where Spring Sandwort (*Minuartia verna*), Thrift (*Armeria maritima*) and Sea Campion (*Silene vulgaris* ssp. *maritima*) are developed on copper-rich soils. Another example is the occurrence of Alpine Penny-cress (*Thalspi alpestre*) on the lead mine tips in the Llanwrst area, and also near Llanafan, to the east of Aberystwyth. It is interesting to note that these anomalous plant communities are commonly used by exploration mining geologists as indicators of possible concealed mineralized zones.

In conclusion, it is worth remarking on some of the fossil plants which have been found in Wales, since they have provided important information on the evolutionary history of the plant kingdom. The earliest known plant fossils from Wales are stromatolites, structures produced by lime-secreting algae occurring in the Gwna Limestone (late Precambrian or possibly very early Cambrian) of Anglesey. Stromatolites have

*B.P. – before present.

also been found in younger strata, including the Wenlock Limestone and Carboniferous Limestone. Present day stromatolites inhabit the shallow intertidal zone of the tropics or subtropics, as for example in Shark Bay, Western Australia, and a similar environment may be inferred for fossil examples. Lime-secreting green and red algae are also found in limestones, for example, the Dolyhir and Nash Scar limestones near Old Radnor, both of Silurian age, and in parts of the Carboniferous Limestone, where they can be important rock builders.

Some of the earliest vascular land plants in the world have been found in late Silurian rocks in Carmarthenshire and Breconshire, where they are represented by the genus *Cooksonia.* This small plant has a simple dichotomously branching stem and terminal sporangia, and has all the important attributes of vascular plants, including tracheids and trilete spores. It probably grew close to the shore of the receding Silurian sea. In late Silurian times, *Cooksonia* was joined both by *Steganotheca* (similar in detail to *Cooksonia*) and by a more complex plant, *Zosterophyllum,* whose sporangia are arranged along the sides of the branches. *Cooksonia, Zosterophyllum* and other simple land plants, such as *Gosslingia,* occur in the Senni Beds, of Devonian (Lower Old Red Sandstone) age. These strata have also yielded plants which are the possible forebears of a Middle Devonian complex of plants from which ferns, gymnosperms and clubmosses probably evolved. The youngest Old Red Sandstone strata at Tongwynlais near Cardiff has yielded the earliest known fossil seed in Britain.

In North Wales, marine strata of early Carboniferous age, which were deposited close to a contemporary shoreline, contain drifted plant remains which include arborescent clubmosses, horsetails and seed ferns — harbingers of the lush Coal Measures flora to follow. This Coal Measures flora comprises gymnosperms (seed-ferns or pteridosperms, cordaitales and conifers), clubmosses, horsetails and true ferns. Presumably because of a unique set of conditions in the low-lying tropical Coal Measures swamps, gigantism occurred in nearly every group, and those which are today represented only by herbaceous forms were represented then by arborescent forms. All of these, however, became extinct by the close of the Palaeozoic, when they were replaced by a different flora, dominated by conifers and by diverse new gymnosperms. The Lias has yielded fronds of *Otozamites,* a gymnosperm belonging to the *Bennettitales,* an extinct group which flourished in the Jurassic and Cretaceous; it probably had a palm-like habit. Plant debris and lignite have been found in the Tertiary sediments in the Mochras borehole. The sequence of floras in the Quaternary, ultimately leading up to that of the present day, is described in the next chapter.

The Effects of Climate, Past and Present, on Plant Distribution in Wales

by J. P. Savidge

An examination of the maps at the end of this book will reveal numerous different patterns of plant distribution in Wales which can be related to differences in the type of rock, the present climate, and changes that have occurred in the flora and fauna since the end of the last Ice Age some 10,000 years ago. In more recent times there have been dramatic changes in the distribution of certain species as a result of man's influence on the natural vegetation through his agricultural and industrial activities. There is a continuing process of both extinction and replacement. Overall there has been an increase in the number of flowering-plants in Wales in recent years with many additional species being brought in both intentionally and accidentally.

The purpose of this chapter is to draw attention to some of the more important climatic factors that have to be taken into consideration when interpreting the distribution maps, especially for those species which are restricted to specific parts of the Principality.

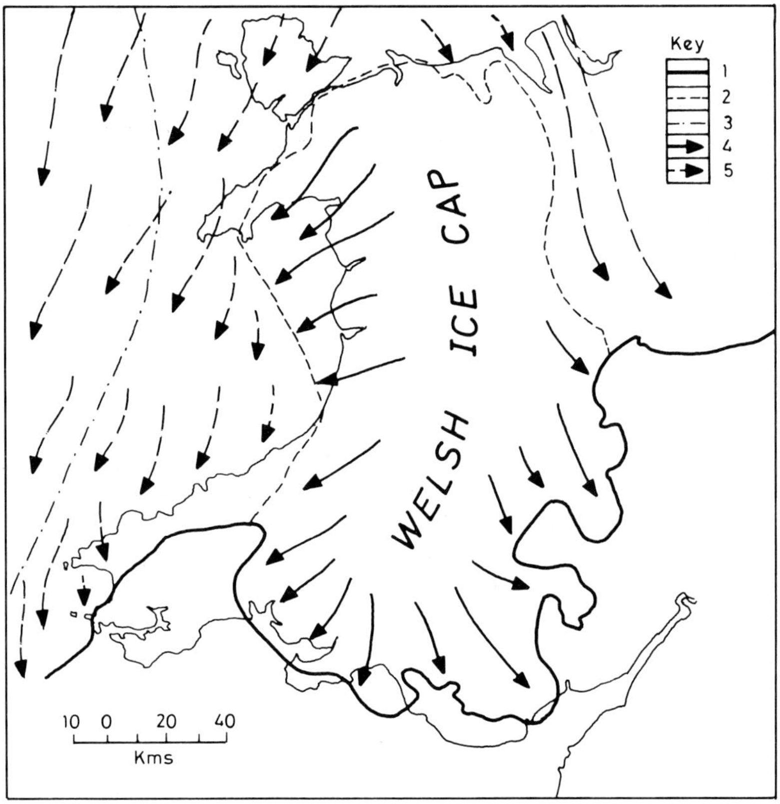

Figure 1. Key to lines: 1 – southern limit of the Devensian ice-sheet; 2 – westward limit of the mainland ice-sheets; 3 – 100 fathom line; 4 – main flow-lines of the Welsh Ice Cap glaciers; 5 – flow-lines of the Irish Sea Glacier. From Bowen (1973) and other sources.

Late Devensian, 26,000 — 13,300 B.P.*

The effects of the last Ice Age were felt throughout the country with the greatest impact occurring in Mid and North Wales which were covered by ice sheets up to 800m in depth (Fig. 1.) A few areas remained vegetated, especially near the southern coasts and on the hills above the ice sheets which would have been covered by alpine grassland and scree similar to that found at the present time near the glaciers of Iceland and northern Norway and Sweden. The plants occurring in these areas would have been small perennials which could withstand the long, cold winters and other features associated with periglacial conditions. The species that were able to survive the Devensian glaciation were important, they formed a nucleus of native species able to migrate back into central and northern areas when conditions became more favourable. This occurred as the glaciers melted at the onset of warmer conditions which set in about 14,000 years ago.

The depth of the ice sheets gradually increased during the Devensian period since more snow and rain became incorporated into the glaciers than melted during the summer thaw. There were two main sources for the ice sheets covering Mid and North Wales: the ice derived from glaciers centred over Snowdonia, Pumlumon and the Brecon Beacons, and the extensive ice sheets which moved southwards down the Irish Sea from the glaciated regions of northern England, Scotland and Northern Ireland. The maximum southward extension of the mainland ice is shown by the bold continuous line in Fig. 1. North of this line there were probably a number of hills and mountains that were above the ice sheets and were covered by alpine plants. In the extreme south and south-west of Pembrokeshire, Carmarthenshire and Glamorgan there may have been a few sheltered valleys containing areas of alpine shrub-heath with small trees and shrubs such as *Betula nana* (Dwarf Birch), *Juniperus* (Juniper) and small species of *Salix* (Willows).

The Late Glacial Interstadial, 13,300 — 11,500 B.P.

About 13,300 years ago there was a rapid rise in temperatures with the estimated mean July temperature rising from about 9°C. to 17°C. between 13,300 and 13,000 B.P. The climate of the extreme south of Wales was not too different from that of today, although there are indications that it was drier and windier; but the landscape would have appeared very different, consisting mainly of herb-rich grassland amongst a range of periglacial landscapes. The climate of the central and northern areas would have remained cooler and damper since it would have taken a long time for the glaciers to completely thaw away. A few of the glaciers in Snowdonia may have survived throughout this period. By 11,500 B.P. most of Wales would have been free from ice sheets. They would have left extensive areas of glacial tills and boulder clay which became gradually colonized by plants migrating into the area from South Wales and the Midlands.

At a number of sites, including Llyn Gwernan, in the Cadair Idris range, and Nant Ffrancon and Glanllyn-nau in Snowdonia, peat deposits have recently been found which contain remains of this period (Lowe, 1981). Up to 2m of peat was formed in the Llyn Gwernan basin, with the base having a radiocarbon date of 13,200 ±120 B.P., and the top layer, a date of 11,160 ±90 B.P. It contained *Betula* (Birch), *Juniperus, Corylus* (Hazel) and *Salix* which are typical species of arctic habitats, several species of grass and sedge and herbs such as *Filipendula* (Meadowsweet) and *Rumex* (Docks and Sorrels).

The warmer conditions of the interglacial were short lived and after a warm spell, of some 800 years, the mean July temperature gradually fell back to a value of about 10°C. (Coope, 1977).

The Younger Dryas Glacial, 11,500 — 10,000 B.P.

With the lower temperatures and associated wetter conditions, many of the former glaciers reformed, especially in Snowdonia and Mid Wales. Favoured areas, particularly hill and mountain slopes facing south, would have kept their mantle of grassland, and peat continued to accumulate in lake basins such as Llyn Gwernan (Lowe, 1981) until about 11,200 B.P. when they became covered by muds and clays from the advancing ice sheets.

This colder period, referred to as either the Younger Dryas Glacial or the Loch Lomond Stadial, was short lived. Over much of Wales the trees and shrub-heath became replaced by plants of more open habitats with sedges dominating in the areas which remained free from the advancing reformed glaciers. In the

*B.P. — before present.

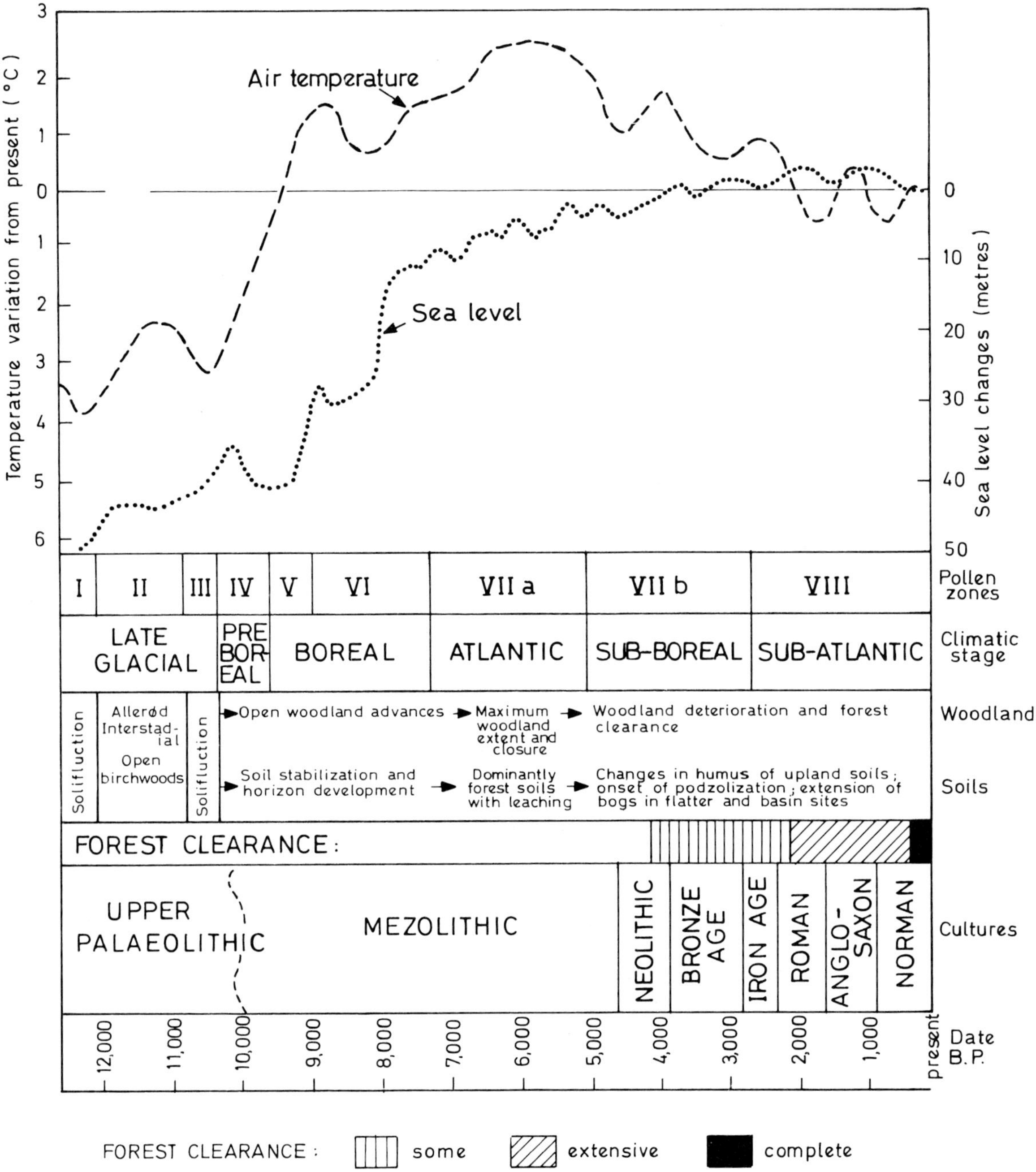

Figure 2. Summary of the main changes in climate and vegetation in Wales since the end of the Devensian glaciation. From various sources.

more southerly regions of Wales, the species-rich grasslands that developed during the previous period continued to survive with the more favoured coastal areas keeping their shrubs and trees such as *Betula, Juniperus* and *Salix.* In a few places, *Corylus* probably managed to keep a foothold until conditions rapidly improved, with the mean July temperature rising from a mean of 10°C. to 15°C. during the 300 years from 10,300 to 10,000 B.P.

As the glaciers retreated towards the end of this period, aquatic plants moved in from the more southerly areas to colonize the numerous lake basins which had formed behind glacial moraines. The remains of these plants, together with pollen from the shrubs and trees colonizing the surrounding hills, can be found in the lacustrine silts on which the more recent peat deposits have been formed. Samples from the base of the newer peats have yielded radiocarbon dates of about 10,000 B.P., with some slightly older dates in a few places such as Cors Tregaron (Hibbert and Switsur, 1976), but most of these samples have a possible radiocarbon date error of some 200 years. From the evidence to date, it would be best to use the date of 10,000 B.P. as indicating the end of the Younger Dryas, although by then the glaciers and ice sheets were in full retreat.

In the Brecon Beacons and other areas of South Wales, where the conditions were not so harsh, peat deposits have been found with radiocarbon dates going as far back as 10,860 ± 70 B.P., at Craig Cerrig-gleisiad and to 11,600 ± 140 B.P. at Traeth Mawr (Walker, 1980).

The Postglacial Period, 10,000 B.P. to the present day.

Climatic conditions improved dramatically from 10,000 to 9,000 B.P. with warmer, drier conditions leading to a rapid melting of the remaining glaciers and permitting the migration of many species into Wales from south-west England and the Midlands. Some came from France before the English Channel formed (southern England was connected to France by land bridges during the main part of the glacial period). This has resulted in many Welsh species having two or more distinct geographical races.

The climate continued to improve with the climatic optimum occurring between 7,500 and 5,000 years ago. The vegetation during this period was in a continual state of flux as the original species-rich alpine grassland was replaced by woodlands which, by 5,000 B.P. covered most of Wales. The next 1,000 years was dominated by the forests; but then man moved into many parts of Wales and started clearing the woodlands during the Late Neolithic and Bronze Ages (Fig. 2.). We can discover how the distribution of the various species changed during the Postglacial period by examination of peat deposits in various parts of Wales. Figure 3 provides information on the changes in the trees and shrubs in the Tregaron area. Details of changes in the herbaceous species, especially those associated with arable farming, can be found in Turner (1970) and Hibbert and Switsur (1976).

The pollen diagram for the south-east raised bog at Cors Tregaron (Fig. 3) provides a good example, especially as it represents a complete record for the whole of the period, with the basal layer having a radiocarbon date of 10,200 ± 200 B.P. *Betula, Corylus, Juniperus* and *Pinus* (Pine) are all present in the basal layer amongst the remains of *Phragmites* (Common Reed) which colonized the margins of the large lake that formed behind the glacial moraine on which the market town of Tregaron is now situated. With the rapid increase in temperature, we find a surprisingly fast northward migration of both *Ulmus* (Elm) and *Quercus* (Oak), both of which arrived by 9,500 B.P. The birch forests gradually gave way to *Pinus sylvestris* (Scots Pine) which eventually became extinct as a native species in Wales. The next major change came just over 7,000 years ago when both *Alnus* (Alder) and *Tilia* (Lime) spread into the area, to be followed by *Fraxinus* (Ash) some 500 years later.

A significant change came about 6,000 years ago when the heathers (*Erica* and *Calluna*), together with species of *Vaccinium* (Bilberry and Cowberry) and *Empetrum* (Crowberry) started to appear as the *Phragmites* peat was replaced by *Sphagnum* (bog moss) peat. This change indicates that the original nutrient rich soils were becoming leached out as the climate became considerably cooler and wetter. The Scots Pine decline set in from 6,000 to 5,000 B.P. and was followed by an Elm decline at about 5,000 B.P. This is just the time when Neolithic man made his appearance and started clearing and cultivating the land. His main effects were in clearing the forests and the introduction of grazing animals. The new grasslands would have

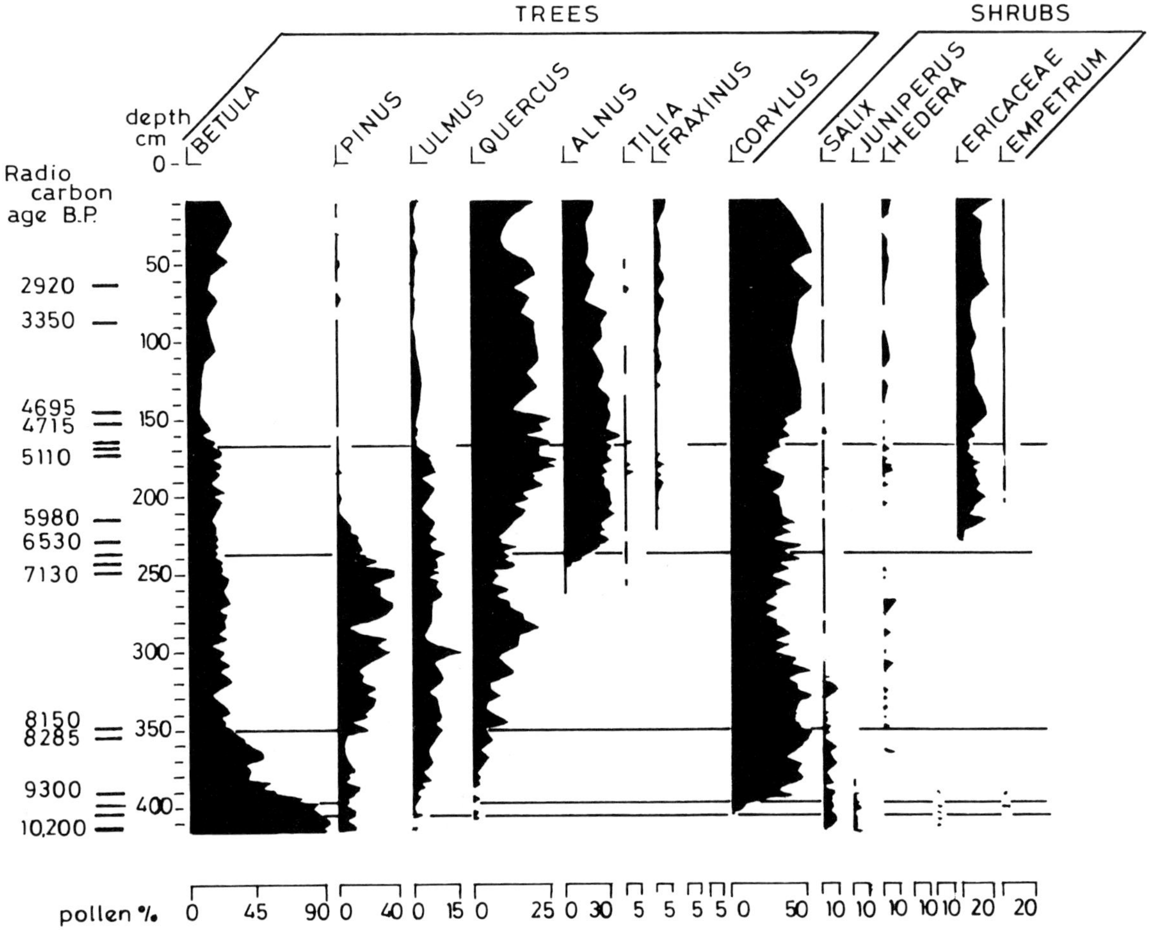

Figure 3. A simplified pollen diagram for trees and shrubs based on a pollen profile obtained at Cors Tregaron by Hibbert and Switsut (1976). The black bands indicate the percentage of pollen found for each of the main species.

been very productive since they occurred on a well developed nutrient rich forest soil; but this was the stage that started the gradual deterioration of upland soils, with the loss of soil structure, loss of nutrients and earthworms, and the formation of the present-day podsols with their iron pans which are now so characteristic of much of upland Wales.

With the decline of the Elm, we start to see the occurrence of a number of typical weed species belonging to such genera as *Plantago* (Plantains), *Rumex* and *Urtica* (Nettles). The traditional type of Welsh tribal pastoral economy continued until the Monastic period (1164-1540 A.D.) in the Tregaron area as in many other parts of Wales. With the coming of the monks, farming methods became more intense, more woodlands were felled, and considerable emphasis was placed on grazing by sheep and cattle. One can pick out this period by the introduction of additional weed species in the peat record. In more recent times there has been a further change to arable farming with the upper layers of peat containing pollen grains from cereal crops and another set of weed species associated with arable farming.

Finally, the last five decades have given rise to further changes with a return to a more pastoral type of farming with more fields devoted to grass, while the surrounding hills have been extensively planted with conifers and once again we can find pine pollen in the top peat horizon. These recent changes have been brought about mainly as a result of a slight deterioration in the climate which has become slightly cooler and wetter over the last 20 years. This has made it difficult for farmers to obtain a good return from cereal crops both in the Tregaron area and in many other parts of western Wales away from the more productive coastal strip. A more detailed account of these changes can be found in papers by Smith (1970), Turner (1965, 1970) and Walker (1982).

Changes in soil chemistry and structure.

The retreating ice sheets left behind them extensive areas of nutrient rich detritus which eventually formed soils containing high levels of chemicals required for good plant growth. Many of the plants which colonized these soils were those which one normally associates with chalk and limestone. From about 10,000 to 7,000 years ago a large part of Wales was probably covered by species-rich alpine meadows of the type that we can see today adjacent to glaciers in Norway. There are still remnants of arctic-alpine meadow vegetation in the more base-rich parts of Cadair Idris and a few mountain slopes in Snowdonia and the Brecon Beacons, but the lower sites are rapidly disappearing as the older methods of hill farming are being replaced by more modern systems which are less labour intensive.

The herb-rich meadows first began to disappear as far back as 7,000 years ago for two main reasons: firstly they were invaded by tree species which soon began to form extensive tracts of woodland; and secondly, those species requiring base-rich soils, or which are unable to compete with species which grow best on neutral to acidic soils, began to decline and were replaced by more invasive species. All these changes were the result of considerable changes in the composition and structure of the soils brought about by the leaching out of nutrients, especially in the high rainfall areas. There was also a loss of nutrients as a result of their becoming temporarily or permanently locked up in the organic and peat deposits. These processes have resulted in a change from a freely drained, crumbly soil full of earthworms and other soil animals, to one in which stratification began to occur to form the characteristic podsol with its associated iron-pan some 20 to 35 cm below the surface. This latter type of soil profile has become widespread over most of upland Wales and has resulted in a species-poor series of plant associations. It is only by deep ploughing to below the iron pan that the Forestry Commission can successfully plant their *Picea sitchensis* (Sitka Spruce) forests in the more exposed hill sites. Without breaking up the iron pan these trees would have a very shallow rooting zone in a nutrient poor soil: a situation which leads to poor growth and a high chance of windblow as the trees age.

Recent changes in soil acidity.

An examination of rain in areas well away from the effects of industrial pollution will show that it is slightly acidic (pH from 5.5 to 5.8) and contains few chemicals present in more than minimal amounts. In areas close to the sea there are usually small quantities of calcium, magnesium and sodium derived from wind blown sea spray which can have beneficial effects on plants and animals. This slightly acidic rainfall would have been typical of that experienced over the whole of Wales until about four hundred years ago and we can still obtain samples of this 'fossil' rain in the present glaciers of Greenland and Iceland which contain a good record of past events, since the different horizons can be dated by radiocarbon techniques.

The composition of rain water began to change with the start of the Industrial Revolution and there has been an even more significant series of changes since the early 1960's. Before 1960, most chemical pollution was localized in or near industrial centres; but in recent years we have found that the effects of chemical pollution from power stations, oil refineries, and the like can travel for many hundreds of miles downwind from the tops of high-rise chimneys. Much of the pollution in southern England and South Wales now comes from the Continent, especially in high summer when it drifts northwards in the light southerly winds. Part of this pollution has been shown to come from internal combustion engines. An investigation of daily rain in West Wales over the last few years has shown that it can be strongly acidic with pH values often below 4.0, although not so acidic as in some parts of the British Isles where values of less than pH 3.0 have been recorded. The main cause of these low pH values is in the increasing amount of sulphur and nitrogenous compounds produced as part of industrial manufacturing processes including the burning of products such as oils and coal which are particularly rich in sulphur.

What is the effect of this increasingly acidic rainfall on the distribution of plant species in Wales? It can seriously damage their growth as seen in the case of lichens which have become almost extinct in industrial areas. Flowering plants are also affected and productivity of grasslands has been reduced in many areas. The main effects are on the germination of seeds and the development of the root system, although the green pigment, chlorophyll, contained in the chloroplasts of the leaves can also be damaged by high levels of sulphur. One of the main problems for plants growing in upland areas is that as the soils become more acidic, more aluminium and aluminium hydroxide ions come into solution. These ions act on seed germination and root growth in the same way as ions of heavy metals such as lead, copper and zinc inhibit plant growth. There are few species that can tolerate high levels of these heavy metal ions and those which have the ability to survive have evolved specialized ecotypes in which a number of the normal biochemical processes have become modified (Bradshaw and McNeilly, 1981). This has been shown to be so in the case of several species such as *Agrostis capillaris* (Common Bent), *Anthoxanthum odoratum* (Sweet Vernal-grass), and *Silene vulgaris* ssp. *maritima* (Sea Campion), which are widespread on old heavy metal sites scattered throughout Wales. These specialized forms tend to be less productive, and have a lowered competitive ability, when compared to more normal forms of the same species.

The effects of the acidic rainfall are not too serious in areas where chalk and limestone occur since the calcium in these rocks can, to some extent, neutralize the acidic products in the rain water; but this feature (the buffering capacity of the soil) is no longer present in most upland soils which, by now, have lost most of their buffering capacity. Some of these upland soils may on analysis be shown to have reasonable amounts of calcium, but this is usually locked up in organic matter which is not breaking down to release calcium and other nutrients required for successful plant growth. The net effect of these processes is for the gradual loss of some of the more interesting species in areas above 200m: they have been able to survive from the initial recolonization of the glacial soils until the start of the industrial revolution, but are now rapidly disappearing from the remaining natural and semi-natural habitats because the present conditions are unsuitable, or they are unable to compete with those species which have been able to adapt themselves to the newer conditions or which have migrated into Wales in more recent times.

The changing coastline of Wales.

Towards the end of the last glaciation, about 12,000 years ago, the coastline of West Wales extended westwards to about 1km east of the present 100m depth line (fig. 1) being separated from Ireland by a narrow strait. In northern England and Scotland there were land-bridges with Ireland at this time, but it appears unlikely that there was a direct land-bridge linking Wales with Ireland. The mean sea-level at this time was about 60m below the present-day coastline as a result of the considerable quantities of water locked up in the glaciers and ice sheets that covered Wales and other parts of northern Europe (Tooley, 1974).

The sea-level gradually rose as the ice melted and this process is still continuing with places in the South-east of England having to be protected by costly sea defence works against flooding during Spring Tides and storm surges. The rise in sea-level has become almost stable in Wales with no obvious increase over the last few centuries. The coastline, however, is still changing as the sea erodes it away, especially in areas where the sea comes into contact with glacial tills left from the last glaciation. The presence of an older coastline beyond the present one can often be deduced by the presence of remnants of submerged forest along various stretches of coastline, as at Borth, Cardiganshire, where one can still see large areas of peat with tree bases and trunks of Pine, Alder and Birch up to 150m from the present-day shingle beach.

The coastal fringe is a particularly important habitat for many of the rarer species since it is often difficult to cultivate, especially on the steeper slopes. As a result, we can usually find areas containing a reasonably natural community of grassland species which form a sub-climax, since grazing usually prevents the establishment of trees; although extensive areas of shrubland can be found along a few sections of coastal cliffs.

The present-day cliff vegetation will normally be found to contain a number of typical meadow species plus the recent addition of the more typical maritime species, which would obviously have been absent in former times when the coastline was further west. The main danger to some of these species is in places where erosion is taking place. This is resulting in the loss of the natural plant communities since, after a certain amount of erosion has taken place, the new coastline comes into contact with the highly cultivated land in which most, if not all, of the native species have been lost through regular ploughing and their inability to compete with the cultivated crops or to tolerate the chemical and other changes that the farmer has carried out to improve the productivity of his land.

Climatic factors affecting plant distribution.

There are numerous climatic factors which determine the distribution of plants and animals. For instance, the combination of factors which enables a colony of *Trichomanes speciosum* (Killarney Fern) to thrive and reproduce itself in Wales is totally different to that required by many of the annual species whose centre of distribution is in the Mediterranean and which occur at their northern limit along the coastal fringe of South Wales.

The more important climatic factors to consider in attempting to correlate the distribution of plants with past and present climate are —

a) temperature: both of the soil and air.
b) growing season: its length and intensity.
c) frost: frequency; intensity; and chance of very early and late killing frosts (minimum less than $-3°C$.)
d) rainfall: amount and seasonal distribution.
e) humidity: especially minimum values.
f) transpiration.
g) snow: cover and duration.
h) wind: degree of exposure; direction and strength.
i) sunshine: duration and intensity.

Just which of the above will be the most important in determining plant distribution will vary from species to species; one must also remember that it is the combination of several climatic factors that are interacting with one another which have to be considered in determining plant distribution. Furthermore, we must not place too much emphasis on temperatures recorded in Stevenson's Screens and rainfall totals recorded at climatological stations (which form the basis for most climatological maps). It is the climate immediately around the plant (including the soil climate around the roots) that is of more significance. For instance, the climate on a south-facing slope of a steep east-west valley is very different to that on the corresponding north-facing slope of the same valley; and the same applies when comparing the vegetation on mountain slopes and, to a lesser extent, on different slopes of a sand-dune. Some flowering plant species are tolerant of a wide range of climatic extremes, but others, including the majority of the lower plants such as mosses, liverworts and lichens, are much more sensitive to subtle variations in climatic factors. It should also be remembered that differences in climate will also affect the structure of the soil and the availability of soil nutrients.

A number of the more important climatic factors listed in the previous paragraph have recently been used to produce a Bioclimatic Map of England and Wales (Bendelow and Hartnup, 1980) which shows that Wales has a far more diverse series of bioclimatic zones than most of England south of the Mersey-Humber line. The part of this map covering Wales (frontispiece) is particularly important for those wishing to correlate the distribution of selected species to one or more of the climatic variables. The map has to be used with care since it incorporates a number of generalizations and, at the scale of reproduction, cannot take into account localized situations such as the differences between north- and south-facing slopes mentioned above.

The Bioclimatic Map of Wales

A full explanation of the various climatic factors and the mapping units used on the bioclimatic map for England and Wales can be found in Bendelow and Hartnup (1980). The purpose of this section is to briefly explain the nature of the mapping units on the Welsh map (frontispiece) and how they relate to the distribution of plants in Wales.

i) **Oceanic subsectors**

The concept of oceanicity is particularly important in studies on plant distribution in Wales. The sheltered west coast valleys, with their extremely oceanic climate, are particularly rich in the moisture dependent species of mosses, liverworts and lichens. The whole of the British Isles falls within the Oceanic Sector of Troll (1965). This Oceanic Sector can be further subdivided into four subsectors which are indicated on the map by the large numbers, 01, 02 and 03 (subsector 04 does not occur in Wales, being confined to the London area and East Midlands). These subsectors are separated by thick light-blue continuous lines. Table 4 summarizes the features of these four oceanic subsectors.

Table 4.

OCEANIC SUBSECTORS					
Subsector		Overall climate			
Zone	Description	Winter	Summer	Growing season	Oceanic Index
01	Hyperoceanic	mild ↑	cool ↑	prolonged, not intense ↑	more than 94
02	Euoceanic	↕	↕	↕	91 – 94
03	Hemioceanic	↕	↕	↕	88 – 91
04	Meioceanic	↓ cold	↓ warm	↓ short, intense	less than 88

The values for the Oceanic Index are based on a formula used by Conrad (1966) in which values for the index range from 100 (extremely oceanic, as in the Faeroe Islands) to 0 (extreme continentality, as in North-east Siberia). The main parameter in this formula is the mean annual range of air temperature.

The extreme west of Caernarvonshire, Merioneth and Anglesey in the north and Pembrokeshire and Cardiganshire in the south, together with parts of Cornwall and Westmorland occur in the Hyperoceanic zone or subsector. In these areas the winters are mild and the summers long and cool. The growing season (days with a soil temperature of 5.6°C. or greater) is from 320 to 365 days a year, but the growth intensity is low since the soil temperatures rarely exceed 16°C. at the height of the summer growing season. These are the areas which are particularly good for the more oceanic mosses, liverworts and lichens and where the best remaining wetland areas occur. Most of Wales is covered by zone 02 (Euoceanic) with slightly colder winters, warmer summers and a shorter growing season of between 240 and 320 days a year. It gradually becomes less humid as one moves eastwards across Wales, especially to the east of the Cambrian Mountains. Zone 03 (Hemioceanic), found along the Welsh/English border, has colder winters, but considerably hotter summers in which the soil temperatures rise to 21°C., thereby allowing farmers to grow cereals and arable crops with expectancy of high yields. It is in this zone that we find the western limit of distribution of many wild plants which have a continental distribution pattern.

ii) **Mapping units**

The colour coded mapping units are based on a combination of thermal and moisture qualities plus exposure categories. The units that occur on the map of Wales are summarised in the following Tables.

Table 5.

THERMAL QUALITIES		
Zone	Description	Accumulated temperature (day °C above 5.6)
A	Moderately cold	Less than 825
B	Slightly cold	825 – 1375
C	Moderately cool	1375 – 1650
D	Slightly cool	1650 – 1925
E	Moderately warm	More than 1925

Day degrees are the accumulated values, in degrees, when the mean temperature for a day is above 5.6°C. The base of 5.6°C. (42°F.) is based on the minimum threshold temperature at which grasses can start to grow.

Table 6.

MOISTURE QUALITIES		
Zone	Description	Moisture deficit (mm)
1	Moderately wet	Less than 40
2	Slightly wet	40 – 60
3	Moderately moist	60 – 100
4	Slightly moist	100 – 180
5	Slightly dry	More than 180

The values for moisture deficit are the calculated values for the average maximum potential soil moisture deficit; that is, actual rainfall minus potential transpiration.

Table 7.

EXPOSURE QUALITIES (based on the effect of heather and tree growth)			
Zone	Description	Approximate average annual windspeed (m/s)	Vegetation effect
m	Unexposed	Less than 4.8	Tree growth moderate to good
p	Exposed	4.8 – 6.6	Tree growth poor
v	Very exposed	More than 6.6	Heather very short, trees absent

Examination of the map will show that the colours are based on 18 different combinations of thermal and moisture qualities, the letter referring to the thermal zone and the number to the moisture zone. A third component, the exposure quality, is represented by the letters m, p, and v. In all this provides us with just over 40 bioclimatic units for Wales. What are the relationships between these units and the distribution of plants? A brief examination of the dependence of plants on a few of the more important climatic factors will show the basis for some relationships.

iii) **Thermal qualities**

All plant growth depends on a supply of radiant energy from the sun for germination, subsequent growth, flowering and successful ripening of fruits and seeds. In cultivated plants it has been shown that some species, such as spring wheat and peas, can germinate at about 4°C., but that potatoes will not grow in the soil unless the soil temperature is 7°C., and runner beans and sweet corn require a more continental temperature of 12°C. or higher for germination. Just how many seeds each plant will produce depends on the total amount of solar radiation that they receive during the summer. In studies on plant distribution it is far more satisfactory to use soil temperatures for mapping, but since there are few places in Wales at which these are measured, we have to use air temperatures recorded in Stevenson's Screens 1.5m above ground level at climatological stations. There are considerable variations in soil and air temperatures throughout Wales: the mean temperatures are up to 2°C. higher on the south coast as compared to the north coast of Wales and it is for this reason that thermal zone E, with accumulated temperature values in excess of 1925 degrees, occur in favoured coastal localities of South Wales. Even greater differences occur when one compares coastal and upland sites since the mean temperature falls 0.6°C. for every 100m rise in altitude: the peaks of Snowdonia being some 6°C. cooler than places such as Porthmadoc just a few miles away. It is for this reason that the growing season is so short in upland parts of West Wales above 300m: over most of the Berwyns and Cambrian Mountains the growing season is from early May to late September, whereas in the Dale Fort area of Pembrokeshire, growth occurs on every day of the year in most years, since it is rare for the soil temperature to fall below 5.6°C. in the extreme south-west of Wales: hence the reason why this is one of the most favoured areas for growing early potatoes.

The same type of responses to germination, growth and productivity can also be found in wild species. Furthermore their growth-form and competitive ability are also affected by thermal factors. Thus, most annual and biennial species are found below 300m in thermal zones C, D and E. A high proportion of the native species occurring in thermal zones A and B are perennials with high chromosome numbers.

iv) **Moisture qualities**

The values for the moisture deficits in Table 6 refer to the moisture status of the soil. It is extremely rare for soils in moisture zone 1 to dry out sufficiently to affect the growth of plants. It is in these areas that we find the greatest development of blanket bogs and wet flushes and valley bogs. The latter, providing the soil water is reasonably rich in soil nutrients such as calcium, can contain a high diversity of plants including some of the rarer species.

Very few parts of Wales are in moisture zone 4, where severe parching of the vegetation occurs during prolonged droughts as in 1976. Droughts in these areas are particularly severe. Areas of natural or semi-natural habitats on the shallower Powys-type soils, which the farmer tends to leave since they are difficult to convert into profitable farm soils, are most sensitive to prolonged droughts. A high proportion of the extreme southern and western coastal fringe occurs in moisture zone 3, but it is rare for soils to dry out sufficiently to really affect the distribution of most native species, although in the hotter summers the lack of soil moisture may restrict the numbers of fruits and seeds.

v) **Exposure qualities**

There are just three categories of exposure quality since it is rather difficult to quantify this factor on a detailed basis mainly because there are few climatological stations that maintain records of wind speed and direction. Strong winds can have a considerable effect on plants since they can uproot trees and break branches as well as causing considerable damage to leaves. Furthermore, strong westerly winds can carry considerable quantitites of salt-spray inland which can cause scorching of foliage as well as increasing the nutrient levels in the soil. For instance, in May 1973 salt-spray damage could be found on both conifers and deciduous trees in the Llangurig and Llanidloes areas of Montgomery some 25 to 30 miles inland from the coast. Frequent winds from the south, west and north maintain a high humidity in adjacent coastal regions; whereas winds from the east tend to be much drier, as well as often carrying harmful pollutants.

Another effect of wind, especially in zone v, is to reduce the size of plants, even those near the ground such as some of the heaths. More conspicuously, shrubs and trees can become lop-sided as growth on the windward side is damaged or suppressed, whereas growth on the leeward side is less stunted.

vi) **Other qualities**
Many other climatic factors, both singly but more usually interacting with one another, can have an influence on the distribution of our native species. The climate is continually changing and one has to consider both past and present climates in order to build up an hypothesis to determine which primary factors are responsible for a particular species or plant association being present in a given area. To-day, one also has to consider how other factors such as man's own influence in modifying the land by drainage and the use of fertilizers, together with the increasing level of atmospheric pollution, can further modify the distributional ranges of our wealth of interesting species.

Geographical elements/components

The two editions of *Welsh Flowering Plants* contained a section on the geographical classification of plants using the system of Watson (1847-59 and 1883) and Matthews (1937, 1955). The appropriate element of both classifications was also given in the systematic list for all those species which were regarded as native in Wales. In this book this information has been confined to Appendix III and the Index. The maps should readily convey the distribution patterns within Wales, although in order to fully appreciate the Welsh distribution, one should be able to compare it with distribution maps for the whole of Europe. Information for Europe is becoming available with the publication of *Atlas Florae Europaea,* Ed. Jalas and Suominen (1972-80), but it will take many years before the full set of maps is published. Maps of a few species have been published in *The Biological Flora of the British Isles* which appears from time to time in *The Journal of Ecology.* The maps in Fitter (1978) for north-west Europe are useful, but lacking in the kind of information one requires for a detailed study.

The elements of Watson (1883) are too generalized since he used just six basic types (British, English, Scottish, Highland, Germanic, and Atlantic) and his original lists are in need of considerable revision in the light of information obtained during the last 100 years. The 16 elements of Matthews (1955) are more useful and still convenient to use in discussions of both past and present distribution patterns in Wales. However, to fully understand why a species occurs in a certain locality one has to obtain information on the type of rock, soil and other environmental parameters of where it is growing; to investigate its methods of reproduction; to examine how it fits into the plant community and whether it is in a stable or transitional state within that community. All areas are in the process of continual change and what used to be an abundant species ten or twenty years ago may have been replaced by another species within the habitat, or by a species coming into the community from outside. In the first instance the change may have been the result of environmental and biotic changes within the existing plant community, whereas in the second case the change may be the result of the new species having recently migrated into the area and once it has come into the community then being able to rapidly increase its numbers because it can replace one or more of the existing species by its greater competitive efficiency.

Two examples illustrate these points. Both *Ammophila arenaria* (Marram), occurring on the younger sand-dunes, and *Spartina* spp. (Cord-grass), a pioneering genus of the lower salt-marshes, are highly successful species, but for only a limited period in any one place. Marram is particularly successful in being the primary colonizer of the embryo and yellow-dunes and in these sites it produces numerous flowers and seeds for up to ten years, but eventually it ceases to reproduce itself sexually. During this period it modifies the surrounding soil environment, making it less suitable for itself, but more suitable for other species such as the small winter annuals and certain mosses and lichens. In the case of *Spartina* we find that it replaces the *Salicornia* (Glasswort) that used to be present and for a time becomes completely dominant and invades hostile regions that were previously not occupied by any other flowering-plants, with the possible exception of *Zostera* (Eel-grass) which has become very rare in Wales in recent years. Like *Ammophila,*

Spartina flowers and fruits for a number of years, but once the original colonies have coalesced, its powers to reproduce itself sexually are reduced. Other species begin to enter the system as the older stems of *Spartina* help to build up the level of the land by the accumulation of mud, silts and plant debris between the stems.

With the fairly rapid reduction in the number of hectares of natural and semi-natural vegetation in Wales, the process of local extinction will continue, especially for those species which are most sensitive to the effects of environmental change. The species in greatest danger of extinction are those which :

a. require special pollination systems involving just one or a few species of pollinating insects;
b. require a certain number of plants in a population to allow for sufficient replacement of the population from new seedlings;
c. because of their rarity are sought after by collectors;
d. are sensitive to the influence of more competitive species which come into Wales as deliberate introductions or on their own accord.

A detailed analysis of the distribution of Welsh species in relation to bioclimatic units is beyond the scope of this volume, although it is hoped that an opportunity will become available to present such an analysis elsewhere. All that is possible in the remaining part of this chapter is to provide a few examples of distribution in relation to Matthews' elements and the recently defined bioclimatic units. All the examples mentioned in the following paragraphs are represented in the Maps Section of this book.

Matthews' **Mediterranean** element is represented in Wales by 19 species whose chief centre of distribution is the Mediterranean basin with extension to the Orient. In Wales they occur mainly in bioclimatic units D3, D4, and E3, as can be seen from the maps of *Lavatera arborea* (Tree-mallow, p.102, map 421), *Linum bienne* (Pale Flax, p.99, map 394) and *Salvia verbenaca* (Wild Clary, p.128, map 613). These contrast to the **Northern-Montane** species which are found in bioclimatic units A1, B1, and B2. Sixteen of these are found in Wales and are characteristic of mountain habitats in northern and central Europe. Examples are *Festuca vivipara* (Viviparous Fescue, p.172, map 864), *Listera cordata* (Lesser Twayblade, p.190, map 1010), *Rubus saxatilis* (Stone Bramble, p.76, map 255) and *Trollius europaeus* (Globe-flower, p.61, map 122).

Species of the **Arctic-Subarctic** element have an even more restricted distribution in Wales being primarily confined to bioclimatic unit A1. These are species which have an exclusively northern distribution. Six species are found in Wales including: *Deschampsia cespitosa* subsp. *alpina* (Alpine Hair-grass, p.178), *Mertensia maritima* (Oysterplant, p.122), *Rubus chamaemorus* (Cloudberry, p.76) and *Saxifraga cespitosa* (Tufted Saxifrage, p.74).

Alpine element species are those which occur in the mountains of central Europe, but which are absent from northern regions. Ten species are found in Wales, examples are: *Arabis scabra* (Bristol Rock-cress, p.68), *Draba aizoides* (Yellow Whitlowgrass, p.69), *Saxifraga hypnoides* (Mossy Saxifrage, p.74, map 242) and *Thlaspi alpestre* (Alpine Penny-cress, p.70), which occur in both lowland and upland areas of Wales and their distribution cannot readily be related to any specific bioclimatic units, although they all occur in very well defined, restricted areas.

The **Arctic-Alpine** element is represented by 26 species in Wales. Overall, they are found mainly in northern and upland areas above the tree line as well as on the mountain ranges of central and southern Europe. Many of them are restricted to bioclimatic unit A1 in Snowdonia; but a few, such as *Empetrum nigrum* (Crowberry, p.115, map 509), *Saxifraga stellaris* (Starry Saxifrage, p.74, map 239), *Rhodiola rosea* (Roseroot, p.74, map 238) and *Vaccinium vitis-idaea* (Cowberry, p.115, map 507), have a greater range extending southwards throughout Wales in units A1, A3, B1, B2 and B3.

In contrast to the above we have several elements whose species occur mainly in coastal areas. *Scilla verna* (Spring Squill, p.166, map 828) is a good example from the **Oceanic West European** element being confined to the Hyperoceanic subsector of Wales where the soil is reasonably nutrient rich. Other examples of this element include *Erodium cicutarium* subsp. *bipinnatum* (Sticky Stork's-bill, p.98, map 392) and *E. maritimum* (Sea Stork's-bill, p.98, map 390), which are also found in the more humid areas of bioclimatic units D3, D4 and E4. There are 96 species of this element in Wales, the overall pattern of distribution being exclusively the western coastal fringe of Europe in areas with a typical 'Atlantic' climate.

As a final example we have the **Continental** element with species characteristic of central Europe and Asia, but thinning out westwards. Of the 56 species of this element in Wales, *Dipsacus pilosus* (Small Teasel, p.138, map 692) and *Silaum silaus* (Pepper-saxifrage, p.112, map 488) are representative examples being restricted to the more continental climates of the Hemioceanic (bioclimatic units C4, D4 and E4).

Readers wishing to explore these aspects of plant distribution in Wales should refer to Matthews (1937, 1955) and Hyde & Wade (1957) for additional background information.

References

Bendelow, V. C. & Hartnup, R. (1980). *Climatic Classification of England and Wales.* Soil Survey Technical Monograph, No. 15. Harpenden.

Bowen, D. Q. (1973). The Pleistocene history of Wales and the borderland. *Geological Journal,* **8**: 207-224.

Bradshaw, A. D. & McNeilly, T. (1981). *Evolution and Pollution.* Studies in Biology No. 130. London.

Conrad, V. (1946). Usual formulas of continentality and their limits of validity. *Trans. Am. Geophys. Un.,* **27**: 663-4.

Coope, G. R. (1977). Quaternary Coleoptera as aids in the interpretation of environmental history. In *British Quaternary Studies : recent advances.* (ed. Shotton, F. W.). Oxford.

Druce, G. C. (1932). *Comital Flora of the British Isles.* Arbroath.

Fitter, A. (1978). *An Atlas of the Wild Flowers of Britain and Northern Europe.* London.

Godwin, H. (1956). *History of the British Flora.* Cambridge.

Hibbert, F. A. & Switsur V. R. (1976). Radiocarbon dating of Flandrian pollen zones in Wales and northern England. *New Phytologist,* **77**: 783-807.

Huntley, B. & Birks, H. J. B. (1982). *An Atlas of Past and Present Pollen Maps for Europe : 0-13,000 Years Ago.* Cambridge.

Hyde, H. A. & Wade, A. E. (1957). *Welsh Flowering Plants,* 2nd Ed. Cardiff.

Jalas J. & Suominen, J. (1972-1980). *Atlas Flora Europaea,* **1-5,** Helsinki.

Lowe, S. (1981). Radiocarbon dating and stratigraphic resolution in Welsh Late-glacial chronology. *Nature,* **293**: 210-212.

Matthews, J. R. (1937). Geographical relationships of the British Flora. *Journal of Ecology,* **25**: 1-90.

— (1955). *Origin and distribution of the British Flora.* London.

Moore, P. D. & Chater, E. H. (1969). Studies in the vegetational history of mid-Wales. I. The post-glacial period in Cardiganshire. *New Phytologist,* **68**: 183-196.

Perring, F. & Walters, S. M. (1962). *Atlas of the British Flora.* London.

Smith, A. G. (1970). The influence on Mesolithic and Neolithic man on British vegetation: a discussion. In *Studies in the Vegetational History of the British Isles.* (eds. Walker, D. & West, R. G.). Cambridge.

Tooley, M. J. (1974). Sea-level changes during the last 9,000 years in North-west England. *Geographical Journal,* **140**: 18-42.

Troll, C. (1965). Seasonal climates of the Earth. In *World maps of Climatology.* (eds. Rodenwaltd, F. & Jusatz, H. J.).

Turner, J. (1965). The *Tilia* decline: an anthropogenic interpretation. *New Phytologist,* **61**: 328-341.

— (1970). Post-Neolithic disturbance of British Vegetation, *Studies in the Vegetational History of the British Isles.* (eds. Walker, D. & West, R. G.). Cambridge.

Walker, D. & West, R. G. (eds.) (1970). *Studies in the Vegetational History of the British Isles.* Cambridge.

Walker, M. C. J. (1980). Late Glacial history of the Brecon Beacons, South Wales. *Nature,* **287**: 133-135.

— (1982). The Late-glacial and early Flandrian deposits at Traeth Mawr, Brecon Beacons, South Wales. *New Phytologist,* **90**: 177-194.

— (1982). Early- and mid-Flandrian environmental history of the Brecons Beacons, South Wales, *New Phytologist,* **91**: 147-165.

Watson, H. C. (1847-59). *Cybele Britannica.* London.

— (1883). *Topographical Botany.* Ed. 2. London.

Plan of the Book

When a new edition of *Welsh Flowering Plants* became necessary the opportunity was taken to enlarge considerably the scope of the book not only by increasing the number of species to be included but also by giving the 10 km square distribution for each.

It was thought desirable to include all taxa down to the rank of subspecies that had been recorded apparently growing wild in Wales, and also those forest trees planted by the Forestry Commission which had become a conspicuous part of our flora.

With our increasing knowledge of the large critical genera; *Rubus, Taraxacum* and *Hieracium,* it also became feasible to include all the micro-species belonging to these genera that had been recorded from Wales. The publication in 1975 of *Hybridization and the Flora of the British Isles* edited by C. A. Stace made it practicable to include all recorded hybrids. Since the identity of many of the varieties included in the previous editions is now in doubt and likely to remain so until the proposed *Flora of Great Britain and Ireland* is published, most varieties have been omitted from the present work but all subspecies known to occur are included.

Once the decision to include all taxa not obviously planted was made, it became clear that a large number of old records of casuals would be involved, especially from Glamorgan. However it was considered better to include details of all published records that could be traced, even if only recorded once 50 or more years ago, rather than to try to discriminate between them. Searches for new records were made through all the relevant Floras, vice-county lists and botanical journals. Lists of county records were sent to all B.S.B.I. recorders and other interested botanists in Wales for their comments and additions.

The maps published in the *Atlas of the British Flora* were used as a starting point in the preparation of the Welsh maps; they were updated by adding records from herbaria, especially the Welsh National Herbarium, through which a complete search was made, from all Floras, lists, journals and from various other sources many of which are listed in Appendix II. All records that were to hand at the end of 1981 are included together with a few 1982 records.

One of the main problems encountered was the lack of any comprehensive gazeteer to Welsh place names, this made it difficult to pinpoint on a map many of the more obscure localities. This was overcome by compiling a list of Welsh place names for each of the Welsh vice-counties with a 1 km square map reference for every name on 1 inch Ordnance Survey Maps of Wales. Copies of these lists may still be available on request.

The updated maps eventually prepared were then circulated to all B.S.B.I. recorders in Wales for checking. This resulted in many doubtful or erroneous records being deleted and new records added.

Arrangement of the Catalogue

The sequence of families, genera and species is that of *Flora Europaea* edited by T. G. Tutin *et. al.* and published in five volumes between 1964 and 1980.

The entry for each species follows the same general pattern : **Latin name; Welsh name; English name; Status; Habitats; Months of flowering; Vice-county distribution; Frequency, and 10 km square distribution.** These are explained in detail below

Latin name : The name adopted here is that given in *Flora Europaea* unless more recent research has shown this to be wrong. Synonyms are only given where the currently accepted name differs from that used for the same taxon in : *Flora Europaea*; *Welsh Flowering Plants*. ed. 2; J. E. Dandy's *List of British Vascular Plants* (1957); Clapham, Tutin and Warburg's *Flora of the British Isles*, ed. 2. (1962) or *The Excursion Flora of the British Isles*, ed. 3. (1981) by the same authors. In the case of species not mentioned in any of these works the name under which the original record was published is always given either as the currently accepted name or as a synonym.

Welsh name : Only one Welsh name is given for each species and has been taken from one of a number of sources. *Enwau Planhigion* published by the University of Wales Press in 1969 sets out to be a recommended list of Welsh names, it is fairly comprehensive and gives a Welsh name for most common species. *Casgliad O Enwau Blodau, Llysiau a Choed* by M. Parry published by the University of Wales Press, 1969, is a more comprehensive work and one or several names are given for each species; a name given there is used if absent from *Enwau Planhigion*. *Welsh Botanology* by H. Davies (1813) and the *Flora of Anglesey and Caernarvonshire* by J. E. Griffith (1895) are other sources of names, as is a manuscript list by Mr. A. Jones which forms the basis of a book planned by the National Museum of Wales on the Welsh names of plants.

English name : Only one English name is given for each species. Names of native plants are taken from *English Names of Wild Flowers* by J. G. Dony, F. H. Perring and C. M. Rob, reprint with corrections published by the B.S.B.I., 1980. Names of some hybrids have been taken from *Hybridization and the Flora of the British Isles* ed. C. A. Stace, 1975. Names of many cultivated plants have been taken from The Reader's Digest *Encyclopaedia of Garden Plants and Flowers*, 1971; the *Pocket Guide to Wild Flowers* by D. M. McClintock and R. S. Fitter, 1956; or its Supplement by D. McClintock, 1957. Other names have been taken from various sources.

Status : The different categories of status here employed are essentially the same as used in the last edition but doubtful natives are now divided into two categories and casuals have been added for the first time. The system followed is adapted from that given in *The recent influx of aliens into the British Flora* by J. E. Lousley in *The Changing Flora of Britain* ed. J. E. Lousley, B.S.B.I. Conference Report, 1953, and is set out below.

NATIVES : Species believed to have been in Wales before man, or to have immigrated without his aid by natural means of dispersal or to have arisen naturally here, e.g. *Betula pubescens* (Downy Birch); *Pyrola rotundifolia* (Round-leaved Wintergreen) and *Senecio cambrensis* (Welsh Groundsel). Hybrids which are believed to have arisen naturally in Wales, even if one or both of the parents are known to be introduced, are termed SPONTANEOUS rather than natives.

DOUBTFUL NATIVES : Species with a long history in Britain but which are suspected of having been introduced by human agency.

DENIZENS : Species growing in natural or semi-natural communities and not dependent for their persistence on human disturbance of the habitat, e.g. *Draba aizoides* (Yellow Whitlowgrass) and *Allium ampeloprasum* (Wild Leek).

COLONISTS : Species which grow only in habitats created and maintained by human activities. These are mainly weeds of cultivation restricted to arable fields and disturbed ground, e.g. *Sinapis arvensis* (Charlock) and *Lamium album* (White Dead-nettle).

ALIENS : Species believed to have been introduced by the intentional or unintentional agency of man.

NATURALIZED ALIENS : Introduced species which are naturalized in natural or semi-natural habitats, e.g. *Epilobium brunnescens* (New Zealand Willowherb) and *Inula helenium* (Elecampane).

ESTABLISHED ALIENS : Introduced species which are established only in man-made habitats, e.g. *Senecio squalidus* (Oxford Ragwort) and *Avena fatua* (Wild-Oat).

CASUALS : Introduced species which do not persist, i.e. are not naturalized or established, e.g. *Lycopersicum esculentum* (Tomato) and *Setaria viridis* (Green Bristle-grass). Some introduced annuals which persist for many years in the same localities either by setting seed or by repeated introduction are classed as Established aliens rather than casuals, e.g. *Lunaria annua* (Honesty) and *Phalaris canariensis* (Canary-grass).

INTRODUCED : This category is used to indicate species known to have been introduced but about which insufficient is known of their persistence for them to be classified further, e.g. *Zantedeschia aethiopica* (Arum-lily) and *Brunnera macrophylla* (Greater Forget-me-not).

For all non-native species the nearest country or area in which that species is known to be native is given in brackets.

When the species or hybrid is known, or believed, to have arisen in cultivation this is indicated by-(garden origin), or for crop plants such as cereals-(cultivated origin), e.g. *Atriplex hortensis* (Garden Orache), *Viola × wittrockiana* (Garden Pansy) and *Avena sativa* (Cultivated Oat).

Habitats : Some species are confined in nature to a particular kind of situation, bog plants for example are found only in wet peaty ground, and salt-marsh plants only in a soil which contains a quantity of common salt. Many plants, however, are able to grow in a variety of situations; woodland species often flourish in the shade of hedgerows, and plants of rocky places are frequently found on gravelly ground or on wall tops. In this book the habitat in which each species is most commonly found is given, followed by a list of many of the habitats (in abbreviated form) in which it has been recorded in Wales.

The abbreviations used are as follows :

Al — Lakes, ponds, still water
Ar — Rivers, streams, flowing water
D — Sand dunes
Ds — Dune slacks
E — Marginal habitats, hedgerows and scrub
F — Farmland
G — Grassland
H — Heath and bog
M — Marshes and wetland
Q — Quarries, walls, rockfaces etc.
R — Ruderals, roadsides, waste ground etc.
Rg — Garden weeds
Rw — Railways
S — Seasides, maritime habitats
Sc — Sea cliffs
W — Woodland

Months of flowering : This information was compiled mainly from herbarium sheets in the Welsh National Herbarium. It is presented in numerical form with 1 representing January and 12, December. Figures in brackets preceding or following the main entry indicate exceptional dates.

Vice-county distribution : Records of plant distribution throughout Great Britain have been kept for many years on the vice-county basis devised by H. C. Watson. He divided the whole country for botanical purposes into 112 divisions, which often coincided exactly or approximately with the political counties, though some large counties had to be sub-divided to give areas of more or less uniform size. Watson called his divisions 'vice-counties' and numbered them serially from west Cornwall (v.c. 1) to Shetland (v.c. 112). The 'census number' of a species, i.e. the number of vice counties in which it occurred in Britain, was thus some measure of its rarity as a British plant.

In Wales the vice-county boundaries follow closely the political boundaries of the 13 Welsh counties as they were before the Local Government reorganization of 1974. No account has been taken of the new political boundaries into which Wales has been divided and apart from being shown on map 2 they are not considered further. The first of the accompanying maps (Map 2) shows the old county boundaries, the new county boundaries and vice-county boundaries. Circles indicate the places where the old county and vice-county boundaries differ. The second (Map 3) shows the vice county boundaries and 10 km squares.

In the paragraph relating to each species the inclusion of the name of a vice-county (in abbreviated form) indicates that the species has been recorded from that vice-county. Some species, while undoubtedly native in certain vice-counties, have been introduced by human agency into others; in such cases the name of the

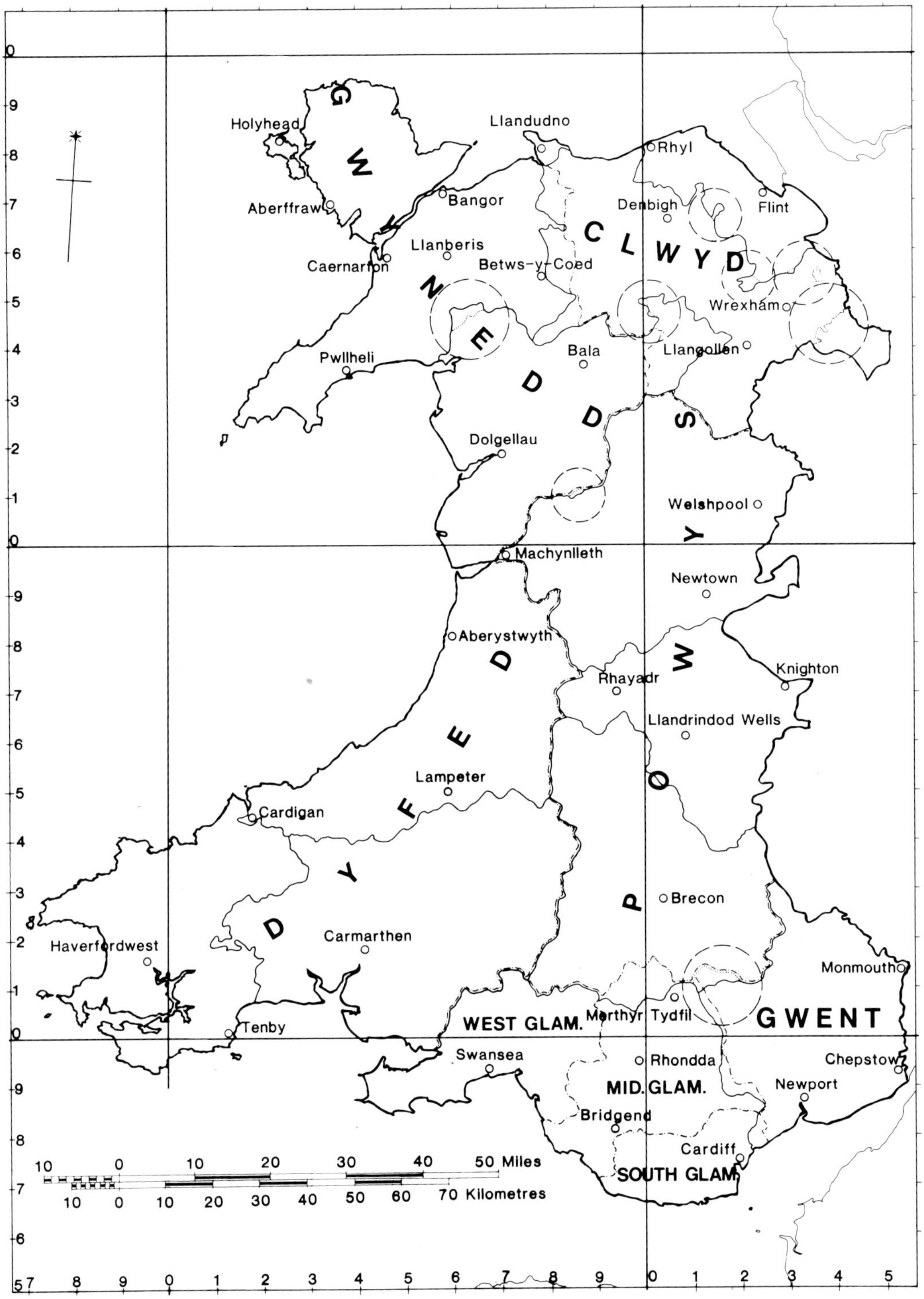

Map 2. Showing old county boundaries (·······), new county boundaries (–·–·–·) and vice-county boundaries (——). The old county boundaries are only given where they differ from the vice-county boundaries (circled).

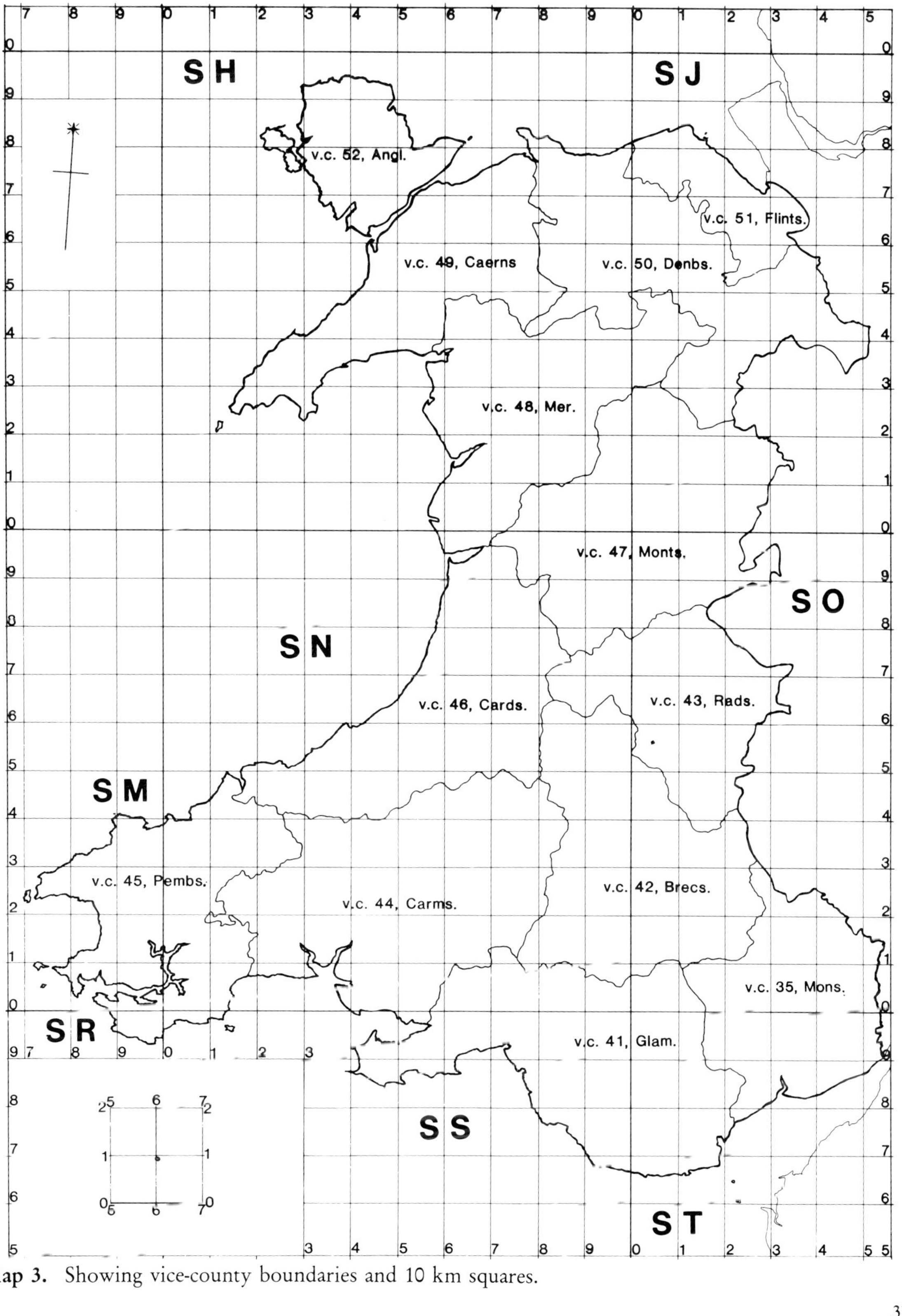

Map 3. Showing vice-county boundaries and 10 km squares.

vice-county is placed in brackets. Roman type indicates that a specimen from the vice-county concerned is in the Welsh National Herbarium, italics show the absence of such a specimen. The names of the vice-counties are arranged in ascending order of their vice-county numbers : Mons. (35, Monmouthshire); Glam. (41, Glamorgan); Brecs. (42, Breconshire); Rads. (43, Radnorshire); Carms. (44, Carmarthenshire); Pembs. (45, Pembrokeshire); Cards. (46, Cardiganshire); Monts. (47, Montgomeryshire); Mer. (48, Merioneth); Caerns. (49, Caernarvonshire); Denbs. (50, Denbighshire); Flints. (51, Flintshire); Angl. (52, Anglesey).

The vice-county distribution given in the previous editions was based upon the records published in the 2nd edition of H. C. Watson's *Topographical Botany* and its two supplements published in the *Journal of Botany* for 1906 and 1929 respectively, together with a considerable number of additional records from many other sources. Vice-county records which have accumulated since the publication of the 2nd edition until the end of 1981 are indicated by an asterisk. Records which are known to be erroneous are enclosed in square brackets.

When a species has been recorded from a county but no locality has been traced, a reference is given to the source of the record in the form of a superscript figure after the county (see p. 39). No attempt has been made to work out the 10 km square distribution of the widely planted forest trees (mainly conifers); for these species a list is provided of vice-counties and references only.

Frequency : Some indication of the frequency of a species within Wales is provided by the number of 10 km squares in which it has been recorded. There are 282 squares in Wales, including some coastal squares which contain only a small amount of land; records from these squares are mapped or listed separately, they are not included with the records from neighbouring squares as was the case in *The Atlas of the British Flora.* The only exception is the Island of Ramsey in Pembs.; part of Ramsey falls in 10 km square **SM**62 and the remainder in **SM**72 but, as no part of the mainland occurs in **SM**62, it was thought convenient to place all Ramsey records in **SM**62.

10 km square distribution : It has long been recognized that the Watsonian method of recording by vice-counties is very imperfect; they are irregularly shaped, few in number, large in size and no account is taken of frequency within the limits of a vice-county. The fact that a given species has been recorded from a particular vice-county may mean that it is common all over it or only that it is known to exist there as a single plant. The adoption by the Ordnance Survey of a National Grid, and its inclusion on their maps from about 1946, provided an opportunity for recording and mapping British plants more precisely and on a uniform plan. The Botanical Society of the British Isles initiated, in 1954, a survey aimed at recording the presence of each species of vascular plant in those 10 km squares of the British Isles in which it occurs. This culminated with the publication, in 1962, of *The Atlas of the British Flora.* As has been mentioned previously (p. 33) these maps were used as a basis for the 10 km square distribution given here. In the *Atlas* pre- and post-1930 records were separated by using different symbols; the same system is adopted here. With the benefit of hindsight there is little doubt that the use of pre- and post-1950 records would have been more appropriate for this book, or even the use of three classes, pre-1930, 1930-1950 and post 1950; with three symbols on the maps. Unfortunately these schemes were thought of too late for the maps to be changed.

The 10 km square distribution is given in the following ways :-
a) for species and subspecies, other than casuals or the microspecies of *Rubus* and *Taraxacum,* recorded in fewer than 10 squares full details, as set out below, are provided for each locality (there may be several localities in one 10 km square, in which case all are treated separately). The entries are listed in vice-county order (see top of page), and 10 km square order within each vice-county (see p. 39).

Locality : The place-name given is usually that of the nearest feature named on the appropriate 1:50,000 O.S. map, but in the case of very rare plants it has been made deliberately vague or omitted altogether. The correct Welsh spelling of place-names is only used when it is unlikely to cause confusion; in all other cases the anglicized spelling is used.

10 km square : For very rare species the correct square is sometimes disguised by listing instead one of the eight adjacent grid squares, or is omitted altogether.

Recorder's initials : A list of these is given in Appendix I (p. 281).

Date : When only an approximate date is known this is indicated by c. (circa). In some cases all that is known of the date of a record is that it was made before or after a certain year; this is indicated by using the prefix pre- and suffix + respectively.

Reference : This takes the form of a superscript number following the date. A list of references is given in Appendix II (p.285).

A complete entry enclosed in round brackets indicates that the species is known to have been introduced in that locality. This is only used when the species is native elsewhere in Wales.

b) For microspecies of *Rubus* recorded in fewer than 10 squares only the 10 km squares are given.

c) For casuals and microspecies of *Taraxacum* recorded in fewer than 10 squares only the 10 km squares and references are given. A square number in italics indicates that the species has not been recorded in that square since before 1930.

d) For species recorded in from 10 to 19 squares, a list of 10 km squares is given. A square number in brackets indicates that the species is known to have been introduced in that square (only used when the species is native elsewhere in Wales). A number in italics indicates a pre-1930 record. There are a few species in this category with records from 20 or more squares; these are listed not mapped because the records which brought the total to over 20 were received after completion of the other maps.

e) Species recorded in from 20 to 250 squares are mapped. On all the maps an open circle (○) indicates a pre-1930 record; a solid dot (●) indicates a post-1930 record and a plus sign (+) indicates an introduction (only used when the species is native elsewhere in Wales).

f) For species recorded in 251 or more squares, a list is provided of the squares from which it has **not** been recorded.

The 10 km squares are always listed in the same order. This is based first on the 100 km square letters starting from the south-west corner of Wales and running west to east and south to north : **SR; SS; ST; SM; SN; SO; SH; SJ.** Within each 100 km square the records are listed in numerical order (see Map 3).

Summary Explanations of the Typography and Abbreviations used in the Main Text.

1. Name of species (see also p. 33).
   Bold type : accepted latin name.
   Italics : synonyms.
   Smaller type is used for all casuals and errors.
   A cross (×) indicates a hybrid.
   Entry enclosed in square brackets [ ] indicates an error.
2. Status (see also p.34)
   The country or region enclosed in brackets following the status indicates the nearest area where the species is native.
3. Habitats (see also p.35)
   The main or typical habitat in Wales is given first, followed by a list of abbreviations enclosed in brackets that indicate the other habitats from which this species has been recorded in Wales. See p. 35 for key to abbreviations.
4. Months of Flowering (see also p.35)
   The months of the year are indicated by numbers; thus Fl.6-8(-10) indicates flowers June to August and occasionally into October.
5. Vice-county distribution (see also p.35)
   Name of vice-county (abbreviated) in roman type : specimens in **NMW.**
   Name of vice-county in italics : species recorded but no specimen in **NMW.**
   Asterisk (*) after the name of the vice-county : record for vice-county published since the appearance of 2nd edition of *Welsh Flowering Plants.*
   A dagger (†) after the name of a vice-county : species believed to be extinct in that vice-county.
   Name of vice-county in round brackets ( ) : species regarded as introduced in that vice-county (only used when the species is native in at least one Welsh vice-county).
   Name of vice-county in square brackets [ ] : error.
   Superscript number (^{00}) after the name of a vice-county : species recorded from that county but without locality; number indicates reference in Appendix II.
6. 10 km square distribution (see also p. 38).
   10 km square or complete locality entry in round brackets : species regarded as introduced in that square or locality (only used when the species is native in at least one Welsh vice-county).
   10 km square record in italics: pre-1930 record.
   Recorders' initials are explained in Appendix I.
   Superscript number (^{00}) after the date or 10 km square : reference to record, see Appendix II.
   A dagger (†) following a record : species believed to be extinct in that locality.

The following abbreviations are employed.

btwn	=	between
c.	=	circa (near or about)
E	=	east
Fl.	=	flowers
km	=	kilometre
m	=	metre
N	=	north
nr	=	near
S	=	south
W	=	west

Catalogue of Welsh Flowering Plants

GYMNOSPERMAE

PINACEAE

ABIES Miller

Abies procera Rehder : Noble Fir. Introduced (N. America), sometimes regenerating. Plantations (acid moorland). *Brecs.*[259], *Rads.*[120], *Carms.*[120], *Cards.*[259], *Monts.*[120], *Mer.*[120], *Caerns.*[87], *Denbs.*[120].

A. grandis (D. Don) Lindley : Giant Fir. Introduced (N. America). Plantations (deep, moist, well-drained soil). *Glam.*[120], *Brecs.*[120], *Rads.*[120], *Carms.*[120], *Cards.*[259], *Monts.*[120], *Caerns.*[87], *Denbs.*[120].

A. alba Miller : Ffynidwydden Arian : European Silver Fir. Introduced (Europe). Parks. *Mons.*[120], *Glam.*[120], *Brecs.*[120], *Rads.*[120], *Carms.*[120], *Monts.*[89], *Mer.*[89], *Caerns.*[87], *Denbs.*[120].

PSEUDOTSUGA Carrière

Pseudotsuga menziesii (Mirbel) Franco : Ffynidwydden Douglas : Douglas Fir. Introduced (N. America), sometimes regenerating. Plantations (well-drained, fertile and firm soil, below 1,200ft.). *Mons.*[88], *Glam.*[86], *Brecs.*[200], *Rads.*[120], *Carms.*[93], *Cards.*[89], *Monts.*[89], *Mer.*[89], *Caerns.*[87], *Denbs.*[120].

TSUGA (Antoine) Carrière

Tsuga heterophylla (Rafin.) Sarg. : Western Hemlock. Introduced (N. America), sometimes regenerating. Plantations (various sites). *Glam.*[120], *Brecs.*[259], *Rads.*[120], *Carms.*[120], Cards.[112], *Monts.*[89], *Caerns.*[87], *Denbs.*[120].

T. canadensis (L.) Carrière : Eastern Hemlock. Introduced (N. America). Parks. *Glam.*[120], *Monts.*[120], *Denbs.*[120].

PICEA A.Dietr.

Picea abies (L.) Karsten : Spriwsen Norwy : Norway Spruce. Introduced (Europe), sometimes regenerating. Plantations (moist, fertile soil). *Mons.*[88], *Glam.*[86], *Brecs.*[102], *Rads.*[93], *Carms.*[255], *Cards.*[89], *Monts.*[89], *Mer.*[89], *Caerns.*[87], *Denbs.*[120], *Flints.*[93].

P. glauca (Moench) Voss : White Spruce. Introduced (N. America). Shelterbelts. *Monts.*[120].

P. engelmannii Parry ex Engelm. : Arizona Spruce. Introduced (N. America). Shelterbelts. *Monts.*[120].

P. sitchensis (Bong.) Carrière : Spriwsen Sitka : Sitka Spruce. Introduced (N. America). Plantations (wet, acid soils). *Mons.*[120], *Glam.*[86], *Brecs.*[102], *Carms.*[120], *Cards.*[89], *Monts.*[89], *Mer.*[89], *Caerns.*[87].

P. omorika (Pančić) Purkyně : Serbian Spruce. Introduced (Yugoslavia). Plantations (mountain peat). *Rads.*[120], *Carms.*[201], *Monts.*[120], *Caerns.*[87].

LARIX Miller

Larix kaempferi (Lamb.) Carrière (*L. leptolepis* (Siebold & Zucc.) Endl.) : Llarwydden Japan : Japanese Larch. Introduced (Japan), sometimes regenerating. Plantations (moist soils). *Mons.*[88], *Glam.*[86], *Brecs.*[200], *Rads.*[120], *Carms.*[201], *Cards.*[89], *Monts.*[89], *Mer.*[89], *Caerns.*[87].

L. decidua Miller : Llarwydden Ewrop : European Larch. Introduced (Europe), occasionally naturalized. Plantations (moist, well-drained, fertile soils). *Mons.*, Glam., *Brecs.*, *Rads.*, *Carms.*, *Pembs.*, *Cards.*, *Monts.*, *Mer.*, *Caerns.*, *Denbs.*, *Flints.*, *Angl.*[6]. 72 squares, map 1.

L. decidua × kaempferi = L. × henryana Rehder (*L.* × *eurolepis* A. Henry) : Hybrid Larch. Introduced and spontaneous. Plantations. Brecs.[112].

CEDRUS Trew

Cedrus libani A. Richard : Cedrwydden Libanus : Cedar-of-Lebanon. Introduced (Asia Minor). Parks. *Carms.*[201], *Monts.*[7], *Mer.*[7].

PINUS L.

Subgen. Pinus

Pinus contorta Douglas ex Loudon : Lodge-pole Pine. Introduced (N. America). Plantations (high exposed moorland, to 2,300ft.). *Mons.*[91], *Glam.*[93], *Brecs.*[259], *Rads.*[259], Cards.[112], *Monts.*[89], *Mer.*[89], *Caerns.*[87], *Denbs.*[120].

P. pinaster Aiton : Pinwydden Arfor : Maritime Pine. Introduced (Mediterranean region). Shelterbelts on dunes. *Mons.*[120], *Glam.*[120], *Carms.*[120], *Pembs.*[120], *Caerns.*[120], *Denbs.*[6], *Angl.*[6].

P. radiata D. Don : Monterey Pine. Introduced (N. America). Plantations near the sea. *Glam.*[120], *Carms.*[120], *Pembs.*[120], *Cards.*[71], *Monts.*[120], *Caerns.*[87], *Denbs.*[120], *Angl.*[120].

P. ponderosa Douglas ex P. & C. Lawson : Western Yellow Pine. Introduced (N. America). Plantations. *Caerns.*[87].

P. nigra Arnold Introduced (Europe). Plantations. *Mons.*[120], *Glam.*[86], *Brecs.*[120], *Rads.*[120], *Carms.*[120], *Pembs.*[120], *Cards.*[93], *Monts.*[89], *Mer.*[89], *Caerns.*[87], *Denbs.*[120], *Flints.*[261], *Angl.*[120].

Subsp. **nigra** : Pinwydden Awstria : Austrian Pine. Introduced (Austria and Hungary). Plantations. *Glam.*[120], *Carms.*[225], *Monts.*[120], *Caerns.*[120].

Subsp. **laricio** (Poiret) Maire (*P. nigra* var *maritima* (Aiton) Melville) : Corsican Pine. Introduced (Mediterranean region). Plantations (sand dunes). *Mons.*[120], *Glam.*[120], *Brecs.*[120], *Rads.*[120], *Carms.*[120], *Monts.*[120], *Mer.*[89], *Caerns.*[120], *Denbs.*[120], *Angl.*[177].

P. sylvestris L. : Pinwydden Sgotland : Scots Pine. Denizen (Scotland). Woods (D,E,H,Q,R,W). Mons., Glam., Brecs., *Rads.*, Carms., Pembs., Cards., Monts., *Mer.*, Caerns., Denbs., Flints., *Angl.*. 102 squares, map 2.

P. mugo Turra : Mountain Pine. Introduced (Europe). Shelterbelts (poor soil to 2,000ft.). *Glam.*[120], *Brecs.*[120], *Carms.*[161].

P. pinea L. : Stone Pine. Introduced (Mediterranean region). Waste ground. *Caerns.*[74].

Subgen. Haploxylon (Koehne) Pilger

P. wallichiana A. B. Jackson : Bhutan Pine. Introduced (Himalaya). Parks. *Mons.*[91], *Glam.*[120], *Rads.*[120], *Carms.*[120], *Monts.*[120], *Denbs.*[120].

P. peuce Griseb. : Macedonian Pine. Introduced (Balkans). Plantations (mountain peats). *Glam.*[120], *Denbs.*[120].

P. strobus L. : White Pine. Introduced (N. America). Plantations (deep well-drained soil). *Glam.*[120], *Caerns.*[120].

TAXODIACEAE

SEQUOIA Endl.

Sequoia sempervirens (Lamb.) Endl. : Coast Redwood. Introduced (N. America). Parks. *Mons.*[120], *Glam.*[120], *Rads.*[120], *Monts.*[120], *Caerns.*[87], *Denbs.*[120].

SEQUOIADENDRON Buchholz

Sequoiadendron giganteum (Lindley) Buchholz (*Sequoia wellingtonia* Seem.) : Wellingtonia. Introduced (N. America). Parks. *Mons.*[120], *Glam.*[120], *Brecs.*[120], *Rads.*[120], *Carms.*[201], *Cards.*[71], *Monts.*[120], *Mer.*[169], *Caerns.*[87], *Denbs.*[120], *Angl.*[7].

CRYPTOMERIA D. Don

Cryptomeria japonica (L.fil.) D. Don : Japanese Cedar. Introduced (Japan). Plantations. *Carms.*[71], Cards.[112].

CUPRESSACEAE

CUPRESSUS L.

Cupressus macrocarpa Hartweg : Cypreswydden Macrocarpa : Monterey Cypress. Introduced (N. America), sometimes regenerating. Plantations (coastal areas). *Glam.*[120], *Carms.*[120], *Pembs.*[94], *Cards.*[259], *Monts.*[120], *Caerns.*[93], *Denbs.*[120], *Angl.*[120].

CHAMAECYPARIS Spach

Chamaecyparis lawsoniana (A. Murray) Parl. : Cypreswydden Lawson : Lawson Cypress. Introduced (N. America), sometimes regenerating. Plantations. *Mons.*[120], *Glam.*[120], *Brecs.*[200], *Rads.*[231], *Carms.*[7], Cards.[112], *Monts.*[89], *Mer.*[7], *Caerns.*[87], *Denbs.*[120].

C. nootkatensis (Lamb.) Spach : Nootka Cypress. Introduced (N. America). Parks. *Mons.*[120], *Monts.*[120], *Caerns.*[120].

CUPRESSUS × CHAMAECYPARIS =
× CUPRESSOCYPARIS Dallimore

Cupressus macrocarpa × Chamaecyparis nootkatensis = × Cupressocyparis leylandii (A. B. Jackson & Dallimore) Dallimore : Leyland Cypress. Spontaneous and introduced. Parks. *Monts.*[120].

THUJA L.

Thuja plicata D. Don ex Lamb. : Western Red Cedar. Introduced (N. America), sometimes regenerating. Plantations. *Mons.*[120], *Glam.*[86], *Brecs.*[120], *Rads.*[259], *Carms.*[120], Cards.[112], *Monts.*[89], *Caerns.*[87], Denbs.[120].

JUNIPERUS L.

Juniperus communis L. Glam., *Brecs.**, Carms., *Pembs.**, Mer., Caerns., *Denbs.**, *Flints.*, *Angl.*. 27 squares, map 3.

Subsp. **communis** : Merywen Gyffredin : Juniper. Native. Rocky hillsides (E,Q,Sc,W). Fl.5-6. Glam., *Brecs.**, Carms., *Pembs.**, Mer., Caerns., *Denbs.**, *Flints.*, [*Angl.*]. 21 squares, map 4.

The Ramsey population (**SM**62) is believed to be intermediate between subsp. *communis* and subsp. *nana*.

Subsp. **nana** Syme (*J. sibirica* Burgsd.) : Merywen Goraidd Fynyddig : Dwarf Juniper. Native. Mountain cliffs (Q,Sc). Fl.3-6. Mer.*, Caerns., *Angl.**. 9 squares.
Mer. : Craig y Gwynt, SH63, WIJ, 1981[3]; Cnicht, SH64, WIJ, 1969[3]; Croesor, SH64, OHB & HHan, 1969[3] and PMB, 1980[3]; Cadair Idris, SH71, JGr, 1932[4] and DMP, 1962[112]. **Caerns. :** Moel Hebog, SH54, LL, 1956[7]; Glyders, SH65, CCo, 1830[112] and ER, 1950+[207]; Cwm Idwal, SH65, JEG, 1883[112] and JHS, 1925[112]; Crib Goch, SH65, HAH, 1934[112]; Y Lliwedd, SH65, CW, 1918[112]; Snowdon, SH65, ER, 1950+[207] and RAB, 1950+[7]; Penyrole-wen, SH66, ER, 1950+[207]; Moel Siabod, SH75, ER, 1946[112]; nr Llyn y Foel, SH75, ER, 1950+[207]; Penllithrig-y-wrach, SH76, ER, 1951[112]; Creigiau Gleision, SH76, ER, 1950+[207]. **Angl.:** Holyhead, SH28, IB, 1978[208].

TAXACEAE

TAXUS L.

Taxus baccata L. : Ywen : Yew. Native but frequently introduced. Woods. (E,G,Q,W). Fl.2-4. Mons., Glam., Brecs., Rads., Carms., (*Pembs.**), *Cards.**, Monts., (Mer.), Caerns., Denbs., Flints., *Angl.*. 135 squares, map 5. Many of the records probably refer to introduced plants.

ANGIOSPERMAE

DICOTYLEDONES

SALICACEAE

SALIX L.

Subgen. Salix

Salix pentandra L. : Helygen Beraroglaidd : Bay Willow. Native. Streamsides (A,E,H,M,R). Fl.5-7. (*Glam.*), Brecs.*, (Carms.), Pembs., (Cards.), Monts.*, (Mer.), *Caerns.*, Denbs., *Flints.*, Angl.. 49 squares, map 6.

S. fragilis L. : Helygen Frau : Crack Willow. Native. Streamsides (Ar,E,R). Fl.4-5. Mons., Glam., Brecs., Rads., Carms., (*Pembs.**), Cards., Monts., (Mer.), Caerns., Denbs., Flints., *Angl.*. 153 squares, map 7.

S. fragilis × pentandra = S × meyerana Rostk. ex Willd. : Pointed-leaved Willow. Introduced (England). Streamsides (Ar,E,M). Fl.4-5. *Brecs.*, Carms., Monts., *Denbs.*, *Angl.*. 7 squares.
Brecs.: —, SN93, PCH, 1956[7]. **Carms.:** Gors-las, SN51, JAW, 1945[112]. **Monts.:** Llanllugan, SJ00, JAW, 1940[112]; Manafon, SJ10, JAW, 1940[112]; Guilsfield, SJ21, JAW, 1940[42]. **Denbs.:** Cerrigydrudion, SH95, PD, 1979[185]. **Angl.:** nr Llangefni, SH47, JEG, pre 1897[101].

[**S. fragilis × triandra = S. × speciosa** Host *Mons.*, *Glam.*, *Cards.*]

S. alba L.

Subsp. **alba** : Helygen Wen : White Willow. Native. Streamsides (Al,Ar,E,M,R). Fl.4-5. (Mons., Glam., *Brecs.*, Rads., Carms., Pembs., Cards., Monts., Mer.), Caerns., Denbs., *Flints.*, (*Angl.*). 89 squares, map 8.

Subsp. **vitellina** (L.) Arcangeli (*S. alba* var. *vitellina* (L.) Stokes) : Helygen Aur : Golden Willow. Introduced (Europe). Streamsides. *Mons.*, *Glam.*, *Brecs.*, Cards., *Mer.*, *Caerns.*, *Angl.*. 10 squares : **SS***69*; **ST***48*; **SN***58*,*69*,74,94; **SO***00*; **SH**47,57,93. Usually planted.

S. alba × fragilis = S. × rubens Schrank Spontaneous or introduced. Streamsides. Fl.4-5. Mons., Glam., *Brecs.*, Rads., *Mer.*, *Denbs.*. 9 squares.
Mons.: Magor, ST48, WAS, 1895[112]. **Glam.:** Swansea, SS69, JWG, c1840[223]; nr Cardiff, ST17, EV, pre 1907[223] and 1933[112]; Aberdare, SO00, HJR, pre 1907[140]. **Brecs.:** nr Hay-on-Wye, SO14, MP, 1970[109]. **Rads.** : Llandrindod Wells, SO06, EV, 1905[112]; —, SO15, FHP, 1956[7]. **Mer.:** Tywyn, SH50, AL, c1886[126]. **Denbs.** : Llansilin, SJ22, EFJ, 1867[144].

S. triandra L. : Helygen Drigwryw : Almond Willow. Native. Streamsides (Ar,D,E,G,M,R). Fl.4-5. Mons., Glam., Brecs., Carms., Cards., Monts., *Caerns.*, Denbs., Flints., *Angl.*, [*Rads.*]. 55 squares, map 9.

S. triandra × viminalis = S. × mollissima Hoffm. ex Elwert : Sharp-stipuled Triandrous Willow. Spontaneous or introduced. River banks. Fl.5. Mons., Glam., *Carms.*. 5 squares.
Mons.: Mounton, ST59, ESM & WAS, 1901[112]; nr Brockweir, SO50, WAS, 1895[112]. **Glam.:** nr Pyle, SS88, WAS, 1901[112]; Cardiff, ST17, EV, 1904[112] & 1934[112]. **Carms.:** Tywyn Burrows, SN30, DSR, 1971[7].

Subgen. Chamaetia Dumort.

[**S. reticulata** L. *Mer.*]

S. herbacea L. : Helygen Leiaf : Dwarf Willow. Native. Mountain cliffs. Brecs., Carms., Cards., Mer., Caerns.. 12 squares : **SN**72,78; **SO**01,02; **SH***55*,61,65, 66,71,*76*,82,84.

Subgen. Caprisalix Dumort.

[**S. phylicifolia** L. *Mons.*]

[**S. nigricans** Sm. *Glam.*]

S. cinerea L.

Subsp. **oleifolia** Macreight (*S. atrocinerea* Brot.; *S. cinerea* subsp. *atrocinerea* (Brot.) P. Silva & Sobrinho) : Helygen Olewydd-ddail : Grey Willow. Native. Streamsides (Ar,D,E,G,H,M,Q,R,Rw,S, Sc,W). Fl.4-5. Mons., Glam., *Brecs.*, Rads., Carms., Pembs., Cards., Monts., Mer., Caerns., Denbs., Flints., Angl.. 268 squares, all except : **SS**38,77; **ST**26; **SM**50,62,73; **SN**36; **SO**30; **SH**29,34,68, 91,98; **SJ**00.

S. cinerea × purpurea = S. × sordida A. Kerner Spontaneous. Streamsides. *Glam.*. 1 square.
Glam.: Taff's Well, ST18, HJR, pre 1907[140].

S. cinerea × purpurea × viminalis = S. × forbyana Sm. Introduced (garden origin?). River bank. Mer.. 1 square.
Mer.: nr Pennal, SN79, PMB, 1973[112].

S. cinerea × repens = S. × subsericea Döll Spontaneous. Fixed dunes. *Mer.*. 1 square.
Mer.: Morfa Dyffryn, SH52, PMB, 1974[3].

S. cinerea × viminalis = S. × smithiana Willd. : Silky-leaved Osier. Spontaneous or introduced. Hedges and streamsides (A,E,G,M,R). Mons., Glam., Brecs., Rads., Carms., Pembs., Cards., Monts., Mer., *Caerns.*, Denbs., Flints., *Angl.*. 65 squares, map 10.

S. aurita L. : Helygen Grynglustiog : Eared Willow. Native. Heaths (E,G,H,M,Q,R,W). Fl.4-5. Mons., Glam., Brecs., Rads., Carms., Pembs., Cards., Monts., Mer., Caerns., Denbs., Flints., Angl.. 198 squares, map 11.

S. aurita × caprea = S. × capreola J. Kerner ex N. J. Andersson Spontaneous. Damp scrub and hedges. Glam., *Brecs.*, Denbs., *Flints.*. 5 squares.
Glam.: Whitchurch, Cardiff, ST18, HJR, pre 1909[142]; Aberdare, SO00, HJR, pre 1907[140] and KA, 1973[112]. **Brecs.:** —, SN94, RCLH, 1956[7]. **Denbs.:** Minera, SJ25, CW, 1919[112]. **Flints.:** nr Lixwm, SJ17, JAWh & AAD, 1906[140].

S. aurita × caprea × viminalis = S. × stipularis Sm. : Helygen Glustenog : Auricled Osier. Spontaneous or introduced. Hedges and damp places. *Glam.*, *Angl.*. 3 squares.
Glam.: Swansea, SS69, WRL, 1891[104]. **Angl.:** nr Dinam, SH46, JEG, pre 1897[101]; nr Treffôs, SH57, JEG, pre 1897[101].

S. aurita × cinerea = S. × multinervis Döll Spontaneous. Woods and hedges (Ds,E,G,H,R,S,W). Fl.3. Mons., Glam., *Brecs.*, *Carms.*, Pembs., *Cards.*, Monts., Mer., *Caerns.*, *Denbs.*, *Flints.*, *Angl.*. 31 squares, map 12.

S. aurita × repens = S. × ambigua Ehrh. : Ambiguous Willow. Spontaneous. Heaths and moors (G,H, M,W). Fl.4-6. *Mons.*, Glam., *Brecs.*, Cards., *Mer.*, Caerns., Angl.. 7 squares.
Mons.: nr Shirenewton, ST49, WAS, pre 1920[216]. **Glam.:** Aberdare, SO00, HJR, 1911[112]. **Brecs.:** nr Cwmgiedd, SN71, MP, 1973[109]. **Cards.:** Llan-non, SN56, JHS, 1930[112] and AOC, 1978[112]. **Mer.:** Llanfair, SH52, WRL, 1894[152]; Morfa Dyffryn, SH52, PMB, 1973[3]. **Caerns.:** nr Rhydyclafdy, SH33, PMB, 1966[112]. **Angl.:** Valley, SH37, WRL & EFL, 1895[107].

S. aurita × viminalis = S. × fruticosa Döll Spontaneous. Scrub. Fl.4. *Mons.*, Glam., *Rads.*. 3 squares.
Mons.: Magor, ST48, WAS, 1895[107]. **Glam.:** Aberdare, SO00, HJR, 1911[112]. **Rads.:** Stanner Rocks, SO25, VG, 1952[7].

S. caprea L. : Helygen Grynddail Fwyaf : Goat Willow. Native. Woods (Ar,D,Ds,E,M,Q,R,Rw,W). Fl.2-5. Mons., Glam., Brecs., Rads., Carms., Pembs., Cards., Monts., Mer., Caerns., Denbs., Flints., *Angl.*. 222 squares, map 13.

S. caprea × cinerea = S. × reichardtii A. Kerner Spontaneous. Woods and streamsides. Fl.4-5. Mons., Glam., *Rads.*, Carms., Cards., *Mer.*, *Caerns.*, *Denbs.*. 19 squares : **SS**58; **ST**08,17,*18*,27,28,59; **SN**61,69,*77*; **SO**00, 10,24,31,50; **SH**12,61; **SJ**24,43.

S. caprea × cinerea × viminalis = S. × calodendron Wimmer : Long-leaved Willow. Spontaneous. Hedges and damp places. *Angl.*. 2 squares.
Angl.: Cors-y-bol, SH38, HD, 1804[107] and JEG, pre 1897[101]; Llangefni, SH47, JEG, pre 1897[101].

[**S. caprea × repens = S. × laschiana** Zahn *Glam.*]

S. caprea × viminalis = S. × sericans Tausch ex. A. Kerner Spontaneous. Hedges and waste ground. *Mons.*, *Brecs.*, *Pembs.*, Cards., Monts., Mer., *Caerns.*, *Denbs.*, *Angl.*, [*Glam.*]. 18 squares : **SM**62; **SN**14,25, 57,68,69,79; **SO**04,21,51; **SH**12,50,*52*,53,*57*,70,85; **SJ**12.

S. repens agg. : Corhelygen : Creeping Willow. Native. Sand dunes and heaths (Ar,D,Ds,G,H,M,Q, R,Rw). Fl.4-5. Mons., Glam., Brecs., Rads.*, Carms., Pembs., Cards., Mer., *Caerns.*, Denbs., Flints., Angl., [*Monts.*]. 134 squares, map 14.

S. repens L. *s.s.* Native. Heaths (D,G,H,M,R,Rw). Fl.4-6. Mons., Glam., Brecs., Rads., Carms., Pembs., Cards., Mer., Caerns., *Denbs.*, Flints., Angl.. 37 squares, map 15. Under-recorded.

S. arenaria L. (*S. repens* var. *argentea* (Sm.) Neilr.; *S. repens* subsp. *argentea* (Sm.) G. & A. Camus). Native. Sand dunes (D,G,R). Fl.3-7. Glam., Carms., Pembs., Mer., Denbs., Flints., Angl.. 17 squares : **SS**19,*49*,*59*,*78*,*87*; **ST**17,18; **SN**30; **SH**36,*37*,*46*, 52,53,71,*87*; **SJ**17,18. Under-recorded.

S. viminalis L. : Helygen Wiail : Osier. Native but frequently introduced. Streamsides (D,Ds,E,G,H,M,R). Fl.3-5. Mons., Glam., Brecs., Rads., Carms., *Pembs.*, *Cards.*, Monts., Mer., Caerns., Denbs., Flints., *Angl.*. 164 squares, map 16.

S. purpurea L. : Helygen Gochlas : Purple Willow. Native but frequently introduced. Streamsides (Ar,E, G,M,R). Fl.4. Mons., Glam., *Brecs.**, Rads., Carms., Cards., Monts., Mer.*, *Caerns.*, Denbs., Flints., Angl.. 51 squares, map 17.

S. purpurea × viminalis = S. × rubra Hudson : Helygen Werdd : Green-leaved Osier. Spontaneous or introduced. Stream and roadsides. Fl.5. *Mons.*, Glam., Cards., Caerns., *Angl.*. 7 squares.
Mons.: Cwmbrân, ST29, MG, 1973[93]; nr Brockweir, SO50, WAS, pre 1920[226]. **Glam.:** Aberdare, SO00, HJR, 1906[112]. **Cards.:** Llanilar, SN67, JHS, 1931[112] and 1941[112]. **Caerns.:** nr Llanberis, SH56, JEG, pre 1894[101]; nr Bangor, SH56, JEG, 1891[112]. **Angl.:** nr Treffôs, SH57, JEG, pre 1894[101]; nr Beaumaris, SH67, JEG, pre 1894[101].

S. daphnoides Vill. : Violet Willow. Introduced (Europe). Sand dunes and hedges. Carms., *Caerns.*. 3 squares.
Carms.: Pembrey, SS49, IMV, 1973[112]. **Caerns.:** Bangor, SH57, JEG, pre 1894[101]; Llanfairfechan, SH67, JEG, pre 1894[101].

POPULUS L.

Populus alba L. : White Poplar. Established alien (Europe). Hedges (D,E,G,M,Sc). Fl.5. Mons., Glam., Brecs., *Rads.*, Carms., Pembs., Cards., Monts., Mer., *Caerns.*, Denbs., Flints., *Angl.*[6]. 63 squares, map 18.

P. alba × **tremula** = **P.** × **canescens** (Aiton) Sm. : Poplysen Wen : Grey Poplar. Established alien (Europe). Hedges (D,E,M,R,W). Fl.3-4. Mons., Glam., *Brecs.**, Rads., Carms., Pembs., *Cards.**, Monts., Caerns., *Denbs.*, Flints., *Angl.*[230]. 55 squares, map 19.

P. tremula L. : Aethnen : Aspen. Native. Woods (E,M,Q,R,S,W). Fl.3-4. Mons., Glam., Brecs., Rads., Carms., Pembs., Cards., Monts., Mer., Caerns., *Denbs.*, Flints., *Angl.*. 164 squares, map 20.

P. gileadensis Rouleau : Balsam-poplar. Established alien (garden origin). Riverbanks (D,E,M). Fl.4. Mons., Glam., *Brecs.*, Rads., Carms., Cards., Monts., Mer., *Caerns.*, Denbs., Flints.. 24 squares, map 21.

P. trichocarpa Torrey & A. Gray ex Hooker : Western Balsam-poplar. Introduced (N. America). Woods and plantations. *Mons.*[45], *Rads.*[6], *Cards.*[6], *Monts.*[6], *Mer.*[6], *Caerns.*[120], *Denbs.*[6], Flints.[112], *Angl.*[6].

P. nigra agg. Mons., Glam., Brecs., Rads., Carms., Pembs., Cards., Monts., Mer., Caerns., Denbs., Flints., Angl.. 105 squares, map 22.

P. nigra L. *s.s.* : Poplysen Ddu : Black Poplar. Established alien (England). Hedges (E,G,M,R,W). Fl.4. Mons., Glam., *Brecs.*, Monts., Mer., Denbs., Flints., [*Caerns.*, *Angl.*]. 26 squares, map 23.

cv. 'Italica' (*P. italica* (Duroi) Moench) : Poplysen Lombardy : Lombardy Poplar. Established alien (garden origin). Hedges. Fl.5. *Mons.*, *Glam.*, *Pembs.*, Monts., Caerns., Denbs., Flints.. 11 squares : **ST**18; **SN**03; **SO**08,19,40; **SH***57*; **SJ**17,20,24,31,36.

P. deltoides × **nigra** = **P.** × **canadensis** Moench : Italian Poplar. Naturalized alien (garden origin). Woods and hedges (E,M,W). Fl.4. Mons., Glam., Brecs., Rads., Carms., Pembs., Cards., Monts., Mer., Caerns., Denbs., Flints.. 50 squares, map 24.

MYRICACEAE

MYRICA L.

Myrica gale L. : Helygen Mair : Bog Myrtle. Native. Bogs (H,M). Fl.4-6. Glam., Carms., Pembs., Cards., Monts., Mer., Caerns., *Denbs.**, *Flints.*, Angl., [*Brecs.*]. 67 squares, map 25.

JUGLANDACEAE

JUGLANS L.

Juglans regia L. : Coeden Cnau Ffrengig : Walnut. Introduced (E. Europe). Hedges and woods. Fl.5. Mons., Glam., *Brecs.*, *Rads.*, *Carms.*, *Pembs.*, Cards., Monts., *Caerns.*, *Denbs.*, Flints., *Angl.*. 35 squares, map 26.

BETULACEAE

BETULA L.

Betula pendula L. (*B. verrucosa* Ehrh.) : Bedwen Arian : Silver Birch. Native. Woods (E,G,H,M,Q,R, Rw,W). Fl.3-5. Mons., Glam., Brecs., *Rads.*, Carms., *Pembs.*, Cards., Monts., Mer., Caerns., Denbs., Flints., Angl.. 206 squares, map 27. Probably over-recorded in mid Wales in mistake for the glabrous form of *B. pubescens*.

B. pendula × **pubescens** = **B.** × **aurata** Borkh. Spontaneous. Woods. Mons., Glam., Brecs., Caerns., [*Mer.*]. 6 squares.
Mons.: Aber-carn, ST29, AEW, 1924[112]; Cwm Big, SO20, AEW, 1925[112]. **Glam.:** Cwm Dimbath, SS98, ARP & AEW, 1974[112]; Taf Fechan, SO00, AEW and RGE, 1969[112]. **Brecs.:** Capel-y-ffin, SO23, MP, 1970. **Caerns.:** Bangor, SH57, JEG, 1890[112].

B. pubescens Ehrh.

Subsp. **pubescens** : Bedwen Gyffredin : Downy Birch. Native. Woods (D,E,G,H,M,Q,R,Rw,W). Fl.4. Mons., Glam., Brecs., Rads., Carms., Pembs., Cards., Monts., Mer., Caerns., Denbs., Flints., *Angl.*. 221 squares, map 28.

Subsp. **carpatica** (Willd.) Ascherson & Graebner (*B. pubescens* subsp. *odorata* (Bechst.) E. F Warburg). Native. Woods, especially in upland areas. Mons., Glam., Brecs., *Rads.*, *Pembs.*, *Monts.*, *Mer.*, *Caerns.*, 20 squares : **SS**78; **ST**29, 49; **SN**10,81,85,90,93,95; **SO**01,02,06,10,21,23,*50*; **SH**51,*71*,92; **SJ**10. Much under-recorded.

ALNUS Miller

Alnus glutinosa (L.) Gaertner : Gwernen : Alder. Native. Streamsides (Ar,D,Ds,E,G,H,M,R,Rw,S,W). Fl.2-4. Mons., Glam., Brecs., Rads., Carms., Pembs., Cards., Monts., Mer., Caerns., Denbs., Flints., *Angl.*. 261 squares, all except : **SR**89; **SS**38,39,77; **ST**26; **SM**50,62,71,73,84,94; **SN**15,25,36,59,87; **SH**29,51,88, 91,98.

A. incarna (L.) Moench : Grey Alder. Established alien (N. Europe). Woods and plantations. Glam., Carms., *Cards.*, Monts.. 6 squares.
Glam.: Porthkerry, ST06, ABP & AEW, 1975[112]. **Carms.:** Pembrey, SN30, IMV, 1975[112]; Llangadog, SN72, IMV, 1976[255]. **Cards.:** Llanafan, SN67, AOC, 1962[71]; Ponterwyd, SN78, JHS, 1930[210] and 1935[211]. **Monts.:** Glasbwll Valley, SN79, JAW, 1941[112].

CORYLACEAE

CARPINUS L.

Carpinus betulus L. : Oestrywydden : Hornbeam. Native or naturalized alien. Woods (E,M,R,W). Fl.4-6. Mons., (Glam., Brecs., Rads., Carms., *Pembs.**, Cards., Monts., Mer., Caerns., Denbs., Flints., Angl.). 103 squares, map 29. Native only in Mons.

CORYLUS L.

Corylus avellana L. : Collen : Hazel. Native. Woods (E,G,M,Q,R,Sc). Fl.2-3 Mons., Glam., Brecs., Rads., Carms., Pembs., Cards., Monts., Mer., Caerns., Denbs., Flints., Angl.. 259 squares, all except : **SR**89; **SS**19,38,39,77; **ST**26; **SM**50,62,70,71,72,73,84,94; **SN**15,25,36; **SH**12,13,24,51,98; **SJ**37.

FAGACEAE

FAGUS L.

Fagus sylvatica L. : Ffawydden : Beech. Native. Woods (E,M,Q,R,W). Fl.4-5. Mons., Glam., Brecs., (Rads., Carms., *Pembs.*, Cards., Monts., Mer., Caerns., Denbs., Flints., *Angl.*). 240 squares, map 30. Native only in S.E. Wales but native and introduced records have not been separated on the map.

CASTANEA Miller

Castanea sativa Miller : Castanwydden : Sweet Chestnut. Naturalized alien (S. Europe). Woods and hedges (E,R,W). Fl.5-8. Mons., Glam., Brecs., Rads., Carms., *Pembs.*, Cards., Monts., Mer., Caerns., Denbs., Flints., *Angl.*. 122 squares, map 31.

QUERCUS L.

Subgen. Erythrobalanus (Spach) Örsted

Quercus rubra L. (*Q. maxima* (Marshall) Ashe; *Q. borealis* Michx) : Red Oak. Introduced (N. America). Woods and plantations. *Glam.*, *Brecs.*, *Rads.*, Carms., *Pembs.*, *Cards.*, *Monts.*, *Mer.*. 10 squares : **SR**99; **SS**79; **ST**17; **SN**61,77,96; **SO**01,09; **SH**61,72. Usually planted but sometimes regenerating.

Subgen. Sclerophyllodrys O. Schwarz

Q. ilex L. : Derwen Fythwyrdd : Evergreen Oak. Naturalized alien (Mediterranean region). Woods (G,Q,S,Sc,W). Mons., Glam., Carms., Pembs., *Cards.*, Monts., Mer., *Caerns.*, *Denbs.*, Flints., *Angl.*[101]. 41 squares, map 32.

Subgen. Cerris (Spach) Örsted

Q. cerris L. : Turkey Oak. Naturalized alien (Europe). Woods and hedges (E,Q,W). Mons., Glam., *Brecs.*, Rads., Carms., Pembs., Cards., Monts., Mer., *Caerns.*, *Denbs.*, Flints., *Angl.*. 66 squares, map 33.

Subgen. Quercus

Q. petraea (Mattuschka) Liebl. : Derwen Ddaildigoes : Sessile Oak. Native. Woods (E,G,H,M,Q,R,W). Fl.4-5. Mons., Glam., Brecs., Rads., Carms., Pembs., Cards., Monts., Mer., Caerns., Denbs., Flints., Angl.. 242 squares, map 34.

Q. petraea × robur = Q. × rosacea Bechst. Spontaneous. Woods (E,W). Mons., Glam., Brecs., Carms., Pembs., *Cards.*, Monts., Mer., *Caerns.*, *Denbs.*, Flints., *Angl.*. 46 squares, map 35.

Q. robur L. : Derwen Goesog : Pedunculate Oak. Native. Woods (E,G,M,R,Rw,W). Fl.4-7. Mons., Glam., Brecs., Rads., Carms., *Pembs.*, *Cards.*, Monts., Mer., Caerns., Denbs., *Flints.*, *Angl.*. 277 squares, map 36.

NOTHOFAGUS Blume

Nothofagus obliqua (Mirbel) Blume : Roblé Beech. Introduced (S. America). Woods and plantations. *Glam.*[120], *Rads.*[120], *Cards.*[6], *Monts.*[6], *Mer.*[6], *Caerns.*[120], *Denbs.*[120], *Flints.*[6].

N. procera (Poeppig & Endl.) Örsted : Rauli Beech. Introduced (S. America). Woods and plantations. *Rads.*[6], *Cards.*[6], *Monts.*[89], *Mer.*[89], *Caerns.*[6], *Denbs.*[6], *Flints.*[6].

ULMACEAE

ULMUS L.

Ulmus glabra Hudson (incl. *U. montana* With.) : Llwyfen Lydanddail : Wych Elm. Native. Woods (E,H,M,Q,R,Rw,W). Fl.3-4. Mons., Glam., Brecs., Rads., Carms., Pembs., Cards., Monts., Mer., Caerns., Denbs., Flints., *Angl.*. 246 squares, map 37.

U. glabra × plotii = U. × elegantissima Horwood Introduced (England). Hedges. Fl.5. Mons., *Glam.*, Cards., Flints.. 5 squares.
Mons.: Chepstow, ST59, WAS, 1909[112]; Monmouth, SO51, AL, 1909[112]. **Glam.:** Gower, SS– –, HAH, c1930[120]. **Cards.:** Llandysul, SN44, AOC, 1976[71]; Trawsgoed, SN67, AOC, 1976[112]. **Flints.:** Prestatyn, SJ08, JAW, 1942[112].

U. procera Salisb. : Llwyfen Gyffredin : English Elm. Native. Hedges (E,G,M,W). Fl.2-3. Mons., Glam., *Brecs.*, *Rads.**, Carms., Pembs., Cards., Monts., Mer., Caerns., Denbs., Flints., Angl.. 158 squares, map 38.

U. angustifolia (Weston) Weston : Cornish Elm. Introduced (England). Hedges. Glam., *Brecs.*, Caerns.. 4 squares.
Glam.: Cardiff, ST17, HJR, 1912[112] and EV, 1931[112]; N. Cardiff, ST18, HAH, 1925[112]. **Brecs.:** Crickhowell, SO21, GCD, 1919[18]. **Caerns.:** Bangor, SH57, HAH, 1925[112].

U. angustifolia × glabra Introduced (S. England). Hedges. Mons., Glam.. 3 squares.
Mons.: Chepstow, ST59, JRe, 1932[112]. **Glam.:** Swansea, SS69, JAW, 1932[112] and 1940[112]; Dinas Powys, ST17, AEW, 1929[112].

U. coritana R. Melville × **plotii** Introduced (England). Hedges. Cards.. 1 square.
Cards.: nr Trawsgoed, SN67, AOC, 1962[112].

U. carpinifolia Gleditsch (*U. minor* auct.) : Pren Llwyf : Small-leaved Elm. Colonist (C. and S. Europe). Hedges. Glam., *Brecs.**, *Carms.**, Pembs.*, Cards.*, *Monts.**, Mer.*, Flints.*. 17 squares : **SS**68,69,78,87,97; **ST**06,28; **SM**80; **SN**41,42,50,66,*80*; **SO***02;* **SH**71,90; **SJ**26.

U carpinifolia × glabra = U. × vegeta (Loudon) A. Ley : Huntingdon Elm. Spontanteous but frequently introduced. River banks. *Mons.*[217], Glam., (Brecs., Rads., Carms., Pembs., Monts.). 13 squares : **SS**68,69, *79*; **ST**17; (**SM**80; **SN**50,61,74,80; **SO***00*,05,12; **SH**80).

U. carpinifolia × glabra × plotii = U. × hollandica Miller : Dutch Elm. Naturalized alien (England). Woods and hedges (E,F,M,Q,R,Sc,W). Fl.4. Mons., Glam., *Brecs.*, Rads., Carms., Pembs., Cards., Monts., Mer., Denbs., Flints.. 45 squares, map 39

U. carpinifolia × plotii = U. × viminalis Loddiges Introduced (S. England). Woods and hedges. Glam., Pembs., Monts., Mer.. 4 squares.
Glam.: Sketty Cwm, SS69, JAW, 1932[112] and 1952[112]. **Pembs.:** Angle, SM80, JBD, 1936[112]. **Monts.:** Cemais, SH80, JAW, 1941[112]. **Mer.:** Dolgellau, SH71, JAW, 1941[112].

[**U. carpinifolia** × **procera** *Glam.*]

U. plotii Druce : Dr. Plot's Elm. Native. Hedges. Mons., *Glam.**, *Brecs.**. 3 squares.
Mons.: Llanfair Cilgedin, SO30, WN, 1941[112]. **Glam.:** nr Swansea, SS69, JAW, 1924-26[221]. **Brecs.:** Crickhowell, SO21, GCD, 1919[18].

U. plotii × procera Introduced (garden origin?). Hedges. *Caerns.*. 1 square.
Caerns.: nr Boduan, SH33, APC, 1980[77].

ZELKOVA Spach

Zelkova carpinifolia (Pallas) C. Koch Introduced (Caucasus). Brecs.. 1 square.
Brecs.: Llanwrtyd Wells, SN84, IMV, 1963[112].

MORACEAE

FICUS L.

Ficus carica L. : Ffigyswydden. : Fig. Naturalized alien (S. Europe). Cliff faces and walls. Fl.7-10. Glam., Brecs., Carms.. 12 squares : **SS**48,68,79,99; **ST**06,08,09, 16,17; **SN**40,51,71.

CANNABACEAE

HUMULUS L.

Humulus lupulus L. : Hopysen : Hop. Native. Hedges (E,Q,R,Rb,W). Fl.(5-)8-9. Mons., Glam., Brecs., Rads., Carms., Pembs., Cards., Monts., Mer., Caerns., Denbs., Flints., Angl.. 174 squares, map 40.

H. japonicus Siebold and Zucc. : Japanese Hop. Casual (Japan). Waste places. Fl.9. Glam.. 1 square. ST27[112].

CANNABIS L.

Cannabis sativa L. : Cywarch : Hemp. Casual (S. and W. Asia). Waste places (R,Rg). Fl.8-10. Mons., Glam., *Cards.*[210], *Mer.*, *Caerns.*, Flints., *Angl.*. 13 squares : **SS**69,88; **ST**17,27, 38,59; **SN**54; **SO***00,10;* **SH**50,*53*,67; **SJ**17.

URTICACEAE

URTICA L.

Urtica dioica L. : Danadl Poethion : Common Nettle. Native. Roadsides, cultivated and waste ground (D,E,F, G,H,M,Q,R,Rg,Rw,S,Sc,W). Fl.4-10. Mons., Glam., Brecs., Rads., Carms., *Pembs.*, Cards., Monts., Mer., Caerns., Denbs., Flints., *Angl.*. 280 squares, all except **SS**39; **SM**50.

U. urens L. : Danhadlen Leiaf : Small Nettle. Native. Waste places (F,G,Q,R,Rg,S). Fl.3-10. Mons., Glam., *Brecs.*, Rads., Carms., Pembs., *Cards.*, Monts., Mer., Caerns., Denbs., Flints., Angl.. 105 squares, map 41.

U. pilulifera L. : Danhadlen Belaidd : Roman Nettle. Established alien (S. Europe). Waste places. *Mons.*[226], Glam., *Mer.*, *Angl.*. 5 squares.
Glam.: Kenfig, SS88, JEBi, c1840[81]; Penarth, ST17, JS, 1876[112]; Cardiff, ST17, JS, c1876[112]. **Mer.:** Barmouth, SH61, WP, c1850[126]. **Angl.:** Baron Hill, SH57, HD, 1757[107]; Beaumaris, SH67, GCD, pre 1930[209]. Long extinct in Wales.

PARIETARIA L.

Parietaria judaica L. (*P. diffusa* Mert. & Koch) . Murlys : Pellitory-of-the-wall. Native. Walls (Q,S,Sc). Fl.5-9. Mons., Glam., Brecs., *Rads.*, Carms., Pembs., Cards., Monts., Mer., Caerns., Denbs., Flints., Angl.. 142 squares, map 42.

SOLEIROLIA Gaud.-Beaup.

Soleirolia soleirolii (Req.) Dandy (*Helxine soleirolii* Req.) : Mam Miloedd : Mind-your-own-business. Established alien (Mediterranean region). Walls (Q,Rg). Fl.5-9. Mons., Glam., *Carms.*, Pembs., Cards., *Monts.*, Mer., Caerns., Flints., *Angl.*. 50 squares, map 43.

SANTALACEAE

THESIUM L.

Thesium humifusum DC. : Geulin y Forwyn : Bastard-toadflax. Casual (S. England). Waste places. Fl.7-10 Glam., 1 square : **ST**16[112].

LORANTHACEAE

VISCUM L.

Viscum album L. : Uchelwydd : Mistletoe. Native. Parasitic on various trees : *Crataegus monogyna*; *Malus sylvestris*; *Populus tremula*; *Quercus* sp.; *Robinia pseudacacia*; *Tilia* × *vulgaris*; *Ulmus glabra*, etc. Fl.2-3. Mons., Glam., *Brecs.*, Rads., *Carms.*, *Cards.*, Monts., (Caerns.), *Denbs.*, *Flints.*, (Angl.). 55 squares, map 44.

ARISTOLOCHIACEAE

ASARUM L.

Asarum europaeum L. : Carn Ebol y Gerddi : Asarabacca. Denizen (Europe). Woods (E,W). Fl.5. *Glam.**, *Brecs.**, Carms., Pembs.*, Caerns., *Denbs.*[230], *Flints.**. 6 squares.
Glam.: Cardiff, ST17, JS, c1876[140]. **Brecs.:** nr Gliffaes, SO12, DEV, 1958[7]. **Carms.:** Dolau Cothi, SN64, MB, 1954[112]. **Pembs.:** Lydstep, SS09, MB, 1965[2]. **Caerns.:** Bangor, SH57, JTCWi, 1876[112]. **Flints.:** Nercwys, SJ26, SF, 1972[181].

ARISTOLOCHIA L.

Aristolochia clematitis L. : Afal Daear : Birthwort. Naturalized alien (C. and S. Europe). Sand dunes and roadsides. Fl.6-7. Glam., *Angl.**. 4 squares.
Glam.: Barry Dock, ST16, CIS, c1936[36]; Merthyr Mawr, SS87, Bry, c1936[36] and MG, 1976[99]; Peterson-super-Ely, ST07, DMHa, 1945[112]. **Angl.:** Rhosneigr. SH37, An, 1972[7].

POLYGONACEAE

POLYGONUM L.

Polygonum maritimum L. : Sea Knotgrass. Casual (Channel Isles). Waste places. Fl.9. Glam., [*Cards.*, *Caerns.*]. 2 squares : **ST***17*[32],*18*[112].

P. patulum Bieb. (*P. bellardii* auct., non All.) : Red Knotgrass. Casual (S. Europe). Waste places. Fl.7-9. Glam., *N. Wales*[16].. 4 squares : **SS***78*[140]; **ST***16*[24],*17*[24],*27*[117].

P. arenarium Waldst. & Kit. subsp. **pulchellum** (Loisel.) D. A. Webb & Chater (*P. pulchellum* Loisel.) : Lesser Red Knotgrass. Casual (Mediterranean region). Waste places. Fl.8-9. Glam., Pembs.. 3 squares : **ST***16*[112],17[112]; **SN**10[112].

P. oxyspermum C. A. Meyer & Bunge ex Ledeb. subsp. **raii** (Bab.) D. A. Webb & Chater (*P. raii* Bab.) : Clymog Eiddil Graban Garw : Ray's Knotgrass. Native. Shingle beaches (R,S). Fl.7-10. Glam., Pembs., Cards., Mer., Caerns., Denbs., Flints., Angl., [*Carms.*]. 43 squares, map 45.

P. aviculare agg. : Canclwm : Knotgrass. Mons., Glam., Brecs., Rads., Carms., Pembs., Cards., Monts., Mer., Caerns., Denbs., Flints., Angl.. 277 squares, all except : **SS**39; **SM**84; **SO**20,50; **SH**51.

P. aviculare L. *s.s.* Native. Cultivated ground and waste places (D,F,Q,R,Rw,S,Sc). Fl.7-10. Mons., Glam., Brecs., Rads., Carms., Pembs., Cards., Monts., Mer., Caerns., Denbs., Flints., Angl.. 126 squares, map 46. Under-recorded.

P. rurivagum Jordan ex Boreau Colonist (S. England). Roadsides and cultivated ground. Fl.9. *Glam.*, Pembs., *Caerns.*, *Angl.*. 6 squares.
Glam.: Pont-lliw, SN60, QONK, 1974[157]. **Pembs.:** Marloes, SM70, TAWD, 1962[112]. **Caerns.:** nr Groeslon, SH45, JEG, pre 1897[101]; nr Bangor, SH57, JEG, pre 1897[101]; Llandudno, SH78, JEG, pre 1897[101]. **Angl.:** Bodorgan, SH37, JEG, pre 1897[101].

P. arenastrum Boreau (*P. aequale* Lindman) : Clymog a Dail Bach : Small-leaved Knotgrass. Native. Roadsides (D,E,G,M,R,S). Fl.7-10. Mons., Glam., Brecs., Rads., Carms., Pembs.*, Cards.*, Monts.*, Mer., Caerns., *Denbs.**, Flints., *Angl.*. 120 squares, map 47. Under-recorded.

P. minus Hudson : Clymog Bychan : Small Water-pepper. Native. Poolsides (Al,G,M). Fl.6-9. Glam., *Carms.*[230], Monts.*, Mer., *Caerns.*, *Denbs.*, Angl., [*Cards.*]. 15 squares : **SS***69*,*78*,*79*,87; **ST**07; **SH**36,*37*,52, 53,77,83,*87*,90,*93*; **SJ**21.

P. minus × **persicaria** = **P.** × **braunianum** F. W. Schultz Spontaneous. Lakeside. Fl.7. Angl.. 1 square. **Angl.:** Llyn Coron, SH36, RHR, 1975[112].

P. mite Schrank : Penboeth Diflas : Tasteless Water-pepper. Native. Streamsides (Al,Ar). Fl.7-8(-10). Mons., Glam., Brecs.*, Carms., Cards., *Monts.**. 13 squares : **SS***69*,*79*,97; **ST**07,17; **SN**34,*40*,81; **SO**41,50,51; **SJ**21,22.

P. mite × **persicaria** = **P.** × **condensatum** (F. W. Schultz) F. W. Schultz Spontaneous. River banks. Fl.8-9. Mons.. 2 squares.
Mons.: nr Monmouth Cap, SO32, SGC, 1945[112]; nr Monmouth, SO51, SGC, 1943[112] and RL, 1944[112].

P. hydropiper L. : Tinboeth : Water-pepper. Native. Marshes (Al,Ar,G,M,R,W). Fl.8-9. Mons., Glam., Brecs., Rads., Carms., Pembs., Cards., Monts., Mer., Caerns., Denbs., *Flints.*, Angl.. 248 squares, map 48.

P. persicaria L. : Dail y Groes : Redshank. Native. Cultivated ground and moist places (F,G,M,Q,R,Rg, Rw,W). Fl.6-10. Mons., Glam., Brecs., Rads., Carms., Pembs., Cards., Monts., Mer., Caerns., Denbs., Flints., Angl.. 266 squares, all except : **SR**89; **SS**38,39,77; **ST**26; **SM**50,62,71,73; **SN**15,34,59,66,87,89; **SH**51.

P. lapathifolium L. (incl. *P. nodosum* Pers.) : Costog y Domen : Pale Persicaria. Native. Cultivated ground and marshy places (Al,Ar,F,G,M,R,Rg,Rw,W). Fl.6-9. Mons., Glam., Brecs., Rads., Carms., Pembs., Cards., Monts., Mer., Caerns., Denbs., Flints., Angl.. 201 squares, map 49.

P. lapathifolium × persicaria = P. × lenticulare Hyl. Spontaneous. Cultivated ground. Fl.8. *Mons.*, *Glam.*. 2 squares.
Mons.: Monmouth, SO51, RL, 1943[42]. **Glam.:** Radyr, ST18, HJR, 1912[12]. These records may refer to one or other of the parent species, both of which are variable.

P. amphibium L. : Canwraidd Goch : Amphibious Bistort. Native. Pools and damp places (Al,Ar,Ds, H,M,R). Fl.6-9. Mons., Glam., Brecs., Rads., Carms., Pembs., Cards., Monts., Mer., Caerns., *Denbs.*, Flints., Angl.. 139 squares, map 50.

P. bistorta L. : Llys y Neidr : Common Bistort. Native. Roadsides (E,G,M,R,Rw,W). Fl.4-7. Mons., Glam., Brecs., Rads., Carms., Pembs.*, *Cards.*, *Monts.*, *Mer.*, Caerns., Denbs., *Flints.*, *Angl.*. 148 squares, map 51.

P. amplexicaule D. Don : Red Bistort. Established alien (Himalayas). Waste places (R,Rg). Fl.9-10. Glam., Mer., Angl.. 4 squares.
Glam.: Gorseinon, SS59, GMR, 1968[112]; Briton Ferry, SS79, CH & HH, 1972[117]. **Mer.:** nr Pennal, SH70, JAW, 1940[112]. **Angl.:** Menai Bridge, SH57, RHR, 1972[112].

P. viviparum L. : Llys y Neidr Fynyddig : Alpine Bistort. Native. Mountain slopes. Fl.6-8. Caerns.. 1 square.
Caerns.: Snowdon, SH65, ER, 1948[112] and 1962[207]; Glyder Fawr, SH65, ER, 1949[112] and 1957[207]; Cwm Idwal, SH65, ER, 1951[112].

P. polystachyum Wallich ex Meissner : Himalayan Knotweed. Established alien (Himalayas). Roadsides (E,H,M,R,Rw,W). Fl.7-10(-12). Mons., *Glam.*, Brecs., *Rads.*, Carms., Pembs., Cards., Monts., Mer., Caerns., Denbs., Flints., Angl.. 54 squares, map 52.

P. campanulatum Hooker fil. : Lesser Knotweed. Established alien (Himalayas). Waste places. Fl.9. Glam., Cards., *Monts.*, Caerns., [*Carms.*]. 5 squares.
Glam.: Bonvilston, ST07, JIL, 1972[112]; Cardiff, ST17, RLS, 1938[112]. **Cards.:** Llanafan, SN67, AOC & APC, 1961[71]. **Monts.:** Powis Castle, SJ20, DHK, 1970+[138]. **Caerns.:** Morfa Dinlle, SH45, LJL, 1977[112].

[**P. lichiangense** W.W. Sm. *Pembs.*, *Mer.*]

FALLOPIA Adans.

Fallopia convolvulus (L.) Á. Löve (*Polygonum convolulus* L.; *Bilderdykia convolulus* (L.) Dumort.) : Taglys yr Ŷd : Black-bindweed. Native. Cultivated ground (D,F,G,M,R,Rg,Rw,S,Sc,W). Fl.6-9. Mons., Glam., *Brecs.*, Rads., Carms., Pembs., Cards., *Monts.*, Mer., Caerns., Denbs., Flints., Angl.. 195 squares, map 53.

F. dumetorum (L.) J. Holub (*Polygonum dumetorum* L.; *Bilderdykia dumetorum* (L.) Dumort.) : Gwenith yr Hydd y Bwch : Copse-bindweed. Casual (England). Hedges. *Caerns.*, *Angl.*, [*Mons.*, *Glam.*]. 4 squares : SH37[7],66[7],*76*[101],*78*[107].

F. aubertii (Louis Henry) J. Holub (*Polygonum baldschuanicum* auct., non Regel; *Bilderdykia aubertii* (Louis Henry) Moldenke) : Russian-vine. Established alien (W. China and Tibet). Roadsides and waste places (E,Q,R). Fl.8-9. *Mons.*, Glam., *Carms.*, Pembs., Cards., *Monts.*, Caerns., Flints.. 16 squares : **SS**68,69; **ST**09,*16*, 17; **SM**81; **SN**10,40,42,50,67; **SO**42; **SH**32; **SJ**11,20,27.

REYNOUTRIA Houtt.

Reynoutria japonica Houtt. (*Polygonum cuspidatum* Siebold & Zucc.) : Pysen Saethwr : Japanese Knotweed. Established alien (Japan). Roadsides and waste places (E,M,R,Rw). Fl.8-10. Mons., Glam., Brecs., Rads., Carms., Pembs., Cards., *Monts.*, Mer., *Caerns.*, Denbs., Flints., *Angl.*. 211 squares, map 54.

R. sachalinensis (Friedrich Schmidt Petrop.) Nakai (*Polygonum sachalinense* Friedrich Schmidt Petrop.) : Giant Knotweed. Established alien (E. Asia). Roadsides (E,G,M,R,W). Fl.8-9. Mons., *Glam.*, Rads., Carms., Pembs., Cards., Mer., Caerns., Denbs.. 20 squares, map 55.

FAGOPYRUM Miller

Fagopyrum esculentum Moench : Gwenith yr Hydd : Buckwheat. Established alien (C. Asia). Waste places (F,R,W). Fl.6-10. Mons., Glam., Carms., *Cards.*, *Monts.*, *Mer.*, *Caerns.*, *Denbs.*, Flints., Angl., [*Pembs.*]. 27 squares, map 56.

F. tataricum (L.) Gaertner Casual (Europe). Waste places (R,W). Fl.7-9. Mons., Glam.. 2 squares : **ST***16*[112],*59*[112].

F. dibotrys (D. Don) Hara (*Polygonum cymosum* Trev.). Established alien (Himalayas). Roadside. Fl.7-8. Pembs.. 1 square.
Pembs.: Dale, SM80, JEL, 1975[112] and TAWD, 1976[112].

OXYRIA Hill

Oxyria digyna (L.) Hill : Suran y Mynydd : Mountain Sorrel. Native. Mountain rocks. Fl.6-8. Mer., Caerns.. 13 squares : **SH**54,55,*62*,*63*,64,65,66,71, 74,75,76,81,82.

RUMEX L.

Subgen. Acetosella Rafin.

Rumex angiocarpus Murb. Native. Heaths (H,R). Fl.5-9. Mons., Glam., Brecs., Rads., Pembs., Cards., Monts., Mer., Caerns., Denbs.. 11 squares : **ST***17*,38; **SM***70*; **SN**68,*84*; **SO***05*,*11*,19; **SH***53*,*65*,*87*. Under-recorded.

R. tenuifolius (Wallr.) Á. Löve Native. Heaths (D,G,H,Q,R). Fl.6-8. Mons., Glam., Rads.*, Mer., Angl.. 11 squares : **SS**88,99; **ST***18*,*27*; **SO**25; **SH**28,37, 50,52,53,61.

Rumex acetosella L. : Suran yr Ŷd : Sheep's Sorrel. Native. Heaths (D,E,G,H,M,Q,R,Rg,Rw,S,W). Fl.5-9. Mons., Glam., Brecs., Rads., Carms., Pembs., Cards., Monts., Mer., Caerns., *Denbs.*, *Flints.*, Angl.. 275 squares, all except : **SR**89; **SS**39; **SM**50; **SO**30,41; **SH**98; **SJ**37.

The distinctions between the three species in the above subgenus are unsatisfactory.

Subgen. Acetosa (Miller) Rech. fil.

R. scutatus L. : French Sorrel. Casual (C. and S. Europe). Walls and railway banks. Fl.7. *Mons.*, *Glam.*, Pembs.. 4 squares : **SS***88*[140]; **ST**59[7]; **SN**10[112]; **SO**50[112].

R. acetosa L. : Suran y Cŵn : Common Sorrel. Native. Pastures (Ar,D,E,G,H,M,Q,R,Rw,S,Sc,W). Fl.4-6. Mons., Glam., Brecs., Rads., *Carms.*, Pembs., Cards., Monts., Mer., Caerns., Denbs., Flints., *Angl.*. 276 squares, all except : **SS**39,77; **SM**50; **SN**59; **SH**51,98.

Subgen. Rumex

R. triangulivalvis (Danser) Rech. fil. : Willow-leaved Dock. Casual (N. America). Waste places. Fl.9. Glam.. 2 squares : **ST***17*[112],*27*[112].

R. frutescens Thouars (*R. cuneifolius* Campd.) : Argentine Dock. Naturalized alien (S. America). Sand dunes. Fl.6-10. Glam.. 5 squares.
Glam.: Oxwich, SS58, GTG, 1955[93]; nr Swansea, SS69, JAW, 1939[112] and HH & CH, 1974[117]; Port Talbot, SS78, MHa, 1961[112]; Kenfig Burrows, SS78, EV, 1934[112] and TGE, 1980[91]; Cardiff, ST17, NYS & GCD, 1925[24]; Cwmbwrla, SN60, JAW, 1951[221].

R. longifolius DC. (*R. domesticus* Hartman; *R. aquaticus* auct., non L.) : Northern Dock. Casual (England). Waste places. *Glam.*, *Caerns.*. 2 squares : **ST**18[21]; **SH**43[7].

R. hydrolapathum Hudson : Tafol y Dŵr : Water Dock. Native. Streamsides (Al,Ar,M,S). Fl.6-8. Mons., Glam., Rads.*, Carms., Pembs., Cards.*, Monts., Mer., *Caerns.*, Denbs., *Flints.*, Angl.. 78 squares, map 57.

R. hydrolapathum × obtusifolius = R. × weberi Fischer-Benzon Spontaneous. Fl.8 *Glam.*, *Pembs.*. 2 squares.
Glam.: Peterston-super-Ely, ST08, HJR, 1907[107].
Pembs.: Manorbier, SS09, HWP, c1924[23].

R. hydrolapathum × conglomeratus or **crispus** Spontaneous. Fl.7-8. Mer.. 1 square.
Mer.: Afon Sgethin, SH52, PMB, 1955[112].

R. cristatus DC. : Greek Dock. Established alien (Mediterranean region). River banks and waste places (E,F,M,R). Fl.6-7. *Mons.**, Glam.. 4 squares.
Mons.: Rumney, ST27, AEW, 1920-70[226] and TGE, 1980[91]. **Glam.:** Sully, ST16, AEW, 1954[112]; Cardiff, ST17, AEW, 1973[227]; E. Cardiff, ST27, JEL, 1938[112] and AEW, 1972[112].

R. cristatus × obtusifolius = R. × lousleyi D. H. Kent Spontaneous. River bank. Fl.7. Glam.. 1 square.
Glam.: E. Cardiff, ST27, GWa, 1982[112].

R. crispus L. : Tafol Crych : Curled Dock. Native. Roadsides (D,G,M,R,Rw,S,Sc,W). Fl.6-10. Mons., Glam., Brecs., Rads., Carms., Pembs., Cards., Monts., Mer., Caerns., Denbs., Flints., Angl.. 263 squares, all except : **SS**39; **SN**75,86,87,97,99; **SO**09,26,30,40; **SH**51,60,72,82,91; **SJ**01,02,15,44.

R. crispus × hydrolapathum = R. × schreberi Hausskn. Spontaneous. Wet places. Fl.7-8. Caerns., [*Mer.*]. 1 square.
Caerns.: nr Bangor, SH57, JTCW, 1876[112].

R. crispus × obtusifolius = R. × pratensis Mert. & Koch : Meadow Dock. Spontaneous. Roadsides. Fl.6-9. Mons., Glam., Brecs., Carms., Cards., Mer., *Caerns.*, *Angl.*. 25 squares, map 58.

R. crispus × pulcher = R. × pseudopulcher Hausskn. Spontaneous. Waste places. Fl.7. *Pembs.*. 1 square.
Pembs.: St David's, SM72, An, 1914[107].

R. crispus × rupestris Spontaneous. Sand dunes. Fl.8. Glam., *Angl.*. 2 squares.
Glam.: Kenfig Burrows, SS78, JSLG, 1932[112] and JEL, 1948[112]. **Angl.:** Newborough Warren, SH46, RHR, 1970+[209] (provisional determination only).

R. crispus × sanguineus = R. × sagorski Hausskn. Spontaneous. River and roadsides. Fl.7-8. *Rads.*, Mer.. 2 squares.
Rads.: nr Hay-on-Wye, SO24, WRL, 1902[107]. **Mer.:** Arthog, SH61, PMB, 1955[112].

R. conglomeratus Murray : Tafol Mair : Clustered Dock. Native. Wet fields (Ar,G,M,R). Fl.6-9. Mons., Glam., Brecs., Rads., Carms., *Pembs.*, *Cards.*, *Monts.**, Mer., *Caerns.*, Denbs., Flints., Angl.. 176 squares, map 59.

R. conglomeratus × crispus = R. × schulzei Hausskn. Spontaneous. Grassland (D,G). Fl.7-8. Mer., *Caerns*.. 3 squares.
Mer.: Mochras, SH52, PMB, 1950+[4]; Harlech, SH53, PMB, 1955[112]; Minfford, SH63, JEL, 1953[112]; **Caerns.:** Portmadoc, SH53, GCD, c1927[27].

R. conglomeratus × obtusifolius = R. × abortivus Ruhmer Spontaneous. Meadows. Fl.8. Mer., *Angl*.. 3 squares.
Mer.: Dolgellau, SH71, PMB, 1955[112]. **Angl.:** Bodorgan, SH36, MSW & GCD, 1918[17]; nr Llanfaelog, SH37, RHR, 1970+[209].

R. conglomeratus × pulcher = R. × murettii Hausskn. Spontaneous. Grassland. Fl.7. Mons., *Pembs*.. 2 squares.
Mons.: nr Chepstow, ST59, WAS, 1903[112]. **Pembs.:** St David's, SM72, AHGA, CIS & NYS, 1931[150].

R. conglomeratus × sanguineus = R. × ruhmeri Hausskn. Spontaneous. Roadside. Mer., *Angl*.[209]. 1 square.
Mer.: Arthog, SH61, PMB, 1950+[4].

R. sanguineus L. : Tafol Coch : Wood Dock. Native. Roadsides and woods (E,G,M,Q,R,Rw,W). Fl.5-8. Mons., Glam., Brecs., *Rads.*, Carms., Pembs., *Cards.*, Monts., Mer., Caerns., Denbs., Flints., Angl.. 209 squares, map 60.

R. rupestris Le Gall : Tafolen y Traeth : Shore Dock. Native. Sand dunes (D,Sc). Fl.6-9. Glam., *Pembs*.*, Angl.*. 5 squares.
Glam.: Three Cliffs Bay, SS58, GB, 1859[37]; Kenfig Burrows, SS78, JSLG, 1932[112] and MG, 1974[99]; Southerndown, SS87, EV, 1934[112]; Merthyr Mawr, SS87, BAM, 1954[53]. **Pembs.:** Lydstep, SS09, EKH, 1957[173]. **Angl.:** Newborough Warren, SH46, An, 1951[108] and RHR, 1977[112].

R. pulcher L. : Tafol Crwth-ddail : Fiddle Dock. Colonist (England). Roadsides (E,G,Q,R,Rg,S). Fl.6-8. Mons., Glam., *Carms*., Pembs., *Monts*., *Mer*., *Caerns*., *Denbs*., *Angl*.*. 19 squares : **SS**19,*59,78*,87; **ST**17,*27*,49, 59; **SM**62,*72,90*; **SN**10,*40*; **SO**19; **SH**57,*67,74,93*; **SJ***24*.

R. obtusifolius L. : Dail Tafol : Broad-leaved Dock Native. Roadsides (E,F,G,M,R,W). Fl.5-8. Mons., Glam., Brecs., Rads., Carms., Pembs., Cards., Monts., Mer., Caerns., Denbs., Flints., Angl.. 269 squares, all except : **SR**89; **SS**38,39; **ST**26; **SM**50,71,73; **SN**21,36, 55,88,89; **SH**51.

R. obtusifolius × sanguineus = R. × dufftii Hausskn. Spontaneous. Roadsides and waste places. Fl.6-8. Glam., *Pembs*., Mer.. 4 squares.
Glam.: Cardiff, ST17, AEW, 1923[112]. **Pembs.:** Haroldston, SM91, AEE, 1930[38]. **Mer.:** Arthog, SH61, PMB, 1955[112]; Dolgellau, SH71, PMB, 1957[112].

R. dentatus L. Casual (S.Europe). Waste places. Fl.7. Glam.. 1 square : **ST**27[112].

R. obovatus Danser Casual (S. America). Waste places. Glam.. 2 squares : **ST***17*[112],*27*[112].

R. palustris Sm. : Marsh Dock. Native. Marshes. Fl.6-9. Mons., (Carms.*), [*Denbs*., *Flints*.]. 2 squares.
Mons.: Magor, ST48, AEW, 1945[112]; Undy, ST48, AEW, 1951[112] and TGE, 1980[112]. (**Carms.:** Cil-y-cwm, SN73, IMV, 1975[112]).

R. maritimus L. : Tafol Arfor : Golden Dock. Native. Marshes (Al,Ds,H,M,R). Fl.7-8. Glam., Brecs., *Pembs*.*, Monts., *Caerns*., Denbs., *Flints*.. 14 squares : **SS***48*,69,78; **ST**(*16,17,27*); **SM**80; **SO**05,12,29; **SH***56*; **SJ***08*,20,35.

R. bucephalophorus L. Casual (Mediterranean region). Waste places. Fl.8. Glam.. 1 square : **ST**16[112].

CHENOPODIACEAE

BETA L.

Beta vulgaris L.

Subsp. **vulgaris** : Betys : Cultivated Beet. Casual (garden origin). Waste places. Fl.7-11. *Mons*., Glam., *Denbs*.. 5 squares : **ST***16*[112],17[112],*27*[112],42[112]; **SH**87[7].

Subsp. **maritima** (L.) Arcangeli (*B. maritima* L.) : Betys Gwyllt : Sea Beet. Native. Sea cliffs and shingle (D,G,M,Rw,S,Sc). Fl.4-9. Mons., Glam., Carms., Pembs., *Cards*., *Mer*., Caerns., Denbs., Flints., Angl.. 92 squares, map 61.

CHENOPODIUM L.

Chenopodium botrys L. Casual (S. Europe). Waste places. Glam., *Flints*.. 2 squares : **ST***17*[112]; **SJ***27*[24].

C. ambrosioides L. Casual (Tropical America). Waste places. Fl.9. Glam., *Flints*.. 3 squares : **ST***16*[24],*27*[112]; **SJ***18*[143].

C. multifidum L. (*Roubieva multifida* (L.) Moq.). Casual (S. America). Waste places. Fl.8. Glam.. 3 squares : **ST***16*[24],*17*[112], *18*[112].

C. bonus-henricus L. : Llys y Gwrda : Good-King-Henry. Established alien (Europe). Roadsides, waste places and cultivated ground (F,Q,R,Rw,S). Fl.5-10. Mons., Glam., Brecs., Rads., Carms., *Pembs., Cards.*, Monts., Mer., Caerns., Denbs., Flints., *Angl*.. 119 squares, map 62.

C. foliosum Ascherson (*C. virgatum* (L.) Ambrosi, non Thunb.). Casual (S. Europe) Waste places. Fl.6-9. Glam., Caerns.. 3 squares : **ST**17[112],27[112]; **SH**78[112].

[**C. virgatum** Thunb. *Glam*., *Caerns*.]

C. capitatum (L.) Ascherson : Strawberry-blite. Casual (origin unknown). Waste places. Fl.8-9. Glam., [*Caerns*.]. 1 square. : **ST***27*[112].

C. glaucum L. : Troed yr Ŵydd Dderw-ddeiliog : Oak-leaved Goosefoot. Casual (Europe). Waste places (R,Rg). Fl.7-9. Mons.*, Glam., *Angl.*. 7 squares : **SS69**[198],*78*[140]; **ST06**[95],*17*[112],*27*[112],38[112]; **SH27**[29].

C. rubrum L. : Troed yr Ŵydd Ruddog : Red Goosefoot. Native. Waste places (Al,D,Ds,F,M,R,Rg,S). Fl.8-10. Mons., Glam., *Brecs.*, Carms., Pembs., *Cards.*, *Monts.**, (*Mer.*), *Caerns.*, Flints., *Angl.*. 51 squares, map 63.

[**C. botryodes** Sm. *Caerns.*]

C. hybridum L. : Troed yr Ŵydd Masarnddail : Maple-leaved Goosefoot. Casual (S. England). Waste places. Fl.9. Glam., Carms.. 4 squares : **ST***17*[112],*27*[112]; **SN31**[180],*80*[195].

C. polyspermum L. : Troed yr Ŵydd Llios-hadog : Many-seeded Goosefoot. Colonist (S. England). Cultivated ground (F,M,R). Fl.7-10. Mons., Glam., Brecs., *Rads.**, Carms., Pembs.*, Cards.*, *Monts.**, Mer.*, Caerns., *Denbs.**, *Flints.**, *Angl.*. 51 squares, map 64.

C. vulvaria L. : Llysgwyn Drewllyd : Stinking Goosefoot. Colonist (England). Waste places. Fl.9. *Mons.*, Glam., *Flints.*. 8 squares.
Mons.: Chepstow, ST59, TW, 1791[226]; Raglan, SO40, BMF, 1928[226]. **Glam.:** Port Talbot, SS78, HJR, pre 1907[140]; nr Sully, ST16, JS, c1876[218]; Cardiff, ST17, HJR, 1907[112]; Radyr, ST18, AEW, 1921[112]; Splott, Cardiff, ST27, AEW, 1925[112] and EV, 1927[112]. **Flints.:** Mostyn Quay, SJ18, AAD, 1909[143].

C. urbicum L. : Troed yr Ŵydd Syth-ddail : Upright Goosefoot. Colonist (Europe). Roadsides and waste places (F,M,R,S). Fl.9. Mons., Glam., *Cards.*, *Caerns.*, *Denbs.*, [*Mer.*]. 8 squares.
Mons.: St Mellons, ST28, AEW, 1949[112]; nr Monmouth SO51, FAR, 1892[226]. **Glam.:** Aberthaw, ST06, EV, 1923[112] and 1933[112]; Cardiff, ST17, JS, c1876[218]; Splott, Cardiff, ST27, EV, 1932[112]. **Cards.:** Tre-saith, SN25, WWB, 1930+[211]. **Caerns.:** nr Llanfairfechan, SH67, AWW & HCG, 1950[197]. **Denbs.:** Llandrillo-yn-Rhos, SH88, WBR, c1911[144].

C. murale L. : Troed yr Ŵydd Ddynad-ddail : Nettle-leaved Goosefoot. Colonist (S. England). Waste places (D,F,G,Q,R,S). Fl.(5-)8-11. Mons.*, Glam., Carms., Pembs., Cards., Caerns., Denbs.*, *Flints.**, *Angl.*, [*Mer.*]. 28 squares, map 65.

C. ficifolium Sm. : Troed yr Ŵydd Ffigys-ddail : Fig-leaved Goosefoot. Established alien (England). Cultivated ground (R,Rg). Fl.6-9. Mons., Glam., Carms., *Cards.*, *Mer.*. 12 squares : **SS78**; **ST***17*,19,27,38,48,*59*; **SN**25,40,*90*; **SO00**; **SH50**.

C. pratericola Rydb. (*C. leptophyllum* auct., non Nutt.) Casual (N. America). Waste places (R,Rg). Fl.7-10. Glam., *Flints.*. 4 squares : **SS***78*[28]; **ST16**[112],*17*[112]; **SJ***36*[18].

C. opulifolium Schrader ex Koch & Ziz : Grey Goosefoot. Established alien (Europe). Waste places (R,Rg). Fl.7-11. Glam., Caerns.. 7 squares.
Glam.: Port Talbot, SS78, HJR, pre 1907[126]; Barry, ST16, AEW, 1924[112] and EV, 1926[112]; Cardiff, ST17, AEW, 1932[112]; Radyr, ST18, RLS, 1920[112]; Splott, Cardiff, ST27, AEW, 1926[112] and EV, 1926[112]; Aberdare, SO00, HJR, 1908[112]. **Caerns.:** Llandudno Junction, SH77, JEG, 1882[112].

C. hircinum Schrader : Foetid Goosefoot. Casual (S. America). Waste places. Fl.9-10. Glam., Carms., *Cards.*, *Caerns.*, Denbs.. 9 squares : **ST***17*[27],27[112]; **SN31**[112],40[110],*58*[210]; **SO***00*[107]; **SH22**[7],*53*[15],97[112].

C. album L. (incl. *C. lanceolatum* Muhl.).
Subsp. **album** : Tafod yr Oen : Fat-hen. Native. Cultivated ground (F,G,Q,R,Rg,W). Fl.6-10. Mons., Glam., *Brecs.*, Rads., Carms., Pembs., *Cards.*, Monts., Mer., Caerns., Denbs., Flints., Angl.. 230 squares, map 66.

Subsp. **album** × subsp. **striatum** (*C. interjectum* J. Murr; *C. pseudostriatum* Zschacke; *C. substriatum* J. Murr). Spontaneous. Waste places. Fl.9-10. *Glam.*. 2 squares.
Glam.: Barry, ST16, GCD, 1926[25]; Cardiff, ST17, GCD, 1924[23].

Subsp. **striatum** (Krašan) J. Murr (*C. striatum* (Krašan) J. Murr Casual (C. and E. Europe). Waste places. Mons., Glam.. 3 squares : **ST***16*[112],*17*[24],38[112].

[**C. album** × **berlandieri** *Mer.*]

C. album × **opulifolium** = **C** × **preissmannii** J. Murr Spontaneous. Waste places. *Glam.*. 2 squares.
Glam.: Barry, ST16, GCD, 1926[17&27]; Leckwith, ST17, AEW, 1922[21].

[**C. album** × **suecicum** = **C.** × **fursajewii** Aellen and Iljin *Glam.*]

C. suecicum J. Murr Casual (N. Europe). Waste places. Fl.9-10. Mons., Glam., Pembs.. 6 squares : **SS***69*[198],*78*[140]; **ST***27*[112],49[112]; **SM91**[112]; **SO***00*[140].

C. berlandieri Moq. subsp. **zschackei** (J. Murr) Zobel : American Goosefoot. Casual (N. America). Waste places. *Glam.*. 2 squares : **SS***78*[107]; **ST17**[29].

C. graveolens Lag. and Rodr. (*C. incisum* Poiret). Casual (C. America). Waste places. Fl.9. Glam.. 1 square : **ST27**[112].

C. probstii Aellen Casual (origin unknown, ? S. America). Waste places. Fl.9-10. Carms.. 1 square : **SN31**[112]. Overlooked, undoubtedly much more frequent than this solitary record indicates.

C. pumilo R. Br. (*C. carinatum* auct., non R. Br.) : Australian Goosefoot. Casual (Australia). Waste places. *Glam.*. 1 square : **ST***17*[27&45].

ATRIPLEX L.

Atriplex hortensis L. : Garden Orache. Casual (garden origin). Waste places (M,R,Rg). Mons., Glam., *Denbs.*. 4 squares : **ST**16[116],38[116]; **SO**21[116]; **SH***87*[210].

A palaestina Boiss. Casual (Middle East). Waste places. *Glam.*. 1 square : **ST**27[25].

A. laciniata L. (*A. sabulosa* Rouy) : Llygwyn Arianaidd : Frosted Orache. Native. Sea-beaches (R,S). Fl.6-8. *Mons.**, Glam., Carms.*, Pembs., Cards., Mer., Caerns., *Denbs.*, Flints.*, Angl.. 46 squares, map 67.

[**A. tatarica** L. *Glam.*]

A. littoralis L. : Llygwyn Arfor : Grass-leaved Orache. Native. Salt-marshes (D,M,R,S). Fl.6-9. *Mons.*, Glam., Carms., Pembs.*, *Mer.*†, Caerns., Denbs., Flints., Angl., [*Cards.*]. 41 squares, map 68.

A. patula L. : Llygwyn Culddail : Common Orache. Native. Cultivated ground (Ar,D,F,M,Q,R,Rg,Rw,S). Fl.6-10. Mons., Glam., Brecs., Rads., Carms., Pembs., Cards., Monts., Mer., Caerns., Denbs., Flints., *Angl.*. 200 squares, map 69.

A. prostrata Boucher ex DC. (*A. hastata* auct. angl., non L.; *A. triangularis* Willd.) : Llygwyn Tryfal : Spear-leaved Orache. Native. Waste places (E,F,G,M, Q,R,S,Sc). Fl.6-9. Mons., Glam., *Brecs.*, Rads., Carms., Pembs., Cards., *Monts.*, Mer., Caerns., Denbs., Flints., *Angl.*. 136 squares, map 70.

A. glabriuscula Edmondston : Llygwyn y Tywod : Babington's Orache. Native. Shingle and sandy beaches (M,S,Sc). Fl.6-9. *Mons.*, Glam., Carms., Pembs., Cards., Mer., Caerns., Denbs., *Flints.*, Angl.. 62 squares, map 71.

A. longipes Drejer Specimens resembling this species have been collected from Glam. (Gower) but await specialist confirmation.

HALIMIONE Aellen

Halimione portulacoides (L.) Aellen : Helys Can : Sea-purslane. Native. Salt-marshes (D,M,S,Sc). Fl.7-10. Glam., Carms., Pembs., (Monts.*), Caerns., Denbs., Flints., Angl.. 46 squares, map 72.

H. pedunculata (L.) Aellen Casual (S. England). Waste places. *Glam.*, [*Carms.*]. 1 square : **SS***78*[140].

AXYRIS L.

Axyris amaranthoides L. : Russian Pigweed. Casual (Russia). Waste places. Fl.8-9. Mons., Glam., *Caerns.*. 3 squares : **ST***27*[112], *39*[112]; **SH**53[16].

KOCHIA Roth

Kochia scoparia (L.) Schrader : Summer-cypress. Casual (S. Russia). Waste places. *Mons.*, Glam., *Caerns.*. 3 squares : **ST**17[112]; **SO**21[159]; **SH**67[18].

SARCOCORNIA A. J. Scott.

Sarcocornia perennis (Miller) A. J. Scott (*Salicornia perennis* Miller; *S. lignosa* Woods; *Arthrocnemum perenne* (Miller) Moss) : Perennial Glasswort. Native. Salt-marshes. Fl.5-8. Mer., [Glam., *Cards.*]. 2 squares. **Mer.:** Pen-sarn, SH52, PMB, 1953[112]; Llanbedr, SH52, EHC, 1953[112]; Barmouth, SH61, PMB, 1952[112]; Fairbourne, SH61, EVW, pre 1955[3] and PMB, 1955[112].

SALICORNIA L.

Salicornia agg. Mons., Glam., Carms., Pembs., Cards., Mer., Caerns., Denbs., Flints., Angl.. 58 squares, map 73.

S. europaea L. (*S. stricta* auct., non Dumort.) : Llyrlys : Glasswort. Native. Salt-marshes. Fl.8-10. Mons., Glam., Carms., Pembs., Cards., Mer., Caerns., Denbs., Flints., Angl.. 40 squares, map 74.

S. ramosissima J. Woods (*S. gracillima* Dumort.; *S. appressa* Dumort.) : Llyrlys Gorweddol : Bushy Glasswort. Native. Salt-marshes. Fl.9-10. Mons., Glam., Carms., Pembs.*, Cards.*, Mer., Angl.*. 25 squares, map 75.

S. pusilla J. Woods (*S. disarticulata* Moss). Native. Salt-marshes. Fl.9-10. Glam., *Carms.*, Pembs.*, *Cards.**, *Caerns.**, *Denbs.**, 14 squares : **SS***49*,58,59,*78*,79; **ST**06; **SM**80,90; **SN**00,14,*40*; **SH**45,77,98.

S. nitens P. W. Ball & Tutin Native. Salt-marshes. Fl.8-10. *Carms.*, Cards.. 2 squares.
Carms.: Llansteffan, SN31, LAE, 1962[7], **Cards.:** Dyfi Estuary, SN69, AOC, 1976[112].

S. fragilis P. W. Ball & Tutin Native. Salt-marshes. Fl.9-10. *Glam.**, Cards., *Caerns.*, *Denbs.*. 6 squares.
Glam.: Llanmadog, SS49, QONK, 1980[157]; Llandimore, SS49, QONK, 1980[157]; Three Cliffs Bay, SS59, QONK 1980[157]; Wern-ffrwd, SS59, QONK, 1980[157]; Jersey Marine, SS79, QONK, 1980[157]. **Cards.:** Teifi Estuary, SN14, AOC, 1976[112]. **Caerns.:** Degannwy, SH77, JMB, 1974[242]. **Denbs.:** Foryd nr Rhyl, SH98, JMB, 1971[242].

S. lutescens P. W. Ball & Tutin Native. Salt-marshes. Fl.9-10. Glam., Pembs., *Mer.*, *Angl.*. 10 squares : **SS**87; **ST**06; **SM**80,90; **SN**00,03,14; **SH**52,57,61.

S. dolichostachya Moss Native. Salt-marshes. Fl.8-10. Mons., Glam., Carms., Pembs.*, Cards., *Mer.**, Denbs.*, *Flints.**. 18 squares : **SS**49,58,59; **ST**06,17,*27*; **SM**80; **SN**00,01,03,14,*40*,69; **SH**52,53,61,77; **SJ**08.

[**S. dolichostachya** × **europaea** *Mer.*]

SUAEDA Forskål ex Scop.

Suaeda vera J. F. Gmelin (*S. fruticosa* auct.) : Shrubby Sea-blite. Denizen (S. England). Salt-marshes. Fl.8-9. Glam., [*Denbs.*]. 1 square.
Glam.: Cardiff, ST17, JS, 1876[140] and ABP & AEW, 1969[112]. Other Glamorgan records from Port Tennant (**SS**69) and Flat Holm (**ST**26) are probably errors.

S. maritima (L.) Dumort : Helys Unflwydd : Annual Sea-blite. Native. Salt-marshes (Ar,M,R,S). Fl.7-10. Mons., Glam., Carms., Pembs., Cards., Mer., Caerns., Denbs., Flints., Angl.. 57 squares, map 76.

S. altissima (L.) Pallas Casual (S.E. Europe). Waste places. Fl.10. *Glam.*. 1 square : **SS***78*[140].

SALSOLA L.

Salsola kali L.

Subsp. **kali** : Helys Ysbigog : Prickly Saltwort. Native. Sea-shores (Ar,D,Rg,S). Fl.5-9. Glam., Carms., Pembs., *Cards.*, Mer., Caerns., Denbs., Flints., Angl.. 53 squares, map 77.

Subsp. **ruthenica** (Iljin) Soó (incl. *S. pestifer* A. Nelson) : Spineless Saltwort. Casual (Europe). Waste places. Fl.9. Glam.. 1 square : **ST**17[112].

AMARANTHACEAE

AMARANTHUS L.

Amaranthus hybridus L. : Prince's feather. Casual (America). Waste places. Fl.9-10. *Mons.*, Glam., *Carms.*. 4 squares : **ST**17[112],*27*[112], 38[245]; **SN**40[7].

A. caudatus L. : Love-lies-bleeding. Casual (S. America). Waste places. *Mons.*, *Carms.*. 2 squares : **ST**38[91]; **SN**40[161].

A. retroflexus L. : Chwyn Moch : Common Amaranth. Casual (N. America). Cultivated ground and waste places (F,R,Rg). Fl.7-10. Mons., Glam., *Brecs.*, Carms., *Cards.*, *Caerns.*. 11 squares : **SS***78*,87; **ST***16*,17,27,38; **SN**51,*58*; **SO***00*,*24*; **SH***53*.

A. quitensis Kunth Casual (S. America). Waste places. *Glam.*. 1 square : **ST***27*[26].

A. spinosus L. Casual (Tropics). Waste places. Fl.9-10. Glam.. 2 squares : **ST***17*[112],*27*[112].

A. acutilobus Uline & Bray Casual (N. America). Waste places. Glam.. 1 square : **ST***27*[112].

[**A. blitoides** S. Watson *Glam.*]

A. standleyanus Parodi ex Covas (*A. vulgatissimus* auct.). Casual (Argentina). Waste places. Fl.10. Mons., Glam.. 2 squares : **ST**27[112],38[112].

A. albus L. : White Pigweed. Casual (N. America). Cultivated ground and waste places. Fl.9. Mons., Glam.. 5 squares : **SS***78*[140]; **ST***16*[24],17[112],*27*[112],*59*[112].

A. graecizans L. (*A. angustifolius* Lam.; *A. sylvestris* Vill.). Casual (S. Europe). Waste places. Fl.9. Glam.. 4 squares : **SS***78*[107]; **ST***16*[112],17[112],27[112].

A. deflexus L. Casual (America). Waste places. *Glam.*. 1 square : **SS***78*[107].

A. lividus L. Casual (S. Europe). Waste places. *Glam.*. 3 squares : **SS***78*[146]; **ST***16*[24],17[112].

PHYTOLACCACEAE

PHYTOLACCA L.

Phytolacca latbenia Maxim. (*P. americana* auct., non L.) : Pokeweed. Casual (E. Asia). Waste places. Fl.9. Mons., Glam.. 2 squares : **ST**17[112],*59*[112].

AIZOACEAE

CARPOBROTUS N.E. Br.

Carpobrotus edulis (L.) N.E. Br. : Hottentot-fig. Naturalized alien (S. Africa). Sea cliffs. Fl.6-7(-10). *Glam.*, Pembs., Caerns., Angl.. 7 squares.
Glam.: Barry, ST16, RLS, 1925[24]. **Pembs.:** Dale, SM80, MGe, c1960[94]; Tenby, SN10, McW, 1954[181] and GEv, 1975[112]. **Caerns.:** Porth Dinllaen, SH24, APC, 1977[74]; Aber-soch, SH32, APC, 1977[74] and RGE, 1978[112]. **Angl.:** Rhoscolyn, SH27, RHR, 1970+[208]; Trearddur Bay, SH27, RHR, 1970+[208]; nr Amlwch, SH49, AJ, 1962[112] and RHR, 1972[208].

MESEMBRYANTHEMUM L.

Mesembryanthemum crystallinum L. (*Cryophytum crystallinum* (L.) N.E. Br.). Casual (Mediterranean region). Sea coast. *Glam.*[170], *Angl.*[119].

LAMPRANTHUS N.E. Br.

Lampranthus falciformis (Haw.) N. E. Br. Naturalized alien (S. Africa). Sea cliffs and quarry. Pembs.. 1 square.
Pembs.: St Ishmael's, SM80, TAWD, 1919-1978[49]; Dale, SM80, TAWD, 1978[49] and BRF, 1978[112].

CLERETUM N.E. Br.

Cleretum bellidiforme (Burm. fil.) Rowley (*Mesembryanthemum criniflorum* L. fil.; *Dorotheanthus bellidiformis* (Burm. fil.) N.E. Br.) : Livingstone-daisy. Casual (S. Africa). Waste places. *Glam.*. 1 square : **SS**87[73].

PORTULACACEAE

MONTIA L.

Montia fontana L. : Gwlyddyn y Ffynnon : Blinks. Native. Moist grassland, spring heads, rills, etc. (Al,Ar, D,Ds,F,G,H,M,Q,R,W). Fl.3-9. Mons., Glam., Brecs., Rads., Carms., Pembs., Cards., Monts., Mer., Caerns., Denbs., *Flints.*, *Angl.*. 207 squares, map 78.

Subsp. **fontana** Native. (Al,Ar,Q,W). Fl.4-8. Pembs., *Cards.*, *Monts.**, Mer., Caerns., *Denbs.*, [*Glam.*, *Brecs.*, *Angl.*]. 29 squares, map 79.

Subsp. **variabilis** Walters Native. (Ar,H,M,R). Fl.5-8. Mons., Glam., Brecs., Rads., Carms., Pembs., Cards., Monts., Mer., Caerns., Denbs., *Flints.*, *Angl.*. 65 squares, map 80.

Subsp. **amporitana** Sennen (*M. fontana* subsp. *intermedia* (Beeby) Walters). Native. (Ar,H,M). Fl.4-9. Mons., Glam., Brecs., Rads., Carms., Pembs., Cards., Monts., Mer., Caerns., *Angl.*. 68 squares, map 81.

Subsp. **chondrosperma** (Fenzl) Walters Native. (Al,Ar,D,Ds,F,G,H,M,R,Sc). Fl.3-6. Mons.*, Glam., Brecs., Rads., *Carms.**, Pembs., Cards., Mer., Caerns., *Denbs.**, Angl., [*Monts.*, *Flints.*]. 51 squares, map 82.

M. perfoliata (Donn ex Willd.) Howell (*Claytonia perfoliata* Donn ex Willd.) : Springbeauty. Established alien (N. America). Waste places (D,M,R,W). Fl.5-7. Glam., *Carms.**, Cards., Mer., *Denbs.**, *Flints.**, [*Mons.*]. 15 squares : **SS**58,68,87,*97*; **ST**06,16,17; **SN**14, 31; **SH**50,52,*53*; **SJ**17,35,36.

M. sibirica (L.) Howell (*Claytonia alsinoides* Sims) : Pink Purslane. Naturalized alien (N. America). Woods and waste places (Ar,E,H,M,R,Rg,Rw,W). Fl.5-8. Mons., Glam., Brecs., Rads.*, Carms.*, *Pembs.**, Cards., Mer., *Caerns.*, Denbs.*, *Flints.*, Angl.*. 33 squares, map 83.

CARYOPHYLLACEAE

ARENARIA L.

Arenaria balearica L. : Mossy Sandwort. Naturalized alien (Mediterranean region). Rocky places and walls. Fl.4. Glam., *Monts.*, *Caerns.*, *Angl.*. 4 squares.
Glam.: Craig Llanishen, ST18, JWD, 1954[112].**Monts.:** nr Llyn Vyrnwy, SH92, DC, 1946[45]. **Caerns.:** nr Llandudno, SH78, An, 1950+[7]. **Angl.:** Holy Island, SH27, An, no date[195].

A. serpyllifolia L. Mons., Glam., Brecs., Rads., Carms., Pembs., Cards., Monts., Mer., Caerns., Denbs., Flints., Angl.. 195 squares, map 84.

Subsp. **serpyllifolia** : Tywodwlydd Gwrywddail : Thyme-leaved Sandwort. Native. Cultivated ground and walls (D,E,G,M,Q,R,Rw,S,Sc,W). Fl.5-11. Mons., Glam., Brecs., Rads., Carms., Pembs., Cards., Monts., Mer., Caerns., Denbs., Flints., Angl.. 151 squares, map 85.

Subsp. **macrocarpa** (Lloyd) F. H. Perring & P. D. Sell (*A. serpyllifolia* var. *macrocarpa* Lloyd). Native. Dunes (D,Sc). Fl.4-6(-8). Glam., Pembs., *Mer.*, Caerns., Denbs., Angl.. 8 squares.
Glam.: Pennard, SS58, HJR, 1903[112]; Kenfig Burrows, SS78, EV, 1927[112]. **Pembs.:** nr Penally, SS19, EV, 1928[112]. **Mer.:** nr Barmouth, SH61, An, 1930+[194]. **Caerns.:** nr Bangor, SH57, JEG, 1878[112]; Gt Ormes Head, SH78, JEG, 1882[112]; Llandudno, SH78, AL, 1901[112]. **Denbs.:** nr Llanddulas, SH97, AW, 1929[112]. **Angl.:** Newborough Warren, SH46, EV, 1937[112].

Subsp. **leptoclados** (Reichenb.) Nyman (*A. leptoclados* (Reichenb.) Guss.) : Slender Sandwort. Native. Cultivated ground and walls (D,E,F,G,Q,R, Rg,Rw,S). Fl.6-11. Mons., Glam., *Brecs.*, *Rads.*, Carms., Pembs., *Cards.*, *Monts.*, Mer., Caerns., *Denbs.*, Flints., *Angl.*. 84 squares, map 86.

MOEHRINGIA L.

Moehringia trinervia (L.) Clairv. : Tywodlys Trinerf : Three-nerved Sandwort. Native. Hedges (E,Q,R,Rw,W). Fl.3-9. Mons., Glam., Brecs., Rads., Carms., Pembs., Cards., Monts., Mer., Caerns., Denbs., Flints., *Angl.*. 216 squares, map 87.

MINUARTIA L.

Minuartia hybrida (Vill.) Schischkin : Tywodlys Deilfain : Fine-leaved Sandwort. Established alien (E. England). Gravelly places (Q,R,Rw). Fl.5. Mons., Glam., [*Rads.*, *Monts.*, *Caerns.*, *Flints.*, *Angl.*]. 7 squares.
Mons.: Tintern, SO50, KEB, 1964[93]; nr Monmouth, SO51, SGC, 1944[112]. **Glam.:** nr Porth-cawl, SS87, JBL, pre 1907[140]; Peterston-super-Ely, ST07, HJR, 1904[112]; Cardiff, ST17, JS, 1876[218]; St Fagans, ST17, EV, 1948[112] and AEW, 1972[112]; Hirwaun, SN90, BAW, pre 1907[223]; Aberdare, SO00, BAW, pre 1907[223].

M. verna (L.) Hiern : Tywodlys y Gwanwyn : Spring Sandwort. Native. Mountain rocks (M,Q,R). Fl.5-8. Brecs., *Cards.**, Mer., Caerns., *Denbs.*, Flints.. 18 squares : **SN**68; **SO**02; **SH**54,*62*,65,71,72,77,78,88; **SJ**07,08,16,17,18,24,26,27.

M. sedoides (L.) Hiern (*Cherleria sedoides* L.) : Eilun Briweg : Cyphel. Native. Mountain rocks. Fl.7. Caerns.. 1 square.
Caerns.: Snowdon, SH65, An, 1870[112] (there is some doubt about whether this specimen was actually collected in North Wales.).

HONKENYA Ehrh.

Honkenya peploides (L.) Ehrh. : Tywodlys Arfor : Sea Sandwort. Native. Sea-shores (Ar,D,S). Fl.5-9.(-11). *Mons.*, Glam., Carms., Pembs., Cards., Mer., *Caerns.*, Denbs., Flints., Angl.. 73 squares, map 88.

STELLARIA L.

Stellaria nemorum L. : Tafod yr Edn y Goedwig : Wood Stitchwort. Native. Woods (M,W). Fl.5-8. Mons., Brecs., Carms., Cards., Monts., Mer., *Caerns.**, Denbs., *Flints.*, [*Glam.*, *Pembs.*]. 24 squares, map 89.

Subsp. **nemorum** Mons., [*Carms.*, *Mer.*]. 3 squares. **Mons.:** Skenfrith, SO42, RGE, 1980[112]; Tintern, SO50, JEL, 1935[112]; Pen-allt, SO51, SGC, 1941[112]; Hadnock, SO51, SGC, 1942[112]. These populations are being, or have been, lost through introgression with subsp. **glochidisperma**.

Subsp. **glochidisperma** Murb. Mons., Brecs.*, Carms.*, Cards., Monts., Mer.. 12 squares : **ST**59; **SN**62,79,83; **SO**03,11,50; **SH**53,61,63,*70*,71.

Subsp. **glochidisperma** × subsp. **nemorum** Spontaneous. Woods. *Mons.*. 1 square. **Mons.:** nr Monmouth, SO51, TGE, 1979[91].

S. media (L.) Vill. : Gwlydd y Gwyddau : Common Chickweed. Native. Cultivated ground (E,F,G,H,M,R, Rg,S,W). Fl.4-10. Mons., Glam., Brecs., Rads., Carms., Pembs., Cards., Monts., Mer., Caerns., Denbs., Flints., Angl.. 278 squares, all except : **SR**89; **SM**50; **SN**36; **SH**51.

S. neglecta Weihe : Brechlys Mwyaf : Greater Chickweed. Native. Hedges (D,E,M,Q,R,W). Fl.4-11. Mons., Glam., Brecs., *Rads.*, Carms.*, Pembs., Cards., Monts., Mer., Caerns., Denbs., Flints., Angl.. 118 squares, map 90.

S. pallida (Dumort.) Piré (*S. apetala* auct., non Ucria) : Gwlydd y Tywod : Lesser Chickweed. Native. Sandy places (Ar,D,G,M,Q,R). Fl.4-7. Mons., Glam., Brecs.*, Carms., Pembs., Cards.*, Monts., Mer., Caerns., Denbs.*, *Flints.*, Angl.. 52 squares, map 91.

S. holostea L. : Serenllys Mawr : Greater Stitchwort. Native. Hedges (E,G,M,R,Rw,W). Fl.3-6(-8). Mons., Glam., Brecs., Rads., Carms., Pembs., Cards., Monts., Mer., Caerns., Denbs., Flints., *Angl.*. 269 squares, all except : **SS**19,38,39,77; **ST**26; **SM**50,62,71,73,84; **SN**86; **SH**29; **SJ**37.

S. uliginosa Murray (*S. alsine* Grimm) : Tafod yr Edn y Gors : Bog Stitchwort. Native. Bogs and marshes (Al,Ar,E,G,H,M,Q,R,W). Fl.4-9. Mons., Glam., Brecs., Rads., Carms., Pembs., Cards., Monts., Mer., Caerns., Denbs., *Flints.*, Angl.. 260 squares, all except : **SR**89; **SS**19,38,39,77,96; **ST**26,58,59; **SM**50,71; **SN**36,59,98; **SO**20,30,41; **SH**29,51,97,98; **SJ**37.

S. palustris Retz. : Tafod yr Edn Llwydlas : Marsh Stitchwort. Native. Marshes (Ar,Mm). Fl.6-8. Pembs., Monts.*, *Caerns.*†, Angl.*, [*Glam.*, *Mer.*, *Denbs.*]. 9 squares.
Pembs.: nr Mathry, SM83, AEW, 1939[112] and TAWD, 1960+[80]; Letterston, SM93, KPe & SD, 1977[112]. **Monts.:** nr Forden, SJ20, VJM, 1950+[116]; Wern, SJ21, VJM, 1956[112] and EDP, 1981[3]. **Caerns.:** Gt Ormes Head, SH78, RT, 1838[107]. **Angl.:** nr Aberffraw, SH36, MHB, 1967[104]; Llyn Coron, SH37, WEH, 1968[112]; Cors Ddyga, SH47, WW, c1800[107] and RHR, 1958[112]; Cors Erddreiniog, SH48, JEG, pre 1895[101] and THB, 1979[208].

S. graminea L. : Tafod yr Edn Lleiaf : Lesser Stitchwort. Native. Pastures and hedges (Ar,D,E,G,M,R,W). Fl.6-10. Mons., Glam., Brecs., Rads., Carms., Pembs., Cards., Monts., Mer., Caerns., Denbs., Flints., Angl.. 267 squares, all except : **SR**89; **SS**38,39,96; **ST**06,26; **SM**50,71,73; **SN**36,59; **SH**51,88,97; **SJ**37.

CERASTIUM L.

Cerastium dichotomum L. Casual (S. Europe). Waste places. Glam., *Mer.*. 2 squares : **ST**27[112]; **SH**60[76].

C. biebersteinii DC. : Snow-plant. Casual (Crimea). Roadsides. Glam.. 2 squares : **ST**06[99],08[252].

C. tomentosum L. : Snow-in-Summer. Established alien (S.E. Europe). Hedges and walls (D,E,F,G,Q, R,Rw,S). Fl.4-7(-10). Glam., Rads., Carms., Pembs., Cards., *Mer.*, Caerns., Denbs., Flints., *Angl.*. 69 squares, map 92.

C. arvense L. : Clust Llygoden y Caeau : Field Mouse-ear. Naturalized alien (England). Sandy and waste places (E,G,R,Rg,Rw). Fl.5-7. *Mons.*, Glam., Brecs.*, Rads., *Pembs.**, *Cards.**, Monts., Mer., *Caerns.*, Denbs., *Flints.*, Angl.. [*Carms.*]. 20 squares, map 93. If native then only in the north-east.

C. alpinum L. : Clust Llygoden Mynyddig : Alpine Mouse-ear. Native. Mountain rocks. Fl.7. Caerns.. 2 squares.
Caerns.: Snowdon, SH65, TH, 1866[113]; Carnedd Llywelyn, SH66, JEG, 1888[112] and ER, 1964[207].

C. alpinum × **arcticum** = **C.** × **blyttii** Baenitz Spontaneous. Mountain rocks. *Caerns.*. 1 square. **Caerns.:** Snowdon, SH65, JKM,1970+[217].

C. arcticum Lange : Clust Llygoden Mynyddig Llydanddail : Arctic Mouse-ear. Native. Mountain rocks. Fl.6-8. Caerns.. 2 squares.
Caerns.: Clogwyn Du'r Arddu, SH55, JEG, 1891[112] and JHS, 1925[112]; Cwm Brwynog, SH55, DPY, 1943[45]; Maen-du'r Arddu, SH55, ER, 1959[207]; Clogwyn y Garnedd, SH65, TB, 1827[7] and ER, 1941[207]; by Llyn Clogwyn, SH65, DGC, 1924[112]; Clogwyn Coch, SH65, ER, 1947[207].

C. arcticum × **fontanum** = **C.** × **richardsonii** Druce Spontaneous. Mountain rocks. *Caerns.*. 1 square. **Caerns.:** Snowdon, SH65, GCD, c1909[11] and JKM, 1950+[217].

C. fontanum Baumg. subsp. **glabrescens** (G. F. W. Meyer) Salman *et.al.* (*C. fontanum* subsp. *triviale* (Link) Jalas; *C. vulgatum* auct.; *C. holosteoides* Fries) : Clust Llygoden Culddail : Common Mouse-ear. Native. Pastures (Ar,D,Ds,E,F,G,H,M,Q,R,Rw,S). Fl.3-10. Mons., Glam., Brecs., Rads., Carms., Pembs., Cards., Monts., Mer., Caerns., *Denbs.*, Flints., Angl.. 278 squares, all except : **SS**39; **SM**50,73; **SH**51.

C. glomeratum Thuill. : Clust Llygoden Llydanddail : Sticky Mouse-ear. Native. Banks and sand dunes (Ar, D,E,F,G,H,M,Q,R,Rg,Rw,S,W). Fl.4-9. Mons., Glam., Brecs., Rads., Carms., Pembs., Cards., Monts., Mer., Caerns., Denbs., Flints., Angl.. 221 squares, map 94.

C. semidecandrum L. : Clust Llygoden Fach : Little Mouse-ear. Native. Dry banks (D,E,G,H,Q,R,S). Fl.4-6. Mons., Glam., *Brecs.**[196], *Rads.*, Carms., *Pembs.*, Cards., Monts., Mer., Caerns., *Denbs.*, Flints., Angl.. 79 squares, map 95.

C. pumilum Curtis : Dwarf Mouse-ear. Native. Limestone grassland. Fl.5. *Pembs.*, *Caerns.*, *Denbs.*, [*Glam.*, *Cards.*]. 3 squares.
Pembs.: Newport, SN03, GCD, 1920[19]. **Caerns.:** Gt Ormes Head, SH78, ERe, 1920[19] and MCFP & JHTa, 1950+[154]. **Denbs.:** Colwyn Bay, SH87, HWD, 1909[21].

C. diffusum Pers. (*C. tetrandrum* Curtis; *C. atrovirens* Bab.) : Clust Llygoden Pedwar-gwryw : Sea Mouse-ear. Native. Sand dunes and rocks (D,G,M,Q,R,Rb,S,Sc). Fl.3-7. Mons., Glam., Brecs.*, Carms., Pembs., Cards., *Monts.**, Mer., Caerns., Denbs., Flints., Angl.. 105 squares, map 96

MOENCHIA Ehrh.

Moenchia erecta (L.) P. Gaertner, B. Meyer & Scherb. : Cornwlyddyn Syth : Upright Chickweed. Native. Gravelly places (E,G,H,Q,R,W). Fl.5-6. Brecs., Rads., *Carms.*, Pembs., Cards., Monts., Mer., Caerns., Denbs., *Flints.*, Angl., [*Mons.*, *Glam.*]. 104 squares, map 97.

MYOSOTON Moench

Myosoton aquaticum (L.) Moench : Llinesg y Dŵr : Water Chickweed. Native. Streamsides (Ar,M). Fl.7-10. Mons., Glam., Brecs., *Rads.*, *Carms.*, *Pembs.**, Monts., *Caerns.*[230], Denbs., *Flints.*, [*Cards.*, *Mer.*]. 44 squares, map 98.

SAGINA L.

Sagina nodosa (L.) Fenzl : Corwlydd Clymog : Knotted Pearlwort. Native. Moist heaths (D,Ds,G,H,M, Q,R,Rw). Fl.5-9. Glam., *Brecs.*, *Rads.*, Carms., Pembs., Cards., Mer., Caerns., Denbs., *Flints.*, Angl., [*Mons.*]. 79 squares, map 99.

S. subulata (Swartz) C. Presl : Corwlyddyn Mynawydaidd : Heath Pearlwort. Native. Heaths and sandy pastures (G,H,Q,Sc). Fl.4-9. Brecs.*, *Carms.*, Pembs., Cards., *Monts.**, Mer., Caerns., *Denbs.*, *Flints.*, Angl.. 60 squares, map 100.

S. procumbens L. : Corwlyddyn Gorweddol : Procumbent Pearlwort. Native. Pastures, paths, etc., (Ds,E,G,H,M,Q,R,Rg,Rw). Fl.4-10. Mons., Glam., Brecs., Rads., Carms., Pembs., Cards., Monts., Mer., Caerns., Denbs., Flints., Angl.. 269 squares, all except : **SR**89; **SS**39; **SM**50; **SN**00,04,12,21,22,87,88; **SH**51,98; **SJ**37.

S. procumbens × subulata = S. × micrantha Boreau ex É. Martin Spontaneous. Dry stony track. *Mer.*. 1 square.
Mer.: Barmouth, SH61, PMB, 1961[4].

S. apetala Ard.

Subsp. **apetala** (*S. ciliata* Fries) : Corwlyddyn Canghenog : Fringed Pearlwort. Native. Heaths (D,G,H,Q,R,Rg,Rw). Fl.5-10. Mons., Glam., *Brecs.*, *Rads.*, Carms., Pembs., Cards., *Monts.*, Mer., Caerns., Denbs., Flints.*, Angl.. 85 squares, map 101.

Subsp. **erecta** (Hornem.) F. Hermann (*S. apetala* auct.) : Corwlyddyn Anaf-flodeuog : Annual Pearlwort. Denizen (England). Walls, railways and paths (D,E,F,Q,R,Rg,Rw,S). Fl.5-9(-12). Mons., Glam., Brecs., Rads., *Carms.*, Pembs., Cards., Monts., Mer., Caerns., Denbs., Flints., Angl.. 156 squares, map 102.

S. maritima G. Don fil. : Corwlyddyn Arfor : Sea Pearlwort. Native. Rocky and sandy places by the sea (D,Q,R,S,Sc). Fl.4-8. Mons., Glam., Carms., Pembs., Cards., Mer., *Caerns.*, Denbs., Flints., Angl.. 72 squares, map 103.

SCLERANTHUS L.

Scleranthus perennis L. : Dinodd Barhaol : Perennial Knawel. Native. Rocky places. Fl.5. Rads., [*Glam.*, *Mer.*]. 1 square.
Rads.: Stanner Rocks, SO25, AHT & EV, 1923[112] and ARP & RGE, 1980[93].

S. annuus L. : Dinodd Flynyddol : Annual Knawel. Native but frequently introduced. Dry pastures and cultivated ground (F,Q,R,Rg,Rw). Fl.5-7. Mons., Glam., Brecs., Rads., Carms., Pembs., Cards., Monts., Mer., Caerns., *Denbs.*, *Flints.*, *Angl.*. 95 squares, map 104.

CORRIGIOLA L.

Corrigiola litoralis L. : Corglymig . Strapwort. Casual (S. England). Waste places. Fl.7-8. Glam.. 1 square : **ST***16*[112].

PARONYCHIA Miller

Paronychia argentea Lam. Casual (S. Europe). Waste places. Fl.7. Glam.. 1 square : **ST***27*[112].

HERNIARIA L.

Herniaria glabra L. : Llys y Fors : Smooth Rupturewort. Casual (S. England). Waste places. Fl.7. Glam.. 2 squares : **ST***16*[112],*27*[24].

H. ciliolata Melderis (*H. ciliata* Bab., non Clairv.) : Fringed Rupturewort. Casual (S. England). Waste places. Fl.7. *Glam.*. 2 squares : **ST***16*[22],*17*[140].

H. hirsuta L. (incl. *H. cinerea* DC.) : Hairy Rupturewort. Casual (C. and S. Europe). Waste places. Fl.8. Glam.. 1 square : **ST***27*[112].

ILLECEBRUM L.

Illecebrum verticillatum L. : Clymogyn Troellog : Coral-necklace. Casual. (S. England). Waste places. Fl.7. Mons., Glam., Mer.. 3 squares : **ST***17*[112],38[112]; **SH**93[112].

POLYCARPON Loefl. ex L.

Polycarpon tetraphyllum (L.) L. : Gorhadog : Four-leaved Allseed. Casual (S. England). Waste places (D,R). Fl.7. Glam., *Mer.*. 5 squares : **SS***88*[82]; **ST***16*[112],*17*[112],*27*[112]; **SH***61*[126].

SPERGULA L.

Spergula arvensis L. : Troellig yr Ŷd : Corn Spurrey. Native. Cultivated ground (F,M,R,Rg). Fl.5-9. Mons., Glam., Brecs., Rads., Carms., Pembs., Cards., Monts., Mer., Caerns., Denbs., Flints., Angl.. 239 squares, map 105.

SPERGULARIA (Pers.) J. & C. Presl

Spergularia rupicola Lebel ex Le Jolis : Tywodwlydd y Morgreigiau : Rock Sea-spurrey. Native. Sea cliffs (G, M,Q,Sc). Fl.5-9. Glam., Carms., Pembs., Cards., Mer., Caerns., Denbs., *Flints.**, Angl., [*Mons.*]. 64 squares, map 106.

S. media (L.) C. Presl (*S. marginata* Kittel) : Troellys Mawr : Greater Sea-spurrey. Native. Salt-marshes (Ar, M,S). Fl.5-10. Mons., Glam., Carms., Pembs., *Cards.*, Mer., Caerns., *Denbs.**, Flints., Angl.. 62 squares, map 107.

S. marina (L.) Griseb. (*S. salina* J. & C. Presl) : Tywodwlydd y Morfa Lleiaf : Lesser Sea-spurrey. Native. Salt-marshes (G,M,S,Sc). Fl.5-9. Mons., Glam., Carms., Pembs., *Cards.*, Monts., Mer., Caerns., Denbs., Flints., Angl.. 77 squares, map 108.

S. marina × rupicola Spontaneous. Sea cliffs (G,Sc). *Pembs.*. 1 square.
Pembs.: Stackpole Head, SR99, JAR, 1971[181].

S. rubra (L.) J. & C. Presl : Troellys Coch : Sand Spurrey. Native. Gravelly places (D,G,H,R,Rg,Rw). Fl.5-10. Mons, Glam., Brecs., Rads., *Carms.*, Pembs., Cards., Monts., Mer., Caerns., Denbs., Flints., Angl.. 119 squares, map 109.

S. nicaeensis Sarato ex Burnat Casual (Europe). Waste places. *Mons.*. 1 square : **ST***38*[24].

S. bocconii (Scheele) Ascherson & Graebner (*S. atheniensis* (Heldr. & Sart.) Ascherson & Schweinf.; *S. campestris* sensu Willk. & Lange, non (L.) Ascherson). Casual (S. and SW. Europe). Waste places. Fl.7-8. Glam.. 2 squares : **ST***16*[112],*27*[112].

LYCHNIS L.

Lychnis chalcedonica L. Casual (Russia). Waste places. Fl.6. Glam.. 1 square : **ST**27[112].

L coronaria (L.) Desr. (*Agrostemma coronaria* L.) : Rose Campion. Casual (S.E. Europe). Waste places (M,Q,R). Fl.6. Mons., *Glam.*, *Monts.*. 6 squares : **SS***68*[19],*69*[222]; **ST**09[227],16[99], 38[112]; **SO***19*[8].

L. flos-cuculi L. : Carpiog y Gors : Ragged-Robin. Native. Marshes (Ar,Ds,G,H,M,R,Sc,W). Fl.5-9. Mons., Glam., Brecs., Rads., Carms., Pembs., Cards., Monts., Mer., Caerns., Denbs., Flints., Angl.. 254 squares, all except : **SS**38,39,77,96; **ST**26,58; **SM**50,71, 73; **SN**04,12,36,86,87,88,89; **SO**30,31,41; **SH**51,68,80, 81,91,97,98; **SJ**02,37.

L. viscaria L. (*Viscaria vulgaris* Bernh.) : Lluglys Gludiog : Sticky Catchfly. Native. Rocks. Fl.5-6. Rads., Monts., [*Mer.*]. 2 squares.
Rads.: Stanner Rocks, SO25, AL, 1874[112] and PMB, 1980[3]. **Monts.:** Breiddin Hill, SJ21, DAJ, 1890[112] and An, 1974[93].

AGROSTEMMA L.

Agrostemma githago L. : Bulwg yr Ŷd : Corncockle. Established alien (Europe). Cornfields and waste places (F,R). Fl.6-9. Mons., Glam., *Brecs.*, *Rads.*, *Carms.*, Pembs., Cards., *Monts.*, *Mer.*, Caerns., Denbs., *Flints.*, *Angl.*. 57 squares, map 110. Rapidly decreasing in numbers, due mainly to changes in agricultural techniques.

SILENE L.

Silene italica (L.) Pers. : Italian Catchfly. Casual (Mediterranean region). Waste places. Fl.6. Mons.. 1 square : **ST**38[112].

S. nutans L. : Gludlys Gogwyddol : Nottingham Catch-fly. Native. Rocks (Q,R,Sc,W). Fl.5-9. (Mons.*, Glam.*), Caerns., Denbs., Flints.. 9 squares.
(**Mons.:** Newport Docks, ST38, JM, 1943[112] and AEW, 1968[112]. **Glam.:** Barry Docks, ST16, RLS & AEW, 1925[24].) **Caerns.:** (Afon-wen, SH43, WAS, 1872[112]); Conway, SH77, JB, 1773[107] and TC, 1848[107]; Degannwy, SH77, HWP, 1921[107] and SMW, 1950[7]; Gt Ormes Head, SH78, RD, 1861[112] and PMB, 1980[3]; nr Colwyn Bay, Little Ormes Head, etc., SH88, CEB, 1836[7] and AWW, 1953[7]; Llangystennin, SH87, AW, 1929[112]. **Denbs.:** Llyn Euryn, SH87, CW, 1900[107]; Bryneuryn, SH87, CW, 1900[112]; Rhos-on-Sea, SH88, VMP, 1929[112]; Llandrillo-yn-Rhos, SH88, JWC, 1933[112]. **Flints.:** Rhuddlan, SJ07, VMP, 1929[112]; nr Prestatyn, SJ08, RB, 1885[140] and TE, 1972[112]; nr Rhyl, SJ08, KP, 1951[112]. Still extant in **SH**77,78,87,88.

S. otites (L.) Wibel : Gludlys Ysbaenaidd : Spanish Catchfly. Casual (E. England). Waste places. *Glam.*. 1 square : **ST***17*[140].

S. vulgaris (Moench) Garcke

Subsp. **vulgaris** (*S. cucubalus* Wibel) : Llys y Poer : Bladder Campion. Native. Cultivated ground, hedges, etc., (E,F,G,M,Q,R,Rg,Rw,W). Fl.4-9. Mons., Glam., Brecs., Rads., Carms., Pembs., *Cards.*, Monts., Mer., Caerns., Denbs., Flints., *Angl.*. 163 squares, map 111.

Subsp. **maritima** (With.) Á. & D. Löve (*S. maritima* With.) : Gludlys Arfor : Sea Campion. Native. Sea shores and mountain rocks (Ar,D,E,M, Q,R,Rw). Fl.(3-)5-10. *Mons.*, Glam., Brecs., Carms., Pembs., Cards., Monts., Mer., Caerns., *Denbs.*, *Flints.*, Angl.. 109 squares, map 112.

Subsp. **maritima** × subsp. **vulgaris** Spontaneous. Waste places (R,S). Fl.7. Glam., Cards., *Mer.*, *Angl.*. 5 squares.
Glam.: Jersey Marine, SS79, MG, 1973[112]. **Cards.:** Aberystwyth, SN58, JHS, 1936[112]; Cwmerfyn, SN68, JPS, 1966[55]. **Mer.:** Llwyngwril, SN50, PMB, 1960[4]. **Angl.:** Cemaes Bay, SH39, RHR, 1960+[208].

S. csereii Baumg. Casual (S.E. Europe). Waste places. *Glam.*. 2 squares : **SS***78*[13]; **SO***00*[13].

S. schafta C. C. Gmelin Casual (Caucasus). Sand dunes. Fl.8. *Flints.*. 1 square : **SJ***08*[107].

S. acaulis (L.) Jacq. : Gludlys Mwsogl : Moss Campion. Native. Mountain rocks. Fl.6-9. Mer.*, Caerns.. 4 squares.
Mer.: Diffwys, SH62, TB, pre 1863[212]; Cadair Idris, SH71, GTG, 1956[112] and PMB, 1962[112]. **Caerns.:** Cwm Idwal, SH65, JHL, 1875[112] and ER, 1964[207]; Glyder Fawr, SH65, ER, 1965[207]; Snowdon, SH65, FHBM, 1927[112] and ER, 1962[112]; Cwm Glas, SH65, ELDa, 1948[112] and ER, 1962[207]; Carnedd Llywelyn, SH66, ER, 1962[207]; Ysgolion Duon, SH66, ER, 1960[207]. Still extant in SH65,66,71.

S. armeria L. : Sweet-William Catchfly. Casual (Europe). Waste places and railways. *Glam.*. 2 squares : **SS***79*[140]; **SN***70*[18].

S. noctiflora L. (*Melandrium noctiflorum* (L.) Fries) : Gludlys Nos-flodeuol : Night-flowering Catchfly. Established alien (England). Cultivated and waste places (F,G,R,Rg). Fl.(7-)10-11. Glam., *Pembs.*, (*Cards.*, *Mer.*, *Caerns.*), *Denbs.*, *Flints.*, [*Mons.*]. 19 squares : **SM**99, **SS**68,69; **ST**07,*16*,17,27,*48*; **SM**82; **SN**03,(58; **SH**53,56,*93*); **SJ**06,07,16,26,36.

S. pratensis (Rafn) Godron & Gren.

Subsp. **pratensis** (*S. alba* (Miller) E. H. L. Krause, non Muhl. ex Britton & A. Br.; *Melandrium album* (Miller) Garcke) : Gludlys Gwyn : White Campion. Native. Cultivated ground and hedges (D,E,F,G,Q, R,Rw). Fl.5-9. Mons., Glam., Brecs., Rads., Carms., Pembs., Cards., Monts., *Mer.*, Caerns., Denbs., Flints., *Angl.*. 156 squares, map 113.

Subsp. **divaricata** (Reichenb.) McNeill & Prentice (*S. alba* subsp. *divaricata* (Reichenb.) Walters; *Lychnis macrocarpa* Boiss. & Reuter; *S. macrocarpa* (Boiss. & Reuter) E. H. L. Krause). Casual (Mediterranean region). Waste places. Fl.5. Glam., *Caerns.*. 3 squares : **ST**16[112],*27*[112]; **SH***53*[15].

S. dioica (L.) Clairv. (*Melandrium rubrum* (Weigel) Garcke; *Lychnis preslii* Sekera) : Blodyn Taranau : Red Campion. Native. Woods and hedges (Ar,D,E,G,M,Q, R,W). Fl.4-9. Mons., Glam., Brecs., Rads., Carms., Pembs., Cards., Monts., Mer., Caerns., Denbs., Flints., Angl.. 270 squares, all except : **SR**89; **SS**38,39; **ST**26; **SM**50; **SN**15,75,99; **SO**10,20; **SH**95,98.

S. dioica × **pratensis** : Pink Campion. Spontaneous. Roadsides and waste places. Fl.(4-)6-9. Mons., Glam., *Brecs.*, Rads., *Carms.*, Pembs., *Cards.*, *Mer.*, Caerns., *Denbs.*, *Flints.*, *Angl.*. 51 squares, map 114.

S. muscipula L. Casual (Mediterranean region). Waste places. Fl.6. Glam., *Carms.*. 3 squares : **ST**17[112],*27*[112]; **SN**30[161].

S. behen L. Casual (Mediterranean region). Waste places. Fl.10. *Glam.*. 1 square : **ST***27*[27].

S. antirrhina L. Casual (N. Africa). Waste places. Fl.6. Glam.. 1 square : **ST**27[112].

S. coeli-rosa (L.) Godron (*Lychnis coeli-rosa* (L.) Desr.) : Rose-of Heaven. Casual (S.W. Europe). Waste places. Fl.6. Glam.. 1 square : **ST**27[112].

S. dichotoma Ehrh. : Forked Catchfly. Casual (E. Europe). Waste places (F,G,R,Rg). Fl.5-9. Mons., Glam., Carms., Pembs., *Angl.*. 10 squares : **SS***59*,69,79,*87*; **ST**17,*27*; **SN***10*; **SO***00*,*51*; **SH***28*.

S. gallica L. (*S. anglica* L.) : Gludlys Brutanaidd : Small-flowered Catchfly. Established alien (Europe). Cultivated ground and roadsides (D,E,F,R,Rg,Rw). Fl.5-10. Mons., Glam., *Brecs.*, *Rads.*, *Carms.*, Pembs., Cards., Mer., Caerns., Denbs., *Flints.*, Angl.. 73 squares, map 115.

S. conica L. : Gludlys Rhesog : Sand Catchfly. Naturalized alien (England). Sandy places (D,R,S). Fl.5-7. Glam., Carms., *Cards.*[4,196], *Denbs.*. 6 squares.
Glam.: Barry Docks, ST16, RLS, 1922[112]; Port Talbot, SS78, HJR, 1906[107]. **Carms.:** nr Pembrey, SS49, RDP, 1973[201] and 1980[201]; Tywyn Burrows, SN30, IMV, SBE & TAWD, 1971[112]; Pembrey, SN40, DH, 1920[19] and JRG, 1968[112]; Burry Port, SN40, JAWhe, 1942[188]; **Denbs.:** Colwyn Bay, SH87, WBR, 1891[144].

S. conoidea L. Casual (Mediterranean region). Waste places. Fl.8. Glam.. 1 square : **ST***27*[112].

[CUCUBALUS L.]

[**Cucubalus baccifer** L. *Angl.*]

GYPSOPHILA L.

Gypsophila paniculata L. Casual (E. Europe). Waste places. *Glam.*. 2 squares : **SS***69*[20]; **ST***17*[140].

G. elegans Bieb. Casual (Ukraine). Waste places. Fl.6. Glam.. 1 square : **ST**27[112].

SAPONARIA L.

Saponaria ocymoides L. Casual (S. Europe). Rubbish tip. *Mons.*. 1 square : **ST**38[184].

S. officinalis L. : Sebonllys : Soapwort. Naturalized alien (Europe). Roadsides and river banks (E,M,Q,R, Rg,Rw,Sc). Fl.5-11. Mons., Glam., Brecs., Rads., Carms., Pembs., Cards., Monts., Mer., Caerns., Denbs., Flints., Angl.. 142 squares, map 116. Often *flore pleno*.

VACCARIA Medicus

Vaccaria hispanica (Miller) Räuschert (*V. pyramidata* Medicus) : Cowherb. Casual (Europe). Waste places. Fl.6-8. *Mons.*, Glam., *Carms.*, *Cards.*, *Caerns.*, *Denbs.*. 14 squares : **SS***78;* **ST***16,17,27,*38; **SN***31,40,42,58,90;* **SO***00;* **SH**33,*66;* **SJ***34.*

PETRORHAGIA (Ser. ex DC.) Link

Petrorhagia saxifraga (L.) Link (*Kohlrauschia saxifraga* (L.) Dandy) : Tunic-flower. Established alien (C. and S. Europe). Waste places (R,Rw). Fl.8-10. Pembs.. 1 square.
Pembs.: Tenby, SN10, JEA, 1908[112] and JED, 1957[236].

P. nanteuilii (Burnat) P. W. Ball & Heywood (*Kohlrauschia prolifera* (L.) Kunth) : Childling Pink. Established alien (S. England). Waste places. Fl.6. Glam.. 1 square.
Glam.: Cardiff Docks, ST27, RLS, 1926[112] and JPC, 1980[75].

DIANTHUS L.

Dianthus barbatus L. : Sweet-William. Casual. (S. Europe). Waste places (E,R,Rw). Fl.8. *Rads.*, Carms.. 2 squares : **SN**50[112]; **SO**16[213].

D. caryophyllus L. : Penigan Rhuddgoch : Clove Pink. Established alien (S. Europe). Walls (Q,S). Fl.5-6. *Mons.*, Glam., *Brecs.**, Carms., *Monts.*, *Flints.*, [*Caerns.*]. 9 squares.
Mons.: Matharn, ST59, An, 1800[226]; Tintern Abbey, SO50, JHC, c1860[226]. **Glam.:** Bridgend, SS88, HTS, 1978[112]; Cardiff, ST17, WT, 1800[82]; Crymlyn, SS79, JAW, 1939[112]. **Brecs.:** Brecon, SO02, JAW, 1930[246]. **Carms.:** Kidwelly, SN40, JAW, 1945[112]. **Monts.:** Llanbryn-mair, SN80, JMo, pre1900[165]. **Flints.:** nr Holywell, SJ17, AAD, c1910[143].

D. plumarius L. : Penigan Cyffredin : Pink. Established alien (S.E. Europe). Walls. Fl.5-7. *Glam.**, *Carms.*, *Mer.**, Caerns., Flints.*. 6 squares.
Glam.: Briton Ferry, SS79, CH & HH, 1973[197]. **Carms.:** Ferryside, SN31, RFM, c1960[161]; Llwchwr Valley, SN50, DH, 1912[161]; Llandybie, SN61, AL, 1883[107]. **Mer.:** Arthog, SH61, AJC, 1930[107]. **Caerns.:** Conwy, SH77, SHB, 1867[107] and WF, 1957[7]. **Flints.:** Shotton, SJ27, TE, 1974[112].

D. deltoides L. : Penigan Gwyryfaidd : Maiden Pink. Native. Dry banks and walls (G,Q,R,Rw,W). Fl.6-9. Glam., Rads., *Carms.*[230], (Monts.), Mer., Caerns., *Denbs.*, Angl.. 19 squares : **SS***69*,78; (**ST***17*,18); **SN**60, 70; **SO**05,06,15; **SH**36,46,53,(61),77,78,(80),87,*93*: **SJ***15*.

D. armeria L. : Penigan y Porfeydd : Deptford Pink. Native. Pastures and banks (E,G,Q,R). Fl.6-8. (*Mons.*, Glam.), Carms., (*Pembs.*), Cards., (Caerns.), *Flints.*, *Angl.*. 14 squares : (**SS**59,79,87; **ST***17,39;* **SM***72*); **SN***30*, 31,54,57; (**SO***51*; **SH***57*),*67*; **SJ***07*.

NYMPHAEACEAE

NYMPHAEA L.

Nymphaea alba L. : Lili-ddŵr Wen : White Water-lily. Native. Lakes. Fl.6-9. *Mons.*, Glam., *Brecs.*, *Rads.*, *Carms.*, *Pembs.*, *Cards.*, *Monts.*, Mer., Caerns., *Denbs.*, *Flints.*, Angl.. 84 squares, map 117.

NUPHAR Sm.

Nuphar lutea (L.) Sibth. & Sm. : Lili-ddŵr Felen : Yellow Water-lily. Native. Lakes, etc. (Al,Ar). Fl.7-8. *Mons.*, Glam., *Brecs.*, Rads., Carms., *Pembs.*, *Cards.*, Monts., Mer., Caerns., *Denbs.*, *Flints.*, *Angl.*. 84 squares, map 118.

N. lutea × pumila = N. × spennerana Gaudin Spontaneous or introduced. Lakes. Fl.7. Mer., Denbs.. 2 squares.
Mer.: Llyn Tal-y-llyn, SH71, PMB, 1961[112]. **Denbs.:** Llyn Tyn-y-llyn, SH87, BS, 1967[112]. Doubt has recently been expressed about the identity of these specimens[217].

[**N. pumila** (Timm) DC. : Bwltys Lleiaf : Least Water-lily. *Mer.*]

CERATOPHYLLACEAE

CERATOPHYLLUM L.

Ceratophyllum demersum L. : Cyrnddail : Rigid Hornwort. Native. Ditches (Al,Ar). Fl.7. Mons., Glam., *Brecs.**, Rads.*, Cards.*, Monts.*, Caerns., Denbs.*, Flints.*, Angl., [*Carms.*]. 24 squares, map 119.

C. submersum L. : Soft Hornwort. Native. Ditches. Fl.7-8. Mons., *Glam.**, Carms.. 6 squares.
Mons.: Marshfield, ST28, AEW, 1921[112]; Peterstone

Wentlooge, ST28, AEW, 1923[112]; Whitson, ST38, TGE, 1960+[226]; Lake's Reen, ST38, MWa, 1975[112]. **Glam.:** nr Oxwich, SS58, QONK, 1970+[157]; Crymlyn Fen, SS79, QONK, 1972[157]. **Carms.:** Llanelli, SS59, DH, 1912[161]; Laugharne, SN20, RFM, 1967[112].

RANUNCULACEAE

HELLEBORUS L.

Helleborus foetidus L. : Llewyg y Llyngyr : Stinking Hellebore. Native. Woods (E,Q,Rw,W). Fl.3-7. Mons., Glam., Rads., (*Carms.*, *Cards.**), Monts., (*Mer.**, Caerns.), Denbs., Flints., (Angl.). 37 squares, map 120.

H. viridis L. subsp. **occidentalis** (Reuter) Schiffner (*H. viridis* var. *occidentalis* (Reuter) Druce) : Crafanc yr Arth : Green Hellebore. Native. Woods and hedges (E,G,M,Q,R). Fl.2-5. Mons., (Glam., Brecs.*, Rads., Carms., Pembs., Cards., *Monts.*, *Caerns.*, Denbs., *Flints.*, Angl.*). 42 squares, map 121.

ERANTHIS Salisb.

Eranthis hyemalis (L.) Salisb. : Bleidd-dag y Gaeaf : Winter Aconite. Naturalized alien (S. Europe). Woods (E,G,W). Fl.2. Glam., *Brecs.*, Monts., *Denbs.*, Angl.*. 8 squares.
Glam.: Widegate, Gower, SS58, JAW, c1940[221]; Llansannor, SS97, RGE & AEW, 1980[95]; Caerau nr Cardiff, ST17, HJR, 1908[112]. **Brecs.:** nr Glasbury, SO13, JAW, c1935[222]. **Monts.:** Llanllugan Church, SJ00, JAW, 1940[112]; nr Crugion, SJ31, An, 1950+[7]. **Denbs.:** Manley Wood nr Erbistock, SJ34, HSM, 1910[146]. **Angl.:** Plas Newydd Estate, SH57, WEH, 1966[112]; Penymynydd, SH57, RHR, 1970+[209].

NIGELLA L.

Nigella damascena L. : Love-in-a-Mist. Casual (S. Europe). Waste places and rubbish tips. Fl.8. Glam., Monts.. 5 squares : **SS***58*[20]; **ST***16*[112],17[95],*27*[112]; **SN**97[112].

TROLLIUS L.

Trollius europaeus L. : Cronnell : Globeflower. Native. Moist upland meadows (G,H,M,Q). Fl.5-7. Mons., Glam., Brecs., Rads., Carms., Cards., Monts., Mer., Caerns., Denbs.. 104 squares, map 122.

CALTHA L.

Caltha palustris L. (incl. *C. minor* Miller and *C. radicans* T. F. Forster) : Gold y Gors : Marsh-marigold. Native. Marshes (Al,Ar,E,G,H,M,Rw,Sc,W). Fl.4-5. Mons., Glam., Brecs., Rads., Carms., Pembs., Cards., Monts., Mer., Caerns., Denbs., Flints., *Angl.*. 258 squares, all except : **SS**09,39,77; **ST**26,27,38,58; **SM**50,62,70,71,84,94; **SN**04,23,25,32,36; **SO**20,30,41; **SH**78,98; **SJ**37.

ACONITUM L.

Aconitum variegatum L. Casual (C. Europe). Waste places. Fl.9. Rads.. 1 square : **SO**06[112].

A. firmum Reichenb. (*A. amoenum* Reichenb.). Casual (C. Europe). Waste places. *Pembs.*. 1 square : **SN**10[15].

A. napellus L. (incl. *A. anglicum* Stapf) : Cwcwll y Mynach : Monk's-hood. Native. Shady streamsides (E,G,M,Q,R,Rg,W). Fl.5-7. Mons., Glam., Brecs., Carms., (Pembs., *Cards.*, Monts., Mer., Caerns., Denbs., Flints.*, Angl.*). 65 squares, map 123.

A. napellus × **variegatum** (*A. stoerckianum* Reichenb.). Casual (garden origin). *Flints.*. 1 square : **SJ**16[7].

A. compactum Reichenb. Naturalized alien (Alps). Woods. *Pembs.*. 1 square.
Pembs.: Dale, SM80, MGe, c1960[94].

CONSOLIDA (DC.) S. F. Gray

Consolida ambigua (L.) P. W. Ball & Heywood (*Delphinium ambiguum* L.) : Larkspur. Casual (Mediterranean region). Waste places (R,Rg,Rw). Fl.6-9(-11). Mons., Glam., *Carms.*, *Angl.*[85]. 10 squares : **SS**59,*78*,87; **ST**17,27,28,38,*49*; **SN**31; **SO***00*.

C. regalis S. F. Gray (*Delphinium consolida* L.) : Yspardun y Marchog : Forking Larkspur. Casual (Europe and W. Asia). Waste places. Fl.7. Glam.. 4 squares : **SS***69*[198],*79*[95]; **ST***17*[24],27[112].

C. pubescens (DC.) Soó (*Delphinium pubescens* DC.). Casual (S. Europe). Garden weed. Fl.9. Glam.. 1 square : **ST***27*[112].

ANEMONE L.

Anemone nemorosa L. : Blodyn y Gwynt : Wood Anemone. Native. Woods (E,G,H,M,Q,R,Rw,W). Fl.2-6. Mons., Glam., Brecs., Rads., Carms., Pembs., Cards., Monts., Mer., Caerns., Denbs., Flints., *Angl.*. 233 squares, map 124.

A. ranunculoides L. : Yellow Anemone. Casual (Europe and W. Asia). Waste places. *Glam.*. 1 square : **SS***69*[140].

A. apennina L. : Blue Anemone. Casual (S. Europe). Woods and hedges. Fl.3. Mons., Glam., *Flints.*. 4 squares : **SS***49*[112],69[111]; **ST**28[112]; **SJ**25[180].

A. coronaria L. Casual (Mediterranean region). Waste places. Glam.. 1 square : **ST***17*[112].

A. hortensis L. × **pavonina** Lam. = **A.** × **fulgens** Gay Casual (garden origin). Grassland and quarry. *Pembs.*. 1 square : **SN***10*[25].

A. × **hybrida** Paxton (*A. japonica* auct.) : Japanese Anemone. Casual (garden origin). Roadsides. *Glam.*, Denbs.. 5 squares : **SS**68[222],*69*[73]; **ST***17*[140]; **SH**85[112]; **SJ**24[112].

CLEMATIS L.

Clematis flammula L. : Virgins-Bower. Established alien (S. Europe). Cliffs. Fl.8-9. Caerns.. 1 square.
Caerns.: Aber-soch, SH32, RGE, 1978[112].

C. vitalba L. : Cudd y Coed : Traveller's Joy. Native. Woods and hedges (D,E,Q,R,Sc,W). Fl.7-9. Mons., Glam., Brecs., (Rads.), Carms., Pembs., (Cards.), Monts., (Mer.), Caerns., Denbs., Flints., Angl.. 137 squares, map 125.

C. montana Buch.-Ham. ex DC. Established alien (Himalaya). *Glam.*. 1 square.
Glam.: nr Swansea, SS79, JAW, 1956[222].

ADONIS L.

Adonis annua L. : Llygad y Goediar Gwanwynol : Pheasant's-eye. Casual (S. Europe and Asia). Waste places. Fl.6-7. Mons., Glam., *Pembs.*, *Denbs.*, *Flints.*. 8 squares : **SS***78*[112],*88*[140]; **ST***17*[140],*48*[226],*59*[112]; **SM**70[193]; **SH***97*[144]; **SJ***07*[141].

A. aestivalis L. Casual (Europe). Allotments. Glam.. 1 square : **ST***27*[112].

RANUNCULUS L.

Subgen. Ranunculus

Ranunculus repens L. : Crafanc y Frân : Creeping Buttercup. Native. Cultivated ground and wet places (Al,D,Ds,E,F,G,H,M,Q,R,Rg,Rw,W). Fl.5-10. Mons., Glam., Brecs., Rads., Carms., Pembs., Cards., Monts., Mer., Caerns., Denbs., Flints., *Angl.*. 279 squares, all except : **SS**39; **SM**50; **SH**51.

R. acris L. : Blodyn Ymenyn : Meadow Buttercup. Native. Pastures (D,Ds,E,F,G,H,M,Q,R,Rg,Rw,W). Fl.5-9. Mons., Glam., Brecs., Rads., Carms., Pembs., Cards., Monts., Mer., Caerns., Denbs., Flints., *Angl.*. 275 squares, all except : **SS**39; **ST**26; **SM**50,71,73,94; **SH**98.

R. bulbosus L. : Chwys Mair : Bulbous Buttercup. Native. Pastures (D,E,F,G,H,Q,R,Rw,W). Fl.4-6. Mons., Glam., Brecs., Rads., Carms., Pembs., Cards., Monts., Mer., Caerns., Denbs., Flints., Angl.. 232 squares, map 126.

R. sardous Crantz : Crafanc y Frân Blewog : Hairy Buttercup. Native. Damp fields (F,G,M,R,Rw). Fl.5-9. *Mons.*, Glam., Brecs.*, *Carms.*, *Pembs.**, *Cards.*, *Monts.*, *Mer.*, *Caerns.*, *Denbs.*, *Flints.*. 31 squares, map 127.

R. muricatus L. : Scilly Buttercup. Casual (Mediterranean region). Waste places. *Glam.*. 1 square : **ST**18[24].

R. arvensis L. : Crafanc yr Ŷd : Corn Buttercup. Established alien (England). Cultivated ground (F,Rg). Fl.5-7. Mons., Glam., *Carms.*, *Cards.*, *Monts.*, *Denbs.*, *Flints.*, [*Mer.*]. 26 squares, map 128.

R. parviflorus L. : Crafanc Y Frân Mânflodeuog : Small-flowered Buttercup. Native. Gravelly places (D,E,F,G,Q,R,Rg,Rw,Sc). Fl.4-8. *Mons.*, Glam., Brecs., Rads.*, Carms., Pembs., Cards., Monts., Caerns., Denbs., *Flints.*[230], Angl., [*Mer.*]. 59 squares, map 129.

R. auricomus L. : Peneuraidd : Goldilocks Buttercup. Native. Woods and hedges (E,M,R,W). Fl.4-5(-8). Mons., Glam., Brecs., Rads., Pembs.*, Monts., Mer., Caerns., Denbs., Flints., Angl.. 77 squares, map 130.

R. sceleratus L. : Crafanc yr Eryr : Celery-leaved Buttercup. Native. Marshes (Al,Ar,Ds,H,M,R,S). Fl.5-7. Mons., Glam., Brecs., Carms., Pembs., Cards., Monts., Mer., *Caerns.*, Denbs., Flints., Angl.. 108 squares, map 131.

R. ficaria L. : Llygad Ebrill : Lesser Celandine. Mons., Glam., Brecs., Rads., Carms., Pembs., Cards., Monts., Mer., Caerns., Denbs., Flints., Angl.. 269 squares, all except : **SS**39,77; **SM**50,94; **SN**59,75,86; **SO**41; **SH**98; **SJ**01,03,12,37.

Subsp. **ficaria** Native. Woods and hedges (E,G,H, M,R,Rw,Sc,W). Fl.3-5(-9). Mons., Glam., Brecs., Rads., Carms., Pembs., Cards., Monts., *Mer.*, Caerns., Denbs., Flints., *Angl.*. 197 squares, map 132.

Subsp. **bulbifer** Lawalrée (*R. ficaria* var. *bulbiferus* Albert). Native. Shady places in woods and hedges (Ds,E,M,R,Rg,W). Fl.4-5. Mons., Glam., Brecs., Rads., Carms., Pembs., Cards., Monts., Mer., *Caerns.*, *Denbs.*, *Flints.*, Angl.. 56 squares, map 133.

Plants resembling subsp. **ficariiformis** Rouy & Fouc. have been recorded from Pembs. and Mer.

R. flammula L. : Llafnlys Bach : Lesser Spearwort. Native. Marshes (Al,Ar,Ds,G,H,M,W). Fl.5-9. Mons., Glam., Brecs., Rads., Carms., Pembs., Cards., Monts., Mer., Caerns., Denbs., Flints., Angl.. 266 squares, all except : **SS**09,38,39; **ST**26,27; **SM**50,71,73,90; **SN**36; **SO**30,41; **SH**51,88,98; **SJ**37.

R. lingua L. : Llafnlys Mawr : Greater Spearwort. Native. Marshes (Al,Ar,Ds,G,M). Fl.6-10. Mons., Glam., Brecs., Rads., Carms.*, Pembs., *Monts.*, Mer., Caerns., Denbs., *Flints.*, Angl.. 47 squares, map 134.

Subgen. Batrachium (DC.) A. Gray

R. hederaceus L. : Egyllt Eiddewddail : Ivy-leaved Crowfoot. Native. Ponds (Al,Ar,M). Fl.4-7. Mons., Glam., Brecs., Rads., Carms., Pembs., Cards., Monts., Mer., Caerns., *Denbs.*, *Flints.*, Angl.. 201 squares, map 135.

R. omiophyllus Ten. (*R. lenormandii* F. W. Schultz) : Egyllt y Rhosdir : Round-leaved Crowfoot. Native. Bogs and marshes (Al,Ar,F,H,M,R). Fl.2-7. Mons., Glam., Brecs., Rads., Carms., Pembs., Cards., Monts., Mer., Caerns., Denbs., *Flints.*, *Angl.*. 203 squares, map 136.

R. omiophyllus × tripartitus Spontaneous. *Glam.*. 1 square.
Glam.: Rhosili, SS48, HJR, pre 1907[140].

R. tripartitus DC. (*R. lutarius* (Revel) Bouvet) : Egyllt y Llynnoedd : Three-lobed Crowfoot. Native. Pools (Al,Ar,H,M). Fl.4-6(-8). Glam., *Brecs.*, Rads.*, *Carms.**, Pembs., Cards., Angl.. 17 squares : **SS**48,49,59,78; **SM**70,71,*91*,92; **SN**25,30,31,40,52; **SO***01*,15; **SH**37,48.

R. baudotii Godron : Egyllt y Mordir : Brackish Water-crowfoot. Native. Pools and ditches near the sea (Al,Ar,Ds,M,Q,S). Fl.4-8. Mons., Glam., Carms., Pembs., Cards., Mer., Caerns., Denbs., Flints., *Angl.*. 49 squares, map 137.

R. baudotii × trichophyllus = R. × segretti A. Felix (*R. × durandii* A. Felix). Spontaneous. *Flints.*[217].

R. aquatilis agg. : Egyllt y Dŵr; Crafanc y Dŵr. Mons., Glam., Brecs., Rads., Carms., Pembs., Cards., Monts., Mer., Caerns., Denbs., Flints., Angl.. 201 squares, map 138.

R. peltatus Schrank (*R. aquatilis* subsp. *peltatus* (Schrank) Syme) : Pond Water-crowfoot. Native. Ponds and slow flowing streams (Al,Ar,H). Fl.4-8. Mons., Glam., Brecs., Rads., Carms., Pembs., Cards., Monts., Mer., Caerns., Denbs., Flints., Angl.. 116 squares, map 139. Under-recorded.

R. peltatus × ?trichophyllus Spontaneous. Rivers. Fl.6. Cards.. 1 square.
Cards.: Afon Cletwr Fawr nr Glan-rhyd-y-dre, SN44, AOC, 1978[112].

R. penicillatus (Dumort.) Bab. (*R. aquatilis* subsp. *pseudofluitans* (Syme) Clapham; *R. pseudofluitans* (Syme) Newbould ex Baker & Foggitt). : Stream Water-crowfoot. Native. Fast flowing rivers. Fl.5-8. Mons.*, Glam., Brecs., Rads., Carms.*, Pembs.*, Cards., Monts., *Mer.*, Caerns., Denbs.*, Flints., Angl.. 86 squares, map 140. Under-recorded.

R. aquatilis L. *s.s.* (*R. aquatilis* subsp. *heterophyllus* (Weber) Syme; *R. aquatilis* subsp. *radians* (Revel) Clapham) : Common Water-crowfoot. Native. Ponds (Al,Ar). Fl.4-8. Mons., Glam., Brecs., *Rads.*, Carms., Pembs., Cards., Monts., Mer., Caerns., *Denbs.*, Flints., Angl.. 100 squares, map 141. Under-recorded.

R. aquatilis × fluitans (or **R. fluitans × trichophyllus**) **= R. × bachii** Wirtgen Spontaneous. Rivers and lakes. *Mons.*, *Denbs.*, *Angl.*. 7 squares.
Mons.: Monmouth, SO51, HHH, 1928[27]. **Denbs.:** River Dee, SJ14,24,34,44,45, NH, 1980[118]. **Angl.:** Llyn Coron, SH37, PJ, 1946[110].

R. trichophyllus Chaix (incl. *R. drouetii* F. W. Schultz ex Godron) : Egyllt Dail Edafaidd : Thread-leaved Water-crowfoot. Native. Ponds (Al,Ar,H,M). Fl.5-8. Mons., Glam., Brecs., Carms.*, Pembs., Cards., *Monts.**, Mer., Caerns., Denbs., Flints., Angl.. 63 squares, map 142.

R. circinatus Sibth. : Egyllt Cylchol-ddail : Fan-leaved Water-crowfoot. Native. Ponds and streams. Fl.6-7. Mons., Glam., *Brecs.*, Carms., *Pembs.**, Monts.*, *Caerns.*, *Denbs.*, *Angl.*, [*Mer.*, *Flints.*]. 25 squares, map 143.

R. fluitans Lam. : Crafanc Hirddail : River Water-crowfoot. Native. Rivers (Al,Ar). Fl.5-8. Mons., Glam., Brecs.*, *Rads.**, *Cards.**, *Caerns.**, *Denbs.**, Flints.*, *Angl.*, [*Carms.*, *Mer.*]. 15 squares : **ST**17,39; **SN**76; **SO**13,21,30,37,41,50,51; **SH**36,66; **SJ**06,07,26.

MYOSURUS L.

Myosurus minimus L. Cynffon Llygoden : Mousetail. Native. Meadow. Fl.5-6. *Denbs.**, [*Caerns.*, *Flints.*]. 1 square.
Denbs.: nr Wrexham, SJ45, AGS, 1973[182] and TE, 1979[112].

AQUILEGIA L.

Aquilegia vulgaris L. : Blodau'r Sipsi : Columbine. Native but frequently introduced. Woods and hedges (D,E,G,M,Q,R,Rw,W). Fl.5-7. Mons., Glam., *Brecs.*, Rads., Carms., Pembs., *Cards.*, Monts., *Mer.*, Caerns., Denbs., Flints., *Angl.*. 141 squares, map 144.

THALICTRUM L.

Thalictrum alpinum L. Arianllys y Mynydd : Alpine Meadow-rue. Native. Damp rocky places. Fl.6. Mer., Caerns.. 3 squares.
Mer.: Cadair Idris, SH71, PMB, 1964[106] and 1975[3].
Caerns.: Snowdon, SH65, CCB, c1831[112]; Twll Du, SH65, JEG, 1880[112]; Cwm Glas, SH65, EV, 1924[112]; Cwm Idwal, SH65, JHS, 1925[112] and ER, c1950[207]; Cwm Cywion, SH66, CML, 1926[112]; Carnedd Dafydd, SH66, ER, c1950[207] and PMB, 1979[3].

T. minus L. (incl. *T. arenarium* Butcher; *T. umbrosum* Butcher; *T. expansum* Jordan) : Arianllys Bach : Lesser Meadow-rue. Native. Rocks and sand dunes (D,E,G, M,Q,R,Rw,Sc,W). Fl.6-7. (*Mons.**), Glam., Brecs., Rads., Carms., Pembs., Monts., Mer., Caerns., Denbs., Flints., Angl.. 73 squares, map 145.

T. flavum L. : Arianllys : Common Meadow-rue. Native. Marshes (Al,Ar,E,M,W). Fl.6-7. Mons., Glam., Brecs., Carms., (Cards., *Mer.*), Caerns., Denbs., *Flints.*, Angl., [*Pembs.*]. 40 squares, map 146.

PAEONIACEAE

PAEONIA L.

Paeonia officinalis L. : Rhosyn Mynydd : Peony. Casual (S. Europe). Waste places. *Monts.*. 1 square : **SO**19[228].

BERBERIDACEAE

BERBERIS L.

Berberis vulgaris L. : Pren Melyn : Barberry. Colonist (Europe). Hedges (E,W). Fl.4-6. Mons., Glam., Brecs., Rads., Carms., Pembs., Cards., Monts., *Mer.*, Caerns., *Denbs.*, Flints.. 87 squares, map 147.

B. glaucocarpa Stapf (*B. aristata* auct.) : Greater Barberry. Casual (Himalaya). Hedges. *Glam.*, Monts.. 2 squares : **SS**69[107]; Vastre[112].

B. darwinii Hooker fil. Naturalized alien (S. America). Woods and scrub. Fl.5-9. Carms., Mer., *Caerns.*. 4 squares.
Carms.: Llandyfaelog, SN40, JAW, 1945[112] and IMV, 1968[112]. **Mer.:** Llanelltud, SH71, JAW, 1941[112]. **Caerns.:** nr Llanbedrog, SH33, APC, 1980[74]; Little Ormes Head, SH88, LL, 1950+[7].

B. hookeri Lemaire Naturalized alien (Himalaya). Woods. Glam.. 1 square.
Glam.: Merthyr Mawr, SS87, JAW, 1943[112].

B. wilsoniae Hemsley Established alien (China). Hedges and scrub. Mer., *Caerns.*. 3 squares.
Mer.: Penmaenpool, SH61, JAW, 1941[112]; Llanelltud, SH71, JAW, 1941[112]. **Caerns.:** Gt Ormes Head, SH78, JGH, 1964[249] and An, 1974[254].

MAHONIA Nutt.

Mahonia aquifolium (Pursh) Nutt. : Oregon-grape. Naturalized alien (N. America). Woodland and scrub (E,Q,R,W). Fl.3-5. Mons., *Glam.*, *Brecs.*, Rads., *Carms.*, *Cards.*, Monts., *Caerns.*, Denbs., *Flints.*. 37 squares, map 148.

LAURACEAE

LAURUS L.

Laurus nobilis L. : Llawrwydden : Bay Laurel. Naturalized alien (Mediterranean region). Woods and cliffs (E,Q,Sc,W). Fl.5. Glam., Carms.. 5 squares.
Glam.: Caswell, SS58, JAW, 1924[220] and MG, 1973[99]; Mumbles, SS68, HH & CH, 1974[117]; Dunraven Park, SS87, MG, 1974[99]; Mountain Ash, ST09, AMP, 1975[99]. **Carms.:** Achddu, SN40, JAW, 1944[112].

PAPAVERACEAE

PAPAVER L.

Papaver somniferum L. : Llysiau'r Cwsg : Opium Poppy. Established alien (Europe and Asia). Waste places (F,R,Rg,S,W). Fl.7-10. Mons., Glam., *Brecs.*, Rads., Carms., Pembs., Cards., Monts., Mer., *Caerns.*, Denbs., Flints., Angl.. 89 squares, map 149.

P. rhoeas L. : Pabi Coch : Common Poppy. Native or colonist. Cultivated ground (F,G,R,Rg,S). Fl.5-10. Mons., Glam., Brecs., Rads., *Carms.*, Pembs., Cards., Monts., Mer., Caerns., Denbs., Flints., *Angl.*. 132 squares, map 150.

P. dubium agg. Mons., Glam., Brecs., Rads., Carms., Pembs., Cards., Monts., Mer., Caerns., Denbs., Flints., Angl.. 130 squares, map 151.

P. dubium L. *s.s.* : Pabi Hirben Llyfn : Long-headed Poppy. Native or colonist. Cultivated ground (F,G,R,Rg,Rw,S). Fl.6-10. Mons., Glam., Brecs., Rads., Carms., Pembs., Cards., Monts., Mer., Caerns., Denbs., Flints., Angl.. 81 squares, map 152. Under-recorded.

P. dubium × rhoeas Spontaneous. *Mer.*. 1 square. **Mer.:** Barmouth, SH61, KMB, 1960[4].

P. lecoqii Lamotte : Pabi Hirben Sudd Melyn : Yellow-juiced Poppy. Colonist (England). Cultivated ground (E,F,R,Rg). Fl.5-7. *Mons.**, Glam., *Brecs.*, *Rads.**, *Pembs.**, Cards., *Mer.**, *Caerns.**, Denbs.. 21 squares, map 153.

P. argemone L. : Pabi Hirben Gwrychog : Prickly Poppy. Established alien (Europe). Cultivated ground (F,R,Rg,S). Fl.5-7. Mons., Glam., *Rads.**, *Carms.*, Pembs., *Cards.*, Monts.*, Mer., Caerns., *Denbs.*, *Flints.*, Angl.. 36 squares, map 154.

P. hybridum L. : Pabi Crynben Pigog : Rough Poppy. Established alien (Europe and Asia). Cultivated ground (F,G,R,Rg). Fl.4-7. Glam., Pembs., Caerns., Denbs., Flints., Angl.. 13 squares : **SR***89*; **ST***17,27*; **SM***72*; **SH***37,58,68,77,78,87,88*; **SJ***05,08*. (Possibly extinct in Wales).

P. atlanticum (Ball) Cosson : Atlas Poppy. Established alien (Morocco). Waste places. Fl.4. Mons., Cards., Mer.. 3 squares.
Mons.: Grosmont, SO42, RGE, 1980[112]. **Cards.:** Llanbadarn Fawr, SN58, RGE, 1977[112]. **Mer.:** Aberdyfi, SN69, PMB, 1981[112].

P. orientale L. (*P. bracteatum* Lindley). Casual (S.W. Asia). Waste places. *Glam.*, *Caerns.*. 2 squares : **SS**69[23]; **SH**57[52].

MECONOPSIS Vig.

Meconopsis cambrica (L.) Vig. : Pabi Cymreig : Welsh Poppy. Native but frequently introduced. Shady rocky places (Ar,E,M,Q,R,W). Fl.4-7. Mons., Glam., Brecs., Rads., Carms., (Pembs.), *Cards.*, *Monts.*, Mer., Caerns., Denbs., *Flints.**, (Angl.*). 86 squares, map 155.

ARGEMONE L.

Argemone mexicana L. : Casual (C. America). Ballast heaps. Glam.. 1 square : **ST***17*[112].

ROEMERIA Medicus

Roemeria hybrida (L.) DC. : Violet Horned-poppy. Casual (C and S. Europe). Waste places (R,Rg). Fl.5-7. Glam., *Flints.*. 3 squares : **ST**17[112],27[112]; **SJ**16[261].

GLAUCIUM Miller

Glaucium flavum Crantz : Pabi Corniog Melyn : Yellow Horned-poppy. Native. Shingle beaches (D,Rw,S,Sc). Fl.6-9. Mons., Glam., Carms., Pembs., *Cards.*, Mer., Caerns., Denbs., Flints., *Angl.*. 62 squares, map 156.

G. corniculatum (L.) J. H. Rudolph (*G. phoeniceum* Crantz; *Chelidonium corniculatum* L.) : Red Horned-poppy. Casual (S. Europe). Waste places. Fl.7. Glam., *Denbs.*. 4 squares : **ST***16*[24],*17*[112], 27[112]; **SJ***23*[144].

CHELIDONIUM L.

Chelidonium majus L. : Dilwydd : Greater Celandine. Established alien (England). Roadsides (E,F,Q,R). Fl.4-8. Mons., Glam., Brecs., Rads., Carms., Pembs., Cards., Monts., Mer., Caerns., Denbs., Flints., *Angl.*. 209 squares, map 157.

ESCHSCHOLZIA Cham.

Eschscholzia californica Cham. : Californian Poppy. Casual (California). Roadsides and waste places. Fl.7-9. Mons., Glam., *Carms.*, *Cards.*, Caerns.. 8 squares : **SS**79[95]; **ST**17[112],27[112]; **SN**31[161],40[161],58[93]; **SH**32[112],*78*[112].

HYPECOUM L.

Hypecoum procumbens L. Casual (S. Europe). Waste places. *Glam.*. 2 squares : **SS***78*[140]; **ST***17*[22].

H. pendulum L. Casual (S. Europe). Waste places. Fl.5. Glam.. 1 square : **ST***27*[112].

DICENTRA Bernh.

Dicentra formosa Walpers Established alien (N. America). Rocky places and streamsides (G,M,R,W). Fl.4-7. Glam., Pembs., Cards., *Caerns.*[119]. 5 squares. **Glam.:** Cardiff, ST17, RD, 1951[112]. **Pembs.:** Henry's Moat, SN02, AOC, 1974[112]; Cwmcneifa, SN23, TAWD, 1972[112]. **Cards.:** nr Capel Dewi, SN44, AOC, 1978[112]; Ysbyty Cynyn, SN77, AOC, 1978[112].

CORYDALIS Vent.

Corydalis claviculata (L.) DC. : Mwg y Ddaear Gafaelgar : Climbing Corydalis. Native. Woods and hedges (E,G,H,M,Q,R,Rw,Sc,W). Fl.5-10. Mons., Glam., Brecs., Rads., Carms., Pembs., Cards., Monts., Mer., Caerns., Denbs., Flints., *Angl.*. 148 squares, map 158.

C. lutea (L.) DC. : Mwg y Ddaear Melyn : Yellow Corydalis. Established alien (S. Europe). Walls. Fl.4-10. Mons., Glam., Brecs., Rads., Carms., Pembs., Cards., Monts., Mer., Caerns., Denbs., Flints., Angl.. 93 squares, map 159.

C. bulbosa (L.) DC. (*C. cava* (L.) Schweigger & Koerte) : Hollow-root. Naturalized alien (S. Europe). Woods. Glam.. 1 square.

Glam.: Stouthall Wood, SS48, JBe, 1981[95] and 1982[112]. The specimen in **NMW** resembles subsp. **marschalliana** (Pallas) Chater.

C. solida (L.) Swartz : Bird-in-a-Bush. Naturalized alien (Europe and Asia). Wood borders. Fl.4. Glam., Pembs.. 4 squares.
Glam.: Penllergaer, SS69, M, c1850[111]; Cottrell, ST07, JS, 1892[112]. **Pembs.:** Cresswell Quay, SN00, WGG, 1943[112]; Amroth, SN10, IMV, SNT & GHT, 1979[245].

FUMARIA L.

Fumaria capreolata L. subsp. **babingtonii** (Pugsley) P. D. Sell (*F. capreolata* var. *babingtonii* Pugsley) : Mwg y Ddaear Afreolus : White Ramping-fumitory. Colonist (England). Cultivated ground and hedge banks (E,F,Q, R). Fl.5-10. Glam., Carms., Pembs., Cards., *Mer.*, Caerns., Denbs., Flints., Angl., [*Brecs.*, *Monts.*]. 62 squares, map 160.

F. purpurea Pugsley : Mwg y Ddaear Glasgoch : Purple Ramping-fumitory. Established alien (England). Cultivated ground (E,F,M,R,Rg,S,Sc). Fl.3-7(-10). Mons., Glam., *Rads.*, Pembs., Caerns., *Denbs.*, Flints., Angl., [*Brecs.*]. 22 squares, map 161.

F. bastardii Boreau : Tall Ramping-fumitory. Established alien (England). Cultivated ground (D,E,F, G,Q,R,Rg). Fl.4-9. Mons.*, Glam., Carms., Pembs., Cards., *Monts.**, Mer., Caerns., Denbs., *Flints.**, Angl., [*Brecs.*]. 89 squares, map 162.

F. muralis Sonder ex Koch

Subsp. **muralis** Established alien (England). Cultivated ground. *Rads.*, *Cards.*, *Caerns.*[145], *Denbs.*, [*Glam.*, *Flints.*]. 3 squares.
Rads.: Boughrood, SO13, GCD, 1929[28]. **Cards.:** Cardigan, SN14, JEL, 1935[211]. **Denbs.:** Colwyn Bay, SH87, CW, 1900[144].

Subsp. **boraei** (Jordan) Pugsley (*F. boraei* Jordan) : Common Ramping-fumitory. Established alien (England). Cultivated ground (E,F,G,Q,R,Rg,Rw). Fl.4-10. Mons., Glam., *Brecs.*, Rads., Carms., Pembs., Cards., Monts., Mer., Caerns., Denbs., Flints., Angl.. 130 squares, map 163.

F. muralis × officinalis Spontaneous. Roadsides. Fl.7. Mer.. 1 square.
Mer.: Pont Dysynni, Bryn-crug, SH50, PMB, 1972[112].

F. densiflora DC. (*F. micrantha* Lag.) : Dense-flowered Fumitory. Established alien (England). Cultivated ground. Caerns.† *Angl.*†, [*Glam.*]. 2 squares.
Caerns.: Conway, SH77, CW, 1917[112]. **Angl.:** nr South Stack, Holyhead, SH28, JEG, 1887[101].

F. officinalis L. : Mwg y Ddaear Cyffredin : Common Fumitory. Established alien (England). Cultivated ground (F,M,R,Rg,Rw,S). Fl.5-8(-11). Mons., Glam., Brecs., Rads., Carms., Pembs., Cards., Monts., Mer.*, *Caerns.*, Denbs., Flints., Angl.. 144 squares, map 164.

F. parviflora Lam. : Mwg y Ddaear Mânflodeuog : Fine-leaved Fumitory. Casual (S. England). Waste places. *Glam..* 2 squares : **ST***16*[24],*17*[218].

CRUCIFERAE

SISYMBRIUM L.

Sisymbrium irio L. : Berwr Caersalem : London-rocket. Casual (Mediterranean region). Waste places (Q,R,Rg,Rw). Fl.7-8. Mons., Glam., Carms., [*Cards.*]. 10 squares : **SS***69*; **ST***16,17,27*; **SN***41*,42,*50*,62; **SO***00*,51.

S. loeselii L. : False London-rocket Casual (S.E. Europe). Waste places. *Glam..* 2 squares : **ST***16*[107],*17*[23].

S. volgense Bieb. ex E. Fourn. : Russian Mustard. Established alien (S.E. European Russia). Waste places near railway lines. *Glam..* 1 square.
Glam.: Barry Docks, ST16, RM, 1930[244].

S. austriacum Jacq. (*S. pyrenaicum* (L.) Vill., non L.). Casual (S. Europe). Waste places. Fl.6 Glam.. 1 square : **ST***17*[112].

S. altissimum L. : Berwr Treigledigol : Tall Rocket. Established alien (E. Europe). Waste places (Q,R,W). Fl.6-7(-10). Mons., Glam., *Brecs.*, *Rads.*[85], Carms., *Pembs.*[85], *Cards.*, *Monts.*[85], *Mer.*, *Caerns.*, Denbs., Flints., Angl.. 40 squares, map 165.

S. orientale L. : Berwr Dwyreiniol : Eastern Rocket. Established alien (S. Europe). Waste places (G,Q,R, Rw). Fl.4-9(-11). Mons., Glam., Rads., Carms., Pembs.*, *Cards.*, Monts., Mer., Caerns., Denbs.*, Flints., *Angl.**. 61 squares, map 166.

[**S. polyceratium** L. (*Chamaeplium polyceratium* (L.) Wallr.). *Glam.*]

S. runcinatum Lag. ex DC. (*Chamaeplium runcinatum* (Lag. ex DC.) Hayek). Casual (S.W. Europe). Waste places. Glam.. 1 square : **ST***27*[25].

S. officinale (L.) Scop. : Cedw'r Berth : Hedge Mustard. Native. Hedgebanks and roadsides (D,E,F,G,M,R). Fl.5-9. Mons., Glam., Brecs., *Rads.*, Carms., Pembs., *Cards.*, Monts., Mer., Caerns., Denbs., Flints., *Angl..* 199 squares, map 167.

DESCURAINIA Webb and Berth.

Descurainia sophia (L.) Webb ex Prantl : Berwr y Fam : Flixweed. Colonist (England). Waste places (G,R,Rg). Fl.6-8. Mons., Glam., *Brecs.*, *Carms.*, Pembs., *Mer.*, *Caerns.*, Denbs., Flints., *Angl..* 29 squares, map 168.

ALLIARIA Scop.

Alliaria petiolata (Bieb.) Cavara & Grande : Garlleg y Berth : Garlic Mustard. Native. Hedges (M,R,Rw,S, W). Fl.4-6(-8). Mons., Glam., Brecs., Rads., Carms., Pembs., Cards., Monts., Mer., Caerns., Denbs., Flints., *Angl..* 220 squares, map 169.

ARABIDOPSIS (DC.) Heynh.

Arabidopsis thaliana (L.) Heynh. : Berwr y Fagwyr : Thale Cress. Native. Walls and cultivated ground (Ar, E,F,G,M,Q,R,Rw,Sc). Fl.3-5. Mons., Glam., Brecs., Rads., Carms., Pembs., Cards., Monts., Mer., Caerns., Denbs., Flints., *Angl..* 173 squares, map 170.

MYAGRUM L.

Myagrum perfoliatum L. Casual (S. Europe). Waste places. *Glam.*, Carms.. 2 squares : **ST***17*[24]; **SN***42*[22].

ISATIS L.

Isatis tinctoria L. : Llysiau'r Lliw : Woad. Casual (C. and S.Europe). Cornfields. Fl.6. Glam.[112], no locality.

BUNIAS L.

[**Bunias erucago** L. *Glam.*]

B. orientalis L. : Warty Cabbage. Casual (E. Europe and W. Asia). Waste places. Fl.7. Glam., *Angl..* 3 squares : **ST**17[31],27[112]; SH37[93].

ERYSIMUM L.

Erysimum decumbens (Schleicher ex Willd.) Dennst. (*E. ochroleucum* DC.). Casual (S. Europe). *Mer..* 1 square : SH*61*[83].

E. repandum L. Casual (Europe). Waste places. *Glam..* 1 square : **ST***17*[24].

E. cheiranthoides L. : Treacle Mustard. Established alien (Europe). Waste places and cultivated ground (M,Q,R,Rg,Sc). Fl.6-10. Mons., Glam., Brecs.*, Rads.*, Carms., *Pembs.*[127], Mer.*, Caerns., Denbs., *Flints..* 34 squares, map 171.

HESPERIS L.

Hesperis matronalis L. : Dame's-violet. Established alien (Europe). Hedges and waste places (E,M,Q,R, Rw,W). Fl.5-11. Mons., Glam., Brecs., Rads., Carms., Pembs., Cards., Monts.*, Mer., Caerns.*, Denbs., Flints.*, Angl.. 111 squares, map 172.

MALCOLMIA R. Br.

[**Malcolmia ramosissima** (Desf.) Thell. (*M. parviflora* (DC.) DC.) *Glam.*]

M. africana (L.) R. Br. (*Wilckia africana* (L.) F. Mueller). Casual (S. Europe). Waste places. Fl.5. Glam.. 1 square : ST*27*[112].

M. maritima (L.) R. Br. (*Wilckia maritima* (L.) Halácsy) : Virginia Stock. Casual (S. Europe). Waste places. Fl.5-9(-11). Mons., Glam., *Pembs.*, *Cards.*, Mer., *Denbs..* 9 squares : **SS**58[222],69[222],*78*[20]; **ST**17[227],*27*[112],38[112]; **SN***10*[18]; **SH***61*[112]; **SJ***12*[144].

CHEIRANTHUS L.

Cheiranthus cheiri L. : Blodyn y Fagwyr : Wallflower. Established alien (Mediterranean region). Walls (Q,R,Sc). Fl.2-6(-11). Mons., Glam., Brecs., Rads., Carms., Pembs., Cards., Monts., Mer., Caerns., Denbs., Flints., Angl.. 87 squares, map 173.

MATTHIOLA R.Br.

Matthiola incana (L.) R. Br. : Hoary Stock. Denizen (England). Sea cliffs. Fl.5. Glam., (*Pembs.*, Cards., *Caerns.*). 6 squares.
Glam.: Mumbles Head, SS68, QONK, 1972[157]; Nash Point, SS96, HJR, 1905[140] and RGE, 1975[112]; Monknash, SS97, AEW, 1970+[227]. (**Pembs.:** Pembroke Dock, SM90, EL, 1853[79]. **Cards.:** Llangrannog, SN35, JHS, pre 1930[210]. **Caerns.:** nr Aberdaron, SH12, APC, 1955[7]).

M. sinuata (L.) R. Br. : Murwyll Tewbannog Arfor : Sea Stock. Native. Sand dunes and rocky places by the sea. Fl.6-8. Glam., Pembs., *Mer.*†, *Caerns.*†, *Denbs.*[85]†, *Flints.*[230]†, *Angl.*†. 16 squares : **SS***09*,19,(48,49),68,69, *78,79*; **SN***10,69*; **SH***12,36,46,52,53,61*.

M. tricuspidata (L.) R. Br. Casual (Mediterranean region). Waste places. Fl.9. Glam.. 1 square : **ST***17*[112].

BARBAREA R.Br.

Barbarea vulgaris R. Br. : Berwr y Gaeaf : Winter-cress. Native. River banks and waste places (Ar,E,G,M, Q,R,Rg,Rw,S). Fl.4-8(-10). Mons., Glam., Brecs., Rads., Carms., *Pembs.*, Cards., Monts., Mer., Caerns., *Denbs.*, Flints., *Angl.*. 155 squares, map 174.

B. stricta Andrz. : Small-flowered Winter-cress. Denizen (England). Streamside. Fl.6. Denbs.. 1 square. **Denbs.:** Rossett, SJ45, TE, 1975[112].

B. verna (Miller) Ascherson : Berwr Tir : American Winter-cress. Established alien (W. Mediterranean region). Cultivated ground and waste places (E,F,M,Q, R,Rg,Rw). Fl.4-7(-11). Mons., Glam., Brecs., Rads., Carms.*, Pembs., Cards., *Monts.**, Mer., *Caerns.*, Denbs., *Flints.*, Angl.. 55 squares, map 175.

B. intermedia Boreau : Medium-flowered Winter-cress. Established alien (W. Europe). Cultivated ground (E,F, G,Q,R,Rw). Fl.4-8. Mons., Glam., Brecs.*, *Rads.*, Carms., *Pembs.*, Cards., Mer., *Caerns.**, Denbs.*. 36 squares, map 176.

RORIPPA Scop.

Rorippa austriaca (Crantz) Besser : Austrian Yellow-cress. Established alien (Europe). Waste places (G,M,R, Rg). Fl.6-8. Glam., Brecs., Carms.. 7 squares.
Glam.: nr Dunvant, SS59, JAW, 1955[112] and QONK, 1972[95]; Swansea SS69, JAW, 1939[112]; Cardiff, ST17, AEW & RLS, 1938[112] and MG, 1974[99]; Leckwith, ST17, JEL, 1960[107] and JPC, 1978[112]; nr Pontardulais, SN50, QONK, 1974[95]; nr Garn-swllt, SN60, QONK, 1974[95]; nr Garn-swllt, SN61, RGE, 1981[112]. **Brecs.:** Tal-y-llyn, SO12, MP, 1972[112]. **Carms.:** nr Ammanford, SN61, AMP, 1981[112].

R. amphibia (L.) Besser : Berwr Melyn Mwyaf y Dŵr : Great Yellow-cress. Native. Riversides (Al,Ar,M). Fl.6-9. Mons., Glam., Monts., Denbs.*. 15 squares : **SS***69*; **ST***07,17,29*,39,48; **SN**60; **SO***00,29*,50,51; **SJ**24, 34,*35*,45.

R. amphibia × sylvestris = R. × anceps (Wahlenb.) Reichenb. Spontaneous. Mons., Brecs.. 2 squares. **Mons.:** Monmouth, SO51, SGC, 1924[112]. **Brecs.:** Three Cocks, SO13, PMH, 1938[112].

R. sylvestris (L.) Besser : Berwr Melyn Ymlusgol y Dŵr : Creeping Yellow-cress. Native. Marshy and waste places (Al,D,F,G,M,Q,R,Rg,Rw,S). Fl.(4-)6-10. Mons., Glam., *Brecs.*, *Rads.*, Carms., *Pembs.**, Cards.*, *Monts.**, Mer., *Caerns.*, Denbs., *Flints.*, *Angl.*. 73 squares, map 177.

R. palustris (L.) Besser (*R. islandica* auct., non (Oeder) Borbás) : Berwr Melyn Blynyddol y Dŵr : Marsh Yellow-cress. Native. Marshy places (Al,Ar,M,Q,R, Rg,Rw). Fl.5-9. Mons., Glam., Brecs., Rads., Carms., Pembs.*, Cards.*, Monts., *Mer.**, *Caerns.*, Denbs., *Flints.*, Angl.. 93 squares, map 178.

R. lippizensis (Wulfen) Reichenb. Casual (Balkans). Waste places. Cards.. 1 square : **SN***58*[112].

ARMORACIA Gilib.

Armoracia rusticana P. Gaertner, B. Meyer & Scherb. : Rhyddygl Poeth : Horse-radish. Established alien (S.E. Europe). Roadsides (Ar,D,G,M,Q,R,Rw,S). Fl.5-7. Mons., Glam., *Brecs.*, *Rads.*, Carms., Pembs., Cards., *Monts.*, *Mer.*, Caerns., *Denbs.*, *Flints.*, *Angl.*. 142 squares, map 179.

NASTURTIUM R. Br.

Nasturtium officinale agg. Mons., Glam., Brecs., Rads., Carms., Pembs., *Cards.*, Monts., Mer., Caerns., Denbs., Flints., Angl.. 224 squares, map 180.

N. officinale R. Br. *s.s.* (*Rorippa nasturtium-aquaticum* (L.) Hayek). : Berwr y Dŵr : Water-cress. Native. Streamsides (Al,Ar,M,W). Fl.5-10. Mons., Glam., Brecs., Rads., Carms., Pembs., Cards.*, Monts., Mer., Caerns., *Denbs.*, Flints.*, *Angl.*. 140 squares, map 181.

N. microphyllum (Boenn.) Reichenb. (*Rorippa microphylla* (Boenn.) Hyl.) : Berwr Dwfr Lleiaf : Narrow-fruited Water-cress. Native. Streamsides (Al,Ar,M). Fl.6-8. Mons., Glam., Brecs., Pembs., *Cards.**, *Monts.**, Mer., Caerns., *Denbs.*, *Flints.**, Angl.. 77 squares, map 182.

N. microphyllum × officinale (*Rorippa × sterilis* Airy Shaw). Spontaneous. Streamsides (Al,Ar,H,M). Fl.6-9. Mons., Glam., Brecs., Rads., *Carms.*, Pembs., Monts., Mer., Caerns., Denbs., *Flints.*, Angl.. 56 squares, map 183.

CARDAMINE L.

Cardamine trifolia L. : Trefoil Cress. Casual (C. Europe). Waste places. Glam.. 1 square : **SS***69*[18].

C. amara L. : Chwerw Mawr : Large Bitter-cress. Native. Streamsides (Ar,M). Fl.5-7. *Mons.*, Glam., *Pembs.**, *Cards.**, Monts., Mer.*, *Caerns.*, Denbs., Flints.. 15 squares : **ST***09*,58,*59*; **SM**81; **SN**54,76; **SH**77; **SJ**04,10,20,21,22,34,35,44.

C. raphanifolia Pourret Naturalized alien (S. Europe). Brookside. Mons.. 1 square.
Mons.: Whitebrook, SO50, STy & TGE, 1981[112].

C. pratensis L. : Blodyn y Gog : Cuckooflower. Native. Moist meadows (Ar,Ds,E,G,H,M,R,W). Fl.3-7. Mons., Glam., Brecs., Rads., Carms., Pembs., Cards., Monts., Mer., Caerns., Denbs., Flints., Angl.. 271 squares, all except : **SS**38,39; **ST**26; **SM**50,71,94; **SN**36; **SH**29,51,68,98.

C. impatiens L. : Chwerw Culddail : Narrow-leaved Bitter-cress. Native. Rocky places on limestone (Q,Rw, W). Fl.5-7. Mons., *Glam.*, Brecs., Rads., Monts., *Mer.*, *Caerns.*[84], Denbs., *Flints.*[230], [*Carms.*]. 21 squares, map 184.

C. flexuosa With. : Chwerw'r Coed : Wavy Bitter-cress. Native. Moist shady places (Al,Ar,E,H,M,R,W). Fl.4-9. Mons., Glam., Brecs., Rads., Carms., Pembs., Cards., Monts., Mer., Caerns., Denbs., Flints., *Angl.*. 251 squares, map 185.

C. flexuosa × hirsuta = C. × zahlbrucknerana O. E. Schulz Spontaneous. Damp places. Fl.7. *Mer.*, [*Monts.*]. 1 square.
Mer.: Tal-y-llyn, SH70, HTM, 1890[107].

C. flexuosa × pratensis = C. × haussknechtiana O. E. Schulz Spontaneous. Mons., Glam.. 3 squares.
Mons.: Pontnewydd, ST29, AEW, 1946[112]; Barnett Woods, ST59, WAS, 1890[112]. **Glam.:** St. Fagans, ST17, JWD, 1972[112].

C. hirsuta L. : Chwerw Blewog : Hairy Bitter-cress. Native. Walls and bare ground (E,G,H,M,Q,R, Rg,Rw,W). Fl.3-9. Mons., Glam., Brecs., Rads., Carms., Pembs., Cards., Monts., Mer., Caerns., Denbs., *Flints.*, *Angl.*. 239 squares, map 186.

CARDAMINOPSIS (C. A. Meyer) Hayek

Cardaminopsis petraea (L.) Hiitonen : Berwr y Cerrig : Northern Rock-cress. Native. Mountain rocks. Fl.6-8. *Mer.*, Caerns.. 3 squares.
Mer.: Cwm Croesor, Moelwyn, SH64, DAJ, 1901[107] and OHB & HHan, 1979[3]; Moel yr Hydd, SH64, JR, c1660[203]; **Caerns.:** Clogwyn Du'r Arddu, SH55, JR, c1660[203] and ER, 1963[207]; Snowdon, SH65, JR, c1660[203] and ER, 1962[207]; Cwm Idwal, SH65, JTCWi, 1876[112] and ER, 1950+[207].

ARABIS L.

Arabis glabra (L.) Bernh. (*Turritis glabra* L.) : Twrged Esmwyth : Tower Mustard. Casual (E. England). Waste places. *Mons.*. 1 square : **ST***38*[226].

A. hirsuta (L.) Scop. : Berwr y Graig : Hairy Rock-cress. Native. Sand dunes and rocky places on limestone (D,E,G,M,Q,R,Rg,Rw,W). Fl.4-8. Mons., Glam., Brecs., Rads., Carms., Pembs., *Cards.*, Monts., Mer., Caerns., Denbs., Flints., Angl.. 105 squares, map 187.

[**A. brownii** Jordan *Pembs.*]

A. verna (L.) R. Br. Casual (Mediterranean region). Waste places. *Glam.*. 1 square : **ST***17*[139].

A. scabra All. (*A. stricta* Hudson) : Bristol Rock-cress. Native. Rocky places. *Rads.*†, [*Denbs.*]. 1 square.
Rads.: nr Llandrindod Wells, SO06, DIGV, 1893[7].

A. caucasica Schlecht. (*A. albida* Steven) : Garden Arabis. Established alien (Mediterranean region). Walls (G,Q,R,S). Fl.4-7(-11). Mons., Glam., Brecs., Rads., Carms., Cards., Monts., Mer., Denbs., Angl.. 22 squares, map 188.

AUBRIETA Adanson

Aubrieta deltoidea (L.) DC. : Aubretia. Established alien (Balkans). Walls. Fl.6. *Mons.*, *Glam.*, Cards., *Monts.*, *Caerns.*, Denbs.. 8 squares.
Mons.: Tintern, SO50, RGE & AEW, 1975[93]. **Glam.:** Mountain Ash, ST09, AMP, 1972[93]; Draethen, ST28, RGE & AEW, 1972[93]. **Cards.:** Llandysul, SN44, AOC, 1978[112]. **Monts.:** Llanfair Caereinion to Melin-y-ddol, SJ00 and SJ10, JAW, 1930+[228]. **Caerns.:** Gt Ormes Head, SH78, JGH, 1964[249]. **Denbs.:** Llanrwst, SH76, JAW, 1943[112].

LUNARIA L.

Lunaria rediviva L. Casual (Europe). Roadsides. *Glam.*. 1 square : **SS***69*[22].

L. annua L. (*L. biennis* Moench) : Swllt Dyn Tlawd : Honesty. Established alien (S.E. Europe). Waste places and roadsides (E,M,Q,R,W). Fl.4-5. Mons., Glam., *Brecs.*, Rads., *Carms.*, *Pembs.*, Cards., *Monts.*, *Caerns.*, *Denbs.*, *Flints.*, *Angl.*. 49 squares, map 189.

ALYSSUM L.

Alyssum saxatile L. : Golden Alison. Casual (C. Europe). Walls. Fl.4-5. Monts., Mer.. 2 squares : **SO**29[112]; **SH**61[112].

A. alyssoides (L.) L. : Small Alison. Casual (Europe). Waste places. Fl.10. *Glam.*. 2 squares : **ST***17*[218]; **SO***00*[107].

BERTEROA DC.

Berteroa incana (L.) DC. : Hoary Alison. Casual (Europe). Cultivated ground and waste places. Fl.7-10. *Mons.*, Glam., *Carms.*, *Cards.*. 11 squares : **SS**69,*78*; **ST**16,17; **SN**25,*40,50, 68,80*; **SO***00*,10.

LOBULARIA Desv.

Lobularia maritima (L.) Desv. : Cyddlin : Sweet Alison. Naturalized alien (Mediterranean region). Sand dunes and waste places (D,G,M,Q,R,Rw,S,Sc). Fl.6-10. Mons.*, Glam., Rads., Carms., *Pembs.**, Cards.*, Monts., Mer., Caerns., Denbs., Flints., *Angl.**. 53 squares, map 190.

DRABA L.

Draba aizoides L. : Llysiau Melyn y Bystwn : Yellow Whitlowgrass. Denizen (Europe). Rocks and walls near the coast. Fl.2-4. Glam.. 3 squares.
Glam.: Worms Head, SS38, TBF, 1887[107] and JAW, 1930+[221]; btwn Worms Head and Port Einon, SS48, MCFP, 1956[7] and QONK, 1970+[157]; Pennard Cliffs, SS58, WT, 1803[107] and TGE, 1980[91].

D. incana L. : Llwydlys y Bystwn : Hoary Whitlowgrass. Native. Mountain rocks. Fl.6-7. Caerns.. 2 squares.
Caerns.: Snowdon, SH65, JEG, 1890[112] and ER, 1964[207]; Penyrole-wen, SH66, DAR, 1950+[207].

D. muralis L. : Magwyrlys y Bystwn : Wall Whitlowgrass. Naturalized alien (England). Walls and limestone rocks. Fl.4-6. *Glam.**, Brecs.*, *Rads.*, *Pembs.*, Cards.*, *Monts.*, Mer., Denbs.*. 14 squares : **SS**58; **SN**11,44,68, 69,84; **SO***25*; **SH**61,*64*,71,*93*; **SJ**15,21,24. Perhaps native in one locality in Denbs.

EROPHILA DC.

Erophila verna (L.) Chevall. Native. Mons., Glam., Brecs., Rads., Carms., Pembs., Cards., Monts., Mer., Caerns., Denbs., Flints., Angl.. 158 squares, map 191.

Subsp. **verna** : Llys y Bystwn : Common Whitlowgrass. Native. Rocks and gravelly places (D,G,Q,R,Rw,W). Fl.3-6. Mons., Glam., Brecs., Rads., Carms., Pembs., Cards., Monts., *Mer.*, Caerns., Denbs., Flints., Angl.. 110 squares, map 192.

[Subsp. **praecox** (Steven) Walters (*E. praecox* (Steven) DC.). *Mons., Glam., Cards., Mer., Caerns., Denbs., Flints.*]

Subsp. **spathulata** (A. F. Láng) Walters (*E. spathulata* A. F. Láng) : Round-podded Whitlowgrass. Native. Walls and rocks (D,G,Q,Rw,Sc). Fl.2-5. Mons., Glam., Carms.*, Pembs., Mer., *Caerns.*, Denbs., Angl.*. 32 squares, map 193.

COCHLEARIA L.

Cochlearia danica L. : Llwylys Denmarc : Danish Scurvygrass. Native. Rocky places near the sea (D,E, G,M,Q,R,S). Fl.1-6(-9). Mons., Glam., Carms., Pembs., Cards., *Monts.*, Mer., Caerns., Denbs., Flints., Angl.. 101 squares, map 194.

C. danica × **officinalis** Spontaneous. Sea cliffs, sand and shingle. Fl.5. Pembs., *Cards.*, *Mer.*, *Caerns.*[217], *Angl.*. 6 squares.
Pembs.: St Ann's Head, SM80, TAWD, 1975[112]. **Cards.:** Llangrannog, SN35, JEH & AJS, 1976[185]; Borth, SN68 and 69, JEH & AJS, 1976[185]. **Mer.:** Llanaber, SH51, PMB, 1961 and 1973[3]. **Angl.:** Lynas Point, SH49, RHR, 1962[208].

C. officinalis L. : Llwylys Cyffredin : Common Scurvygrass. Native. Sea-shores (E,M,Q,R,S,Sc). Fl.3-8. Mons., Glam., *Carms.*, Pembs., Cards., Monts., Mer., Caerns., *Denbs.*, *Flints.*, Angl.. 106 squares, map 195.

C. alpina (Bab.) H. C. Watson (*C. officinalis* subsp. *alpina* (Bab.) Hooker): Môrlwyau Creigiau Mynyddig : Alpine Scurvygrass. Native. Mountain rocks. Fl.6-9. Mer., Caerns., [*Pembs.*, *Denbs.*]. 6 squares.
Mer.: Diffwys, SH62, TB, pre 1863[212]; Cadair Idris, SH71, AL, pre 1875[126] and PMB & KS, 1966[3]; nr Dinas Mawddwy, SH81, MR, 1938[3]. **Caerns.:** Clogwyn Du'r Arddu, SH55, JLOL, 1924[112] and ER, c1958[207]; Snowdon, SH65, ELD, 1948[112] and ER, 1965[207]; Glyders, SH65, JTCW, 1876[112] and ER, 1956[207]; Carneddau, SH66, HAH, 1926[112] and ER, 1960[207].
The relationship between this and *C. pyrenaica* DC. is uncertain, some authors refer Welsh plants to the latter species.

C. anglica L. : Llwylys Lloegr : English Scurvygrass. Native. Muddy sea-shores and estuaries (Ar,M,S). Fl.2-7. Mons., Glam., Carms., Pembs., Cards., Monts., Caerns., Denbs., Flints., Angl., [*Mer.*]. 59 squares, map 196.

KERNERA Medicus

Kernera saxatilis (L.) Reichenb. Casual (Europe). Waste places. *Glam.*. 1 square : **SO***00*[140].

CAMELINA Crantz

Camelina sativa (L.) Crantz (*Myagrum sativum* L.) : Gold-of-pleasure. Casual (E. Europe). Cultivated ground and waste places (F,R,Rg,Rw). Fl.6-10. Mons., Glam., *Carms.*, Pembs., Cards., *Mer.*, *Caerns.*, *Denbs.*, *Flints.*, *Angl.*[85]. 18 squares : **SS***59,78*: **ST**16,*17,18*,*27*,38; **SM**80; **SN***42*,58,*69*; **SO***00,30,51*; **SH**33,*67*, **SJ***24,26*.

C. microcarpa Andrz. ex DC. Casual (Europe). Waste places. *Carms.*. 1 square : **SS***59*[161].

C. alyssum (Miller) Thell. Casual (Europe). Waste places. *Glam.*. 1 square : **ST***27*[69].

NESLIA Desv.

Neslia paniculata (L.) Desv. (*Myagrum paniculatum* L.; *Vogelia paniculatum* (L.) Hornem.) : Ball Mustard. Casual (Near East). Waste places (R,Rw). Fl.8-9. Glam., *Flints*.. 5 squares : **SS***79*[140]; **ST***27*[112]; **SN***90*[107]; **SO***00*[107]; **SJ***26*[143].

CAPSELLA Medicus

Capsella bursa-pastoris (L.) Medicus : Pwrs y Bugail : Shepherd's-purse. Native. Cultivated ground and waste places (D,E,F,G,M,R,Rg). Fl.4-11. Mons., Glam., Brecs., Rads., Carms., Pembs., Cards., Monts., Mer., Caerns., Denbs., Flints., *Angl*.. 272 squares, all except : **SR**89; **SS**39; **SM**50,73; **SN**35,53,59; **SO**28; **SH**51; **SJ**00.

C. rubella Reuter Casual (S. Europe). Waste places. Fl.6. Mons.. 1 square : **ST**38[112].

HORNUNGIA Reichenb.

Hornungia petraea (L.) Reichenb. : Beryn Creigiog : Hutchinsia. Native. Limestone rocks and sand dunes (D,G,Q,S,Sc). Fl.2-6. Glam., Brecs., Pembs., (*Mer.*), Caerns., Denbs., *Angl*.. 23 squares, map 197.

TEESDALIA R. Br.

Teesdalia nudicaulis (L.) R. Br. : Beryn Coesnoeth : Shepherd's Cress. Native. Sandy or gravelly places (D,G,Q,R). Fl.(1-)4-6. Rads., *Cards*., *Monts*., Mer., Caerns., Denbs., *Flints*., Angl., [*Glam*., *Brecs*.]. 29 squares, map 198.

THLASPI L.

Thlaspi arvense L. : Codywasg : Field Penny-cress. Colonist (England). Cultivated ground (E,F,G,R,Rg). Fl.5-9. Mons., Glam., Brecs.*, Rads., *Carms*., Pembs., *Cards*., Monts., Mer.*, *Caerns*., Denbs., *Flints*., Angl.. 64 squares, map 199.

[**T. alliaceum** L. *Denbs*.]

T. alpestre L. : Codywasg Creigiog Mynyddog : Alpine Penny-cress. Native. Mountain rocks (M,Q,R, Rw). Fl.4-8. *Glam*., *Rads*.[232], Cards., Caerns., [*Denbs*.]. 11 squares : **SN**56,57,*58*,67,68,78,*80*; **SH**53,*66*,75,76. The *Glam*. record (**SN**80) is regarded as doubtful.

IBERIS L.

Iberis sempervirens L. Casual (S. Europe). Waste places. *Glam*., *Caerns*.. 2 squares : **ST**09[95]; **SH**57[7].

I. amara L. : Beryn Cherw. Wild Candytuft. Casual (S. England). Waste places (E,R). Fl.7-8. *Mons*., Glam., *Brecs*., *Carms*., *Monts*., *Caerns*.[85], Denbs., *Flints*.[85], *Angl*.. 14 squares : **SS***88*; **ST***08,09,16,18*,48; **SO**00,*13*; **SN**31,*40*,42,79; **SH***67*; **SJ**53.

I. umbellata L. : Garden Candytuft. Casual (S. Europe). Waste places (R,Rg,Rw). Fl.7-8. Mons., Glam., *Denbs*., Flints.. 9 squares : **SS**58[248],79[157],*98*[18]; **ST**09[112],*17*[112],*39*[112]; **SO***50*[112]; **SJ**06[90],36[112].

LEPIDIUM L.

Lepidium campestre (L.) R. Br. : Codywasg y Maes : Field Pepperwort. Native. Dry pastures (E,F,G,Q,R, Rg,Rw,W). Fl.4-6(-8). Mons., Glam., *Brecs*., Rads., *Carms*., Pembs., *Cards*.*, Monts., *Caerns*., Denbs., Flints., *Angl*.*, [*Mer*.]. 58 squares, map 200.

L. heterophyllum Bentham (*L. smithii* Hooker) : Pupurlys : Smith's Pepperwort. Native. Fields and roadsides (E,G,M,Q,R,Rw,W). Fl.5-10. Mons., Glam., Brecs., Rads., Carms., Pembs., Cards., Monts., Mer., Caerns., Denbs., Flints., Angl.. 178 squares, map 201.

L. sativum L. : Berwr Gardd : Garden Cress. Casual (Egypt and W. Asia). Waste places (F,R). Fl.6-10. Mons., Glam., Brecs., Carms., Cards., Monts., *Mer*., Caerns., Denbs.. 21 squares, map 202.

L. virginicum L. : Poor-man's Pepper. Casual (N. America). Waste places. Fl.6-8(-10). Mons., Glam., *Rads*., Cards., *Caerns*., *Flints*.. 10 squares : **SS***69*; **ST***17,18,27*,38,*59*; **SN***58*; **SO**25; **SH***33*; **SJ***36*.

L. densiflorum Schrader : American Pepperwort. Casual (N. America). Waste places. Fl.8. *Mons*., *Glam*., *Rads*., *Pembs*., *Cards*.. 9 squares : **SS***69*[222]; **ST***16*[23],*17*[27],*18*[25],*27*[27],*59*[107]; **SN**10[30],*58*[26]; **SO**06[229].

L. ramosissimum A. Nelson Casual (N. America). Waste places. *Glam*.. 1 square : **ST***27*[27].

L. neglectum Thell. : Least Pepperwort. Casual (N. America). Waste places. Fl.6-10. Mons., Glam., Rads., Cards., *Caerns*.. 6 squares : **ST***06*[112],*16*[112],27[112]; **SN**58[112]; **SO**51[112]; **SH**33[18].

L. ruderale L. : Pupurlys Culddail : Narrow-leaved Pepperwort. Established alien (England). Waste places (M,R,Rw,S). Fl.5-7. Mons., Glam., *Brecs*., *Carms*., *Pembs*., *Cards*.*, *Mer*., Caerns., *Flints*.. 28 squares, map 203.

L. perfoliatum L.: Perfoliate Pepperwort. Casual (E. Europe and W. Asia). Waste places (M,R,Rw). Fl.5-7. Glam., *Brecs*., *Carms*., Cards., Caerns., *Denbs*.. 7 squares : **ST***17*[112],*27*[112]; **SN***42*[20],*68*[112],*90*[107]; **SH***66*[112],*87*[22].

L. latifolium L. : Berwr Gwyllt : Dittander. Denizen (England). Sea-shores, roadsides and river banks (E,M, R,Rw,S). Fl.7-9. Mons., Glam., *Carms*., Denbs., *Flints*., *Angl*.. 12 squares : **SS**79; **ST***08*,16,17,27,38; **SN**42; **SH***36,38*,97; **SJ***08*,18. Possibly native in Glam. and Denbs.

L. graminifolium L. : Tall Pepperwort. Casual (Mediterranean region). Waste places. Fl.7-9. Glam., Caerns.. 6 squares : **SS**69[112],*78*[140]; **ST***16*[112],17[112],27[95]; **SH**53[112].

CARDARIA Desv.

Cardaria draba (L.) Desv. : Pupurlys Llwyd : Hoary Cress. Established alien (Mediterranean region and W. Asia). Waste places (D,G,M,R,Rg,Rw,S). Fl.4-8. Mons., Glam., Brecs., Carms., Pembs., Cards., Monts., Mer., Caerns., Denbs., Flints., Angl.. 65 squares, map 204.

CORONOPUS Haller

Coronopus squamatus (Forskål) Ascherson : Olbrain Dafadennog : Swine-cress. Native. Roadsides (G,M,R, Rg,Rw,S). Fl.5-9. Mons., Glam., *Brecs.*, *Rads.*, Carms., Pembs., *Cards.*, *Monts.*, *Mer.**, Caerns., Denbs., *Flints.*, Angl.. 96 squares, map 205.

C. didymus (L.) Sm. : Olbrain Lleiaf : Lesser Swine-cress. Established alien (S. America). Roadsides (E,F,M, R,Rg,Rw,S). Fl.4-10. Mons., Glam., Brecs., Rads., Carms., Pembs., Cards., Monts.*, Mer., Caerns., *Flints.**, *Angl.*. 108 squares, map 206.

SUBULARIA L.

Subularia aquatica L. : Mynawydlys Dyfrdig : Awlwort. Native. Lakes (Al,M). Fl.7-9. *Carms.*, Cards., Mer., Caerns., *Denbs.*, *Angl.*. 18 squares : **SN**66,*76*,82,86; **SH**_38_,44,48,55,56,62,63,64,65,66,70,71,*85*,*94*.

CONRINGIA Adanson

Conringia orientalis (L.) Dumort. : Hare's-ear Mustard. Casual (Mediterranean region). Waste places (G,R,Rg,Rw). Fl.6-8. Mons., Glam., *Carms.*, *Cards.*, *Mer.*, *Caerns.*, *Denbs.*. 19 squares : **SS**69,*78*,*87*; **ST**08,16,*17*,*18*,*27*,*48*; **SN**_40_,*58*; **SO**_00_,*51*; **SH**_66_; **SJ**_03_,*06*,*07*,*23*,*34*.

C. austriaca (Jacq.) Sweet Casual (S.E. Europe). Waste places. *Glam.*, *Brecs.*. 2 squares : **SO**_00_[140],*05*[22].

DIPLOTAXIS DC.

Diplotaxis tenuifolia (L.) DC. : Cedw Meindwf y Tywod : Perennial Wall-rocket. Denizen (England). Cliffs, walls, etc. (D,M,Q,R,Rw). Fl.5-10. Mons., Glam., *Brecs.*, Carms., Pembs., Mer., Caerns., *Denbs.*, *Flints.*, *Angl.**. 45 squares, map 207.

D. erucoides (L.) DC. : White Rocket. Casual (S.W. Europe). Waste places. Fl.8-10. Glam.. 2 squares : **ST**_16_[112],17[112].

[**D. viminea** (L.) DC. *Glam.*]

D. muralis (L.) DC. : Mwstard y Tywod : Annual Wall-rocket. Established alien (C. and S. Europe). Roadsides and cultivated ground (D,F,Q,R,Rg,Rw,S). Fl.4-10. Mons., Glam., Brecs.*, Carms., Pembs., Cards., Monts., Mer., Caerns., Denbs., Flints., *Angl.*. 70 squares, map 208.

BRASSICA L.

Brassica elongata Ehrh. Casual (S. Europe). Waste places. Fl.6-10. *Glam.*, *Brecs.*. 3 squares : **SS**_78_[140]; **SN**_90_[107]; **SO**_00_[110].

B. oleracea L. : Bresych Gwyllt : Wild Cabbage. Native. Sea cliffs (M,Q,R,Rw,Sc). Fl.4-8. Mons., Glam., (*Rads.**), *Carms.**, Pembs., Caerns., *Denbs.*, (Flints.*), *Angl.**, [*Mer.*]. 34 squares, map 209.

B. napus L. : Bresych yr Ŷd : Rape. Established alien (origin unknown). Waste places (E,F,G,M,Q,R,W). Fl.5-7(-9). Mons., Glam., Brecs., Rads.*, *Carms.**, Pembs.*, *Cards.*, *Monts.**, *Mer.**, Caerns., Denbs., Flints., Angl.[209]. 84 squares, map 210.
Due to confusion between this and the following species some of the records may be referable to the latter.

B. rapa L. : Meipen : Wild Turnip. Naturalized alien (origin unknown). River banks (E,F,G,H,R,Rg,Rw). Fl.5-10. Mons., Glam., Brecs., Rads., Carms., Pembs., Cards., Monts., Mer., *Caerns.*, Denbs., Flints., Angl.. 161 squares, map 211.

B. juncea (L.) Czern. Casual (Asia). Waste places (R,S). Fl.8-10. Glam., *Carms.*, *Angl.*. 6 squares : **SS**69[112]; **ST**16[112],17[112],*27*[27]; **SN**_40_[19]; **SH**_68_[18].

B. fruticulosa Cyr. Casual (S. Europe). Waste places. Fl.7. Glam.. 1 square : **ST**_17_[112].

B. tournefortii Gouan : Pale Cabbage. Casual (S. Europe and Asia). Waste places. *Glam.*. 1 square : **ST**_16_[23].

B. nigra (L.) Koch : Cedw Du : Black Mustard. Denizen (Europe). Sea cliffs and roadsides (Ar,E,F,G, M,Q,R,S,Sc). Fl.5-9. Mons., Glam., *Brecs.*, *Rads.**, Carms., Pembs., Cards., *Monts.*, Mer., Caerns., Denbs., Flints., Angl.. 86 squares, map 212.

SINAPIS L.

Sinapis arvensis L. : Cedw Gwyllt : Charlock. Colonist (England). Cultivated ground (D,F,R). Fl.4-8(-10). Mons., Glam., Brecs., Rads., Carms., Pembs., Cards., Monts., Mer., Caerns., Denbs., Flints., *Angl.*. 213 squares, map 213.

S. alba L.

Subsp. **alba** : Cedw Gwyn : White Mustard. Established alien (Mediterranean region). Waste places and cultivated ground (F,R,Rw). Fl.6-10. Mons., Glam., Brecs., Rads., *Carms.*, *Pembs.*, *Cards.*, Monts., Mer., *Caerns.*, Denbs., Flints., *Angl.*. 60 squares, map 214.

Subsp. **dissecta** (Lag.) Bonnier (*S. dissecta* Lag.). Casual (Mediterranean region). Waste places. Fl.7. Glam.. 3 squares : **SS**69[95],*78*[112]; **SN**80[95].

ERUCA Miller

Eruca vesicaria (L.) Cav. subsp. **sativa** (Miller) Thell. (*E. sativa* Miller) : Garden Rocket. Casual (Mediterranean region). Waste places. Fl.6-7. Glam., Caerns., *Flints.*. 4 squares : **ST**_27_[112]; **SH**_53_[16],*75*[112]; **SJ**_36_[18].

ERUCASTRUM C. Presl

Erucastrum gallicum (Willd.) O. E. Schulz : Hairy Rocket. Casual (W. and C. Europe). Waste places (R,Rg). Fl.7. Glam., *Carms.*, [*Pembs.*]. 3 squares, : **SS**69[112]; **ST**27[112]; **SN**_40_[19].

HUTERA Porta

Hutera cheiranthos (Vill.) Gómez-Campo (*Rhynchosinapis cheiranthos* (Vill.) Dandy; *R. erucastrum* (L.) Dandy ex Clapham) : Berwr Murwyll y Môr : Wallflower Cabbage. Established alien (Europe). Waste places (D,R,Rw). Fl.6-10. Mons.*, Glam., Carms., *Cards.**, Denbs.*. 24 squares, map 215.

H. monensis (L.) Gómez-Campo (*Rhynchosinapis monensis* (L.) Dandy ex Clapham) : Berwr Môn a Manaw : Isle of Man Cabbage. Native. Sea cliffs (D,G,R,Sc). Fl.5-9. Glam., (Carms.*, Caerns., *Angl.*). 4 squares.
Glam.: nr Pennard Castle, SS58, JWG, 1841[198] and MPe, 1971[112]; Three Cliffs Bay, SS58, AL, 1903[106] and MG, 1976[99] (all other records for vc 41 are errors for *H. cheiranthos*). (**Carms.:** nr Llanelli, SS59, IMV, 1971[181]. **Caerns.:** nr Bangor, SH57, JTCWi, c1876[112]. **Angl.:** Abermenai, SH46, HD, c1800[78] and JEG, c1890[101]).

H. hispida (Cav.) Gómez-Campo (*Rhynchosinapis hispida* (Cav.) Heywood; *Eruca hispida* Cav.). Casual (Iberian Peninsula). Waste places. Fl.6. Glam.. 1 square : **ST***27*[112].

HIRSCHFELDIA Moench

Hirschfeldia incana (L.) Lagrèze-Fossat : Cedw Penllwyd : Hoary Mustard. Established alien (Mediterranean region). Waste places (D,G,R,Rw). Fl.7-9. Mons., Glam., *Rads.**, Carms., Pembs.*, *Caerns.*. 26 squares, map 216.

CARRICHTERA DC.

Carrichtera annua (L.) DC. Casual (Mediterranean region). Waste places. Fl.7. Glam.. 1 square : **ST***27*[112].

ERUCARIA Gaertner

Erucaria hispanica (L.) Druce Casual (Aegean region). Waste places. Fl.9. Glam.. 2 squares : **ST**17[112],27[112].

CAKILE Miller

Cakile maritima Scop. : Hegydd Arfor : Sea Rocket. Native. Sea beaches (Ar,D,S). Fl.5-11. Glam., Carms., Pembs., Cards., Monts., Mer., Caerns., Denbs., Flints., Angl.. 72 squares, map 217.

RAPISTRUM Crantz

Rapistrum perenne (L.) All. : Steppe Cabbage. Casual (Europe). Waste places. Fl.6. Glam., *Caerns.*. 5 squares : **SS**48[222],*78*[140]; **ST***17*[140],27[112]; **SH**78[7].

R. rugosum (L.) All. : Bastard Cabbage. Casual (Mediterranean region). Waste places. Fl.6-10. Mons., Glam., Carms., *Pembs.*, Mer., Caerns., Flints.. 17 squares : **SS**68,*78*; **ST***07,16,27*,38,48,58; **SM**93; **SN***10*,31,*40,42*; **SH**50,*56*,78; **SJ**17.

Subsp. **rugosum** Mons., Glam., *Carms.*, Mer., *Caerns.*. 10 squares : **SS**68,*78*; **ST***16,27*,38; **SN***40,42*; **SH**50,*56*,78.

Subsp. **orientale** (L.) Arcangeli (*R. orientale* (L.) Crantz). Glam., Carms., *Pembs.*, Caerns., Flints.. 8 squares : **ST***16*[26], *27*[112]; **SM**93[53]; **SN***10*[18],31[112],40[161]; **SH**78[112]; **SJ**17[112].

Subsp. **linnaeanum** Rouy & Fouc. (*R. hispanicum* (L.) Crantz). Glam., Carms., Caerns.. 3 squares : **ST***27*[112]; **SN**31[112]; **SH**78[112]

CRAMBE L.

Crambe maritima L. : Bresych y Môr : Sea-kale. Native. Shingle beaches and cliffs. Fl.5-7. *Glam.*, Pembs., Mer., Caerns., Denbs., Angl.. 28 squares, map 218.

RAPHANUS L.

Raphanus raphanistrum L.

Subsp. **raphanistrum** : Rhuddygl Gwyllt : Wild Radish. Established alien (England). Cultivated ground (E,F,R). Fl.6-10. Mons., Glam., *Brecs.*, Rads., *Carms.*, Pembs., Cards., Monts.*, Mer., *Caerns.*, *Denbs.*, Flints., *Angl.*. 104 squares, map 219.

Subsp. **maritimus** (Sm.) Thell. (*R. maritimus* Sm.) : Bysedd yr Iâr Arforol : Sea Radish. Native. Seashores (D,R,S). Fl.7-9. *Mons.**, Glam., Carms., Pembs., *Mer.*, Caerns., Angl.*, [*Cards.*, *Denbs.*]. 33 squares, map 220.

Subsp. **landra** (Moretti ex DC.) Bonnier and Layens (*R. landra* Moretti ex DC.). Casual (Mediterranean region). Waste places. Mons.. 1 square : **ST**38[112].

R. sativus L. : Rhuddygl : Garden Radish. Casual (origin unknown). Waste places (F,R). Fl.8-10. Mons., Glam., *Carms.*, Monts., Mer., Denbs., Flints.. 13 squares : **SS**,69,*78*; **ST***16*,17,*27*; **SN**50,61; **SO***00*,19,42; **SH**71; **SJ**17,21.

RESEDACEAE

RESEDA L.

Reseda luteola L. : Melyn-gu : Weld. Native. Waste places, roadsides, etc. (D,E,G,Q,R,Rg,Rw,W). Fl.5-8(-10). Mons., Glam., Brecs., Rads., Carms., Pembs., Cards., Monts., Mer., Caerns., Denbs., Flints., Angl.. 144 squares, map 221.

R. alba L. : Melyngu Gwyn Unionsyth : White Mignonette. Established alien (Mediterranean region). Waste places. Fl.6-8(-10). Mons.*, Glam., Carms., *Pembs.*, *Caerns.**. 11 squares : **SS**48,*69,87*: **ST***07*,16,17, 27,38; **SN***10,40*: **SH**78.

R. phyteuma L. : Corn Mignonette. Casual (Europe). Waste places (R,Rg). Glam., *Mer.*. 2 squares : **ST***17*[112]; **SN**69[177].

R. odorata L. Casual (N. Africa). Waste places. *Glam.*. 1 square : **SS**69[222].

[**R. stricta** Pers. *Glam.*]

R. lutea L. : Perllys : Wild Mignonette. Native. Sandy and waste places (D,F,Q,R,Rg,Rw). Fl.5-9. Mons., Glam., Brecs., *Rads.**, Carms., Pembs., Cards.*, Monts.*, *Mer.*, Caerns., Denbs., Flints., *Angl.**. 71 squares, map 222.

CAYLUSEA St-Hil.

Caylusea hexagyna (Forskål) M. L. Green (*C. canescens* St-Hil.). Casual (N. Africa and Middle East). Waste places. *Glam.*. 1 square : **ST***27*[26].

DROSERACEAE

DROSERA L.

Drosera rotundifolia L. : Gwlithlys : Round-leaved Sundew. Native. Bogs (Ar,H,M). Fl.6-8. Mons., Glam., Brecs., Rads., Carms., Pembs., Cards., Monts., Mer., Caerns., Denbs., Flints., *Angl.*. 188 squares, map 223.

D. anglica Hudson : Gwlithlys Mawr : Great Sundew. Native. Bogs. Fl.6-9. Glam.†, Cards., Mer., Angl., [*Monts., Caerns., Flints.*]. 5 squares.
Glam.: Crymlyn Reservoir nr Swansea, SS69, LWD, c1840[112]†; Townhill nr Swansea, SS69, JWG, c1840[198]†. **Cards.:** Tregaron Bog, SN66, HAH, 1936[112]; Cors Fochno, SN69, HWP & FDr, 1905[107] and JMB, 1966[112]. **Mer.:** nr Llyn Barfog, SN69, AOC, c1955[4], nr Aberdyfi, SN69, PMB, 1965[177]; Arthog, SH61, LHP, 1920[107] and PMB, 1980[3]. **Angl.:** Llanfairpwllgwyngyll, SH57, JTCWi, 1876[112] and JEG, 1882[112]. All other records require confirmation.

D. anglica × rotundifolia = D. × obovata Mert. & Koch Spontaneous. Bogs. Fl.5-8. Cards., *Mer.*. 1 square.
Cards.: Cors Fochno, SN69, JHS, 1937[112] and PMB, 1960[174]. **Mer.:** nr Aberdyfi, SN69, PMB, 1965[177].

D. intermedia Hayne : Gwlithlys Hirddail : Oblong-leaved Sundew. Native. Bogs (Ar,G,H,M). Fl.6-7. Glam., Carms., Pembs., Cards., *Monts.**, Mer., Caerns., *Denbs.*[230], *Flints.*[230], *Angl.*. 38 squares, map 224.

D. intermedia × rotundifolia = D. × beleziana Camus. Spontaneous. Bogs. Cards.. 1 square.
Cards.: Cors Fochno, SN69, JHS, 1940[112].

CRASSULACEAE

UMBILICUS DC.

Umbilicus rupestris (Salisb.) Dandy : Deilen Gron : Navelwort. Native. Rocks and walls (E,Q,Sc,W). Fl.4-8. Mons., Glam., Brecs., Rads., Carms., Pembs., Cards., Monts., Mer., Caerns., Denbs., *Flints.*, Angl.. 233 squares, map 225.

SEMPERVIVUM L.

Sempervivum tectorum L. : Bywfyth : House-leek. Established alien (S. Europe). Walls and roofs. Fl.6-7. *Mons.*, Glam., Brecs., *Rads.**, *Carms.*, *Pembs.*, *Cards.*[210], Monts., *Mer.*[47], Caerns., *Denbs.*, *Flints.*, *Angl.*[101]. 53 squares, map 226.

SEDUM L.

Sedum hybridum L. Casual (N. Asia). Waste places. *Glam.*. 1 square : **ST***07*[140].

S. telephium L. : Berwr Taliesin : Orpine. Native. Hedgebanks and rocky places (E,F,G,Q,Rw,W). Fl.6-9. Mons., Glam., Brecs., Rads., Carms., Pembs., Cards., Monts., Mer., Caerns., Denbs., *Flints.*, Angl.. 174 squares, map 227.

Subsp. **telephium** (*S. telephium* subsp. *purpurascens* (Koch) Syme). Mons., Glam., Brecs., Rads., Carms., Pembs.*, Cards.*, Monts., Mer., Caerns.*, *Denbs.*[230], *Flints.*[230], *Angl.*. 25 squares, map 228.

Subsp. **fabaria** (Koch) Kirschleger Mons., Glam., Brecs., Rads., Carms., Pembs., Cards., Monts., Mer., Caerns., Denbs., Angl.. 53 squares, map 229.

S. spurium Bieb. : Caucasian Stonecrop. Established alien (Caucasus). Walls (Q,R,Rg,S). Fl.6-8(-10). Mons., Glam., *Brecs.*, Rads., Carms., *Pembs.*, Cards., Monts., *Caerns.*, *Denbs.*, *Flints.*, *Angl.*. 50 squares, map 230.

S. reflexum L. : Llwynau'r Fagwyr : Reflexed Stonecrop. Established alien (S. Europe). Walls (M,Q,R). Fl.6-10. Mons., Glam., Brecs., Rads., Carms., Pembs., Cards., Monts., Mer., Caerns., Denbs., Flints., *Angl.*. 117 squares, map 231.

S. forsteranum Sm. : Bywydog Cymreig : Rock Stonecrop. Mons., Glam., Brecs., Rads., Carms., (Pembs.), Cards., Monts., Mer., Caerns., Denbs., Flints., (*Angl.**). 108 squares, map 232.

Subsp. **forsteranum** Native. Rocky places (E,M, Q,Sc). Fl.6-8. Mons., Glam., Brecs., Rads., Carms., (Pembs.), Cards., Monts., Mer., Caerns., Denbs.. 31 squares, map 233.

Subsp. **elegans** (Lej.) E. F. Warburg (*S. rupestre* auct.; *S. elegans* Lej.). Native. Rocky places and walls (E,Q,R,Rw,S). Fl.6-7. Mons., (Glam.), Brecs., Rads., Carms., Cards., *Monts.*, Mer., Caerns., Denbs., Flints.. 41 squares, map 234.

S. acre L. : Bywydog Boeth : Biting Stonecrop. Native but frequently introduced. Limestone rocks, walls and sand dunes (D,G,Q,R,Rw,S). Fl.6-7. Mons., Glam., Brecs., Rads., Carms., Pembs., Cards., Monts., Mer., Caerns., Denbs., Flints., Angl.. 208 squares, map 235.

S. sexangulare L. : Tasteless Stonecrop. Established alien (Europe). Walls. Fl.6-7. Glam., Rads., Carms., *Cards.**, Caerns.. 13 squares : **SS**59,79; **SN**44,70,73, 74,*80*; **SO**04,08,14; **SH***76,77,78*.

S. album L.

Subsp. **album** : Gwenith y Gwylanod : White Stonecrop. Naturalized alien (Europe). Walls and shingle beaches (D,E,G,Q,Rw,S). Fl.6-8. Mons., Glam., Brecs., Rads., Carms., Pembs., *Cards.*, Monts., Mer., Caerns., Denbs., Flints., *Angl.**. 127 squares, map 236.

Subsp. **micranthum** (Bast.) Syme (*S. micranthum* Bast.) : Small-flowered Stonecrop. Established alien (S. Europe). Walls. Glam.. 2 squares.
Glam.: Penrice, SS48, JAW, 1930[220]; Pontalun, SS87, EV, 1937[112].

S. anglicum Hudson : Briweg y Cerrig : English Stonecrop. Native. Rocky banks near the sea (E,M,Q, Sc). Fl.6-8. Glam., Brecs., Rads., Carms., Pembs., Cards., Monts., Mer., Caerns., Denbs., *Flints.*, Angl., [*Mons.*]. 182 squares, map 237.

S. dasyphyllum L. : Thick-leaved Stonecrop. Established alien (S. Europe). Walls. Fl.6-8. Glam., Monts., Mer., Caerns., Denbs.. 17 squares : **SS***48*,*49*,79; **SO**39; **SH**52,*56*,*57*,*61*,*77*,78,87,88; **SJ**05,06,15,16,25.

S. villosum L. : Briweg Blewog : Hairy Stonecrop. Casual (England). Waste places. *Glam.*, *Caerns.*[119]. 1 square : **SS***89*[218].

S. spathulifolium Hooker : Colorado Stonecrop. Casual (N. America). Walls. *Monts.*. 2 squares : **SN**99[228]; **SO**09[228].

RHODIOLA L.

Rhodiola rosea L. (*Sedum rosea* (L.) Scop.) : Pren y Ddannoedd : Roseroot. Native. Mountain and coastal rocks. Fl.5-9. Glam., Brecs., Carms., *Pembs.*, Mer., Caerns., *Angl.**. 25 squares, map 238.

SAXIFRAGACEAE

SAXIFRAGA L.

Saxifraga nivalis L. : Tormaen Mynyddig Sypiog : Alpine Saxifrage. Native. Mountain rocks. Fl.5-7. Caerns., [*Mer.*]. 3 squares.
Caerns.: Cwm Meillionen, SH54, JEG, pre 1897[101] and ER, 1967[207]; Clogwyn Du'r Arddu, SH55, JEG, pre 1897[101]; Snowdon, SH65, CCB, 1832[112] and ER, 1965+[207]; Cwm Idwal, SH65, JEG, 1883[112] and PMB 1979[3].

S. stellaris L. : Tormaen Serennog : Starry Saxifrage. Native. Sides of mountain rills. Fl.6-8. Cards., *Monts.*, Mer., Caerns., Denbs., *Flints.*[230], [*Brecs.*]. 30 squares, map 239.

S. cuneifolia L. Casual (S. Europe). Waste places. *Glam.*, *Carms.*. 3 squares : **SS**79[117]; **SN**51[7],70[157].

[**S. hirsuta** L. *Caerns.*]

S. hirsuta × **spathularis** = **S.** × **polita** (Haw.) Link Established alien (Ireland). Waste places. Fl.6. Glam., *Cards.*, Caerns.. 4 squares.
Glam.: Neath, SS79, EMT, 1944[112]. **Cards.:** Aberystwyth, SN58, HDr, c1920[83]. **Caerns.:** Plas Lodwig, SH– –, JTCWi, 1876[112]; nr Bethesda, SH66, MRo, 1967[179].

S. spathularis Brot. × **umbrosa** L. = **S.** × **urbium** D. A. Webb : Balchder Llundain : London pride. Established alien (garden origin). Roadsides (E,Q,R, Rw,W). Fl.5-6. Glam., Brecs., *Rads.*, *Carms.*, *Cards.*, Monts., Caerns., *Denbs.*, Flints., Angl.. 34 squares, map 240. All records previously referred to **S. umbrosa** L. are included here.

S. cymbalaria L. var. **huetiana** (Boiss.) Engler & Irmscher (*S. huetiana* Boiss.) : Celandine Saxifrage. Established alien (W. Asia). Damp walls. Fl.5-8. Mons., Glam., Rads., Mer., *Flints.*. 5 squares.
Mons.: Chepstow, ST59, TGE, 1979-1980[112]. **Glam.:** Ewenni Priory, SS97, MHa, 1963[112]. **Rads.:** Glasbury, SO13, EB, 1954[112]. **Mer.:** Llandderfel, SH93, JAWhe, 1940[112]. **Flints.:** Mold, SJ26, WBY, 1929[191].

S. tridactylites L. : Tormaen Tribys : Rue-leaved Saxifrage. Native. Walls and rocks (D,G,Q,R,Rw,W). Fl.3-7. Mons., Glam., Brecs., Rads., Carms., Pembs., Cards., Monts., Mer., Caerns., Denbs., Flints., Angl.. 98 squares, map 241.

S. aizoides L. : Tormaen Melyn Mynyddig : Yellow Saxifrage. Native. Mountain rocks. Fl.7. (*Glam.*), Caerns.†. 2 squares.
(**Glam.:** Maesteg, SS89, M, c1840[140]). **Caerns.:** N. Wales, –, JEG, c1900[112].

S. cespitosa L. : Tormaen Siobynnog : Tufted Saxifrage. Native. Mountain rocks. Fl.4-8. Caerns.. 1 square.
Caerns.: Cwm Idwal, SH65, WW, 1825[112] and ER, 1965[207].

S. rosacea Moench : Tormaen Mynyddig Ymylgrych : Irish Saxifrage. Native. Mountain rocks. Fl.5-6. (Rads.*), Caerns.. 2 squares.
(**Rads.:** Llandrindod Wells, SO06, JAW, 1938[112]). **Caerns.:** Snowdon, SH65, ESM, c1918[112] and RHR, 1978[208].

S. hypnoides L. : Tormaen Llydandroed : Mossy Saxifrage. Native. Mountain rocks (G,Q,R,Rw,W). Fl.5-7. Mons., Glam., Brecs., Rads., Carms., Cards., Monts., Mer., Caerns., *Denbs.*. 48 squares, map 242.

S. granulata L. : Tormaen Gwyn : Meadow Saxifrage. Native. Meadows and stream banks (E,G,M,Q,R,W). Fl.4-6. Mons., Glam., Brecs., Rads., Carms., Cards., Monts., Mer., Caerns., Denbs., Flints., Angl.. 57 squares, map 243.

S. oppositifolia L. : Tormaen Cyferbynddail : Purple Saxifrage. Native. Mountain rocks. Fl.4-5. Brecs., Mer., Caerns.. 11 squares : **SN***91*,92; **SO**01,02; **SH**55,65,66, 71,75,76,81.

TOLMIEA Torrey & A. Gray

Tolmiea menziesii (Pursh) Torrey & A. Gray : Pick-a-back-plant. Established alien (N. America). Damp waste places. Fl.5. Mons., *Glam.*, Caerns.. 4 squares. **Mons.:** Llydart, SO40, PC, 1971[112]. **Glam.:** Llanrhidian, SS49, MG, 1974[99]. **Caerns.:** Llanbedrog, SH33, HHu, 1976[112]; Pont-rug, SH56, WEH, 1967[112].

TELLIMA R. Br.

Tellima grandiflora (Pursh) Douglas ex Lindley : Fringe-cups. Established alien (N. America). Woods and hedges. Fl.4-5. Rads., Caerns., Denbs.. 4 squares. **Rads.:** btwn Nash and Walton, SO25, EMRu, 1961[112]. **Caerns.:** Degannwy, SH77, WEH, 1967[112]. **Denbs.:** Trefor, SJ24, MM, 1966[112]; Gresford, SJ35, RLB, 1977[112].

BERGENIA Moench

Bergenia cordifolia (Haw.) Sternb. Casual (Asia). Glam.. 1 square : **ST**06[99].

TIARELLA L.

Tiarella wherryi Lakela Casual (N. America). *Carms.*. 1 square : **SN**40[201].

CHRYSOSPLENIUM L.

Chrysosplenium alternifolium L. : Eglyn Cylchddail : Alternate-leaved Golden-saxifrage. Native. Wet shady places (Ar,M,Q,W). Fl.3-6. Mons., Glam., Brecs., Rads., Carms., Monts., Denbs., Flints.. 62 squares, map 244.

C. oppositifolium L. : Eglyn Cyferbynddail : Opposite-leaved Golden-saxifrage. Native. Wet shady places (Ar,E,G,M,Q,R,Sc,W). Fl.3-6. Mons., Glam., Brecs., Rads., Carms., Pembs., Cards., Monts., Mer., Caerns., Denbs., Flints., *Angl.*. 234 squares, map 245.

ASTILBE Buch.-Ham. ex D. Don

Astilbe japonica (Morren & Decne) A. Gray Casual (Japan). Roadsides. Fl.6. Mons.. 1 square : **SO**50[112].

BOYKINIA Nutt.

Boykinia major A. Gray Casual (N. America). Waste places. Fl.6. Mons.. 1 square : **ST**38[112].

RODGERSIA A. Gray

Rodgersia podophylla A. Gray Introduced (China and Japan). Damp grassland. Fl.6. *Monts.*. 1 square. **Monts.:** Gregynog Hall, SO09, DM, 1978[107].

PARNASSIACEAE

PARNASSIA L.

Parnassia palustris L. : Brial y Gors : Grass-of-Parnassus. Native. Marshes (Ds,H,M). Fl.8-9. *Glam.*†, Caerns., *Denbs.*[230], Flints., Angl., [*Mons.*, *Cards.*, *Mer.*]. 22 squares, map 246.

HYDRANGEACEAE

PHILADELPHUS L.

Philadelphus coronarius L. : Mock-Orange. Established alien (S. Europe). Roadsides. Fl.4-6. *Glam.*, 1 square.
Glam.: nr Oxwich, SS58, JAW, c1930[221].

ESCALLONIACEAE

ESCALLONIA Mutis ex L. fil.

Escallonia macrantha Hooker & Arnott : Gum-box. Naturalized alien (Chile). Sea cliffs and hedges. Fl.6-10. Glam., Pembs., Mer., Flints.. 8 squares.
Glam.: nr Horton, SS48, JAW, c1940[221]; Rhosili, SS48, DWB, 1958[7]; Langland Cliffs, SS58, AEW, 1947[112]; Langland SS68, JAW, 1941[112]. **Pembs.:** Tenby, SS19, AMP, 1978[112]; Amroth, SN10, AMP, 1978[112]. **Mer.:** Aberdyfi, SN69, RMe, 1929[107] and JAW, 1941[112]; Harlech, SH53, JAW, 1940[112]. **Flints.:** Caerwys, SJ17, JAW, 1942[112].

GROSSULARIACEAE

RIBES L.

Ribes rubrum L. (*R. sylvestre* (Lam.) Mert. & Koch) : Rhyfon Coch : Red Currant. Denizen (Europe?). Woods and shady stream banks (E,M,R,W). Fl.4-5. Mons., Glam., Brecs., Rads.*, Carms., *Pembs.*, Cards., Monts., Mer., Caerns., Denbs., Flints., Angl.*. 125 squares, map 247.

[**R. spicatum** Robson *Mons.*]

R. nigrum L. : Rhyfon Duon : Black Currant. Denizen (England). Wood borders (Ar,E,M,R,W). Fl.4-5. Mons., Glam., Brecs., Rads., Carms., *Pembs.*, *Cards.*, *Monts.**, *Mer.*, Caerns., *Denbs.*, *Flints.*, *Angl.*. 83 squares, map 248.

R. sanguineum Pursh : Flowering Currant. Established alien (N. America). Roadsides (E,R,W). Fl.4-8. Glam., *Brecs.*, *Rads.*, *Carms.*, Cards., Monts., Mer.. 20 squares, map 249.

R. uva-crispa L. : Eirin Mair : Gooseberry. Denizen (England). Hedges and woods (D,E,M,Q,R,Sc,W). Fl.4-5(-9). Mons., Glam., Brecs., Rads., Carms., Pembs., Cards., Monts., Mer., Caerns., Denbs., Flints., Angl.. 176 squares, map 250.

R. alpinum L. : Rhyfon Mynydd : Mountain Currant. Native or denizen. Bushy places and hedges (E,R,W). Fl.4-5. (*Mons.**, *Brecs.*, Rads., *Cards.*, Monts., Mer., *Caerns.*), Denbs., *Flints.*, (*Angl.*), [*Glam.*, *Carms.*]. 27 squares, map 251.

ROSACEAE

SPIRAEA L.

Spiraea salicifolia agg. : Erwain Helygddail : Bridewort. Established alien (C. Europe). Hedges (E,M,R,W). Fl.6-10. Mons., Glam., Brecs., Rads., Carms., Pembs., Cards., Monts., Mer., Caerns., Denbs., Flints., *Angl.*. 87 squares, map 252.

S. salicifolia L. *s.s.* Established alien (C. Europe). Hedges and streamsides. Fl.6-9. Carms., Pembs., Cards., Mer., Caerns., Denbs.. 11 squares : **SN***13*,53, *58*,64,84; **SH**71,73,*75*,*86*,93,94.

S. alba Duroi × **salicifolia** = **S.** × **rosalba** Dippel Established alien (garden origin). Hedges. Fl.7. Cards.. 1 square.
Cards.: nr Rhydowen, SN44, AOC & DGJ, 1979[112].

S. douglasii Hooker : Steeplebush. Casual (N. America). Waste places (R,Rw). Fl.8-9. Glam., *Caerns.*. 3 squares : **SS**79[18]; **ST**17[112]; **SH**32[29].

S. douglasii × **salicifolia** = **S.** × **billiardii** Hérincq. Established alien (garden origin). Roadsides (D,E,G,R,Rw). Fl.7-10. Mons., Glam., Brecs., Rads., Carms., Cards., Monts., Mer., Denbs., Flints.. 20 squares, map 253.

S. japonica L. fil. (*S. bumaldi* Hort.). Casual (Japan). Waste places. *Caerns.*. 1 square : **SH***54*[16].

S. canescens D. Don Casual (India). Waste places. Fl.7. *Pembs.*. 1 square : **SM**93[47].

FILIPENDULA Miller

Filipendula vulgaris Moench : Crogedyf : Dropwort. Native. Limestone pastures (E,G,M,Q,R,Rw,W). Fl.6-9. Mons., Carms., Pembs., (Cards.), *Monts.**, Caerns., *Denbs.*, *Flints.*, *Angl.*, [*Glam.*, *Brecs.*, *Mer.*]. 38 squares, map 254.

F. ulmaria (L.) Maxim. : Bugeiles-y-weirglodd : Meadowsweet. Native. Marshes (Ar,E,G,H,M,Q,R, Rw,W). Fl.6-9. Mons., Glam., Brecs., Rads., Carms., Pembs., Cards., Monts., Mer., Caerns., Denbs., Flints., Angl.. 271 squares, all except : **SS**38,39,77; **ST**26; **SM**50,62,71,73; **SN**15,36; **SH**98.

KERRIA DC.

Kerria japonica (L.) DC. : Bachelor's-buttons. Casual (Japan). Hedges. Fl.5. Mons.. 1 square : **SO***50*[112].

RUBUS L.

Subgen. Chamaemorus (Hill) Focke

Rubus chamaemorus L. : Mwyaren Doewan : Cloudberry. Native. Peaty moors. Fl.6-8. *Monts.*, Mer., Denbs., [*Pembs.*, *Caerns.*]. 3 squares.
Monts.: nr Pen-y-bont-fawr, SO02, FHP, 1956[7]; Moel Sych, SJ03, VJM, 1960+[116] and RGW, 1977[259]. **Mer.:** Cadair Fronwyn, SJ03, WP, c1860[198] and An, 1960+[4]; Berwyn, SJ03, AL, 1898[126] and An, 1960+[4]; Moel Sych, SJ03, MR, 1957[112]. **Denbs.:** Craig Berwyn, SJ03, EJ, c1867[144]; Moel Sych; SJ03, AJS, 1957[112]; Mynydd Tarw, SJ13, TWH, c1870[144].

Subgen. Cylactis (Rafin.) Focke

R. saxatilis L. : Corfwyaren : Stone Bramble. Native. Rocky banks (M,Q,Rw,W). Fl.4-6. Mons., Glam., Brecs., Rads.*, Carms., *Cards.**[259], *Monts.*, Mer., Caerns., Denbs., *Flints.*. 38 squares, map 255.

Subgen. Idaeobatus Focke

R. idaeus L. : Mafonen : Raspberry. Native. Woods (E,H,M,R,Rw,W). Fl.5-7. Mons., Glam., Brecs., Rads., Carms., *Pembs.*, Cards., Monts., *Mer.*, Caerns., *Denbs.*, *Flints.*, *Angl.*. 218 squares, map 256.

R. phoenicolasius Maxim. : Japanese Wineberry. Established alien (Japan). Hedges. Fl.7-8. Glam., *Brecs.*. 2 squares.
Glam.: Church Village, ST08, CMD, 1938[112].
Brecs.: Crickhowell, SO21, MP, 1981[200].

Subgen. Rubus

R. fruticosus agg.: Mwyaren Ddu : Bramble. Native. Woods and hedges (Al,D,E,F,G,H,M,R,Rg,Rw,S, Sc,W). Fl.5-9. Mons., Glam., Brecs., Rads., Carms., Pembs., Cards., Monts., Mer., Caerns., Denbs., Flints., Angl.. 278 squares, all except : **SS**39; **SM**50,73,94.

The following account of the **Rubus** microspecies recorded from Wales is based largely on records compiled by Mr. A. Newton whose help is gratefully acknowledged. In view of the provisional nature of this account, no attempt has been made to provide a complete treatment for each microspecies, details of their vice-county and 10km square distribution only are given. The sequence follows that of W. C. R. Watson's *Handbook of the Rubi of Great Britain and Ireland* (1958), with subsequently described species inserted in their correct position, but all microspecies erroneously recorded from Wales are listed together at the end of each section.

Sect. Suberecti P. J. Mueller

R. nessensis W. Hall (*R. suberectus* G. Anderson ex Sm.). Mons., Glam., Brecs., Carms., Pembs., Cards., *Monts.*, *Mer.*, *Denbs.*. 24 squares, map 257.

R. scissus W. C. R. Watson (*R. fissus* auct., non Lindley). Mons., Glam., Brecs., Rads., *Carms.*, Cards., *Monts.*, Mer., Caerns., Denbs., Flints.. 49 squares, map 258.

R. bertramii G. Braun (*R. opacus* auct., non Focke). Mons., Glam., Brecs., Rads., Carms., Pembs., Cards., *Monts.*, Mer., Caerns., *Denbs.*, *Flints.*. 42 squares, map 259.

R. plicatus Weihe & Nees *Brecs.*, *Rads.*, *Cards.*, *Mer.*, *Caerns.*, *Denbs.*. 9 squares : **SN**44,55,69,94,95; **SO**05; **SH**54,71; **SJ**43.

R. fissus Lindley (*R. rogersii* E. F. Linton). Glam., *Brecs.*, *Monts.*. 6 squares : **ST**07,08; **SN**70,81; **SO**00; **SJ**02.

R. arrheniiformis W. C. R. Watson Mons., *Brecs.*, Mer., *Caerns.*. 8 squares : **SN***92*; **SO***50*; **SH***52*,55,61, 71,72,*82*.

R. affinis Weihe & Nees Mons., Glam., Rads., [*Pembs.*, *Caerns.*, *Angl.*]. 5 squares : **ST**07,48,49; **SO**16,26.

R. briggsianus (Rogers) Rogers Pembs., Caerns., Angl., [*Mons.*, *Mer.*]. 10 squares : **SM***72,93*; **SH***27,36, 37,54*,56,*57*,58,*67*.

[**R. nobilissimus** (W. C. R. Watson) Pearsall *Glam.*]

[**R. divaricatus** P. J. Mueller (*R. nitidus* Weihe & Nees). *Brecs.*, *Rads.*, *Pembs.*, *Mer.*, *Caerns.*, *Angl.*]

Sect. Triviales P. J. Mueller

R. conjungens (Bab.) Rogers Mons., *Brecs.*, Pembs., *Cards.*. 15 squares : **ST**28,*49,59*; **SM**72,82,91; **SN**03,04, 14,24,35; **SO**02,12,13,*51*.

R. eboracensis W. C. R. Watson *Denbs.*, *Flints.*. 3 squares : **SJ**07,36,43.

R. sublustris Lees *Glam.*, *Brecs.*, *Rads.*, *Denbs.*, *Flints.*, *Angl.*. 19 squares : **SS**58,59,68,87,97; **ST**07,16,18,28; **SN**90,91; **SO**02,26; **SH**37,47; **SJ**15,16,17,25.

R. nemorosus Hayne & Willd. (*R. balfourianus* Bloxam ex Bab.). Mons., *Mer.*. 4 squares : **ST**28,38,39; **SH**61.

R. warrenii Sudre *Denbs.*, [*Mons.*]. 1 square : **SJ**05.

R. halsteadensis W. C. R. Watson (*R. raduliformis* (A. Ley) W. C. R Watson, non Sudre). Mons., [*Glam.*, *Flints.*]. 2 squares : **SO**40,50.

R. tuberculatus Bab. Mons., Brecs., *Rads.*, *Carms.*, Pembs., *Cards.*, *Monts.*, *Mer.*, *Caerns.*, *Denbs.*, *Flints.*, *Angl.*. 44 squares, map 260.

[**R. babingtonianus** W. C. R. Watson (*R. altheaefolius* auct.). *Mons.*]

[**R. rubriflorus** Purchas *Mons.*]

[**R. scabrosus** auct. *Mons.*, *Brecs.*]

[**R. britannicus** Rogers *Glam.*, *Brecs.*, *Rads.*, *Carms.*, *Caerns.*]

[**R. tenuiarmatus** Lees *Mons.*, *Cards.*]

Sect. Sylvatici P. J. Mueller

R. gratus Focke Glam., *Brecs.*, [*Mer.*]. 4 squares : **SN**70,80,91,92.

R. monensis W. C. Barton & Riddelsd. Mer., *Caerns.*, Angl., [*Cards.*]. 15 squares : **SH***27,37*,44,47,52,54, 56,*57,63*, 64,65,71,74,*83*,93.

R. riparius W. C. Barton ex A. Newton *Cards.*, *Monts.*, *Mer.*, Caerns., *Denbs.*, *Angl.*. 16 squares : **SN***69*,79; **SH**44,52,53,54,55,56,*57*,62,63,64,*67*,71,*72*,85.

R. dasycoccus W. C. R. Watson Mons., [*Cards.*]. 2 squares : **SO**50,51.

R. sciocharis Sudre Mons., Brecs., *Denbs.*. 7 squares : **ST***59*; **SO**21,22,*23*,32,*50*; **SJ**43.

R. ludensis W. C. R. Watson Brecs., Rads., [*Glam.*, *Caerns.*]. 15 squares : **SN**81,*83,84*,85,91,93,94,95,96; **SO**03,04,05,06,13,14.

R. platyacanthus P. J. Mueller & Lefèvre (incl. *R. laxissimus* A. Newton; *R. carpinifolius* auct.). Mons., Brecs., *Rads.*, [*Caerns.*]. 9 squares : **ST***49,59*; **SN***83,93*; **SO**05,06,*36,37,50*.

R. polyoplus W. C. R. Watson (*R. salteri* auct.). Glam., *Brecs.*. 8 squares : **SS***98*; **ST***07,08*,17,18; **SN**82,92,93.

R. nemoralis P. J. Mueller (*R. selmeri* Lindeb.; *R. pistoris* W. C. Barton & Riddelsd.). Mons., Glam., *Brecs.*, *Rads.*, *Carms.*, *Pembs.*, Cards., *Monts.*, *Mer.*, Caerns., *Denbs.*, *Flints.*, *Angl.*. 66 squares, map 261.

R. laciniatus Willd. Glam., *Rads.*. 2 squares : **ST**17; **SO**05.

R. cambrensis W. C. R. Watson *Mons.*, [*Brecs.*, *Mer.*, *Caerns.*]. 2 squares : **ST**59; **SO**50.

R. lindleianus Lees Mons., Glam., Brecs., *Rads.*, *Carms.*, Pembs., *Cards.*, *Monts.*, Mer., Caerns., Denbs., *Flints.*, *Angl.*. 95 squares, map 262.

R. robii (W. C. R. Watson) A. Newton *Denbs.*. 2 squares : **SJ**43,53.

R. macrophyllus Weihe & Nees *Denbs.*. 2 squares : **SJ**43,53.

R. subinermoides Druce *Glam.*. 3 squares : **SS**69; **ST**08,18.

R. perdigitatus A. Newton (*R. digitatus* A. Newton, non Sprengel). Glam., Brecs., *Rads.*, *Carms.*, Pembs., Cards.. 37 squares, map 263.

R. silvaticus Weihe & Nees *Mons.*, Caerns.. 3 squares : **SO***50*; **SH***57*,66.

R. chrysoxylon (Rogers) Rogers Caerns., *Denbs.*, [*Mons., Brecs., Angl.*]. 4 squares : **SH**56,*57*,66; **SJ**43.

R. amplificatus Lees Mons., *Glam.*, Cards., *Denbs., Flints.*. 11 squares : **ST***49,59*; **SN**25,70; **SO***40,41*; **SH**85,*87*; **SJ**15,22,36.

R. pyramidalis Kaltenb. Mons., Glam., Brecs., *Rads., Carms.*, Cards., *Monts., Mer., Caerns., Denbs., Angl.*. 39 squares, map 264.

R. albionis W. C. R. Watson (*R. macrophyllus* subsp. *schlechtendalii* Rogers *p.p.; R. schlechtendalii* var. *anglicus* Sudre). *Glam.*, Brecs., *Carms.*, Pembs., *Cards., Monts.*. 15 squares : **SS**98; **ST**07,18; **SM**80; **SN**00,24,33, 34,35,69,70,79,80,82,92.

R. septentrionalis W. C. R. Watson (incl. *R. insularis* sensu W. C. R. Watson, non F. Aresch.). Brecs., Rads.. 4 squares : **SN**83; **SO**03,05,12.

R. incurvatus Bab. Glam., Brecs., *Rads., Carms.*, Pembs., *Cards., Monts.*, Mer., Caerns., Denbs., *Flints.*, Angl.. 65 squares, map 265.

R. incurvatiformis Edees *Flints.*. 1 square : **SJ**25.

R. pampinosus Lees (*R. favonii* W. C. R. Watson; *R. buttii* W. C. Barton & Riddelsd.). Brecs., *Cards.*, [*Mons., Caerns.*]. 4 squares : **SN**68,*83*; **SO***02,05*.

R. riddelsdellii Rilstone Mons., Glam., *Cards.*. 8 squares : **SS***48,58*,59,68; **ST**08,18,*59*; **SN**14.

R. polyanthemus Lindeb. (*R. pulcherrimus* Neuman, non Hooker). Mons., Glam., Brecs., *Rads., Carms., Pembs.*, Cards., *Monts., Mer., Caerns.*, Denbs., *Flints., Angl.*. 78 squares, map 266.

R. rubritinctus W. C. R. Watson (*R. cryptadenes* Sudre, non Dumort). Mons., Glam., *Brecs., Rads.*, Carms., Pembs., *Cards., Monts.*, [*Mer.*]. 83 squares, map 267.

R. rhombifolius Weihe ex Boenn. (*R. incurvatus* var. *subcarpinifolius* (Rogers) Riddelsd.). Glam., *Brecs.*, Rads., [*Mons., Flints.*]. 6 squares : **SS***96*; **SN**83,*96*; **SO***04*,05,16.

R. acclivitatum W. C. R. Watson (*R. clivicola* (A. Ley) Druce, non Sudre). *Mons.*, Brecs., *Rads., Carms.*, [*Monts.*]. 24 squares. map 268.

R. prolongatus Boulay & Letendre (*R. griseoviridis* W. C. Barton & Riddelsd.). Mons., Glam., Brecs., Rads., *Carms.*, Pembs., Cards., *Monts., Mer., Caerns.*, [*Angl.*]. 48 squares, map 269.

R. viridescens (Rogers) T. A. W. Davis (*R. thyrsoideus* Wimmer var. *viridescens* Rogers). Pembs.. 4 squares : **SM**70,*72*,80,92.

R. cardiophyllus P. J. Mueller & Lefèvre (*R. rhamnifolius* auct.). Mons., Glam., *Brecs., Cards., Monts., Mer., Caerns.*, Denbs., *Flints., Angl.*. 46 squares, map 270.

R. dumnoniensis Bab. Glam., *Brecs., Rads., Carms.*, Pembs., *Cards., Mer.*, Caerns.. 25 squares, map 271.

R. altiarcuatus W. C. Barton & Riddelsd. Glam., Brecs., *Carms.*, Pembs., Cards.. 68 squares, map 272.

R. lindebergii P. J. Mueller Mons., *Brecs., Rads., Caerns., Denbs., Flints.*. 24 squares, map 273.

R. errabundus W. C. R. Watson (*R. scheutzii* auct.). *Denbs.*, [*Caerns., Angl.*]. 1 square : **SH**96.

R. imbricatus Hort Mons., *Pembs.*. 4 squares : **SM**80,82; **SO**50,*51*.

R. silurum (A. Ley) Druce Mons., Glam., *Brecs., Rads.*, Carms., Pembs., Cards., *Monts., Mer., Caerns., Denbs., Angl.*. 91 squares, map 274.

[**R. calvatus** Lees ex Bloxam *Denbs.*]

[**R. oxyanchus** Sudre *Glam.*]

[**R. egregius** Focke *Mer., Caerns.*]

[**R. leucandrus** Focke *Mons.*]

[**R. mercicus** Bagnall *Caerns.*]

[**R. schlechtendalii** Weihe ex Link *Caerns., Denbs.*]

[**R. hirtifolius** P. J. Mueller & Wirtgen *Glam.*]

[**R. ramosus** Bloxam ex T. R. A. Briggs *Caerns.*]

[**R. rhodanthus** W. C. R. Watson (*R. rhombifolius* auct.). *Mons.*]

[**R. alterniflorus** P. J. Mueller & Lefèvre *Mons., Mer.*]

[**R. obvallatus** Boulay & Gillot *Mer.*]

Sect. Discolores P. J. Mueller

R. ulmifolius Schott (*R. discolor* auct.; *R. rusticanus* Merc.). Mons., Glam., *Brecs., Rads., Carms.*, Pembs., Cards., *Monts., Mer.*, Caerns., Denbs., *Flints., Angl.*. 163 squares, map 275.

R. armipotens W. C. Barton ex A. Newton (*R. pseudobifrons* auct. angl., non Sudre). Mons., *Brecs.*. 3 squares : **SO**14,*22,24*.

R. rossensis A. Newton (*R. propinquus* auct., non P. J. Mueller; *R. crassifolius* auct., non Genev.). *Mons., Glam., Brecs., Pembs.*. 11 squares : **SS***69,79*; **ST***59*; **SM**90; **SN***70,81*,92; **SO***00*,01,50,*51*.

R. procerus P. J. Mueller (*R. armeniacus* Focke; *R. hedycarpus* Focke) : Himalayan-giant Bramble. Mons., Glam., *Brecs., Pembs.*. 14 squares : **SR**99; **SM**91; **SS**59, 69; **ST**06,27,29; **SN**72,80; **SO**02,05,12,21,51.

[**R. winteri** P. J. Meuller ex Focke *Mons.*]

[**R. anglocandicans** A. Newton (*R. falcatus* auct., non Kaltenb.). *Angl.*]

[**R. thyrsanthus** Focke *Mons.*]

Sect. Sprengeliani (Focke) W. C. R. Watson

R. sprengelii Weihe Mons., *Glam.*, *Brecs.*, Cards., *Mer.*, *Caerns.*, Denbs., Flints.. 16 squares : **ST***07*,49,*59*; **SN**45,70; **SO***05*,41,*50*; **SH**56,64,72,75; **SJ**16,25,36,43.

R. lentiginosus Lees (*R. cambricus* Focke). Caerns., *Denbs.*, *Flints.*, Angl., [*Mons.*, *Mer.*]. 15 squares : **SH**54,*55*,56,57,58,64,*66*,*67*,75,85; **SJ**08,25,35,36,43.

Sect. Appendiculati (Genev.) Sudre

R. fuscicortex Sudre (*R. podophyllus* auct.). Mer., Caerns., Denbs., [*Brecs.*]. 11 squares : **SH**54,56,63, 64,71,72,74,75,*76*,85; **SJ**43.

R. bartonii A. Newton *Brecs.*, Rads., *Carms.*, Cards., Monts., *Mer.*, Caerns., *Denbs.*, Flints.. 63 squares, map 276.

R. lanaticaulis Edees & A. Newton (*R. hebecaulis* sensu W. C. R. Watson, non Sudre). *Mons.*, *Brecs.*, *Monts.*, *Mer.*, *Caerns.*, *Denbs.*, Angl., [*Glam.*, *Carms.*, *Pembs.*]. 24 squares, map 277.

R. thurstonii Rilstone *Pembs.*. 1 square : **SM**90.

R. plymensis (Focke) Edees & A. Newton *Brecs.*. 1 square : **SO**12.

R. orbus W. C. R. Watson *Glam.*, *Brecs.*. 2 squares : **ST**08; **SN***71*.

R. vestitus Weihe & Nees (*R. diversifolius* Lindley; *R. leucostachys* auct., non Sm.; *R. leightonianus* Bab.). Mons., Glam., Brecs., Rads., *Carms.*, Pembs., Cards., *Monts.*, *Mer.*, *Caerns.*, *Denbs.*, *Flints.*, *Angl.*. 113 squares, map 278.

R. longus (Rogers & A. Ley) A. Newton (*R. lasioclados* var. *longus* Rogers & A. Ley). *Mons.*, Glam., Brecs., *Rads.*, *Carms.*, *Pembs.*, *Cards.*. 19 squares, map 279.

R. ordovicum A. Newton (*R. leucostachys* var. *macrothyrsus* N.E. Br. and var. *gymnostachys* sensu Griffith in *Fl. Angl. & Caern.*, et Rogers *Handbook Brit. Rubi*). *Mer.*, Caerns., *Denbs.*, *Flints.*. 11 squares : **SH**52,53,54,*55*,56,*57*,61,63,64,65; **SJ**36.

R. melanocladus Sudre Mons., Glam. 3 squares : **ST**18,*48*; **SO***00*.

R. wirralensis A. Newton Brecs., Carms., *Cards.*, *Caerns.*, *Denbs.*, *Flints.*, *Angl.*. 15 squares : **SN**25,45, *71*,83,92; **SO**02,*03*,13,14; **SH***28*,56,97; **SJ**07,25,26.

R. celticus A. Newton (*R. obvallatus* auct., non Boulay & Gillot). Cards., *Monts.*, Mer., *Caerns.*. 13 squares : **SN***69*; **SH**52,53,54,61,*62*,63,64,70,71,72,74,82

R. leyanus Rogers Glam., Brecs., Rads., Carms., Pembs., *Cards.*, *Flints.*, [*Caerns.*]. 68 squares, map 280.

R. hibernicus (Rogers) Rogers Mons., Glam., Brecs.. 10 squares : **SS**98; **ST***07*,08,*17*,18,28; **SN**90,91; **SO***00*,*01*.

R. mucronatoides A. Ley *Rads.*. 3 squares : **SO***16*,*26*,*36*.

R. trelleckensis Edees & A. Newton (*R. orthocladus* A. Ley, non Boulay). Mons.. 2 squares : **ST**28; **SO**50.

R. dentatifolius (Briggs) W. C. R. Watson (*R. vectensis* W. C. R. Watson). Mons., Glam., *Brecs.*, Carms., Pembs., Cards., Monts., *Denbs.*, [*Mer.*]. 23 squares, map 281.

R. infestus Weihe ex Boenn. (*R. taeniarum* Lindeb.). *Brecs.*, Caerns., [*Mer.*]. 3 squares : **SN**81; **SH***56*,*66*.

R. radula Weihe ex Boenn. *Brecs.*, *Caerns.*, [*Mons.*, *Angl.*]. 2 squares : **SO**03; **SH***53*.

R. echinatus Lindley (*R. discerptus* P. J. Mueller; *R. rudis* auct.). Mons., Glam., Rads., Monts., Caerns., Denbs., [*Brecs.*]. 11 squares : **ST***18*,*59*; **SO**14,*27*,*41*, *50*,*51*; **SH**57; **SJ***20*,24,35.

R. echinatoides (Rogers) Dallman *Mons.*, Caerns., *Denbs.*, Flints., Angl.. 8 squares : **SO**50; **SH***36*,*37*, *53*,*66*; **SJ**16,17,23.

R. rudis Weihe & Nees Mons.. 1 square : **SO***51*.

R. radulicaulis Sudre (*R. radula* var. *sertiflorus* P. J. Mueller). Mons., Brecs.. 3 squares : **ST**49; **SO**02,50.

R. troiensis A. Newton Mons., Glam.. 5 squares : **ST**08,09,18,*49*; **SO**51.

R. effrenatus A. Newton *Cards.*, *Monts.*, *Mer.*, *Caerns.*. 18 squares : **SN***68*,69,79; **SH***23*,52,53,54,56,61,63,64,65, *66*,70,71,72,*74*,80.

R. flexuosus P. J. Mueller & Lefèvre Mons., Glam., *Brecs.*, Pembs., *Mer.*, *Caerns.*. 14 squares : **ST**29,*49*,*59*; **SM**91; **SN**00,01,02,10,70,71; **SO***00*,*50*; **SH***56*,70.

R. rubristylus W. C. R. Watson *Denbs.*, [*Mons.*, *Glam.*, *Brecs.*, *Caerns.*]. 1 square : **SJ**54.

R. adenanthoides A. Newton *Denbs.*, *Flints.*.2 squares: **SJ**35,36.

R. bloxamii Lees Mons., [*Carms.*, *Caerns.*]. 1 square : **ST**49.

R. largificus W. C. R. Watson *Cards.*. 1 square : **SN**24.

R. cavatifolius P. J. Mueller Mons.. 4 squares : **ST***59*; **SO***40*,50,*51*.

R. fuscicaulis Edees *Mons.*, *Glam.*, *Brecs.*, *Denbs.*. 9 squares : **SS***79*; **ST***49*; **SN**71,*80*,81,90; **SO***00*,51; **SJ**24.

R. hirsutissimus Sudre & A. Ley *Mons.*. 1 square : **SO***51*.

R. aequalidens A. Newton *Glam.*, Brecs., *Carms.*, Pembs.. 13 squares : **SR**99; **SS**87; **SM**70,80,82,90; **SN**00,01,02,10,11,21,84.

R. pallidus Weihe & Nees Mons., *Glam.*, Brecs., Rads., Carms., [*Caerns.*]. 16 squares : **ST***18*,*27*,29; **SN**71,*74*,81,*83*,*84*,*90*,91,*92*,93,95; **SO**03,05,06.

R. morganwgensis W. C. Barton & Riddelsd. Glam., Brecs., [*Caerns.*]. 12 squares : **SS**79; **ST**07,08,09; **SN**71, 80,90,92,94; **SO**00,01,02.

R. glareosus Rogers *Brecs.*, *Rads.*, *Carms.*, *Cards.*, [*Mons.*, *Angl.*]. 6 squares : **SN**46,74,83,84,93,96.

R. acutifrons A. Ley *Mons.*, *Brecs.*. 6 squares : **ST**29, 49,*59*; **SO***03*,41,50.

R. euryanthemus W. C. R. Watson Mons., *Rads.*. 2 squares : **SO**06,50.

R. insectifolius P. J. Mueller & Lefèvre *Glam.*. 1 square : **ST***08*.

R. scaber Weihe & Nees Mons.. 1 square : **SO***50*.

R. longithyrsiger Lees ex Baker (*R. pyramidalis* (Bab.) Bab., non Kaltenb.). Mons., *Brecs.*, *Rads.*, Carms., Caerns.. 10 squares : **SN**23,33; **SO**24,*36*,*50*,51; **SH**54, *55*,56,*67*.

R. botryeros (Focke ex Rogers) Rogers (*R. cenomanensis* Sudre; *R. oigocladus* auct.). Mons., Brecs., [*Rads.*, *Flints.*]. 3 squares: **ST***49*,*59*; **SN***83*.

R. rufescens P. J. Mueller & Lefèvre (*R. rosaceus* var. *infecundus* Rogers). Mons., *Glam.*, Brecs., *Mer.*, *Caerns.*. 24 squares, map 282.

R. purchasianus Rogers Mons., *Rads.*. 6 squares : **SO***14*,*30*,40,41,50,51.

R. micans Gren. & Godron (*R. anglosaxonicus* Gelert; *R. apiculatus* auct., non Weihe & Nees). Brecs.. 5 squares : **SN**92; **SO**02,03,13; **SJ**05.

R. raduloides (Rogers) Sudre *Mons.*, Glam., Brecs., *Rads.*, *Denbs.*, *Flints.*, [*Mer.*, *Caerns.*]. 29 squares, map 283.

R. pascuorum W. C. R. Watson Brecs., *Rads.*, [*Mons.*, *Mer.*]. 3 squares : **SO***02*,03,26.

R. melanodermis Focke Glam., *Brecs.*, [*Caerns.*]. 3 squares : **ST**18,28; **SN**84.

R. griffithianus Rogers Caerns., *Denbs.*, *Flints.*, *Angl.*, [*Mons.*]. 7 squares : **SH***37*,56,57,66,86; **SJ***25*,26.

R. leightonii Lees ex Leighton (*R. radula* var. *anglicanus* Rogers). Denbs., *Flints.*, [*Mons.*, *Brecs.*]. 3 squares : **SJ**36,43,53.

R. moylei W. C. Barton & Riddelsd. (*R. disjunctus* auct. angl., non P. J. Mueller & Lefèvre). *Mons.*, Glam., Brecs., *Rads.*. 26 squares, map 284.

R. diversus W. C. R. Watson (*R. hirtus* subsp. *kaltenbachii* Rogers *p.p.*). Mons., [*Glam.*, *Brecs.*]. 1 square : **SO**50.

R. angusticuspis Sudre (*R. apiculatus* var. *setulosus* (Rogers) Riddelsd.). Mons., Glam., *Brecs.*. 17 squares : **ST**07,17,18,49,59; **SN**80,83,*90*,91,93; **SO**00,05,*22*,*40*,*41*, 50,51.

R. hastiformis W. C. R. Watson (*R. thyrsiger* Bab., non Banning & Focke). Glam., *Cards.*. 3 squares : **ST***07*,08; **SN**46.

R. breconensis W. C. R. Watson (*R. lejeunei* auct.; *R. wedgewoodiae* W. C. Barton & Riddelsd. var. *sabrinae* W. C. Barton & Riddelsd.). Glam., Brecs., [*Rads.*, *Pembs.*]. 4 squares : **SS**69; **SN***81*,82; **SO***00*.

[**R. podophyllus** P. J. Mueller *Mons.*, *Caerns.*]

[**R. conspicuus** P. J. Mueller *Mons.*]

[**R adscitus** Genev. (*R. hypoleucus* P. J. Mueller & Lefèvre, non Vest). *Mons.*, *Mer.*]

[**R. boraeanus** Genev. *Mons.*, *Mer.*, *Caerns.*]

[**R. macrothyrsus** Lange *Mons.*]

[**R. cinerosus** Rogers *Mons.*, *Brecs.*, *Rads.*]

[**R. sectiramus** W. C. R. Watson *Mons.*]

[**R. genevieri** Boreau *Mons.*]

[**R. crispus** P. J. Mueller & Lefèvre *Mer.*, *Caerns.*]

[**R. aspericaulis** P. J. Mueller & Lefèvre *Mer.*]

[**R. granulatus** P. J. Mueller & Lefèvre *Mons.*, *Pembs.*]

[**R. acutipetalus** P. J. Mueller & Lefèvre *Mons.*]

[**R. trichodes** W. C. R. Watson *Mons.*, *Brecs.*]

[**R. chlorocaulon** Sudre *Caerns.*]

[**R. curtiglandulosus** Sudre *Cards.*]

[**R. sprengeliiflorus** Sudre *Mons.*]

[**R. heterobelus** Sudre (*R. penhallowensis* Rilstone; *R. praeruptorum* auct.). *Rads.*, *Mer.*, *Caerns.*, *Angl.*]

[**R. melanoxylon** P. J. Mueller & Wirtgen *Mons.*]

[**R. phaeocarpus** W. C. R. Watson (*R. babingtonii* auct.). *Mons.*, *Rads.*, *Carms.*]

[**R. euanthinus** W. C. R. Watson (*R. apiculatus* var. *vestitiformis* (Rogers) Riddelsd.). *Mons.*, *Carms.*, *Caerns.*]

[**R. furvicolor** Focke (*R. melanoxylon* auct.). *Carms.*]

[**R. turneri** W. C. R. Watson (*R. glandulosus* auct.). *Pembs.*, *Caerns.*]

Sect. Glandulosi P. J. Mueller

R. murrayi Sudre (*R. rosaceus* subsp. *adornatus* sensu Rogers). *Mons.*, *Denbs.*, [*Caerns.*]. 2 squares : **SO***50*; **SJ**53.

R. babingtonii Bell Salter (*R. ochrodermis* A. Ley). Mons., Brecs., Rads., [*Cards., Caerns.*]. 11 squares : **ST**59; **SN***92*,93,94,95; **SO***02,03*,04,13,*16,22*.

R. bercheriensis (Druce ex Rogers) Rogers *Brecs., Carms.*. 4 squares : **SN***70*,74,*81*,83.

R. hylocharis W. C. R. Watson Glam., Brecs., Rads., *Carms.*, Pembs., Cards., *Monts., Caerns., Flints.*. 28 squares, map 285.

R. semiglaber (Rogers) W. C. R. Watson *Mer., Caerns., Denbs., Flints.*, [*Mons., Angl.*]. 14 squares : **SH**56,*57*,64,*66,67*,76,93,96,97; **SJ**04,05,06,07,25.

R. milesii A. Newton *Mons.*. 1 square : **SO**50.

R. dasyphyllus (Rogers) E. S. Marshall Mons., Glam., *Brecs.*, Rads., *Carms., Pembs., Cards.*, Monts., *Mer.*, Caerns., *Denbs., Flints., Angl.*. 122 squares, map 286.

R. aristisepalus (Sudre) W. C. R. Watson (*R. velatus* auct). Mons., [*Brecs., Carms.*]. 1 square : **SO***50*.

R. bellardii Weihe & Nees *Brecs., Mer., Caerns.*. 5 squares : **SN***83*; **SH***56*,64,*74*,75.

R. pallidisetus Sudre (*R. divexiramus* auct.). Mons., [*Caerns., Angl.*]. 1 square : **SO**51.

R. hylonomus P. J. Mueller & Lefèvre Carms., Cards.. 2 squares : **SN**46,74.

[**R. adornatus** P. J. Mueller ex Wirtgen *Mons.*]

[**R. spinulifer** P. J. Mueller & Lefèvre *Caerns.*]

[**R. scabripes** Genev. (*R. rosaceus* auct.). *Mons.*]

[**R. hystrix** Weihe & Nees *Mons., Glam., Mer., Caerns.*]

[**R. adenolobus** W. C. R. Watson *nom. nud. Glam., Caerns., Angl.*]

[**R. fuscoater** Weihe & Nees *Caerns.*]

[**R. oigocladus** P. J. Mueller & Lefèvre *Mons.*]

[**R. elegans** P. J. Mueller (*R. angustifrons* Sudre; *R. serpens* auct.). *Mons.*]

Sect. Glaucobatus (Dumort.) W. C. R. Watson

R. caesius L. : Mwyaren Mair : Dewberry. Native. Hedges, woods, banks and sand dunes on calcareous soils (D,E,G,R,Sc). Fl.4-8. Mons., Glam., Brecs., *Rads.*, Carms., Pembs., *Cards.*, Monts., *Mer.*, Caerns., *Denbs.*, Flints., Angl.. 123 squares, map 287.

R. caesius × **fruticosus** agg. (*Rubi Corylifolii*). Hybrids between **R. caesius** and the other sections of the subgenus **Rubus** are of fairly frequent occurrence but have been reliably recorded only from Mons., Glam., *Brecs.*, Carms., *Denbs.*, Angl..

ROSA L.

Rosa arvensis Hudson : Rhosyn Gwyn Gwyllt : Field Rose. Native. Hedges (D,E,H,M,Q,R,W). Fl.6-8. Mons., Glam., Brecs., Rads., Carms., Pembs., Cards., Monts., Mer., Caerns., Denbs., Flints., Angl.. 228 squares, map 288.

R. arvensis × **canina** = **R.** × **wheldonii** Wolley-Dod Spontaneous. Hedges. Fl.8-9. Glam., *Brecs.*, Carms., Monts., *Angl.*. 8 squares.
Glam.: Llandough, ST17, AEW, 1920[112]. **Brecs.:** Heolsenni, SN92, MP, 1975[200]; Aberduhonw, SO05, MP, 1979[245]. **Carms.:** Ferryside, SN31, IMV, 1977[112]; Rhandir-mwyn, SN74, IMV, 1966[112]. **Monts.:** Carno, SN99, JAW, 1940[112]; Abermule, SO19, IMV, 1968[112]. **Angl.:** Rhoscefn-hir, SH57, GGG, 1979[245].

R. arvensis × **dumetorum** Spontaneous. Hedges. Fl.8. *Mons.*[217], Glam., *Brecs.*, Caerns.. 3 squares.
Glam.: St-y-nyll, ST07, HJR, 1908[112]. **Brecs.:** Llan-y-wern, SO13, MP, 1980[200]. **Caerns.:** Bangor, SH57, RHR, 1979[112].

R. arvensis × **rubiginosa** Spontaneous. Hedges. Carms.. 1 square.
Carms.: btwn Ferryside and St Ishmael, SN30, IMV, 1977[112].

R. arvensis × **sherardii** Spontaneous. Hedges. *Brecs.*. 1 square.
Brecs.: Grwyne Fechan, SO22, MP, 1979[245].

R. arvensis × **stylosa** = **R.** × **bibracteoides** Wolley-Dod Spontaneous. Hedges. Fl.9. Mons.. 1 square.
Mons.: Tintern, SO50, WAS, 1881[112].

R. pimpinellifolia L. (*R. spinosissima* L.) : Rhosyn Burnet : Burnet Rose. Native. Sand dunes and heaths (D,E,G,H,Q,R,Rw,Sc,W). Fl.4-8. *Mons.*, Glam., Brecs., *Rads.*, Carms., Pembs., Cards., Monts., Mer., Caerns., Denbs., Flints., Angl.. 107 squares, map 289.

R. pimpinellifolia × **sherardii** = **R.** × **involuta** Sm. (incl. *R.* × *wilsonii* Borrer). Spontaneous. Hedges. Fl.5-7. *Brecs.*, Caerns., *Denbs.*[217], Angl.. 4 squares.
Brecs.: Dyffryn Crawnon, SO01, AL, 1898[200]; Lower Chapel, SO03, MP, 1979[245]; nr Hay-on-Wye, SO23, AL, 1899[200]. **Caerns.:** nr Bangor, SH57, CCo, c1830[112] and GHB, 1900[112]. **Angl.:** nr Rhoscefn-hir, SH57, RHR, 1978[184].

R. pimpinellifolia × **villosa** = **R.** × **sabinii** J. Woods Spontaneous. Woods and hedges. Fl.7. Denbs.. 1 square.
Denbs.: nr Llaneilian-yn-Rhos, SH87, WBR, pre 1911[144]; Bryn-y-maen, SH87, CEB, 1925[112].

R. majalis J. Herrmann (*R. cinnamomea* sensu L. 1759, non L. 1753). Casual (N. and C. Europe). River-banks. Fl.8. Carms.. 1 square : **SN**71[112] (provisional determination).

R. rugosa Thunb.: Japanese Rose. Established alien (E. Asia). Waste places (D,E,R,W). Fl.6. Mons., Glam., *Brecs., Rads.*, Carms., *Mer., Caerns., Angl.*. 16 squares : **SS***48*,69,99; **ST**28,38; **SN**30,40,97; **SO**02; **SH**22,23,27, 32,33,52,71.

R. blanda Aiton Casual (N. America). Waste places. Fl.7. Glam.. 2 squares : **SS**48[221],69[112].

R. virginiana J. Herrmann Casual (N. America). Hedges. Glam., Cards., *Flints.*. 4 squares : **ST**17[112]; **SN***34*[47],*78*[112]; **SJ**16[261].

R. carolina L. Casual (N. America). Hedges. Fl.7. Cards.. 1 square : **SN**68[112].

R. stylosa Desv. : Rhosyn Ungolofn : Close-styled Rose. Native. Hedges (E,Sc,W). Fl.5-7. Mons., Glam., Brecs.*, Carms., Pembs.*, *Cards.**[210], *Denbs., Flints.**. 23 squares, map 290.

R. canina agg. Mons., Glam., Brecs., Rads., Carms., Pembs., Cards., Monts., Mer., Caerns., Denbs., Flints., Angl.. 262 squares, all except : **SR**89; **SS**38,39,77,99; **ST**38, **SM**50,71,73,84; **SN**59,66,77,86,87; **SH**29,51,98; **SJ**37,44.

R. canina L. *s.s.* : Rhosyn Coch Gwyllt : Dog Rose. Native. Hedges (E,G,M,Q,R,Rg,Rw,Sc,W). Fl.5-7. Mons., Glam., Brecs., Rads., Carms., Pembs., Cards., Monts., Mer., Caerns., Denbs., Flints., Angl.. 148 squares, map 291.

R. canina × coriifolia Spontaneous. Hedges. *Brecs., Angl.*. 5 squares.
Brecs.: nr Merthyr Cynog, SN93, MP, 1978[245]; nr Defynnog, SN92, MP, 1978[200]; nr Llangorse, SO12, MP, 1979[200]; Cwm Clydach, SO21, MP, 1979[200]. **Angl.:** nr Foel, Llangeinwen, SH46, RHR, 1980[209].

R. canina × micrantha = R. × toddii Wolley-Dod Spontaneous. Hedges. *Brecs.*. 1 square.
Brecs.: Talgarth, SO13, MP, 1979[200].

R. canina × pimpinellifolia Spontaneous. Hedges. *Angl.*. 1 square.
Angl.: nr Talgwynedd, Dwyran, SH46, RHR, 1978[184].

R. canina × rubiginosa = R. × latens Wolley-Dod Spontaneous. Roadsides and hedges. Fl.6-9. Mons., Brecs., Rads., Monts., Caerns., [*Mer.*]. 7 squares.
Mons.: Dingestow, SO40, NYS, 1946[112]; nr Monmouth, SO51, HJR, 1924[116]. **Brecs.:** Cwm Cilieni, SN93, JAW, 1941[112]; nr Llansanffraid, SO12, MP, 1975[245]. **Rads.:** btwn Llandrindod Wells and Newtown, SO07, IMV, 1962[112]. **Monts.:** nr Machynlleth, SH70, JAW, 1940[112]. **Caerns.:** Deganwy, SH77, JEG, 1885[112].

R. canina × sherardii Spontaneous. Hedges (E,R,Rw). *Brecs., Carms.*, Monts., *Caerns.*, Angl.. 8 squares.
Brecs.: nr Pontarhydfer, SN82, MP, 1978[245]; nr Llywel, SN83, IMV, 1962[109]; Penderyn, SN90, MP, 1978[200]; Nant Senni, SN92, MP, 1978[200]; nr Builth, SO05, MP, 1978[200]. **Carms.:** Ystrad-ffin, SN74, IMV, 1960+[161]. **Caerns.:** nr Bangor, SH57, RHR, 1974 and 1976[184]. **Angl.:** nr Foel, Llangeinwen, SH46, RHR, 1980[245]; Penmynydd, SH57, RHR, 1978[112].

R. canina × tomentosa = R. × curvispina Wolley-Dod (*R. sherardii* var. *woodsiana* (Groves) Wolley-Dod). Spontaneous. Roadsides (E,Q,R). Fl.6-7. Brecs., Carms., Pembs., Cards., Flints.. 7 squares.
Brecs.: nr Llwynderw, SN85, IMV, 1966[112]; nr Llandewi'r-cwm, SO04, MP, 1973[200]; Builth Wells, SO05, IMV, 1962[112]. **Carms.:** Rhandir-mwyn, SN74, IMV, 1966[112]. **Cards.:** Upper Tywi Gorge, SN85, IMV, 1966[112]. **Pembs.:** Whitchurch, SN13, TAWD, 1973[112]. **Flints.:** Cilcain, SJ16, TE, 1973[116]. All these records were determined as var. *woodsiana*.

R. canina × villosa Spontaneous. Hedges. *Brecs.*. 1 square.
Brecs.: Merthyr Cynog, SN93, MP, 1977[245].

R. afzeliana Fries (*R. dumalis* Bechst. *p.p.*) : Rhosyn Llwydwyrdd : Glaucous Rose. Native. Hedges (E,Q,R,Rw). Fl.6-7. Mons., Glam., Brecs., Rads., Carms.*, Pembs.*, Cards.*, Monts., Mer., Caerns., Denbs., *Flints.*, Angl.. 28 squares, map 292.

R. afzeliana × canina Spontaneous. Hedges. *Brecs., Caerns., Angl.*. 5 squares.
Brecs.: Sennybridge, SN92, MP, 1978[200]; nr Merthyr Cynog, SN93, MP, 1973[200]; Llan-y-wern, SO02, MP, 1979[200]. **Caerns.:** nr Bangor, SH57, RHR, 1979[208]. **Angl.:** btwn Moel y Don and Porthaml, SH56, RHR, 1978[209]; nr Penmynydd, SH57, RHR, 1978[184].

R. afzeliana × sherardii Spontaneous. Hedges. *Brecs., Angl.*. 2 squares.
Brecs. Defynnog, SN92, MP, 1978[200]. **Angl.:** nr Penmynydd, SH57, RHR, 1978[184].

R. coriifolia Fries (*R. dumalis* Bechst. *p.p.*) : Leathery-leaved Rose. Native. Hedges (E,Q,R). Fl.6-8. Mons., Brecs., Rads., Carms.*, Mer., Caerns., Denbs., Flints.. 17 squares : **ST***59*; **SN**73, 74,82,92,93,94; **SO***02*,03,05,*06*; **SH**66,71,83; **SJ**15, 16,17.

R. coriifolia × sherardii Spontaneous. Hedges. *Brecs.*. 1 square.
Brecs.: nr Newbridge-on-Wye, SN95, MP & RGW, 1979[200].

R. obtusifolia Desv. : Rhosyn Deilen Llawban : Felt-leaved Rose. Native. Hedges. Fl.7-8. Mons., Glam., *Brecs., Rads., Carms.**, Pembs., *Angl.*. 22 squares, map 293.

R. dumetorum Thuill. : Rhosyn Dryslwyn : Thicket Rose. Native. Hedges (E,G,M,Q,R). Fl.6-7. Mons., Glam., Brecs., Rads., Carms., Pembs., *Cards.*, Monts., Mer., Caerns., Denbs., Flints., *Angl.*. 73 squares, map 294.

R. dumetorum × **sherardii** Spontaneous. Banks. *Brecs.*. 2 squares.
Brecs.: nr Defynnog, SN92, MP, 1978[245]; nr Merthyr Cynog, SN93, MP, 1978[200].

R. tomentosa agg. Mons., Glam., Brecs., Rads., Carms., Pembs., Cards., Monts., Mer., Caerns., Denbs., Flints., *Angl.*. 189 squares, map 295.

R. tomentosa Sm. *s.s.* : Rhosyn Lled-wlanog : Downy Rose. Native. Hedges (D,E,Q,R,Sc,W). Fl.6-7. Mons., Glam., Brecs., Rads., Carms., Pembs., Cards., Monts., Mer., Caerns., Denbs., Flints., *Angl.*. 59 squares, map 296.

R. sherardii Davies : Rhosyn Sherard : Sherard's Rose. Native. Hedges (E,G,Q,R,Rw). Fl.6-8. Mons., Glam., Brecs., Rads., Carms.*, Pembs.*, Cards., Monts., Mer.*, Caerns., Denbs., Flints.*, Angl.. 128 squares, map 297.

R. sherardii × **villosa** = **R.** × **shoolbredii** Wolley-Dod Spontaneous. *Brecs.*. 1 square.
Brecs.: Merthyr Cynog, SN93, MP, 1978[245].

R. villosa L. : Rhosyn Gwlanog : Soft-leaved Rose. Native. Hedges (E,H,Q,R). Fl.6-7. Mons., Glam., Brecs., Rads., Carms., *Pembs.**, Cards., Monts., Mer., *Caerns.*, Denbs., Flints., Angl.. 30 squares, map 298.

R. rubiginosa agg. Mons., Glam., Brecs., *Rads.*, Carms., Pembs., Cards., *Monts.*, Mer., Caerns., Denbs., *Flints.*, *Angl.*. 77 squares, map 299.

R. rubiginosa L. *s.s.* : Drysen Bêr : Sweet Briar. Native. Hedges (D,E,G,H,R). Fl.6-7. *Mons.*[230], Glam., Brecs.*, *Carms.*[230], Pembs.*, *Monts.*, (*Mer.*), Caerns., Denbs., *Flints.*, *Angl.*. 25 squares, map 300.

R. rubiginosa × **sherardii** = **R.** × **burdonii** Wolley-Dod (*R. sherardii* var. *suberecta* (A. Ley) Wolley-Dod). Spontaneous. Hedges. Fl.6-9. Brecs., Rads., Carms., Monts., Caerns.[74], Flints., Angl.. 11 squares : **SN**73,74,82,83,92,98; **SO**04,07,12; **SH**27; **SJ**08. Most specimens were determined as var. *sub-erecta.*

R. rubiginosa × **tomentosa** Spontaneous. Hedges. Fl.7. Carms., *Pembs.*[217]. 1 square.
Carms.: Cilycwm, SN74, IMV, 1963[112].

R. agrestis Savi : Miaren Culddail : Small-leaved Sweetbriar. Native. Bushy places. Fl.7-8. Caerns., [*Denbs.*]. 2 squares.
Caerns.: Gt Ormes Head, SH78, CB, 1895[112] and WAS, 1911[112]; nr Gloddaith, SH88, JEG, 1886[112] and CB, 1887[112].

R. micrantha Borrer ex Sm. : Rhoslwyn Pêr : Small-flowered Sweetbriar. Native. Hedges (E,F,G,Q,R,Sc,W). Fl.6-7. Mons., Glam., Brecs., *Rads.*[47], Carms., *Pembs.*, Cards., Monts., Mer., Caerns., *Denbs.*, *Flints.*, Angl.. 50 squares, map 301.

R. micrantha × **sherardii** Spontaneous. Hedges. *Brecs.*. 1 square.
Brecs.: N.E. Llech-faen, SO02, MP, 1979[200].

AGRIMONIA L.

Agrimonia eupatoria L. : Llysiau'r Dryw : Agrimony. Native. Fields and hedge borders (D,E,G,H,M,Q,R, Rw,W). Fl.5-9. Mons., Glam., Brecs., Rads., Carms., Pembs., Cards., Monts., Mer., Caerns., Denbs., Flints., Angl.. 204 squares, map 302.

A. procera Wallr. (*A. odorata* auct., non Miller) : Llysiau'r Dryw Peraroglus : Fragrant Agrimony. Native. Fields and roadsides (D,E,G,M,Q,R,W). Fl.7-9. Mons., Glam., Brecs., Rads., Carms., Pembs., Cards., Monts., Mer., Caerns., Denbs., Flints., *Angl.*. 105 squares, map 303.

SANGUISORBA L.

Subgen. Sanguisorba

Sanguisorba officinalis L. : Llysyrllys : Great Burnet. Native. Damp meadows (Ar,E,F,G,M,R,Rw). Fl.6-10. Mons., Glam., Brecs., Rads., Carms., Pembs., Cards., Monts., Mer., *Caerns.*, Denbs., Flints., *Angl.*. 107 squares, map 304.

S. canadensis L. Casual (N. America). Damp grassland. Fl.9. Mons.. 1 square : **ST**28[112].

Subgen. Poterium (L.) A. Braun & Bouché

S. minor Scop.
Subsp. **minor** (*Poterium sanguisorba* L.) : Gwyddlwdn Cyffredin : Salad Burnet. Native. Limestone pastures (D,E,G,Q,R,Rw,Sc). Fl.4-8. Mons., Glam., Brecs., *Rads.*, Carms., Pembs., Monts., Caerns., Denbs., Flints., Angl., [*Cards.*, *Mer.*]. 99 squares, map 305.

Subsp. **muricata** Briq. (*Poterium polygamum* Waldst. & Kit.) : Fodder Burnet. Established alien (Mediterranean region). Waste places (E,R,Rw,W) Fl.5-10. Mons., Glam., Brecs., *Pembs.*, Cards., *Monts.**, Caerns.*, [*Rads.*, *Carms.*]. 11 squares : **SS***78,96*; **ST***06*,17,*27,59*; **SN**10,14,79; **SO**21; **SH**12.

ACAENA Mutis ex L.

Acaena anserinifolia (J. R. & G. Forster) Druce (*A sanguisorbae* auct., *p.p.*) : Pirri-pirri-bur. Established alien (New Zealand). Garden weed. Fl.6. Glam.. 1 square.
Glam.: Llandaff, ST17, MCr, 1964[112].

A. novae-zelandiae Kirk (*A. anserinifolia* auct., non (J. R. & G. Forster) Druce; *A. sanguisorbae* auct. *p.p*). Established alien (New Zealand). Walls. Fl.6. Pembs.. 1 square.
Pembs.: St David's, SM72, CLW, 1951[112].

DRYAS L.

Dryas octopetala L. : Derig : Mountain Avens. Native. Mountain ledges. Fl.5-7. Caerns.. 2 squares.
Caerns.: Y Gribin, SH65, JEG, 1885[112] and HAH, 1924[112]; Cwm Idwal, SH65, ER, 1960[207]; Creigiau Gleision, SH76, ER, 1946[112] & 1965[207].

GEUM L.

Geum rivale L. : Mapgoll Glan y Dŵr : Water Avens. Native. Moist woods (Ar,E,H,M,R,Rw,W). Fl.4-8. Mons., Glam., Brecs., Rads., Carms., Pembs., Cards., Monts., Mer., Caerns., Denbs., Flints., Angl.. 129 squares, map 306.

G. rivale × urbanum = G. × intermedium Ehrh. Spontaneous. Roadsides and river banks (E,M,R,W). Fl.4-6. Mons., Glam., Brecs., Rads., Carms., Pembs., Cards., *Monts.*, Mer., Caerns., *Denbs.*, Flints., *Angl.*. 37 squares, map 307.

G. urbanum L. : Mapgoll : Wood Avens. Native. Hedges (Ar,E,G,M,Q,R,W). Mons., Glam., Brecs., Rads., Carms., *Pembs.*, Cards., Monts., Mer., Caerns., Denbs., Flints., *Angl.*. 259 squares, all except : **SR**89; **SS**38,39,77,89; **ST**26; **SM**50,62,71,73,84,94; **SN**36,59,75, 78,86; **SH**24,29,36,51,95,98.

POTENTILLA L.

Subgen. Trichothalamus (Lehm.) Reichenb.

Potentilla fruticosa L. : Shrubby Cinquefoil. Casual (N. England). Waste places. *Glam.*, *Caerns.*[166]. 1 square : **ST***18*[218].

Subgen. Comarum (L.) Syme

P. palustris (L.) Scop. : Pumdalen y Gors : Marsh Cinquefoil. Native. Marshes and fens (Ar,Ds,G,H,M). Fl.4-7. *Mons.*, Glam., Brecs., Rads., Carms., Pembs., Cards., Monts., Mer., Caerns., Denbs., Flints., Angl.. 176 squares, map 308.

Subgen Potentilla

P. anserina L. : Tinllwyd : Silverweed. Native. Roadsides (D,Ds,E,F,G,M,R,Rw,S,W). Fl.4-9. Mons., Glam., Brecs., Rads., Carms., Pembs., Cards., Monts., Mer., Caerns., Denbs., Flints., *Angl.*. 268 squares, all except : **SS**38,39; **ST**26; **SM**50; **SN**36,75,78,86,87; **SO**20; **SH**51,92; **SJ**02,13.

P. rupestris L. : Rock Cinquefoil. Native. Rocky places (M,Q). Fl.6-8. Rads., Monts., [*Mer.*]. 2 squares.
Rads.: nr Boughrood, AL, 1886[116] and RGW, 1980[259].
Monts.: Craig Breiddin, SJ21, JEB, 1830[112] and VJM, 1960+[116].

P. argentea L. : Pumbys Arian-ddail : Hoary Cinquefoil. Native. Dry gravelly places (G,R,Rw). Fl.5-8. (Glam.), *Monts.*, *Denbs.*[230], *Flints.*, [*Mons.*, *Caerns.*, *Angl.*]. 6 squares.
(**Glam.:** Barry, ST16, RM, 1924[112] and ABP & AEW, 1974[112]; Cardiff, ST17, RM and RLS, 1924[23]). **Monts.:** nr Churchstoke, SO29, PN, 1958[7]; nr Llandysilio, SJ21, VJM, 1956[7]; Breiddin Hill, SJ21, VJM, 1950+[116]; Bellan Bank, SJ31, JDB and PCH, 1961[107] and EDP, 1979[185]. **Flints.:** nr Tremeirchion, SJ07, RB, 1886[140] and JAG, 1977[185].

[**P. inclinata** Vill. (*P. canescens* Besser). *Glam.*]

P. norvegica L. : Ternate-leaved Cinquefoil. Established alien (N. and C. Europe). Waste places (R,Rg,Rw). Fl.6-9. Mons., Glam., *Brecs.**, *Carms.**, Pembs.*, *Caerns.*, *Denbs.*, *Flints.*. 13 squares : **SS**69; **ST***16*,*17*,18,*27*; **SM**90; **SN***50*,*90*; **SO***00*,51; **SH***53*; **SJ***16*,*23*.

P. intermedia L. : Russian Cinquefoil. Casual (N. and C. Russia). Waste places. Fl.6-7. Glam.. 1 square : **ST**17[112].

P. recta L. : Sulphur Cinquefoil. Casual (C. and S. Europe). Waste places (F,G,Q,R,Rw,S). Fl.6-9. *Mons.*, Glam., Pembs., Monts., Caerns., *Denbs.*, *Angl.*. 10 squares : **SS**87; **ST**16; **SM**80,91; **SN**98; **SO**50; **SH**57,77,78,87.

P. hirta L. Casual (Mediterranean region). River banks. Fl.7. *Mons.*. 1 square : **SO***30*[107].

P. crantzii (Crantz) G. Beck ex Fritsch : Pumdalen Gwyrdd : Alpine Cinquefoil. Native. Mountain rocks. Fl.5-8. *Monts.*, Caerns., [*Cards.*]. 4 squares.
Monts.: Craig Breiddin, SJ21, FAL, 1871[107]. **Caerns.:** Moel yr Ogof, SH54, JEG, pre 1895[101] and PMB, 1979[3]; Snowdon, SH65, JEG, 1891[112] and ER, 1962[207]; Ysgolion Duon, SH66, JEG, pre 1895[101]; Carnedd Llywelyn, SH66, ER, 1964[207] and ER & PMB, 1968[3].

P. tabernaemontani Ascherson : Pumdalen y Gwanwyn : Spring Cinquefoil. Native. Limestone pastures and rocks (G,Sc). Fl.4-7. Glam., Monts.*, Caerns., Denbs., Flints.*, *Angl.**, [*Mons.*]. 11 squares : **SS**48,58; **SH**58,78,87,88,97; **SJ**06,07,08,22.

P. erecta (L.) Räuschel

Subsp. **erecta** : Tresgl y Moch : Tormentil. Native. Heaths and moors (Ar,D,Ds,E,G,H,M,Q,R,Rw,W). Fl.4-9. Mons., Glam., Brecs., Rads., Carms., Pembs., Cards., Monts., Mer., Caerns., Denbs., Flints., Angl.. 270 squares, all except : **SS**39,77; **ST**06,26,27; **SM**50; **SN**36; **SH**51; **SJ**33,37,45,54.

Subsp. **strictissima** (Zimmeter) A. J. Richards Native. Upland heaths. Fl.7. *Brecs.*, *Mer.*. 2 squares.
Brecs.: Llanwrtyd Wells, SN84, GCD, pre 1930[241].
Mer.: nr Barmouth, SH61, HWM, 1890[241].

P. anglica Laicharding : Tresgl Ymlusgol : Trailing Tormentil. Native. Woods and hedgebanks (E,G,M, Q,R,Rw,S,W). Fl.5-10. Mons., Glam., Brecs., Rads., Carms., Pembs., *Cards.*, Monts., Mer., Caerns., *Denbs.*, Flints., *Angl.*. 184 squares, map 309.

P. anglica × **erecta** = **P.** × **suberecta** Zimmeter Spontaneous. Hedgebanks (E,G,M,R,S,W). Fl.5-9. Mons., Glam., *Brecs.*, Carms., Pembs., Cards., Monts., Mer., Caerns., Angl.. 35 squares, map 310.

P. anglica × **reptans** = **P.** × **mixta** Nolte ex Reichenb. (including *P.* × *italica* Lehm.). Spontaneous. Hedgebanks and roadsides (D,E,G,R,Sc,W). Fl.6-9. Mons., Glam., Brecs., Pembs., *Mer.*, *Caerns.*, Flints., *Angl.*. 32 squares, map 311.

P. reptans L. : Pumdalen Ymlusgol : Creeping Cinquefoil. Native. Waysides (D,E,G,M,R,Rg,Rw,S, Sc,W). Fl.4-8(-10). Mons., Glam., Brecs., Rads., Carms., Pembs., Cards., Monts., Mer., Caerns., Denbs., Flints., Angl.. 260 squares, all except : **SS**39; **SM**50,73; **SN**59, 65,75,77,86,87; **SH**51,60,62,63,64,72,74,82,85,95; **SJ**01,02,03.

Subgen. Fragariastrum (Heister ex Fabr.) Reichenb.

P. sterilis (L.) Garcke : Coegfefusen : Barren Strawberry. Native. Hedgebanks (D,E,G,M,Q,R,W). Fl.2-5(-9). Mons., Glam., Brecs., Rads., Carms., Pembs., Cards., Monts., Mer., Caerns., Denbs., Flints., Angl.. 260 squares, all except : **SR**89; **SS**38,39,77; **ST**26,27;**SM**50,62,71,73,84; **SN**15,34,40,43,59,86,88; **SH**51,98; **SJ**33,37.

FRAGARIA L.

Fragaria vesca L. : Mefusen y Goedwig : Wild Strawberry. Native. Woods and hedges (D,E,M,Q, R,Rw,W). Fl.4-7. Mons., Glam., Brecs., Rads., Carms., Pembs., Cards., Monts., Mer., Caerns., Denbs., Flints., Angl.. 259 squares, all except : **SR**89; **SS**38,39,77; **ST**26,58; **SM**50,62,70,71,73,84; **SN**59,78,86,87; **SH**13, 27,28,29,51,98; **SJ**37.

F. moschata Duchesne : Mefusen Fawr : Hautbois Strawberry. Naturalized alien (C. Europe). Hedgebanks (E,R,Rw). Fl.4-6(-10). Mons., Glam., Brecs., *Rads.**, Carms.*, Cards., Denbs.. 21 squares, map 312.

F. chiloensis (L.) Duchesne × **virginiana** Duchesne = **F.** × **ananassa** Duchesne : Mefusen : Garden Strawberry. Established alien (garden origin). Roadsides and waste places (R,Rw). Fl.5-7. Mons., Glam., *Brecs.**, *Rads.*, Carms., *Pembs.**, Cards., *Denbs.*, Angl., [*Monts.*]. 21 squares, map 313.

DUCHESNEA Sm.

Duchesnea indica (Andrews) Focke (*Fragaria indica* Andrews) : Yellow-flowered Strawberry. Casual (Asia). Waste places. Glam., *Pembs.*. 2 squares : **ST**18[112]; **SS**19[170].

ALCHEMILLA L.

Alchemilla conjuncta Bab. Naturalized alien (Scotland). Rocky places. Fl.6-8. Caerns.. 1 square.
Caerns.: Cwm Idwal, SH65, CWW, 1876[112]; Snowdon, SH65, KMG, 1949[112].

A. vulgaris agg. : Mantell Fair Gyffredin : Lady's-mantle. Mons., Glam., Brecs., Rads., Carms., *Pembs.*, Cards., Monts., Mer., Caerns., Denbs., Flints., *Angl.**. 178 squares, map 314.

A. xanthochlora Rothm. Native. Grassland (G,M,R,Rw,W). Fl.5-9. Mons., Glam., Brecs., Rads., Carms., Cards.*, Monts., Mer., Caerns., *Denbs.*, *Flints.**, *Angl.**. 107 squares, map 315.

A. filicaulis Buser
Subsp. **filicaulis** Native. Calcareous upland pastures (G,Q,R). Fl.5-7. *Mons.*, Brecs., Carms.*, Mer., *Caerns.**, *Denbs.**, *Flints.**. 19 squares : **SN**61,62,71; **SO**02,23; **SH**56,61,62,65,71,72, 73,82,83,84,95; **SJ**04,16,*17*.

Subsp. **vestita** (Buser) M. E. Bradshaw (*A. vestita* (Buser) Raunk.). Native. Grassland (E,G,M,R,Rw,W). Fl.(3-)5-8. Mons., Glam., Brecs., Rads., Carms., Monts.*, Mer., Caerns., Denbs.*, Flints.*, Angl.*. 92 squares, map 316.

A. wichurae (Buser) Stefánsson Casual or established alien (N. England). Roadsides. Fl.4. *Mer.*. 1 square.
Mer.: Pontdolgefeiliau, SH72, RSF & SMW, 1959[7].

A. glabra Neygenf. Native. Streamsides, cliffs and pastures (E,G,H,M,Q,R,Rw,W). Fl.4-9. Mons., Glam., Brecs., Rads., Carms., *Pembs.**, Cards., Monts., Mer., Caerns., Denbs., Flints.. 104 squares, map 317.

A. mollis (Buser) Rothm. Casual (E. Carpathians). Waste places. Fl.7-8. Glam., *Carms.*, Denbs.. 5 squares : **ST**08[95]; **SN**40[245],90[112]; **SO**10[95]; **SJ**33[112].

APHANES L.

Aphanes arvensis agg. : Troed y Dryw : Parsley-piert. Mons., Glam., Brecs., Rads., Carms., Pembs., Cards., Monts., Mer., Caerns., Denbs., Flints., Angl.. 219 squares, map 318.

A. arvensis L. *s.s.* : Parsley-piert. Native. Dry pastures and cultivated ground (D,E,F,G,Q,R,Rg, Rw,S,W). Fl.4-9. Mons., Glam., Brecs., Rads., Carms.*, Pembs., Cards.*, Monts., Mer.*, Caerns., Denbs., Flints.*, *Angl.*. 126 squares, map 319.

A. microcarpa (Boiss. & Reuter) Rothm. : Slender Parsley-piert. Native. Dry acidic pastures (D,E,F,G,H,Q,R). Fl.4-8(-10). Mons.*, Glam.*, Brecs., Rads., Carms., Pembs., Cards.*, *Monts.**, Mer., Caerns., Denbs., *Flints.**, Angl.. 119 squares, map 320.

CYDONIA Miller

Cydonia oblonga Miller (*Pyrus cydonia* L.) : Quince. Introduced (Asia). Woods. *Glam.*. 1 square.
Glam.: nr Cardiff, ST18, LP, 1937[36].

PYRUS L.

[**Pyrus cordata** Desv. *Mons.*]

P. pyraster Burgsd. (*P. communis* auct., non L.) : Gellygen : Wild Pear. Naturalized alien (Europe). Hedges (D,E,Q,R,Rw,W). Fl.4-5. Mons., Glam., *Brecs.*, *Rads.*, Carms., *Pembs.**, Monts., *Mer.*, Caerns., Denbs., *Flints.*, Angl.. 37 squares, map 321.

P. amygdaliformis Vill. Casual (Mediterranean region). Woods. Mons.. 1 square : **ST***59*[112].

ARONIA Medicus

Aronia melanocarpa (Michx) Elliott Introduced (N. America). Boggy places. Fl.5. Caerns..
Caerns.: nr Llyn Hafod-y-llyn, SH64, JSy, 1980[112].

MALUS Miller

Malus sylvestris agg. : Coeden Afalau Surion : Crab Apple. Mons., Glam., Brecs., Rads., Carms., Pembs., Cards., Monts., Mer., Caerns., Denbs., Flints., *Angl.*. 228 squares, map 322.

M. sylvestris Miller *s.s.* Native. Woods and hedges (E,Q,W). Fl.4-5. Mons., Glam., Brecs., Rads., Carms., Pembs., Cards., Monts., Mer., Caerns., Denbs., Flints., *Angl.*. 102 squares, map 323.

M. domestica Borkh. (*M. sylvestris* subsp. *mitis* (Wallr.) Mansfeld). Established alien (garden origin). Hedges (E,R,W). Fl.4-6. Mons., Glam., Carms., *Pembs.**, *Cards.**, Monts.[112], Mer., Caerns., Flints., *Angl.*. 51 squares, map 324.

SORBUS L.

[**Sorbus domestica** L. *Mons.*]

S. aucuparia L. : Criafolen : Rowan. Native. Mountain rocks and woods (E,G,H,M,Q,R,W). Fl.5-6. Mons., Glam., Brecs., Rads., Carms., Pembs., Cards., Monts., Mer., Caerns., Denbs., Flints., Angl.. 238 squares, map 325.

S. aucuparia × **intermedia** (*S.* × *pinnatfida* auct.). Introduced (origin unknown). *Mer.*[217]. Reported to be spontaneous but **S. intermedia** has yet to be recorded from *Mer.*

S. torminalis (L.) Crantz : Criafolen Wyllt : Wild Service-tree. Native. Woods (E,M,Q,R,Sc,W). Fl.5-6. Mons., Glam., Brecs., Rads., Carms., Pembs., *Cards.*, *Monts.*, Mer., Caerns., Denbs., Flints., *Angl.*. 45 squares, map 326.

S. aria agg. : Criafolen Wen : Common Whitebeam. Mons., Glam., Brecs., Rads., Carms., Pembs., *Cards.*, Monts., Mer., Caerns., Denbs., Flints., *Angl.*. 67 squares, map 327.

S. aria (L.) Crantz Native. Woodland and limestone rocks (E,Q,R,W). Fl.5-6. Mons., (*Glam.**, Rads.*, Carms.*, Pembs.*, *Cards.**, Monts., Mer., Caerns., *Denbs.**, *Flints.**, *Angl.*). 38 squares, map 328.

S. aria × **aucuparia** = **S.** × **thuringiaca** (Ilse) Fritsch Spontaneous. Woods. Mons.. 2 squares.
Mons.: (nr Newport, ST38, CT, 1974[91]); Piercefield Cliffs, ST59, TGE & CT, 1972[112].

S. aria × **torminalis** = **S.** × **vagensis** Wilmott Spontaneous. Woods (E,Q,W). Fl.5-6. Mons.. 1 square.
Mons.: Wynd Cliff, ST59, WAS, 1878[112] and TGE, 1972[91]; Mounton, ST59, WAS, 1891[112]; nr Chepstow, ST59, WAS, 1909[112].

S. leptophylla E. F. Warburg Native. Limestone rocks. Fl.5-6. Brecs., Monts.. 5 squares.
Brecs.: Craig y Rhiwarth, SN81, BAM, 1956[112] and MP, 1970[112]; Craig y Cilau, SO11, HAH & AEW, 1926[112] and RGW, 1977[108]; Cwm Clydach, SO21, DPMG, 1950+[202]. **Monts.:** Kerry, SO19, JAW, 1939[112]; Craig Breiddin, SJ21, An, 1955[108] and RGW, 1977[108].

S. leptophylla × **rupicola** Spontaneous. Limestone rocks. *Brecs.*. 1 square.
Brecs.: Mynydd Llangatwg, SO11, An, no date[217](may be abnormal **S. aria** × **rupicola**[217]).

S. eminens E. F. Warburg Native. Limestone rocks. Woods. Fl.6. Mons.. 3 squares.
Mons.: Itton, ST49, FJH, 1852[108]; Wynd Cliff, ST59, ESM, 1895[108]; nr Tintern, ST59, WAS, pre 1920[226]; nr Monmouth, SO51, SGC, 1943[226].

S. porrigentiformis E. F. Warburg Native. Rocky places usually on limestone (E,Q,Sc,W). Fl.6-7. Mons., Glam., Brecs., Rads., *Carms.*, *Monts.**, Caerns., *Flints.**. 19 squares : **SS**48,58; **ST**18,59; **SN**61,81,90,92; **SO**00,01,04,11,21,22,23; **SH**57,77; **SJ**17,21.

S. rupicola (Syme) Hedl. : Rock Whitebeam. Native. Limestone cliffs and rocks (E,Q,Sc,W). Fl.6. Mons.*, Glam., Brecs., *Carms.**, Monts., Caerns., Denbs., Flints., [*Rads.*]. 18 squares : **ST**59; **SN**61, 90; **SO**00,01,11; **SH***57*,67,77,*78*,87,*95*,97; **SJ**08,16,21, 22,24.

S. rupicola × **torminalis** Spontaneous. Limestone rocks. Fl.5-6. *Brecs.*, [*Glam.*]. 2 squares.
Brecs.: Cefncoedycymer, SO00, AL, 1899[15]; Dan y Graig, SO01, AL & WAS, 1896[15].

S. intermedia agg.

S. minima (A. Ley) Hedl. : Cerdin Wen Lleiaf : Lesser Whitebeam. Native (endemic). Limestone rocks. Fl.5-6. Brecs.. 1 square.
Brecs.: Craig y Cilau, SO11, AL, 1893[112] and TGE, 1980[91]; Cwm Cleisfer, SO11, DPMG, 1951[112].

S. leyana Wilmott : Cerdin Darren Fach : Ley's Whitebeam. Native (endemic). Limestone rocks. Fl.5-6. Brecs.. 2 squares.
Brecs.: Penmoelallt, SO00, PH, 1963[112] and PDS, 1975[108]; Darren Fach, SO01, AL, 1899[112] and TGE, 1980[91].

S. anglica Hedl. : English Whitebeam. Native. Limestone rocks (E,Q,Rw,W). Fl.5-7. Mons., Brecs., *Rads.**, Monts., Denbs.. 8 squares.
Mons.: nr Chepstow, ST59, WAS, 1893[112] and TGE, 1978[112]; Wynd Cliff, ST59, WAS, 1878[112] and TGE, 1980[91]. **Brecs.:** Elan Valley, SN96, RGW, 1969[108] and 1979[259]; Craig y Cilau, S011, AL, 1894[108] and TGE, 1980[91]; nr Gilwern, SO21, DPMG, 1951[112] and MP, 1970+[200]. **Rads.:** Elan Valley, SN96, RGW, 1979[245]. **Monts.:** Vaynor, SJ10, VJM, 1960+[116]; Craig Breiddin, SJ21, WAL, 1836[108] and IMV, 1962[112]; Llanymynech Hill, SJ22, An, 1955[108]. **Denbs.:** Cefn Fedw, SJ24, AL, 1901[112]; Eglwyseg Rocks, SJ24, JAW, 1943[112] and TE, 1974[112].

S. intermedia (Ehrh.) Pers. : Swedish Whitebeam. Established alien (N. Europe). Woods and hedges (E,H,Q,R,W). Fl.6-7. Glam., Rads., Pembs., Cards., Caerns., Flints.. 12 squares : **ST**17,18; **SM**72,91; **SN**58; **SO**06; **SH**56,66,78; **SJ**08,26,27.

S. latifolia (Lam.) Pers. *s.l.* : Broad-leaved Whitebeam Naturalized alien (Europe). Woods and roadsides. Fl.5. Carms., *Cards.*, Flints., [*Mons., Brecs., Denbs.*]. 3 squares.
Carms.: Carreg Cennen, SN61, IMV, 1970[112]. **Cards.:** Rhosgellan Fawr, SN58, AOC, 1980[112]. **Flints.:** Caergwrle, SJ35, JAW, 1942[112].

Recent research by P. D. Sell has indicated the presence of two previously undescribed species of **Sorbus** in Wales. The names of these had not been published when this book went to press (Dec. 1982). One, belonging to **S. aria** agg., has been recorded from Carms., *Cards., Caerns.*, Flints., Angl. and the other belonging to **S. intermedia** agg., from *Caerns., Denbs.*

COTONEASTER Medicus

Cotoneaster horizontalis Decne : Wall Cotoneaster. Established alien (W. China). Walls (Q,Rw,Sc). Fl.6. Mons., Glam., *Carms., Pembs., Caerns.*, Denbs., *Flints., Angl.*[209]. 24 squares, map 329.

C. simonsii Baker : Himalayan Cotoneaster. Naturalized alien (E. India). Woods and hedges (E,Q,R,Sc,W). Fl.6-8. Mons., Glam., Brecs., Rads., Carms., Pembs., Cards.*, Monts., Mer., Caerns., Denbs., Flints., Angl.. 51 squares, map 330.

C. dielsianus E. Pritzel ex Diels Established alien (China). Rocky places and walls. Fl.6-7. Glam.. 3 squares.
Glam.: nr Coity, SS98, ABP & AEW, 1974[112]; Rhoose, ST06, ABP & RGE, 1971[112]; Cardiff, ST18, HMS, 1980[112].

C. integerrimus Medicus : Wild Cotoneaster. Native. Sea cliffs (limestone). Fl.4-5. Caerns.. 1 square.
Caerns.: Gt Ormes Head, SH78, CCo, c1830[112] and MM, 1979[197].

C. bullatus Boiss. Established alien (China). Garden weed. Fl.6. Glam.. 1 square.
Glam.: Cardiff, ST18, HMS, 1980[112].

C. microphyllus Wallich ex Lindley : Small-leaved Cotoneaster. Naturalized alien (Himalaya). Walls and rocky places (E,G,Q,R,Rw,Sc). Fl.3-6. *Mons.*, Glam., Brecs., Rads., Carms., Pembs., *Cards.**, Monts., Mer., Caerns., Denbs., Flints., Angl.. 73 squares, map 331.

C. frigidus Wallich ex Lindley Established alien (Himalaya). Walls and waste places (Q,R,Rg,W). Fl.6-8(-11). Glam., Rads., Carms., Monts., *Caerns.*[119], *Angl.*[119]. 7 squares.
Glam.: Liliput, SS68, JAW, 1923[112]; Coltshill, SS68, JAW, c1940[221]; nr Swansea, SS69, JAW, c1918[221] & 1941[112]; Mountain Ash, ST09, AMP, 1970[112]; Cwm-pennar, SO00, AMP, 1973[112]. **Rads.:** Llandrindod Wells, SO06, JAW, 1939[112]; **Carms.:** Llanelli, SS59, JAW, 1944[112]. **Monts.:** Pont Llyfnant, SN79, HWil, 1981[112].

C. salicifolius Franchet Casual (China). Waste places. Fl.4. Glam.. 1 square : **ST**17[112].

PYRACANTHA M. J. Roemer

Pyracantha coccinea M. J. Roemer : Firethorn. Introduced (S. Europe). Roadsides (E,R,Sc). Fl.6-8. Mons., Glam., *Denbs.*. 3 squares.
Mons.: nr Monmouth, SO51, RL & VG, 1951[112]. **Glam.:** nr Penarth, ST17, PM, 1953[112]. **Denbs.:** nr Ruthin, SJ15, AAD, 1910[144].

MESPILUS L.

Mespilus germanica L. : Merysbren : Medlar. Introduced (S.E. Europe). Woods. Glam.. 4 squares.
Glam.: Stouthall Woods, SS48, JBe, 1981[197]; Swansea, SS69, JAW, 1939[112]; Ewenni, SS97, HJR, c1906[110]; Radyr, ST18, HJR, c1906[140].

CRATAEGUS L.

Crataegus intricata Lange Casual (N. America). Hedges. Fl.7. *Mons.*, Glam.. 4 squares : **SS***58*[221],69[225],78[112]; **SO***51*[27].

C. crus-galli L. Casual (N. America). Hedges. Mons., Glam.. 4 squares : **SS**59[17],69[17],79[112]; **ST**49[112].

C. laevigata (Poiret) DC. (*C. oxyacanthoides* Thuill.) : Draenen Ysbyddaden : Midland Hawthorn. Denizen (England). Hedges. Fl.4-5. Mons., Glam., Pembs.*, Caerns.*, *Denbs., Angl.,* [*Carms.*]. 10 squares : **SS**58,87; **ST**09,38,39; **SN**13; **SO***00*; **SH**47,66; **SJ***22*. Some records may refer to **C. laevigata** × **monogyna** = **C.** × **media** Bechst.

C. monogyna Jacq. : Draenen Wen : Hawthorn. Native. Hedges and woods (D,E,G,H,M,Q,R,Sc,W). Fl.4-6(-12). Mons., Glam., Brecs., Rads., Carms., Pembs., Cards., Monts., Mer., Caerns., Denbs., Flints., Angl.. 275 squares, all except : **SR**89; **SS**39,77; **SM**50, 62,71,73.

C. laciniata Ucria (*C. orientalis* Pallas ex Bieb.). Casual (S.E. Europe). Waste places. *Glam..* 1 square : **SS***69*[17].

CRATAEGUS × MESPILUS = × CRATAEMESPILUS Camus

Crataegus monogyna × **Mespilus germanica** = × **Crataemespilus grandiflora** (Sm.) Camus Introduced (garden origin). *Mer..* 1 square.
Mer.: nr Tal-y-llyn, SH71, IMR, 1931[30].

PRUNUS L.

Subgen. Prunus

Prunus cerasifera Ehrh. : Cherry Plum. Established alien (Balkans). Roadsides (E,R,Sc,W). Fl.1-4. Glam., Carms., *Cards.*[210], Mer., *Denbs., Angl.*†. 6 squares. **Glam.:** Colwinston, SS97, ABP, 1978[112]. **Carms.:** Laugharne, SN31, JFT, 1957[112]. **Mer.:** nr Corwen, SJ04, EDP, 1980[6] and PMB, 1981[112]. **Denbs.:** Trefnant, SJ07, DS, 1976[183]; nr Gresford, SJ35, BF, 1979[67]. **Angl.:** nr Britannia Bridge, SH57, PDe, 1970+[209]†.

P. spinosa L. : Draenen Ddu : Blackthorn. Native. Woods and hedges (E,G,M,Q,R,Rw,Sc,W). Fl.4-5(-7). Mons., Glam., Brecs., Rads., Carms., Pembs., Cards., Monts., Mer., Caerns., Denbs., Flints., Angl.. 274 squares, all except : **SR**89; **SS**39,77; **SM**50; **SN**59; **SO**10; **SH**51; **SJ**37.

P. domestica L. Mons., Glam., Brecs., Rads., Carms., Pembs., Cards., Monts., Mer., *Caerns.,* Denbs., Flints., *Angl..* 137 squares, map 332.

Subsp. **domestica** : Eirinen : Wild Plum. Established alien (Caucasus). Hedges (E,W). Fl.3-5. Mons., Glam., Brecs., Rads., Carms., Pembs., Cards., Monts., Mer., *Caerns.,* Denbs., Flints., *Angl..* 93 squares, map 333.

Susp. **insititia** (L.) C. K. Schneider (*P. insititia* L.; incl. *P. domestica* subsp. *italica* (Borkh.) Hegi) : Eirin Bwlas : Bullace. Naturalized alien (Caucasus). Woods and hedges (E,Q,W). Fl.3-5. Mons., Glam., *Brecs., Rads.**, Carms., *Pembs.,* Cards., Monts., *Mer., Caerns.,* Denbs., *Flints., Angl..* 75 squares, map 334.

Subgen. Cerasus (Miller) Focke

P. avium L. : Ceiriosen Ddu : Wild Cherry. Native. Woods (E,M,Q,R,Rw,W). Fl.3-5. Mons., Glam., Brecs., Rads., Carms., Pembs., Cards., Monts., Mer., Caerns., *Denbs.,* Flints., *Angl..* 156 squares, map 335.

P. cerasus L. : Ceiriosen : Dwarf Cherry. Established alien (S.W. Asia). Hedges (E,M,W). Mons., Glam., Brecs., *Rads.,* Carms., Pembs., Cards., Monts., Mer., *Caerns.,* Denbs., Flints., Angl.. 95 squares, map 336.

Subgen. Padus (Miller) Focke

P. padus L. : Ceiriosen yr Aderyn : Bird Cherry. Native. Woods and hedges (E,M,Q,W). Fl.4-6. Mons., Glam., Brecs., Rads., Carms., *Pembs.**, Cards., Monts., Mer., Caerns., Denbs., Flints.. 118 squares, map 337.

P. serotina Ehrh. : Rum Cherry. Casual (N. America). Woods. *Glam..* 1 square : **SS**59[45].

Subgen. Laurocerasus (Duh.) Rehder

P. lusitanica L. : Portugal Laurel. Introduced (Iberian Peninsula). Hedges and waste places. *Mons., Glam.,* Mer., Denbs., Flints.. 5 squares.
Mons.: nr Magor, ST48, TGE, 1970+[91]. **Glam.:** nr Ystradmynach, ST19, RGE, 1978[95]. **Mer.:** nr Dolgellau, SH71, JAW, 1941[112]. **Denbs.:** Fron, SJ25, JAW, 1943[112]. **Flints.:** nr Ysceifiog, SJ17, GW, 1970+[261].

P. laurocerasus L. : Llawr-sirianen : Cherry Laurel. Established alien (Balkans). Hedges and waste places (E,M,R,Rw,W). Fl.4. Mons., Glam., *Rads.,* Carms., Cards., Monts., *Mer., Caerns.,* Denbs., Flints.. 35 squares, map 338.

LEGUMINOSAE

CASSIA L.

Cassia nictitans L. Casual (N. America). Waste places. *Glam..* 1 square : **ST***16*[26].

LABURNUM Fabr.

Laburnum anagyroides Medicus : Tresi Aur : Laburnum. Established alien (C. and S. Europe). Hedgerows and waste places. *Glam., Rads., Carms., Pembs.,* Cards., Monts., *Denbs..* 45 squares, map 339.

L. alpinum (Miller) Berchtold & J. Presl : Scotch Laburnum. Casual (C. and S. Europe). Waste places. Fl.6-8. *Rads.,* Carms., Monts.. 5 squares : **SN**32[112]; **SO**05[229],07[229]; **SJ**11[112],22[228].

CYTISUS L.

Cytisus multiflorus (L' Hér.) Sweet : White Broom. Introduced (Iberian Peninsula). Roadside bank. *Mer..* 1 square.
Mer.: btwn Bala and Corwen, SH93, GK, 1977[257].

C. striatus (Hill) Rothm. Established alien (Iberian Peninsula). Roadside verges. Fl.6-7. *Brecs.*[98], Rads., Cards., Mer.. 3 squares.
Rads.: nr Llanwrthwl, SN96, AOC, 1977[112]; Rhayader, SN96, AOC, 1978[71]. **Cards.:** Penparcau, SN58, AOC, 1977[112]. **Mer.:** nr Maentwrog, SH64, RGE, 1978[112].

C. scoparius (L.) Link (*Sarothamnus scoparius* (L.) Wimmer ex Koch) : Banhadlen : Broom. Native. Heaths (E,H,M,Q,R,Rw,W). Fl.3-9. Mons., Glam., Brecs., Rads., Carms., Pembs., Cards., Monts., Mer., Caerns., Denbs., Flints., *Angl.*. 228 squares, map 340. Prostrate forms occur in *Pembs.*, *Cards.*, *Caerns.*

GENISTA L.

Genista tinctoria L.

Subsp. **tinctoria** : Melynog y Waun : Dyer's Greenweed. Native. Pastures (E,F,G,H,M,Q,R, Rw,Sc). Fl.6-10. Mons., Glam., Brecs., Rads., Carms., *Pembs.*, *Cards.*, Monts., (*Mer.*), Caerns., Denbs., Flints., *Angl.*. 109 squares, map 341.

Subsp. **littoralis** (Corb.) Rothm. Native. Sea cliffs. *Pembs.*. 1 square.
Pembs.: St David's Head, SM72, An, pre 1930[79]†.

G. pilosa L. : Aurfanadl Blewog : Hairy Greenweed. Native. Heaths (G,H,Sc). Fl.5-9. Pembs., Mer.. 7 squares.
Pembs.: St David's, SM72, EA, 1902[112] and TAWD, 1962[112]; Tremynydd, SM73, TAWD, 1967[92]; Porthmynawyd, SM82, TAWD, 1963[92] and SGH, 1965[112]; St Elvis, SM82, TAWD, 1969[112]; Strumble Head, SM83 & 84, TAWD, 1960+[79] & 1976[201]; Llanwrda, SM94, TAWD, 1966[79]. **Mer.:** Cadair Idris, SH71, JHS, 1938[112] and PMB, 1978[3].

G. anglica L. : Cracheithin : Petty Whin. Native. Moist heaths (G,H,M,R,W). Fl.5-7. Mons., Glam., Brecs., Rads., Carms., Pembs., *Cards.*, Monts., Mer., Caerns., Denbs., Flints., Angl.. 129 squares, map 342.

G. hispanica L. subsp. **occidentalis** Rouy : Spanish Gorse. Naturalized alien (Spain). Fl.4-7. Cards.. 1 square.
Cards.: Aberystwyth, SN58, A, 1927[107] and AOC, 1977[112].

SPARTIUM L.

Spartium junceum L. : Spanish Broom. Established alien (S. Europe). Waste places and roadsides (E,R,Sc). *Mons.*, Glam., Mer.. 4 squares.
Mons.: Newport, ST38, TGE, 1979[91]. **Glam.:** St Athan, ST06, AEW, 1977[227]; Cardiff Docks, ST17, AEW, 1926[112]. **Mer.:** Harlech, SH53, JAW, 1941[112].

ULEX L.

Ulex europaeus L. : Eithinen Ffrengig : Gorse. Native. Heaths and roadsides (D,E,G,H,M,Q,R,Rw, Sc,W). Fl.(9-)3-5(-7). Mons., Glam., *Brecs.*, Rads., Carms., Pembs., Cards., Monts., Mer., Caerns., Denbs., Flints., *Angl.*. 274 squares, all except : **SR**89; **SS**39; **ST**26,48,58; **SM**50; **SH**98; **SJ**37.

U. europaeus × **galli** = **U.** × **douieae** Druce Spontaneous. Heaths (E,H,Q,Sc,W). Pembs., *Monts.*, Mer., *Caerns.*, *Angl.*. 12 squares : **SM**72,82,93; **SH**28, 32,48,49,57,58,61,62; **SJ**22.

U. minor Roth : Eithinen Goraidd : Dwarf Gorse. Established alien (England). Roadsides. *Monts.*, *Mer.*, *Caerns.*, *Denbs.*, [*Angl.*]. 11 squares : **SN**69; **SH**32, 50,61,63,64,72,82,93; **SJ**14,*15*.

U. gallii Planchon : Eithinen Mân : Western Gorse. Native. Heaths and hill pastures (Ar,D,E,G,H,Q, R,Rw). Fl.7-11. Mons., Glam., Brecs., Rads., Carms., Pembs., Cards., Monts., Mer., Caerns., Denbs., Flints., Angl.. 234 squares, map 343.

LUPINUS L.

[**Lupinus luteus** L. *Glam.*]

L. angustifolius L. : Annual Lupin. Casual (S. Europe). Waste places. Fl.8. Glam., *Carms.*, *Caerns.*. 4 squares : **ST***18*[112]; **SN***40*[19]; **SH***56*[18],78[52].

L. micranthus Guss. (*L. hirsutus* L. 1796 *p.p.* et auct., non L. 1753). Casual (Mediterranean region). Waste places and cultivated ground). Fl.9-10. Glam.. 1 square : **ST***27*[112].

L. nootkatensis Donn ex Sims : Bys y Blaidd : Nootka Lupin. Casual (N. America). Waste tips. *Glam.*. 1 square : **SS***69*[23].

L. polyphyllus Lindley : Garden Lupin. Casual (N. America). Waste places (R,Rw). Fl.5. Mons., *Glam.*, *Brecs.*, Rads.. 4 squares : **SS**79[117]; **ST**27[112]; **SN**97[112]; **SO**04[206].

L. arboreus Sims : Coeden Bys y Blaidd : Tree Lupin. Naturalized alien (N. America). Sand dunes and waste places. Fl.6-8. Mons., Glam., *Carms.*, Cards.*, *Mer.**, *Caerns.*, *Flints.*, *Angl.*. 14 squares : **SS**48,58,69,79,(96); **ST**48; **SN**14,31; **SH**33,46,53,76; **SJ**08,18.

ROBINIA L.

Robinia pseudacacia L. : Acacia. Naturalized alien (N. America). Sand dunes and woods (D,E,W). Fl.6-8. *Mons.*, Glam., Carms.. 5 squares.
Mons.: St Pierre, ST59, TGE, 1960+[91]; Trelleck, SO50, TGE, 1960+[91]. **Glam.:** nr Swansea, SS69, MHS, c1942[222]; Michaelston-le-Pit, ST17, AEW, 1934[112]; St Fagans, ST17, RGE, 1976[95]. **Carms.:** Pembrey Forest, SN30, RFM, 1959[161] and RDP, 1980[201].

GALEGA L.

Galega officinalis L. : Goat's-rue. Established alien (Europe). Roadsides and waste places (E,M,R,Rg,Rw). Fl.6-10. Mons., Glam., *Carms., Pembs.*. 8 squares. **Mons.:** Lamby, nr Cardiff, ST27, RGE, 1979[112]; nr Newport, ST28, GP, 1957[112]; nr Llangstone, ST38, AEW, 1942[112]; nr Milton, ST38, AEW, no date[226]. **Glam.:** Horton SS48, BJ, 1972[95]; Barry, ST16, RLS, 1921[112] and JPC, 1980[112]; Penarth, ST17, JS, 1876[21] and THT 1902[112]; Cardiff, ST17, RLS, 1923[112]; St Fagans, ST17, EV, 1941[112]; Leckwith, ST17, SGH, 1967[112]; Splott, Cardiff, ST27, AEW, 1937[112] and RGE & AEW, 1973[95]. **Carms.:** Pembrey, SN40, IMV, 1970+[225]. **Pembs.:** Tenby, SN10, RFM, 1959[79].

COLUTEA L.

Colutea arborescens L. Bladder-senna. Casual (Mediterranean region). Waste places. Glam.. 1 square : **ST17**[112].

ASTRAGALUS L.

Subgen. Trimeniaeus Bunge

Astragalus boeticus L. Casual (Mediterranean region). Waste places (R,Rg). Fl.9. Glam.. 1 square : **ST27**[112].

A. stella Gouan Casual (Mediterranean region). Waste places. Glam.. 1 square : **ST*17***[112].

Subgen. Epiglottis (Bunge) Willk.

A. hamosus L. Casual (S. Europe). Waste places. Glam.. 2 squares : **ST*18***[112],*27*[112].

Subgen. Hypoglottis Bunge

A. danicus Retz. : Llaethwyg Mynyddig Rhuddlas : Purple Milk-vetch. Casual (England). Roadsides. *Mer.*. 1 square : **SH*53***[126].

Subgen. Phaca (L.) Bunge

A. glycophyllos L. : Llaethwyg : Wild Liquorice. Native. Hedges (E,G,Q,R,W). Fl.6-7. Mons., Glam., Caerns., Denbs., Flints.. 13 squares : **SS**78,79; **ST**16, 17,18,48,49; **SO**51; **SH***78,87*,88; **SJ***24*,35.

PSORALEA L.

Psoralea bituminosa L. Casual (S. Europe). Waste places. Fl.7-9. Glam.. 2 squares : **ST*17***[112],*18*[112].

P. americana L. Casual (S. Europe). Waste places. *Mons.*. 1 square : **ST38**[184].

PHASEOLUS L.

Phaseolus vulgaris L. : Ffaen Ffrengig : Kidney Bean. Casual (S. America). Waste places. *Mons.*. 1 square : **ST38**[42].

GLYCINE Willd.

Glycine soja Siebold and Zucc. : Soya Bean. Casual (E. Asia). Waste places. Fl.9. Mons.. 1 square : **ST38**[112].

CICER L.

Cicer arietinum L. Casual (S.W. Asia). Waste places. Fl.8. Glam., Rads.. 4 squares : **ST*16***[112],*17*[69],*27*[112]; **SO27**[112].

VICIA L.

Vicia orobus DC. : Pysen y Coed : Wood Bitter-vetch. Native. Pastures (E,F,G,H,M,Q,R,Rw). Fl.5-8. Mons., Glam., Brecs., Rads., Carms., Pembs., Cards., Monts., Mer., Caerns., Denbs.. 90 squares, map 344.

V. cracca L. : Tagwyg Bysen : Tufted Vetch. Native. Hedges (Ar,D,Ds,E,G,M,Q,R,Rw). Fl.6-9. Mons., Glam., Brecs., Rads., Carms., Pembs., Cards., Monts., Mer., Caerns., Denbs., Flints., Angl.. 273 squares, all except : **SS**38,39,77; **ST**26; **SM**50,71; **SN**36; **SH**51; **SJ**02.

V. tenuifolia Roth : Fine-leaved Vetch. Casual (Europe). Waste places. *Glam.*. 1 square : **SS69**[35].

V. cassubica L. : Danzig Vetch. Casual (Europe). Waste places. Fl.5. Glam.. 1 square : **ST*18***[112].

V. sylvatica L. : Ffugbysen y Wig : Wood Vetch. Native. Woods and hedges (E,R,S,Sc,W). Fl.5-8. Mons., Glam., Brecs.*, *Rads., Carms.**, Pembs., Cards., Caerns., Denbs., Flints., Angl., [*Mer.*]. 45 squares, map 345.

V. villosa Roth : Fodder Vetch.

Subsp. **villosa** Casual (Europe). Waste places. Fl.6. Mons., Glam., *Cards.*. 7 squares : **SS*69***[23],*78*[140]; **ST*16***[112],17[112], *18*[21],38[112]; **SN*45***[210].

Subsp. **varia** (Host) Corb. (*V. varia* Host, non Brot.; *V. dasycarpa* auct.). Casual (Europe). Waste places and cultivated ground. Fl.6-10. Mons., Glam.. 9 squares : **SS78**[112],79[157]; **ST*16***[112],17[112],18[22],27[112],28[112]; **SN*90***[107]; **SO*00***[112].

Subsp. **eriocarpa** (Hausskn.) P. W. Ball (*V. eriocarpa* (Hausskn.) Halácsy). Casual (S. Europe). Waste places. *Glam.*. 1 square : **ST17**[37].

Subsp. **pseudocracca** (Bertol.) P. W. Ball (*V. pseudocracca* Bertol.; *V. elegantissima* R. J. Shuttlew.). Casual (S.W. Europe). Waste places. Glam.. 1 square : **ST*18***[27].

V. benghalensis L. Casual (Mediterranean region). Waste places. Fl.6-9. Glam.. 2 squares : **ST16**[112],*18*[112].

V. hirsuta (L.) S. F. Gray : Corbysen Flewog : Hairy Tare. Native. Roadsides and cultivated ground (D,E,F, G,M,R,Rw,S,Sc). Fl.5-9. Mons., Glam., Brecs., Rads., Carms., Pembs., Cards., Monts., Mer., Caerns., Denbs., Flints., Angl.. 186 squares, map 346.

[**V. tenuissima** (Bieb.) Schinz and Thell. : Slender Tare. *Mons., Glam.*]

V. tetrasperma (L.) Schreber : Corbysen Lefn Bedwarronynnog : Smooth Tare. Native. Grassy roadsides (Ar,E,F,G,M,Q,R,Rw,W). Fl.5-9. Mons., Glam., Brecs., *Rads.*, Carms., Pembs., Cards., *Monts.*, Mer., Caerns., Denbs., *Flints.*, *Angl.*. 78 squares, map 347.

V. sepium L. : Ffugbysen y Cloddiau : Bush Vetch. Native. Hedges and woods (E,G,M,R,Rw,W). Fl.5-9. Mons., Glam., Brecs., Rads., Carms., Pembs., Cards., Monts., Mer., Caerns., Denbs., Flints., *Angl.*. 250 squares, map 348.

V. pannonica Crantz Casual (Europe). Waste places and cultivated ground. Fl.7-8. *Mons.*, Glam., Pembs., Mer.. 4 squares : **ST**59[91]; **SN**10[112],*90*[112]; **SH**70[112].

V. sativa L.

Subsp. **sativa** : Ffugbysen Faethol : Common Vetch. A rare casual or relic of cultivation. Due to confusion with *V. angustifolia* var. *segetalis* it has proved impossible to work out the vice-county distribution or to map this taxon.

Subsp. **nigra** (L.) Ehrh. (*V. sativa* subsp. *angustifolia* (L.) Gaudin; *V. angustifolia* L.; *V. angustifolia* var. *segetalis* Koch) : Ffugbysen Gulddail Ruddog : Narrow-leaved Vetch. Native. Pastures and banks (D,Ds,E,F,G,H,Q,R,Rw,W). Fl.5-7(-10). Mons., Glam., Brecs., Rads., Carms., Pembs., Cards., Monts., Mer., Caerns., Denbs., *Flints.*, Angl.. 227 squares, map 349.

V. lathyroides L. : Ffugbysen y Gwanwyn : Spring Vetch. Native. Sandy places (D,G,Q,R,Rw,Sc,W). Fl.4-8. Glam., *Rads.*, Carms., *Pembs.*, Cards., *Monts.*[230], Mer., Caerns., *Denbs.**, Flints., *Angl.*. 43 squares, map 350.

V. peregrina L. Casual (S. Europe). Waste places. Glam.. 2 squares : **ST***17*[112],*27*[25].

V. lutea L.

Subsp. **lutea** (*V. laevigata* Sm.) : Ffugbysen Felen Arw-godog : Yellow-vetch. Established alien (England). Waste places and cultivated ground (E,F,R, Rw). Fl.7-10. Glam., Pembs., Cards., Flints.. 10 squares : **SS**69; **ST***07*,*09*,16,17,*27*; **SN**10,68; **SO***00*; **SJ***36*.

Subsp. **vestita** (Boiss.) Rouy (*V. vestita* Boiss.). Casual (S. W. Europe). Waste places. Glam.. 1 square : **ST***27*[112].

V. hybrida L. : Ffugbysen Gymysgryw : Hairy Yellow-vetch. Casual (S. Europe). Waste places. Fl.5. Glam.. 1 square : **ST**16[112].

V. bithynica (L.) L. : Ffugbysen Rhuddlas Arw-godog : Bithynian Vetch. Native. Hedges (E,W). Fl.5-7. Glam., *Brecs.*, (Caerns.*), Denbs., *Flints.*, [*Cards.*]. 7 squares.
Glam.: (Barry, ST16, MC, 1853[7]); Penarth, ST17, HL, 1837[107] and EV, 1936[112]; (Cardiff, ST17, RLS, 1925[112]; Radyr, ST18, RLS, 1922[112]). **Brecs.:** nr Brecon, SO02, Fr, 1884[129]. (**Caerns.:** Llandudno Junction, SH77, RL, 1957[112]). **Denbs.:** Llanddulas, SH97, SHB, 1891[7]; Trefnant, SJ07, DN, 1973[112]. **Flints.:** nr St Asaph, SJ07, An, 1888[107]; nr Rhuddlan, SJ07, DS, 1976[183].

V. narbonensis L. Casual (S. Europe). Waste places. Glam.. 2 squares : **ST***16*[112],*27*[112].

V. faba L. : Broad Bean. Casual (garden origin). Waste places. Glam., *Monts.*. 7 squares : **SS***69*[221],*78*[140], 79[157]; **ST***09*[140],*17*[112]; **SO***00*[140],*08*[228].

LENS Miller

Lens culinaris Medicus (*L. esculenta* Moench) : Lentil. Casual (garden origin). Waste places. Glam.. 1 square : **ST***16*[112].

LATHYRUS L.

Lathyrus japonicus Willd. subsp. **maritimus** (L.) P. W. Ball (*L. maritimus* Bigelow) : Pysen Gochlas Arfor : Sea Pea. Native. Shingle beaches and sandy shores. Glam.†, *Caerns.**. 3 squares.
Glam.: Aberthaw, ST06, EV, 1933[112]. **Caerns.:** Bardsey Island, SH12, APC, 1978[74]; nr Criccieth, SH43, An, c1966[74]†).

L. montanus Bernh. : Pysen y Coed Gnapwreiddiog : Bitter-vetch. Native. Wood borders (E,G,H,M,Q,R, Rw,W). Fl.4-9. Mons., Glam., Brecs., Rads., Carms., Pembs., Cards., Monts., Mer., Caerns., Denbs., Flints., *Angl.*. 215 squares, map 351.

L. pratensis L. : Ytbysen y Ddôl : Meadow Vetchling. Native. Grassland and roadsides (D,E,G,M,Q,R, Rw,W). Fl.5-9. Mons., Glam., Brecs., Rads., Carms., Pembs., Cards., Monts., Mer., Caerns., Denbs., Flints., Angl.. 272 squares, all except : **SS**39,77; **ST**26; **SM**50,71,73; **SN**59,75; **SH**51,65.

L. palustris L. : Ytbysen Las y Morfa : Marsh Pea. Native. Marshes (G,M). Fl.5-7. Carms.*, *Mer.*†, *Caerns.*†, [*Glam.*]. 4 squares.
Carms.: Pembrey, SN30, RDP, 1977[201] & 1981[201]; Pembrey, SN40, IMV, 1971[112] and RDP, 1980[201]. **Mer.:** Bala, SH93, JEB, c1830[126]. **Caerns.:** nr Llandudno, SH78, JEB, 1837[107] and JEG, c1879[167].

L. tuberosus L. : Tuberous Pea. Established alien (Europe). Waste places. Fl.5-9. Glam., *Pembs.**, *Flints.*, [*Mons.*]. 6 squares.
Glam.: Jersey Marine, SS79, VMP, 1932[112] and QONK, 1980[157]; Llanharry, ST08, JL, 1968[99]; Barry, ST16, PHH, 1916[112] and RGE, 1980[112]; Cardiff Docks, ST17, HJR, 1902[107] and AEW, 1956[112]; Penarth, ST17, ABP, 1963[112]. **Pembs.:** nr Haverfordwest, SM91, Car, 1852[107]. **Flints.:** Rhyl, SJ08, AAD, 1927[26].

L. grandiflorus Sibth. & Sm. Established alien (S. Europe). Hedges. Fl.6. Mons., Glam., *Monts.*. 5 squares.
Mons.: Dinham, ST49, TGE, 1978[91]; Penpergwm Station, SO31, TGE, 1978[112]; Monmouth, SO51, TGE, 1981[91]. **Glam.:** Mountain Ash, ST09, AMP, 1973[112]. **Monts.:** Glascoed, —, An, 1967[250].

L. sylvestris L. : Ytbysen Barhaus Gulddail : Narrow-leaved Everlasting-Pea. Native. Hedges, sea cliffs and railways (D,E,G,Q,R,Rg,E,Sc,W). Fl.5-9. Mons., Glam. Brecs.*, *Rads.*, Carms., Pembs., Cards., Monts., Mer., Caerns., *Denbs.*, *Flints.**, *Angl.*. 69 squares, map 352.

L. latifolius L. : Ytbysen Barhaus Lyndanddail : Broad-leaved Everlasting Pea. Established alien (S. Europe). Hedges (D,E,R,Rw,S). Fl.7-9. *Mons.**, Glam., *Rads.**, Carms.*, Pembs., Monts., Mer., Caerns., Denbs., Flints., Angl.*. 33 squares, map 353.

L. odoratus L. Casual (S. Europe). Waste places. *Mons.*, *Glam.*. 2 squares : **ST***07*[26],38[91].

L. sphaericus Retz. Casual (S. Europe). Waste places. Fl.5. Glam.. 1 square : **ST***16*[112].

L. cicera L. Casual (S. Europe). Waste places. Fl.7-9. Glam.. 3 squares : **ST***16*[23], 18[112], 27[112].

L. clymenum L. Casual (Mediterranean region). Waste places. Fl.7. Glam.. 2 squares : **ST***18*[24],27[112].

L. ochrus (L.) DC. Casual (S. Europe). Waste places. Glam.. 3 squares : **ST***16*[24],*17*[24],27[112].

L. nissolia L. : Ytbysen Goch : Grass Vetchling. Native. Pastures (E,G,Q,R,Rw,Sc). Fl.6-7. Mons., Glam., (Brecs.*, *Pembs.**, *Cards.**), Denbs.. 16 squares : **ST**06,07,16,17,*27*,*28*,38,48,49,*59*; (**SM**90; **SN**14; **SO**12), *22*; **SJ**34,35.

L. aphaca L. : Ytbysen Felen : Yellow Vetchling. Established alien (Europe). Waste places and cultivated ground (D,F,R,Rg,Rw). Fl.6-7. *Mons.*, Glam., Caerns., Denbs., *Flints.*, [*Angl.*]. 21 squares, map 354.

PISUM L.

Pisum sativum L. (incl. *P. arvense* L.) : Pysen : Garden Pea. Casual (S. Europe). Cultivated ground and waste places. Fl.8-9. Mons., Glam., Rads., Pembs., Monts.. 12 squares : **SS**68,*78*,79; **ST**09,*16*,*17*,27,38; **SN**10; **SO***00*,19,36.

ONONIS L.

Ononis natrix L. Casual (S. and W. Europe). Waste places. Glam.. 1 square : **ST***27*[112].

O. reclinata L. : Small Restharrow. Native. Dry grassy places near the sea (G,Sc). Fl.7. Glam., *Pembs.**. 4 squares.
Glam.: Port Einon, SS48, JW, 1828[140] and QONK, 1979[157]; Overton Mere, SS48, MHB, BAM & JAW, 1956[112]; nr Salthouse Point, SS59, HRW, 1912[221]. **Pembs.:** Stackpole, SR99, DSR & SBE, 1971[184]; nr Tenby, SS19, NJ, 1959[79].

O. spinosa L. : Cas Gan Arddwr : Spiny Restharrow. Native. Pastures (G,M,Q,R,Rw,S,Sc). Fl.5-9. Mons., Glam., *Pembs.*, *Mer.*, Caerns., Denbs., Flints.. 32 squares, map 355.

O. repens L. : Tagaradr : Common Restharrow. Native. Sand dunes and dry pastures (D,G,M,Q,R, Rw,S,Sc). Fl.5-9. Mons., Glam., Brecs., *Rads.*, Carms., Pembs., *Cards.*, Monts., Mer., Caerns., Denbs., Flints., Angl.. 170 squares, map 356.

O. alopecuroides L. Casual (S. Europe). Cultivated ground. Fl.9. *Cards.*. 1 square : **SN**68[104].

MELILOTUS Miller

Melilotus altissima Thuill. : Meillionen y Ceirw : Tall Melilot. Established alien (Europe). Roadsides and waste places (E,M,Q,R,Rw). Fl.6-9. Mons., Glam., Brecs.*, Carms., Pembs., *Cards.*, *Monts.*, *Mer.*, Caerns., *Denbs.*, *Flints.*, *Angl.*, [*Rads.*]. 65 squares, map 357.

M. alba Medicus : Meillion Tair Dalen Gwyn : White Melilot. Established alien (Europe). Waste places (D,F,R,Rw). Fl.6-10. Mons., Glam., *Brecs.**, *Rads.**, Carms., *Pembs.**, Cards., Monts.*, Mer., Caerns., Denbs., *Flints.*, Angl.*. 51 squares, map 358.

M. officinalis (L.) Pallas : Gwydro Rhesog : Ribbed Melilot. Established alien (Europe). Roadsides and waste places (D,E,F,Q,R,Rw). Fl.5-8(-10). Mons., Glam., *Brecs.**, Carms., Pembs., *Cards.*, *Monts.**, Mer., Caerns., *Denbs.*, Flints., *Angl.**. 62 squares, map 359.

M. indica (L.) All. : Gwydro Blodau Bach : Small Melilot. Introduced (Mediterranean region). Waste places (R,Rw). Fl.6-10. Mons.*, Glam., Pembs.*, *Cards.*, Monts., Caerns., *Denbs.*, Flints., [*Mer.*]. 24 squares, map 360.

[**M. sulcata** Desf. *Glam.*]

M. messanensis (L.) All. (*M. sicula* (Turra) B. D. Jackson). Casual (Mediterranean region). Waste places. Glam.. 1 square : **ST***17*[112].

TRIGONELLA L.

Subgen. Trigonella

Trigonella corniculata (L.) L. Casual (Mediterranean region). Waste places. Fl.10. Mons., Glam.. 3 squares : **ST**16[112],*27*[24], 38[112].

T. polyceratia L. (*T. pinnatifida* Cav.). Casual (S.W. Europe). Waste places. Fl.9. Glam.. 1 square : **ST***27*[112].

T. monspeliaca L. Casual (C. and S. Europe). Waste places. Glam.. 1 square : **ST***17*[112].

T. hamosa L. Casual (Egypt). Waste places. Fl.8. Glam.. 2 squares : **ST***17*[112],*27*[112].

T. laciniata L. Casual (Egypt). Waste places. Glam.. 1 square : **ST***27*[112].

Subgen. Trifoliastrum (Moench) G. Beck

T. caerulea (L.) Ser. agg. (incl. *T. procumbens* (Besser) Reichenb.). Casual (Europe). Waste places. Fl.9. Glam., *Carms.*, *Flints.*. 6 squares : **SS***78*[140]; **ST***17*[112],*27*[112]; **SN***40*[15]; **SO***00*[140]; **SJ***26*[143].

MEDICAGO L.

Subgen. Medicago

Medicago lupulina L. : Maglys : Black Medick. Native. Limestone pastures, also widespread as an introduction on railways, roadsides and cultivated ground (D,E,F,G, M,Q,R,Rw,S,Sc). Fl.4-10. Mons., Glam., *Brecs.*, Rads., Carms., Pembs., *Cards.*, Monts., Mer., Caerns., Denbs., Flints., Angl.. 205 squares, map 361.

M. sativa L.

Subsp. **sativa** : Maglys Rhuddlas : Lucerne. Established alien (Mediterranean region). Waste places (F,G,R,Rw,S). Fl.4-10. Mons., Glam., Brecs.*, Carms., Pembs., *Cards.*, *Monts.**, Mer., Caerns., Denbs., Flints., *Angl.*. 56 squares, map 362.

Subsp. **falcata** (L.) Arcangeli (*M. falcata* L.) : Meillionen Gorniog : Sickle Medick. Established alien (S. England). Waste places (R,Rw). Fl.7-9. Mons., Glam., Pembs., *Cards.**, *Mer.*, *Caerns.*, Flints., [*Denbs.*, *Angl.*]. 15 squares : **SS***48*,69, *78*,*79*,*87*; **ST**16,*17*,*27*,38; **SN**10,*69*,*80*; **SH***53*,*78*; **SJ***08*.

Subsp. **varia** Martyn (*M. sativa* subsp. *falcata* × subsp. *sativa*). Established alien (Europe). Waste places. Fl.7. Glam., Caerns.. 7 squares.
Glam.: nr Port Einon, SS48, EF, c1800[140]; Swansea, SS69, JAW, 1956[112]; Porth-cawl, SS87, VCB, 1973[112]; Cardiff, ST17, PHH, 1926[112] and EV, 1929[28]; Cardiff Docks, ST27, ECu, 1978[95] and ABP, 1980[112]. **Caerns.:** Morfa Nevin, SH24, FLS, 1926[26]; Llandudno, SH78, MHS, 1953[112].

M. ciliaris (L.) All. Casual (Mediterranean region). Waste places. Glam.. 3 squares : **ST***16*[112],*17*[112],*27*[112].

Subgen. Cymatium (Pospichal) Gams

M. soleirolii Duby Casual (Crimea). Waste places. *Glam.*. 1 square : **ST***16*[24].

M. tornata (L.) Miller (*M. italica* (Miller) Steudel; *M. orbicularis* All., non (L.) Bartal.). Casual (Mediterranean region). Waste places. Fl.5-8. Glam.. 4 squares : **ST***16*[112], *17*[112],*18*[112],*27*[112].

M. marina L. Casual (Mediterranean region). Waste places. Glam.. 1 square : **ST***27*[112].

M. truncatula Gaertner (*M. tribuloides* Desr.). Casual (Mediterranean region). Waste places. Fl.5-7. Glam.. 3 squares : **ST***16*[112],*17*[22],*27*[112].

M. rigidula (L.) All. (*M. gerardii* Waldst. & Kit. ex Willd.). Casual (S. Europe). Waste places. Fl.9. Glam.. 1 square : **ST***27*[112].

M. littoralis Rohde ex Loisel. Casual (Mediterranean region). Waste places. Fl.7. Glam.. 3 squares : **ST***16*[112],*17*[112],*27*[112].

M. doliata Carmign. (*M. aculeata* Gaertner; *M. turbinata* Willd., non (L.) All.). Casual (Mediterranean region). Waste places. Glam.. 3 squares : **ST***16*[112],*17*[22],*27*[112].

M. turbinata (L.) All. (*M. tuberculata* (Retz.) Willd.). Casual (Mediterranean region). Waste places. *Glam.*. 1 square : **ST***17*[25].

M. murex Willd. Casual (Mediterranean region). Waste places. Glam.. 3 squares : **ST***16*[23],*18*[22],*27*[112].

M. arabica (L.) Hudson : Maglys Amrywedd : Spotted Medick. Naturalized alien (England). Waste places (D,F,G,M,R,S). Fl.4-8. Mons., Glam., *Carms.*, Pembs., Cards., Mer., Caerns., *Denbs.*, Angl.. 40 squares, map 363.

M. polymorpha L. (*M. hispida* Gaertner) : Maglys Eiddiog : Toothed Medick. Established alien (S. England). Waste places. Fl.6-9. Mons., Glam., *Brecs.*[85], *Pembs.*, *Mer.*, *Caerns.*, *Denbs.*, *Flints.*. 15 squares : **SS**19, *78*; **ST***16*,17,18,27,*59*; **SN**10,*90*; **SO***00*,51; **SH***53*,*93*; **SJ***24*,*26*.

M. minima (L.) Bartal. : Bur Medick. Casual (S. England). Waste places and cultivated ground (R,Rg). Fl.5-7. Glam., Carms.. 5 squares : **SS***58*[140]; **ST***16*[112],17[112],*27*[112]; **SN**31[112].

TRIFOLIUM L.

Subgen. Falcatula (Brot.) D. E. Coombe

Trifolium ornithopodioides L. (*Trigonella ornithopodioides* (L.) DC.) . Corfeillionen Wen : Fenugreek. Native. Pastures and heaths (D,E,G,H,M,R,Rg,Rw,S,Sc). Fl.5-8. Mons., Glam., Brecs., Rads.*, Carms.*, Pembs., *Cards.*, Mer., Denbs., *Flints.*, Angl.. 27 squares, map 364.

Subgen. Lotoidea Pers.

T. strictum L. : Meillionen Unionsyth : Upright Clover. Native. Dry grassy places. Fl.6-7. Rads., *Angl.*. 2 squares.
Rads.: Stanner Rocks, SO25, WHH, 1936[112] and QC, 1980[71]. **Angl.:** nr Aberffraw, SH37, FD, 1837[108].

T. repens L. : Meillionen Wen : White Clover. Native. Pastures (D,E,G,H,M,Q,R,Rw,S,W). Fl.4-8(-10). Mons., Glam., Brecs., Rads., Carms., Pembs., Cards., Monts., Mer., Caerns., Denbs., Flints., Angl.. 278 squares, all except : **SS**39; **SM**50,73; **SH**51.

T. hybridum L.

Subsp. **hybridum** : Alsike Clover. Established alien (origin unknown). Cultivated ground and roadsides (E,F,G,M,R,Rw). Fl.6-9. Mons., Glam., Brecs., Rads., Carms., *Pembs.,* Cards., Monts., Mer., Caerns., Denbs., Flints., Angl.. 155 squares, map 365.

Subsp. **elegans** (Savi) Ascherson & Graebner (*T. elegans* Savi) : Lesser Alsike Clover. Casual (S. Europe). Roadsides and waste places. Mons., *Glam.*, Brecs., *Caerns.*. 6 squares : **SS***78*[140]; **ST***16*[140],*49*[112]; **SO**04[112]; **SH***67*[101],*78*[101].

T. nigrescens Viv. Casual (S. Europe). Waste places. Fl.6. Glam.. 2 squares : **ST***17*[112],*18*[112].

T. michelianum Savi Casual (S. Europe). Waste places. *Glam.*. 1 square : **ST***16*[24].

T. glomeratum L. : Meillionen Gwastadwedd : Clustered Clover. Casual (S. England). Dry places. *Mons., Glam., Carms.*, [*Caerns., Denbs.*]. 5 squares : **SS***69*[140]; **ST***17*[218],58[91]; **SN***40*[161]; **SO***00*[140].

T. suffocatum L. : Meillionen Fygiedig : Suffocated Clover. Native. Sandy places. Fl.6. (Glam.), Angl.†, [*Cards., Flints.*]. 4 squares.
(**Glam.:** Rhosili, SS48, EMW, 1910[17]; Barry, ST16, JS, c1876[218]; Penarth, ST17, JS, 1876[112]). **Angl.:** Beaumaris, SH67, HD, 1807[112] and SAN, 1839[7].

T. fragiferum L. : Meillionen Fefusaidd : Strawberry Clover. Native. Roadsides and grassy places near the sea (D,G,M,R,Rw,S). Fl.5-9. Mons., Glam., Carms., Pembs., Cards., Mer., Caerns., Denbs., *Flints.*, Angl.. 53 squares, map 366.

T. resupinatum L. : Reversed Clover. Established alien (Asia). Waste places. Fl.6-9. Mons., Glam.. 11 squares : **SS***68,69,87*; **ST***06,16,17,18,27*,38,*49*; **SN***90*.

T. tomentosum L. : Wool Clover. Casual (Mediterranean region). Waste places. Fl.5. Glam.. 2 squares : **ST***16*[112],*27*[112].

T. aureum Pollich : Large Hop Trefoil. Casual (Europe). Waste places. Fl.6-7. Glam.. 2 squares : **ST***17*[112],18[112].

T. campestre Schreber : Meillionen Hopys : Hop Trefoil. Native. Banks and walls (D,G,M,Q,R,Rw,S). Fl.5-8. Mons., Glam., Brecs., Rads., Carms., Pembs., Cards., Monts., Mer., Caerns., Denbs., Flints., *Angl.*. 213 squares, map 367.

T. dubium Sibth. : Meillionen Felen Fechan : Lesser Trefoil. Native. Pastures (D,E,F,G,H,M,Q,R,Rw). Fl.4-10. Mons., Glam., Brecs., Rads., Carms., Pembs., Cards., Monts., Mer., Caerns., Denbs., Flints., Angl.. 265 squares, all except : **SS**38; **SM**50,71; **SN**45,57,63, 64,75,78,87; **SO**09,41; **SH**51,82; **SJ**01,02,13.

T. micranthum Viv. : Meillionen Felen Eiddil : Slender Trefoil. Native. Pastures (D,G,H,Q,R, Rg,Rw,W). Fl.5-8. Mons., Glam., Brecs., *Rads.*, Carms., Pembs., Cards., *Monts.*, Mer., *Caerns., Denbs., Flints.*, Angl.. 106 squares, map 368.

T. ciliatum Nutt. Casual (N. America). Waste places. Fl.8. Caerns.. 1 square : **SH***56*[18].

Subgen. Trifolium

T. striatum L. : Meillionen Rychiog : Knotted Clover. Native. Dry pastures (G,M,Q,R,S,Sc). Fl.5-7. Mons., Glam., Brecs., Rads., Carms., Pembs., Cards., *Monts.**, Mer., Caerns., Denbs., Flints., Angl.. 87 squares, map 369.

T. arvense L. : Meillionen Gedenog : Hare's-foot Clover. Native. Sand dunes and gravelly places (D,G, Q,R,Rw,S). Fl.6-9. (Mons.), Glam., *Brecs.*, Rads., Carms., Pembs., Cards., Monts., Mer., Caerns., *Denbs.*, Flints., Angl.. 104 squares, map 370.

T. scabrum L. : Meillionen Ger y Môr : Rough Clover. Native. Dry pastures (D,E,G,M,Q,R,S,Sc). Fl.5-9. Mons., Glam., *Rads.**, Carms., Pembs., Cards., *Monts.**, Mer., Caerns., *Denbs., Angl.*, [*Flints.*]. 57 squares, map 371.

T. stellatum L. : Starry Clover. Casual (Mediterranean region). Waste places. Fl.5. Glam.. 4 squares : **SS***69*[20],*78*[140]; **ST***16*[112],*17*[112].

T. incarnatum L. : Meillionen Ysgarlad : Crimson Clover. Casual (Mediterranean region). Fields and waste places (F,G,R). Fl.4-9. Mons., Glam., Mer., Caerns., *Denbs.*. 15 squares : **SS**48,*69*,78,*97*; **ST***06*,07,*16,17*,48,*49*; **SO***00*; **SH**50,*57*; **SJ***05,25*.

T. pratense L. : Meillionen Goch : Red Clover. Native. Grassland (D,E,F,G,M,Q,R,Rw). Fl.4-10. Mons., Glam., Brecs., Rads., Carms., Pembs., Cards., Monts., Mer., Caerns., Denbs., Flints., Angl.. 276 squares, all except: **SS**39; **SM**50,73,94; **SH**51; **SJ**02.

T. diffusum Ehrh. Casual (Europe). Waste places. Glam.. 1 square : **ST***17*[112].

T. lappaceum L. Casual (S. Europe). Waste places. Fl.6-9. Glam.. 4 squares : **ST***16*[112],*17*[112],*27*[24]; **SO***00*[140].

T. cherleri L. Casual (S. Europe). Waste places. Fl.6. Glam.. 1 square : **ST***18*[112].

T. medium L. : Meillionen Wyrgam : Zigzag Clover. Native. Roadsides (D,E,G,H,M,R,Rw,W). Fl.6-9. Mons., Glam., Brecs., Rads., Carms., Pembs., Cards., Monts., Mer., Caerns., Denbs., Flints., Angl.. 184 squares, map 372.

T. angustifolium L. : Narrow Clover. Casual (S. Europe). Waste places. Fl.6-7. Glam.. 3 squares : **ST***16*[112],*17*[112],*18*[112].

T. ochroleucon Hudson : Sulphur Clover. Casual (E. England). Waste places (D,R). Fl.7. Glam., [*Caerns., Flints.*]. 2 squares : **SS***79*[244]; **ST***27*[112].

T. pannonicum Jacq. Casual (S. Europe). Waste places. *Glam..* 1 square : **ST***17*[218].

T. constantinopolitanum Ser. Casual (S. Europe). Waste places. Glam.. 1 square : **ST***27*[112].

T. squamosum L. : Meillionen y Morfa : Sea Clover. Native. Salt-marshes (G,M,R,Rw,S). Fl.6-11. Mons., Glam., Carms., *Caerns., Angl.*, [*Mer.*]. 17 squares : **SS**69,*79*,*87*; **ST**06,*16*,*17*,*27*,*28*,38,*39*,59; **SN**21,*41*,*42*; **SO***50*; **SH**49,*78*.

T. subterraneum L. : Meillionen Wen Ymgudd : Subterranean Clover. Native. Sandy grassland (G,M,Q, R,S,Sc). Fl.4-8. Mons., Glam., Carms., Pembs., Cards., Caerns., Denbs., *Flints.*, Angl.. 23 squares, map 373.

DORYCNIUM Miller

Dorycnium hirsutum (L.) Ser. Casual (Mediterranean region). Waste places. Fl.7. Glam.. 3 squares : **ST***16*[24],*17*[25],*18*[112].

D. rectum (L.) Ser. Casual (Mediterranean region). Waste places. Fl.8. Glam.. 1 square : **ST**16[112].

D. pentaphyllum Scop. subsp. **herbaceum** (Vill.) Rouy (*D. herbaceum* Vill.) : Greater Badassi. Casual (C. and S.E. Europe). Waste places. Fl.8-10. Glam.. 1 square : **ST***16*[112].

LOTUS L.

Lotus tenuis Waldst. & Kit. ex Willd. : Troed Aderyn Culddail : Narrow-leaved Bird's-foot-trefoil. Native. Pastures (G,R). Fl.6-8. (Mons.), Glam., Carms., *Monts.*[231], *Flints., Angl.*[230], [*Rads., Mer.*]. 8 squares. (**Mons.:** Newport Docks, ST38, WW, 1881[226] and TGE, 1978[91]; Caerleon, ST39, EV, 1937[112]). **Glam.:** Sully, ST16, AEW, 1950[112]; (Cardiff, ST17, AEW, 1920[112]; Pengam Cardiff, ST27, HJR, 1905[112]). **Carms.:** Kidwelly Quay, SN30, RFM, 1965[112]; nr Carmarthen, SN32, TWB, 1905[161]. **Flints.:** nr Harwarden Bridge, SJ36, AAD, 1909[143].

L. corniculatus L. : Pys y Ceirw : Common Bird's-foot-trefoil. Native Pastures (D,E,G,H,M,Q,R, Rw,S,Sc,W). Fl.4-9. Mons., Glam., Brecs., Rads., Carms., Pembs., Cards., Monts., Mer., Caerns., Denbs., Flints., Angl.. 280 squares, all except : **SS**39; **SM**50.

L. uliginosus Schkuhr (*L. pedunculatus* auct., non Cav.) : Pysen y Ceirw Mwyaf : Greater Bird's foot-trefoil. Native. Marshes (Ar,Ds,E,G,H,M,Q,R,Rw,W). Fl.5-9. Mons., Glam., Brecs., Rads., Carms., Pembs., Cards., Monts., Mer., Caerns., Denbs., *Flints.*, Angl.. 269 squares, all except : **SS**38,39,77; **ST**58,59; **SM**50,71,73; **SN**87; **SH**51,91,98; **SJ**37.

L. subbiflorus Lag. (*L. hispidus* Desf. ex DC. 1815, non 1805) : Pysen y Ceirw Flewog : Hairy Bird's-foot-trefoil. Native. Grassy cliff tops. Fl.6-9. (*Glam.* *), Pembs.. 4 squares.
(**Glam.:** Barry, ST16, RLS, 1927[26]). **Pembs.:** Westdale Bay, SM70, CDP, 1953[53]; St David's, SM72, GCL, 1977[112]; Watwick Bay, SM80, FWA, 1953[53],

L. angustissimus L. : Pysen y Ceirw Eiddilaidd : Slender Bird's-foot-trefoil. Denizen or naturalized alien (England). Grassy places. *Caerns.*, [*Cards.*]. 1 square. **Caerns.:** btwn Tremadoc and Tan-y-bwlch, SH64, SHB, 1863[53]†.

L. edulis L. Casual (Mediterranean region). Waste places. Fl.6. Glam.. 2 squares : **ST***17*[24],*18*[112].

L. conimbricensis Brot. Casual (Mediterranean region). Waste places. Fl.6. Glam.. 1 square : **ST***27*[112].

L. cytisoides L. Casual (Mediterranean region). Waste places. Glam.. 1 square : **ST***16*[112].

L. creticus L. Casual (S. Europe). Waste places. Fl.9. Glam.. 1 square : **ST***16*[112].

L. ornithopodioides L. Casual (S. Europe). Waste places. Fl.8. Glam.. 1 square : **ST***18*[112].

TETRAGONOLOBUS Scop.

[**Tetragonolobus maritimus** (L.) Roth *Glam.*]

T. purpureus Moench (*Lotus tetragonolobus* L.). Casual (S. Europe). Waste places. Glam., *Caerns..* 2 squares : **ST***17*[112]; **SH**22[184].

SECURIGERA DC.

Securigera securidaca (L.) Degen & Dörfler (*S. coronilla* DC.; *Bonaveria securidaca* (L.) Desv.). Casual (S. Europe). Waste places. Glam.. 1 square : **ST***17*[112].

ANTHYLLIS L.

Anthyllis vulneraria L. : Plucen Felen : Kidney Vetch.
Subsp. **vulneraria** Native. Grassland, duneland and rocky places on limestone (D,E,G,Q,R,Rw,S,Sc). Fl.4-10. Mons., Glam., Brecs., Rads., Carms., Pembs., Cards., *Monts.*, Mer., Caerns., *Denbs.*, Flints., Angl.. 139 squares, map 374.

Subsp. **corbierei** (Salmon and Travis) Cullen Native. Sea cliffs. Fl.3. Angl.. 4 squares.
Angl.: South Stack, Holyhead, SH28, JCu, 1958[110]; W. side of Holy Island, SH27 & 28, RHR, 1970+[209]; Aberffraw Common, SH36, RHR, 1970+[209]; Porth Trecastell and Porth Nobla, SH37, RHR, 1970+[209]; N. Rhosneigr, SH37, JCu, 1961[110] and RHR, 1970+[209].

Subsp. **carpatica** (Pant.) Nyman (*A. vulneraria* subsp. *vulgaris* (Koch) Corb.). Casual or naturalized alien (Europe). Dry waste places and cliffs. Fl.6. Pembs.. 4 squares.
Pembs.: Marloes, SM70, TAWD, 1956[112]; —, SM80, An, 1950+[194]; Tre-fin, SM83, TAWD, 1959[7]; Ratford Bridge, SM91, TAWD, 1965[112].

Subsp. **iberica** (Becker) Jalas Native. Sea cliffs. Fl.5-6. Pembs.. 4 squares.
Pembs.: Caldey Island, SS19, TAWD, 1905[112]; St Brides, SM70, TAWD, 1962[112]; Marloes, SM70, TAWD, 1963[112]; St David's Head, SM72, TAWD, 1962[112]; Porthmynawyd, SM72, TAWD, 1963[112]; Tre-fin, SM83, TAWD, 1963[112]. All these specimens have been identified by JCu as approaching subsp. **iberica**.

A. tetraphylla L. Casual (Mediterranean region). Waste places. Glam.. 2 squares : **ST***16*[112],*18*[112].

ORNITHOPUS L.

Ornithopus compressus L. Casual (S. Europe). Waste places. Fl.7-8. Glam.. 1 square : **ST***16*[112].

O. sativus Brot. (*O. roseus* Dufour). Casual (S.W. Europe). Waste Places. Glam.. 1 square : **ST**17[112].

O. perpusillus L. : Troed yr Aderyn : Bird's-foot. Native. Gravelly places (D,E,G,H,M,Q,R,Rg,Sc,W). Fl.4-9. Mons., Glam., Brecs., Rads., Carms., Pembs., Cards., Monts., Mer., Caerns., Denbs., Flints., Angl.. 134 squares, map 375.

O. pinnatus (Miller) Druce : Orange Bird's-foot. Casual (Scilly Is.). Waste places. Fl.7-8. Glam.. 2 squares : **ST***16*[112],*17*[140].

CORONILLA L.

Coronilla varia L. : Ffugbysen Goronog : Crown Vetch. Established alien (C. and S. Europe). Waste places (G,M,Q,R,Rw). Fl.7-9. Mons., Glam., Carms., Pembs.*, *Cards.*, Monts., *Mer.*, *Caerns.**, Flints.. 16 squares : **SR**99; **SS***78*,88; **ST**16,17,18,*27*,*28*,*29*; **SM**90; **SN**41,*69*; **SO***00*,51; **SH**57; **SJ***20*.

C. scorpioides (L.) Koch Casual (S. Europe). Waste places. Glam.. 4 squares : **SS***78*[140]; **ST***17*[112],*27*[112]; **SO***00*[140].

HIPPOCREPIS L.

Hippocrepis comosa L. : Pedol y March : Horseshoe Vetch. Native. Grassy cliff tops on limestone (D,G,Sc). Fl.5-7. *Mons.*, Glam., *Caerns.*. 11 squares : **SS**48,58,68,69, 78,87,96,97; **ST**17,*59*; **SH**78.

H. unisiliquosa L. Casual (S. Europe). Waste places. Glam.. 1 square : **ST***18*[112].

SCORPIURUS L.

Scorpiurus muricatus L. (incl. *S. sulcatus* L.; *S. subvillosus* L.). Casual (S. Europe). Waste places. Fl.6-9. Glam.. 5 squares : **ST**07[112],*16*[24],*17*[112],*18*[112],*27*[112].

HEDYSARUM L.

Hedysarum coronarium L. Casual (Mediterranean region). Waste places. Fl.7. Glam.. 4 squares : **ST***16*[112],*17*[112],*18*[24],*27*[112].

H. glomeratum F. G. Dietrich (*H. capitatum* Desf., non Burm. fil.). Casual (Mediterranean region). Waste places. Glam.. 1 square : **ST***17*[112]

ONOBRYCHIS Miller

Onobrychis viciifolia Scop. : Codog : Sainfoin. Denizen (S. England). Limestone grassland (G,R,Rw). Fl.6-7. Mons., Glam., *Brecs.**, *Carms.*, *Cards.**, *Mer.*, *Denbs.*, *Flints.*, *Angl.*. 24 squares, map 376.

ARACHIS L.

Arachis hypogaea L. : Peanut. Casual (S. America). Waste places. Glam.. 2 squares : **ST**17[112],*27*[24].

LIMNANTHACEAE

LIMNANTHES R. Br.

Limnanthes douglasii R. Br. : Poached-egg-flower. Casual (N. America). Waste places. Fl.5. Glam., *Rads.*. 3 squares : **ST***17*[112]; **SN**71[157]; **SO***07*[229].

OXALIDACEAE

OXALIS L.

Oxalis corniculata L. : Suran-y-Coed Felen Orweddol : Procumbent Yellow-sorrel. Established alien (S. Europe). Roadsides (Q,R,Rg,Rw). Fl.6-8. Mons., Glam., *Brecs.**, Rads., Carms., Pembs., Cards., Monts., Mer., *Caerns.**, *Flints.*, *Angl.*. 47 squares, map 377.

[**O. stricta** L. (*O. dillenii* Jacq.). *Mons.*, *Glam.*, *Brecs.*, *Carms.*, *Cards.*, *Monts.*, *Mer.*, *Caerns.*, *Angl.*]

O. europaea Jordan (*O. stricta* auct., non L.) : Suran-y-Coed Felyn Unionsyth : Upright Yellow-sorrell. Established alien (N. America). Roadsides (E,R, Rg,Rw). Fl.6-9. Mons.*, Glam., Brecs.*, *Rads.**, Carms., *Pembs.**, *Cards.*, Monts.*, Mer., Caerns.*, *Angl.**. 21 squares, map 378.

O. articulata Savigny : Pink Oxalis. Established alien (S. America). Waste places and roadsides (E,R,S,Sc). Fl.(4-)6-7(-9). Glam., Carms., Pembs., *Mer.*, Caerns., *Angl.*. 19 squares : **SS**69,87,88; **SM**80,82,93; **SN***31*,40; **SH**12,23,24,28,32,33,49,58,61,68,78.

O. acetosella L. : Suran-y-Coed : Wood-sorrel. Native. Woods (E,H,M,Q,R,Sc,W). Fl.3-6. Mons., Glam., Brecs., Rads., Carms., Pembs., Cards., Monts., Mer., Caerns., Denbs., Flints., *Angl.*. 255 squares, all except : **SR**89; **SS**09,19,38,39,77; **ST**26,58; **SM**50,62,70,71,73, 83,84,94; **SN**15,25,36,59; **SO**41; **SH**13,29,36,51,98; **SJ**37.

O. corymbosa DC. : Small Pink Oxalis. Established alien (S. America). Waste places. *Glam.*. 2 squares. **Glam.:** nr Port Einon, SS48, QONK, 1974[157]; nr Gowerton, SS59, QONK, 1974[157].

[**O. violacea** L. *Carms.*]

O. pes-caprae L. : Bermuda-buttercup. Casual (S. Africa). Waste places. *Cards.*. 1 square : **SN**56[95].

O. incarnata L. : Pale Oxalis. Established alien (S. Africa). Roadsides. Fl.8. Mons.. 1 square. **Mons.:** Grosmont, SO42, RGE, 1980[112].

GERANIACEAE

GERANIUM L.

Geranium sanguineum L. : Pig yr Aran Rhuddgoch : Bloody Crane's-bill. Native. Rocky places and sand dunes (D,E,Q,Sc,W). Fl.5-8. Mons., Glam., Brecs., Rads., Carms., (Pembs.*, Cards.*), *Monts.*, Mer., Caerns., Denbs., Flints., Angl.. 57 squares, map 379.

G. pratense L. : Pig yr Aran y Weirglodd : Meadow Crane's-bill. Native. Pastures and roadsides (E,G, M,Q,R,Rg,Rw). Fl.5-9. Mons., Glam., Brecs., Rads., Carms., (*Pembs.*, *Cards.*), *Monts.*, (Mer., Caerns.), Denbs., Flints., (*Angl.*). 107 squares, map 380.

G. sylvaticum L. : Pig yr Aran y Goedwig : Wood Crane's-bill. Native. Woods (M,Q,W). Fl.6-8. Mons., *Brecs.*, Rads.*, (*Monts.*), [*Glam.*, *Pembs.*]. 7 squares. **Mons.:** Grwyne Valley, SO22, Pa, 1916[112]. **Brecs.:** nr Capel-y-ffin, SO23, MP, 1970+[200]. **Rads.:** Water-break-its-neck, SO16, AL, 1885[107]; Radnor Forest, SO16, ACP, 1971[112]. (**Monts.:** nr Cemais, SH80, VJM, 1950+[116]; nr Llanwddyn, SJ01, IP, 1979[61]; Pennant Melangell, SJ02, WWh, 1866[107]; Llanfair Caereinion, SJ10, JAW, 1931[228]).

G. endressii Gay : Troedrudd Ffrengig : French Crane's-bill. Established alien (S. Europe). Hedges (E,G,R,Rw). Fl.6-8(-10). Mons., Glam.*, *Brecs.**, *Carms.**, Pembs., Monts., Caerns.*, Flints., *Angl.**. 14 squares : **SS***19*; **ST**18; **SN**21,80; **SO**05,50,51; **SH**22,27, 32,75,90; **SJ**11,17.

G. endressii × versicolor Casual (garden origin). Hedges. Mons.. 2 squares : **ST**19[112]; **SO**42[112].

G. versicolor L. : Pig yr Aran Llinellgoch : Pencilled Crane's-bill. Established alien (S. Europe). Hedges (E,G,R,Rw). Fl.5-9. Mons., Glam., *Brecs.**, Carms., Pembs., *Cards.*, Monts., *Caerns.*, Flints., Angl.*. 25 squares, map 381.

G. nodosum L. : Pig yr Aran Clymog : Knotted Crane's-bill. Established alien (S. Europe). Hedges. Fl.6-7. Glam., Brecs., Carms., *Monts.*, [*Pembs.*]. 7 squares.
Glam.: Baglan, SS79, JAW, 1932[220]; Cwrtyrala, ST17, EV, 1931[112]. **Brecs.:** nr Brecon, SO02, DPW, 1945[112]. **Carms.:** Gelli Aur, SN51, RDP, 1980[201]; Taliaris, SN62, MB, 1950[112]. **Monts.:** Bontdolgadfan, SH80, JAW, c1940[228]; Bwlchycibau, SJ11, JAW, c1940[228].

G. ibericum Cav. × **platypetalum** Fischer & C. A. Meyer = **G. × magnificum** Hyl. (*G. ibericum* auct.) : Purple Crane's-bill. Established alien (garden origin). Waste places (E,R,Rw). Fl.6. Mons., Glam.. 9 squares. **Mons.:** Cwmyoy, SO22, SGC, 1942[112]; Abergavenny, SO31, GCD, 1924[23]. **Glam.:** West Cross, SS68, JAW, 1942[112]; Jersey Marine, SS79, JAW, 1939[112]; Cowbridge, SS97, AEW & ABP, 1975[95]; Rhyd Lafr, ST18, RGE & AEW, 1974[112]; nr Garn-swllt, SN61, RGE, 1981[112]; Llwydcoed, SN90, AEW & ABP, 1974[95]; Aberdare, SO00, AMP, 1974[112].

G. phaeum L. : Pig yr Aran Dulwyd : Dusky Crane's-bill. Established alien (C. and S. Europe). Hedges (E,R,W). Fl.4-8. Mons., Glam., Brecs., Rads., Carms., *Pembs.**, Cards., Monts., Mer., *Denbs.*, Flints., *Angl.*. 67 squares, map 382.

G. pyrenaicum Burm. fil. : Pig yr Aran y Gwrych : Hedgerow Crane's-bill. Denizen (Europe). Hedges and roadsides (D,E,G,M,Q,R,W). Fl.5-8. Mons., Glam., Brecs., *Rads.**, Carms.*, Pembs., Cards., Monts., Mer., Caerns., Denbs., Flints., Angl.. 80 squares, map 383.

G. rotundifolium L. : Pig yr Aran Crynddail : Round-leaved Crane's-bill. Denizen (S. England). Roadsides and waste places (E,G,Q,R,Rw). Fl.5-10. Mons., Glam., Brecs.*, Carms., Pembs.*, (Mer.*, Caerns.*, *Flints.*), [*Denbs.*]. 24 squares, map 384.

G. molle L. : Troed y Golomen : Dove's-foot Crane's-bill. Native. Roadsides and dry pastures (D,E,F,G,M,Q,R,Rw,S). Fl.4-10. Mons., Glam., Brecs., Rads., Carms., Pembs., Cards., Monts., Mer., Caerns., Denbs., Flints., *Angl.*. 242 squares, map 385.

G. pusillum L. : Pig yr Aran Mânflodeuog : Small-flowered Crane's-bill. Native. Limestone pastures, also widespread as an introduction in cultivated ground and waysides (E,F,G,R,Rw,W). Fl.5-9. Mons., Glam., Brecs.*, *Rads.**, *Carms.*, Pembs., *Cards.*, *Monts.*, Mer., Caerns., *Denbs.*, *Flints.*, *Angl.*. 65 squares, map 386.

G. columbinum L. : Pig yr Aran Hirgoesog : Long-stalked Crane's-bill. Native. Hedges and banks (E,G,M, Q,R,Rw,Sc,W). Fl.5-10. Mons., Glam., Brecs., Rads., Carms., Pembs., Cards., Monts., Mer., Caerns., Denbs., Flints., Angl.. 150 squares, map 387.

G. dissectum L. : Pig yr Aran Llarpiog : Cut-leaved Crane's-bill. Native. Roadsides (E,G,H,M,Q,R, Rg,Rw,W). Fl.5-8. Mons., Glam., Brecs., Rads., Carms., Pembs., *Cards.*, Monts., Mer., Caerns., Denbs., Flints., Angl.. 245 squares, map 388.

G. lucidum L. : Pig yr Aran Disglair : Shining Crane's-bill. Native. Walls and hedge banks (E,Q, R,W). Fl.4-8. Mons., Glam., Brecs., Rads., Carms., Pembs., Cards., Monts., Mer., Caerns., Denbs., Flints., Angl.. 155 squares, map 389.

G. robertianum L.

Subsp. **robertianum** : Llys y Llwynog : Herb-Robert. Native. Woods and hedges (D,E,G,M,Q, R,Rw,W). Fl.3-10. Mons., Glam., Brecs., Rads., Carms., Pembs., Cards., Monts., Mer., Caerns., Denbs., Flints., Angl.. 274 squares, all except : **SS**39; **SM**50,62,71,73; **SN**59; **SH**51; **SJ**37.

Subsp. **celticum** Ostenf. (*G. robertianum* var. *celticum* (Ostenf.) Druce). Native. Limestone rocks (Q,Sc). Fl.5-6. *Glam.*, *Brecs.*, Carms., Pembs.. 7 squares.
Glam.: nr Oxwich, SS48, EFW, 1961[7]; Three Cliffs Bay, SS58, HGB, 1946[235]; nr Mountain Ash, ST09, AMP, 1970+[95]. **Brecs.:** Erwood, SO04, MSW, 1926[235]. **Carms.:** Llyn Llech Owen, SN51, DH, 1922[23]; Carmel, SN61, IMV, 1963[112]. **Pembs.:** Penally, SS19, MB, 1963[112].

Subsp. **maritimum** (Bab.) H. G. Baker Native. Shingle beaches (S,Sc). Fl.6-7. Glam., *Carms.**, Pembs., *Cards.*, Mer., *Denbs.*. 15 squares : **SS**19, *48*,*58*,*59*,*68*,*78*,87; **ST**06,*16*,26; **SN**31,58; **SH**50, 61,97.

G. purpureum Vill. : Llys Robert Bychan : Little Robin. Native. Hedges. *Carms.*, [*Glam.*, *Brecs.*]. 1 square.
Carms.: nr Newcastle Emlyn, SN34, JFGC & TGP, 1934[33]. This record may be referable to **G. purpureum** × **robertianum.**

ERODIUM L'Hér.

Erodium chium (L.) Willd. Casual (Mediterranean region). Waste places. Glam.. 1 square : **ST***27*[112].

E. laciniatum (Cav.) Willd. Casual (Mediterranean region). Waste places. *Glam.*. 1 square : **ST***27*[25].

E. malacoides (L.) L'Hér. Casual (Mediterranean region). Waste and cultivated ground. Glam.. 1 square : **ST***27*[112].

E. maritimum (L.) L'Hér. : Pig y Crëyr Arfor : Sea Stork's-bill. Native. Grassy cliff tops and fixed dunes (D,E,G,S,Sc). Fl.5-9. Glam., *Carms.*, Pembs., Cards., Mer., Caerns., *Denbs.**, *Flints.*, Angl., [*Mons.*, *Monts.*]. 54 squares, map 390.

E. botrys (Cav.). Bertol. : Mediterranean Stork's-bill. Casual (Mediterranean region). Waste places. *Glam.*. 1 square : **ST***16*[24].

E. cicutarium (L.) L'Hér.

Subsp. **cicutarium** (incl. *E. cicutarium* subsp. *dunense* Andreas) : Pig y Crëyr Cegidaidd : Common Stork's-bill. Native. Sand dunes, grassland and cultivated ground (D,E,F,G,Q,R,Rg,Rw). Fl.4-9. Mons., Glam., *Brecs.*, Rads., Carms., Pembs., Cards., Monts., Mer., Caerns., Denbs., Flints., *Angl.*. 136 squares, map 391.

Subsp. **bipinnatum** Tourlet (*E. glutinosum* Dumort.) : Sticky Stork's-bill. Native. Sand dunes (D,G,R,Rw,S). Fl.6-9. Glam., Carms., *Pembs.*, Cards.*, Mer., Caerns., Angl.. 28 squares, map 392.

Subsp. **anaristatum** Andreas (subsp. *bipinnatum* × subsp. *cicutarium*). Spontaneous. Sand dunes (D,W). Fl.6-9. Glam., Carms., Mer., *Caerns.*, Angl.. 7 squares.
Glam.: Newton Burrows, SS87, HJR, 1904[107] and AEW, 1925[112]. **Carms.:** Tywyn Burrows, SN30, IMV, 1960[112] & 1970[112]. **Mer.:** Morfa Dyffryn, SH52, PMB, 1961[4]; Morfa Harlech, SH53, PMB, 1961[112]. **Caerns.:** Morfa Dinlle, SH45, RHR, c1961[176]. **Angl.:** Aberffraw Common, SH36, RHR, 1970+[209]; Newborough Warren, SH46, RHR, 1970+[209].

E. moschatum (L.) L'Hér. : Pig y Crëyr Mwsgaidd : Musk Stork's-bill. Native. Sea cliffs (D,E,G,Q,R,S,Sc). Fl.4-8. (Mons.), Glam., Carms., Pembs., Cards., *Mer.*, Caerns., (*Denbs.*), Flints., *Angl.*, [*Rads.*]. 39 squares, map 393.

E. acaule (L.) Becherer and Thell. (*E. romanum* (Burm. fil.) L'Hér.). Casual (Mediterranean region). Waste places. Fl.6. Glam.. 1 square : **ST***16*[112].

[**E. cygnorum** Nees *Glam.*]

TROPAEOLACEAE

TROPAEOLUM L.

Tropaeolum majus L. : Nasturtium. Casual (S. America). Waste places. *Mons.*, *Glam.*, *Rads.*, Pembs., *Monts.*, *Caerns.*, *Angl.*. 11 squares : **SS***69*; **ST***17*,38; **SM**93; **SO**09,26; **SH**23,*67*; **SJ**10,19,21.

T. minus L. Casual (S. America). Waste places. *Glam.*. 2 squares : **SS**97[95]; **ST**08[95].

ZYGOPHYLLACEAE

TRIBULUS L.

Tribulus terrestris L. Casual (S. Europe). Waste places. Glam.. 1 square : **ST***27*[112].

LINACEAE

LINUM L.

Linum bienne Miller : Llin Culddail : Pale Flax. Native. Banks and cliff tops on limestone (F,G,R,Sc). Fl.5-7. Mons., Glam., *Rads.*, Carms., Pembs., *Cards.*, *Caerns.*, Denbs., Flints., Angl.. 54 squares, map 394.

L. usitatissimum L. : Llin Amaeth : Flax. Casual (garden origin). Waste places and cultivated ground (E,F,R,Rw). Fl.6-9. Mons., Glam., *Brecs.*, *Rads.*, Carms., *Pembs.*, *Cards.*, *Monts.*, Mer., Caerns., *Denbs.*, *Flints.*, *Angl.*. 53 squares, map 395.

L. catharticum L. : Llin y Tylwyth Teg : Fairy Flax. Native. Pastures and heaths (D,Ds,G,H,M,Q,R,Rw,Sc). Fl.5-9. Mons., Glam., Brecs., Rads., Carms., Pembs., Cards., Monts., Mer., Caerns., *Denbs.*, Flints., Angl.. 241 squares, map 396.

RADIOLA Hill

Radiola linoides Roth : Gorhilig : Allseed. Native. Damp heaths (D,Ds,G,H,M,R). Fl.5-9. Glam., *Carms.*, Pembs., Cards., *Monts.*, Mer., Caerns., *Denbs.*, Flints., Angl., [*Brecs.*]. 57 squares, map 397.

EUPHORBIACEAE

MERCURIALIS L.

Mercurialis annua L. : Clais yr Hydd Blynyddol : Annual Mercury. Established alien (S. Europe). Cultivated ground (Q,R,Rg,Rw). Fl.(4-)6-10(-12). Mons., Glam., Brecs., Rads., Carms., Pembs., Cards., Mer., Caerns., *Flints.**, Angl.. 44 squares, map 398.

M. perennis L. : Bresych y Cŵn : Dog's Mercury. Native. Woods (E,M,Q,W). Fl.2-5. Mons., Glam., Brecs., Rads., Carms., Pembs., Cards., Monts., Mer., Caerns., Denbs., Flints., *Angl.*. 247 squares, map 399.

RICINUS L.

Ricinus communis L. : Castor-oil-plant. Casual (Tropics). Waste places. *Mons.*, Glam.. 3 squares : **ST***16*[112],*17*[21],38[42].

EUPHORBIA L.

Subgen. Chamaesyce Rafin.

Euphorbia peplis L. : Fflamgoed Ruddlas : Purple Spurge. Native. Shingle beaches. *Cards.*†, [*Glam.*, *Mer.*]. 1 square.
Cards.: Aberystwyth, SN58, E, 1805[140].

E. maculata L. Established alien (S. Europe). Cultivated ground. Fl.7. Carms.†. 1 square.
Carms.: Bynea, SS59, RDP, 1976[112].

Subgen. Esula Pers.

E. villosa Waldst. & Kit. ex Willd. (*E. pilosa* auct. eur., non L.) : Hairy Spurge. Casual (Europe). Waste places. Fl.8. Glam.. 1 square : **ST**17[112].

E. corallioides L. : Coral Spurge. Casual (Mediterranean region). Waste places (R,Rg). Mons., Glam.. 2 squares : **SS**87[112]; **ST**59[91].

[**E. palustris** L. *Glam.*]

E. ceratocarpa Ten. Casual (Mediterranean region). Waste places. Fl.5-6. Glam.. 1 square : **ST***16*[112].

E. dulcis L. : Sweet Spurge. Established alien (Europe). Roadsides. Fl.4-9. Brecs., Caerns., Denbs.. 5 squares.
Brecs.: nr Llangorse, SO12, EV, 1945[112] and ARP, 1972[112]. **Caerns.:** Llanberis, SH55, HDr, c1910[83]; Penisa'r-waun, SH56, JAWhe & JRa, 1946[112] and LJL, 1977[184]; Capel y Graig, SH56, JRa, 1946[112]; Bangor, SH57, RHR, 1972[208]. **Denbs.:** nr Llansilin, SJ22, EJ, 1864[107] and 1888[112].

E. platyphyllos L. : Fflamgoed Lydanddail : Broad-leaved Spurge. Established alien (England). Cultivated ground (F,Rg,Sc). Fl.6-8. *Mons.*, Glam.. 4 squares.
Mons.: Caldicot, ST48, CT, 1979[91]; Chepstow, ST59, TGE, 1980[91]. **Glam.:** Penarth, ST16, AEW, 1940[112]; Lavernock, ST16, AEW, 1949[112] and JPC, 1978[112]; Ely, Cardiff, ST17, RLS, 1947[112].

E. serrulata Thuill. (*E. stricta* L., nom. illegit.) : Upright Spurge. Native. Woods (E,Q,R,W). Fl.5-8. Mons., [*Glam.*]. 5 squares.
Mons.: Slade Wood, ST48, CT, 1979[91]; Black Cliff & Wynd Cliff, ST59, RL, 1944[112] and TGE, 1980[112]; Chepstow, ST59, TGE, 1980[91]; Mitchel Troy, SO41, ST, 1980[91]; nr Bigswier, SO50, WAS, 1878[112]; Llandogo, SO50, WAS, 1901[112]; nr Tintern, SO50, CR, 1943[108]; nr Monmouth, SO51, SGC, 1957[7].

E. helioscopia L. : Llaeth Ysgyfarnog : Sun Spurge. Colonist (England). Cultivated ground (F,G,R,Rg). Fl.7-9. Mons., Glam., Brecs., Rads., Carms., Pembs., Cards., *Monts.*, Mer., Caerns., Denbs., Flints., Angl.. 220 squares, map 400.

E. lathyris L. : Fflamgoed Gaperol : Caper Spurge. Naturalized alien (S. Europe). Woods (E,G,R,Rg,W). Fl.4-7. Mons., Glam., Brecs.*, Rads., Carms., Pembs., *Cards.*, *Monts.*, *Mer.**, Caerns., *Denbs.*, *Flints.*, Angl.*. 45 squares, map 401.

E. exigua L. : Fflamgoed Eiddil Flaenfain : Dwarf Spurge. Colonist (England). Cultivated ground (F,G,Q,R). Fl.6-10. Mons., Glam., Brecs., *Rads.*, *Carms.*, Pembs., *Cards.*, Monts., *Caerns.*, Denbs., Flints., Angl., [*Mer.*]. 63 squares, map 402.

E. falcata L. Casual (Europe). Waste places. *Glam.*. 1 square : **ST***17*[22].

E. peplus L. : Llaeth y Cythraul : Petty Spurge. Colonist (England). Cultivated ground (D,E,F,Q,R, Rg). Fl.3-10. Mons., Glam., Brecs., Rads., Carms., Pembs., Cards., Monts., Mer., Caerns., Denbs., Flints., Angl.. 185 squares, map 403.

E. portlandica L. : Llaethlys Portland : Portland Spurge. Native. Sand dunes (D,S,Sc). Fl.5-8. Glam., Carms., Pembs., Cards., Mer., *Caerns.*, *Denbs.*, Flints., Angl.. 40 squares, map 404.

E. paralias L. : Llaethlys y Môr : Sea Spurge. Native. Sand dunes (Ar,D,S). Fl.5-8. Glam., Carms., Pembs., Cards., Mer., Caerns., Denbs., Flints., Angl.. 51 squares, map 405.

E. paralias × **portlandica** Spontaneous. Sand dunes. *Glam.*, *Mer.*, *Angl.*. 4 squares.
Glam.: Margam Moors, SH78, HJR, pre 1907[140]. **Mer.:** Morfa Dyffryn, SH52, PMB, 1950+[4]; Morfa Harlech, SH53, PMB, 1950+[4]. **Angl.:** Newborough Warren, SH46, PMB, 1950+[217].

[**E. esula** L. *Mons.*, *Glam.*, *Brecs.*, *Carms.*, *Flints.*]

E. esula × **uralensis** = **E.** × **pseudovirgata** (Schur) Soó : Hungarian Spurge. Established alien (Europe). Waste places. Fl.5-9. Mons., Glam., Brecs., *Carms.*. 9 squares.
Mons.: nr Monmouth, SO51, AL, 1904[112]. **Glam.:** Aberafan, SS79, HJR, 1909[107]; Aberthaw, ST06, RGE, 1979[112]; Barry, ST16, RLS, 1922[112] and AEW & ABP, 1974[112]; Cardiff, ST17, PHH, 1926[112]; Radyr, ST18, RLS, 1923[22]; Cardiff Docks, ST27, JPC, 1981[75]. **Brecs.:** Taf Fechan, SO01, MG, 1971[112]. **Carms.:** Llanelli, SS59, DH, 1911[161].
There is some confusion over the identity of this taxon. It is usually found as an escape from gardens.

[**E. uralensis** Fishcher ex Link (*E. esula* subsp. *tommasiniana* (Bertol.) Nyman). *Glam.*]

E. cyparissias L. : Fflamgoed Gyprysol : Cypress Spurge. Established alien (Europe). Roadsides (D,E,R, Rg,Rw). Fl.5-7. Glam., *Brecs.**, Carms., Pembs., *Cards.*, Monts., Caerns.*, *Denbs.*, Flints.. 22 squares, map 406.

E. cyparissias × **esula** = **E.** × **pseudoesula** Schur Introduced (garden origin). Garden weed. Fl.5-6. Brecs., Pembs.. 2 squares.
Brecs.: Dyffryn Callwen, SN81, AL, 1904[112]. **Pembs.:** Tenby, SN10, FLR, 1946[112].

E. amygdaloides L. : Llaethlys y Coed : Wood Spurge. Native. Woods (E,Q,Rw,Sc,W). Fl.4-6. Mons., Glam., *Brecs.*, Carms., Pembs., *Cards.*, *Monts.*, *Mer.*, Caerns., Denbs.*, *Flints.**, *Angl.**. 63 squares, map 407.

RUTACEAE

RUTA L.

Ruta graveolens L. : Rue. Casual (Mediterranean region). Limestone rocks. Fl.8. *Denbs.*. 1 square : **SJ***15*[107].

POLYGALACEAE

POLYGALA L.

Polygala vulgaris L. (incl. *P. oxyptera* Reichenb.) : Llysiau Crist : Common Milkwort. Native. Pastures, hedgebanks and sand dunes (D,Ds,E,G,H,Q,R,Rw, S,Sc). Fl.4-9. Mons., Glam., Brecs., Rads., Carms., Pembs., Cards., *Monts.*, Mer., Caerns., *Denbs.*, Flints., *Angl.*. 215 squares, map 408.

P. serpyllifolia J. A. C. Hose : Llysiau'r Groes : Heath Milkwort. Native. Heaths (D,E,F,G,H,M,Q,W). Fl.5-7(-9). Mons., Glam., Brecs., Rads., Carms., Pembs., Cards., Monts., *Mer.*, Caerns., Denbs., Flints., *Angl.*. 227 squares, map 409.

ANACARDIACEAE

RHUS L.

Rhus typhina L. : Stag's-horn Sumach. Casual (N. America). Railways and waste places. *Mons.*, *Glam.*. 2 squares : **ST**27[112],38[91].

ACERACEAE

ACER L.

Acer platanoides L. : Norway Maple. Introduced (Europe), sometimes becoming naturalized. Woods. Fl.4-5. *Glam.*, *Carms.*, *Pembs.*, Cards., Monts., *Mer.*, *Caerns.*, *Denbs.*, Flints.. 28 squares, map 410.

A. campestre L. : Masarnwydd Lleiaf : Field Maple. Native. Woods and hedges (E,G,M,Q,R,W). Fl.5-6. Mons., Glam., Brecs., Rads., Carms., *Pembs.*, Cards., Monts., Mer., Caerns., Denbs., Flints., (Angl.). 160 squares, map 411.

A. pseudoplatanus L. : Masarnen : Sycamore. Naturalized alien (C. and S. Europe). Woods (D,E,G,H,M, Q,R,Sc,W). Fl.4-5. Mons., Glam., Brecs., Rads., Carms., Pembs., Cards., Monts., Mer., Caerns., Denbs., Flints., *Angl.*. 270 squares, all except : **SR**89; **SS**38,39,77; **SM**50,62,73,84,94; **SN**15; **SH**51; **SJ**37.

A. negundo L. : Box-elder. Casual (N. America). Woods. *Glam.*. 1 square : **ST**18[95].

SAPINDACEAE

KOELREUTERIA Medicus

Koelreuteria paniculata Laxm. Casual (E. Asia). Waste places. Mons.. 1 square : **SO**50[112].

HIPPOCASTANACEAE

AESCULUS L.

Aesculus hippocastanum L. : Castanwydden y Meirch : Horse-chestnut. Established alien (S. Europe). Roadsides (E,F,R,W). Fl.4-6. *Mons.*, *Glam.*, Brecs., *Rads.*, *Carms.*, *Pembs.*, Cards., Monts., Mer., *Caerns.*, Denbs., *Flints.*, *Angl.*. 126 squares, map 412.

BALSAMINACEAE

IMPATIENS L.

Impatiens noli-tangere L. : Ffromlys : Touch-me-not Balsam. Native. Moist woods (E,W). Fl.8-9. (Mons.), *Monts.*, Mer., *Caerns.*[230], *Denbs.*, *Flints.*[230], [*Glam.*, *Pembs.*]. 9 squares.
Mons.: Piercefield Woods, ST59, WAS, 1888[112]. **Monts.:** Marrington Dingle, SO29, VJM, 1950+[116] and RGW, 1976[259]; Leighton Woods, SJ20, VJM, 1950+[116]. **Mer.:** Barmouth, SH61, WW, 1833[112]; Dolgellau, SH71, CW, 1879[112]; Hengwrt, SH71, JAW, 1940[112] and PMB, 1978[3]; Nannau Hall, SH72, EV, 1908[112] and PMB, 1954[3]. **Denbs.:** nr Chirk, SJ23, AAD, c1910[144]; Llangollen, SJ24, TR, 1906[146]; Erbistock, SJ34, AAD, c1910[144].

I. capensis Meerb. : Orange Balsam. Naturalized alien (N. America). River banks. Fl.7-9. Mons., Glam., *Flints.*. 5 squares.
Mons.: nr Dyffryn, ST28, AEW, 1921[112]; Coedkernew, ST28, AEW, 1934[112]; Newport, ST38, AMc, 1924[227]. **Glam.:** nr Newton, SS97, GRW, 1931[112]; nr Llansanwyr, SS97, GRW, 1931[112]; nr Cowbridge, SS97, ABP & AEW, 1969[112]; Gigman, ST07, EV, 1908[112] and ABP & AEW, 1969[112]; Ffynon Deilo nr Pendoylan, ST07, RJHL, 1959[112] and ABP, 1981[112]. **Flints.:** nr Rhyl, SJ08, WH, c1910[144].

I. parviflora DC. : Ffromlys Melyn Bychanflodeuog : Small Balsam. Naturalized alien (C. Asia). Shady river banks (E,M,R,W). Fl.8-9. Glam., *Rads.**, Carms.*, Monts.*, Mer., Denbs., *Angl.*. 12 squares : **SS**79; **SN**50, 60,61,79; **SO**24,26,36; **SH***57*,70,*71;* **SJ**35.

I. glandulifera Royle : Ffromlys Chwarennog : Indian Balsam. Naturalized alien (Himalaya). River banks (Ar,G,M,Q,R). Fl.7-10. Mons., Glam., Brecs., Rads., Carms., Pembs., Cards., Monts., Mer., Caerns., Denbs., Flints., *Angl.*. 114 squares, map 413.

AQUIFOLIACEAE

ILEX L.

Ilex aquifolium L. : Celynnen : Holly. Native. Woods and hedges (E,G,H,M,Q,R,Sc,W). Fl.3-5. Mons., Glam., Brecs., Rads., Carms., Pembs., Cards., Monts., Mer., Caerns., Denbs., Flints., Angl.. 257 squares, all except : **SR**89; **SS**38,39,77; **ST**26; **SM**50, 62,70,71,72,73,84,94; **SN**15,36,59,86,87; **SO**17; **SH**13, 24,29,51,98; **SJ**37.

CELASTRACEAE

EUONYMUS L.

Euonymus europaeus L. : Piswydden : Spindle. Native. Woods and hedges (D,E,Q,R,Sc,W). Fl.4-7. Mons., Glam., Brecs., Rads., Carms., Pembs., Cards., Monts., Mer., Caerns., Denbs., Flints., Angl.. 153 squares, map 414.

E. japonicus L. fil. Casual (Japan). Waste places (D,E,Q,R,Sc). *Glam.*, Carms.. 8 squares : **SS**58[99],68[99],69[222],87[94]; **ST**09[95],17[112]; **SN**40[112],50[201].

STAPHYLEACEAE

STAPHYLEA L.

Staphylea pinnata L. : Dagrau Addaf : Bladder-nut. Casual (C. and S. Europe). Hedges and scrub. Fl.6. *Glam.*, Cards., Denbs.. 3 squares : **ST***18*[218]; **SN**34[112]; **SJ**24[112].

BUXACEAE

BUXUS L.

Buxus sempervirens L. : Bocyswydden : Box. Established alien (S. England). Hedges and scrub (E,R,W). Fl.3-5. *Mons.*, Glam., Brecs., Rads., Carms., Pembs., *Cards.*, Monts., Mer., Caerns., Denbs., Flints., Angl.. 44 squares, map 415.

RHAMNACEAE

RHAMNUS L.

Rhamnus alaternus L. Naturalized alien (Mediterranean region). Woods and sea cliffs. Fl.3. Glam., Caerns., *Denbs.*. 3 squares.
Glam.: Penarth, ST17, AEW, 1943[112]. **Caerns.:** Gt Ormes Head, SH78, EE, 1980[264]; nr Llandudno Junction, SH87, RKB, 1960[112] and DM, 1974[107]. **Denbs.:** nr Colwyn Bay, SH87, JMB, 1970+[67].

R. catharticus L. : Rhafnwydden : Buckthorn. Native. Hedges (D,E,R,Sc,W). Fl.5-6. Mons., Glam., Brecs., *Rads.**, Carms., *Monts.*, Caerns., Denbs., Flints., *Angl.*. 46 squares, map 416.

FRANGULA Miller

Frangula alnus Miller : Breuwydd : Alder Buckthorn. Native. Marshy thickets (E,H,M,R,Rw,W). Fl.6-9. Mons., Glam., Brecs., *Rads.**, Carms., Cards., Mer., *Caerns.*, Denbs., *Flints.*, [*Angl.*]. 62 squares, map 417.

VITACEAE

VITIS L.

Vitis vinifera L. : Gwinwydden : Grape-vine. Casual (Europe). Waste places. *Mons.*, *Glam.*. 4 squares : **ST**09[95],*16*[24],*27*[24],38[245]. Cultivated now or formerly in *Glam.*, *Pembs.*, *Cards.*, and perhaps elsewhere.

PARTHENOCISSUS Planchon

Parthenocissus quinquefolia (L.) Planchon : True Virginia-creeper. Casual (N. America). Waste places. Glam., Carms., Cards., Caerns.[119]. 3 squares : **SS**69[22]; **SN**53[225],58[7].

P. inserta (A. Kerner) Fritsch : Common Virginia-creeper. Casual (N. America). Roadsides and sand dunes (D,E,G,Q,R). Glam., Mer.. 8 squares : **SS48**[112],58[112],68[117], 69[117],99[95]; **ST09**[95], 16[112]; **SH71**[112].

P. tricuspidata (Siebold & Zucc.) Planchon : Virginia-creeper. Casual (China & Japan). Waste ground (M,R). Glam., *Pembs.*, *Monts.*, [*Mer.*]. 6 squares : **SS58**[7],69[221]; **SM91**[7]; **SN02**[93]; **SO09**[33],19[33].

TILIACAEAE

TILIA L.

Tilia platyphyllos Scop. : Pisgwydden Deilen Fawr : Large-leaved Lime. Denizen (Europe). Woods (E,M,Q,R,W). Fl.6-9. Mons., Glam.*, Brecs., Rads., *Carms.*, Cards.*, Monts., Mer., Caerns., Denbs., Flints.. 33 squares, map 418. (Possibly native in eastern parts of *Mons.* and *Brecs.*).

T. cordata Miller : Pisgwydden Deilen Fach : Small-leaved Lime. Native. Woods (E,M,Q,R,W). Fl.7-8. Mons., Glam., Brecs., Rads., Carms., Cards., Monts., Mer., Caerns., Denbs., Flints., (*Angl.*). 92 squares, map 419.

T. cordata × platyphyllos = T. × vulgaris Hayne (*T.* × *europaea* L. *p.p.*) : Gwaglwyfen : Lime. Naturalized alien (Europe). Woods and hedges (E,M,Q,R,W). Fl.7.Mons., Glam., Brecs., Rads., Carms., *Pembs.*, Cards., Monts., Mer., Caerns., Denbs., Flints., Angl.. 111 squares, map 420.

MALVACEAE

SIDA L.

Sida spinosa L. Casual (S. America). Waste places. *Glam.*. 1 square : **ST***17*[218].

MALVA L.

Malva hispanica L. Casual (S. Europe). Waste places (R,Rg). Fl.8. Glam.. 1 square : **ST***27*[112].

M. moschata L. : Hocysen Fws : Musk Mallow. Native. Hedge banks (D,E,G,Q,R,Rw,W). Fl.5-9. Mons., Glam., Brecs., Rads., Carms., Pembs., *Cards.*, Monts., Mer., Caerns., Denbs., Flints., *Angl.*. 192 squares, map 421.

M. sylvestris L. : Hocysen Gyffredin : Common Mallow. Denizen (England). Waste places (D,E,G, M,Q,R,Rg,Sc). Fl.5-10. Mons., Glam., Brecs., Rads., Carms., Pembs., Cards., Monts., Mer., Caerns., Denbs., Flints., Angl.. 196 squares, map 422.

M. nicaeensis All. Casual (Mediterranean region). Waste places (R,Rg). Fl.7-9. Glam.. 2 squares : **ST***16*[24],*27*[112].

M. parviflora L. : Least Mallow. Casual (Mediterranean region). Waste places. Fl.8. Glam., *Caerns.*. 4 squares : **ST***16*[26],17[112],27[112]; **SH***67*[107].

M. pusilla Sm. : Hocys Blodau Bychan : Small Mallow. Introduced (Europe). Waste places (R,Rg). Fl.8-10. Mons., Glam., *Pembs.*. 7 squares.
Mons.: Newport Docks, ST38, JH, 1975[112]; Rogiet, ST48, WAS, 1910[112]. **Glam.:** St Hilary, ST07, ISN 1908[112]; Cardiff, ST17, RLS, 1935[112]; Splott, Cardiff, ST27, RLS, 1926[112] and EV, 1929[112]; Aberdare, SO00, HJR, 1901[112]. **Pembs.:** Tenby, SN10, EST, 1929[28].

M. neglecta Wallr. : Hocys Bychan : Dwarf Mallow. Denizen (England). Waste places (E,F,R,Rw,S). Fl.6-10. Mons., Glam., Brecs., Rads.*, Carms., Pembs., *Cards.*, *Monts.*, Mer., Caerns., Denbs., Flints., *Angl.*. 87 squares, map 423.

M. verticillata L. : Chinese Mallow. Casual (E. Asia). Cultivated ground. *Carms.*. 1 square : **SS***59*[107].

LAVATERA L.

Lavatera cretica L. : Môr Hocysen Bychan : Smaller Tree-mallow. Established alien (S. England). Waste places near the sea (G,R,Sc). Fl.7-9(-10). Glam., Pembs., *N. Wales*[107]. 5 squares.
Glam.: Llantwit Major, SS96, MG, 1974[99]; Cardiff, ST17, HJR, c1909[142] and RLS, 1938[112]†; Splott, Cardiff, ST27, RLS & AEW, 1929[37]. **Pembs.:** Nolton Haven, SM81, CCJ, 1970[181]; Tenby, SN10, EV, 1941[112] and JRP, 1970[181]. **N. Wales.:** no locality, JEG, 1888[107].

L. arborea L. : Hocyswydden : Tree-mallow. Native or denizen. Sea cliffs (E,Q,R,Sc). Fl.5-9. *Mons.**, Glam., Carms., Pembs., Cards., Mer., *Caerns.*, *Denbs.*, *Flints.*, Angl.. 71 squares, map 424.

L. punctata All. Casual (S. Europe). Waste places. Fl.8. Glam.. 1 square : **ST***27*[112].

ALTHAEA L.

Althaea hirsuta L. : Hispid Mallow. Casual (Europe). Waste places (F,M,Q,R,Rw,W). Fl.6-9. Glam., *Carms.*, *Pembs.*, *Denbs.*. 6 squares : **ST***17*[112],27[112]; **SN***10*[18],62[228],*90*[112]; **SJ***23*[144].

A. officinalis L. : Hocys y Morfa : Marsh-mallow. Native. Salt-marshes (Ar,D,E,M,Q,S). Fl.7-9. Mons., Glam., Carms., Pembs., *Caerns.*, (*Denbs.*), [*Angl.*]. 27 squares, map 425.

ALCEA L.

Alcea rosea L. (*Althaea rosea* (L.) Cav.) : Hocysen Bendigaid : Hollyhock. Casual (garden origin). Waste places (R,Rg,Rw). Fl.7. *Mons.*, Glam.. 5 squares : **SS***68*[19],69[222]; **ST16**[112],17[112], 59[99].

ABUTILON Miller

Abutilon theophrasti Medicus (*A. avicennae* Gaertner) : Velvetleaf. Casual (S.E. Europe). Waste places (E,F,R). Fl.10. Mons., Glam.. 3 squares : **ST08**[112],49[184]; **SO40**[112].

HIBISCUS L.

Hibiscus trionium L. Casual (Europe). Waste places. Fl.9. *Mons.*, Glam.. 2 squares : **ST***27*[112],38[245].

SIDALCEA A. Gray

Sidalcea malvaeflora (DC.) A. Gray ex Bentham Casual (N. America). Roadsides. Fl.6. Glam.. 1 square : **ST**08[112].

MALVASTRUM A. Gray

Malvastrum campanulata Nicholson Casual (Chile). Ditches. Fl.9. Denbs.. 1 square : **SJ**04[112].

THYMELAEACEAE

DAPHNE L.

Daphne mezereum L. : Bliwlys : Mezereon. Native. Woods (E,Q,W). Fl.2-4. Mons., (*Glam.*, Carms.*, *Monts.**, Caerns.). 6 squares.
Mons.: Hardwick, ST48, CT, 1961[91]; nr Chepstow, ST59, WAS, pre 1920[225]. (**Glam.:** Draethen, ST28, Gr, pre 1909[142]. **Carms.:** Crwbin, SN41, IMV & MH, 1971[112]. **Monts.:** Leighton, SJ20, JAW, c1940[228]. **Caerns.:** nr Bangor, SH57, JTCW, 1877[112] and JEG, 1878[112]).

D. laureola L. : Clust yr Ewig : Spurge-laurel. Native. Woods (E,W). Fl.1-4. Mons., Glam., *Rads.**, Carms.*, Pembs., (*Cards.*), Monts.*, (*Mer.*), Caerns., Denbs., Flints., Angl.. 51 squares, map 426.

ELAEAGNACEAE

HIPPOPHAE L.

Hippophae rhamnoides L. : Mŷr-rhafnwydd : Sea-buckthorn. Naturalized alien (England). Sand dunes (D,E,M,Q,R). Fl.3-5. Glam., Brecs.*, Carms., Pembs., Cards., Monts.*, Mer., Caerns., *Denbs.**, Flints.*, Angl.. 30 squares, map 427. Often planted originally.

ELAEAGNUS L.

Elaeagnus umbellata Thunb. : Oleaster. Introduced (Temperate Asia). Hedges. Fl.8-9. Pembs.. 1 square. **Pembs.:** Amroth, SN10, AMP, 1978[112].

GUTTIFERAE

HYPERICUM L.

Hypericum calycinum L. : Rhosyn Saron : Rose-of-Sharon. Established alien (S.E. Europe). Plantations, woods, etc. (E,Q,Rw,W). Fl.5-11. *Mons.**, Glam., *Brecs.**, Rads., *Carms.*, Pembs., Cards., Monts., Mer., Caerns., Denbs., Flints., *Angl.*. 44 squares, map 428.

H. hircinum L. : Stinking Tutsan : Naturalized alien (Europe). River banks and woods (E,M,R,Rw,S,W). Fl.7-9. Mons., *Glam.**†, *Cards.**, *Mer.**†, Caerns., *Flints.*[121], Angl.. 7 squares.
Mons.: nr Pontypool, ST29, JND, 1975[112]; The Coombe, ST49, WAS, c1920[226]. **Glam.:** Pen-maen, SS58, JAW, 1927[220]. **Cards.:** Llanbadarn Fawr, SN68, AOC, 1977[71]. **Mer.:** Tywyn, SH50, DAJ, 1886[126]. **Caerns.:** Llanfairfechan, SH67, JEG, 1884[112] and JDM, 1929[112]. **Angl.:** nr Menai Bridge, SH57, FCG, 1952[112] and RHR, 1970+[209].

H. androsaemum L. : Dail y Fendigaid : Tutsan. Native. Woods and hedges (E,G,M,Q,R,Rw,W). Fl.5-10. Mons., Glam., Brecs., Rads., Carms., Pembs., Cards., Monts., Mer., Caerns., Denbs., Flints., Angl.. 208 squares, map 429.

H. androsaemum × hircinum = H. × inodorum Miller (*H. elatum* Aiton) : Tall Tutsan. Established alien (Madeira, Canaries). Hedges (E,Q,Rg). Fl.7-9. Glam., Carms., *Pembs.**, Cards., Flints.. 11 squares : **SS**59,69; **ST***16*,17; **SM**91; **SN**25,35,42,46,77; **SJ**17.

H. hirsutum L. : Eurinllys Blewog : Hairy St. John's-wort. Native. Woods on limestone (E,M,Q,R, Rw,Sc,W). Fl.6-9. Mons., Glam., Brecs., Rads., *Pembs.*, (*Cards.**), Monts., Mer., *Caerns.*, Denbs., Flints., Angl.*. 67 squares, map 430.

H. pulchrum L. : Eurinllys Mân Syth : Slender St. John's-wort. Native. Heaths and woods (E,H,M,Q,R, Rw,W). Fl.6-9. Mons., Glam., Brecs., Rads., Carms., Pembs., Cards., Monts., Mer., Caerns., Denbs., Flints., Angl.. 263 squares, all except : **SS**38,39,77,78; **ST**27,49; **SM**50,73; **SN**36,59; **SO**30,31; **SH**50,51,98; **SJ**18,36, 37,45.

H. montanum L. : Eurinllys Mynyddig : Pale St. John's-wort. Native. Woods on limestone (E,G,Q,R, Rw,W). Fl.6-8. Mons., Glam., *Brecs.*[231], Rads., *Carms.*, Pembs., Monts., Caerns., Denbs., Flints., Angl., [*Mer.*]. 45 squares, map 431.

H. elodes L. : Eurinllys y Gors : Marsh St. John's-wort. Native. Bogs (Al,H,M). Fl.6-9. Mons., Glam., *Brecs.*, Rads., Carms., Pembs., Cards., Monts., Mer., Caerns., Denbs., *Flints.*[230], Angl.. 159 squares, map 432.

H. linarifolium Vahl : Eurinllys Culddeiliog : Toadflax-leaved St. John's-wort. Native. Pastures and heaths (E,G,H,Q). Fl.7-8(-11). *Rads.*, *Mer.*, Caerns., Angl.. 4 squares.
Rads.: nr Llandrindod Wells, SO05, NYS, 1945[45]. **Mer.:** Barmouth, SH61, JHP, 1907[126]. **Caerns.:** Pwllheli, SH33, JAWhe, 1940[112] and APC & PMB, 1979[3]. **Angl.:** Trearddur Bay, SH27, G, 1915[112].

H. humifusum L. : Eurinllys Mân Ymdaenol : Trailing St. John's-wort. Native. Heaths and banks (D,E, G,H,M,Q,R,Rg,Rw,W). Fl.6-10. Mons., Glam., Brecs., Rads., Carms., Pembs., Cards., Monts., Mer., Caerns., Denbs., *Flints.*, *Angl.*. 225 squares, map 433.

H. tetrapterum Fries : Eurinllys Pedrongl : Square-stalked St. John's-wort. Native. Marshes (Ar,Ds,G, H,M, R,Rw,W). Fl.6-10. Mons., Glam., Brecs., Rads., Carms., Pembs., *Cards., Monts.,* Mer., Caerns., Denbs., *Flints.,* Angl.. 230 squares, map 434.

H. undulatum Schousboe ex Willd. : Eurinllys Tonnog-ddail : Wavy St. John's-wort. Native. Marshes (Ar,D,E,G,M,R,Sc). Fl.7-8. Pembs., Cards., Mer.. 13 squares : **SM**72,81,82,83,84,91,92,93,94; **SN**01,57,67; **SH**61.

H. maculatum Crantz subsp. **obtusiusculum** (Tourlet) Hayek (*H. dubium* Leers) : Eurinllys Mawr : Imperforate St. John's-wort. Native. Hedges (Ds,E,G,M,Q, R,Rw,W). Fl.6-10. Mons., Glam., Brecs., Rads., Carms., Pembs., Cards., Monts., Mer., Caerns., Denbs., Flints., Angl.. 233 squares, map 435.

H. maculatum × perforatum = H. × desetangsii Lamotte Spontaneous. Hedges and railways (E,G, Q,R,Rw). Fl.7. Glam., Brecs., *Rads.*, Cards., Monts., *Mer., Caerns.,* Denbs., *Angl.*. 11 squares : **ST**18; **SN**68; **SO**13,21; **SH**48,53,57,61,66,85; **SJ**21.

H. perforatum L. : Eurinllys Trydwll : Perforate St. John's-wort. Native. Open woods and roadsides (D,E, G,M,Q,R,Rg,Rw,W). Fl.6-10. Mons., Glam., Brecs., Rads., Carms., Pembs., Cards., Monts., Mer., Caerns., Denbs., Flints., *Angl.*. 228 squares, map 436.

VIOLACEAE

VIOLA L.

Viola odorata L. : Crinllys Aroglys : Sweet Violet. Native. Woods and hedges (E,M,R,Rw,W). Fl.(12-)2-4. Mons., Glam., *Brecs.,* Rads., Carms., Pembs., (*Cards.*), Monts., (*Mer.*), Caerns., Denbs., Flints., Angl.. 124 squares, map 437.

V. hirta L.

Subsp. **hirta** : Gwiolydd Flewog : Hairy Violet. Native. Hedges and pastures (D,E,G,R,W). Fl.3-5(-9). Mons., Glam., Rads., *Carms.,* Pembs., *Monts.,* Caerns., Denbs., Flints., Angl.. 60 squares, map 438.

Subsp. **calcarea** (Bab.) E. F. Warburg (*V. calcarea* (Bab.) E. S. Gregory). Native. Limestone grassland. Glam.. 1 square.
Glam.: Ogmore Down, SS87, AEW, 1933[112] and BAM, 1951[7].

V. hirta × odorata = V. × permixta Jordan Spontaneous. Woodland. Fl.4. Mons., *Glam., Denbs.,* [*Cards.*]. 4 squares.
Mons.: Portskewett, ST48, WAS, 1904[112]; Trap Hill, Mounton, ST59, WAS, 1891[112]; Wynd Cliff, ST59, WAS, c1900[226]; Thornwell, ST59, WAS, c1900[226]. **Glam.:** Coedrhiglan, ST17, GTr, c1900[223]. **Denbs.:** nr Colwyn Bay, SH87, JMB, c1970[67].

[**V. rupestris** F. W. Schmidt *Glam.*]

V. reichenbachiana Jordan ex Boreau : Gwiolydd y Goedwig : Early Dog-violet. Native. Woods and hedges (E,M,Q,R,W). Fl.4-5. Mons., Glam., Brecs.*, *Rads.**, Carms., Pembs., Cards., *Monts.**, Caerns., *Denbs.*, Flints., *Angl.,* [*Mer.*]. 69 squares, map 439.

V. reichenbachiana × riviniana Spontaneous. Woods and hedges. Fl.5-6. Mons., *Glam., Brecs., Caerns., Angl.*. 9 squares.
Mons.: Itton, ST49, WAS, 1916[112]; Wynd Cliff, ST59, ESM & WAS, 1901[226]. **Glam.:** Taff's Well, ST18, HJR, c1900[140]; Merthyr Tydfil, SO00, HJR, c1900[140]. **Brecs.:** Cwm Cleisfer, SO11, MP, 1968[179]; Partrishow, SO22, MP, 1970[200]; Llanigon, SO23, HSo, 1969[200]. **Caerns.:** nr Bangor, SH57, RHR, 1961[177]. **Angl.:** Plas Llugwy, SH48, RHR, 1979[245].

V. riviniana Reichenb. : Gwiolydd Cyffredin : Common Dog-violet. Native. Woods and hedges (D,E,G,H,Q,R,Rw,W). Fl.4-6. Mons., Glam., Brecs., Rads., Carms., Pembs., Cards., Monts., Mer., Caerns., Denbs., Flints., Angl.. 278 squares, all except : **SS**39; **SM**50; **SN**59; **SJ**45. Var. **minor** (Murb. ex E. S. Gregory) Valentine (subsp. *minor* (Murb. ex E. S. Gregory) Valentine) has been recorded from all the Welsh vice-counties.

[**V. riviniana × rupestris = V. × burnatii** Greml. *Angl.*]

V. canina L. : Fioled y Cŵn : Heath Dog-violet. Native. Sand dunes and heaths (D,Ds,E,G,H,Q,R,Rw). Fl.4-7. *Mons.*, Glam., Brecs., *Rads.*, Carms., Pembs., Cards., Mer., Caerns., *Denbs.*, Flints., Angl., [*Monts.*]. 67 squares, map 440.

V. canina × lactea = V. × militaris Savouré Spontaneous. Heaths and scrub. Fl.5. *Glam., Pembs., Angl.*. 4 squares :
Glam.: Rhosili, SS48, HJR, c1900[140]; Gower, —, AD, c1930[29]. **Pembs.:** Red Cliff, Dale, SM70, DMo, 1956[154]. **Angl.:** nr Four Mile Bridge, SH27, JEG, 1885[101]; Penrhos nr Holyhead, SH28, JEG, 1885[101] and GCD, 1925[24]. It is probable that most if not all of these records refer to **V. lactea × riviniana.**

V. canina × riviniana Spontaneous. Woods and sand dunes (D,Ds,M,W). Fl.5-8. Mons., Glam., Pembs., Cards., Mer., *Caerns., Angl.*. 17 squares : **SS***48*: **ST***38*; **SM**70,73; **SN**04,*46*,59,69; **SO***50*; **SH**27,33,36,37,51, 52,53,61.

V. lactea Sm. : Millyn Welw Grugog : Pale Dog-violet. Native. Heaths and moors (E,G,H,M). Fl.5-8. Mons., Glam., Pembs., Cards., *Caerns.*, Angl., [*Mer., Denbs.*]. 34 squares, map 441.

V. lactea × riviniana Spontaneous. Heaths and moors. Fl.6. Glam., Pembs., *Caerns., Angl.*. 10 squares : **SS**97; **ST***18*; **SM**70,72; **SH**,27,28,33,48,57,58.

[**V. persicifolia** Schreber (*V. stagnina* Kit.). *Glam.*]

V. palustris L. : Fioled y Gors : Marsh Violet.
Subsp. **palustris** Native. Bogs and marshes (G,H,M,Q,W). Fl.4-5(-8). Mons., Glam., *Brecs.*, *Rads.*, Carms., Pembs., Cards., *Monts.*, Mer., Caerns., *Denbs.*, *Flints.*, *Angl.*. 196 squares, map 442.

Subsp. **juressi** (Link ex K. Wein) Coutinho (*V. juressi* Link ex K. Wein). Native. Bogs. Mons., Glam., *Rads.*[85], *Pembs.*, Mer., *Caerns.*. 19 squares : **SS**79; **ST***49,59*; **SM**82,92,93; **SN**01,02,03,13; **SH**44, 62,63,64,66,71,72,73,85.

V. cornuta L. Casual (Pyrenees). Waste places. Fl.6. *Glam.*, Flints.[112]. 1 square : **SS**69[222].

V. lutea Hudson : Fioled y Mynydd : Mountain Pansy. Native. Mountain pastures (G,H,R). Fl.5-9. *Mons.*, Glam., Brecs., Rads.*, Carms., *Pembs.*, Cards., Monts., Mer., Caerns., Denbs., Flints., Angl.. 99 squares, map 443. Doubt has been expressed whether the specimen from Angl. (SH36) in **NMW** was actually collected there.

V. × wittrockiana Gams (*V. hortensis* auct.) : Garden Pansy. Casual (garden origin). Cultivated ground and waste places. Fl.5-8(-10). Mons., Glam., Brecs., Rads., Pembs., Flints.. 8 squares : **ST***07*[112],38[245]; **SO***12*[112],*30*[112],36[112],42[93]; **SJ**07[112],16[7].

V. tricolor L. : Trilliw : Wild Pansy.
Subsp. **tricolor** (incl. *V. saxatilis* auct.). Native. Cultivated ground (D,F,G,R,Rg). Fl.4-8(-10). Mons., Glam., Brecs., Rads., *Carms.*, Pembs., Cards., Monts., Mer., Caerns., *Denbs.*, Flints., *Angl.*. 126 squares, map 444.

Subsp. **curtisii** (E. Forster) Syme (*V. curtisii* E. Forster). Native. Sand dunes. Fl.6-9(-11). Glam., Carms., Pembs., Mer., *Caerns.*, Flints., Angl.. 31 squares, map 445.

V. arvensis Murray : Ofergaru : Field Pansy. Native. Cultivated ground (D,E,F,R,Rg,Rw,W). Fl.5-10. Mons., Glam., Brecs., Rads., Carms., Pembs., Cards., Monts., Mer., Caerns., Denbs., Flints., Angl.. 193 squares, map 446.

V. arvensis × tricolor Spontaneous. Cultivated ground. Fl.6-8(-10). Mons., *Pembs.*, *Cards.*, Caerns., Flints.. 7 squares.
Mons.: Pen-how, ST49, FV, 1931[112]; nr Monmouth, SO51, RL, 1945[112]. **Pembs.:** Skokholm Island, SM70, GTG & MG, c1950[153]. **Cards.:** nr Aberystwyth, SN58, AP, 1950+[238]. **Caerns.:** Bangor, SH57, JEG, 1884[112]. **Flints.:** Cwm, SJ07, JAWh, 1909[112]; Pantasaph, SJ07, JAW, 1942[112]; Caerhys to Llyn Helyg, Lloc and nr Holywell, SJ17, JAW, 1942[112].

CISTACEAE

TUBERARIA (Dunal) Spach

Tuberaria guttata (L.) Fourr. (*Helianthemum guttatum* (L.) Miller; incl. subsp. *breweri* (Planchon) Syme) : Cor-rosyn Rhuddfannog : Spotted Rock-rose. Native. Sea cliffs (H,Sc). Fl.5-7. Caerns., Angl.. 6 squares.
Caerns.: Braich y Pwll, SH12, BMCM, 1941[112] and 1947[112]; nr Aberdaron, SH12, APC, 1971[74] and 1980[74]. **Angl.:** Trearddur Bay, SH27, CCT, 1953[7] and An, 1967[7]; South Stack, Holyhead, SH28, VMP, 1929[112]; Holyhead Mountain, SH28, CCT, 1952[7] and AR & MCFP, 1956[112]; nr Church Bay, SH28, RHR, 1960[112]; Church Bay, SH29, MCFP, 1955[7]; nr Llanfair-yng-nghornwy, SH38, JEG, 1885[112] and CWi, 1877[112]; Amlwch, SH49, HD, c1800[135].

HELIANTHEMUM Miller

Subgen. Helianthemum

Helianthemum nummularium (L.) Miller
Subsp. **nummularium** (*H. chamaecistus* Miller) : Cor-rosyn Cyffredin : Common Rock-rose. Native. Rocky places and pastures on limestone (D,G,M, Q,Sc,W). Fl.4-9. Mons., Glam., Brecs., Rads., Carms., *Pembs.*, (Cards.*), Monts., Mer., Caerns., Denbs., Flints., Angl.. 65 squares, map 447.

Subsp. **grandiflorum** (Scop.) Schinz & Thell. (*H. grandiflorum* (Scop.) Lam.). Casual (Europe). Waste places. Flints.. 1 square : **SJ**17[112].

H. apenninum (L.) Miller : Cor-rosyn Gwyn y Mynydd : White Rock-rose. Naturalized alien (S. England). Sea cliffs. Fl.5. Glam.. 1 square.
Glam.: nr Pennard, SS58, DRH, 1969[112]. Originally introduced c.1960 but not seen recently.

Subgen. Plectolobum Willk.

H. canum (L.) Baumg. : Cor-rosyn Lledlwyd : Hoary Rock-rose. Native. Sea cliffs. Fl.5-6. Glam., Pembs.*, Caerns., *Denbs.*, Flints., Angl., [*Brecs.*]. 15 squares : **SR**99; **SS**38,48,58,68,*69*; **SH**58,68,77,78,87,88,97; **SJ**07, 08.

[**H. canum × nummularium = H. × bickhamii** E. S. Marshall *Caerns.*]

TAMARICACEAE

TAMARIX L.

Tamarix gallica L. (incl. *T. anglica* Webb) : Grugbren : Tamarisk. Naturalized alien (Europe). Sand dunes (D,Q). Fl.7-9. Glam., *Pembs.*, Cards., Mer., *Caerns.*, *Denbs.*, *Flints.*, *Angl.*. 26 squares, map 448. Often planted originally.

FRANKENIACEAE

FRANKENIA L.

Frankenia laevis L. : Grugeilyn Llyfn : Sea-heath. Denizen (S. England). Salt-marshes. Fl.7-9. Glam., Angl.. 2 squares.
Glam.: Merthyr Mawr, SS87, SW, 1981[112]. **Angl.:** Rhosneigr, SH37, EW, 1965[209] and LJL, 1979[112].

ELATINACEAE

ELATINE L.

Elatine hydropiper L. : Gwybybyr Wythochrog : Eight-stamened Waterwort. Native. Lakes (Al,Ar). Fl.(6-)8-9. *Denbs.*[232], Angl., [*Cards.*]. 4 squares.
Angl.: Llyn Coron, SH36 & 37, JEG, 1884[112] and RHR, 1958[173]; Llyn Llywenan, SH38, WEH, 1968[112] and RHR, 1975[112]; Cefni Reservoir, SH47, WEH, 1968[112].

E. hexandra (Lapierre) DC. : Gwybybyr Chweochrol : Six-stamened Waterwort. Native. Lakes (Al,Ar,M). Fl.6-9. Glam., Rads.†, Carms., Cards., Mer., Caerns., Angl., [*Denbs.*]. 30 squares, map 449.

CUCURBITACEAE

ECBALLIUM A. Richard

Ecballium elaterium (L.) A. Richard : Squirting Cucumber. Casual (S. Europe). Waste places. Glam.. 2 squares : **ST***16*[24], *27*[112].

BRYONIA L.

Bryonia cretica L. subsp. **dioica** (Jacq.) Tutin (*B. dioica* Jacq.) : Bloneg y Ddaear : White Bryony. Native. Hedges (E,G,R). Fl.5-9. Mons., Glam., Brecs., *Rads.*, *Carms.**, *Pembs.*, *Monts.**, Caerns., Denbs., Flints., [*Mer.*, *Angl.*]. 54 squares, map 450.

CITRULLUS Schrader

Citrullus lanatus (Thunb.) Mansfeld (*C. vulgaris* Schrader) : Water-melon. Casual (S. Africa). Waste places. Mons., Glam.. 2 squares : **ST***27*[112],38[112].

LAGENARIA Ser.

Lagenaria siceraria (Molina) Standley (*L. vulgaris* Ser.) : Bottle-gourd. Casual (Tropics). Waste places. Glam.. 1 square : **ST***27*[112].

CUCURBITA L.

Cucurbita pepo L. : Marrow. Casual (C. America). Waste places. Mons., Glam.. 3 squares : **ST**17[112],27[112],38[91].

LYTHRACEAE

LYTHRUM L.

Lythrum salicaria L. : Llys y Milwr : Purple-loosestrife. Native. Marshes (Al,Ar,D,Ds,E,H,M,R). Fl.7-9. Mons., Glam., Brecs., *Rads.*, Carms., Pembs., Cards., Monts., Mer., Caerns., *Denbs.*, Flints., Angl.. 174 squares, map 451.

L. junceum Banks & Solander (*L. graefferi* Ten.; *L. acutangulum* auct., non Lag.; *L. flexuosum* auct., non Lag.; *L. meonanthum* Link ex Steudel). Casual (S. W. Europe). Waste places. Fl.7-9. Mons., Glam., Pembs., Mer.. 4 squares : **ST**27[112],38[112]; **SM**91[112]; **SH**50[112].

L. hyssopifolia L. : Gwyarllys Isop-ddail : Grass-poly. Casual (S. England). Waste places. Fl.7. Glam., *Carms.*. 3 squares : **ST***16*[112],*17*[112]; **SN***42*[82].

L. portula (L.) D. A. Webb (*Peplis portula* L.) : Troed y Gywen : Water-purslane. Native. Marshy places (Al,Ar,M,R). Fl.6-9. Mons., Glam., Brecs., Rads., Carms., Pembs., Cards., *Monts.*, Mer., Caerns., *Denbs.*, *Flints.*, Angl.. 172 squares, map 452. Var. **longidentata** Gay (subsp. *longidentata* (Gay) P. D. Sell) has been recorded from Mons., Glam., Brecs., Rads., Pembs., *Cards.*, *Mer.*, Caerns.

ONAGRACEAE

FUCHSIA L.

Fuchsia magellanica Lam. : Ffwsia : Fuchsia. Established alien (S. America). Hedges (E,Sc). Fl.8-11. *Glam.*, Pembs., *Cards.*, Mer., *Caerns.*, Denbs.. 14 squares : **SS**19,68; **ST**17; **SM**70,80,93; **SN**46; **SH**12, 22,32,34,50,72,87.

GAURA L.

Gaura coccinea Nutt. ex Pursh Casual (N. America). Sand dunes. Fl.7. Carms.. 1 square : **SN**40[112].

CIRCAEA L.

Circaea lutetiana L. : Llysiau Steffan : Enchanter's-nightshade. Native. Woods (E,M,Q,R,Rg,W). Fl.6-9. Mons., Glam., Brecs., Rads., Carms., Pembs., Cards., Monts., Mer., Caerns., Denbs., Flints., Angl.. 242 squares, map 453.

C. alpina L. : Llysiau Steffan Mynyddig : Alpine Enchanter's-nightshade. Native. Shady rocky places. Fl.5-7. *Glam.*, Brecs., Mer., Caerns., [*Mons.*, *Rads.*, *Carms.*, *Pembs.*, *Cards.*, *Monts.*, *Denbs.*]. 6 squares.
Glam.: Rhondda, SS99, AL, 1890[94]. **Brecs.:** Craig y Cilau, SO11, DPMG, 1952[112] and RGW, 1980[259]. **Mer.:** Corris, SH70, DMP, 1964[112] and PMB, 1980[3]; Llanymawddwy, SH82, WCo & MR, 1955[112] and PMB, 1980[3]. **Caerns.:** Dolbadarn, SH55, An, 1832; nr Capel Curig, SH75, ER, 1948[107] and RHR, 1979[208].

C. alpina × lutetiana = C. × intermedia Ehrh. : Gwynlys Mynydd-dir : Upland Enchanter's-nightshade. Spontaneous or naturalized alien. Roadsides and rocky woods (E,Q,R,W). Fl.6-10. Mons., *Brecs.*, *Rads.*, Carms., Cards., Monts., Mer., Caerns., *Denbs.*, [*Pembs.*]. 47 squares, map 454.

OENOTHERA L.

Subgen. Oenothera

Oenothera biennis L. : Melyn yr Hwyr : Common Evening-primrose. Naturalized alien (Europe). Sand dunes and waste places (D,R,Rw). Fl.6-8. *Mer.*, Caerns., [*Mons.*, *Glam.*, *Brecs.*, *Carms.*, *Pembs.*, *Cards.*, *Monts.*, *Denbs.*, *Flints.*]. 3 squares.
Mer.: Barmouth, SH61, MAL, 1867[107] and PMB, 1960[109]. **Caerns.:** Morfa Bychan, SH53, JEG, 1890[112]; Bangor, SH57, JTCWi, 1876[112].

O. fallax Renner em. Rostański (*O. lamarkiana* auct.). Naturalized alien (Europe). Sand dunes and waste places. Fl.9. Glam.. 2 squares.
Glam.: Whitford Burrows, SS49, RGE, 1977[112]; Swansea, SS69, KRo, 1977[184]. This species arose from the hybrid **O. biennis × erythrosepala.**

[**O. rubricaulis** Klebahn (*O. muricata* L. nomen confusum). *Mons.*, *Glam.*, *Brecs.*, *Carms.*, *Pembs.*, *Cards.*, *Monts.*, *Denbs.*]

O. erythrosepala Borbás (*O. glazioviana* Micheli) : Briallu yr Hwyr Mwyaf : Large-flowered Evening-primrose. Naturalized alien (N. America). Sand dunes and waste places (D,E,G,R,Rw). Fl.6-9. Mons.*, Glam., *Brecs.**, Carms.*, Pembs.*, Cards., Mer., Caerns.*, *Denbs.**, Flints., *Angl.**. 72 squares, map 455.

O. perangusta Gates Casual (N. America). Waste places. Fl.8. Glam.. 3 squares : **ST**09[112],16[112],*18*[112].

O. cambrica Rostański (*O. parviflora* auct., non L.; *O. ammophila* auct., non Focke; *O. biennis* auct., non L.; *O. muricata* auct., non L.) : Briallu yr Hwyr Cymreig : Welsh Evening-primrose. Naturalized alien (Canada). Sand dunes and waste places (D,E,R,Rg,Rw,W). Fl.(3-)5-9. Mons., Glam., Brecs., Carms., Pembs., Cards., Monts., Mer., Caerns., Denbs.. 59 squares, map 456.

O. cambrica × erythrosepala = O. × britannica Rostański Spontaneous. Sand dunes and waste places. Fl.7-9. Glam., *Carms.*. 5 squares.
Glam.: Oxwich Burrows, SS58, KRo & RGE, 1977[184]; Swansea, SS69, KRo & RGE, 1977[184]; Kenfig Burrows, SS78, RMS, 1963[104]; Cardiff, ST17, JWD, 1956[112].
Carms.: Pembrey Burrows, SS49, FHW, 1980[201].

O. salicifolia Desf. ex G. Don Casual (N. America). Waste places. Fl.9. Glam.. 1 square : **ST***27*[112].

O. parviflora L. : Small-flowered Evening-primrose. Casual (N. America). Waste places. Fl.10. *Glam.*. 2 squares : **SS***78*[113]; **SO***00*[107].

O. rubricuspis Renner ex Rostański Casual (N. America). Waste places. Fl.7-8. Glam.. 1 square : **ST***17*[112].

[**O. ammophila** Focke *Carms.*, *Pembs.*]

Subgen. Raimannia (Rose) Munz

O. stricta Ledeb. ex Link : Briallu yr Hwyr Perar-oglys : Fragrant Evening-primrose. Naturalized alien (N. America). Sand dunes and waste places (D,E,R,Rg,Rw). Fl.5-10. Mons.*, Glam., Carms.*, Cards.*, Mer.. 11 squares : **SS**49,69,*78*,87; **ST***17*,*27*,38; **SN**14,40,59,69.

O. laciniata Hill Casual (N. America). Waste places. Fl.8-10. Glam., Flints.. 2 squares : **ST***17*[112]; **SJ***08*[112].

Subgen. Hartmannia (Spach) Munz

O. rosea L'Hér. ex Aiton Casual (N. America). Waste places. Fl.9. Glam.. 1 square : **ST**17[112].

CLARKIA Pursh

Clarkia unguiculata Lindley (*C. elegans* Douglas, non Poiret). Casual (N. America). Waste places. *Mons.*, Glam.. 2 squares : **ST**16[112],38[91].

C. pulchella Pursh Casual (N. America). Waste places. Glam.. 1 square: **ST***17*[112].

C. purpurea (Curtis) A. Nelson & Macbride subsp. **viminea** (Douglas) F. H. & M. E. Lewis (*Godetia viminea* Spach.) Casual (N. America). Rubbish tip. Fl.8. Glam.. 1 square : **ST**17[112].

C. tenella (Cav.) F. H. & M. E. Lewis (*Oenothera tenella* Cav.; *Godetia tenella* (Cav.) Spach ex Steudel). Casual (Chile). Glam.. 1 square : **ST***17*[112].

EPILOBIUM L.

Epilobium angustifolium L. (*Chamaenerion angustifolium* (L.) Scop.; *Chamerion angustifolium* (L.) J. Holub.) : Helyglys Hardd : Rosebay Willowherb. Native on mountain cliffs but widespread as a naturalized alien in wood clearings and along railway banks (D,E,G,H,M,Q,R,Rw,W). Fl.7-9. Mons., Glam., Brecs., Rads., Carms., Pembs., Cards., Monts., Mer., Caerns.. Denbs., Flints., *Angl.*. 263 squares, all except : **SS**38,39,77; **ST**26; **SM**50,62,70,71,73,84; **SN**15,36,59,75,89,99; **SH**24,29,51.

E. hirsutum L. : Helyglys Pêr : Great Willowherb. Native. Marshes (Al,Ar,Ds,E,G,H,M,R,W). Fl.7-9. Mons., Glam., Brecs., Rads., Carms., Pembs., Cards., Monts., Mer., Caerns., Denbs., Flints., Angl.. 213 squares, map 457.

E. hirsutum × montanum = E. × erroneum Hausskn. Spontaneous. Waste places (R,W). Fl.6-9. Glam.. 3 squares.
Glam.: Cardiff, ST17, AEW, 1929[112]; Castell Coch, ST18, EV, 1935[35]; Aberdare, SO00, HJR, 1903[140].

E. hirsutum × obscurum = E. × anglicum E. S. Marshall Spontaneous. Marshes. Fl.9. Pembs.. 1 square.
Pembs.: Marloes, SM70, TAWD, 1963[112].

E. hirsutum × parviflorum Spontaneous. Ditches and river banks. Fl.8. *Glam.*, *Mer.*, *Angl.*. 3 squares.
Glam.: Cardiff, ST17, AHT, c1905[223] and HJR, 1913[107]. **Mer.:** Mochras Estuary, SH52, PMB, 1958[4]. **Angl.:** nr Afon Ffraw, SH36, RHR, 1978[209].

E. hirsutum × roseum = E. × goerzii Rubner Spontaneous. Roadsides. *Mons.*.
Mons.: nr Matharn, ST59, WAS, pre 1920[216].

E. parviflorum Schreber : Helyglys Lledlwyd : Hoary Willowherb. Native. Streamsides and marshes (Ar,D, E,G,H,M,Q,R,Rg,Rw,W). Fl.6-9. Mons., Glam., Brecs., Rads., Carms., Pembs., Cards., Monts., Mer., Caerns., *Denbs.*, Flints., *Angl.*. 201 squares, map 458.

E. parviflorum × roseum = E × persicinum Reichenb. Spontaneous. Roadsides and waste places. *Mons.*, Glam., *Rads.*. 3 squares.
Mons.: nr Matharn, ST59, WAS, 1909[107]. **Glam.:** Aberdare, SO00, HJR, c1903[140]. **Rads.:** Old Radnor, SO25, CIS & NYS, 1945[44].

E. parviflorum × tetragonum = E. × weissenburgense F. W. Schultz Spontaneous. Ditches. Cards.. 1 square.
Cards.: nr New Quay, SN35, JHS, 1937[112] (provisional determination).

E. montanum L. : Helyglys Llydanddail : Broad-leaved Willowherb. Native. Woods and hedgebanks (D,E,M,Q,R,Rg,Rw,W). Fl.6-9. Mons., Glam., Brecs., Rads., Carms., Pembs., Cards., Monts., Mer., Caerns., Denbs., Flints., *Angl.*. 267 squares, all except : **SR**89; **SS**38,39,77; **ST**26; **SM**50,71,73,84,94; **SN**36; **SH**51,98; **SJ**37,44.

E. montanum × obscurum = E × aggregatum Čelak. Spontaneous. Waste places (Ar,E,Q,R). Fl.7-9. Mons., *Glam.*, Brecs., *Rads.*, Cards., *Monts.*[217], *Mer.*, Caerns., *Angl.*. 11 squares : **SS***69*; **SN**68,*92*; **SO***00*,23, *50*; **SH**36,37,*66*,71,*75*.

E. montanum × parviflorum = E. × limosum Schur Spontaneous. Waste places (R,Rw). Fl.7-9. Mons., Glam., *Caerns.*. 6 squares.
Mons.: Portskewett, ST58, ESM & WAS, 1902[112]; nr Matharn, ST59, WAS, 1909[107]. **Glam.:** nr Cadoxton, ST16, AEW, 1921[112]; Cardiff, ST17, HJR, 1912[107] and AEW, 1951[112]; Aberdare, SO00, HJR, 1903[140]. **Caerns.:** Bangor, SH57, JEG, c1890[101].

E. montanum × roseum = E. × mutabile Boiss. & Reuter Spontaneous. Waste places (E,Q,R,Rg). Fl.7. *Mons.*, *Glam.*. 6 squares.
Mons.: Shirenewton, ST49, ESM, 1895[107]; Chepstow, ST59, WAS, pre 1920[216]; Tintern, SO50, WAS, pre 1920[216]. **Glam.:** Swansea, SS69, EFL, 1892[107]; Dyffryn, ST07, EV, 1942[233]; Aberdare, SO00, HJR, 1903[107].

E. lanceolatum Sebastiani & Mauri : Helyglys Gwaywddail : Spear-leaved Willowherb. Native. Banks and walls (Ar,D,M,Q,R,Rg,Rw). Fl.6-9. Mons., Glam., Brecs., Rads.*, Carms.*, *Pembs.**, Cards., Mer.*, Denbs.. 26 squares, map 459.

E. lanceolatum × montanum = E. × neogradense Borbás Spontaneous. Waste places (E,R,Rb). Fl.7. Mons.. 2 squares.
Mons.: Portskewett, ST58, WAS, pre 1920[216]; Dixton, SO51, HJR, 1911[112].

E. tetragonum L.

Subsp. **tetragonum** (*E. adnatum* Griseb.) : Helyglys Pedrongl : Square-stalked Willowherb. Native. Damp places (H,M,Q,R,Rg,Rw,W). Fl.6-8. Mons., Glam., *Brecs.*, Rads., Carms., Pembs., *Cards.*, *Monts.**, *Caerns.*, *Denbs.*, *Flints.**, *Angl.*. 91 squares, map 460.

Subsp. **lamyi** (F. W. Schultz) Nyman (*E. lamyi* F. W. Schultz) : Helyglys Llwydwyrdd : Glaucous Willowherb. Native. Damp woods. *Mons.*, Glam., *Brecs.*. 6 squares.
Mons.: Portskewett, ST58, HSR, 1922[107] and EC, 1955[7]; Chepstow, ST59, AL, c1900[226] and HSR, 1922[107]; nr Monmouth, SO51, AL, c1900[226]. **Glam.:** nr Cardiff, ST17, EV, 1935[112]; Whitchurch Canal, ST18, MG, 1971[99]. **Brecs.:** Tal-y-llyn, SO12, AL, 1884[130].

E. obscurum Schreber : Helyglys Rhedegydd Tenau : Short-fruited Willowherb. Native. Marshy places (Ar,H, M,Q,R,Rg,W). Fl.6-10. Mons., Glam., Brecs., Rads., Carms., Pembs., Cards., Monts., Mer., Caerns., Denbs., Flints., *Angl.*. 222 squares, map 461.

E. obscurum × palustre = E. × schmidtianum Rostk. Spontaneous. Marshy ground (Ar,H,M). Fl.7-9. Glam., *Brecs.*, Rads., Carms., *Pembs.*, *Mer.*, *Caerns.*. 15 squares : **SS***58*,59,98; **ST**59; **SM***72*; **SN**61,71,*96*; **SO***14*, *50*; **SH***33*,*57*,61,71,*75*.

E. obscurum × parviflorum = E. × dacicum Borbás Spontaneous. Damp places (D,G,M,Q,W). Fl.6-9. Mons., Glam., *Cards.*, *Monts.*, Mer., Denbs., *Flints.*. 8 squares.
Mons.: Itton, ST49, WAS, 1895[112]; Bigswier, SO50, WAS, 1897[112]. **Glam.:** St Fagans, ST17, EV, 1941[112]. **Cards.:** New Quay, SN35, JHS, c1940[211]. **Monts.:** nr Crugion, SJ31, An, 1950+[1]. **Mer.:** Morfa Harlech, SH53, PMB, 1954[112]. **Denbs.:** nr Llangollen, SJ24, CW, 1915[112]. **Flints.:** Nant y Ffrith, SJ25, VG, 1957[7].

E. obscurum × roseum = E. × brachiatum Čelak. Spontaneous. Waste places. Fl.7. Mons.. 1 square.
Mons.: Monmouth, SO51, SGC, 1936[112].

E. obscurum × tetragonum = E. × thuringiacum Hausskn. Spontaneous. Waste places (Q,R). Fl.7. Mons., *Glam.*. 4 squares.
Mons.: Tintern, SO50, WAS, 1891[112]. **Glam.:** Port Talbot, SS78, HJR, pre 1907[140]; Cardiff, ST17, HJR, 1908[142]; Aberdare, SO00, HJR, pre 1907[140].

E. roseum Schreber : Helyglys Coesig : Pale Willowherb. Native or colonist. Roadsides. (Ar,E,M,Q,R, Rg,Rw,W). Fl.6-9. Mons., Glam., Brecs., *Rads.*, Carms., Pembs., Cards., Monts., Mer., *Caerns.*, Denbs., Flints., *Angl.*. 95 squares, map 462.

E. palustre L. : Helyglys Culddail y Fawnog : Marsh Willowherb. Native. Marshes (Al,Ar,Ds,E,G,H,M, Rw,W). Fl.7-9. Mons., Glam., Brecs., Rads., Carms., Pembs., Cards., Monts., Mer., Caerns., Denbs., Flints., Angl.. 233 squares, map 463.

E. palustre × parviflorum = E. × rivulare Wahlenb. Spontaneous. Marshes. Fl.7. Pembs., *Mer.*, *Angl.*. 5 squares.
Pembs.: Marloes, SM70, TAWD, 1959[112]. **Mer.:** Morfa Dyffryn, SH52, PMB, 1950+[4]; Morfa Harlech, SH53, PMB, 1950+[4]. **Angl.:** Trearddur, SH27, RHR, 1970+[209]; nr Aberffraw, SH36, RHR, 1979[245].

E. alsinifolium Vill. : Helyglys Gwlyddyn-ddail : Chickweed Willowherb. Native. Mountain cliffs. Fl.7. Caerns.. 3 squares.
Caerns.: Cwm Idwal, SH65, WAS, 1912[112]; Cwm Glas, SH65, LL, 1959[112]; Clogwyn y Garnedd, SH65, ER, 1950+[207]; Ysgolion Duon, SH66, JEG, 1892[112]; Carnedd Llywelyn, SH66, ER, 1950+[207]; Craig Dulyn, SH66, ER, 1950+[207]; Braich-ty-du, SH66, ER, 1963[207]; nr Aber falls, SH67, JEG, c1895[101].

E. ciliatum Rafin. (*E. adenocaulon* Hausskn.) : American Willowherb. Established alien (N. America). Cultivated ground and waste places (E,M,Q,R,Rg,Rw,W). Fl.6-8. Mons., Glam., Brecs.*, *Rads.**, Carms.*, Pembs., Cards.*, *Monts.**, Mer., Caerns.*, Denbs.*, Flints.*, Angl.*. 145 squares, map 464.

E. ciliatum × hirsutum Spontaneous. Waste places. Fl.8. Glam.. 1 square.
Glam.: Cardiff, ST17, JWD, 1975[112].

E. ciliatum × montanum Spontaneous. Waste places. Fl.7-9. Glam., *Mer.*, *Caerns.*. 3 squares.
Glam.: Treherbert, SS99, RGE, & AEW, 1970[112]. **Mer.:** Dolgellau, SH71, PMB, 1955[4]. **Caerns.:** Llangain, SH22, APC, 1980[74].

E. ciliatum × obscurum Spontaneous. Waste places. Fl.9. Carms., *Mer.*, *Caerns.*. 3 squares.
Carms.: Capel Hendre, SN51, RFM, 1963[112]. **Mer.:** Dolgellau, SH71, PMB, 1961[4]. **Caerns.:** nr Pwllhelli, SH33, APC, 1970+[74].

E. ciliatum × palustre Spontaneous. Marshy meadows. Fl.7. *Caerns.*. 2 squares.
Caerns.: Cors Lleferin, SH32, APC, 1980[74]; nr Llanbedrog, SH33, APC, 1980[74].

E. ciliatum × parviflorum Spontaneous. Garden weed. Fl.7. Glam.. 1 square.
Glam.: Radyr, ST18, MSP, 1951[112].

E. ciliatum × roseum Spontaneous. Waste places. Fl.6-8. Glam., Flints.. 2 squares.
Glam.: Mountain Ash, ST09, AMP, 1970[112]. **Flints.:** Saltney, SJ36, TE, 1969[112]. Provisional determinations only.

E. brunnescens (Cockayne) P. H. Raven & Engelhorn (*E. pedunculare* auct. brit., non A. Cunn.; *E. nerteroides* A. Cunn.) : Helyglys Gorweddol : New Zealand Willowherb. Naturalized alien (New Zealand). Walls and gravelly places (Ar,Ds,E,G,M,Q,R,Rg,Rw,W). Fl.5-10. Mons., Glam., Brecs.*, Rads.*, Carms., Pembs.*, Cards.*, Monts.*, Mer., Caerns., Denbs.*, *Flints.**, *Angl.**. 108 squares, map 465.

HALORAGACEAE

GUNNERA L.

Gunnera tinctoria (Molina) Mirbel : Giant Rhubarb. Casual (N. America). Damp waste places. *Glam.*. 2 squares : **SS**69[221]; **ST**17[37].

G. manicata Linden ex André : Prickly Rhubarb. Casual (S. America). River banks in woodland. Fl.3. Mons., *Pembs.*, *Angl.*. 3 squares : **ST**39[112]; **SM**81[80]; **SH**57[208].

MYRIOPHYLLUM L.

Myriophyllum verticillatum L. : Myrdd-ddail Troellog : Whorled Water-milfoil. Native. Lakes and ditches. Fl.6-8. Mons., Glam.*, Brecs.*, *Rads.**, *Monts.*, Denbs., *Angl.*, [*Pembs.*, *Mer.*, *Flints.*]. 12 squares : **SS**79; **ST**16, 28,29,48; **SN**92; **SO**04,05; **SH**37; **SJ**14,21,22.

M. spicatum L. : Myrdd-ddail Tywysennaidd : Spiked Water-milfoil. Native. Lakes and ditches (Al,Ar,Ds). Fl.5-8. Mons., Glam., Brecs., *Rads.*, Carms., *Pembs.*, *Cards.*, Monts., Mer., *Caerns.*, Denbs., Flints., Angl.. 90 squares, map 466.

M. alterniflorum DC. : Myrdd-ddail Cylchynol : Alternate Water-milfoil. Native. Lakes and ditches (Al,Ar,Ds,H). Fl.6-8. Glam., Brecs., Rads., Carms., Pembs., Cards., Monts.*, Mer., Caerns., Denbs., Flints., Angl.. 124 squares, map 467.

HIPPURIDACEAE

HIPPURIS L.

Hippuris vulgaris L. : Rhawn y Gaseg : Mare's-tail. Native. Pond and lake margins (Al,Ar,H,M). Fl.5-9. Mons., Glam., Brecs.*, Carms., *Pembs.**, Mer., *Caerns.*, Denbs., *Flints.*, Angl., [*Rads.*, *Cards.*, *Monts.*]. 52 squares, map 468.

CORNACEAE

CORNUS L.

Cornus sanguinea L. (*Thelycrania sanguinea* (L.) Fourr.; *Swida sanguinea* (L.) Opiz) : Cwyros : Dogwood. Native. Hedges and woods (E,M,Q,Sc,W). Fl.5-7(-12). Mons., Glam., Brecs., Rads., Carms., Pembs., (Cards.*), Monts., (*Mer.*), Caerns., Denbs., Flints., *Angl.*. 116 squares, map 469.

C. alba L. (*Thelycrania alba* (L.) Pojark.) : Cornel. Established alien (N.E. Asia). Hedges. *Cards., Mer., Denbs., Flints.*. 9 squares.
Cards.: nr Mydroilyn, SN45, JHS, c1930[210]; Lampeter, SN54, JHS, c1930[210]; Llangeitho, SN65, JHS, c1930[210]; Nanteos, SN67, AOC, 1955[71]; nr Devil's Bridge, SN77, JHS, c1930[210]. **Mer.:** Tywyn, SH50, AL, 1875[14]. **Denbs.:** Llanfihangel Glyn Myfyr, SH94, AAD, pre 1911[144]. **Flints.:** Coed Du, SJ16, AAD, pre 1910[143]; Rhyd-y-mwyn, SH26, AAD, pre 1910[143].

C. sericea L. (*C. stolonifera* Michx; *Thelycrania sericea* (L.) Dandy; *Swida sericea* (L.) J. Holub) : Red-osier Dogwood. Established alien (N. America). Woodland and scrub (E,R,Rw,W). Fl.5-8. Glam., *Brecs., Rads.*, Carms., *Monts.*, Denbs., Flints.. 21 squares, map 470.

ARALIACEAE

HEDERA L.

Hedera helix L.

Subsp. **helix** : Iorwg : Ivy. Native. Woods (E,M,Q,R,Sc,W). Fl.8-11. Mons., Glam., Brecs., Rads., Carms., Pembs., Cards., Monts., Mer., Caerns., Denbs., Flints., Angl.. 278 squares, all except : **SR**89; **SS**39; **SM**50; **SN**15.

Subsp. **canariensis** (Willd.) Coutinho (*H. canariensis* Willd.). Casual (Canary Isles). Woods. Mer.. 1 square : **SH**61[112].

Recent research by H. A. McAllister has indicated the presence of two species within the **Hedera helix** complex : **H. helix** L. s.s., diploid, with a more easterly distribution and **H. hibernica,** tetraploid, with a more westerly distribution in Wales; the detailed distribution of both species has yet to be worked out.

H. colchica (C. Koch) C. Koch Established alien (Caucasus). Walls. Glam.. 1 square.
Glam.: Aberdare, SO00, AMP, 1981[112].

UMBELLIFERAE

HYDROCOTYLE L.

Hydrocotyle vulgaris L. : Cron y Gweunydd : Marsh Pennywort. Native. Marshes (Al,D,Ds,G,H,M,W). Fl.5-9. Mons., Glam., Brecs., Rads., Carms., Pembs., Cards., *Monts.,* Mer., Caerns., Denbs., Flints., Angl.. 239 squares, map 471.

SANICULA L.

Sanicula europaea L. : Clust yr Arth : Sanicle. Native. Woods (E,Q,R,W). Fl.4-8. Mons., Glam., Brecs., Rads., Carms., Pembs., Cards., Monts., Mer., Caerns., Denbs., Flints., Angl.. 213 squares, map 472.

ASTRANTIA L.

Astrantia major L. : Astrantia. Established alien (S. Europe). Roadsides (E,M,R,Rg,W). Fl.6-7. Mons., Glam., Monts., *Caerns.*. 6 squares.
Mons.: Pontnewynydd, SO20, HJF, 1952[112]. **Glam.:** Ton-teg, ST08, DMH, 1974[112]; Cardiff, ST17, HES, 1926[112]. **Monts.:** Leighton Woods, SJ20, VJM, 1950+[116]; Long Mountain, SJ20, DMP, 1969[240]; Guilsfield, SJ21, HW, 1964[112]. **Caerns.:** Llanllechid, SH66, RHR, c1960[208].

ERYNGIUM L.

Eryngium maritimum L. : Celyn y Môr : Sea-holly. Native. Sea shores (D,S). Fl.7-9. Glam., Carms., Pembs., Cards., Mer., Caerns., Denbs., Flints., Angl.. 62 squares, map 473.

E. planum L. Casual (Europe). Rubbish tip. Fl.9. *Mons.*. 1 square : **ST**38[184].

E. amethystinum L. Established alien (Balkans). Waste places. Fl.9. Glam., Caerns.. 2 squares.
Glam.: Penarth, ST17, LE, 1979[112]. **Caerns.:** Llandudno, SH78, EP, 1980[245].

E. campestre L. : Boglynnon : Field Eryngo. Established alien (S. England). Waste places. Fl.8-9. *Mons.*, Glam., Caerns.. 6 squares.
Mons.: Tintern, SO50, TWG, 1857[226]. **Glam.:** Port Talbot, SS78, HJR, c1904[107] and RLB, 1933[32]; Barry, ST16, RLS, 1926[112] and DM, 1974[93]; Cardiff, ST17, FB, 1848[107]; Roath, Cardiff, ST27, JS, c1876[218]. **Caerns.:** Portmadoc, SH53, ABC & MCo, 1916[15] and MHA, 1951[112].

CHAEROPHYLLUM L.

Chaerophyllum aureum L. : Bulbous Chervil. Casual (Europe). Farmyard. Fl.1. Pembs.. 1 square : **SM**93[112].

C. temulentum L. : Perllys y Perthi : Rough Chervil. Native. Hedgebanks (E,M,R,S). Fl.5-10. Mons., Glam., Brecs., Rads., Carms., Pembs., Cards., Monts., Mer., Caerns., Denbs., Flints., *Angl.*. 198 squares, map 474.

ANTHRISCUS Pers.

Anthriscus sylvestris (L.) Hoffm. : Gorthyfail Llyfn : Cow Parsley. Native. Hedge banks (D,E,G,H,M,Q,R, Rw,S,W). Fl.4-5(-8). Mons., Glam., Brecs., Rads., Carms., Pembs., Cards., Monts., Mer., Caerns., Denbs., *Flints.*, Angl.. 258 squares, all except : **SR**89; **SS**39; **ST**26; **SM**50,62,71,73,84,94; **SN**15,75,77,86,87,91; **SO**10,20,41; **SH**29,60,62,91,92; **SJ**02.

A. cerefolium (L.) Hoffm. : Gorthyfail y Gerddi : Garden Chervil. Casual (E. Europe). Waste places. *Glam.*, *Carms.*, *Mer.*. 3 squares : **SS***87*[224]; **SN***42*[161]; **SH***93*[126].

A. caucalis Bieb. (*A. neglecta* Boiss. & Reuter) : Gorthyfail Cyffredin : Bur Chervil. Native or colonist. Hedgebanks and sandy waste places (D,E,M,Q,R,Sc). Fl.5-6. *Mons.*, Glam., Carms., Pembs., *Cards.*[231], (Mer.), Caerns., *Denbs.*, Flints., Angl., [*Brecs.*]. 39 squares, map 475.

SCANDIX L.

Scandix pecten-veneris L. : Nodwydd y Bugail : Shepherd's-needle. Established alien (England). Cultivated ground (F,G,Q,R,Rg). Fl.4-7. Mons., Glam., *Brecs.*[130], *Rads.*[230], *Carms.*, Pembs., Cards., *Monts.*, *Mer.*, Caerns., Denbs., *Flints.*, Angl.. 47 squares, map 476.

MYRRHIS Miller

Myrrhis odorata (L.) Scop. : Cegiden Bêr : Sweet Cicely. Colonist (Europe). Roadsides (Ar,E,F,G, Q,R,Rg,Rw). Fl.4-7. Mons., Glam., Brecs., Rads., Carms., Pembs.*, Cards., Monts., Mer., Caerns., Denbs., Flints., *Angl.*. 106 squares, map 477.

CORIANDRUM L.

Coriandrum sativum L. : Brwysgedlys : Coriander. Casual (Mediterranean region). Waste places (F,R,Rw,S). Fl.6-9. Mons., Glam., Carms., *Flints.*, Angl.. 12 squares : **SS***78*; **ST***16*,17,*18*,*27*,38; **SN**30,40,*90*; **SO**51; **SH**46; **SJ***26*.

SMYRNIUM L.

Smyrnium olusatrum L. : Dulys : Alexanders. Naturalized alien (S.W. Europe). Roadsides and cliffs (E,M,R,Rw,Sc). Fl.3-6. Mons., Glam., *Brecs.*, Carms., Pembs., Cards., Mer., Caerns., Denbs., Flints., Angl.. 113 squares, map 478.

CONOPODIUM Koch

Conopodium majus (Gouan) Loret : Cnau'r Ddaear : Pignut. Native. Woods and pastures (E,G,M,Q,R,W). Fl.5-7. Mons., Glam., Brecs., Rads., Carms., Pembs., Cards., Monts., Mer., Caerns., Denbs., Flints., Angl.. 266 squares, all except : **SR**89; **SS**39,77; **ST**26; **SM**50, 71,73,94; **SN**59; **SO**20; **SH**13,51,98; **SJ**03,13,37.

PIMPINELLA L.

Pimpinella major (L.) Hudson : Gwreiddiriog Mawr : Greater Burnet-saxifrage. Casual (England). Waste places. Fl.6-7. *Glam.**, Monts.*, [*Carms.*, *Mer.*, *Caerns.*]. 2 squares : **SS***87*[140]; **SJ**10[112].

P. saxifraga L. : Tormaen Burnet : Burnet-saxifrage. Native. Pastures (D,E,G,M,Q,R,Rg,Rw,Sc). Fl.5-10. Mons., Glam., Brecs., Rads., Carms., Pembs., Cards., Monts., Mer., Caerns., Denbs., Flints., Angl.. 231 squares, map 479.

AEGOPODIUM L.

Aegopodium podagraria L. : Llys y Gymalwst : Ground-elder. Denizen (Europe). Roadsides (D,E, G,Q,R,Rw,W). Fl.5-7. Mons., Glam., Brecs., *Rads.*, Carms., Pembs., *Cards.*, Monts., Mer., Caerns., Denbs., Flints., *Angl.*. 232 squares, map 480.

[SIUM L.]

[**Sium latifolium** L. *Mons.*, *Glam.*, *Mer.*, *Caerns.*, *Flints.*, *Angl.*]

BERULA Koch

Berula erecta (Hudson) Coville : Dyfrforonen Gulddail : Lesser Water-parsnip. Native. Streams (Al,Ar, M,S). Fl.7-10. Mons., Glam., *Brecs.*, Carms., Pembs., Monts., Mer., *Caerns.**, *Denbs.*, *Flints.*, Angl.. 81 squares, map 481.

CRITHMUM L.

Crithmum maritimum L. : Corn Carw'r Môr : Rock Samphire. Native. Sea cliffs (M,Q,S,Sc). Fl.5-9. Mons., Glam., Carms., Pembs., Cards., Mer., Caerns., *Denbs.**, Angl.. 72 squares, map 482.

OENANTHE L.

Oenanthe fistulosa L. : Dibynlor Pibellaidd : Tubular Water-dropwort. Native. Streamsides (Al,Ar,G,M, S,W). Fl.6-9. Mons., Glam., Brecs., Rads., Carms., Pembs.*, Monts., Mer., Caerns., Denbs., Flints., Angl.. 62 squares, map 483.

[**O. pimpinelloides** L. *Glam.*, *Angl.*]

[**O. silaifolia** Bieb. *Mons.*]

O. lachenalii C. C. Gmelin : Dibynlor Perllysddail : Parsley Water-dropwort. Native. Marshes near the sea (Ar,M,S). Fl.6-9. Mons., Glam., *Brecs.**, Carms., Pembs., Cards., Monts., Mer., Caerns., Denbs., Flints., Angl.. 64 squares, map 484.

O. crocata L. : Cegid y Dŵr : Hemlock Water-dropwort. Native. Marshes and streamsides (Al,Ar,E, G,H,M,S,W). Fl.5-8. Mons., Glam., Brecs., Rads., Carms., Pembs., Cards., *Monts.*, Mer., Caerns., Denbs., *Flints.*, Angl.. 265 squares, all except : **SS**38,39,77; **ST**26; **SM**50,71,73; **SN**36,*78*,*87*,88,89; **SH**34,51,95,98; **SJ**35.

O. fluviatilis (Bab.) Coleman : River Water-dropwort. Native. Ponds. *Denbs.*, [*Caerns.*, *Angl.*]. 1 square. **Denbs.:** nr Wrexham, SJ34, AGS, 1973[182].

O. aquatica (L.) Poiret : Cegid Manddail y Dŵr : Fine-leaved Water-dropwort. Native. Ditches (Al, Ar,M). Fl.6-8. Mons., Glam.*, Monts., *Caerns.*, Denbs., Flints.. 20 squares, map 485.

AETHUSA L.

Aethusa cynapium L. : Persli'r Ffŵl : Fool's Parsley. Denizen (England). Cultivated ground (D,E,F,M,R, Rg,Rw,W). Fl.6-9. Mons., Glam., Brecs., Rads., Carms., Pembs., Cards., Monts., Mer., Caerns., Denbs., Flints., *Angl.*. 149 squares, map 486.

FOENICULUM Miller

Foeniculum vulgare Miller : Ffenigl Cyffredin : Fennel. Naturalized alien (Mediterranean region). Sea cliffs and waste places inland (D,E,Q,R,Rw,S,Sc,W). Fl.7-9. Mons., Glam., *Brecs.**, Carms., Pembs., Cards., Mer., Caerns., Denbs., Flints., Angl.. 84 squares, map 487.

ANETHUM L.

Anethum graveolens L. (*Peucedanum graveolens* (L.) Bentham & Hooker fil.) : Dill. Casual (Europe). Waste places. Fl.7. Glam., Pembs.. 2 squares : **ST***27*[112]; **SM**91[112].

SILAUM Miller

Silaum silaus (L.) Schinz & Thell. : Ffenigl yr Hwch : Pepper-saxifrage. Native. Pastures and roadsides (G,M,R). Fl.7-9. Mons., Glam., Brecs., *Rads.*, *Carms.*, *Monts.**, Denbs., Flints., [*Cards.*, *Mer.*, *Caerns.*, *Angl.*]. 42 squares, map 488.

MEUM Miller

Meum athamanticum Jacq. : Ffenigl Helen Lueddog : Spignel. Denizen (England). Pastures. Fl.6-7. Mer., Caerns., Denbs.. 7 squares.
Mer.: Barmouth, SH61, TR, c1840[126]; Diffwys, SH62, An, 1960+[5]; Cadair Idris, SH71, JHS, 1930[112] and PMB, 1968[3]; Brithdir, SH71, PMB, 1953[112] & 1975[3]; btwn Bala and Dolgellau, SH82, An, c1800[82] and WCo & PMB, 1975[3]; Llangower, SH93, An, c1800[82]. **Caerns.:** nr Bethesda, SH66, CWi, 1876[112]. **Denbs.:** nr Pentrefoelas, SH95, EPE, 1952[112].

CONIUM L.

Conium maculatum L. : Cegiden : Hemlock. Native or colonist. Roadsides (Ar,E,G,M,Q,R,Sc). Fl.6-10. Mons., Glam., Brecs., Rads., Carms., Pembs., Cards., Monts., Mer., Caerns., Denbs., Flints., Angl.. 187 squares, map 489.

PRANGOS Lindley

Prangos uechtritzii Boiss. & Hausskn. Casual (Turkey). Waste places. Fl.8. *Glam.*. 1 square : **ST**16[30].

BUPLEURUM L.

Bupleurum rotundifolium L. : Paladr Trwyddo : Thorow-wax. Casual (Europe). Cornfields and waste places (E,F,R). Fl.7-8. Glam., *Cards.*, *Mer.*, *Caerns.*[121], *Denbs.*, *Flints.*. 9 squares : **SS***69*[247]; **ST***17*[112]; **SN***57*[210]; **SO***00*[107]; **SH***53*[126], *61*[126],*93*[126]; **SJ***18*[143],*35*[146]. Some of these records doubtless refer to **B. subovatum** with which this species is often confused.

B. subovatum Link ex Sprengel (*B. lancifolium* auct., non Hornem.; *B. intermedium* (Loisel. ex DC.) Steudel) : False Thorow-wax. Established alien or casual (Mediterranean region). Waste places (E,R,Rg). Fl.6-7. *Mons.*, Glam., *Rads.*, Carms., Pembs., *Cards.*, Monts., Denbs., *Angl.*. 16 squares : **SS**19; **ST**06,*16*, 17,*27*,49; **SM**90,91; **SN**11,31,68; **SO**06; **SH**28; **SJ**10,12,34.

B. tenuissimum L. : Paladr Trwyddo Eiddilddail : Slender Hare's-ear. Native. Salt-marshes. Fl.7-9. Mons., Glam., *Caerns.**, *Flints.*[85], *Angl.*[231], [*Mer.*]. 8 squares.
Mons.: Rumney, ST27, AEW, 1951[112]; Magor Pill, ST48, WAS, 1894[112]; Caldicot Pill, ST48, TGE, 1972[181]; Blackrock, ST58, TGE, 1972[112]; nr Chepstow, ST59, MG, 1960+[99]. **Glam.:** Aberthaw, ST06, EV, 1935[112]; Cardiff, ST17, EV, 1903[224] and AEW, 1956[112]; E. Cardiff, ST27, HJR, 1908[112] and AEW, 1926[112]. **Caerns.:** nr Bangor, SH57, An, 1938[107].

B. falcatum L. : Sickle-leaved Hare's-ear. Casual (Europe). Waste places. Glam.. 1 square : **ST***27*[112].

APIUM L.

Apium graveolens L. : Mers : Wild Celery. Native. Salt-marshes (Ar,E,G,H,M,S). Fl.6-8. Mons., Glam., Carms., Pembs., *Cards.*, Mer., *Caerns.*, Denbs., Flints., Angl., [*Monts.*]. 70 squares, map 490.

A. nodiflorum (L.) Lag. : Dyfrforonen Sypflodeuog : Fool's Water-cress. Native. Ditches (Al,As,Ds,E,G,H, M,Q,R,Sc,W). Fl.6-9. Mons., Glam., Brecs., Rads., Carms., Pembs., Cards., Monts., Mer., Caerns., Denbs., Flints., Angl.. 213 squares, map 491.

A. inundatum (L.) Reichenb. fil. : Dyfrforonen Nofiadwy : Lesser Marshwort. Native. Pool margins (Al,Ar,H,M). Fl.5-9. Mons., Glam., Brecs., Rads., Carms., Pembs., Cards., *Monts.*, Mer., Caerns., Denbs., Flints., Angl.. 105 squares, map 492.

A. leptophyllum (Pers.) F. Mueller ex Bentham (*A. tenuifolium* Thell.). Casual (America). Waste places. *Glam.*. 1 square : **ST***17*[22].

PETROSELINUM Hill

Petroselinum crispum (Miller) A. W. Hill : Persli : Garden Parsley. Naturalized alien (S. Europe). Banks and sea walls (E,Q,R,Rw,Sc). Fl.5-9. Mons., Glam., Brecs.*, Rads., Carms., Pembs., Cards., Mer., Caerns., Denbs., *Flints.*, Angl.. 36 squares, map 493.

P. segetum (L.) Koch : Eilunberllys : Corn Parsley. Native. Dry banks (E,F,G,Q,R,Sc). Fl.7-9. Mons., Glam., Pembs., Cards., [*Denbs.*]. 20 squares, map 494.

SISON L.

Sison amomum L. : Githran : Stone Parsley. Native. Hedgebanks (Ar,E,G,M,R,Rw,W). Fl.7-9. Mons., Glam., *Brecs.*, *Rads.*[85], *Carms.*[85], *Pembs.*, *Monts.**, Caerns., Denbs., Flints., Angl.. 38 squares, map 495.

CICUTA L.

Cicuta virosa L. : Buladd : Cowbane. Native. Pond margins (Al,Ar,M). Fl.8-9. *Rads.*†, *Monts.*, *Denbs.*, [*Mons.*, *Mer.*, *Caerns.*, *Flints.*]. 7 squares.
Rads.: Carneddau, SO05, HNR, 1880[128]; Rhos-goch Common, SO14, HNR, 1880[128]. **Monts.:** nr Llandrinio, SJ31, An, 1950+[7]. **Denbs.:** Hafod Wood, SJ34, AGS, 1978[197]; Llyn Bedydd, SJ43, AGS, 1977[197] and JMB, 1978[184]; nr Bronington, SJ43, PD, 1979[197]; nr Penley, SJ43 & 44, PD, 1979[197]; nr Eglwys Cross, SJ44, PD, 1979[197]; nr Fenn's Bank, SJ53, PD, 1979[197].

AMMI L.

Ammi visnaga (L.) Lam. : Toothpick-plant. Casual (Mediterranean region). Waste places. *Carms.*. 1 square : **SN***42*[19].

A. majus L. Casual (Mediterranean region). Waste places. Fl.8-10. *Mons.*, Glam., Carms.. 7 squares : **ST***27*[112],38[91], 48[93],59[91]; **SN**31[112],*42*[161]; **SO***00*[140].

FALCARIA Fabr.

Falcaria vulgaris Bernh. : Longleaf. Established alien (Europe). Waste places. Fl.7-9. Glam., *Mer.*. 3 squares.
Glam.: Port Talbot, SS78, HJR, 1909[112]; Barry Docks, ST16, EV, 1931[112]. **Mer.:** Barmouth, SH61, An, 1901[107] and PMB, 1980[3].

CARUM L.

Carum carvi L. : Carwy : Caraway. Established alien (N. and C. Europe). Waste places (E,R). Fl.4-6. Mons., Glam., *Brecs.**, *Carms.**, *Cards.*, *Caerns.*, Angl.. 12 squares : **SS***69,78,87*; **ST***17,27,29*; **SN***42,58,81*; **SO***00*; **SJ***27,67*.

C. verticillatum (L.) Koch : Carwas Troellog : Whorled Caraway. Native. Meadows (Ar,G,H,M). Fl.6-8. Glam., Brecs., Rads., Carms., Pembs., Cards., Monts.*, Mer., Caerns.. 84 squares, map 496.

ANGELICA L.

Angelica sylvestris L. : Llys yr Angel : Wild Angelica. Native. Marshes (Al,Ar,E,G,H,M,Q,R,Rw,W). Fl.7-9. Mons., Glam., Brecs., Rads., Carms., Pembs., *Cards.*, Monts., Mer., Caerns., Denbs., Flints., Angl.. 271 squares, all except : **SS**38,39,77; **ST**26; **SM**50,71,73,84; **SN**36,59; **SH**98.

A. archangelica L. : Llys yr Angel Peraroglaidd : Garden Angelica. Casual (N. Europe). Waste places. *Carms.*, *Angl.*. 2 squares : **SN**64[161]; **SH***57*[101].

LEVISTICUM Hill

Levisticum officinale Koch : Lovage. Established alien (Iran). Roadsides. *Cards.*. 1 square.
Cards.: Llanrhystud, SN56, JHS, c1930[210].

PEUCEDANUM L.

Peucedanum ostruthium (L.) Koch : Llysiau'r Ddannoedd : Masterwort. Naturalized alien (S. Europe). Moist meadows. Fl.7-8. *Carms.*, *Cards.*, *Mer.*, *Caerns.*, Denbs., *Flints.*. 11 squares : **SS***59*; **SN***45*; **SH***56,70, 78,85,93,95*; **SJ***23,35,36*. Apparently extinct in Wales.

PASTINACA L.

Pastinaca sativa L. : Moronen y Moch : Wild Parsnip. Native. Roadsides and waste places (D,E,G,M,R,Rw,S). Fl.6-9. Mons., Glam., (*Brecs.**, *Rads.**), Carms., *Pembs.*, (Cards., *Mer.*, Caerns., *Denbs.*, Flints.). 68 squares, map 497.

HERACLEUM L.

Heracleum sphondylium L. : Panasen y Cawr : Hogweed. Native. Roadsides (D,E,G,H,M,Q,R,Rw,S,W). Fl.5-9. Mons., Glam., Brecs., Rads., Carms., Pembs., Cards., Monts., Mer., Caerns., Denbs., Flints., Angl.. 279 squares, all except : **SS**39; **SM**50; **SH**51.

H. mantegazzianum Sommier & Levier : Giant Hogweed. Established alien (Caucasus). River banks and roadsides (E,M,R). Fl.4-7. Mons., Glam., Brecs., *Rads.*, Carms., *Pembs.*, *Monts.*, *Caerns.*, *Denbs.*, *Flints.*, Angl.. 27 squares, map 498.

TORILIS Adanson

Torilis nodosa (L.) Gaertner : Troed-y-cyw Clymog : Knotted Hedge-parsley. Native. Dry banks (Ar,G,M,R, S,Sc). Fl.5-8. Mons., Glam., Brecs., *Rads.*[230], Carms., Pembs., *Cards.*, Monts., (*Mer.*), Caerns., *Denbs.*, Flints., Angl.. 69 squares, map 499.

T. arvensis (Hudson) Link : Troed-y-cyw Ymdaenol : Spreading Hedge-parsley. Established alien (England). Cultivated ground. Fl.7-8. *Mons.**, *Glam.*, *Brecs.*[130], *Rads.*[230], *Carms.*[230], *Monts.*[230], *Mer.*, *Caerns.*[230], *Denbs.*, *Flints.*, [*Angl.*]. 5 squares.
Mons.: nr Raglan, SO40, An, 1807[226]. **Glam.:** Swansea, SS69, TBF & EL, c1840[198]; Aberdare, SO00, HJR, c1905[140]. **Mer.:** Llandderfel, SH93, TR, c1840[126]. **Flints.:** nr Prestatyn, SJ08, RB, 1885[140].

T. japonica (Houtt.) DC. : Troed-y-cyw Syth : Upright Hedge-parsley. Native. Hedges (D,E,G,M,R,W). Fl.5-9. Mons., Glam., Brecs., Rads., Carms., Pembs., Cards., Monts., Mer., Caerns., Denbs., Flints., *Angl.*. 253 squares, all except : **SR**89; **SS**38,39; **ST**26; **SM**50,71,73 84; **SN**36,53,54,75,77,78,86,87; **SO**30,39; **SH**27,29,51, 55,65,73,74,91,92,95; **SJ**13.

T. leptophylla (L.) Reichenb. fil. (*Caucalis leptophylla* L.). Casual (S. Europe). Waste places. *Glam.*. 1 square : **ST***27*[224].

CAUCALIS L.

Caucalis platycarpos L. : Eilunberllys Bychan : Small Bur-parsley. Casual (Europe). Waste places (R,Rg,Rw). Fl.5-6. Glam., *Brecs., Denbs., Flints.*. 7 squares : **SS***68*[112]; **ST**16[112], *17*[112]; **SN***90*[107]; **SO***00*[140]; **SH***87*[144]; **SJ**36[144].

TURGENIA Hoffm.

Turgenia latifolia (L.) Hoffm. (*Caucalis latifolia* L.) : Great Bur-parsley. Casual (C. and S. Europe). Waste places. Fl.5-8. Mons., Glam., *Carms.*. 6 squares : **SS***78*[140]; **ST***16*[112],17[112]; **SN***50*[161]; **SO***00*[140],*51*[112].

DAUCUS L.

Daucus carota L.

Subsp. **carota** : Moronen y Maes : Wild Carrot. Native. Pastures (D,E,F,G,Q,R,Rw,S,Sc). Fl.5-9. Mons., Glam., Brecs., Rads., Carms., Pembs., Cards., Monts., Mer., Caerns., Denbs., Flints., Angl.. 181 squares, map 500.

Subsp. **gummifer** Hooker fil. (*D. gingidium* auct.) : Moronen y Môr : Sea Carrot. Native. Sea cliffs. Fl.6-9. Glam.*, Pembs., *Cards.*, *Caerns.**, *Angl.*, [*Carms.*]. 20 squares, map 501.

PYROLACEAE

PYROLA L.

Pyrola minor L. : Glesyn y Gaeaf : Common Wintergreen. Native. Woods. Fl.5-7. Mons., *Glam.*, *Carms.**, *Angl.*, [*Caerns.*, *Flints.*]. 8 squares.
Mons.: Chepstow Park Wood, ST49, RWR, 1923[112]; Wynd Cliff, ST59, WAS, 1878[112] and TGE, 1971[181]; Blorenge, SO21, SH, pre 1900[226]; Trelleck, SO50, SGC, c1930[226]; nr Monmouth, SO51, SGC, 1929[112] and AEW, 1930+[226]. **Glam.:** Sgŵd Einion Gam, SN80, JHo, c1830[81] (this locality may be in *Brecs.*). **Carms.:** Tywyn Burrows, SN30, IMV, 1975[112] and RDP, 1981[93]; Trimsaron, SN40, RMd & RDP, 1982[201]. **Angl.:** Llugwy Woods, SH48, WM, c1760[78].

[**P. media** Swartz *Mons.*, *Flints.*]

P. rotundifolia L.

[Subsp. **rotundifolia** *Glam.*]

Subsp. **maritima** (Kenyon) E. F. Warburg : Coedwyrdd Crynddail : Round-leaved Wintergreen. Native. Sand dunes. Fl.8-9. Glam., Carms.*, Mer.*, Flints., Angl.*. 13 squares : **SS**49,58,*78*,79,87; **SN**20, 30; **SH**36,46,52; **SJ**07,*08*,18.

ORTHILIA Rafin.

Orthilia secunda (L.) House (*Ramischia secunda* (L.) Garcke) : Coedwyrdd Bylchog : Serrated Wintergreen. Native. Rocky woods and slopes. Fl.4-8. *Mons.*†, *Glam.*†, Brecs., Mer.*. 5 squares.
Mons.: Wynd Cliff, ST59, BMW, 1879[107]. **Glam.:** Craig y Llyn, SN90, HJR, 1906[107]. **Brecs.:** Fan Nedd, SN91, WLB, 1921[107]; Cwm Senni, SN92, HHK, 1906[107]; Craig Cerrig-gleisiad, SN92, BAM, 1955[112] and MP, 1979[200]. **Mer.:** Llanuwchllyn, SH82, DAR, 1961[112].

MONOTROPA L.

Monotropa hypopitys L. (incl. *M. hypophegea* Wallr.) : Cyd-dwf : Yellow Bird's-nest. Native. Dune slacks and woods. Fl.7-8. Mons., Glam., *Brecs.*, *Carms.**, *Mer.**, *Denbs.**, Flints., *Angl.*, [*Caerns.*]. 19 squares : **SS**49,78, 87,88;**ST***17*,18,59; **SN**30,40,41,51,61,62; **SO**21; **SH**46, 53; **SJ**16,25,26. Recent work suggests that **M. hypopitys** and **M. hypophegea** are not taxonomically distinct.

ERICACEAE

ERICA L.

Erica ciliaris L. : Dorset Heath. Casual (S. England). Angl., [*Glam.*]. 1 square : **SH***48*[112].

E. tetralix L. : Grug Deilgroes : Cross-leaved Heath. Native. Wet moorland (Ds,G,H,M). Fl.6-10. Mons., Glam., Brecs., Rads., Carms., Pembs., Cards., Monts., Mer., Caerns., Denbs., Flints., Angl.. 225 squares, map 502.

E. cinerea L. : Clychau'r Grug : Bell Heather. Native. Heaths (Ar,G,H,W). Fl.6-10. Mons., Glam., *Brecs.*, Rads., Carms., Pembs., Cards., Monts., Mer., Caerns., Denbs., Flints., Angl.. 206 squares, map 503.

E. vagans L. : Cornish Heath. Introduced (S. England). River banks. *Denbs.*, [*Glam.*]. 1 square.
Denbs.: nr Cefn Mawr, SJ24, VG, 1952[52].

[**E. erigena** R. Ross (*E. mediterranea* auct., non L.). *Glam.*]

CALLUNA Salisb.

Calluna vulgaris (L.) Hull : Grug Cyffredin : Heather. Native. Heaths and moors (Al,Ar,Ds, E,G,H,M,Q,R,Rw,Sc,W). Fl.8-10. Mons., Glam., Brecs., Rads., Carms., Pembs., Cards., Monts., Mer., Caerns., Denbs., Flints., Angl.. 263 squares, all except : **SR**89; **SS**19,38,39; **ST**06,16,17,27,38,58; **SM**50; **SN**30,59; **SO**30,41; **SH**51,98; **SJ**33,37.

RHODODENDRON L.

Rhododendron luteum Sweet : Yellow Azalea. Naturalized alien (Asia Minor). Woods. Fl.6. Cards., *Mer.*. 2 squares.
Cards.: Devil's Bridge, SN77, HJMB, 1956[53]; Hafod, SN77, AOC, 1976[112]. **Mer.:** nr Ganllwyd, SH72, TT, 1956[7].

R. ponticum L. : Rhododendron. Naturalized alien (S. Europe). Woods (D,E,M,W). Fl.5-6. *Mons.*, Glam., Brecs., *Rads.*, Carms., *Pembs.*, Cards., Monts., Mer., Caerns., *Denbs.*, *Flints.*, *Angl.*. 136 squares, map 504.

ARBUTUS L.

Arbutus unedo L. : Mefusbren : Strawberry-tree. Naturalized alien (Ireland). Limestone cliffs. Fl.8-11. Mons., Glam., Caerns.. 3 squares.
Mons.: Chepstow, ST59, TGE, 1978[112]. **Glam.:** Caswell Bay, SS58, JAW, 1927[45] & 1952[112]. **Caerns.:** Gt Ormes Head, SH78, DPY, 1941[54] and JMB, 1977[67].

PERNETTYA Gaud.-Beaup.

Pernettya mucronata (L. fil.) Gaud.-Beaup. ex Sprengel : Prickly Heath. Established alien (Chile). Hedges (E,H,W). Monts., Mer., Caerns., *Angl.*. 4 squares.
Monts.: Bont Dolgadfan, SH80, JAW, 1941[112]. **Mer.:** Penmaenpool, SH61, JAW, 1941[112]. **Caerns.:** Madryn Castle, SH23, FCG, 1962[112]. **Angl.:** Holyhead Mountain, SH28, IB, 1978[209].

ANDROMEDA L.

Andromeda polifolia L. : Rhosmari Gwyllt : Bog-rosemary. Native. Bogs. Fl.4-9. *Glam.*†, Brecs.*, Carms., Cards., *Monts.**, Mer., Denbs., [*Caerns.*, *Flints.*]. 26 squares, map 505.

VACCINIUM L.

Subgen. Oxycoccus (Hill) A. Gray

Vaccinium oxycoccus L. (*Oxycoccus palustris* Pers.) : Llygaeron : Cranberry. Native. Bogs (H,M). Fl. 5-7. Mons., Glam., Brecs., Rads., Carms., Pembs., Cards., Monts., Mer., Caerns., Denbs., *Flints.*, Angl.. 119 squares, map 506.

V. macrocarpum Aiton : American Cranberry. Naturalized alien (N. America). Bogs. Fl.4. Pembs., Flints.. 2 squares.
Pembs.: Cwm Dewi, Dinas, SN03, SBE & RAL, 1981[112]. **Flints.:** Soughton nr Mold, SJ26, EPo, 1859[143]. A vegetative specimen resembling this species has been collected from Glam. (Crymlyn Fen, SS69, RMd, 1981[112]).

Subgen. Vaccinium

V. vitis-idaea L. : Llus y Geifr : Cowberry. Native. Moors (H,Q,W). Fl.6. Mons., Glam., Brecs., Rads., Carms., Cards., *Monts.*, Mer., Caerns., Denbs., *Flints.*. 64 squares, map 507.

V. myrtillus L. : Llus : Bilberry. Native. Heaths and moors (E,G,H,M,Q,R,W). Fl.4-6. Mons., Glam., Brecs., Rads., Carms., Pembs., Cards., Monts., Mer., Caerns., *Denbs.*, Flints., *Angl.*. 215 squares, map 508.

V. myrtillus × vitis-idaea = V. × intermedium Ruthe Spontaneous. Heath. Fl.7. *Mer.*. 1 square.
Mer.: nr Llanfachreth, SH72, EV, 1908[27].

EMPETRACEAE

EMPETRUM L.

Empetrum nigrum L.
Subsp. **nigrum** : Creiglys : Crowberry. Native. Moors (Ar,H,M). Fl.4. Mons., Glam., Brecs., Rads., Carms., Cards., Monts., Mer., Caerns., Denbs., *Flints.*. 107 squares, map 509.

Subsp. **hermaphroditum** (Hagerup) Böcher (*E. hermaphroditum* Hagerup) : Mountain Crowberry. Native. Moors. Fl.4. Mer.*, Caerns.. 3 squares.
Mer.: Mynydd Moel, Cadair Idris, SH71, DAR, 1963[3] and PMB, 1965[112]. **Caerns.:** Glyders, SH65, CGT, 1930[112] and ER, 1957[207]; Snowdon, SH65, ER, 1950+[207]; Craig yr Ysfa, SH66, ER, 1950+[207]; Pen-yr-helgi-ddu, SH66, ER, 1950+[207].

PRIMULACEAE

PRIMULA L.

Primula vulgaris Hudson : Briallu : Primrose. Native. Woods (E,G,Q,W). Fl.1-5. Mons., Glam., Brecs., Rads., Carms., Pembs., Cards., Monts., Mer., Caerns., Denbs., Flints., Angl.. 272 squares, all except : **SR**89; **SS**39,77; **ST**19, **SM**50; **SN**36,59,89,96; **SO**10.

P. veris L. : Briallu Mair : Cowslip. Native. Pastures (D,G,M,Q,R,Rw,Sc,W). Fl.4-6. Mons., Glam., Brecs., Rads., Carms., Pembs., *Cards.*, Monts., Mer., Caerns., Denbs., Flints., *Angl.*. 187 squares, map 510.

P. veris × vulgaris = P. × tommasinii Gren. & Godron Spontaneous. Pastures and roadsides (E,G,M,R,Rw,Sc,W). Fl.4-5. Mons., Glam., *Brecs.*, Rads., Carms., Pembs., Cards., *Monts.*, *Mer.*, Caerns., *Denbs.*, Flints., Angl.. 66 squares, map 511.

P. japonica A. Gray : Japanese Primrose. Casual (Japan). Waste places. Fl.7. Carms.. 1 square : **SN**72[112].

HOTTONIA L.

Hottonia palustris L. : Fioled y Dŵr : Water-violet. Native. Ponds and ditches (Al,Ar,M). Fl.6. *Glam.*, *Carms.*†, *Monts.*, Caerns., *Denbs.*, *Flints.*, Angl., [*Mer.*]. 11 squares : **ST**16; **SN**42; **SH**37,47,68,78; **SJ**17,20,22,44,53.

CYCLAMEN L.

Cyclamen hederifolium Aiton (*C. neapolitanum* Ten.) : Bara'r Hwch : Cyclamen. Naturalized alien (S. Europe). Hedges (E,G,W). Fl.8-9(-11). Mons., *Glam.*, Rads.*, *Pembs.**, *Cards.**, Angl.*. 8 squares.
Mons.: St. Arvans, ST59, WAS, 1904[112] and HSR, 1922[107]. **Glam.:** Fairyhill Wood, SS49, JAW, 1931[221]; Coedrhiglan, ST17, JS, c1885[218]; N. Cardiff, ST18, V, 1979[95]. **Rads.:** Clyro, SO24, ACP, 1974[182] & 1980[112]. **Pembs.:** Brownslade Wood, SR99, FLR, c1950[204]. **Cards.:** Llangybi Common, SN55, IWC, 1977[245]. **Angl.:** Plasnewydd, SH56, WEH, 1966[112].

C. repandum Sibth. & Sm. Established alien (S. Europe). Waste places. Fl.4. *Angl.*. 1 square.
Angl.: Penrhos, nr Holyhead, SH28, GCD, 1924[24].

C. coum Miller Introduced (S.E. Europe). *S. Wales*[251].

C. persicum Miller (*C. latifolium* Sibth. & Sm.). Casual (Greece). Waste places. Fl.5. Caerns.. 1 square : **SH**57[112].

LYSIMACHIA L.

Lysimachia nemorum L. : Gwlydd Melyn Mair : Yellow Pimpernel. Native. Woods (E,G,H,R,Rw,W). Fl.5-9. Mons., Glam., Brecs., Rads., Carms., Pembs., Cards., Monts., Mer., Caerns., Denbs., Flints., Angl.. 230 squares, map 512.

L. vulgaris L. : Trewynyn : Yellow Loosestrife. Native. Marshes (Ds,H,M,W). Fl.6-9. Mons., Glam., *Brecs.,* Rads., Carms., *Pembs., Cards.,* Monts., *Mer., Caerns.,* Denbs., *Flints.,* Angl.. 124 squares, map 513.

L. ciliata L. : Fringed Loosestrife. Established alien (N. America). Lakesides. Fl.9. *Rads., Caerns.*. 2 squares.
Rads.: Tyn-yr-ynn, SO17, RCLH, 1956[53]. **Caerns.:** Afon-wen, SH43, HCL, 1884[101].

L. nummularia L. : Siani Lusg : Creeping-Jenny. Native. Damp pastures and marshes (Ar,E,G,M, R,Rw,W). Fl.6-8. Mons., Glam., Brecs., Rads., Carms., Pembs., *Cards.,* Monts., Mer., Caerns., *Denbs., Flints.,* (*Angl.**). 153 squares, map 514.

L. punctata L. : Dotted Loosestrife. Naturalized alien (Europe). Marshy places (Ar,M,R). Fl.6-7. Glam., Brecs.*, *Rads.*,* Carms., Cards.*, *Monts.*,* Caerns., Angl.*. 29 squares, map 515.

GLAUX L.

Glaux maritima L. : Glas yr Heli : Sea-milkwort. Native. Salt-marshes (Ds,S). Fl.5-6. Mons., Glam., Carms., Pembs., Cards., Monts., Mer., Caerns., Denbs., *Flints.,* Angl., [*Brecs., Rads.*]. 90 squares, map 516.

ANAGALLIS L.

Anagallis minima (L.) E. H. L. Krause (*Centunculus minimus* L.) : Bril-lys Coraidd : Chaffweed. Native. Damp shady places (D,Ds,E,G,H,M,R,Rw,S). Fl.6-8. Glam., Brecs.*, *Carms., Pembs.,* Cards., Mer., Caerns., *Denbs.,* Flints., Angl., [*Monts.*]. 56 squares, map 517.

A. tenella (L.) L. : Gwlyddyn Mair y Gors : Bog Pimpernel. Native. Bogs (Ar,D,Ds,E,G,H,M,R,W). Fl. 6-9. Mons., Glam., Brecs., Rads.*, Carms., Pembs., Cards., Monts., Mer., Caerns., Denbs., *Flints.,* Angl.. 206 squares, map 518.

A. arvensis L. : Gwlydd Mair : Scarlet Pimpernel. Native. Cultivated and sandy ground (D,Ds,F,G, Q,R,Rg,Rw,S). Fl.6-9. Mons., Glam., Brecs., Rads., Carms., Pembs., Cards., Monts., Mer., Caerns., Denbs., Flints., Angl.. 239 squares, map 519.
The blue flowered f. **azurea** has been recorded from Mons., Glam., *Brecs., Carms.,* Pembs., Cards., *Mer., Denbs., Flints.*

A. foemina Miller (*A. arvensis* subsp. *foemina* (Miller) Schinz & Thell.) : Gwlyddyn Mair Benyw : Blue Pimpernel. Established alien (S. England). Cultivated ground. Fl.6-11. Mons.*, Glam., *Carms.**, Flints.*, [*Mons., Carms., Pembs., Cards., Mer., Denbs., Angl.*]. 6 squares.
Mons.: Newport, ST38, TGE, 1980[112]. **Glam.:** Port Talbot Docks, SS78, HJR, c1906[140]; Lavernock, ST16, CTV, 1894[112]; Barry, ST16, MTH, 1929[112]; Cardiff, ST17, HJR, 1907[112] and LB, 1962[112]; Dinas Powis, ST17, GMB, 1953[112]. **Carms.:** Cwmffrwd, SN41, TWB, 1905[201]. **Flints.:** Cilcain, SJ16, HGH, 1960[112].

SAMOLUS L.

Samolus valerandi L. : Claerlys : Brookweed. Native. Wet places, chiefly near the sea (Ar,D,Ds,M,S,Sc). Fl. 6-8. Mons., Glam., Carms., Pembs., Cards., Monts., Mer., Caerns., *Denbs.,* Flints., Angl.. 87 squares, map 520.

PLUMBAGINACEAE

ARMERIA Willd.

Armeria maritima (Miller) Willd. : Clustog Mair : Thrift. Native. Salt-marshes and sea cliffs, occasionally on mountain cliffs inland (Ar,D,E,G,M,Q,R,S,Sc). Fl.4-10. Mons., Glam., Carms., Pembs., Cards., Monts., Mer., Caerns., Denbs., Flints., Angl.. 107 squares, map 521.

LIMONIUM Miller

Limonium vulgare Miller : Lafant-y-Môr : Common Sea-lavender. Native. Salt-marshes (D,S). Fl.7-9. Mons.*, Glam., Carms., Mer., Caerns., Denbs., Angl., [*Pembs., Flints.*]. 32 squares, map 522.

L. humile Miller : Llemyg Blodau Rhydd : Lax-flowered Sea-lavender. Native. Salt-marshes. Fl.7-8. Pembs., Mer., Angl., [*Carms., Cards., Denbs.*]. 13 squares : **SR**99; **SM**80,90,91; **SN**00,01; **SH**27,*28*,37, 38,52,53,61.

L. humile × vulgare = L. × neumanii Salmon Spontaneous. Salt-marshes. Fl.8. Mer., *Angl.*. 4 squares.
Mer.: Mochras, SH52, PMB, 1964[57]; Harlech, SH53, PMB, 1972[3]; Arthog, SH61, PMB, 1962[112]. **Angl.:** nr Four Mile Bridge, SH27, JEG, 1885[107].

L. binervosum (G.E.Sm.) Salmon : Llemyg y Morgreigiau : Rock Sea-lavender. Native. Sea cliffs (D,M, S,Sc). Fl.6-10. Glam., Carms., Pembs., Cards., Caerns., *Flints.,* Angl., [*Mer.*]. 44 squares, map 523.

[**L. binervosum × paradoxum** *Pembs.*]

L. transwallianum (Pugsley) Pugsley Native. Sea cliffs. Fl.7-8. Pembs.. 2 squares.
Pembs.: Bosherston, SR99, TAWD, 1971[112]; nr Penally, SS19, HT, 1867[107] and TAWD, 1960+[79].

L. paradoxum Pugsley Native. Sea cliffs. Fl.8. Pembs.. 1 square.
Pembs.: St David's Head, SM72, HWP, 1931[112] and TAWD, 1960+[79].
Recent research by M. J. Ingrouille on the **L. binervosum** complex has indicated the presence of several new segregates in Wales but details were not available at the time of going to press (December 1982).

OLEACEAE

JASMINUM L.

Jasminum officinale L. : White Jasmine. Casual (S.W. Asia). Hedges. *Glam.*. 1 square : **ST**09[95].

FORSYTHIA Vahl

Forsythia suspensa (Thunb.) Vahl × **viridissima** Lindley = **F. × intermedia** Zabel Casual (garden origin). Woods. *Glam.*. 1 square : **SS**97[99].

FRAXINUS L.

Fraxinus excelsior L. : Onnen : Ash. Native. Woods (E,G,H,M,Q,R,Rw,Sc,W). Fl.4. Mons., Glam., Brecs., Rads., Carms., Pembs., Cards., Monts., Mer., Caerns., Denbs., Flints., *Angl.*. 269 squares, all except : **SR**89; **SS**38,39,77; **ST**26; **SM**50,62,73,84; **SN**15; **SH**51,98; **SJ**37.

SYRINGA L.

Syringa vulgaris L. : Lelog : Lilac. Established alien (S.E. Europe). Hedges. Fl.4-5. Mons., Glam., *Brecs., Rads., Carms., Pembs., Cards.,* Monts., *Caerns., Flints., Angl.*. 36 squares, map 524.

LIGUSTRUM L.

Ligustrum vulgare L. : Yswydden : Wild Privet. Native but frequently introduced. Woods and hedges (E,R,Sc,W). Fl.6-8. Mons., Glam., Brecs., Rads., Carms., Pembs., Cards., Monts., Mer., Caerns., Denbs., Flints., Angl.. 190 squares, map 525.

L. ovalifolium Hassk. : Garden Privet. Established alien (Japan). Hedges (E,M,W). Fl.7-8. *Mons.,* Glam., *Brecs., Rads.,* Carms., *Pembs.,* Cards., Monts., *Mer., Caerns., Denbs.,* Flints., Angl.. 41 squares, map 526. Many records of **L. vulgare**, especially from hedges, probably refer to this species.

DESFONTAINIACEAE

DESFONTAINIA Ruiz & Pavón

Desfontainia spinosa Ruiz & Pavón Introduced (S. America). *Caerns.*. 1 square.
Caerns.: Madryn Castle, SH23, JEG, 1917[16].

GENTIANACEAE

CICENDIA Adanson

Cicendia filiformis (L.) Delarbre : Canri Felen Eiddil : Yellow Centaury. Native. Damp peaty places. Fl.7-8. Pembs., *Caerns.**. 8 squares.
Pembs.: Penally, SS19, CCB, 1851[79]; Ramsey Island, SM62, TAWD, 1970+[79]; St David's Head, SM72, EA, 1902[112] and TAWD, 1970+[79]; nr Nolton, SM81, TAWD, 1974[80]; nr Mathry Station, SM83, BL, 1930[112]; Trefaser, SM83, TAWD, 1970+[79]; Tre-rhos, SM92, TAWD, 1970+[79], nr Goodwick, SM93, MHB, 1967[104].
Caerns.: nr Tudweiliog, SH23, APC, 1956[171] and An, 1981[74].

BLACKSTONIA Hudson

Blackstonia perfoliata (L.) Hudson : Canri Felen : Yellow-wort. Native. Limestone pasture and cliffs (D,Ds,E,G,H,Q,R,Rw,S,Sc,W). Fl.6-10. Mons., Glam., Carms., Pembs., *Cards.**, *Monts.*, (*Mer.**), Caerns., Denbs., Flints., Angl.. 69 squares, map 527.

CENTAURIUM Hill

Centaurium scilloides (L.fil.) Samp. (*C. portense* (Brot.) Butcher) : Canri Tryflwyddol : Perennial Centaury. Native. Grassy sea cliffs. Fl.6-7. Pembs.. 2 squares.
Pembs.: nr Newport, SN03, GCD, 1919[112] and ARP, 1975[93]; nr Newport, SN04, TAWD, 1967[7] and TGE, 1975[91].

C. erythraea Rafn (*C. minus* auct.; incl. *C. capitatum* (Willd.) Borbás) : Bustl y Ddaear : Common Centaury. Native. Dry pastures and sand dunes (D,Ds,E,F,G,Q, R,Rw,W). Fl.6-10. Mons., Glam., Brecs., Rads., Carms., Pembs., Cards., Monts., Mer., Caerns., Denbs., Flints., Angl.. 196 squares, map 528.

C. erythraea × littorale Spontaneous. Dune slacks. Fl.7. Mer., *Flints.*, Angl.. 4 squares.
Mer.: Mochras, SH52, PMB, 1956[112]. **Flints.:** Talacre, SJ18, KGA, 1961[108]. **Angl.:** Aberffraw Common, SH36, RHR, 1979[112]; Newborough Warren, SH46, RU, 1971[243]. Plants similar to this hybrid have been reported from *Pembs.*, but **C. littorale** has yet to be recorded from that vice-county.

C. littorale (D. Turner) Gilmour : Canri Goch Arfor : Seaside Centaury. Native. Sand dunes (D,Ds,G). Fl. 6-9. Carms., *Cards.,* Mer., *Caerns., Flints.,* Angl., [*Glam., Denbs.*]. 17 squares : **SN**15,20,30,*69*, **SH**27, 28,*32*,36,37,45,46,52,*61*,68,*78*,*88*; **SJ***08*.

C. pulchellum (Swartz) Druce : Canri Leiaf : Lesser Centaury. Native. Damp sandy places (D,Ds,F,R, S,Sc,W). Fl.6-9. Mons., Glam., Carms., Pembs., Cards., Mer., *Caerns.*, Flints., Angl.. 43 squares, map 529.

GENTIANA L.

Gentiana pneumonanthe L. : Crwynllys y Rhos : Marsh Gentian. Native. Moist heaths. Fl.8-9. *Caerns.*, Angl., [*Carms.*, *Pembs.*]. 5 squares.
Caerns.: nr Llyn Crafnant, SH76, JEG, c1895[101]. **Angl.:** Trearddur Bay, SH27, CCL, 1905[107]; Holyhead, SH28, Kn, 1812[107] and RHR, 1927[208]; nr Llanddaniel-fab, SH46, RHR, 1970+[208]; Penrhosllugwy, SH48, TR, 1891[112] and TGE, 1977[91]; nr Llanbedr-goch, SH48, RHR, 1972[208].

GENTIANELLA Moench

Gentianella campestris (L.) Börner : Crwynllys y Maes : Field Gentian. Native. Pastures (D,G,H,R). Fl 7-10. Mons., Glam., *Brecs.*, *Rads.*, *Carms.*, Pembs., Cards., *Monts.*, Mer., Caerns., *Denbs.*, *Flints.*, Angl.. 63 squares, map 530.

C. amarella (L.) Börner : Crwynllys Chwerw : Autumn Gentian. Native. Limestone pastures and sand dunes (D,Ds,G,H,Q,R). Fl.7-9. Mons., Glam., Brecs., Carms., Pembs., *Monts.*, Caerns., Denbs., Flints., Angl., [*Mer.*]. 58 squares, map 531.

G. amarella × uliginosa Spontaneous. Sand dunes. *Glam.*, *Carms.*, *Pembs.*. 4 squares.
Glam.: Whitford Burrows, SS49, NMP, 1950+[236]; Oxwich Burrows, SS58, NMP, 1950+[236]. **Carms.:** Pembrey, SN30, NMP, 1950+[217]. **Pembs.:** nr Tenby, SN10, NMP, 1950+[217]. It is probable that most colonies of **G. uliginosa** show signs of hybridization.

G. uliginosa (Willd.) Börner : Crwynllys y Tywod : Dune Gentian. Native. Damp sandy ground (D,Ds,G). Fl.7-9. Glam., Carms., Pembs.. 5 squares.
Glam.: Llanmadog, SS49, AGSt, 1934[45] and JEL, 1948[112]; Whitford Burrows, SS49, NMP, 1956[7] and QONK, 1970+[157]; Llangennith, SS49, MG, 1970[99]; Oxwich, SS58, JEL, 1948[112] and QONK, 1970+[157]. **Carms.:** Laugharne, SN20, JFT, 1930+[161] and IMV et al 1960+[181]; Tywyn Burrows, SN30, RFM, 1961[112] and RDP, 1980[201]. **Pembs.:** Tenby Burrows, SS19, HWP, 1923[107] and An, 1954[79].

[**G. germanica** (Willd.) E. F. Warburg *Pembs.*, *Flints.*]

MENYANTHACEAE

MENYANTHES L.

Menyanthes trifoliata L. : Ffa'r Gors : Bogbean. Native. Bogs and marshes (Al,Ar,H,M). Fl.4-7. Mons., Glam., Brecs., Rads., Carms., Pembs., Cards., Monts., Mer., Caerns., Denbs., Flints., Angl.. 200 squares, map 532.

NYMPHOIDES Séguier

Nymphoides peltata (S. G. Gmelin) O. Kuntze : Ffâen Cors Eddïog : Fringed Water-lily. Naturalized alien (England). Lakes and streams. Fl.6-9. Glam., Brecs., *Denbs.**. 9 squares.
Glam.: Broad Pool, SS59, HDa, 1960[112] and RHH, 1970[112]; nr Swansea SS69, JAW, c1925[221]; Crymlyn, SS79, QONK, 1972[157]; Pysgodlyn Mawr, ST07, MG, 1972[99]; nr Ferndale, ST09, JT, 1975[95]; Cardiff, ST17, AEW, 1942[95]; Whitchurch, Cardiff, ST18, EV, 1943[43]. **Brecs.:** Llangorse Lake, SO12, EB, 1936[112] and DEV, 1951[112]. **Denbs.:** nr Abergele, SH97, HWil, 1976[67].

APOCYNACEAE

VINCA L.

Vinca minor L. : Perfagl : Lesser Periwinkle. Denizen or naturalized alien (Europe). Woods and hedges (E,R,W). Fl.4-5(-9). Mons., Glam., *Brecs.*, Rads., Carms., Pembs., *Cards.*, Monts., Mer., Caerns., Denbs., *Flints.*, *Angl.*. 100 squares, map 533.

V. major L. : Perfagl Mwyaf : Greater Periwinkle. Established alien (Europe). Hedges (E,M,R,Rg,Rw). Fl. 3-5. Mons., Glam., *Brecs.*, Rads., Carms., Pembs., Cards., Monts., Mer., Caerns., *Denbs.*, Flints., Angl.. 119 squares, map 534.

[**V. major** subsp. **hirsuta** (Boiss.) Stearn (*V. herbacea* auct. angl., non Waldst. & Kit.). *Glam.*. The specimen from *Glam.* is now regarded as **V. major** cv. **'Oxyloba'.**]

RUBIACEAE

SHERARDIA L.

Sherardia arvensis L. : Mandon Las yr Ŷd : Field Madder. Native. Cultivated ground (E,F,Q,R). Fl.5-9. Mons., Glam., Brecs., Rads., Carms., Pembs., Cards., Monts., *Mer.*, Caerns., Denbs., Flints., *Angl.*. 159 squares, map 535.

PHUOPSIS (Griseb.) Hooker fil.

Phuopsis stylosa (Trin.) B. D. Jackson : Caucasian Crosswort. Established alien (Persia). Hedgerow (E,R). Fl.5-7. Glam., Carms., *Pembs.*, Caerns., Flints.. 7 squares.
Glam.: St Donats, SS96, EV, 1927[26] and JMT, 1970[112]; nr Creigiau, ST08, MD, 1952[112]. **Carms.:** Llansteffan, SN31, HSo, 1958[112]. **Pembs.:** Camros, SM91, CFH, 1955[107]. **Caerns.:** Tre-garth, SH56, RHR, 1955[112]. **Flints.:** nr Diserth, SJ07, JPAJ, 1950[7]; Prestatyn, SJ08, BA, 1927[26].

ASPERULA L.

Asperula cynanchica agg. : Mandon Fechan : Squinancywort. Native. Dry calcareous pastures (D,G,Sc). Fl.5-7. Glam., Carms., Pembs., *Cards.*, *Denbs.*[230]. 23 squares, map 536.

A. occidentalis Rouy Native. Sand dunes. Fl.5-7. Glam., Carms., Pembs.. 6 squares.
Glam.: Rhosili, SS48, HJR, 1904[107] and VMP, 1929[112]; Pennard Castle, SS58, EMT, 1920[112]; Porthcawl, SS87, WAS, 1901[112] and EKH, 1953[242]. **Carms.:** Pendine, SN20, JAW, 1952[112]. **Pembs.:** Penally Burrows, SS19, HT, 1869[107]; Tenby, SN10, JSi, 1881[107] and RMB, 1975[255].

A. cynanchica L. *s.s.* Native. Dry calcareous pastures (D,G,Sc). Fl.6-8. Glam., Carms., Pembs.. 16 squares : **SR**99; **SS**09,48,49,*68*,78,88,96,*97*; **ST***16*; **SM**72,80,91; **SN**10,40,61.

A. arvensis L. Casual (Europe). Waste places (F,R,Rw). Fl. 6-9. Mons., Glam., *Flints*.. 7 squares : **SS***78*[140]; **ST**17[112], *27*[25], 49[184]; **SO***00*[140], *51*[42]; **SJ***16*[143].

A. orientalis Boiss. & Hohen. Casual (S.W. Asia). Waste places. Fl.6. Glam.. 1 square : **ST**27[112].

GALIUM L.

Galium boreale L. : Briwydden Mynyddog-creigiog : Northern Bedstraw. Native. Mountain rocks. Fl.6-8. *Glam.*, Brecs., Carms., Mer., Caerns., [*Denbs.*]. 15 squares : **SN**72,81,82,*90*,92; **SO***01*; **SH**54,64,65,66, 71,73,74,75,*78*.

G. odoratum (L.) Scop. (*Asperula odorata* L.) : Briwydden Bêr : Woodruff. Native. Woods (E,G,M,Q,W). Fl.5-6. Mons., Glam., Brecs., Rads., Carms., Pembs., Cards., Monts., Mer., Caerns., Denbs., Flints., *Angl*.. 187 squares, map 537.

G. uliginosum L. : Briwydden y Fign : Fen Bedstraw. Native. Marshes and bogs (Al,Ar,G,H,M). Fl.6-9. Mons., Glam., Brecs.*, Rads., Carms.*, Pembs.*, Cards., *Monts.*, Mer., Caerns., *Denbs.*, *Flints.*, Angl.. 101 squares, map 538.

G. palustre agg. (incl. *G. elongatum* C. Presl) : Briwydden y Gors : Common Marsh-bedstraw. Native. Marshes (Al,Ar,D,Ds,G,H,M,R,W). Fl.5-9. Mons., Glam., Brecs., Rads., Carms., Pembs., Cards., Monts., Mer., Caerns., Denbs., Flints., Angl.. 270 squares, all except : **SS**38,39; **ST**26; **SM**50,71,73,94; **SN**36; **SO**30; **SH**51,98; **SJ**02.

G. palustre L. *s.s.* Native. Marshes (Al,Ar,D,Ds, G,H,M,R). Fl.5-9. Mons., Glam., Brecs., Rads., Carms., Pembs., Cards., Monts., Mer., Caerns., Denbs., Flints., Angl.. 63 squares, map 539.

G. elongatum C. Presl (*G. palustre* subsp. *elongatum* (C. Presl) Lange). Native. Marshes and swamps (Al,Ar,M). Fl.6-9. Mons., Glam., *Brecs.**, Carms.*, Pembs.*, *Monts.**, Mer.*, *Caerns.**, *Angl.**. 27 squares, map 540.

Both segregates are very much under-recorded owing to the difficulty in separating them, at least in Wales.

G. verum L. : Briwydden Felen : Lady's Bedstraw. Native. Pastures (D,E,G,H,M,Q,R,Rw,S,Sc,W). Fl.6-9. Mons., Glam., Brecs., Rads., Carms., Pembs., Cards., Monts., Mer., Caerns., *Denbs.*, Flints., Angl.. 210 squares, map 541.

G. mollugo agg. : Llysiau'r Pannwr. Native. Hedges and grassy places (D,E,F,G,M,Q,R,Rw,S,W). Fl.6-9. Mons., Glam., *Brecs.*, *Rads.*, Carms., Pembs., Cards., *Monts.*, Mer., *Caerns.*, Denbs., Flints.. 105 squares, map 542.

G. mollugo L. *s.s.* : Hedge-bedstraw. Native. Hedges (Ar,E,G,M,Q,R,Rw,S,Sc). Fl.6-9. Mons., Glam., *Brecs.*, *Rads.*[230], Carms., Pembs., Cards., Monts., (Mer.), *Caerns.*, Denbs., Flints.. 54 squares, map 543.

G. album Miller (*G. mollugo* subsp. *erectum* Syme; *G. erectum* Hudson, 1778 non 1762) : Upright Hedge-bedstraw. Native. Grassy places (E,F,G,R, Rw,S). Fl.5-9. Mons., Glam., *Brecs.**, *Rads.*, Carms.*, Cards., *Monts.**, *(Mer.)*, *Caerns.*, *Denbs.*, Flints.*. 18 squares : **SS**59; **ST**06,18,*49*,*59*; **SN**40,56, 57,67,*86*,95; **SO**02; **SH**(61),88; **SJ**10,16,*24*,26.

G. mollugo agg. × **verum** = **G.** × **pomeranicum** Retz. Spontaneous. Sand dunes. Fl.7. Glam.. 2 squares.
Glam.: Cwm Ivy, SS49, QONK, 1980[157]; Kenfig Burrows, SS78, AEW, 1956[112].

[**G. pumilum** Murray *Mons., Glam., Brecs., Rads., Caerns.*]

G. sterneri Ehrend. : Limestone Bedstraw. Native. Calcareous grassland and basic rocks. Fl.6-8(-10). Mons., Glam., Brecs., Caerns., *Flints.*, [*Rads.*, *Monts.*]. 12 squares : **SN**81,90,91,92; **SO**00,01,02,11; **SH**54, 65,66; **SJ**07.

G. saxatile L. (*G. hercynicum* auct.) : Briwydden Wen : Heath Bedstraw. Native. Heaths and pastures (E,G,H, M,Q,R,W). Fl.5-9. Mons., Glam., Brecs., Rads., Carms., Pembs., Cards., Monts., Mer., Caerns., Denbs., *Flints.*, *Angl.*. 260 squares, all except: **SR**89; **SS**39; **ST**16,17,27,48,58; **SM**50,73,90; **SN**00,11,59; **SO**30,40,41; **SH**36,51,98; **SJ**33,34,37.

G. saxatile × **sterneri** Spontaneous. Rocky places. *Caerns.*. 1 square.
Caerns.: Ysgolion Duon, SH66, KG, 1960+[197]; Llech Ddu, SH66, KG, 1960+[197].

G. spurium L. : False Cleavers. Casual (Europe). Waste places. Fl.10. Glam.. 1 square : **ST**27[112].

G. aparine L. : Llau'r Offeriad : Cleavers. Native. Hedges (E,Q,R,Rw,S,Sc,W). Fl.5-7. Mons., Glam., Brecs., Rads., Carms., Pembs., Cards., Monts., Mer., Caerns., Denbs., *Flints.*, Angl.. 279 squares, all except : **SS**38; **SM**50; **SH**51.

G. tricornutum Dandy : Corn Cleavers. Established alien (Mediterranean region). Waste places (E,F,R,Rg). Fl.6. Mons., Glam., *Carms.*, [*Mer.*]. 10 squares : **SS***69*,*78*; **ST***06*,*07*,16,*17*,*49*; **SN***41*; **SO***00*,*51*.

[**G. verrucosum** Hudson *Glam.*]

G. parisiense L. : Wall Bedstraw. Native. Walls and cornfield. *Carms.*†. 2 squares.
Carms.: Llanelli, SS59 M, 1859[81]; Llanlleidi, SN50, M, c1850[81].

CRUCIATA Miller

Cruciata laevipes Opiz (*C. chersonensis* auct.; *Galium cruciata* (L.) Scop.) : Briwydden Groes : Crosswort. Native. Hedges (D,E,M,Q,R,Rw,W). Fl.4-8. Mons., Glam., Brecs., Rads., Carms., Pembs., Cards., Monts., Mer., Caerns., Denbs., Flints., (Angl.*). 129 squares, map 544.

RUBIA L.

Rubia peregrina L. : Cochwraidd Gwyllt : Wild Madder. Native. Woods and hedges (E,Q,Sc,W). Fl.6-7. Mons., Glam., Carms., Pembs., *Cards., Mer.,* Caerns., *Denbs., Flints., Angl.*. 55 squares, map 545.

POLEMONIACEAE

POLEMONIUM L.

Polemonium caeruleum L. : Ysgol Jacob : Jacob's-ladder. Naturalized alien (England). Copses and roadsides (M,R,Rw). Fl.5-8. Mons., Glam., Brecs., Rads., *Carms.**, Pembs., Cards., *Mer.**, Caerns., Denbs., Flints.. 22 squares, map 546.

GILIA Ruiz & Pavón

Gilia capitata Sims Casual (N. America). Waste places (R,Rw). Fl.6. Glam., *Carms.*. 3 squares : **ST***16*[112], *17*[112]; **SN***42*[22].

G. achilleifolia Bentham Casual (N. America). Waste places. Fl.8. *Caerns.*. 1 square : **SH***67*[107].

CONVOLVULACEAE

CUSCUTA L.

Subgen. Grammica (Lour.) Yuncker

Cuscuta australis R. Br. subsp. **tinei** (Insenga) Feinbrun (*C. tinei* Insenga). Casual (S. Europe). Parasitic on *Aster* sp. Fl. 7-10. Mons., Glam.. 3 squares : **SS**69[112]; **ST***17*[141]; **SO**21[112].

C. campestris Yuncker Casual (N. America). Parasitic on various cultivated species. Glam.. 1 square : **ST**17[112].

C. suaveolens Ser. Casual (Chile). Parasitic on various cultivated species. Fl.8-9. Mons., Glam.. 2 squares : **ST**17[112]; **SO**20[112].

Subgen. Cuscuta

C. europaea L. : Llindag Mwyaf : Greater Dodder. Status unknown (England). *Carms.*, [*Glam.*]. 2 squares. **Carms.:** Machynys, SS59, RVI, c1840[161]; nr Furnace, SN50, RVI, c1840[161]. These two records may refer to **C. epilinum.**

C. epilinum Weihe : Flax Dodder. Casual (Asia). Parasitic on *Linum usitatissimum*. Fl.7. *Mons.*, Glam., Monts., Angl.. 4 squares : **ST**17[112]; **SO***30*[226]; **SH***47*[112]; **SJ***21*[112].

C. epithymum (L.) L. (incl. *C. trifolii* Bab.) : Llindag Lleiaf : Dodder. Native. Parasitic on *Ulex* sp., *Erica* sp., *Trifolium* sp. etc. (D,E,G,H,R,Sc). Fl.7-9. Mons., Glam., *Rads.*[230], (*Carms.**), Pembs., Cards., (Mer.), Caerns., *Denbs., Angl.,* [*Brecs.*]. 28 squares, map 547.

C. racemosa C. F. P. Mart. (*C. hassiaca* Pfeiffer). Casual (S. America). Parasitic on *Nepeta mussinii*. Glam.. 1 square : **ST**17[24].

C. americana L. Casual (America). Parasitic on *Pelargonium* sp. *Caerns.*. 1 square : **SH***56*[107].

CALYSTEGIA R. Br.

Calystegia soldanella (L.) R. Br. : Cynghafog Arfor : Sea Bindweed. Native. Sand dunes (D,S,Sc). Fl.6-7. Glam., Carms., Pembs., *Cards., Mer.,* Caerns., *Denbs.,* Flints., Angl.. 57 squares, map 548.

C. sepium (L.) R. Br. : Clych y Perthi : Hedge Bindweed.

Subsp. **sepium** Native. Hedges (Ar,D,E,F,G,M, Q,R,Rg,Rw,S,W). Fl.6-9. Mons., Glam., Brecs., Rads., Carms., Pembs., Cards., Monts., Mer., Caerns., Denbs., *Flints., Angl.*. 222 squares, map 549.

Subsp. **roseata** Brummitt Native. Sandy places near the sea (Ar,D,G,M,Rw,S). Fl.7-8. *Glam.,* Carms., *Monts., Mer., Caerns.,* Denbs.. 18 squares : **SS**69,*79*; **SN**30,31,40,50,51,*79*; **SH**33,50,52,53,*54*,61,65,77,78,87.

Subsp. **spectabilis** Brummitt Established alien (Siberia or garden origin). Waste places. Fl.8. *Mer.*. 1 square.
Mer.: nr Arthog, SH61, RKB, PMB et al, 1961[109].

C. sepium × silvatica = C. × lucana (Ten.) G. Don Spontaneous. Hedges and waste places. Fl.6-10. Glam., Rads., Carms., Cards., Mer., *Caerns.,* Flints.. 8 squares. **Glam.:** Llanmadog, SS49, JAW, 1952[112]. **Rads.:** New Radnor, SO26, JAW, 1938[112]. **Carms.:** Llandybie, SN61, JAW, 1944[112]. **Cards.:** Aber-porth, SN25, JAW, 1951[112]. **Mer.:** Aberdyfi, SN69, EM, 1960[56]; Llanbedr, SH52, JAW, 1953[112]. **Caerns.:** nr Conwy, SH77, RKB, 1963[109]. **Flints.:** Talacre, SJ18, JAW, 1942[112]. Expert opinion differs on the identity of many of these putative hybrids.

C. silvatica (Kit.) Griseb. (*C. sepium* subsp. *silvatica* (Kit.) Batt.) : Cynghafog Fawr : Large Bindweed. Naturalized alien (S. Europe). Hedges (E,R,Rg). Fl. 6-9. Mons.*, Glam., Brecs.*, Rads., Carms., Pembs., Cards., *Monts.**, Mer., *Caerns.**, Denbs., Flints., *Angl.**. 147 squares, map 550.

C. pulchra Brummit & Heywood (*C. sepium* subsp. *pulchra* (Brummitt & Heywood) Tutin; *C. dahurica* auct., non (Herbert) G. Don) : Hairy Bindweed. Established alien (N.E. Asia or garden origin). Hedges (E,G,Q,R,Rw). Fl.7-9. Mons., Glam., Brecs., Rads., Carms., Cards., Monts., Mer., Caerns., *Denbs.,* Flints.. 72 squares, map 551.

C. pulchra × sepium Spontaneous or introduced. *Monts.*[217].

CONVOLVULUS L.

Convolvulus tricolor L. Casual (Mediterranean region). Waste places. Fl.7. Glam.. 1 square : **ST***27*[112].

C. arvensis L. : Cwlwm y Cythraul : Field Bindweed. Native. Cultivated ground and waysides (E,F,M,Q, R,Rg,Rw,S,W). Fl.7-9. Mons., Glam., Brecs., Rads., Carms., Pembs., Cards., Monts., Mer., Caerns., Denbs., Flints., Angl.. 194 squares, map 552.

HYDROPHYLLACEAE

PHACELIA Juss.

Phacelia ciliata Bentham Casual (N. America). Waste places. Glam.. 2 squares : **SS**87[112]; **ST**17[112].

BORAGINACEAE

HELIOTROPIUM L.

Heliotropium europaeum L. Casual (Europe). Waste places. Glam.. 5 squares : **SS***87*[135]; **ST***16*[112], *17*[24], *18*[112], *27*[112].

LITHOSPERMUM L.

Lithospermum officinale L. : Maenhad Meddygol : Common Gromwell. Native. Thickets (Ar,D,E,H, Q,R,Rw,W). Fl.5-7. Mons., Glam., Carms., Pembs., Cards., *Mer.*, Caerns., Denbs., Flints., Angl.. 58 squares, map 553.

NEATOSTEMA I. M. Johnston

Neatostema apulum (L.) I. M. Johnston (*Lithospermum apulum* (L.) Vahl). Casual (S. Europe). Waste places. Fl.7. Caerns.. 1 square : **SH***78*[112].

BUGLOSSOIDES Moench

Buglossoides purpurocaerulea (L.) I. M. Johnston (*Lithospermum purpurocaeruleum* L.) : Y Gromandi Gwyrddlas : Purple Gromwell. Native. Limestone woods. Fl.4-7(-9). Mons., Glam., *Denbs.*. 11 squares : **SS**48,58,87,*88,96,97*; **ST**06,16,*17*,48; **SJ**06.

B. arvensis (L.) I. M. Johnston (*Lithospermum arvense* L.) : Maenhad yr Âr : Field Gromwell. Established alien (England). Cultivated and waste ground (F,Q, R,Rg). Fl.5-7(-9). Mons., Glam., Carms., *Pembs., Cards., Monts.**, *Mer.*, Caerns., *Denbs., Flints., Angl.*. 33 squares, map 554.

ECHIUM L.

[**Echium humile** Desf. *Glam.*]

E. italicum L. Casual (S. Europe). Waste places. Glam.. 1 square : **SS***78*[112].

E. vulgare L. : Tafod y Bwch : Viper's-bugloss. Native. Sand dunes and waste places (Ar,D,G,M, R,Rw). Fl.6-9. Mons., Glam., *Brecs., Rads.*[230], Carms., Pembs., *Cards.*, Monts., Mer., Caerns., Denbs., *Flints.*, Angl.. 93 squares, map 555.

E. plantagineum L. (*E. lycopsis* L.) : Purple Viper's-bugloss. Casual (S. England). Waste places. Fl.5-6. Glam., *Cards.*. 4 squares : **SS***78*[107]; **ST***16*[112], *17*[112]; **SN**58[210].

E. rosulatum Lange Established alien (Iberian Peninsula). Waste places. Fl.8. Glam.. 2 squares.
Glam.: Barry, ST16, RLS, 1927[112] and JPC, 1980[112]; Cardiff, ST17, JHS, 1930[112].

[**E. arenarium** Guss. *Glam.*]

PULMONARIA L.

Pulmonaria officinalis L. : Llys yr Ysgyfaint : Lungwort. Naturalized alien (Europe). Woods (E,G,M, Q,R,W). Fl.4-5. Mons., Glam., *Brecs.**, *Rads.**, *Carms.*, Pembs.*, Cards., Monts., *Mer.**, Caerns., *Denbs.**, Flints.*, Angl.*. 43 squares, map 556. Some of these records may refer to **P. obscura** Dumort.

P. longifolia (Bast.) Boreau : Narrow-leaved Lungwort. Naturalized alien (S. England). Woods. Fl.2-5. *Glam.*, Denbs., Angl., [*Flints.*]. 4 squares.
Glam.: Ilston, SS59, JAW, 1929[221]; Swansea, SS69, An, c1891[140]. **Denbs.:** Coed Coch, SH87, JEG, 1879[112]. **Angl.:** Plasnewydd, SH56, WEH, 1966[112].

NONEA Medicus

Nonea lutea (Desr.) DC. Established alien (Russia). Lawn and garden weed. Fl.4-5. Carms., *Caerns.*. 2 squares.
Carms.: Gilfach, SN41, RDP, 1982[112]. **Caerns.:** Bangor, SH57, RHR, 1977[107].

SYMPHYTUM L.

Symphytum officinale L. : Dail Cwmffri : Common Comfrey. Native. River banks and waste places (E,M,R, Rw). Fl.5-9. Mons., Glam., Brecs., Rads., Carms., *Pembs., Cards.*, Monts., Caerns., *Denbs.*, Flints., *Angl.*, [*Mer.*]. 125 squares, map 557. Many of these records doubtless refer to **S. × uplandicum.**

S. asperum Lepechin : Rough Comfrey. Established alien (S.W. Asia). Roadsides (E,G,M,R,Rw). Fl.6-7. Mons., Glam., Brecs.*, Carms., Cards., *Caerns., Flints., Angl.*, [*Monts.*]. 13 squares : **ST***59*; **SN**54,62,72,77,90; **SO**02; **SH**32,*36,44,57*,77; **SJ**26. Most populations exhibit introgression with **S. officinale.**

S. asperum × officinale = S. × uplandicum Nyman (*S. peregrinum* auct., non Ledeb.) : Cyfardwf Glas : Russian Comfrey. Naturalized alien (garden origin or, rarely, of spontaneous local origin). River banks and roadsides (E,M,R,W). Fl.5-9(-11). Mons., Glam., Brecs., *Rads.**, Carms., Pembs., Cards., Monts., Mer., Caerns.*, Denbs., Flints., Angl.. 141 squares, map 558.

S. tuberosum L. : Cyfardwf Oddfynog : Tuberous Comfrey. Naturalized alien (England). Woods and hedges (E,G,M,R,Rw,W). Fl.5-8. Mons.*, Glam.*, Brecs.*, Carms., Monts., *Mer.*, Caerns., Denbs., *Flints.**, [*Cards.*]. 20 squares : **SS**48; **SN**31; **ST**49; **SO**05,29; **SH***44*,56,57,66,67,76,77,*78*,87,*93*,97; **SJ**05,16, 17,36.

S. ibiricum Steven (*S. grandiflorum* auct., non DC.) : Dwarf Comfrey. Established alien (Caucasus). Hedges. *Denbs..* 1 square.
Denbs.: nr Nebo, SH85, JMB, 1979[67].

[**S. orientale** L. *Flints.*]

S. caucasicum M. Bieb. : Crimaean Comfrey. Casual (Caucasus). Waste places. Fl.5. Flints.. 1 square : **SJ08**[112].

S. bulbosum C. Schimper Established alien (S. Europe). Scrub. Fl.2-3. Cards.. 1 square.
Cards.: Llanbadarn Fawr, SN58, VGE, 1975[112] and AOC, 1980[112].

BRUNNERA Steven

Brunnera macrophylla (Adams) I. M. Johnston : Great Forget-me-not. Introduced (Caucasus). Woodland. *Mons..* 1 square.
Mons.: Wynd Cliff, ST59, MDT, 1980[264].

ANCHUSA L.

Anchusa officinalis L. : Alkanet. Casual (Europe). Waste places (D,Rw). Fl.7. Mons., Glam., Carms., Mer.. 7 squares : **SS59**[112],*78*[112],*87*[135]; **ST***17*[93]; **SO***30*[260],51[112]; **SH61**[112].

A. azurea Miller (*A. italica* Retz.) : Garden Anchusa. Casual (S. Europe). Waste places (R,W). Fl.6-8. Mons., Glam., Mer.. 6 squares : **ST***17*[112],*38*[112]; **SN***90*[112]; **SO***00*[112],42[112]; **SH52**[112].

A. arvensis (L.) Bieb. (*Lycopsis arvensis* L.) : Bleidd Drem : Bugloss. Denizen or colonist (England). Cultivated ground and sand dunes (D,F,G,Q,R). Fl. 5-9. Mons., Glam., *Brecs.*, Carms., Pembs., Cards., *Monts.**, Mer., Caerns., *Denbs.*, Flints., Angl.. 88 squares, map 559. May be native in duneland.

PENTAGLOTTIS Tausch

Pentaglottis sempervirens (L.) Tausch ex L. H. Bailey : Llys y Gwrid : Green Alkanet. Denizen (W. Europe). Hedges (E,M,Q,R,Rw,W). Fl.4-11. Mons., Glam., Brecs.*, *Rads.**, Carms., Pembs., Cards., Monts., Mer., Caerns., Denbs., Flints., Angl.. 95 squares. map 560.

BORAGO L.

Borago officinalis L. : Tafod yr Ych : Borage. Established alien (S. Europe). Waysides (E,R). Fl.6-8. Mons., Glam., Carms., Pembs., *Cards.*, *Monts.**, Mer., Caerns., *Denbs.*, *Flints..* 54 squares, map 561.

B. pygmaea (DC.) Chater & W. Greuter (*B. laxiflora* Willd.) : Slender Borage. Naturalized alien (Mediterranean region). Damp pathsides. Fl.6-7. *Mer.*, Caerns.. 2 squares.
Mer.: Llanfendigaid, SH50, PMB, 1953[3]. **Caerns.:** Porth Dinllaen, SH24, APC, c1975[112].

TRACHYSTEMON D. Don

Trachystemon orientalis (L.) G. Don fil. : Abraham-Isaac-Jacob. Naturalized alien (E. Mediterranean region). Woods and roadsides. Fl.3-5. Caerns., Angl.. 3 squares.
Caerns.: Penmaen-mawr, SH77, AEB, 1955[112]. **Angl.:** Plasnewydd, SH56, WEH, 1967[112]; nr Moel y Don, SH56, WEH, 1960+[209]; Llanddona, SH57, RHR, 1970+[209].

MERTENSIA Roth

Mertensia maritima (L.) S. F. Gray : Llysiau'r Ysgyfaint Arfor : Oysterplant. Native. Shingle beaches. Fl.7. *Mer.**†, Caerns.†, Denbs.*, Angl.†, [*Cards.*]. 6 squares.
Mer.: Aberdyfi, SN69, AAi, c1797[71]. **Caerns.:** Dinas Dinlle, SH45, JEG, 1884[112]; Gt Ormes Head, SH78, AAi, c1880[82]. **Denbs.:** Locality withheld, DWC, 1979[245] and HWil, 1980[112]. **Angl.:** Cemlyn, SH39, JEG, 1914[112]; nr Trefarthen, SH46, An, c1800[82].

AMSINCKIA Lehm.

Amsinckia lycopsoides (Lehm.) Lehm. (*Benthamia lycopsoides* Lindley). Casual (N. America). Cultivated ground and waste places. Fl.5-7. *Mons.*, Glam., *Pembs.*, *Caerns.*, *Angl..* 8 squares : **SS***19*[107],*78*[140]; **ST***07*[112],26[99],*59*[226]; **SN***90*[107]; **SO***00*[140]; **SH***57*[112].

A. calycina (Moris) Chater (*A. hispida* I. M. Johnston; *A. angustifolia* Lehm.; *Benthamia hispida* (Ruiz & Pavón) Druce; *Benthamia angustifolia* Lehm.). Casual (S. America). Waste places. Fl.8. Glam., *Caerns.*, *Denbs..* 4 squares: **ST17**[112],*18*[112]; **SH***56*[18]; **SJ***24*[146].

A. intermedia Fischer & C. A. Meyer (*Benthamia intermedia* (Fischer & C. A. Meyer) Druce) : Fiddleneck. Casual (N. America). Waste places. *Mer.*, [*Glam.*] 1 square : **SH50**[253].

A. menziesii (Lehm.) A. Nelson & Macbride (*A. retrorsa* Suksdorf). Casual (N. America). Waste places. Fl.8. Glam.. 1 square : **ST17**[112].

A. tessellata A. Gray Casual (N. America). Waste places. Fl. 6-8. Glam.. 2 squares : **ST***16*[107],17[112].

ASPERUGO L.

Asperugo procumbens L. : Cynghafan Mwyaf : Madwort. Casual (Europe). Waste places. Fl.7-8. Glam., *Denbs..* 5 squares : **SS98**[112]; **ST***17*[24],*27*[28]; **SO***00*[107]; **SH***87*[144].

MYOSOTIS L.

Myosotis arvensis (L.) Hill : Ysgorpionllys y Meysydd : Field Forget-me-not. Native. Cultivated ground, hedges and woods (D,E,F,G,Q,R,Rg,Rw,S,W). Fl.4-9. Mons., Glam., Brecs., Rads., Carms., Pembs., Cards., Monts., Mer., Caerns., Denbs., Flints., *Angl..* 247 squares, map 562.

M. ramosissima Rochel (*M. hispida* Schlecht.) : Ysgorpionllys Cynnar : Early Forget-me-not. Native. Limestone rocks and sandy places (D,E,G,Q,R,Rw,S). Fl.4-6(-8). Mons., Glam., *Brecs.**, *Rads.**, Carms., Pembs., Cards., *Monts.**, (Mer.), Caerns., Denbs., Flints., Angl.. 88 squares, map 563.

M. discolor Pers. : Ysgorpionllys Amryliw : Changing Forget-me-not. Native. Grassland and waste places (D,E, F,G,H,Q,R,Rg,Rw). Fl.4-6. Mons., Glam., Brecs., Rads., Carms., Pembs., Cards., Monts., Mer., Caerns., *Denbs., Flints.,* Angl.. 164 squares, map 564.

M. sylvatica Hoffm. : Ysgorpionllys y Coed : Wood Forget-me-not. Established alien (England). Woods and hedges (D,E,M,R,Rw,W). Fl.4-5. Mons., Glam., Brecs., Rads., Carms., *Pembs.**, Cards.*, Monts., Caerns., Denbs., Flints., Angl.. 88 squares, map 565.

[**M. alpestris** F. W. Schmidt *Glam.*]

M. secunda A. Murray : Ysgorpionllys Ymlusgaidd : Creeping Forget-me-not. Native. Bogs and marshes (Al,Ar,M). Fl.5-9. Mons., Glam., Brecs., Rads., Carms., Pembs., Cards., Monts., Mer., Caerns., Denbs., *Flints.,* Angl.. 200 squares, map 566.

M. laxa Lehm. subsp. **caespitosa** (C. F. Schultz) Hyl. ex Nordh. (*M. caespitosa* C. F. Schultz) : Ysgorpionllys Siobynnog : Tufted Forget-me-not. Native. Marshes (D,E,G,M,Q,R,Rg,Rw,W). Fl.5-7(-9). Mons., Glam., Brecs., Rads., Carms., Pembs., Cards., Monts., Mer., Caerns., *Denbs.,* Flints., Angl.. 194 squares, map 567.

M. laxa × scorpioides Spontaneous. Marshes. Fl.6-9. Mer., Caerns., *Angl..* 7 squares.
Mer.: Minffordd, SH53, PMB, 1962[112]; Penrhyndeudraeth, SH63, PMB, 1957[54]; Dolgellau, SH71, PMB, 1956[112]; nr Llanfor, SH93, PMB, 1959[112]. **Caerns.:** Pwllheli, SH33, PMB, 1979[112]. **Angl.:** Llyn Coron, SH36, RHR, 1970+[209]; Llyn Cerrig Bach, SH37, RHR, 1975[183].

M. scorpioides L. : Ysgorpionllys y Gors : Water Forget-me-not. Native. Streamsides (Al,Ar,H,M,W). Fl.6-10. Mons., Glam., Brecs., Rads., Carms., Pembs., *Cards.,* Monts., Mer., Caerns., Denbs., *Flints.,* Angl.. 194 squares, map 568. This is often confused with **M. secunda** and many of the records may refer to that species.

ERITRICHIUM Schrader ex Gaudin

Eritrichium nanum (L.) Schrader ex Gaudin (*E. terglovense* (Hacq.) A. Kerner). Casual (Alps). Cultivated ground. *Caerns..* 2 squares : **SH***57*[101],*78*[101].

LAPPULA Gilib.

Lappula squarrosa (Retz.) Dumort. (*L. echinata* Fritsch; *L. myosotis* Moench; *Echinospermum lappula* (L.) Lehm). : Bur Forget-me-not. Casual (Europe). Waste places (R,Rw). Fl. 6-9. Mons., Glam., *Cards., Caerns., Flints..* 10 squares : **SS***78*; **ST**16,*17,27*; **SN***67,90*; **SO***00,51*; **SH***56*; **SJ***08*.

OMPHALODES Miller

Omphalodes verna Moench : Blue-eyed-Mary. Naturalized alien (Europe). Woods (Q,R,W). Fl.3-5. *Glam.,* Carms., Pembs., Cards., *Denbs.,* Angl.. 9 squares.
Glam.: Penrice, SS48, LWD, c1840[81]; nr Reynoldstone, SS48, JAW, c1927[221]; Fairyhill Wood, SS49, JAW, c1925[24]. **Carms.:** Llanegwad, SN52, CBe, 1927[112]. **Pembs.:** Orielton, SR99, RML, 1959[79]; Ludchurch, SN11, ECH, 1942[112]. **Cards.:** Cwm Woods nr Aberystwyth, SN58, JHS, 1925[112]. **Denbs.:** Pennant Hall, SH87, WBR, pre 1900[146]; nr Llangollen, SJ24, JAWhe, 1943[42]. **Angl.:** Treiorwerth, SH38, LKS, 1928[28] and RHR, 1972[208]; Bodedern, SH38, FCG, 1955[112].

CYNOGLOSSUM L.

Cynoglossum officinale L. : Tafod y Bytheiad : Hound's-tongue. Native. Sand dunes and waysides (D,G,Q,R,S). Fl.5-7. Mons., Glam., *Brecs.,* Rads.*, Carms., Pembs., Cards., Monts., Mer., Caerns., Denbs., Flints., Angl.. 73 squares, map 569.

C. germanicum Jacq. : Tafod y Bytheiad Gwyrdd-ddail : Green Hound's-tongue. Casual (England). Waste places. Glam.. 1 square : **ST***17*[140].

VERBENACEAE

VERBENA L.

Verbena officinalis L. : Briw'r March : Vervain. Native. Roadsides (G,M,Q,R). Fl.7-10. Mons., Glam., Brecs., *Rads.,* Carms., Pembs., *Cards., Monts.,* Mer., Caerns., Denbs., Flints., *Angl..* 116 squares, map 570.

CALLITRICHACEAE

CALLITRICHE L.

Callitriche hermaphroditica L. (*C. autumnalis* L.) : Brigwlydd Cynauafol : Autumnal Water-starwort. Native. Lakes and streams (Al,Ar). Fl.7-9. *Brecs.**, Monts., Angl., [*Glam., Cards.*]. 7 squares.
Brecs.: Above Usk Reservoir, SN82, NH, 1980[118]. **Monts.:** nr Berriew, SJ10, An, 1950+[195]; nr Welshpool, SJ20, VJM, 1958[7]; Pool Quay, SJ21 JAW, 1940[112] and RGW, 1974[259]; Welshpool Canal, SJ21, VJM, 1950+[116]. **Angl.:** Llyn Coron, SH36, BS, 1962[112] and RHR, 1975[112]; Llyn Maelog, SH37, JHS, 1925[112] and RHR, 1974[208]; Llyn Padrig, SH37, BS, 1962[112]; Llyn Dinam, SH37, BS, 1962[97]; Cefni Reservoir, SH47, WEH, 1968[112].

C. stagnalis agg. Mons., Glam., *Brecs.,* Rads., Carms., Pembs., Cards., Monts., Mer., Caerns., *Denbs.,* Flints., Angl.. 219 squares, map 571.

C. stagnalis Scop. *s.s.* : Brigwlydd y Dŵr : Common Water-starwort. Native. Ponds and ditches (Al,Ar,G,H,M,R). Fl.5. Mons., Glam., *Brecs.,* Rads., Carms., Pembs., Cards., Monts., Mer., *Caerns., Denbs., Flints.,* Angl.. 152 squares, map 572.

C. platycarpa Kütz. (*C. verna* auct.) : Brigwlydd y Gwanwyn : Various-leaved Water-starwort. Native. Ponds and ditches (Al,Ar,G). Fl.5-8. Mons.*, Glam., *Rads.*[231], *Carms.**, *Pembs.**, Cards.*, *Monts.**, Mer.*, Caerns.*, *Denbs.**, Flints.*, Angl.*. 67 squares, map 573.

C. obtusangula Le Gall : Blunt-fruited Water-starwort. Native. Ponds and ditches (Al,Ar,G,H,M,Rw). Fl.5-8. Mons., Glam., *Brecs., Rads., Carms.,* Pembs.*, Cards., Mer., Caerns., *Denbs.,* Flints.*, Angl.. 67 squares, map 574.

C. intermedia agg. : Intermediate Water-starwort. Native. Lakes, ponds and ditches (Al,Ar,G,H,M). Fl.4-8. Mons., Glam., Brecs., Rads., Carms., Pembs., Cards., Monts., Mer., Caerns., Denbs., Flints., Angl.. 152 squares, map 575.

C. hamulata Kütz. ex Koch (*C. intermedia* Hoffm.; *C. intermedia* subsp. *hamulata* (Kütz. ex Koch) Clapham). Native. Ponds and ditches. Fl.4-5. Mons., Glam., Brecs., Rads.*, Carms., Pembs.*, Cards., Monts., Mer., Caerns., Denbs., Flints., *Angl..* 96 squares, map 576.

C. brutia Petagna (*C. intermedia* subsp. *pedunculata* (DC.) Clapham). Native. Ponds, streams and ditches. Fl.5-8. Mons., Glam., Rads., Carms.*, Pembs., *Cards.,* Monts.*, Caerns.*, Angl.. 25 squares, map 577.

LABIATAE

AJUGA L.

Ajuga genevensis L. : Cornish Bugle. Casual (Europe). Sea cliff. Fl.5. Pembs.. 1 square : **SS**09[112].

A. reptans L. : Glesyn y Coed : Bugle. Native. Woods (D,E,G,H,M,Q,R,W). Fl.4-7. Mons., Glam., Brecs., Rads., Carms., Pembs., Cards., Monts., Mer., Caerns., Denbs., Flints., Angl.. 265 squares, all except : **SR**89; **SS**39,77; **ST**26; **SM**50,71,73,94; **SN**15,36,59; **SO**30; **SH**29,50,98; **SJ**02,13.

[**A. chamaepitys** (L.) Schreber *Pembs., Monts.*]

TEUCRIUM L.

Teucrium scorodonia L. : Chwerwlys yr Eithin : Wood Sage. Native. Woods and hedges (E,G,Q,R, Sc,W). Fl.6-9. Mons., Glam., Brecs., Rads., Carms., Pembs., Cards., Monts., Mer., Caerns., Denbs., Flints., Angl.. 273 squares, all except : **SR**89; **SS**39,77; **ST**38, 58; **SM**50; **SN**59; **SH**51,98.

T. chamaedrys L. : Chwerwlys y Mur : Wall Germander. Naturalized alien (C. and S. Europe). Rocky places (E,Q). Fl.7-8. Glam., *Carms.*[161], *Pembs., Denbs.,* [*Mons.*]. 5 squares.
Glam.: South Cornelly, SS88, HJR, 1904[112] and QONK, 1970+[157]; Penyrheolgerrig, SO00, An, c1890[140]. **Pembs.:** Nevern, SN04, HNR, 1882[107]; Tenby, SN10, HM, 1835[107]. **Denbs.:** Ruthin Castle, SJ15, Gri, c1800[82].

T. flavum L. Casual (Mediterranean region). Waste places. Fl.7. Mons.. 1 square : **ST***59*[112].

SCUTELLARIA L.

Scutellaria galericulata L. : Cycyllog : Skullcap. Native. Streamsides (Al,Ar,D,H,M,W). Fl.7-9. Mons., Glam., Brecs., Rads., Carms., Pembs., Cards., Monts., Mer., Caerns., Denbs., *Flints.,* Angl.. 160 squares, map 578.

S. minor Hudson : Cycyllog Bach : Lesser Skullcap. Native. Wet heaths (Ar,G,H,M,W). Fl.6-9. Mons., Glam., Brecs., Rads., Carms., Pembs., Cards., Monts., Mer., Caerns., Denbs., Flints., Angl.. 177 squares, map 579.

MARRUBIUM L.

Marrubium vulgare L. : Llwyd y Cŵn : White Horehound. Colonist or established alien (England). Roadsides (D,E,G,R). Fl.6-10. *Mons.,* Glam., Carms., Pembs., *Cards.,* Monts., Mer., Caerns., Denbs., Flints., *Angl..* 56 squares, map 580. Perhaps native in Gower.

SIDERITIS L.

Sideritis montana L. Casual (S. Europe). Waste places. Fl. 7-10. Glam.. 4 squares : **SS***78*[140]; **ST***17*[112],18[112],*27*[69].

MELITTIS L.

Melittis melissophyllum L. : Gwenynog : Bastard Balm. Native. Woods and hedges. Fl.6-9. (*Glam.*), Carms., Pembs., Cards., (*Denbs.),* [*Mer.*]. 10 squares : (**SS***69*;**ST***17*); **SM**82,90,91; **SN**13,14,24,34; (**SJ***24*).

PHLOMIS L.

Phlomis samia L. Casual (Greece). Waste places. Glam.. 1 square : **ST**18[112].

P. fruticosa L. : Jerusalem Sage. Casual (Mediterranean region). Waste places. *Caerns..* 1 square : **SH**88[67].

GALEOPSIS L.

Subgen. Ladanum Reichenb.

Galeopsis segetum Necker (*G. dubia* Leers) : Downy Hemp-nettle. Established alien (Europe). Cultivated ground. Fl.7-9. *Carms.,* Caerns.. 4 squares.
Carms.: Llangennech, SN50, DH, 1912[7]. **Caerns.:** nr Port Dinorwic, SH56, JEB, 1830[107] and WBW, 1957[7]; Bangor, SH57, JEG, 1895[112] and EV, 1926[112]; Aber, SH67, WW, 1830[112].

G. ladanum L. Casual (Europe). Waste places. *Glam., Carms., Denbs., Flints.,* [*Pembs.*]. 6 squares : **SS***79*[157]; **SN***50*[157]; **SJ**08[90], *12*[107], *24*[146],*35*[146].

G. angustifolia Ehrh. ex Hoffm. : Penboeth Gulddail : Red Hemp-nettle. Native. Cultivated ground and shingle beaches (E,F,R,Rg,Rw,S). Fl.7-9. (*Mons.*), Glam., Brecs.*, *Rads., Carms.,* Pembs., Monts.*, *Mer.,* Caerns., *Denbs., Flints.,* [*Angl.*]. 36 squares, map 581.

Subgen. Galeopsis

G. speciosa Miller : Penboeth Amryliw : Large-flowered Hemp-nettle. Established alien (England). Cultivated ground (F,R,Rg). Fl.7-9. Glam., *Brecs.*, Rads., Carms., Pembs., Cards., Monts., Mer., Caerns., Denbs., Flints., Angl., [*Mons.*]. 84 squares, map 582.

G. tetrahit L. : Penboeth Cyffredin : Common Hemp-nettle. Native. Wood borders (E,F,Rw,W). Fl. 7-9. Mons., Glam., Brecs., Rads., Carms., Pembs., Cards., Monts., Mer., Caerns., Denbs., Flints., *Angl.*. 263 squares, all except : **SR**89; **SS**38,39,77; **ST**26; **SM**62, 71,73,84; **SN**00,15,36,66,87; **SH**88,97,98; **SJ**15,18. Some of these records may refer to the following species which has not always been separated.

G. bifida Boenn. (*G. tetrahit* var. *bifida* (Boenn.) Lej. & Court.). Native. Scrub and roadsides (E,F,R,Rw). Fl. 6-8. Mons., Glam., Carms.*, Pembs., Cards., Mer.*, Caerns., Denbs.*, *Angl.*. 73 squares, map 583.

[**G. bifida × speciosa = G. × sulfurea** Druce *Pembs.*]

G. bifida × tetrahit = G. × ludwigii Hausskn. Spontaneous Roadsides. Fl.8-9. Mer.. 3 squares. **Mer.:** Peniarth, SH60, PMB, 1960[217]; Arthog, SH61, PMB, 1963[217]; Maentwrog, SH64, WCB, 1922[112].

LAMIUM L.

Lamium maculatum L. : Marddanhadlen Fraith : Spotted Dead-nettle. Established alien (Europe). Hedges (E,R). Fl.4-11. Mons., Glam., Brecs., Rads., Carms., *Pembs.*[47], *Cards.*[210], Monts., Mer.*, Caerns., *Denbs.*, Flints., *Angl.*. 56 squares, map 584.

L. album L. : Marddanhadlen Wen : White Dead-nettle. Colonist (England). Hedges and roadsides (E,F,G,M,Q,R,Rg,Rw,W). Fl.3-10. Mons., Glam., Brecs., Rads., Carms., Pembs., Cards., Monts., Mer., Caerns., Denbs., Flints., Angl.. 178 squares, map 585.

L. purpureum L. : Marddanhadlen Goch : Red Dead-nettle. Native. Cultivated ground (Ds,E,F,G,R,Rg,Rw). Fl.1-12. Mons., Glam., Brecs., Rads., Carms., Pembs., Cards., Monts., Mer., Caerns., Denbs., Flints., *Angl.*. 221 squares, map 586.

L. hybridum Vill. : Marddanhadlen Rwygddail : Cut-leaved Dead-nettle. Colonist or established alien (England). Cultivated ground (E,F,G,R,S). Fl.3-9. Mons.*, Glam., Carms., Pembs., Cards., Caerns., *Denbs.*, Flints., Angl.. 66 squares, map 587.

L. moluccellifolium Fries : Northern Dead-nettle. Established alien (Scotland). Arable land. Monts.*, [*Glam., Denbs.*]. 1 square.
Monts.: nr Welshpool, SJ20, CASm, 1980[112].

L. amplexicaule L. : Marddanhadlen Goch Gron : Henbit Dead-nettle. Established alien (England). Cultivated ground (F,R,Rg,Rw). Fl.3-8(-10). *Mons.*, Glam., *Carms., Pembs.*, Cards., *Monts., Mer.*, Caerns., Denbs., Flints., Angl.. 65 squares, map 588.

LAMIASTRUM Heister ex Fabr.

Lamiastrum galeobdolon (L.) Ehrend. & Polatschek subsp. **montanum** (Pers.) Ehrend. & Polatschek (*Galeobdolon luteum* Hudson) : Marddanhadlen Felen : Yellow Archangel. Native. Woods (E,G,M,Q,R,Rw,W). Fl.4-7. Mons., Glam., Brecs., Rads., Carms., Pembs., *Cards.*, Monts., Mer., Caerns., Denbs., Flints.. 142 squares, map 589.

LEONURUS L.

Leonurus cardiaca L. : Mamlys : Motherwort. Established alien (Europe). Roadsides (E,F,R,Rg). Fl. 6-9. Mons., Glam., Brecs.*, *Carms.**, Pembs., Cards., *Monts.*, Mer., Caerns., Denbs., Flints., *Angl.*. 47 squares, map 590.

BALLOTA L.

Ballota nigra L. subsp. **foetida** Hayek : Marddanhadlen Ddu : Black Horehound. Colonist (England). Roadsides (F,Q,R,Rw,W). Fl.6-9. Mons., Glam., Brecs., *Rads.*, Carms., Pembs., *Cards.*, Monts., Mer., Caerns., Denbs., Flints., Angl.. 112 squares, map 591.

STACHYS L.

Stachys officinalis (L.) Trevisan (*Betonica officinalis* L.) : Cribau San Ffraid : Betony. Native. Woods (E,G,H, M,Q,R,Sc,W). Fl.7-10. Mons., Glam., Brecs., Rads., Carms., Pembs., Cards., Monts., Mer., Caerns., Denbs., Flints., *Angl.*. 243 squares, map 592.

S. alpina L. : Briwlys Calchfaen : Limestone Woundwort. Native. Hilly woods. Fl.8. Denbs.. 2 squares. **Denbs.:** nr Cerrigydrudion, SH94, AW, 1927[112] and LWF, 1954[7†]; nr Ruthin, SJ15, BI, 1973[183] and JMB, 1981[67].

[**S. germanica** L. *Denbs.*]

S. cretica L. (*S. italica* Miller). Casual (Mediterranean region). Waste places. Glam.. 1 square ; ST27[112].

S. byzantina C. Koch (*S. lanata* Jacq., non Crantz) : Lambsear. Introduced (S.W. Asia). Waste places. Fl.6. Rads., *Pembs.*. 2 squares.
Rads.: nr Llanelwedd, SO05, RGE, 1982[112]. **Pembs.:** nr Tenby, SN10, JEA, c1917[16].

S. sylvatica L. : Briwlys y Goedwig : Hedge Woundwort. Native. Woods and hedges (F,R,Rg,S,W). Fl.4-9. Mons., Glam., Brecs., Rads., Carms., Pembs., Cards., Monts., Mer., Caerns., Denbs., *Flints., Angl.*. 270 squares, all except : **SS**38,39,77; **ST**26; **SM**50,71,73,84; **SN**87; **SH**29,51,98.

S. palustris L. : Briwlys y Gors : Marsh Woundwort. Native. River banks (Ar,E,F,G,M,R,S,W). Fl.7-9. Mons., Glam., Brecs., *Rads.*, Carms., Pembs., Cards., Monts., Mer., Caerns., Denbs., Flints., Angl.. 236 squares, map 593.

S. palustris × sylvatica = S. × ambigua Sm. : Briwlys Amheus. Spontaneous. Roadsides and waste places (Ar,E,R,S). Fl.7-8. Mons., Glam., Brecs., Rads., Carms., *Pembs., Cards., Monts.**, Mer., Caerns., Flints., *Angl.*. 68 squares, map 594.

S. recta L. : White Woundwort. Established alien (Europe). Waste places. Fl.6-7. Glam.. 1 square.
Glam.: Barry Dock, ST16, AEW & RLS, 1923[112] and JPC, 1980[112].

S. annua (L.) L. Casual (C. and S. Europe). Waste places (R,Rg). Fl.9. Mons., Glam., *Pembs.*. 5 squares : **ST**17[112], *18*[24],27[112],38[112]; **SM**70[153].

S. arvensis (L.) L. : Briwlys yr Ŷd : Field Woundwort. Colonist (England). Cultivated ground (E,F,G,R,Rg,W). Fl.(4-)7-11. Mons., Glam., *Brecs., Rads.*, Carms., Pembs., Cards., Monts., Mer., Caerns., Denbs., Flints., *Angl.*. 188 squares, map 595.

NEPETA L.

Nepeta cataria L. : Mintys y Gath : Cat-mint. Native. Hedges (D,E,R). Fl.7-9. Mons., Glam., *Pembs., Monts.*, Caerns., *Denbs., Flints., Angl.*, [*Carms., Mer.*]. 37 squares, map 596.

N. mussinii Sprengel ex Henckel Casual (Caucasus). Waste places. Fl.8-9. Mer., Denbs.. 2 squares : **SN**69[112]; **SJ**04[112].

N. mussinii × nepetella L. = **N. × faassenii** Bergmans ex Stearn (*N. mussinii* auct., non Sprengel). Casual (garden origin). Roadside. Fl.8. Mons.. 1 square : **SO**42[112].

GLECHOMA L.

Glechoma hederacea L. : Eidral : Ground-ivy. Native. Hedges (Ar,D,E,G,H,M,Q,R,Rw,Sc,W). Fl.4-9. Mons., Glam., Brecs., Rads., Carms., Pembs., Cards., Monts., *Mer.*, Caerns., Denbs., Flints., Angl.. 273 squares, all except : **SS**39; **SM**50; **SN**15,59,75,86; **SH**29,51,98.

DRACOCEPHALUM L.

Dracocephalum parviflorum Nutt. Casual (N. America). Waste places. Fl.7. Glam.. 1 square : **ST***27*[112].

PRUNELLA L.

Prunella laciniata (L.) L. : Cut-leaved Selfheal. Naturalized alien (Europe). Grassland. Fl.6. *Mons.*. 1 square.
Mons.: Common Coed, ST48, TGE, 1971[181].

P. vulgaris L. : Craith Unnos : Selfheal. Native. Pastures (Ar,Ds,G,H,M,Q,R,Rw,W). Fl.6-9. Mons., Glam., Brecs., Rads., Carms., Pembs., Cards., Monts., Mer., Caerns., Denbs., Flints., *Angl.*. 275 squares, all except : **SS**38,39; **SM**50,73; **SN**36; **SH**51,98.

MELISSA L.

Melissa officinalis L. : Gwenynddail : Balm. Established alien (S. Europe). Roadsides (E,M,R,Sc). Fl.6-9. Mons., Glam., *Brecs., Rads.**, Carms., *Pembs., Cards.*, Monts., Mer., Caerns., Denbs., Flints., *Angl.*. 65 squares, map 597.

SATUREJA L.

[**Satureja montana** L. *Carms.*]

S. hortensis L. : Summer Savory. Casual (Mediterranean region). Waste places. Fl.8. Glam., *Carms.*. 5 squares : **SS***68*[112]; **ST***17*[112],*27*[112]; **SN***42*[13],*90*[140].

ACINOS Miller

Acinos arvensis (Lam.) Dandy : Brenhinllys : Basil Thyme. Native. Cliffs (E,G,Q,R,Rw,Sc). Fl.5-8. Mons., Glam., *Carms.**, Pembs., *Monts.*, Caerns., *Denbs., Flints., Angl.*. 25 squares, map 598.

CALAMINTHA Miller

Calamintha sylvatica Bromf. subsp. **ascendens** (Jordan) P. W. Ball (*C. ascendens* Jordan) : Erbin Cyffredin : Common Calamint. Native. Hedgebanks (E,G,Q,R, Rw,W). Fl.7-9. Mons., Glam., Brecs., *Carms.*, Pembs., *Cards.*, Monts., Caerns., Denbs., *Flints., Angl.*, [*Mer.*]. 65 squares, map 599.

C. nepeta (L.) Savi : Erbin Bach : Lesser Calamint. Established alien (S. England). Waysides (E,R). Fl. 9-10. *Mons.*, Glam., [*Rads., Pembs., Caerns., Flints.*]. 2 squares.
Mons.: nr Monmouth, ST51, SGC, 1957[7] (requires confirmation). **Glam.:** Barry, ST16, RLS, 1921[112].

CLINOPODIUM L.

Clinopodium vulgare L. : Brenhinllys Gwyllt : Wild Basil. Native. Hedges (E,G,M,Q,R,W). Fl.7-10. Mons., Glam., Brecs., Rads., Carms., Pembs., Cards., Monts., Mer., Caerns., Denbs., Flints., Angl.. 135 squares, map 600.

HYSSOPUS L.

Hyssopus officinalis L. : Isop : Hyssop. Casual (S. Europe). Rubbish tip. Mons.. 1 square : **ST**38[112].

ORIGANUM L.

Origanum vulgare L. : Penrhudd : Marjoram. Native. Hedgebanks (E,Q,R,Rw,Sc,W). Fl.7-11. Mons., Glam., *Brecs.*, Rads., Carms., Pembs., Cards., *Monts.*, Mer., Caerns., Denbs., Flints., Angl.. 144 squares, map 601.

O. majorana L. (*Majorana hortensis* Moench) : Sweet Marjoram. Casual (N. Africa). Waste places. Fl.8. Mons.. 1 square : **ST**38[112].

THYMUS L.

Thymus praecox Opiz subsp. **arcticus** (E. Durand) Jalas (*T. drucei* Ronniger) : Teim Gwyllt : Wild Thyme. Native. Heaths (D,E,G,H,Q,R,S,Sc,W). Fl. 6-9. *Mons.*, Glam., Brecs., Rads., Carms., Pembs., Cards., *Monts.*, Mer., Caerns., Denbs., Flints., Angl.. 220 squares, map 602.

[**T. praecox × pulegioides = T. × henryi** Ronniger *Glam., Mer.*]

T. pulegioides L. : Gruwlys Gwyllt Mwyaf : Large Thyme. Native. Heaths (G,H,R). Fl.7-10. Mons., Glam., Brecs.*, *Carms.*, (*Caerns.*), Flints., [*Cards.*]. 15 squares : **ST**17,*29*,48,49,59; **SN***82*; **SO**12,21,22,*41*,50,51; (**SH***53*); **SJ***07*,08.

[**T. pulegioides × serpyllum = T. × oblongifolius** Opiz *Caerns.*]

T. pulegioides × vulgaris L. = **T. × citriodorus** (Pers.) Schreber : Lemon Thyme. Casual (garden origin). Waste places. Fl.6. Caerns.. 1 square : **SH***78*[112].

LYCOPUS L.

Lycopus europaeus L. : Llys y Sipsiwn : Gipsywort. Native. Streamsides (Ar,G,H,M). Fl.7-9. Mons., Glam., Brecs., Rads., Carms., Pembs., Cards., Monts., *Mer.*, Caerns., Denbs., *Flints.*, Angl.. 170 squares, map 603.

MENTHA L.

Mentha requienii Bentham : Corsican Mint. Established alien (Corsica and Sardinia). Damp places (G,R,Re,Rg). Fl.7-8. Mons., Glam.. 3 squares. **Mons.:** Usk, SO30, DTP, 1965[112]. **Glam.:** Penlle'r-gaer, SS69, HJR, 1906[112]; Cardiff, ST17, LSP, 1965[112].

M. pulegium L. : Brymlys : Pennyroyal. Native. Wet heaths (Al,H,M,R). Fl.7-9. *Mons.*, Glam., Brecs.*, Carms., Pembs., (*Mer.*), Caerns., *Flints.*, Angl.. 24 squares, map 604.

M. arvensis L. : Mintys yr Ŷd : Corn Mint. Native. Cultivated ground (Ar,E,F,G,H,M,R,W). Fl.7-10. Mons., Glam., Brecs., *Rads.*, Carms., Pembs., Cards., Monts., *Mer.*, Caerns., Denbs., *Flints.*, Angl.. 188 squares, map 605.

M. arvensis × spicata = M. × gentilis L. (incl. *M. × gracilis* Sole) : Mintys Culddail : Bushy Mint. Spontaneous or introduced. Streamsides (M,R,Rg). Fl.7. Mons., Glam., Rads., Carms., *Pembs.**, Cards., Monts., Mer., Caerns., Denbs., *Flints.*, Angl.*. 40 squares, map 606.

M. aquatica L. : Mintys y Dŵr : Water Mint. Native. Streamsides (Ar,D,Ds,G,H,M,R,S,W). Fl.7-10. Mons., Glam., Brecs., Rads., Carms., Pembs., Cards., Monts., Mer., Caerns., Denbs., Flints., Angl.. 265 squares, all except : **SS**38,39; **SM**50,71; **SN**04,25,32,36,59,86,87,88; **SO**20; **SH**92,94,98; **SJ**02.

M. aquatica × arvensis = M. × verticillata L. Mintys Troellaidd : Whorled Mint. Spontaneous. Streamsides (Ar,D,E,G,H,M,R). Fl.7-10. Mons., Glam., Brecs., Rads., Carms., Pembs., Cards., Monts., Mer., Caerns., Denbs., Flints., Angl.*. 127 squares, map 607.

M. aquatica × arvensis × spicata = M. × smithiana R. A. Graham : Mintys Coch : Red Mint. Spontaneous. Wet grassy places (Ar,M,R). Fl.7-10. Mons., Glam., Brecs., Rads., Carms., Cards., Monts., Mer., *Caerns., Angl.**. 46 squares, map 608.

M. aquatica × spicata = M. × piperita L. : Pupurfintys : Peppermint. Naturalized alien (garden origin) or sometimes spontaneous. Wet places (Ar,E,G,H, M,R,Rw). Fl.7-10. Mons., Glam., Brecs., Rads., Carms., Pembs., Cards., *Monts.*, Mer., *Caerns.*, Denbs., *Flints.*, Angl.. 113 squares, map 609.

M. suaveolens Ehrh. (*M. rotundifolia* auct., non (L.) Hudson) : Mintys Deilgrwn : Round-leaved Mint. Colonist (S. England). Roadsides (E,M,R,Rw). Fl.7-10. Mons., Glam., Brecs., *Rads.**, Carms., Pembs., Cards., Monts., Mer., Caerns., *Denbs.*, *Flints.*, Angl.. 103 squares, map 610. Some of these records may refer to **M. × villosa.**

[**M. longifolia** (L.) Hudson *Mons., Glam., Brecs., Rads., Carms., Pembs., Cards., Monts., Mer., Caerns., Denbs., Flints., Angl.*. All records refer to **M. spicata** L.]

M. longifolia × spicata Casual (garden origin). Waste places. Glam., Monts., *Denbs., Angl.*. 4 squares : **SN**61[112]; **SH***28*[17], 70[112]; **SJ**25[7].

M. spicata L. (incl. *M. longifolia* auct., non (L.) Hudson) : Mintys Ysbigog : Spear Mint. Established alien (garden origin). Roadsides and streamsides (M,R,Rw). Fl.6-10. Mons., Glam., Brecs., Rads.*, Carms., Pembs., Cards., Monts., Mer., Caerns., Denbs., Flints., *Angl.**. 121 squares, map 611.

M. spicata × suaveolens = M. × villosa Hudson (*M. × cordifolia* Opiz; *M. × niliaca* auct.; incl. *M. alopecuroides* Hull) : Mintys Lled Crynddail : Large Apple Mint. Established alien (garden origin). Roadsides and waste places (D,M,Q,R). Fl.6-10. Mons., Glam., Brecs., Rads., Carms., Pembs., Cards., Mer., *Caerns.*, Denbs., Flints., *Angl.*. 47 squares, map 612. Some of these records may refer to varieties of **M. spicata.**

SALVIA L.

Salvia officinalis L. : Sage. Casual (S. Europe). Waste places (R,S). Fl.7. *Monts., Angl.*. 2 squares : **SH**28[56]; **SJ**00[104].

S. sclarea L. : Clary. Casual (S. Europe). Waste places. Fl.7. Glam.. 1 square : **SS**78[112].

S. argentea L. Casual (S. Europe). Waste places. Fl.7. *Glam.*. 1 square : **SS***78*[140].

S. aethiopis L. Casual (S. Europe). Waste places. Fl.9. Glam.. 1 square : **ST***17*[112].

S. pratensis L. : Saets y Waun : Meadow Clary. Native. Pastures. Fl.6-7. Mons., (*Glam.*), [*Angl.*]. 5 squares.

Mons.: Rogiett, ST48, EV, 1903[112] and TGE, 1978[91]; Caldicot, ST48, SH, c1909[226]; Abergavenny Castle, SO20, An, 1807[226]. (**Glam.:** Port Einon, SS48, WT, 1802[82]; Merthyr Mawr, SS87, HJR, pre 1907[140]; Cardiff, ST17, JS, c1876[219]).

S. nemorosa L. (*S. sylvestris* auct., non L.) Casual (S. Europe). Waste places. Fl.6-8. Glam., *Brecs.*. 5 squares : **SS***78*[18],*87*[135]; **ST***16*[112],*27*[112]; **SN***90*[107].

S. verbenaca L. (*S. horminoides* Pourret) : Saets Gwyllt : Wild Clary. Native but possibly introduced in some of its old localities in S. Wales. Roadsides (E,G,R,Rw). Fl.4-7. Mons., Glam., Carms., Pembs., (*Mer.*), Caerns., *Denbs., Flints., Angl.*. 32 squares, map 613.

S. verticillata L. : Troellenog : Whorled Clary. Established alien (S. Europe). Roadsides (D,E,R,Rw). Fl.7-9. Mons.*, Glam., *Carms.**, Denbs.. 13 squares : **SS**48,*78,87*; **ST***06*,09,16,17,18,*27*,38; **SN***40*; **SO***00*; **SJ**35.

SOLANACEAE

NICANDRA Adanson

Nicandra physalodes (L.) Gaertner : Apple-of-Peru. Casual (Peru). Cultivated ground and waste places. Fl.8-10. Mons, Glam., *Brecs.*, Pembs., *Caerns.*. 8 squares : **SS***48*[112],*87*[112],97[112]; **ST**38[112]; **SM**92[112]; **SO**02[181]; **SH**33[68],78[177].

LYCIUM L.

Lycium barbarum L. (*L. halimifolium* Miller) : Duke of Argyll's Teaplant. Established alien (S. Europe). Hedges (E,Q,R,S). Fl.4-9. Mons., Glam., Brecs., *Rads.**, Carms., Pembs., Cards.*, Monts., Caerns., Denbs., Flints., Angl.*. 90 squares, map 614.

L. chinense Miller : Ysbeinwydd : China Teaplant. Established alien (E. Asia). Hedges (E,Rw,S). Fl.9. *Mons.**, Carms., *Pembs.**, Cards., *Flints.**, [*Glam.*]. 16 squares: **ST**58; **SN**10,11,15,20,24,25,31,35,40,58; **SJ**08, 16,17,25,27.

ATROPA L.

Atropa bella-donna L. : Codwarth : Deadly Nightshade. Native. Woods and waste places (E,Q,R,Rg, Rw,S,W). Fl.6-7. Mons., Glam., (*Brecs., Carms., Pembs.*), *Monts.*, Caerns., Denbs., Flints.. 23 squares, map 615.

HYOSCYAMUS L.

Hyoscyamus niger L. : Llewyg yr Iâr : Henbane. Native but frequently introduced. Sand dunes and waste places (D,F,Q,R,Rg,S,Sc). Fl.6-8. Mons., Glam., *Brecs.**, Carms., Pembs., *Cards.*, Monts., Mer., Caerns., Denbs., Flints., Angl.. 67 squares, map 616.

H. albus L. Casual (S. Europe). Waste places. Fl.9. Glam.. 2 squares : **ST***16*[112],*17*[218].

PHYSALIS L.

Physalis alkekengi L. : Suran Godog : Cape-gooseberry. Casual (C. and S. Europe). Waste places. Fl.8. *Mons.*, Carms.. 2 squares : **ST**38[91]; **SN**- Hendy[112].

P. peruviana L. Casual (S. America). Waste places. Glam.. 1 square : **ST**17[112].

SOLANUM L.

Solanum nigrum L. : Codwarth Du : Black Nightshade. Colonist (England). Cultivated ground and waste places (Ar,F,G,H,M,R,Rg,Rw). Fl.7-10. Mons., Glam., Brecs.*, *Rads.**, Carms., Pembs., *Cards., Monts.**, Mer., Caerns., *Denbs.*, Flints., *Angl.*. 87 squares, map 617.

S. chenopodioides Lam. (*S. sublobatum* Willd. ex Roemer & Schultes; *S. gracile* Dunal, non Sendtner). Casual (S. America). Waste places. Fl.9. Glam.. 1 square : **ST**16[112].

S. villosum Miller

Subsp. **villosum** (*S. luteum* Miller subsp. *luteum*) Casual (Europe). Waste places. Fl.9. Glam.. 1 square : **ST***18*[112].

Subsp. **puniceum** (Kirschleyer) Edmonds (*S. villosum* subsp. *alatum* (Moench) Edmonds; *S. luteum* subsp. *alatum* (Moench) Dostál; *S. miniatum* Bernh. ex Willd.) Casual (Europe). Waste places. Fl.8. Glam.. 1 square : **ST***16*[112].

S. sarrachoides Sendtner : Green Nightshade. Casual (Europe). Waste places. Fl.8. *N. Wales*[21].

S. triflorum Nutt. : Small Nightshade. Casual (N. America). Waste places. Fl.8. Glam., *Caerns.*. 3 squares : **ST***17*[112],*27*[112]; **SH***45*[107].

S. pseudocapsicum L. : Jerusalem-cherry. Casual (S. America). Waste places. Fl.10. Glam.. 1 square : **ST**27[112].

S. capsicastrum Link ex Schauer : Winter-cherry. Casual (S. America). Waste places. Fl.10. Glam.. 1 square : **ST***27*[112].

S. dulcamara L. : Codwarth Caled : Bittersweet. Native. Woods and hedges (Al,D,E,G,H,M,R, Rg,Rw,S,W). Fl.6-8. Mons., Glam., Brecs., Rads., Carms., Pembs., Cards., Monts., Mer., Caerns., Denbs., Flints., Angl.. 229 squares, map 618.

S. tuberosum L. : Tatws : Potato. Casual (S. America). Cultivated ground and waste places. Fl.7. *Mons.*, Glam., *Brecs.*, Rads., *Carms.*, Cards., *Monts., Caerns., Flints.*. 22 squares, map 619.

S. sisymbrifolium Lam. Casual (S. America). Cultivated ground (R,Rg). Fl.6-9. Glam., Pembs.. 2 squares : **ST**17[112]; **SM**92[112].

S. cornutum Lam. (*S. rostratum* Dunal). Casual (C. and N. America). Waste places. Fl.8. *Mons.*, Glam., Carms., *Flints.*. 5 squares: **ST***16*[112]17[112],38[49]; **SN**84[112]; **SJ***18*[143].

S. ciliatum Lam. Casual (S. America). Waste places. Glam.. 3 squares : **SS**87[36]; **ST**17[112],*27*[112].

LYCOPERSICON Miller

Lycopersicon esculentum Miller : Tomato. Casual (C. and S. America). Cultivated ground and waste places (Ar,F,M,R, Rw,S). Fl.8. *Mons.*, Glam., Brecs., *Carms.*, *Cards.*, *Monts.*, *Caerns.*, *Flints.*. 30 squares, map 620.

DATURA L.

Datura stramonium L. : Meiwyn : Thorn-apple. Casual or established alien (America). Cultivated ground and waste places (F,M,R,Rg). Fl.7-9. Mons., Glam., Brecs., *Rads.*, Carms., Pembs., *Cards.*, *Monts.*, *Mer.*, *Caerns.*, *Denbs.*, *Flints.*, *Angl.*. 68 squares, map 621.

D. ferox L. : Angels-trumpets. Casual (E. Asia). Waste places. Fl.9. Glam., Brecs., Pembs.. 4 squares : **ST**17[112],18[112]; **SO**23[112]; **SM**91[112].

NICOTIANA L.

Nicotiana alata Link & Otto Casual (S. America). Waste places. Fl.8. Glam.. 1 square : **ST**17[112].

N. longiflora Cav. Casual (S. America). Waste places. Fl.9. Glam.. 1 square : **ST**17[112].

PETUNIA Juss.

Petunia axillaris (Lam.) Britton, E. E. Stearns & Poggenb. Casual (S. America). Roadsides. Glam.. 1 square : **ST**18[112].

P. axillaris × integrifolia (Hooker) Schinz & Thell. = **P. × hybrida** Vilm.- Andr. Casual (garden origin). Waste places. *Glam.*. 1 square : **SS**79[117].

BUDDLEJACEAE

BUDDLEJA L.

Buddleja davidii Franchet (incl. *B. spectabilis* auct.) : Bwdleia : Butterfly-bush. Established alien (China). Walls and waste places (E,Q,R,Rw). Fl.6-9. Mons., Glam., *Brecs.*, *Rads.*, Carms., Pembs., Cards., *Monts.*, Mer., Caerns., *Denbs.*, Flints., *Angl.*. 86 squares, map 622.

B. globosa Hope Introduced (S. America). Roadsides Fl.6. *Angl.*. 1 square.
Angl.: Holyhead, SH28, FHo, 1976[256].

SCROPHULARIACEAE

LIMOSELLA L.

Limosella aquatica L. : Lleidlys : Mudwort. Native. Pool margins. Fl.7-9. Glam., Brecs., *Carms.*, *Monts.**, Mer.*, *Flints.*, *Angl.*, [*Caerns.*]. 10 squares : **SS**78,88; **SN***42*,*80*,92; **SO**04; **SH***37*,83; **SJ***08*,20.

L. aquatica × australis Spontaneous. Pools. Glam.. 1 square.
Glam.: Morfa Pools, SS78, EV, 1935[112] and 1945[224] (not seen recently).

L. australis R. Br. (*L. subulata* Ives) : Lleidlys Cymreig : Welsh Mudwort. Native. Pool margins (Al,Ar). Fl.6-8. Glam., Mer., Caerns.. 5 squares.
Glam.: Crymlyn Fen, SS69, HJR, 1907[140]; Kenfig Pool, SS78, ESM & WAS, 1901[111] and MRS, 1970[95]; Morfa Pools, SS78, EV, 1935[224] and JAW, 1942[112]. **Mer.:** Dysynni Broadwater, SH50, PMB, 1950+[7] & 1979[3]; Tywyn, SH50, JEL, 1953[112]; Glaslyn Estuary, SH53, DAJ, 1921[112] and PMB, 1979[3]; Minffordd, SH53, JEL, 1935[107]; Morfa Harlech, SH53, An, 1965[7]; Bryn-crug, SH60, CMG, 1952[112] and PMB, 1979[3]. **Caerns.:** Glaslyn Estuary, SH53, DAJ, 1921[112]; Portmadoc, SH53, APC, c1955[74] and PMB, 1979[3].

MIMULUS L.

Mimulus guttatus DC. : Blodyn y Mwnci : Monkeyflower. Naturalized alien (N. America). Shallow streams (Ar,M). Fl.6-10. Mons., Glam., Brecs., Rads., Carms., Pembs., *Cards.*, Monts., Mer., Caerns., Denbs., Flints., *Angl.**. 164 squares, map 623.

M. guttatus × luteus Naturalized alien (America). Streamsides (Ar,M). Mons., *Glam.*, *Brecs.*, *Rads.*, *Carms.*, Pembs., *Cards.*, *Monts.*, Mer., Caerns., *Denbs.*, *Angl.*. 58 squares, map 624.

M. guttatus × tigrinus Hort. Naturalized alien (garden origin). Streamsides. Caerns.. 1 square.
Caerns.: Bethesda, SH66, RHR & EDM, 1971[242]; Llanllechid, SH66, RHR & EDM, 1971[242].

[**M. luteus** L. *Mons.*, *Brecs.*, *Carms.*, *Pembs.*, *Mer.*, *Caerns.* All records refer to **M. guttatus × luteus.**]

M. cupreus Dombrain × **guttatus × luteus** Naturalized alien (garden origin). Streamsides. Fl.6. Rads..
Rads.: nr Llanbadarn Fynydd, SO17, ACP, 1981[112].

M. moschatus Douglas ex Lindley : Mwsg : Musk. Naturalized alien (N. America). Streamsides (M,Q, Rg,Rw). Fl.6-9(-12). Mons.*, Glam., Brecs., Rads., (Carms.*), *Pembs.**, Cards., Monts., Mer., Caerns., Denbs.. 36 squares, map 625.

VERBASCUM L.

Verbascum ovalifolium Donn ex Sims Casual (S. E. Europe). Waste places. *Glam.*. 1 square: **ST***17*[22].

V. blattaria L. : Gwyfynnog : Moth Mullein. Established alien (Europe). Waste places (G,M,R). Fl.7-8. Mons., Glam., *Carms.*, Pembs., Cards.*, *Mer.*, *Caerns.*, *Denbs.*, *Flints.*, *Angl.**. 25 squares, map 626.

V. virgatum Stokes : Tewbannog : Twiggy Mullein. Established alien (Europe). Waste places (E,Q,R,Rw) Fl.7-9. Glam., Brecs.*, Carms.*, Pembs., Cards., *Monts.**, *Mer.*, Caerns., Denbs., Flints.*. 19 squares : **SS**79; **ST**16,17; **SM**82; **SN**10,30,31,40,58,67,*68*,*69*; **SO**02,21; **SH***57*; **SJ**16,20,25,35.

V. phoeniceum L. : Pannog Brithgoch : Purple Mullein. Casual (Europe). Waste places. Fl.8. *Glam., Brecs..* 2 squares : **SS***79*[35]; **SN***90*[107].

V. phlomoides L. : Orange Mullein. Established alien (S. Europe). Waste places (E,Q,R,Rg). Fl.7-8(-10). Mons., Glam., Carms., Pembs., Cards., *Denbs.,* Flints.. 11 squares : **SS**69,79; **ST**17,28,38; **SN**10,58,71; **SH**85; **SJ**07,25.

V. thapsus L. : Pannog Melyn : Great Mullein. Native. Quarries and waste places (D,E,G,Q,R,Rw,W). Fl.6-9. Mons., Glam., Brecs., Rads., Carms., Pembs., Cards., Monts., *Mer.,* Caerns., Denbs., Flints., Angl.. 199 squares, map 627.

[**V. speciosum** Schrader *Glam.*]

V. sinuatum L. Casual (S. Europe). Waste places. Fl.9. Glam., *Caerns..* 2 squares : **ST***16*[112]; **SH***53*[17].

V. pulverulentum Vill. : Pannog Blawrwyn : Hoary Mullein. Established alien (England). Waste places (M,R,Rw). Fl.7-8. Mons., Glam., Cards.*. 6 squares. **Mons.:** Redbrook, SO51, SGC, 1936[112]. **Glam.:** Crymlyn Burrows, SS79, AEW, 1956[112] and CH & HH, 1976[117]; nr Neath, SS79, EV & EMT, 1937[112]; Barry, ST16, RLS, 1922[112]; Cardiff, ST17, RM & RLS, 1924[23]; nr Clydach, SN60, QONK, 1974[157]. **Cards.:** Cardigan, SN14, MPa, 1981[71] and LF & SBE, 1981[112].

V. lychnitis L. : Hanner Pan : White Mullein. Naturalized alien (England). Waste places and rocky woods (G,Q,R,W). Fl.6-7. *Mons.**, *Glam., Carms.**, *Cards.**, *Mer.,* Denbs., *Flints..* 15 squares : **SS**78; **ST***17, 38*; **SN**40,55,58,62,74; **SH**61,72; **SJ**08,*24*,27,35,36.

[**V. lychnitis × nigrum = V. × schiedeanum** Koch *Pembs.*]

V. lychnitis × thapsus = V. × thapsi L. Spontaneous. Quarries and waste places (Q,R,Rw). Fl.6-8. Cards., Denbs., Flints., *Angl..* 4 squares.
Cards.: Aberystwyth, SN58, AOC, 1977[112]. **Denbs.:** Marford, SJ35, TE, 1971[112]. **Flints.:** Flint, SJ27, TE, 1973[112]. **Angl.:** Friars, SH67, WWM, 1926[25].

V. chaixii Vill.

Subsp. **chaixii** : Nettle-leaved Mullein. Established alien (S. Europe). Waste places. Fl.7. Mons., Glam.. 5 squares.
Mons.: Pontnewydd, ST29, TGE, 1981[112]. **Glam.:** Port Talbot, SS78, HJR, 1905[107]; Porth-cawl, SS87, HJR, 1904[107]; Canton, Cardiff, ST17, RLS, 1923[22]; Aberdare, SO00, HJR, 1902[107].

Subsp. **austriacum** (Schott ex Roemer & Schultes) Hayek (*V. austriacum* Schott ex Roemer & Schultes). Casual (C. Europe). Waste places. Fl.8. *Glam..* 1 square : **SN***90*[140].

V. nigrum L. : Pannog Tywyllddu : Dark Mullein. Colonist (England). Waste places (D,G,M,R,Rw,Sc). Fl. 7-9. Mons., Glam., *Brecs.**, *Carms.**, Pembs., *Cards.*[210], *Monts.**, *Mer.,* Caerns., Denbs., *Flints..* 39 squares, map 628.

[**V. nigrum × phlomoides = V. × brockmuelleri** Ruhmer *Pembs.*]

V. nigrum × thapsus = V. × semialbum Chaub. Spontaneous. Quarries. Fl.6-7. Pembs.. 1 square. **Pembs.:** Tenby, SN10, FLR, 1942[112] & 1943[112].

(*V. albiflorum × nigrum* recorded in *List of Pembrokeshire Plants* F. L. Rees (1950), is an error for **V. nigrum** f. **albiflorum).**

SCROPHULARIA L.

Scrophularia vernalis L. : Gornerth Felen : Yellow Figwort. Naturalized alien (Europe). Hedgebanks (E,R,W). Fl.3-8. Cards.*, Caerns., Denbs., Angl., [*Mer.*]. 7 squares.
Cards.: Llanbadarn Fawr, SN58, RGE, 1977[112]. **Caerns.:** nr Conway, SH77, RL, 1956[7]; Gt Ormes Head, SH78, LWF, 1956[7]; Llandudno, SH78, JMB, 1965[67]; Colwyn Bay, SH87, JMB, 1965[67]; Gloddaith, SH88, CB, 1876[112] and PMB, 1981[3]; Little Ormes Head, SH88, DC, 1956[7]; Llandrillo-yn-Rhos, SH88, JMB, c1965[67]. **Denbs.:** Llansanffraid Glan Conwy, SH87, RL, 1957[112]; Colwyn Bay, SH87, JMB, 1971[180]; Abergele, SH97, JAG, 1971[112]. **Angl.:** nr Bethel, SH36, JEG, 1890[112].

S. scorodonia L. : Gornerth Gwenynddail : Balm-leaved Figwort. Naturalized alien (S. England). Waste places and thickets (D,E,R,Rw). Fl.5-8. *Mons.**, Glam., *Carms.**, Pembs.*, Cards.*. 7 squares.
Mons.: Newport Docks, ST38, TGE & CT, 1972[182] & 1980[91]. **Glam.:** Porth-cawl, SS87, PWR, 1920[112]; Cardiff, ST17, AEW, 1924[112] & 1956[112]; Cardiff Docks, ST27, JPC, 1981[112]. **Carms.:** nr Pembrey, SN40, CSe & JOM, 1978[245]. **Pembs.:** Ramsey Island, SM72, JWDo 1974[112]. **Cards.:** Llanbadarn Fawr, SN58, VGE, 1970[112] & 1978[112].

S. scopolii Hoppe Casual (Europe). Waste places. *Glam..* 1 square : **ST***17*[22].

S. nodosa L. : Melyn y Gwanwyn : Common Figwort. Native. Woods (E,G,M,Q,R,Rw,W). Fl.5-9. Mons., Glam., Brecs., Rads., Carms., Pembs., Cards., Monts., Mer., Caerns., Denbs., Flints., Angl.. 262 squares, all except : **SR**89; **SS**38,39; **SM**50,71,73,84; **SN**15,59,74,75,86,87; **SH**13,29,51,98; **SJ**03,45,54.

S. auriculata L. (*S. aquatica* auct., non L.) : Gornerth y Dŵr : Water Figwort. Native. Streamsides (Ar,F,M, R,Rg,Rw,W). Fl.6-9. Mons., Glam., Brecs., *Rads.,* Carms., Pembs., *Cards.,* Monts., *Mer., Caerns.,* Denbs., *Flints.,* Angl.. 136 squares, map 629.

S. umbrosa Dumort. : Gornerth Gorllewinol : Green Figwort. Native. Damp woods (Ar,M,W). Fl.8. *Mons.,* Monts.*, *Denbs.,* [*Mer.*]. 3 squares.
Mons.: Bigswier, SO50, AL, pre 1900[226]. **Monts.:** Stalloe Farm, SO29, VJM, 1972[112]. **Denbs.:** nr Ruthin, SJ15, H, 1916[15].

S. canina L.

Subsp. **canina** : French Figwort. Established alien (S. Europe). Waste places. Fl.6-7(-11). Mons., Glam.. 5 squares.

Mons.: Newport, ST38, TGE & CT, 1973[112]. **Glam.:** Barry, ST16, An, 1930+[72]; Cardiff, ST17, JS, 1876[112]; Cardiff Docks, ST27, JS, 1876[112] and DM, c1955[162]; Radyr, ST18, RLS, 1920[112].

Subsp. **hoppii** (Koch) P. Fourn. (*S. hoppii* Koch). Casual (S. Europe). Waste places. *Glam.*. 1 square : **ST***18*[21].

CALCEOLARIA L.

Calceolaria chelidonioides Humboldt, Bonpland & Kunth. : Slipper-flower. Casual (S. America). Cultivated ground. Fl.9. Mons., *Caerns.*. 2 squares : **SO**10[112]; **SH***54*[17].

[**C. tripartita** Ruiz & Pavón (*C. gracilis* Humboldt, Bonpland & Kunth.). *Caerns.*]

ALONSOA Ruiz & Pavón

Alonsoa incisifolia Ruiz & Pavón : Mask-flower. Casual (S. America). Cultivated ground. Fl.8. Brecs.. 1 square : **SO**12[112].

ANTIRRHINUM L.

Antirrhinum majus L. : Trwyn y Llo Mwyaf : Snapdragon. Established alien (S.W. Europe). Walls (Q,R). Fl.6-8(-10). Mons., Glam., *Brecs.**, Rads., *Carms.*, *Pembs.*, *Cards.*, Monts., Mer., Caerns., Denbs., Flints., *Angl.*. 64 squares, map 630.

MISOPATES Rafin.

Misopates orontium (L.) Rafin. (*Antirrhinum orontium* L.) : Trwyn y Llo Bychan : Lesser Snapdragon. Established alien (England). Cultivated ground (F,R, Rg,Rw). Fl.6-11. Mons., Glam., Carms., Pembs., Cards., Monts., Mer., Caerns., Denbs., Angl.. 56 squares, map 631.

CHAENORHINUM (DC.) Reichenb.

Chaenorhinum minus (L.) Lange : Gingroen Bychan : Small Toadflax. Established alien (England). Railway tracks and cultivated ground (F,Q,R,Rg,Rw). Fl.6-10. Mons., Glam., Brecs., Rads., Carms., Pembs., Cards., Monts., Mer., *Caerns.*, Denbs., Flints., *Angl.*. 140 squares, map 632.

LINARIA Miller

Linaria reflexa (L.) Desf. Casual (Mediterranean region). Waste places. *Glam.*. 1 square : **ST***17*[22].

L. triphylla (L.) Miller Casual (Mediterranean region). Cultivated ground (R,Rg). Glam.. 1 square : **ST***27*[112].

L. genistifolia (L.) Miller subsp. **dalmatica** (L.) Maire & Petitmengin (*L. dalmatica* (L.) Miller) : Balkan Toadflax. Casual (S. Europe). Waste places. Fl.8. Glam.. 1 square : **ST***17*[112].

L. viscosa (L.) Dum.-Courset (incl. *L. heterophylla* Desf.; *L. stricta* (Sibth. & Sm.) Guss., non Hornem.). Casual (S. Europe). Waste places. Fl.8. Glam.. 1 square : **ST***18*[112].

L. incarnata (Vent.) Sprengel (*L. bipartita* auct. eur., non (Vent.) Willd.). Casual (N.W. Africa). Waste places. Fl.7. Glam.. 1 square : **ST**17[112].

L. purpurea (L.) Miller : Gingroen Cochlas : Purple Toadflax. Established alien (S. Europe). Walls (E,G,Q,R,Rw,S). Fl.6-8. Mons., Glam., *Brecs.**, *Rads.**, Carms.*, Pembs.*, *Cards.**, *Monts.**, Mer.*, Caerns.*, Denbs., Flints., Angl.*. 66 squares, map 633.

L. purpurea × repens = L. × dominii Druce Spontaneous. Waste places (Q,R,Rg). Fl.7-8. Mons., Glam., Carms., *Pembs.*, Caerns., *Angl.*. 8 squares.

Mons.: Rumney, ST27, ARP, 1978[112]; Severn Tunnel Junction, ST48, TGE, 1978[184]. **Glam.:** nr Gowerton, SS59, QONK, 1974[157]; nr Port Talbot, SS79, HH & CH, 1974[117]; Barry Docks, ST16, RGE & TGE, 1980[112]. **Carms.:** Llandybie, SN61, PHD, 1976[112]. **Pembs.:** Fishguard, SM93, DM, 1956[7]. **Caerns.:** nr Bangor, SH57, RHR, 1973[112]. **Angl.:** Menai Bridge, SH57, RHR, 1979[209].

L. repens (L.) Miller : Gingroen Porffor Gwelw : Pale Toadflax. Colonist (England). Railway banks and road sides (Ar,D,E,G,M,Q,R,Rg,Rw,S). Fl.6-9. Mons., Glam., Brecs., Rads., Carms., Pembs., Cards., Monts., Mer., Caerns., Denbs., Flints., *Angl.**. 119 squares, map 634.

L. repens × vulgaris = L. × sepium Allman Spontaneous. Railway banks and roadsides (E,Q,R,Rg, Rw,S,Sc). Fl.6-10. Mons., Glam., Brecs., Carms., Pembs., Cards., Mer., *Caerns.*, Flints.. 40 squares, map 635.

L. pelisseriana (L.) Miller : Jersey Toadflax. Casual (Jersey). Waste places. *Glam.*, *Cards.*. 3 squares : **SS***69*[18]; **ST***17*[140]; **SN***58*[210].

L. vulgaris Miller : Llin y Llyffant : Common Toadflax. Native. Hedgebanks and waste places (D,E, G,Q,R,Rg,Rw,S). Fl.6-10. Mons., Glam., Brecs., Rads., Carms., Pembs., Cards., Monts., Mer., Caerns., Denbs., Flints., Angl.. 232 squares, map 636.

L. aeruginea (Gouan) Cav. (*L. reticulata* auct. eur., non (Sm.) Desf.). Casual (Iberian Peninsula). Waste places. *Cards.*. 1 square : **SN***58*[19].

L. caesia (Pers.) DC. ex Chav. Casual (Iberian Peninsula). Waste places. *Glam.*. 1 square : **ST***17*[218].

L. supina (L.) Chaz. : Prostrate Toadflax. Established alien (S.W. Europe). Railways and waste places. Fl.6-8 *Glam.*, Carms.. 2 squares.

Glam.: Cardiff, ST17, JS, c1876[140] (casual only). **Carms.:** nr Burry Port, SN40, TWB, pre 1900[141] and RDP, 1980[201].

L. arvensis (L.) Desf. Casual (Europe). Waste places. Fl.10. *Glam.*. 1 square : **ST***18*[107].

CYMBALARIA Hill

Cymbalaria muralis P. Gaertner, B. Meyer & Scherb. : Llin y Fagwyr : Ivy-leafed Toadflax. Established alien (S. Europe). Walls (Ar,E,G,Q,R,Rw,Sc). Fl.3-10. Mons., Glam., Brecs., Rads., Carms., Pembs., Cards., Monts., Mer., Caerns., Denbs., Flints., Angl.. 221 squares, map 637.

C. pallida (Ten.) Wettst. : Greater Ivy-leaved Toadflax. Casual (S. Italy). Walls and waste places. Fl.7-9. *Glam.*[107], Pembs., *Mer., Caerns., Flints.*. 4 squares : **SM**80[112]; **SH**54[7],61[7]; **SJ**27[261]. The *Glam.* record is from the Gower but the exact locality is unknown.

KICKXIA Dumort.

Kickxia elatine (L.) Dumort. : Trwyn y Llo Blaen-feinddail : Sharp-leaved Fluellen. Established alien (England). Cultivated ground (F,R,Rg,Rw,Sc). Fl.7-10. Mons., Glam., *Rads.**, Carms., Pembs., Cards., Monts., Mer.*, Caerns., Denbs., *Flints.*, Angl.. 77 squares, map 638.

K. spuria (L.) Dumort. : Llysiau Llywelyn : Round-leaved Fluellen. Established alien (England). Cultivated ground (F,R,Rg,S,W). Fl.7-9. Mons., Glam., *Carms.**, *Pembs.*, [*Cards.*]. 18 squares : **SS***48,69*,87,*96;* **ST**06,07, 08,16,17,*27,38*,48,49,*59*; **SM**70,72,80; **SN***50.*

DIGITALIS L.

Digitalis purpurea L. : Bysedd y Cŵn : Foxglove. Native. Hedges and copses (D,E,G,H,M,Q,R,Rw,W). Fl.5-9(-11). Mons., Glam., Brecs., Rads., Carms., Pembs., Cards., Monts., Mer., Caerns., Denbs., Flints., Angl.. 276 squares, all except : **SS**38,39,77; **ST**58; **SM**50; **SH**98.

ERINUS L.

Erinus alpinus L. : Fairy Foxglove. Established alien (Europe). Walls. Fl.4-8(-10). Mons., Glam., *Brecs.*, Carms., *Caerns.*, Denbs., Flints., *Angl.*. 16 squares : **SS**97; **ST**07,17,*59*; **SN**63; **SO**12,21,*51*; **SH**67,75,*97*; **SJ***06*,07,16,17,*27*.

VERONICA L.

Veronica serpyllifolia L. : Rhwyddlwyn Gwrywddail : Thyme-leaved Speedwell.

Subsp. **serpyllifolia** Native. Pastures (E,F,G,H,M, Q,R,Rg). Fl.4-9. Mons., Glam., Brecs., Rads., Carms., Pembs., Cards., Monts., Mer., Caerns., Denbs., Flints., Angl.. 245 squares, map 639.

Subsp. **humifusa** (Dickson) Syme Native. Damp places in mountain areas. Fl.7-8. Mons.*, Brecs., *Carms.**, Caerns., *Denbs.*[230]. 8 squares.
Mons.: nr Monmouth, SH51, SGH, 1962[112]. **Brecs.:** nr Ystradfellte, SN91, An, 1930+[195]; Merthyr Tydfil, SO01, MG, 1970[99]; Pen y Fan, SO02, AEW, 1926[112] and MP, 1970+[200]. **Carms.:** Cwm Mihertach, SN72, DPMG, 1950+[106]. **Caerns.:** nr Beddgelert, SN54, FCR, pre 1895[101]; Snowdon, SH65, ER, 1950+[207]; Carnedd Ddafydd, SH66, JLW, 1893[112] and JHS, 1925[112].

V. austriaca L. subsp. **teucrium** (L.) D. A. Webb (*V. teucrium* L.). Casual (Europe). Waste places. *Caerns.*. 1 square : **SH**32[74].

V. prostrata L. Casual (Europe). Waste places. *Cards.*. 1 square : **SN***14*[19].

V. officinalis L. : Rhwyddlwyn Meddygol : Heath Speedwell. Native. Pastures (D,Ds,E,G,H,M,Q,R,W). Fl.5-10. Mons., Glam., Brecs., Rads., Carms., Pembs., Cards., Monts., Mer., Caerns., Denbs., Flints., Angl.. 258 squares, all except : **SR**89; **SS**39,77,96; **ST**06,26,27; **SM**50,73; **SN**02,15,22,32,36,46,59,87; **SO**30,41,50; **SH**51,60,98; **SJ**44.

V.chamaedrys L. : Llygad Doli : Germander Speedwell. Native. Hedgebanks (D,Ds,E,G,M,Q,R,Rg, Rw,W). Fl.4-7. Mons., Glam., Brecs., Rads., Carms., Pembs., Cards., Monts., Mer., Caerns., Denbs., Flints., Angl.. 275 squares, all except : **SS**39,77; **SM**50,94; **SN**59; **SH**51,98.

V. montana L. : Rhwyddlwyn Mynyddol : Wood Speedwell. Native. Woods (E,M,W). Fl.4-6. Mons., Glam., Brecs., Rads., Carms., Pembs., *Cards., Monts.*, Mer., Caerns., Denbs., Flints., *Angl.*. 192 squares, map 640.

V. scutellata L. : Rhwyddlwyn Culddail y Gors : Marsh Speedwell. Native. Marshes (G,H,M). Fl.6-9. Mons., Glam., Brecs., Rads., Carms., Pembs., Cards., Monts., Mer., Caerns., Denbs., Flints., Angl.. 188 squares, map 641.

V. beccabunga L. : Gorferini : Brooklime. Native. Streams and ditches (Al,Ar,H,M,W). Fl.5-8. Mons., Glam., Brecs., Rads., Carms., Pembs., Cards., Monts., Mer., Caerns., Denbs., Flints., Angl.. 247 squares, map 642.

V. anagallis-aquatica agg. Native. Streamsides (Al,Ar, D,Ds,G,M,S). Fl.5-9. Mons., Glam., Brecs., Rads.*, Carms., Pembs., *Cards.**, *Monts.*, Mer., *Caerns.*, Denbs., Flints., Angl.. 91 squares, map 643.

V. anagallis-aquatica L. *s.s.* : Graeanllys y Dŵr : Blue Water-speedwell. Native. Streamsides (Al,Ar, Ds,G,M). Fl.6-8. Mons., Glam., Brecs.*, *Rads.**, *Carms.**, *Pembs.*, *Cards.**, *Monts.**, Mer.*, *Caerns.**, *Denbs.**, *Flints.**, Angl.. 50 squares, map 644.

V. anagallis-aquatica × catenata = V. × lackschewitzii Keller Spontaneous. *Glam.*. 1 square. **Glam.:** Gower, SS -, NGM, 1970[93].

V. catenata Pennell : Graeanllys y Dŵr Rhudd-goch : Pink Water-speedwell. Native. Streamsides (Al,Ar,D,Ds,M,S). Fl.5-9. Mons., Glam., Brecs., Carms., Pembs., *Monts.**, Mer., *Caerns.*, Denbs., Flints., Angl.. 61 squares, map 645.

V. arvensis L. : Mur-rwyddlwyn : Wall Speedwell. Native. Walls and dry fields (D,F,G,H,Q,R,Rg,Rw,S). Fl.4-7(-11). Mons., Glam., Brecs., Rads., Carms., Pembs., Cards., Monts., Mer., Caerns., Denbs., Flints., Angl.. 239 squares, map 646.

V. peregrina L. : American Speedwell . Casual (N. America). Cultivated ground. *Monts.*, [*Glam.*]. 1 square : **SJ**10[180].

V. agrestis L. : Rhwyddlwyn Gorweddol : Green Field-speedwell. Colonist (England). Cultivated ground (F,Q,R,Rg,Rw). Fl.(3-)6-9. Mons., Glam., Brecs., *Rads.*, Carms., Pembs., *Cards.*, Monts., Mer., Caerns., Denbs., Flints., *Angl.*. 145 squares, map 647.

V. polita Fries : Rhwyddlwyn Llwyd : Grey Field-speedwell. Colonist (England). Cultivated ground (D,E,F,Q,R,Rg). Fl.5-10. Mons., Glam., Brecs., *Rads.*, Carms., Pembs., *Cards.*, *Monts.*, Mer., Caerns., Denbs., *Flints.*, Angl.. 101 squares, map 648.

V. persica Poiret : Rhwyddlwyn y Gerddi : Common Field-speedwell. Established alien (Europe). Cultivated ground (E,F,G,M,Q,R,Rg,Rw,S). Fl.3-10. Mons., Glam., Brecs., Rads., Carms., Pembs., Cards., Monts., Mer., Caerns., Denbs., Flints., Angl.. 211 squares, map 649.

V. filiformis Sm. : Rhwyddlwyn Crwn-ddail : Slender Speedwell. Naturalized alien (Caucasus). River banks and waysides (E,G,M,Q,R). Fl.4-6. Mons., Glam., *Brecs.**, Rads.*, Carms., *Pembs.**, Cards.*, *Monts.*, Mer., *Caerns.**, *Denbs.**, Flints., *Angl.**. 131 squares, map 650.

V. hederifolia L. : Rhwyddlwyn Eiddewddail : Ivy-leaved Speedwell. Native. Cultivated ground and hedgebanks (E,F,M,Q,R,Rg,Rw,W). Fl.3-6. Mons., Glam., Brecs., Rads., Carms., Pembs., Cards., Monts., Mer., Caerns., Denbs., Flints., Angl.. 179 squares, map 651.

Subsp. **lucorum** (Klett & Richter) Hartl (*V. sublobata* M. Fischer). Native. Woods and roadsides (E,M,Q,R,Rg,Rw,W). Fl.3-5. Mons., Glam., Brecs., Rads., Carms., Pembs., Cards., Monts., Mer., Caerns., Denbs., Angl.. 70 squares, map 652. Under-recorded.

Subsp. **hederifolia** Denizen (England). Cultivated ground (F,M,R,Rg). Fl.3-6. Mons., Glam., Pembs., Cards., Monts., Mer., Caerns., Denbs., Flints., Angl.. 25 squares, map 653.

V. crista-galli Steven Casual (Caucasus). Waste places. Fl.5. Cards.. 1 square : **SN**68[112].

V. longifolia L. Casual (Europe). Waste places (E,M,R,Rw). Fl.7-9. Glam., Carms., Cards., *Monts.*. 7 squares : **SS**59[157], 79[117],87[95]; **ST**17[112]; **SN**51[161],68[112]; **SJ**31[7].

V. paniculata L. (*V. spuria* auct., non L.). Casual (Europe). Waste places. *Glam.*. 1 square : **SN**70[157].

V. spicata L. (incl. *V. hybrida* L.) : Spiked Speedwell. Native. Limestone and volcanic rocks (G,H,Q,R,Sc). Fl.6-9. Glam., Rads., Carms., Pembs., Monts., Mer., Caerns., Denbs., *Flints.*. 13 squares : **SS**48; **SM**83; **SN**61,69; **SO**25; **SH**61,78,87,88; **SJ**06,07,08,21.

HEBE Commerson

Hebe speciosa (R. Cunn. ex A. Cunn.) Andersen Naturalized alien (New Zealand). Sea cliffs. Fl.6-10. *Glam.*, *Mer.*. 2 squares.
Glam.: Langland, SS68, JAW, 1920[19]. **Mer.:** nr Barmouth, SH61, GCD, 1920[19]. These records may refer to the following hybrid.

H. elliptica (G. Forster) Pennell × **speciosa** = **H. × franciscana** (Eastwood) Souster (*H. × lewisii* auct.). Naturalized alien (garden origin). Sea cliffs. Fl.6-9. Pembs., *Cards.*, Mer., Caerns., *Angl.*. 9 squares.
Pembs.: Goodwick, SM93, NDS, 1962[107]; Tenby, SN10, AM, 1974[112]. **Cards.:** Aberystwyth, SN58, AOC, 1975[183]. **Mer.:** Llanaber, SH61, JAW, 1941[112]. **Caerns.:** , SH22, APC, c1978[74]; Aber-soch, SH32, RGE, 1978[112]; Gt Ormes Head, SH78, JMB, 1970+[67]. **Angl.:** Moelfre, SH58, RHR, 1974[208]; Trwyn Du, SH68, RHR, 1974[208].

SIBTHORPIA L.

Sibthorpia europaea L. : Ceinioglys : Cornish Moneywort. Native. Moist shady places (Ar,H,M,R). Fl.6-8. Glam., Carms., Pembs.*, Cards.. 12 squares : **ST**08,09,18,19; **SM**80; **SN***31*,33,*34*,*42*,44,52,53.

MELAMPYRUM L.

Melampyrum pratense L. : Gliniogai Cyffredin : Common Cow-wheat. Native. Woods (E,H,R,Rw,W). Fl.5-9. Mons., Glam., Brecs., Rads., Carms., Pembs., Cards., Monts., Mer., Caerns., Denbs., Flints., *Angl.*. 204 squares, map 654.

EUPHRASIA L.

Euphrasia officinalis agg. : Effros : Eyebright. Native. Pastures and heaths (D,Ds,E,F,G,H,M,Q,R,Rw,W). Fl.5-10. Mons., Glam., Brecs., Rads., Carms., Pembs., Cards., Monts., Mer., Caerns., Denbs., Flints., Angl.. 266 squares, all except: **SS**39; **ST**27,58; **SM**50,73; **SN**50; **SO**09,30,31,40,41; **SH**51,98; **SJ**36,44,45.

This account of **Euphrasia,** especially of the hybrids, is largely based on the work of P. F. Yeo, who kindly lent his notes and records. It is important to remember that positive identification of hybrids in this genus is difficult and most determinations should be regarded as provisional only.

E. rostkoviana Hayne

[Subsp. **montana** (Jordan) Wettst. (*E. montana* Jordan). *Mons., Glam., Brecs., Carms., Cards., Mer., Caerns.*]

Subsp. **rostkoviana** (incl. *E. hirtella* auct. brit., non Jordan ex Reuter *p.p.*). Native. Meadows and upland pastures (F,G,H,M,R,W). Fl.7-9. Mons., Glam., Brecs., Rads.*, Carms., *Pembs.**, Cards., *Monts.*, Mer., Caerns., *Denbs.**. 87 squares, map 655.

E. rivularis Pugsley Native. Mountain turf near streams (G,M). Fl.6-9. *Mer.*, Caerns.. 6 squares.
Mer.: Cadair Idris, SH71, WAV, 1905[262]; nr Llanfrachreth, SH72, JBM, 1939[107]. **Caerns.:** Moel yr Ogof, SH54, DGC, 1924[262]; Llanberis, SH55, JAWh, 1919[112]; Clogwyn Du'r Arddu, SH55, JEL, 1935[112]; btwn Llanberis and Cwm-y-glo, SH56, JLOL, 1924[262]; Snowdon, SH65, DM, 1955[262] and ER, 1965[207]; Cwm Idwal, SH65, PFY, 1955[262] and ER, 1965[207].

E. anglica Pugsley (incl. *E. hirtella* auct. brit., non Jordan ex Reuter *p.p.*). Native. Heaths and pastures (Ds,G,H,M,R). Fl.6-10. Mons., Glam., Brecs., *Rads.**, Carms.*, Pembs., Cards., Monts., Mer., *Caerns.**, Denbs., *Angl.**. 88 squares, map 656.

E. anglica × confusa Spontaneous. Upland pastures. Fl.7. *Brecs., Rads., Carms.*. 3 squares.
Brecs.: Craig Cerrig-gleisiad, SN92, PS, 1963[107]. **Rads.:** nr Gladestry, SO25, NYS, 1945[109]. **Carms.:** Ystradffin, SN74, IMV, 1960[112].

E. anglica × micrantha Spontaneous. *Pembs.*. 1 square.
Pembs.: St David's Head, SM72, PFY, 1960+[217].

E. anglica × nemorosa = E. × glanduligera Wettst. Spontaneous. Damp places (E,M). Fl.7-8. *Rads.*, Pembs.. 2 squares.
Rads.: nr Newbridge-on-Wye, SO05, LW, 1971[197].
Pembs.: Freni Fach, SN23, TAWD, 1971[112].

E. anglica × rostkoviana Spontaneous. Heaths and bogs. Fl.8. *Cards., Monts.*. 2 squares.
Cards.: nr Aber-arth, SH46, ESM, 1899[262]. **Monts.:** Llyfnant Valley, SN79, EKH, 1956[262].

E. arctica Lange ex Rostrup subsp. **borealis** (Townsend) Yeo (*E. borealis* (Townsend) Wettst.; *E. brevipila* auct., non Burnat & Gremli). Native. Rough pastures and meadows (D,Ds,E,G,H,Q,R). Fl.5-9. Mons., Glam., Brecs., Rads., Carms., Pembs., Cards., Monts., Mer., Caerns., *Denbs., Angl.*. 108 squares, map 657.

E. arctica × micrantha Spontaneous. Railway line. *Mer.*. 1 square.
Mer.: Tan-y-bwlch, SH64, EKH, pre 1930[262].

[**E. arctica × rostkoviana = E. × notata** Townsend *Cards., Mer.*]

E. tetraquetra (Bréb.) Arrondeau (*E. occidentalis* Wettst.) Native. Grassy sea cliffs (D,Ds,G,Sc). Fl.5-9. Glam., *Carms.**, Pembs., Cards., *Monts.**, Mer., Caerns., *Denbs.**, *Flints.**, *Angl.*. 60 squares, map 658.

E. tetraquetra × confusa *or* **nemorosa** Spontaneous. Sea cliffs. Fl.7. Glam., Pembs.. 2 squares.
Glam.: Oxwich, SS58, TAWD, 1961[112]. **Pembs.:** Marloes, SM70, TAWD, 1963[112].

E. nemorosa (Pers.) Wallr. Native. Heaths and pastures (D,E,G,H,M,Q,R,Rw,Sc,W). Fl.6-9. Mons., Glam., Brecs., Rads., Carms., Pembs., Cards., Monts., Mer., Caerns., Denbs., Flints., Angl.. 165 squares, map 659.

E. nemorosa × tetraquetra Spontaneous. Grassy places near the sea (D,Ds,G,H). Fl.7-8. *Pembs., Mer., Caerns.*. 5 squares.
Pembs.: nr Skrinkle Haven, SS09, EKH, 1957[262]; Tenby, SN10, CCB, 1848[262]. **Mer.:** Aberdyfi, SN69, EM, 1958[109]; Morfa Harlech, SH53, MR, 1964[262]. **Caerns.:** Pwllheli, SH33, APC, 1961[262].

E. pseudokerneri Pugsley Native. Limestone grassland. Fl.9. *Flints.*, [*Glam., Rads., Carms.*]. 1 square.
Flints.: Holywell, SJ17, VG, 1962[7].

E. confusa Pugsley Native. Heaths and grassy cliffs (D,E,G,H,M,Q,R,Sc). Fl.7-9. Mons., Glam., Brecs., *Rads.**, Carms., Pembs., Cards., *Monts.**, Mer., *Caerns.*, Denbs., Flints.*, Angl.. 110 squares, map 660.

E. confusa × micrantha Spontaneous. Grassland. Fl. 6-9. Brecs., Carms., Pembs., *Cards., Caerns.*. 9 squares.
Brecs.: Heolsenni, SN92, TAWD, 1961[112]. **Carms.:** Blaen Cennen, SN61, IMV, 1972[201]; Mynydd Mallaen, SN74, IMV, 1963[112]. **Pembs.:** Trefeiddan Moor, SM72, TAWD, 1967[112]; St Ishmael's, SM80, TAWD, 1963[112]; Maenclochog, SN12, TAWD, 1963[112]; Mynydd Preseli, SN13, TAWD, 1961[112]. **Cards.:** nr Ponterwyd, SN77, FW, 1959[262]. **Caerns.:** Capel Curig, SH75, PFY, 1952[262].

E. confusa × nemorosa Spontaneous. Grassy places (E,G,R). Fl.7-9. *Glam.*, Pembs., *Caerns., Flints., Angl.*. 9 squares.
Glam.: Pontalun, SS87, EKH, 1953[262]. **Pembs.:** Solva, SM72, EFL, 1905[262]; Canaston Wood, SN01, EKH, 1957[262]; Henry's Moat, SN02, TAWD, 1959[112]; nr Kilgetty, SN10, EKH, 1959[262]. **Caerns.:** Llanberis, SH55, ABC, 1919[262]; Gt Ormes Head, SH78, PFY, 1955[262] and JHR, 1965[262]. **Flints.:** Prestatyn, SJ08, RIG, 1978[261]. **Angl.:** Arthur's Seat, SH58, AWW, 1959[262].

E. confusa × ostenfeldii Spontaneous. Grassy places. Fl.6-8. *Caerns.*. 1 square.
Caerns.: Cwm Idwal, SH65, PFY, 1956[262].

E. confusa × scottica Spontaneous. Marshy ground (G,H,M). Fl.6-9. Mons., *Brecs.*, Carms., *Pembs.*, Cards., *Mer., Caerns., Flints.*. 13 squares : **SN**65,74,*77,92*; **SO***10*; **SH**34,61,65,66,71,72,76; **SJ**17.

E. confusa × tetraquetra Spontaneous. Cliffs and sand dunes (D,Ds,G,Sc). Fl.7-9. *Glam.*, Carms., Pembs., Cards., *Mer., Caerns., Angl.*. 15 squares : **SR**99; **SS**58,78,87,88; **SM**80,82,83; **SN**03,15,30,57; **SH**23,36,53.

E. cambrica Pugsley Native. Mountain slopes (G,Q). Fl.6-8. Mer.*, Caerns., [*Brecs., Carms.*]. 4 squares.
Mer.: Cadair Idris, SH71, PMB, 1965[262] and PFY, 1970+[262]. **Caerns.:** Snowdon, SH65, JAWh, 1919[112] and DM, 1955[262]; Cwm Idwal, SH65, PFY, 1952[262] and 1970+[262]; Capel Curig, SH75, AHW, 1914[262]; Llyn Crafnant, SH76, ESM, 1912[262].

E. cambrica × scottica Spontaneous. Cliff ledges. Fl. 8. *Mer.,* Caerns.. 2 squares.
Mer.: Cadair Idris, SH71, PMB, 1965[262]. **Caerns.:** Clogwyn Du'r Arddu, SH65, JAWh, 1919[112].

E. ostenfeldii (Pugsley) Yeo (*E. curta* auct., non (Fries) Wettst.). Native. Grassy cliff ledges (G,H,M,Q) Fl.6-8. *Caerns..* 5 squares.
Caerns.: Beddgelert, SH54, RLB, 1930+[262]; Moel Hebog, SH54, An, 1930+[262]; Clogwyn Du'r Arddu, SH55, CCT, 1956[262]; Snowdon, SH65, MR, 1936[262] and DM, 1955[262]; Cwm Idwal, SH65, PFY, 1956[262]; Ysgolion Duon, SH66, WRL, 1900[262]; Capel Curig, SH75, AHW, 1914[262].

E. ostenfeldii × scottica Spontaneous. Bogs. Fl.6-9. *Mer., Caerns..* 3 squares.
Mer.: Brithdir, SH71, PMB, 1955[262]. **Caerns.:** Snowdon, SH65, DM, 1955[262]; Cwm Idwal, SH65, PFY, 1956[262]; Llyn Crafnant, SH76, ESM, 1912[262].

E. micrantha Reichenb. Native. Heaths (E,G,H,M). Fl.5-9. *Brecs., Rads.**, Carms.*, Pembs., Cards., *Mer.,* Caerns., Denbs., Flints., *Angl.,* [*Monts.*]. 52 squares, map 661.

E. micrantha × nemorosa Spontaneous. Pastures. Fl. 8. Cards.. 2 squares.
Cards.: nr Aber-arth, SN46, ESM, 1899[262]; Goginan, SN68, AOC, 1970[112].

E. micrantha × tetraquetra Spontaneous. Cliff grassland. Fl.7. *Angl..* 1 square.
Angl.: Holyhead, SH28, NYS, 1943[109].

E. scottica Wettst. Native. Streamsides and damp upland grassland (G,H,M,R). Fl.6-8. *Brecs.**, *Rads.**, Carms.*, Cards., Mer., *Caerns., Flints., Angl..* 19 squares : **SN**72,74,*76*,77,78,86,*87*,92; **SO**11; **SH**55,*56*, 58,*61*,65,*66*,71,72,*76*; **SJ**17.

ODONTITES Ludwig

Odontites verna (Bellardi) Dumort. : Gorudd : Red Bartsia. Native but also widespread as a casual weed of cultivation. Pastures (D,E,F,G,R,Rw). Fl.6-9. Mons., Glam., Brecs., Rads., Carms., Pembs., Cards., Monts., Mer., Caerns., Denbs., Flints., Angl.. 213 squares, map 662.

Subsp. **verna** Mons., Glam., Pembs., *Caerns.**, Denbs.*. 15 squares : **ST**06,07,16,17,27,38,48,*49*; **SM**83; **SO**00,30; **SH**78,87; **SJ**25,35. Probably under-recorded.

Subsp. **litoralis** (Fries) Nyman *Mer..* 1 square.
Mer.: Mochras, SH52, ECW, 1930[107].

Subsp. **serotina** (Dumort.) Corb. Mons., Glam., Brecs., Rads., Carms., Pembs., Cards., Monts., Mer., Caerns., Denbs., Flints., Angl.. 88 squares, map 663.

PARENTUCELLIA Viv.

Parentucellia viscosa (L.) Caruel : Gorudd Melyn : Yellow Bartsia. Native. Wet grassy places (D,Ds,G, H,R). Fl.6-8. Glam., Carms., Pembs., Mer.*, (*Caerns.*). 16 squares : **SS**58,59,*68*,69,79,*98*; **SM**82,83,84,91,93,94; **SN**03,30; **SH**(32),52.

PEDICULARIS L.

Pedicularis palustris L. : Melog y Waun : Marsh Lousewort. Native. Marshes (G,H,M). Fl.5-9. Mons., Glam., Brecs., Rads., Carms., Pembs., Cards., *Monts.,* Mer., Caerns., Denbs., *Flints.,* Angl.. 173 squares, map 664.

P. sylvatica L. : Melog y Cŵn : Lousewort. Native. Damp heaths (E,G,H,M,Q,W). Fl.4-6. Mons., Glam., Brecs., Rads., Carms., Pembs., Cards., Monts., Mer., Caerns., Denbs., Flints., Angl.. 238 squares, map 665.

RHINANTHUS L.

Rhinanthus minor L. (incl. *R. borealis* (Sterneck) Druce) : Cribell Felen : Yellow-rattle. Native. Pastures (D,E,F,G,H,M,Q,R,Rw,S,Sc). Fl.5-8. Mons., Glam., Brecs., Rads., Carms., Pembs., Cards., Monts., Mer., Caerns., Denbs., Flints., Angl.. 259 squares, all except : **SS**39,77; **ST**26; **SM**50,62,71,73,94; **SN**22,36,59,88; **SO**08, 20,26,29,31; **SH**51,98; **SJ**01,43,53,54.

var. **minor** (*R. minor* subsp. *minor*) is now regarded as an ecotypic variant of the 'vernal' type and has been recorded from Mons., Glam., Brecs., Rads., Carms., Pembs., Cards., Monts., Mer., Caerns., Denbs., Flints., Angl..

var. **stenophyllus** Schur (*R. stenophyllus* (Schur) Druce; *R. minor* subsp. *stenophyllus* (Schur) O. Schwarz) is now regarded as an ecotypic variant of the 'autumnal' type and has been recorded from Mons.*, Glam.*, Brecs., Carms.*, Pembs.*, Cards., *Monts.**, *Mer., Caerns., Denbs.**, Flints., *Angl..*

R. angustifolius C. C. Gmelin (*R. major* auct., non L.; *R. serotinus* (Schönheit) Oborny) : Greater Yellow-rattle. Native. Marshy places. *Angl..* 1 square.
Angl.: Llyn Coron, SH36, GCD, 1900[136] (probably extinct).

LATHRAEA L.

Lathraea squamaria L. : Deintlys Cennog : Toothwort. Native. Woods, parasitic on roots of various trees (G,M,Q,R,W). Fl.3-5. Mons., Glam., Brecs., Rads., Carms.*, Cards., Monts., (*Mer.*), Caerns., *Denbs., Flints..* 65 squares, map 666.

ACANTHACEAE

ACANTHUS L.

Acanthus mollis L. : Bear's-breech. Established alien (S. Europe). Waste places (G,R,Rg,S). Fl.6-8. Glam., Pembs.. 4 squares.
Glam.: Aberafan, SS79, LT, 1950[112]; Cardiff, ST18, WMD, 1950[112]. **Pembs.:** Lydstep, SS09, JESJ, 1955[112]; St. Ishmael's, SM80, TAWD, 1910[79] & 1960[112].

[GESNERIACEAE]

[HABERLEA Friv.]

[**Haberlea rhodopensis** Friv. *Pembs.*]

OROBANCHACEAE

OROBANCHE L.

Orobanche purpurea Jacq. : Gorfanc Glasgoch : Yarrow Broomrape. Native. Grassland, parasitic on *Achillea millefolium*. Fl.7. Mons., Pembs., [*Glam., Mer.*]. 3 squares.
Mons.: nr Chepstow, ST59, FJH, 1852[226] and WAS, 1923[112]; Matharn, ST59, GCF, 1912[112]. **Pembs.:** Bosherston, SR99, FLR, 1950[79]; Tenby, SN10, JEA, 1926[112].

[**O. alba** Stephan ex Willd. *Glam.*]

[**O. reticulata** Wallr. *Brecs.*]

O. loricata Reichenb. (incl. *O. picridis* F. W. Schultz ex Koch) : Oxtongue Broomrape. Native. Grassland, parasitic on *Picris* sp. and *Crepis* sp. *Brecs.**, *Pembs.*. 2 squares.
Brecs.: nr Llandovery, SN83, HHK, 1905[107]. **Pembs.:** nr Giltar Point, SS19, CCB, 1851[127].

O. minor Sm. (incl. *O. maritima* Pugsley) : Gorfanc Lleiaf : Common Broomrape. Native. Grassy places, parasitic on *Trifolium* and other plants (D,F,G,M, Q,R,Rw). Fl.6-8. Mons., Glam., *Brecs.*, Carms., Pembs., *Cards.*, Mer.*, Caerns., Flints., Angl.. 43 squares, map 667.

O. hederae Duby : Gorfanc Eiddew : Ivy Broomrape. Native. Shrubby places, parasitic on *Hedera helix* (E,G,Q,Rg,Rw,Sc,W). Fl.6-8. Mons., Glam., *Brecs.*[230], Carms., Pembs., Cards., Mer., Caerns., *Denbs., Angl.*. 45 squares, map 668.

[**O. caryophyllacea** Sm. *Glam.*]

O. elatior Sutton : Gorfanc Hir : Knapweed Broomrape. Native. Limestone grassland, parasitic on *Centaurea scabiosa*. Glam., *Brecs.*[232], *Caerns.*[232], [*Mer.*]. 4 squares.
Glam.: Penrice, SS48, JWG, c1840[198]; Port Talbot, SS78, HJR, pre 1907[140]; Porthkerry, ST06, AEW, 1957[93] and AWC, 1961[112]; Cwrtyrala, ST17, TCh, pre 1876[218].

O. rapum-genistae Thuill. : Gorfanc Mwyaf : Greater Broomrape. Native. Shrubby places, parasitic on *Cytisus* sp., *Ulex* sp., etc. (D,E,Rg,W). Fl.4-7(-9). Mons., Glam., *Brecs., Rads.*, Carms., Pembs., *Cards., Monts.**, Mer., Caerns., *Denbs., Flints., Angl.*. 46 squares, map 669.

LENTIBULARIACEAE

PINGUICULA L.

Pinguicula lusitanica L. : Pale Butterwort. Native. Bogs. Fl.6-8. Pembs., [*Glam.*]. 4 squares.
Pembs.: Hayscastle, SM92, ECH, 1937[112]; Tre-rhos, SM92, TAWD, 1960+[79]; Ambleston, SM92, TAWD, 1960[112]; Mathry, SM93, BL, 1930[112]; Carn Ingli, SN03, AHGA, 1931[150]; Brynberian, SN13, TAWD, 1960+[79].

P. vulgaris L. : Toddyn Cyffredin : Common Butterwort. Native. Bogs (Ar,Ds,H,M,Q,W). Fl.5-8. Mons., Glam., Brecs., Rads., Carms., Pembs., Cards., Monts., Mer., Caerns., Denbs., *Flints.*, Angl.. 60 squares, map 670.

UTRICULARIA L.

Utricularia minor L. : Chwysigenwraidd Lleiaf : Lesser Bladderwort. Native. Pools (Al,Ar,H). Fl.6-9. Glam., Brecs., Rads., Carms., Pembs., Cards., Monts., Mer., Caerns., Denbs., *Flints.*[85], Angl.. 47 squares, map 671.

U. intermedia Hayne : Chwysigenwraidd Canolig : Intermediate Bladderwort. Native. Pools (Al,H,M). Fl. 7-9. Caerns.. 3 squares.
Caerns.: nr Tudweiliog, SH23, APC, 1958[7]; Nevin, SH33, JAWhe, 1940[112]; Cors Geirch, SH33, PWR, 1950[112] and PMB, 1981[3]; Cors Graianwg, SH44, WEH & CSi, 1968[112].

U. vulgaris L. Native. Pools. Fl.7. Pembs.*, *Caerns.*, [*Mons., Glam., Rads., Carms., Cards., Monts., Mer., Flints., Angl.*]. 2 squares.
Pembs.: Dowrog, SM72, RMd, 1979[245]; Waun Llechell, SM72, IRS, 1979[245] and SBE, 1981[112]. **Caerns.:** Llandudno, SH78, WHP, 1868[104]. All other records refer to **U. australis.**

U. australis R. Br. (*U. neglecta* Lehm.) : Chwysigenwraidd Cyffredin : Greater Bladderwort. Native. Pools (Al,Ar,Ds,H,M). Fl.7-9. Mons., Glam., Rads., Carms., Cards., Monts., Mer., Caerns., Denbs.*, *Flints.**, Angl.. 47 squares, map 672.

PLANTAGINACEAE

PLANTAGO L.

Subgen. Plantago

Plantago major L. : Llwynhidydd Mawr : Greater Plantain. Native. Roadsides (Ds,E,G,M,Q,R,Rg,Sc,W). Fl.6-9. Mons., Glam., Brecs., Rads., Carms., Pembs., Cards., Monts., Mer., Caerns., Denbs., Flints., Angl.. 280 squares, all except : **SS**39; **SH**51.

P. coronopus L. : Llwynhidydd Corn y Carw : Buck's-horn Plantain. Native. Grassy places near the sea (D,Ds,G,Q,R,Rw,S,Sc). Fl.5-9. Mons., Glam., Carms., Pembs., Cards., Mer., Caerns., Denbs., Flints., Angl.. 113 squares, map 673.

P. maritima L. : Llwynhidydd Arfor : Sea Plantain. Native. Salt-marshes (Ds,G,M,R,S,Sc). Fl.5-9(-11). Mons., Glam., Carms., Pembs., Cards., Monts., Mer., Caerns., Denbs., Flints., Angl.. 109 squares, map 674.

P. media L. : Llwynhidydd Blewog : Hoary Plantain. Native. Pastures (E,G,M,R,Rw). Fl.6-9. Mons., Glam., *Brecs.**, *Rads.**, Carms., *Pembs.*, *Cards.*, *Monts.*, (*Mer.*), *Caerns.*, Denbs., Flints., *Angl.**. 70 squares, map 675.

P. lanceolata L. : Llwynhidydd : Ribwort Plantain. Native. Pastures (D,Ds,E,F,G,H,M,Q,R,Rg,Rw,S,W). Fl.3-8(-9). Mons., Glam., Brecs., Rads., Carms., Pembs., Cards., Monts., Mer., Caerns., Denbs., Flints., Angl.. 280 squares, all except : **SS**39; **SM**50.

P. lagopus L. Casual (S. Europe). Waste places. Fl.7. Glam., *Caerns.*. 2 squares : **ST***27*[112]; **SH***56*[18].

Subgen. Psyllium (Miller) Harms

P. arenaria Waldst. & Kit. (*P. indica* L., *nom. illegit.*) Casual (S. Europe). Waste places (R,Rg). Fl.6-8. Mons., Glam., Cards.. 5 squares : **ST***17*[112],*27*[112],28[112]; **SO**00[140]; **SN**58[112].

P. afra L. Casual (S. Europe). Waste places. Fl.6. Glam.. 2 squares : **ST***17*[112],*27*[112].

LITTORELLA Bergius

Littorella uniflora (L.) Ascherson : Beistonnell : Shoreweed. Native. Lakes (Al,Ds,M). Fl.6-9. Mons.*, Glam., Brecs., Rads., Carms., Pembs., Cards., Monts., Mer., Caerns., Denbs., Flints., Angl.. 112 squares, map 676.

CAPRIFOLIACEAE

SAMBUCUS L.

Sambucus ebulus L. : Ysgawen Fair : Dwarf Elder. Naturalized alien (Europe). Waysides (D,E,G,M,Q, R,Rw,W). Fl.5-9. Mons., Glam., Brecs., Rads., Carms., *Pembs.*, *Cards.*, *Monts.*, Mer., Caerns., Denbs., Flints., *Angl.*. 72 squares, map 677.

S. nigra L. : Ysgawen : Elder. Native. Hedges (D,E, M,Q,R,Rw,W). Fl.5-9. Mons., Glam., Brecs., Rads., Carms., Pembs., Cards., Monts., Mer., Caerns., Denbs., Flints., Angl.. 274 squares, all except: **SR**89; **SS**38,39,77; **SM**50,73,84; **SH**51.

S. racemosa L. : Red-berried Elder. Established alien (Europe). Woods. Fl.9-10. *Brecs.*, *Rads.*, *Cards.*, Monts.. 3 squares.
Brecs.: nr Newbridge-on-Wye, SO05, FMS, 1980[200]. **Rads.:** Kerry Hills, SO18, AJB, 1957[7]. **Cards.:** nr Trawsgoed, SN67, AOC, 1975[71]. **Monts.:** Kerry Hills, SO18, MGBE, 1941[112].

VIBURNUM L.

Viburnum opulus L. : Corswigen : Guelder-rose. Native. Woods (E,M,Q,Rw,W). Fl.5-7. Mons., Glam., Brecs., Rads., Carms., Pembs., *Cards.*, Monts., Mer., Caerns., Denbs., Flints., *Angl.*. 189 squares, map 678.

V. lantana L. : Gwifwrnwydd : Wayfaring-tree. Native. Woods on limestone (E,G,Q,W). Fl.4-6. Mons., Glam., *Carms.*[161], Pembs., (Monts., *Mer.**, Caerns.), *Flints.*. 29 squares, map 679.

V. tinus L. : Laurustinus. Introduced (S. Europe). Sea cliff. Glam. 1 square.
Glam. : Caswell, SS58, JAW, 1941[112].

SYMPHORICARPOS Duh.

Symphoricarpos albus (L.) S. F. Blake (*S. rivularis* Suksdorf) : Llus Eira : Snowberry. Established alien (N. America). Hedges (E,M,Q,R,Rw,W). Fl.6-9. Mons., Glam., Brecs., Rads., Carms., *Pembs.*, Cards., Monts., Mer., *Caerns.**, Denbs., Flints., *Angl.**. 166 squares, map 680.

LEYCESTERIA Wall.

Leycesteria formosa Wall. : Bachgen Llwm : Himalayan Honeysuckle. Established alien (India & China). Hedges (E,Q,Rw,W). Fl.7-10. Mons., Glam., Carms., Pembs., *Cards.*, Mer., Caerns., *Denbs.*, Angl.. 29 squares, map 681.

LONICERA L.

Lonicera xylosteum L. : Gwyddfid Syth : Fly Honeysuckle. Established alien (England). Hedges (E,R). Fl.(5-)7-9. Glam., Brecs., Carms., *Pembs.**, *Cards.*, Caerns., *Denbs.*, *Flints.**, *Angl.*, [*Mer.*]. 20 squares, map 682.

L. nitida E. H. Wilson : Chinese Honeysuckle. Introduced (China). Hedges (E,M,Q,W). Fl.6-9. *Mons.*, Glam., Carms., *Pembs.*, Cards., *Caerns.*, Denbs.. 20 squares, map 683.

L. japonica Thunb. : Japanese Honeysuckle. Established alien (E. Asia). Hedges and sand dunes (D,E,Q). Fl. 6-7. Glam., Carms., *Pembs.*. 6 squares.
Glam.: nr West Cross, SS68, JAW, 1943[112]; Swansea, SS69, JAW, 1957[112]; Derwen-fawr, SS69, JAW, 1957[112]; Baglan, SS79, JAW, 1941[112]; Rhoose, ST06, ABP & RGE, 1971[112]. **Carms.:** Laugharne Burrows, SN30, IMV, 1974[112]. **Pembs.:** Fishguard, SM93, DM, 1956[7].

L. ledebourii Eschsch. : Californian Honeysuckle. Introduced (California). Parkland. Fl.5. *Flints.*. 1 square.
Flints.: nr St Asaph, SH97, PWa, 1952[107].

L. caprifolium L. : Perfoliate Honeysuckle. Established alien (Europe). Hedges (E,Q). Fl.6. Glam., *Rads.**, *Denbs.**. 4 squares.
Glam.: Stonemill, SS58, JAW, 1956[112]; Newton Nottage, SS87, JBL, pre 1907[140]. **Rads.:** Presteigne, SO36, JAW, 1948[229]. **Denbs.:** Llansilin, SJ22, EJo, 1871[7].

L. periclymenum L. : Gwyddfid : Honeysuckle. Native. Hedges (E,G,H,M,R,Rw,W). Fl.4-10. Mons., Glam., Brecs., Rads., Carms., Pembs., Cards., Monts., Mer., Caerns., Denbs., Flints., Angl.. 270 squares, all except : **SR**89; **SS**38,39,77; **ST**26; **SM**50,71,73; **SN**15, 36; **SH**29,98.

ADOXACEAE

ADOXA L.

Adoxa moschatellina L. : Mysglys : Moschatel. Native. Hedgebanks (E,M,Q,R,Rg,Rw,W). Fl.3-5. Mons., Glam., Brecs., Rads., Carms., Pembs., Cards., Monts., Mer., Caerns., Denbs., Flints., *Angl.*. 206 squares, map 684.

VALERIANACEAE

VALERIANELLA Miller

Valerianella discoidea (L.) Loisel. Casual (Mediterranean region). Waste places and marshes (G,H,M,R). *Glam.*, *Caerns.*. 3 squares : **SO**00[140]; **SH**12[7],44[7].

V. locusta (L.) Laterrade : Llysiau'r Oen : Common Cornsalad. Native. Cultivated ground and cliffs (Ar,D, E,F,Q,R,Rg,Rw,S,Sc). Fl.4-6. Mons., Glam., Brecs., *Rads.**, Carms., Pembs., Cards., Monts., Mer., Caerns., Denbs., Flints., *Angl.*. 100 squares, map 685.
A taxon from sand dunes sometimes called subsp. **dunensis** (D. E. Allen) P. D. Sell has been recorded from Glam., Pembs., Cards., *Mer.*, *Angl.*.

V. carinata Loisel. : Keeled-fruited Cornsalad. Native. Banks and walls (E,G,Q,R,Rg). Fl.4-6. Mons., Glam., Brecs.*, Rads.*, Pembs., Cards.*, Monts.*, Mer.*, Caerns.*, Denbs.. 24 squares, map 686.

V. dentata (L.) Pollich : Gwylaeth yr Oen Deintiog : Narrow-fruited Cornsalad. Established alien (England). Cultivated ground (F,R,Rg). Fl.5-9. Mons., Glam., *Brecs.*, *Rads.*, *Carms.*, Pembs., Cards., Monts., Caerns., Denbs., Flints., *Angl.*, [*Mer.*]. 61 squares, map 687. Possibly native on Gt Ormes Head.

V. rimosa Bast. : Broad-fruited Cornsalad. Established alien (England). Cultivated ground (F,Rg). Fl.7-8. *Mons.*, Glam., *Brecs.*, *Carms.*[230], Pembs., Caerns., *Denbs.*. 12 squares : **SS***68*,*69*,87,97; **ST**06; **SM***72*; **SN**03; **SO**05,*40*,51; **SH***78*,*87*.

[**V. eriocarpa** Desv. : Hairy-fruited Cornsalad. *Glam.*, *Pembs.*, *Caerns.*. All probably errors for **V. dentata** var **mixta** (L.) Dufresne]

VALERIANA L.

V. officinalis L. : Triaglog : Common Valerian. Native. Woods and marshes (Ar,E,G,H,M,Q,R,W). Fl. 6-9. Mons., Glam., Brecs., Rads., Carms., Pembs., Cards., Monts., Mer., Caerns., Denbs., *Flints.*, Angl.. 235 squares, map 688.
Plants resembling subsp **sambucifolia** (Mikan fil.) Celak. have been recorded from Mons., Glam., Brecs., Pembs., Mer..

V. phu L. Naturalized alien (N. Anatolia). Roadsides. *Cards.*. 1 square.
Cards.: Mydroilyn SN45, JHS, c1930[210] and AOC, 1958[71].

V. pyrenaica L. : Pyrenean Valerian. Naturalized alien (Pyrenees). Waysides (Ar,E,M,R). Fl.4-5. Caerns.. 3 squares.
Caerns.: nr Penllyn, SH56, WEH, 1967[112]; Bangor, SH57, GC, 1914[112] and JO, 1960[174]; Bethesda, SH66, JEG, 1879[112]; Carneddau, SH66, LJL, 1956[195].

V. dioica L. : Triaglog y Gors : Marsh Valerian. Native. Marshes (Ar,G,H,M,Q,W). Fl.4-7. Mons., Glam., Brecs., Rads., Carms., *Cards.**, Monts., Mer., *Caerns.*, Denbs., Flints., Angl., [*Pembs.*]. 102 squares, map 689. Possibly over-recorded in error for **V. officinalis**.

CENTRANTHUS DC.

Centranthus ruber (L.) DC. (*Kentranthus ruber* (L.) DC.). : Triaglog Coch : Red Valerian. Naturalized alien (Mediterranean region). Cliffs and walls (E,Q, R,Sc). Fl.5-10. Mons., Glam., Brecs., Rads., Carms., Pembs., Cards., Monts., Mer., Caerns., Denbs., Flints., Angl.. 151 squares, map 690.

DIPSACACEAE

DIPSACUS L.

Dipsacus sativus (L.) Honckeny (*D. fullonum* subsp. *sativus* (L.) Thell.) : Fuller's Teasel. Casual (origin unknown). Waste places. *Mons.*. 1 square : **ST**38[245].

D. fullonum L. : Crib y Pannwr : Teasel. Native. Thickets (Ar,E,M,Q,R,W). Fl.7-9. Mons., Glam., *Brecs.**, *Rads.*, Carms., Pembs., *Cards.*, Monts., Mer., Caerns., Denbs., *Flints.*, Angl.. 112 squares, map 691.

D. pilosus L. : Ffon y Bugail : Small Teasel. Native. Woods (E,M,Q,R,Rw,W). Fl.7-10. Mons., (Glam.), *Brecs.*, Rads., Monts., Denbs., *Flints.*, [*Caerns.*]. 25 squares, map 692.

SUCCISA Haller

Succisa pratensis Moench : Clafrllys Gwreidd-don : Devil's-bit Scabious. Native. Pastures (Ds,E,G,H, M,Q,W). Fl.7-9. Mons., Glam., Brecs., Rads., Carms., Pembs., Cards., Monts., Mer., Caerns., Denbs., Flints., Angl.. 269 squares, all except : **SS**38,39,77; **ST**26,58; **SM**50,71; **SN**36,59; **SO**30,31; **SH**51,98.

KNAUTIA L.

Knautia arvensis (L.) Coulter : Clafrllys : Field Scabious. Native. Pastures and waysides (D,E,F,G,H,R,Rw). Fl.6-9. Mons., Glam., Brecs., *Rads.*, Carms., Pembs., *Cards.*, Monts., Mer., Caerns., Denbs., Flints., *Angl.*. 182 squares, map 693.

SCABIOSA L.

Scabiosa atropurpurea L. : Sweet scabious. Casual (S. Europe). Waste places. *Caerns.*. 2 squares : **SH***33*[101],*57*[101].

S. columbaria L. : Clafrllys Bychan : Small Scabious. Native. Calcareous pastures (D,G,Q,Sc). Fl.7-10. Mons., Glam., Brecs., Carms., *Pembs.*, (*Cards.*), *Monts.*, Caerns., Denbs., Flints., *Angl.*, [*Mer.*]. 46 squares, map 694.

CAMPANULACEAE

CAMPANULA L.

Campanula patula L. : Clychlys Lledaenol : Spreading Bellflower. Native. Woods (E,Rw,W). Fl.6-8. Mons., Brecs., Rads., (Mer.), [*Glam.*]. 15 squares : **ST***39*; **SO**01,04,13,14,23,24,25,26,*30*,*32*,*42*,50,51; (**SH**71).

C. rapunculus L. : Clychlys Erfin : Rampion Bellflower. Casual (England). Waste places. *Angl.*. 1 square : **SH**37[7].

C. persicifolia L. : Clychlys Peatws-ddail : Peach-leaved Bellflower. Established alien (Europe). Roadsides. Fl.7. *Brecs.*, *Carms.*, *Caerns.*. 3 squares.
Brecs.: Brecon, SO02, SIL, 1970[200]. **Carms.:** nr Carmarthen, SN41, DH, 1919[18]; Llangunnor, SN41, RFM, 1950+[161] **Caerns.:** nr Pwllheli, SJ33, JBy, 1981[74].

C. medium L. : Clychlys Caer-gaint : Canterbury-bells. Casual (Italy). Waste places (E,W). Fl.6-7. Mons., Glam., *Brecs.*, *Carms.*, *Pembs.*, *Caerns.*. 8 squares : **SS**79[117]; **ST**16[117],27[112], *29*[112]; **SO***21*[18]; **SN***10*[18],40[201]; **SH**33[74].

C. glomerata L. : Clychlys Clwster : Clustered Bellflower. Native. Limestone grassland. Fl.6-8. Glam., (Rads.*), [*Mons.*, *Caerns.*]. 6 squares.
Glam.: Southerndown, SS87, JS, 1876[112] and FHP, 1971[197]; Dunraven, SS87, EV, 1896[112] and BAM, 1953[93]; Marcross, SS96, JS, c1876[218]; Llantwit Major, SS96, JS, c1876[218]; Nash Point, SS96, HJR, 1907[112] and ARP, 1976[197]; Monkash, SS97, HAH, 1934[112]; Cwm Nash, SS97, RGE & AEW 1972[95]; Penarth, ST17, JS, 1876[218]; (Resolven, SN80, MG, c1970[99]. **Rads.**: nr Old Radnor, SO25, ACP, 1980[112].)

C. portenschlagiana Schultes : Adria Bellflower. Casual (Jugoslavia). Walls and roadsides. Fl.6-11. Glam., *Carms.*, Cards. 8 squares : **SS**97[95],99[95]; **ST**08[95],09[95],16[95],17[112]; **SN**31[161],36[112].

C. poscharskyana Degen : Trailing Bellflower. Established alien (Jugoslavia). Rocky river bank and walls. *Glam.*, *Rads.*. 3 squares.
Glam.: Cardiff, ST17, AEW, 1972[95]; Llanedern, ST28, AEW & ABP, 1975[95]. **Rads.**: Boughrood, SO13, GMK, 1978[50].

C. latifolia L. : Clychlys Mawr : Giant Bellflower. Native. Woods (E,M,R,Rw,W). Fl.7-9. Mons., Glam., Brecs., *Rads.*, Carms., *Monts.**, *Mer.*, *Caerns.*[230], Denbs., Flints., *Angl.*. 54 squares, map 695.

C. trachelium L. : Clychlys Danadl : Nettle-leaved Bellflower. Native. Woods and roadsides (E,R,Rw,W). Fl.7-9. Mons., Glam., Brecs., Rads., *Carms.*, *Monts.*, *Caerns.*[230], Denbs., Flints., *Angl.*. 45 squares, map 696.

C. rapunculoides L. : Clychlys Llusg : Creeping Bellflower. Naturalized alien (Europe). River banks and roadsides (F,M,R,Rw). Fl.9. Glam., *Rads.**, Carms.*, *Monts.**, Caerns.. 5 squares.
Glam.: Merthyr Mawr, SS87, JAW, 1939[112]. **Rads.**: Downton Farm, SO26, JFH & PMH, 1956[53]. **Carms.**: Glanaman, SN61, RFM, 1958[112]. **Monts.**: Newtown, SO19, AJB, 1957[7]. **Caerns.**: Bangor, SH57, CW, 1876[112].

C. latiloba DC. (*C. grandis* Fischer & C. A. Meyer). Casual (Asia Minor). Waste places. Fl.6. Glam.. 1 square : **ST**17[112].

C. rotundifolia L. : Clychlys Deilgrwn : Harebell. Native. Pastures and heaths (E,G,H,M,Q,R,Sc,W). Fl. 7-10. Mons., Glam., Brecs., Rads., Carms., Pembs., Cards., Monts., Mer., Caerns., Denbs., Flints., Angl.. 234 squares, map 697.

LEGOUSIA Durande

Legousia hybrida (L.) Delarbre : Drych Gwener : Venus's-looking-glass. Casual (England). Waste places. *Glam.*. 1 square : **SS***87*[140].

L. speculum-veneris (L.) Chaix Casual (Europe). Waste places. Fl.7. Glam.. 1 square : **SS***97*[112].

PHYTEUMA L.

Phyteuma spicatum L. subsp. **coeruleum** R. Schulz : Cyrnogyn Pigfain : Spiked Rampion. Casual (S. Europe). Grassland. Fl.6. *Mer.*, Caerns.. 2 squares : **SH***57*[112],*71*[21].

[**P. orbiculare** L. (incl. *P. tenerum* R. Schulz). *Glam.*]

WAHLENBERGIA Schrader ex Roth

Wahlenbergia hederacea (L.) Reichenb. : Clychlys Eiddew : Ivy-leaved Bellflower. Native. Damp peaty places (Ar,E,G,H,M,Rg,W). Fl.7-10. Mons., Glam., Brecs., Rads., Carms., Pembs., Cards., Monts., Mer., Caerns., Denbs., *Flints.*. 120 squares, map 698.

JASIONE L.

Jasione montana L. : Clefryn : Sheep's-bit. Native. Heaths and sea cliffs (D,E,G,H,M,Q,R,W). Fl.5-9. Mons., Glam., Brecs., Rads., Carms., Pembs., Cards., Monts., Mer., Caerns., Denbs., Flints., Angl.. 234 squares, map 699.

LOBELIA L.

Lobelia dortmanna L. : Lobelia'r Dŵr : Water Lobelia. Native. Lakes. Fl.7-9. Glam., *Rads., Carms.,* Cards., *Monts.,* Mer., Caerns., *Denbs., Angl..* 32 squares, map 700.

L. erinus L. Casual (S. Africa). Waste places. Fl.9. Glam., Denbs.. 2 squares : **ST**27[112]; **SJ**33[112].

COMPOSITAE

EUPATORIUM L.

Eupatorium cannabinum L. : Byddon Chwerw : Hemp-agrimony. Native. Moist woods and river banks (D,E,G,H,M,Q,R,Rw,Sc,W). Fl.7-8. Mons., Glam., Brecs., *Rads.,* Carms., Pembs., Cards., Monts., Mer., Caerns., Denbs., Flints., Angl.. 226 squares, map 701.

GRINDELIA Willd.

Grindelia squarrosa (Pursh) Dunal : Gumplant. Casual (N. America). Waste places. Fl.7. Glam.. 2 squares : **ST***17*[24],*27*[112].

SOLIDAGO L.

Subgen. Solidago

Solidago virgaurea L. : Eurwialen : Goldenrod. Native. Dry woods (D,E,G,H,M,Q,R,Rw,W). Fl.6-10. Mons., Glam., Brecs., Rads., Carms., Pembs., Cards., Monts., Mer., Caerns., Denbs., Flints., Angl.. 228 squares, map 702.

S. canadensis L. : Canadian Goldenrod. Established alien (N. America). Waste places (D,M,R,Rw). Fl.7-9. Mons., Glam., *Brecs., Rads.,* Carms., Cards., *Monts., Caerns.,* Denbs., Flints.. 31 squares, map 703.

S. altissima L. : Garden Goldenrod. Casual (N. America). Waste places. *Carms..* 1 square : **SN**31[161].

S. gigantea Aiton subsp. **serotina** (O. Kuntze) McNeill (*S. serotina* Aiton, non Retz.) : Early Goldenrod. Established alien (N. America). Roadsides. Fl.8. Mons., Glam., Carms., Cards.. 10 squares : **SS**69, 79; **ST**16,17,38; **SN**50,58,61,99; **SJ**36.

Subgen. Euthamia Nutt.

S. graminifolia (L.) Salisb. Casual (N. America). Waste places. *Glam..* 1 square : **ST***17*[24].

BELLIS L.

Bellis perennis L. : Llygad y Dydd : Daisy. Native. Grassy places (Al,D,Ds,E,F,G,H,M,Q,R,Rg,Rw,S, Sc,W). Fl.3-11. Mons., Glam., Brecs., Rads., Carms., Pembs., Cards., Monts., Mer., Caerns., Denbs., Flints., Angl.. 279 squares, all except : **SS**39; **SM**50,71.

ASTER L.

Aster novae-angliae L. Casual (N. America). Waste places (D,R). Fl.8-10. Glam., *Carms., Pembs..* 9 squares : **SS**78[112], 57[7]; **ST***17*[112]; **SN**02[7],*31*[19],*40*[19],*42*[19],60[7],70[7].

A. laevis L. × **novi-belgii** = **A.** × **versicolor** Willd. (*A. novi-belgii* subsp. *laevigatus* (Lam.) Thell.). Casual (garden origin). Waste places. *Glam..* 1 square : **SN***90*[18].

A. novi-belgii L. (incl. *A. longifolius* Lam.) : Blodyn Mihangel : Michaelmas-daisy. Established alien (N. America). Marshes and waste places (D,G,M,R,Rw). Fl.8-11. Mons., Glam., *Brecs., Rads.,* Carms., Pembs., Cards., Monts., Mer., Caerns., Denbs., Flints., *Angl..* 64 squares, map 704.

A. lanceolatus Willd. (*A. tradescantii* auct. eur., non L.). Casual (N. America). Waste places (M,R). Fl.8-11. Glam.. 4 squares : **SS**69[104],*87*[140]; **ST**09[112],17[112].

A. lanceolatus × **novi-belgii** = **A.** × **salignus** Willd. Casual (garden origin). Waste places (M,R). Fl.10. Glam., Caerns.. 6 squares : **SS***49*,[140],*78*[140],87[99]; **SN***90*[112]; **SO***00*[140]; **SH**55[112].

[**A. tradescantii** L. *Glam.*]

A. adscendens Lindley Casual (N. America). Waste places. Fl.9. Glam.. 1 square : **ST***27*[112].

A. tripolium L. : Seren y Morfa : Sea Aster. Native. Salt-marshes (D,S,Sc). Fl.7-10. Mons., Glam., Carms., Pembs., Cards., Monts., Mer., Caerns., Denbs., Flints., Angl., [*Brecs., Rads.*]. 74 squares, map 705.

A. linosyris (L.) Bernh. (*Linosyris vulgaris* Cass. ex DC.; *Crinitaria linosyris* (L.) Less.) : Gold y Môr : Goldilocks Aster. Native. Sea cliffs on limestone (S,Sc). Fl.8-9. *Glam.,* Pembs.*, Caerns., [*Denbs., Angl.*]. 3 squares.
Glam.: Gower Coast, SS48, CDPi & SMW, c1950[153] and QONK, 1970+[157]. **Pembs.**: Castle Martin, SR99, RLa, 1981[112]. **Caerns.**: Gt Ormes Head, SH78, HWi, 1870[112] and PMB, 1979[3].

BOLTONIA L'Hér

Boltonia asteroides (L.) L'Hér. : False Chamomile. Casual (N. America). Waste places. Cards.. 1 square : **SN***58*[112].

ERIGERON L.

Erigeron karvinskianus DC. (*E. mucronatus* DC.) : Mexican Fleabane. Established alien (Mexico). Walls and cliffs (D,E,Q,Sc,W). Fl.4-9(-12). Glam., Pembs., *Monts.,* Caerns., Denbs., *Angl..* 9 squares.
Glam.: Caswell Bay, SS58, CH, 1978[117]; Newton Nottage, SS87, JAW, 1932[221]; Dinas Powys, ST17, DM, 1958[55] and AEW, 1974[227]; St Fagans, ST17, SGH, 1966[112]; Llandaf, ST17, MG 1971[99]; Lisvane, ST18, JEB, 1947[112]. **Pembs.**: Tenby, SN10, MB, 1963[112]. **Monts.:** Powis Castle, SJ20, DHK, 1970+[197]. **Caerns.**: Llandudno, SH78, MM, 1961[112] and JMB, 1965[67]. **Denbs.**: Colwyn Bay, SH87, JMB, 1965[112]. **Angl.**: Menai Bridge, SH57, FJ, 1974[57].

E. acer L. : Cedowydd Glas : Blue Fleabane. Native. Limestone banks and calcareous duneland (D,Ds,Q, R,Rw). Fl.5-9(-11). Mons., Glam., Brecs., *Rads.**,

Carms., Pembs., Cards., *Monts.*, Mer., Caerns., Denbs., Flints., *Angl.*. 81 squares, map 706.

E. philadelphicus L. : Robin's-plantain. Established alien (N. America). Walls (Q,R). Fl.6-7. *Mons.*, Glam.. 2 squares.
Mons.: Tintern, SO50, EL, 1917[16]. **Glam.**: Llandaf, ST17, SJ, 1944[112] and AEW, 1973[112].

CONYZA Less.

Conyza canadensis (L.) Cronq. (*Erigeron canadensis* L.) : Amrhydlwyd Canada : Canadian Fleabane. Naturalized alien (N. America). Roadsides and sand dunes (D,E,M,Q,R,Rg,Rw). Fl.7-9. Mons., Glam., Brecs.*, Carms., *Pembs.**, Cards.*, Mer.*, *Caerns.*, Denbs.*, *Flints.**. 47 squares, map 707.

C. bonariensis (L.) Cronq. (*C. ambigua* DC.; *Erigeron bonariensis* L.; *Erigeron linifolius* Willd.) : Horseweed. Casual (Tropical America). Waste places. Fl.9. Mons., Glam.. 2 squares : **ST***17*[112],38[112].

FILAGO L.

Filago vulgaris Lam. (*F. germanica* L., non Hudson) : Edafeddog : Common Cudweed. Native. Dry pastures (D,Q,R,Rg,Rw,W). Fl.6-8. Mons., Glam., Brecs., Rads., Carms., Pembs., Cards., Monts., Mer., Caerns., *Denbs., Flints., Angl.*. 80 squares, map 708.

LOGFIA Cass.

Logfia minima (Sm.) Dumort. (*Filago minima* (Sm.) Pers.) : Edafeddog Leiaf : Small Cudweed. Native. Sandy and gravelly places (D,H,M,R,Rw). Fl.5-9(-11). Mons.*, Glam., *Brecs.*, Rads., Carms., Pembs., *Cards.*, Monts., Mer., Caerns., Denbs., *Flints.*, Angl.. 82 squares, map 709.

L. gallica (L.) Cosson & Germ. (*Filago gallica* L.) : Edafeddog Culddail : Narrow-leaved Cudweed. Casual (Europe). Hayfield. *Mer.*. 1 square : **SH***93*[18]. This record may be a misidentification of **Filaginella uliginosa.**

OMALOTHECA Cass.

Omalotheca sylvatica (L.) Schultz Bip. & F. W. Schultz (*Gnaphalium sylvaticum* L.) : Edafeddog y Goedwig : Heath Cudweed. Native. Heaths (D,F,G, H,R,W). Fl.8-10(-12). Mons., Glam., Brecs., Rads., *Carms., Pembs.**, *Cards.*, Monts.*, *Mer.*, Caerns., Denbs., *Flints., Angl.*. 46 squares, map 710.

FILAGINELLA Opiz

Filaginella uliginosa (L.) Opiz (*Gnaphalium uliginosum* L.) : Edafeddog y Gors : Marsh Cudweed. Native. Damp waysides (F,G,M,R,W). Fl.7-9. Mons., Glam., Brecs., Rads., Carms., Pembs., Cards., Monts., Mer., Caerns., Denbs., Flints., Angl.. 225 squares, map 711.

GNAPHALIUM L.

Gnaphalium luteo-album L. : Edafeddog Melynwyn : Jersey Cudweed. Casual (Channel Isles). Waste places. Fl.7-9. Glam.. 3 squares : **ST***16*[26],*17*[24],*27*[112].

HELIPTERUM DC.

Helipterum manglesii (Lindley) F. Mueller ex Bentham : Australian Everlasting. Casual (W. Australia). Garden weed. Fl.7. Mer. 1 square : **SH**83[112].

HELICHRYSUM Miller

Helichrysum stoechas (L.) Moench Casual (Europe). Waste places. Glam.. 1 square : **ST***16*[112].

H. italicum (Roth) G. Don fil. subsp. **microphyllum** (Willd.) Nyman (*H. microphyllum* (Willd.) Camb.). Casual (Mediterranean region). Waste places. Fl.6. *Monts.*, [*Glam.*]. 1 square : **SH**80[107].

ANTENNARIA Gaertner

Antennaria dioica (L.) Gaertner : Edafeddog y Mynydd : Mountain Everlasting. Native. Heaths (G,H,Q,R,Sc). Fl.5-8. *Mons.*, Glam., Brecs., *Rads.*, Carms., *Pembs.*, Cards., *Monts.**, Mer., Caerns., Denbs., Flints., Angl.. 59 squares, map 712.

ANAPHALIS DC.

Anaphalis margaritacea (L.) Bentham : Edafeddog Dlysog : Pearly Everlasting. Naturalized alien (N. America). Railway banks etc. (D,E,G,H,M,Q,R, Rw,W). Fl.6-10. Mons., Glam., Brecs., Rads., Carms., Pembs., Cards., Monts., Mer., *Caerns.**, Flints.*, [*Angl.*]. 80 squares, map 713.

INULA L.

Inula helenium L. : Marchalan : Elecampane. Naturalized alien (S.E. Europe). Pastures and thickets (E,G,Q,R,Rg,Sc,W). Fl.7-8. Mons., Glam., Brecs., Rads., Carms., Pembs., Cards., *Monts.*, Mer., Caerns., Denbs., Flints., *Angl.*. 107 squares, map 714.

I. conyza DC. : Meddyg Mair : Ploughman's-spikenard. Native. Dry banks (D,E,G,M,Q,R,Rw,W). Fl.7-10. Mons., Glam., Brecs., Rads., Carms., Pembs., *Cards.*, Monts., Mer., Caerns., Denbs., Flints., *Angl.*. 100 squares, map 715.

I. crithmoides L. : Cedowys Sugnol : Golden-samphire. Native. Sea cliffs. Fl.7-10. Glam., Carms.*, Pembs., *Caerns.*, Angl.. 26 squares, map 716.

DITTRICHIA W. Greuter

Dittrichia viscosa (L.) W. Greuter (*Inula viscosa* (L.) Aiton). Casual (S. Europe). Waste places. Fl.9-10. Glam., *Caerns.*. 4 squares : **ST***16*[112],*17*[23],*27*[112]; **SH***53*[15].

PULICARIA Gaertner

Pulicaria dysenterica (L.) Bernh. : Cedowydd : Common Fleabane. Native. Marshy places (Ar,D,Ds,E, G,M,R,S). Fl.7-9. Mons., Glam., Brecs., *Rads.*, Carms., Pembs., Cards., Monts., Mer., Caerns., Denbs., Flints., Angl.. 180 squares, map 717.

P. vulgaris Gaertner : Cedowydd Bach : Small Fleabane. Casual (England). Waste places. *Mons.**, Glam., [*Monts., Mer.*]. 2 squares : **ST***17*[112],*38*[218].

P. paludosa Link (*P. uliginosa* Hoffmans. & Link, non S. F. Gray). Casual (Iberian Peninsula). Waste places. *Glam..* 1 square : **ST***17*[22].

TELEKIA Baumg.

Telekia speciosa (Schreber) Baumg. (*Buphthalmum speciosum* Schreber) : Yellow Ox-eye. Casual (E. and C. Europe). Waste places. Fl.9. Mer.. 1 square : **SH**83[112].

ASTERISCUS Miller

Asteriscus aquaticus (L.) Less. (*Bubonium aquaticum* (L.) Hill; *Odontospermum aquaticum* (L.) Schultz Bip.). Casual (Mediterranean region). Waste places. *Glam..* 1 square : **ST***17*[33].

A. maritimus (L.) Less. (*Odontospermum maritimum* (L.) Schultz Bip.). Casual (Mediterranean region). Waste places. Glam.. 1 square : **ST***17*[112].

GUIZOTIA Cass.

Guizotia abyssinica (L. fil.) Cass. : Niger. Casual (E. Africa). Waste places. Fl.8-10. Mons., Glam., *Flints..* 6 squares : **ST**17[112],18[112],27[20],38[112]; **SO**00[140]; **SJ**18[23].

COREOPSIS L.

Coreopsis tinctoria Nutt. : Tickseed. Casual (N. America). Waste places. Fl.9. Glam.. 1 square : **ST***27*[112].

BIDENS L.

Bidens tripartita L. : Graban Deiran : Trifid Bur-marigold. Native. Marshy places (Al,Ar,E,H,M,R,W). Fl.7-10. Mons., Glam., Brecs., *Rads.**, Carms., Pembs., Cards., *Monts., Mer.,* Caerns., *Denbs., Flints.,* Angl.. 119 squares, map 718.

B. cernua L. : Graban Ogwydd : Nodding Bur-marigold. Native. Marshy places (Al,Ar,G,H,M,R,Rw). Fl.6-10. Mons., Glam., Brecs., *Rads.**, Carms., Pembs., Cards., Monts., Mer., Caerns., Denbs., *Flints.,* Angl.. 84 squares, map 719.

B. frondosa L. : Beggarticks. Established alien (America). Wet places and waste ground (Ar,G,M,R). Fl.8-9. Mons., Glam.. 9 squares.
Mons.: Pontnewydd, ST29, RMo, 1977[112]; Newport, ST38, DM, 1954[107] and TGE, 1975[112]; nr Chepstow, ST59, TGE, 1980[91]. **Glam.**: Swansea, SS69, JEL, 1948[107] and QONK, 1972[157]; Briton Ferry, SS79, CH & HH, 1973[117]; Welsh St Donat's, ST07, AEW, 1969[112]; Pysgodlyn Mawr, ST07, AEW, 1969[227]; Mountain Ash, ST09, AMP, 1972[95]; Cardiff, ST17, RLS & AEW, c1925[24] and AEW, 1947[112]; Penarth, ST17, PHH, 1916[112]; Nantgarw, ST18, EV, 1934[112]; Taff's Well, ST18, EV, 1936[112].

COSMOS Cav.

Cosmos bipinnatus Cav. : Mexican Aster. Casual (Mexico). Waste places. *Mons.,* Glam.. 2 squares : **ST**27[112], 38[184].

RUDBECKIA L.

Rudbeckia hirta L. : Black-eyed-Susan. Casual (N. America). Rubbish tip. Fl.8. Mons.. 1 square : **ST**38[112].

R. laciniata L. : Coneflower. Casual (N. America). Waste places. Fl.7. Glam.. 1 square : **ST**17[112].

HELIANTHUS L.

Helianthus annuus L. : Blodau'r Haul : Common Sunflower. Casual (N. America). Waste places. Fl.7-8. Mons., Glam., *Carms., Angl..* 6 squares : **ST**17[99],27[112],38[91]; **SN***40*[107],*90*[140]; **SH***67*[18].

H. tuberosus L. : Artisiog Jerwsalem : Jerusalem-artichoke. Casual (N. America). Waste places. *Glam..* 1 square : **SS***78*[140].

H. rigidus (Cass.) Desf. (*H. diffusus* Sims) : Prairie Sunflower. Casual (N. America). Waste places. *Glam., Monts..* 5 squares : **SS***69*[23]; **ST***17*[24]; **SN***90*[140]; **SO**09[228],19[228]. Some or all of these records may be referable to **H.** × **laetiflorus.**

H. rigidus × **tuberosus** = **H.** × **laetiflorus** Pers. (*H. scaberrimus* Elliott) : Showy Sunflower. Established alien (N. America). Waste places (D,R). Fl.8-9. Mons., Glam., Carms.. 6 squares.
Mons.: Lamby nr Cardiff, ST27, RGE, 1979[112]; Llanwern, ST38, AEW, 1968[112]. **Glam.**: Newton Burrows, SS87, JAW, 1953[112]; Cardiff Docks, ST17, PHD, 1969[112]; Cardiff, ST17, MG, 1974[99]. **Carms.**: Pwll, SN40, JAW, 1944[112]; Lougher Valley, SN50, DH, c1912[12].

H. hirsutus Rafin. Established alien (N. America). Roadside. Fl.8. *Flints..* 1 square.
Flints.: Bagillt, SJ27, TE, 1971[197].

IVA L.

Iva xanthifolia Nutt. (*Cyclachaena xanthifolia* (Nutt.) Fresen. Casual (N. America). Waste places. Fl.8. Glam.. 2 squares : **ST**08[112],17[112].

AMBROSIA L.

Ambrosia maritima L. Casual (Mediterranean region). Waste places. Fl.10. Glam.. 2 squares : **SS***78*[140]; **ST***17*[112].

A. artemisiifolia L. : Ragweed. Casual (N. America). Waste places. Fl.8-10. Mons., Glam., Carms., Caerns., *Flints..* 10 squares : **SS***78*; **ST***16*,17,*18*,38; **SN***50*; **SO**40; **SH***53*,75; **SJ***17.*

A. coronopifolia Torrey & A. Gray (*A. psilostachya* auct., non DC.). Casual (N. America). Waste places. Fl.8-9. Glam.. 2 squares : **ST***16*[112], *27*[25].

A trifida L. : Great Ragweed. Casual (N. America). Waste places. Fl.9. *Mons., Glam.,* Brecs.. 3 squares : **ST***17*[107],*28*[107]; **SN**95[112].

XANTHIUM L.

Xanthium strumarium L.

Subsp. **strumarium** : Cacamwci Lleiaf : Rough Cocklebur. Casual (Europe). Waste places. Fl.7-10. Glam.. 4 squares : **SS**69[112]; **ST**16[107],*17*[112],*27*[224].

Subsp. **italicum** (Moretti) D. Löve (*X. echinatum* Murray). Casual (America). Waste places. *Glam*.. 1 square : **ST***17*[22].

X. spinosum L. : Spiny Cocklebur. Casual (S. America). Waste places. Fl.8-10. Glam., *Caerns*.. 5 squares : **SS**78[140]; **ST***16*[25],*17*[112],*27*[112]; **SH***53*[16].

GALINSOGA Ruiz & Pavón

Galinsoga parviflora Cav. : Gallant Soldier. Established alien (S. America). Waste places (F,R,Rg). Fl.6-10. Mons., Glam., *Cards., Mer., Caerns., Denbs.,* [*Flints.*]. 14 squares : **SS**69,*78*,87,88; **ST**08,17,27,38; **SN**54,58; **SH***43*,52,61,87.

G. quadriradiata Ruiz & Pavón (*G. ciliata* (Rafin.) S. F. Blake) : Shaggy Soldier. Established alien (S. America). Waste places (R,Rg). Fl.8-10. Mons., Glam., *Brecs.,* Carms., Mer., Caerns., *Flints*.. 13 squares : **SS**87,98; **ST**17,27,59; **SN**50,62; **SO**13; **SH**24,33,50,78; **SJ**17.

MADIA Molina

Madia capitata Nutt. (*M. savita* Molina) : Tarweed. Casual (America). Waste places. Glam., *Monts*.. 2 squares : **ST***17*[112]; **SJ**20[108].

HEMIZONIA DC.

Hemizonia kelloggii E. L. Greene Casual (N. America). Waste places. Fl.9. Glam., Flints.. 2 squares : **ST***17*[112]; **SJ***36*[112].

H. pungens Torrey & A. Gray : Spikeweed. Casual (N. America). Waste places. Glam., *Flints*.. 3 squares : **ST***16*[24], 17[112], **SJ***36*[26].

BAERIA Fischer & C. A. Meyer

Baeria platycarpha (A. Gray) A. Gray (*Lasthenia platycarpha* (A. Gray) E. L. Greene). Casual (N. America). Waste places. Caerns.. 1 square : **SH***67*[112].

SANTOLINA L.

Santolina chamaecyparissus L. : Llwyn Cotymog : Lavender-cotton. Casual (Mediterranean region). Waste places. Glam., *Carms., Caerns*.. 4 squares : **SS**49[161], 69[112]; **SN**30[19]; **SH***56*[18].

ANTHEMIS L.

Subgen. Anthemis

Anthemis punctata Vahl subsp. **cupaniana** (Tod. ex Nyman) R. Fernandes (*A. cupaniana* Tod. ex Nyman) : Italian Mayweed. Established alien (Sicily). Cliffs (Q,R). Fl.4-9. Pembs.. 1 square.
Pembs.: nr Fishguard, SM93, GCD, 1923[112] and ECH, 1975[112].

A. arvensis L. : Camri'r Ŷd : Corn Chamomile. Colonist (England). Cultivated ground (E,F,R,Rg). Fl.6-9. *Mons.,* Glam., Brecs., Carms., Pembs.*, *Cards., Monts.**, *Mer.,* Caerns., *Denbs., Flints., Angl*.. 34 squares, map 720.

A. tomentosa L. (*A. peregrina* L.). Casual (Mediterranean region). Waste places. *Glam*.. 1 square : **ST***17*[22].

A. cotula L. : Camri'r Cŵn : Stinking Chamomile. Colonist (England). Cultivated ground (F,R,Rg). Fl.6-8. Mons., Glam., Brecs., *Carms.,* Pembs., *Cards., Monts., Mer.*[230], *Caerns., Denbs., Flints.,* Angl., [*Rads.*]. 65 squares, map 721.

Subgen. Cota (Gay ex Guss.) Rouy

A. tinctoria L. : Camri Melyn : Yellow Chamomile. Casual (Europe). Waste places (R,Rw). Fl.6-9. Mons., Glam., *Brecs., Carms.,* Caerns.. 7 squares : **ST**16[112],17[112],38[112]; **SN***42*[161],*91*[9]; **SH**45[112],*57*[27].

[**A. macrantha** Heuffel *Pembs.*]

ACHILLEA L.

Achillea ptarmica L. : Ystrewlys : Sneezewort. Native. Pastures (Ar,E,F,G,H,M,R,Rw,W). Fl.7-10. Mons., Glam., Brecs., Rads., Carms., Pembs., Cards., Monts., Mer., Caerns., Denbs., Flints., Angl.. 257 squares, all except : **SR**89; **SS**09,38,39,77,96; **ST**26,49,58; **SM**50, 62,71,73,83,84,94; **SN**04,36,59; **SO**20,36,41; **SH**97, 98; **SJ**18.

[**A. distans** Waldst. & Kit. ex Willd. subsp. **tanacetifolia** Janchen (*A. tanacetifolia* All., non Miller). *Glam.*]

A. millefolium L. : Milddail : Yarrow. Native. Pastures (D,E,G,H,M,Q,R,Rg,Rw,S,W). Fl.6-10. Mons., Glam., Brecs., Rads., Carms., Pembs., Cards., Monts., Mer., Caerns., Denbs., Flints., Angl.. 280 squares, all except : **SS**39; **SM**50

A. nobilis L. Casual (S. Europe). Waste places. *Glam*.. 2 squares : **SS***78*[140]; **ST***17*[23].

A. ligustica All. Established alien (Mediterranean region). Waste places. Fl.6-9(-11) Mons., Glam.. 4 squares.
Mons.: Newport Docks, ST38, JM, 1953[112] and TGE, 1980[91]. **Glam.**: Barry Docks, ST16, RM, 1924[233] and GCD, 1930[233]; Cardiff Docks, ST17, AEW, 1924[112]; Radyr, ST18, RLS, 1920[107].

A. cretica L. Casual (Aegean region). Waste places. *Glam.*. 1 square : **ST***17*[140].

CHAMAEMELUM Miller

Chamaemelum nobile (L.) All. (*Anthemis nobilis* L.) : Camri : Chamomile. Native. Sandy pastures. Fl.7-9(-11). *Mons.**, Glam., *Carms.*, Pembs., *Cards.*, *Mer.*, Caerns., *Flints.**, Angl., [*Monts.*]. 36 squares, map 722.

C. mixtum (L.) All. (*Anthemis mixta* L.). Casual (S. Europe). Waste places. Fl.7. Glam.. 2 squares : **ST***16*[27],*17*[112].

TRIPLEUROSPERMUM Schultz Bip.

Tripleurospermum maritimum agg. Mons., Glam., Brecs., Rads., Carms., Pembs., Cards., Monts., Mer., Caerns., Denbs., Flints., Angl.. 228 squares, map 723.

T. maritimum (L.) Koch *s.s.* (*Matricaria maritima* L.) : Sea Mayweed. Native. Shingle beaches (G,S,Sc). Fl. 7-8(-11). Glam.*, Carms.*, Pembs., Cards.*, Monts.*, Mer., Caerns., *Denbs.**, *Flints.**, Angl.*. 68 squares, map 724.

T. inodorum Schultz Bip. (*T. maritimum* subsp. *inodorum* (C. Koch) Hyl. ex Vaarama; *Matricaria inodora* L.; *M. perforata* Mérat) : Ffenigl y Cwn : Scentless Mayweed. Native. Sea cliffs, waste places, etc. (F,M,R,Rw,S,Sc). Fl.7-10(-12). Mons., Glam., Brecs., Rads., Carms., Pembs., Cards., Monts., Mer., Caerns., Denbs., Flints., Angl.. 82 squares, map 725.

T. inodorum × maritimum Spontaneous. Roadsides and waste places. *Glam.*. 1 square. **Glam.**: Lower Swansea Valley, SS69, QONK, 1970+[217].

MATRICARIA L.

Matricaria recutita L. (*M. chamomilla* auct.; *Chamomilla recutita* (L.) Rauschert) : Amranwen : Scented Mayweed. Established alien (England). Cultivated ground (E,F,R). Fl.6-10. Mons., Glam., Brecs., Rads., Carms., Pembs., Cards., Monts., Mer., Caerns., Denbs., Flints., Angl.. 133 squares, map 726.

M. matricarioides (Less.) Porter (*Chamomilla suaveolens* (Pursh) Rydb.) : Chwyn Afal Pinwydd : Pineappleweed. Established alien (N.E. Asia). Roadsides and cultivated ground (E,F,G,R,Rg). Fl. 5-10. Mons., Glam., Brecs., Rads., Carms., Pembs., Cards., Monts., Mer., Caerns., Denbs., Flints., Angl.. 273 squares, all except : **SR**89; **SS**39; **ST**26; **SM**73, 84,94; **SN**15,78; **SH**51.

ANACYCLUS L.

Anacyclus clavatus (Desf.) Pers. (*A. tomentosus* DC.). Casual (Mediterranean region). Waste places. Fl.7-9. Glam.. 3 squares : **ST***16*[24],*17*[112],*27*[112].

A. radiatus Loisel. Casual (Mediterranean region). Waste places. Fl.9-10. Glam.. 3 squares : **ST***16*[22],*17*[112],*27*[112].

A. homogamos (Maire) Humphries × **radiatus** = **A. × valentinus** L. (*A. dissimilis* Pomel). Casual (S. Europe). Waste places. Fl.6. Glam.. 2 squares : **ST***17*[23],*27*[112].

OTANTHUS Hoffmanns. & Link

Otanthus maritimus (L.) Hoffmanns. & Link : Edafeddog y Môr : Cottonweed. Native. Sandy shores. *Caerns.*†, *Angl.*†, [*Glam.*, *Mer.*]. 3 squares.
Caerns.: nr Glynllifon, SH45, TG, c1630[124]; nr Llandwrog, SH45, TG, c1639[125]. **Angl.**: Llanfaelog, SH37, HD, c1800[107] and HWS, 1894[101]; Abermenai, SH46, JR, 1662[203].

CHRYSANTHEMUM L.

Chrysanthemum segetum L. : Gold yr Ŷd : Corn Marigold. Established alien (Aegean region). Cultivated ground (D,F,G,R). Fl.(4-)6-11. Mons., Glam., Brecs., *Rads.*, Carms., Pembs., Cards., Monts., Mer., Caerns., *Denbs.*, *Flints.*, *Angl.*. 165 squares, map 727.
This species has decreased greatly in numbers since the 1930's due mainly to changes in agricultural techniques.

C. coronarium L. Casual (Mediterranean region). Waste places. Fl.7-9(-11). Glam.. 4 squares : **ST***07*[112],*17*[112],*27*[112]; **SO***00*[140].

TANACETUM L.

Tanacetum vulgare L. (*Chrysanthemum vulgare* (L.) Bernh., non (Lam.) Gaterau) : Tanclys : Tansy. Native. River banks, roadsides and waste places (E,F,G,M,Q, R,Rw). Fl.7-10. Mons., Glam., Brecs., Rads., Carms., Pembs., Cards., Monts., Mer., Caerns., Denbs., Flints., Angl.. 186 squares, map 728.

T. corymbosum (L.) Schultz Bip. (*Chrysanthemum corymbosum* L.; *Leucanthemum corymbosum* (L.) Gren. & Godron; *Pyrethrum corymbosum* (L.) Scop.). Casual (Europe). Waste places. Caerns.. 1 square : **SH***57*[112].

T. parthenium (L.) Schultz Bip. (*Chrysanthemum parthenium* (L.) Bernh.) : Wermod Wen : Feverfew. Established alien (Balkan Peninsula). Roadsides (E,Q, R,Rg,W). Fl.7-10. Mons., Glam., Brecs., Rads., Carms., Pembs., Cards., Monts., Mer., Caerns., Denbs., Flints., *Angl.*. 187 squares, map 729.

LEUCANTHEMELLA Tzvelev

Leucanthemella serotina (L.) Tzvelev (*Chrysanthemum uliginosum* Pers.; *C. serotinum* L.; *Tanacetum serotinum* (L.) Schultz Bip.; *Pyrethrum uliginosum* Waldst. & Kit. ex Willd.) : Moon Daisy. Casual (E. Europe). Waste places. *Rads.*. 1 square : **SO***25*[12].

BALSAMITA Miller

Balsamita major Desf. (*Chrysanthemum balsamita* (L.) Baillon, non L.) : Costmary. Casual (S.W. Asia). Waste places. *Glam.*. 2 squares : **ST***16*[24],*17*[21].

COLEOSTEPHUS Cass.

Coleostephus myconis (L.) Reichenb. fil. (*Chrysanthemum myconis* L.). Casual (S. Europe). Waste places. Glam.. 1 square : **ST***27*[112].

LEUCANTHEMUM Miller

Leucanthemum vulgare Lam. (*Chrysanthemum leucanthemum* L.) : Aspygan : Oxeye Daisy. Native. Meadows (D,E,G,M,Q,R,Rw,Sc). Fl.4-10. Mons., Glam., Brecs., Rads., Carms., Pembs., Cards., Monts., Mer., Caerns., Denbs., Flints., Angl.. 271 squares, all except : **SR**89; **SS**39; **ST**39; **SM**50; **SN**75,79,89; **SH**51, 92,98; **SJ**01.

L. maximum (Ramond) DC. (*Chrysanthemum maximum* Ramond) : Shasta Daisy. Established alien (Pyrenees). Waste places. Fl.7-8. Mons., Glam., Rads., *Pembs.*, Cards., *Monts.*, Caerns., *Denbs.*, Flints., *Angl.*. 25 squares, map 730.

L. lacustre (Brot.) Samp. (*Chrysanthemum lacustre* Brot.). Casual (Portugal). Waste places. *Glam.*. 1 square : **SS***68*[19].

ARTEMISIA L.

Artemisia vulgaris L. : Beidiog Lwyd : Mugwort. Native. Roadsides and waste places (Ar,D,E,G,M, Q,R,Rw,S). Fl.5-10. Mons., Glam., Brecs., Rads., Carms., Pembs., Cards., Monts., Mer., Caerns., Denbs., Flints., Angl.. 199 squares, map 731.

A. verlotiorum Lamotte : Chinese Mugwort. Established alien (S.W. China). Waste places. Fl.9. Caerns.. 2 squares.
Caerns.: Aber-soch, SH32, PMB & APC, 1960[74] and RGE, 1978[112]; Port Dinllaen, SH24, APC, 1978[184].

A. absinthium L. : Wermod Lwyd : Wormwood. Denizen (Europe). Roadsides and waste places (As,E, F,M,Q,R,Rw). Fl.6-10. Mons., Glam., Brecs., Rads., Carms., Pembs., *Cards.*, Monts., Mer., Caerns., Denbs., Flints., Angl.. 139 squares, map 732.

A. maritima L. : Wermod y Môr : Sea Wormwood. Native. Salt-marshes (S,Sc). Fl.8-9. Mons., Glam., Carms., Pembs., *Cards.*, *Caerns.*, Denbs., *Flints.*[230], Angl., [*Mer.*]. 31 squares, map 733.

A. abrotanum L. : Southernwood. Casual (unknown origin). Rubbish tip. *Caerns.*. 1 square : **SH**54[7].

A. pontica L. : Roman Wormwood. Casual (C. and E. Europe). Waste places. Fl.8. Glam.. 1 square : **SS**69[112].

A. biennis Willd. : Lesser Mugwort. Casual (N. America). Waste places. Fl.7-9. Mons., Glam., *Carms.*. 6 squares : **ST**17[112],18[112],*27*[112],38[112], **SN***30*[19]; **SO**21[104].

A. ludoviciana Nutt. (incl. *A. gnaphaloides* Nutt.) : White-sage. Casual (N. America). Waste places. Fl.6-10. Glam., *Angl.*. 3 squares : **ST***17*[112],*18*[112]; **SH***67*[18].

A. annua L. Casual (S. E. Europe). Waste places. *Glam.*, *Carms.*. 2 squares : **ST**27[37]; **SN***42*[18].

A. campestris L. : Llysiau'r Corff : Field Wormwood. Established alien (England). Sand dunes and waste places. Fl.7-9. Glam., (*Mer.*). 4 squares.
Glam.: Crymlyn Burrows, SS79, AEW, 1956[112] and QONK et al, 1981[93]; (Blaenrhondda, SS99, HHa, 1903[103]; Cardiff, ST27, JS, 1876[112]. **Mer.**: Barmouth, SH61, RMe, 1929[107]†).

TUSSILAGO L.

Tussilago farfara L. : Carn yr Ebol : Colt's-foot. Native. Stony riversides but widespread as an introduction in cultivated ground and waste places (E,M,Q,R, Rg,Rw,S,Sc,W). Fl.2-4. Mons., Glam., Brecs., Rads., Carms., Pembs., Cards., Monts., *Mer.*, Caerns., Denbs., Flints., *Angl.*. 265 squares, all except : **SS**38,39; **ST**26; **SM**50,62,71,73,84,92; **SN**00,36,59,87,89; **SH**51,60,62.

PETASITES Miller

Petasites albus (L.) Gaertner : White Butterbur. Established alien (Europe). Roadsides (E,R). Fl.1-3. Glam.*, Carms.*, Cards., *Mer.*, *Caerns.**, *Denbs.*, Flints.. 9 squares.
Glam.: Stouthall Wood, SS48, JBe, 1981[197] & 1982[112]. **Carms.**: Cil-y-cwm, SN73, JRG, 1972[112]. **Cards.**: nr Aberystwyth, SN68, JHS, 1925[112]. **Mer.**: nr Dolgellau, SH71, PMB, 1960+[3] & 1981[3]. **Caerns.**: nr Tre-garth, SH66, MJM, 1956[7]; nr Conwy, SH77, JMB, 1960+[67]. **Denbs.**: Pentre'r-felin, SH87, AAD, 1911[146]; Llangollen, SJ24, JAWhe, 1940[42]. **Flints.**: Ffynnongroyw, SJ18, RB, 1898[140] and GW, 1978[59].

P. hybridus (L.) P. Gaertner, B. Meyer & Scherb. : Alan Mawr : Butterbur. Native. Streamsides (E,M,R, Rw,W). Fl.3-4. Mons., Glam., Brecs., Rads., Carms., Pembs., Cards., Monts., Mer., Caerns., Denbs., *Flints.*, Angl.. 157 squares, map 734.

P. fragrans (Vill.) C. Presl : Winter Heliotrope. Established alien (Mediterranean region). Roadsides (E,G,R,Rw). Fl.11-2. Mons., Glam., Brecs., Rads., Carms., Pembs., Cards., Monts., Mer., Caerns., *Denbs.*, *Flints.*, *Angl.*. 124 squares, map 735.

P. japonicus (Siebold & Zucc.) Maxim. : Giant Butterbur. Naturalized alien (Japan). Streamsides (Ar,M,W). Fl.2-4. Glam., *Monts.*, *Mer.*, Denbs., *Flints.*. 8 squares.
Glam.: Margam, SS88, DMC, 1966[112] and MGC, 1977[112]; nr Pont-sarn, ST07, IR, 1974[112]; Cardiff, ST17, JPS, 1952[112] and LPe, 1979[112]. **Monts.**: Meifod, SJ11, HW, 1970[180]. **Mer.**: Dolgellau, SH71, KS, 1969[197]. **Denbs.**: Tal-y-cafn, SH77, FTJ, 1941[112]; nr Conwy, SH77, Co, 1946[112]; Bodidris, SJ25, AGS, 1960-80[197]. **Flints.**: Nannerch, SJ16, VG, 1970+[261].

DORONICUM L.

Doronicum orientale Hoffm. (*D. caucasicum* Bieb.). Casual (S.E. Europe). Hedges. Fl.4-5. Glam., Caerns.. 2 squares : **ST**18[112]; **SH**57[112].

D. plantagineum L. : Plantain-leaved Leopard's-bane. Naturalized alien (S.W. Europe). Woods and roadsides. Fl.4-5. *Carms., Pembs.,* Cards., Angl., [*Mons.*]. 5 squares.
Carms.: Carmel, SN63, IMV, 1974[225]; Llangadog, SN72, IMV, 1974[225]. **Pembs.:** Caldey Island, SS19, FNH, c1950[52]. **Cards.:** Lovesgrove, SN68, AOC, 1980[112]. **Angl.:** Llanfairpwllgwyngyll, SH57, WEH, 1966[112].

D. pardalianches L. : Llewpard-dag : Leopard's-bane. Naturalized alien (Europe). Hedges and woods (E,R, Rg,Rw,W). Fl.4-7. Mons., Glam., *Brecs.,* Rads., Carms., Pembs., Cards., Monts., Mer., *Caerns.*,* Denbs., Flints.. 45 squares, map 736.

SENECIO L.

Senecio bicolor (Willd.) Tod. subsp. **cineraria** (DC.) Chater (*S. cineraria* DC.) : Llys y Lludw : Silver Ragwort. Established alien (Mediterranean region). Sea cliffs and waste places (M,Q,R,Rw,Sc). Fl.7-8. Glam., *Pembs., Mer.,* Caerns.. 8 squares.
Glam.: Langland Bay, SS68, JRS, 1913[107]; Aberafan, SS78, RGE, 1981[95]; Cwmfelin, SS88, ABP, 1975[95]; Barry Docks, ST16, RGE, 1980[112]; Leckwith, ST17, RLS, 1922[112]. **Pembs.**: Lydstep, SS09, AW, 1916[146]. **Mer.**: Tywyn, SH50, An, 1975[7]. **Caerns.**: Llandudno, SH78, PMB, 1978[3]. Some of these records refer to cultivated forms.

S. bicolor subsp. **cineraria** × **jacobea** = **S.** × **albescens** Burbidge & Colgan. Established alien (garden origin). Sea cliffs and waste places. Fl.7. Mons., Glam., Brecs., Carms., Cards., Mer., *Caerns., Angl..* 13 squares : **SS***48*; **ST**16,18,38,59; **SN**50,58,61; **SO**21; **SH**50,57,61,78.

S. fluviatilis Wallr. : Broad-leaved Ragwort. Naturalized alien (Europe). Streamsides (E,G,M). Fl.6-7. Monts., Denbs., Flints.*, *Angl.*,* [*Caerns.*]. 7 squares.
Monts.: Pool Quay, SJ21, JAW, c1940[112]. **Denbs.**: Llansilin, SJ22, EJo, 1888[112]; Five Fords, SJ34, AGS, 1964[177]; Trefalun, SJ35, AGS, 1973[182]; Sutton Green, SJ44, AGS, 1964[177]. **Flints.**: Hawarden, SJ36, VG, 1956[7]. **Angl.**: Bodedern, SH38, EGW, 1945[45].

S. integrifolius (L.) Clairv. subsp. **maritimus** (Syme) Chater (*S. spathulifolius* auct. angl.; *S. integrifolius* var. *maritimus* (Syme) Clapham) : Chweinllys Arfor : Field Fleawort. Native. Sea cliffs. Fl.6-9. Angl.. 2 squares.
Angl.: Porth Dafarch, SH27, JEG, c1890[101] and RHR, 1970+[209]; Trearddur Bay, SH27, WSL, 1955[7] and RLB, 1959[7]; South Stack, Holyhead, SH28, JEG, 1901[112] and RHR, 1970+[209].

S. congestus (R. Br.) DC. (*S. palustris* (L.) Hooker, non Velloso) : Marsh Fleawort. Native. Marshes. *Glam.*†. 1 square.
Glam.: Aberafan, SS78, JR, 1662[203].

S. jacobaea L. : Creulys Iago : Common Ragwort. Native. Pastures and roadsides (D,E,G,H,M,Q,R, Rw,W). Fl.7-10. Mons., Glam., Brecs., Rads., Carms., Pembs., Cards., Monts., Mer., Caerns., Denbs., Flints., Angl.. 276 squares, all except : **SS**39; **SM**50,73; **SN**98; **SO**09; **SH**51.

[**S. jacobaea** × **vulgaris** *Mer.*]

S. aquaticus Hill : Creulys y Gors : Marsh Ragwort. Native. Marshes (Al,G,H,M,Q,R,S,W). Fl.5-9. Mons., Glam., Brecs., Rads., Carms., Pembs., Cards., Monts., Mer., Caerns., Denbs., *Flints.,* Angl.. 228 squares, map 737.

S. aquaticus × **jacobaea** = **S.** × **ostenfeldii** Druce Spontaneous. Damp pastures (G,M,R). Fl.8-9. *Carms.,* Cards., Mer., *Caerns., Angl..* 22 squares, map 738.

S. erucifolius L. : Creulys Llwyd : Hoary Ragwort. Native. Pastures and roadsides (E,G,M,Q,R,Rw,S, Sc,W). Fl.7-10. Mons., Glam., *Brecs.*,* Carms., Pembs., *Cards.,* Monts., Caerns., Denbs., Flints., Angl., [*Mer.*]. 78 squares, map 739.

S. squalidus L. : Creulys Rhydychen : Oxford Ragwort. Established alien (C. and S. Europe). Walls and waste ground (Ar,D,E,Q,R,Rg,Rw,S). Fl.4-11. Mons., Glam., Brecs., Rads.*, Carms., Pembs., *Cards.,* Monts., Mer., Caerns., Denbs., Flints., *Angl.*.* 117 squares, map 740.

S. squalidus × **viscosus** = **S.** × **londinensis** Lousley Spontaneous. Waste places (R,Rw). Fl.6. *Glam.,* Mer., *Denbs..* 3 squares.
Glam.: Ynystawe, SS69, QONK, 1970+[157]. **Mer.**: Barmouth Junction, SH61, PMB, 1960[112] & 1973[3].
Denbs.: Llay, SJ35, TE, 1972[112].

S. squalidus × **vulgaris** = **S.** × **baxteri** Druce Spontaneous. Waste places. Fl.4-8. Glam., *Monts., Denbs..* 7 squares.
Glam.: Gowerton, SS59, QONK, 1974[157]; nr Swansea, SS69, CH & HH, 1974[117]; Barry, ST16, GCD, 1927[26]; Cardiff, ST17, HJR, 1906[10]; E. Cardiff, ST27, AEW, 1975[112]. **Monts.**: Welshpool, SJ20, DHK, 1975[197].
Denbs.: Gwersyllt, SJ35, HHJ, 1926[25].

S. cambrensis Rosser : Creulys Cymreig : Welsh Groundsel. Native (endemic). Roadsides (E,R). Fl. 5-7(-10), Denbs., Flints.. 4 squares.
Denbs.: Colwyn Bay, SH87, JMB, 1966[179] & 1970[112]; nr Chirk, SJ23, PMB, 1980[112]; Ffrith, SJ25, EPAJ, 1948[235] and AS, 1967[112]. **Flints.**: Ffrith, SJ25, EMR & LWF, 1954[112]; Queensferry, SJ36, EPAJ, 1958[7].
This species is the naturally occurring fertile allohexaploid of **S.** × **baxteri** and, so far as is known, does not occur elsewhere.

S. nebrodensis L. Casual (Spain). Waste places. Fl.8. *Glam.*. 1 square : **ST***17*[107].

S. leucanthemifolius Poiret (*S. crassifolius* Willd.). Casual (Mediterranean region). Waste places. *Glam.*. 1 square : **ST***17*[140].

[**S. vernalis** Waldst. & Kit. *Glam.*]

[**S. vernalis** × **vulgaris** = **S.** × **pseudovernalis** Zabel *Glam.*]

S. sylvaticus L. : Creulys y Coed : Heath Groundsel. Native. Heaths (E,H,R,Rg,W). Fl.5-9. Mons., Glam., Brecs., Rads., Carms., Pembs., Cards., Monts., Mer., Caerns., Denbs., *Flints.*, *Angl.*. 176 squares, map 741.

S. viscosus L. : Creulys Gludiog : Sticky Groundsel. Colonist (Europe). Roadsides (M,Q,R,Rw). Fl.7-11. Mons., Glam., Brecs., Rads., Carms., Pembs., Cards., Monts., Mer., Caerns., Denbs., Flints., *Angl.*. 152 squares, map 742.

S. vulgaris L. : Creulys Cyffredin : Groundsel.
Subsp. **vulgaris.** Native. Cultivated ground (D,E, F,Q,R,Rw,S). Fl.3-10. Mons., Glam., Brecs., Rads., Carms., Pembs., Cards., Monts., Mer., Caerns., Denbs., Flints., Angl.. 275 squares, all except : **SR**89; **SM**50,71,73,84,94; **SH**51.

Var. **hibernicus** Syme (var. *radiatus* auct. angl.) has been recorded from all the Welsh counties. Recent research suggests that this variety may have arisen as a result of introgression from **S. squalidus** into **S. vulgaris.**

Subsp. **denticulatus** (O. F. Mueller) P. D. Sell Native. Sand dunes. *Angl.*. 1 square.
Angl.: Newborough Warren, SH46, DEA, 1950+[238].

LIGULARIA Cass.

Ligularia tangutica (Maxim.) Bergmans (*Senecio tanguticus* Maxim.). Established alien (China). Damp ground. *Angl.*. 1 square.
Angl.: Beaumaris, SH57, NC, 1979[59].

CALENDULA L.

Calendula officinalis L. : Melyn Mair : Pot Marigold. Casual (unknown origin). Waste places (G,R,Rw,S). Fl.7-12. Mons., Glam., Carms., *Pembs.*, *Monts.*, Mer., *Caerns.*, Denbs., Flints., *Angl.*. 27 squares, map 743.

C. arvensis L. : Field Marigold. Casual (S. Europe). Waste places (R,Rg). Fl.10. Glam.. 3 squares : **ST**06[99],*17*[112],*27*[112].

ARCTOTHECA Wendl.

Arctotheca calendula (L.) Levyns (*Cryptostemma calendulacea* (Hill) R. Br.). Casual (S. Africa). Waste places. Fl.7. Glam.. 2 squares : **ST***16*[112],*17*[112].

CARLINA L.

Carlina vulgaris L. : Ysgallen Siarl : Carline Thistle. Native. Limestone grassland and sea cliffs (D,G,H,Q,R,Rw,S,Sc,W). Fl.6-9. Mons., Glam., Brecs., Rads., Carms., Pembs., *Cards.*, Monts., Mer., Caerns., Denbs., *Flints.*, Angl.. 162 squares, map 744.

ECHINOPS L.

Echinops sphaerocephalus L. : Globe Thistle. Casual (C. and S. Europe). Waste places. *Glam.*, *Carms.*. 2 squares : **ST**17[75]; **SN***42*[19].

E. exaltatus Schrader Casual (C. Europe). Hedgerow. *Mons.*. 1 square : **SO**21[91].

ARCTIUM L.

Arctium lappa L. : Cyngaf Mawr : Greater Burdock. Native. Roadsides (Ar,E,G,M,R,Rw,Sc,W). Fl.7-9. Mons., Glam., *Brecs.**, *Rads.*, Carms., Pembs.*, Monts., *Caerns.*, *Denbs.*, *Flints.**, *Angl.*. 49 squares, map 745. Some of the early records may refer to **A. minus** agg.

A. minus agg. Mons., Glam., Brecs., Rads., Carms., Pembs., Cards., Monts., Mer., Caerns., Denbs., Flints., Angl.. 269 squares, all except : **SS**39; **SM**50,73; **SN**25, 36,59,78,86,87,98; **SO**09; **SH**51; **SJ**13.
The distributions given for the three segregate species are provisional only.

A. pubens Bab. (*A. minus* subsp. *pubens* (Bab.) J. Arènes). Plants resembling this species have been recorded from *Glam.*, *Brecs.*, *Pembs.*, *Flints.*. but no specimens have been positively identified as such.

A. minus Bernh. *s.s.* (*A. minus* subsp. *minus*) : Cyngaf Bychan : Lesser Burdock. Native. Roadsides and woods (E,G,M,R,Rw,Sc,W). Fl.6-8. Mons., Glam., Brecs., Rads., Carms., Pembs., Cards., Monts., Mer., Caerns., *Denbs.*, *Flints.*, *Angl.*. 118 squares, map 746.

A. nemorosum Lej. (*A. minus* subsp. *nemorosum* (Lej.) Syme; *A. vulgare* auct.) : Cyngaf y Coed : Intermediate Burdock. Native. Roadsides (E,F,G,R, Sc,W). Fl.8-9. *Mons.*, Glam., *Brecs.*, *Rads.*, Carms., Pembs., *Cards.*, Monts., *Mer.*, *Caerns.*, Denbs., *Flints.*, *Angl.*. 85 squares, map 747.

Hybrids occur between all four species of **Arctium** but so far it has not proved practicable to separate them in Wales. They have been recorded from *Mons.*, *Glam.*, *Brecs.*, *Pembs.*, *Cards.*, *Mer.*, *Caerns.*, *Denbs.*, *Flints.*, *Angl.*

SAUSSUREA DC.

Saussurea alpina (L.) DC. : Lliflys Mynyddig : Alpine Saw-wort. Native. Mountain rocks. Fl.6-8. Mer.*, Caerns.. 3 squares.
Mer.: Llyn Aran, SH71, SCC, 1962[112] and PMB, 1965[112]. **Caerns.**: Snowdon, SH65, WI, 1895[112] and ER, 1962[228]; Twll Du, SH65, JEG, 1884[112]; Glyders, SH65, ER, 1965[228]; Carnedd Llywelyn, SH66, AL, 1888[112]; Ysgolion Duon, SH66, ER, 1950+[228].

CARDUUS L.

Carduus nutans L. : Ysgallen Ogwydd : Musk Thistle. Native. Limestone grassland and calcareous dunes (D,F,G,H,M,R). Fl.6-11. Mons., Glam., Brecs., Rads., Carms., Pembs., Cards., Monts., Mer., *Caerns.*, Denbs., Flints., Angl.. 106 squares, map 748.

C. acanthoides L. (*C. crispus* auct., non L.) : Ysgallen Grych : Welted Thistle. Native. Roadsides (Ar,D,E, G,M,R,W). Fl.6-10. Mons., Glam., Brecs., Rads., Carms., *Pembs.*, Cards., Monts., Mer., Caerns., Denbs., *Flints.*, Angl.. 120 squares, map 749.

C. acanthoides × nutans = C. × orthocephalus Wallr. Spontaneous. Waste places. Fl.7-10. Mons., Glam., Denbs.. 4 squares.
Mons.: Dinham, ST49, WAS, 1924[112]. **Glam.**: Bishopston, SS58, JAW, 1943[112]. **Denbs.**: Llanrhaeadr, SJ06, JAW, 1943[112]; Marford, SJ35, TE, 1972[112].

C. tenuiflorus Curtis : Ysgallen Flodfain : Slender Thistle. Native. Waste places near the sea (D,G,R,S). Fl.5-11. Mons., Glam., Carms., Pembs., *Cards.*, Mer., Caerns., Denbs., Flints., Angl.. 75 squares, map 750.

C. pycnocephalus L. : Plymouth Thistle. Casual (S. Europe). Waste places. *Glam.*. 1 square : **ST**17[30].

CIRSIUM Miller

Cirsium eriophorum (L.) Scop. : Ysgallen Benwlanog : Woolly Thistle. Native. Limestone pastures (F,G,R). Fl.7-9. Mons., Glam., *Rads.**, (*Carms.*), *Pembs.*[230], *Monts.*[230], (*Flints.*), [*Brecs.*, *Mer.*]. 21 squares, map 751.

C. vulgare (Savi) Ten. : March Ysgallen : Spear Thistle. Native. Pastures (D,E,F,G,H,M,Q,R,Rw,S,W). Fl.7-11. Mons., Glam., *Brecs.*, Rads., Carms., Pembs., Cards., Monts., Mer., Caerns., Denbs., Flints., *Angl.*. 280 squares, all except : **SS**39; **SM**50.

C. dissectum (L.) Hill : Ysgallen Mignwern : Meadow Thistle. Native. Marshes (D,Ds,G,H,M,R,W). Fl.6-9. Mons., Glam., Brecs., Rads., Carms., Pembs., Cards., Mer., [*Denbs.*]. 97 squares, map 752.

C. dissectum × palustre = C. × forsteri (Sm.) Loudon Spontaneous. Moorland. Fl.6-7. Carms., Pembs.. 2 squares.
Carms.: nr Trimsaran, SN40, RDP, 1982[112]. **Pembs.**: Dowrog Common, SM72, FBo, 1979[112].

[**C. dissectum × vulgare** *Cards.*]

C. tuberosum (L.) All. : Ysgallen Oddfynog : Tuberous Thistle. Native. Limestone pastures (G,R). Fl.7-8. Glam.. 5 squares.
Glam.: Southerndown, SS87, An, 1920[112]; Nash Point, SS96, TWe, c1840[198] and RGE & AEW, 1979[95]; Monknash, SS97, HAH, 1934[112]; Cwm Nash, SS97, RGE, 1972[105]; (Barry, ST16, AEW, 1927[112]; Cardiff Docks, ST17, HJR, 1909[142]). Most of the native populations show signs of introgression with **C. acaule.**

C. acaule Scop. (*C. acaulon* (L.) Scop.) : Ysgallen Ddigoes : Dwarf Thistle. Native. Limestone pastures (D,Ds,F,G,H,Q,Sc). Fl.7-10. Mons., Glam., Brecs.*, *Pembs.**, Monts., *Denbs.*, *Flints.**. 28 squares, map 753.

C. acaule × arvense = C. × boulayi Camus Spontaneous. *Mons.*. 1 square.
Mons.: nr St Arvans, ST59, ESM & WAS, 1903[137].

[**C. acaule × dissectum = C. × woodwardii** (H. C. Watson) Nyman *Glam.*]

C. acaule × tuberosum = C. × zizianum Koch Spontaneous. Limestone grassland. Fl.5-8. Glam.. 2 squares.
Glam.: Nash Point, SS96, AEW, 1976[95]; Monknash, SS97, JLB, 1935[112] and MCK, 1954[7]; Cwm Nash, SS97, EV, 1938[112].

C. helenioides (L.) Hill (*C. heterophyllum* (L.) Hill) : Ysgallen Bwyth : Melancholy Thistle. Native. Moist pastures (E,G,H,M,Rw). Fl.6-9. Rads., (Monts.*), Mer., Caerns., Denbs., *Flints.*, [*Glam.*, *Cards.*, *Angl.*]. 19 squares : (**SN**99); **SO**07,08,17,18; **SH***65*,73,74,83,84, 85,*92*,*93*,94,95; **SJ**(04),07,*15*,25.

C. palustre (L.) Scop. : Ysgallen y Gors : Marsh Thistle. Native. Wet pastures and woods (Ar,E,G, H,M,Q,R,Rw,W). Fl.5-9. Mons., Glam., Brecs., Rads., Carms., Pembs., Cards., Monts., Mer., Caerns., Denbs., *Flints.*, *Angl.*. 275 squares, all except : **SS**38,39; **ST**26; **SM**50,71,73; **SN**98.

C. palustre × tuberosum = C. × semidecurrens Richter Spontaneous. Limestone grassland. *Glam.*. 1 square.
Glam.: Nash Point, SS96, JLB, 1932[32].

C. arvense (L.) Scop. : Ysgallen Gyffredin : Creeping Thistle. Native. Cultivated ground and pastures (D,E, F,G,M,Q,R,Rw,S,Sc,W). Fl.6-10. Mons., Glam., Brecs., Rads., Carms., Pembs., Cards., Monts., Mer., Caerns., Denbs., Flints., Angl.. 280 squares, all except : **SS**39; **SM**50.

C. arvense × palustre = C. × celakovskianum Knaf fil. Spontaneous. Waste places. Fl.9. Glam., *Mer.*. 4 squares.
Glam.: Cardiff, ST17, EV, 1922[112]. **Mer.**: Arthog, SH61, PMB, 1971[59]; Cilfor, SH63, PMB, 1980[3]; Glyndyfrdwy, SJ14, PMB, 1974[59].

GALACTITES Moench

Galactites tomentosa Moench (*Lupsia galactites* (L.) O. Kuntze; *G. galactites* (L.) Druce). Casual (Mediterranean region). Waste places. Fl.8. Glam.. 2 squares : **ST***16*[27],*18*[112].

ONOPORDUM L.

Onopordum acanthium L. : Ysgallen Gotymog : Cotton Thistle. Established alien (Europe). Fields and waste places (D,E,Q,R). Fl.7. Mons., Glam., *Carms.*[230], *Pembs., Cards., Monts.*[230], *Mer.* *, Caerns., *Denbs., Flints., Angl.*. 26 squares, map 754.

SILYBUM Adanson

Silybum marianum (L.) Gaertner : Ysgallen Fair : Milk Thistle. Established alien (S. Europe). Waste places (D,R,Rg). Fl.6-9. Mons., Glam., Carms., *Pembs., Monts.* *, *Mer.*[85], Caerns., *Denbs., Flints., Angl.*. 34 squares, map 755.

SERRATULA L.

Serratula tinctoria L. : Dant y Pysgodyn : Saw-wort. Native. Pastures (E,F,G,H,M,R,Rw,Sc,W). Fl.6-10. Mons., Glam., Brecs., Rads., Carms., Pembs., Cards., Monts., Mer., Caerns., Denbs., Flints., Angl.. 191 squares, map 756.

MANTISALCA Cass.

Mantisalca salmantica (L.) Briq. & Cavillier (*Centaurea salmantica* L.; *Microlonchus salmanticus* (L.) DC.). Casual (Mediterranean region). Waste places. Fl.8. Glam.. 1 square : **ST***27*[112].

CENTAUREA L.

Subgen. Lopholoma (Cass.) Dobrocz.

Centaurea orientalis L. Casual (S.E. Europe). Waste places. Glam.. 1 square : **ST**17[112].

C. scabiosa L. : Pengaled Fawr : Greater Knapweed. Native. Dry grassland (D,E,F,G,H,Q,R,Rw). Fl.6-9. Mons., Glam., Carms., Pembs., Cards., Monts., (Mer.), Caerns., Denbs., Flints., Angl.. 107 squares, map 757.

Subgen. Acrolophus (Cass.) Dobrocz.

[**C. paniculata** L. *Glam.*]

C. leucophaea Jordan Casual (S. Europe). Waste places. Fl.8. Glam.. 1 square : **ST***27*[112]

C. diffusa Lam. Casual (S. Europe). Waste places. Glam.. 1 square : **ST**27[112].

Subgen. Calcitrapa (Heister ex Fabr.) Hayek

C. iberica Trev. ex Sprengel Casual (S. Europe). Waste places. Fl.9. Glam.. 1 square : **ST**16[112].

C. calcitrapa L. : Red Star-thistle. Casual (S. Europe). Waste places. Fl.6-10. *Mons.*, Glam., *Carms., Denbs., Flints.*. 7 squares : **SS***78*[140]; **ST***16*[112],*17*[112],27[112],*38*[226]; **SN***40*[161]; **SJ**07[195].

Subgen. Seridia (Juss.) Czerep.

C. aspera L. : Rough Star-thistle. Established alien (S. W. Europe). Waste places. Fl.9. Glam., Carms.. 4 squares.
Glam.: Port Talbot, SS78, RLB, 1933[32]; Barry Docks, ST16, RLS, 1921[112] and DM, 1974[197]; Cardiff, ST17, JS, c1876[218]. **Carms.**: Pembrey, SN30, SP & IMV, 1977[107].

Subgen. Solstitiaria (Hill) Dobrocz.

C. solstitialis L.

Subsp. **solstitialis** : Yellow Star-thistle. Casual (S. Europe). Waste places (F,R,Rg). Fl.8. Glam., Brecs., *Mer., Flints.*. 8 squares : **SS***48*[17],*78*[17],*87*[112]; **ST**17[112],*27*[112]; **SO***21*[112]; **SH***61*[199]; **SJ***08*[26].

Subsp. **adamii** (Willd.) Nyman (*C. adamii* Willd.). Casual (Mediterranean region). Waste places. Fl.8. *Glam.*. 1 square : **SO***00*[107].

C. melitensis L. : Maltese Star-thistle. Casual (S. Europe). Waste places. Fl.6-8. Glam., *Carms., Caerns.*. 7 squares : **SS**48[112],78[112]; **ST***17*[112],27[112]; **SN***31*[161]; **SO***00*[140]; **SH***56*[18].

C. nicaeensis All. Casual (Mediterranean region). Waste places. Fl.9. Glam.. 1 square : **ST***27*[112].

C. diluta Aiton : Lesser Star-thistle. Casual (N. W. Africa). Waste places. Fl.7-10. Mons., *Glam.*, Carms.. 4 squares : **SS***78*[107]; **ST**27[112],38[112]; **SN**31[112].

Subgen. Jacea (Miller) Hayek

C. jacea L. : Pengaled Lwytgoch : Brown Knapweed. Casual (Europe). Waste places. Fl.9. *Glam.*. 2 squares : **SS**69[104],*78*[17].

C. jacea × nigra agg. = **C. × drucei** C. E. Britton Status unknown. Waste places. Glam., *Denbs.*. 2 squares.
Glam.: Barry, ST16, AEW, 1922[112]. **Denbs.**: Ruabon, SJ34, EMJ & WBT, pre 1930[160]. **C. jacea** has not been recorded from either locality.

C. nigra agg. : Pengaled : Knapweed. Native. Pastures (D,E,G,M,Q,R,Rw,S,W). Fl.6-9. Mons., Glam., Brecs., Rads., Carms., Pembs., Cards., Monts., Mer., Caerns., Denbs., Flints., Angl.. 276 squares, all except : **SS**38,39, 77; **SM**50,62,73.
It is very difficult to separate the following segregates in Wales, hence the few definite records.

C. debeauxii Gren. & Godron subsp. **nemoralis** (Jordan) Dostál (*C. nemoralis* Jordan; *C. nigra* subsp. *nemoralis* (Jordan) Gremli) : Pengaled Cochddu : Slender Knapweed. Native. Pastures (E,G,R,Rw). Fl.7-9. Mons., Glam., Brecs., *Rads.* *, Carms., Pembs., Cards., Monts., Mer., Caerns., Denbs., Flints.. 60 squares, map 758.

C. debeauxii × nigra. Intermediates between these two taxa have been reported from *Mons., Pembs., Cards., Mer.*

C. nigra L. *s.s.* : Pengaled Benddu : Common Knapweed. Native. Pastures (G,R,Rw,S). Fl.6-9. Mons., Glam., Brecs., Rads., Carms., Pembs., Cards., Monts., Mer., Caerns., *Denbs.*, Flints., *Angl.*. 187 squares, map 759.

Subgen. Cyanus (Miller) Hayek

C. montana L : Perennial Cornflower. Established alien (Europe). Waste places (E,G,Q,R,Rw). Fl.4-8. Mons., Glam., Brecs., Rads., Carms., Cards., *Monts.*. 27 squares, map 760.

C. cyanus L. : Penlas yr Ŷd : Cornflower. Colonist (England). Cultivated ground and waste places (F,R,Rg). Fl.6-10. Mons., Glam., *Brecs.**, *Rads.*, *Carms.**, Pembs.*, *Cards.*, *Monts.**, *Mer.*, Caerns., Denbs., Flints., *Angl.*. 59 squares, map 761.

Subgen. Tetramorphaea (DC.) Czerep.

C. bruguierana (DC.) Hand.-Mazz. subsp. **belangerana** (DC.) Bornm. (*Tetramorphaea belangeriana* DC.). Casual (Asia). Waste places. Fl.7. Glam.. 1 square : **ST**27[112].

C. algeriensis Cosson & Durieu Casual (N. Africa). Waste places. Fl.6-10. Glam.. 1 square : **ST***27*[112].

CARTHAMUS L.

Carthamus tinctorius L. : Safflower. Casual (W. Asia). Waste places. Fl.7-9. Mons., Glam., Brecs., Cards., *Caerns.*, Flints.. 8 squares : **ST**17[112],18[112],38[112]; **SO**00[112]; **SN***58*[112]; **SH***53*[16]; **SJ**17[112],*18*[23].

C. lanatus L. : Downy Safflower. Casual (S. Europe). Waste places. Fl.8. Glam.. 4 squares : **SS***78*[140]; **ST***17*[112],*18*[20],*27*[112].

SCOLYMUS L.

Scolymus maculatus L. Casual (S. Europe). Waste places. *Glam.*. 1 square : **ST***27*[26].

S. hispanicus L. : Spanish Oysterplant. Casual (S. Europe). Waste places. Fl.6-9. Mons., Glam.. 4 squares : **ST***16*[112], *17*[218],*27*[112],38[112].

CICHORIUM L.

Cichorium intybus L. : Ysgellog : Chicory. Colonist (England). Cultivated ground and waste places (F,G,R, Rg,Rw). Fl.6-9. Mons., Glam., Brecs., *Rads.*, *Carms.*, *Pembs.*, *Cards.*, *Monts.*, Mer., Caerns., Denbs., *Flints.*, Angl.. 108 squares, map 762.

C. endivia L. : Endive. Casual (S. Europe). Waste places. Fl.6. Glam.. 2 squares : **ST***18*[112],*27*[112].

TOLPIS Adanson

Tolpis barbata (L.) Gaertner : Yellow Hawkweed. Casual (S. Europe). Waste places. Fl.8. Glam.. 1 square : **ST***16*[112].

HEDYPNOIS Miller

Hedypnois cretica (L.) Dum.-Courset (*Rhagadiolus hedypnois* (L.) All.). Casual (S. Europe). Waste places. *Glam.*. 1 square : **ST***27*[25].

RHAGADIOLUS Zinn

Rhagadiolus stellatus (L.) Gaertner (*R. edulis* Gaertner). Casual (S. Europe). Waste places. Glam.. 1 square : **ST***27*[112].

HYPOCHOERIS L.

Hypochoeris maculata L. : Melynydd Brych : Spotted Cat's-ear. Native. Sea cliffs. Fl.6-7. *Mer.*†, Caerns.. 2 squares.
Mer.: nr Barmouth, SH61, TGa, 1863[212]. **Caerns.**: Gt Ormes Head, SH78, RD, 1861[112] and PMB, 1979[3].

H. glabra L. : Melynydd Moel : Smooth Cat's-ear. Native. Dry sandy places (D,G). Fl.5-9. Glam., (*Brecs.*), Carms., *Pembs.*, *Mer.*, Caerns., *Flints.*, Angl.. 19 squares : **SR**99; **SS**78,87,88; **SN***10*,30; (**SO**02); **SH**36,37,45,46, 50,52,53,*57*,58,61; **SJ**08,*14*.

H. glabra × radicata Spontaneous. Sandy grassland. Fl.7. *Mer.*. 2 squares.
Mer.: Mochras, SH52, PMB, 1959[4]; Morfa Harlech, SH53, PMB, 1959[4].

H. radicata L. : Melynydd : Cat's-ear. Native. Pastures (D,E,F,G,H,M,Q,R,Rw,S). Fl.6-10(-12). Mons., Glam., Brecs., Rads., Carms., Pembs., Cards., Monts., Mer., Caerns., Denbs., Flints., Angl.. 279 squares, all except : **SR**89; **SS**39; **SM**50.

LEONTODON L.

Leontodon autumnalis L. : Peradyl yr Hydref : Autumn Hawkbit. Native. Pastures (D,E,F,G,H,M,Q, R,S). Fl.7-10. Mons., Glam., Brecs., Rads., Carms., Pembs., Cards., Monts., Mer., Caerns., Denbs., Flints., *Angl.*. 270 squares, all except : **SS**09,39; **SM**50,62,73; **SN**11,21,36,46; **SO**36,42; **SH**98.

L. hispidus L. : Peradyl Garw : Rough Hawkbit. Native. Pastures (D,G,Q,R,Rg,Rw). Fl.6-10. Mons., Glam., Brecs., Rads., Carms., Pembs., Cards., *Monts.*, Mer., *Caerns.*, Denbs., *Flints.*, *Angl.*. 193 squares, map 763.

L. taraxacoides (Vill.) Mérat (*L. leysseri* G. Beck) : Peradyl Blewog : Lesser Hawkbit. Native. Gravelly fields and heaths (D,Ds,E,G,Q,R,S). Fl.6-10. Mons., Glam., Brecs., *Rads.*, Carms., Pembs., *Cards.*, Monts., Mer., Caerns., Denbs., *Flints.*, Angl.. 164 squares, map 764.

PICRIS L.

Picris echioides L. : Gwylaeth Chwerw : Bristly Ox-tongue. Native. Roadsides (E,F,G,Q,R,Rg,Rw,W). Fl.7-10. Mons., Glam., Brecs., Rads., Carms., *Pembs.*, *Cards.*, Monts., Mer., Caerns., Denbs., Flints., *Angl.*. 83 squares, map 765.

P. hieracioides L. : Gwylaeth yr Hebog : Hawkweed Oxtongue. Native. Roadsides (D,E,G,Q,R,Rw,W). Fl.7-9. Mons., Glam., Brecs., *Rads.*, Carms., Pembs., Monts., (Mer.), *Caerns.**, Denbs., *Flints.**. 71 squares, map 766.

TRAGOPOGON L.

Tragopogon porrifolius L. : Barf yr Afr Gochlas : Salsify. Naturalized alien (Mediterranean region). Pastures and railway banks (G,M,R,Rw). Fl.5-7. Mons., Glam., *Carms., Pembs., Cards.*, Mer.*, Flints., Angl.**. 16 squares : **SS***59,89*; **ST***16,17*,18,27,28,*39,59*; **SN***10,40*,46,*69*; **SJ***07,17*.

T. crocifolius L. Established alien or casual (S. Europe). Grassy and waste places. Fl.6. Mons., *Glam.*. 2 squares : **ST***27*[25],*59*[112].

T. pratensis L.

Subsp. **pratensis** Casual (Europe). Waste places. Fl.5-6. Mons., Glam., Brecs.. 5 squares : **ST***16*[112],17[112],*48*[226]; **SO***01*[112],*20*[112].

Subsp. **minor** (Miller) Wahlenb. : Barf yr Afr Felen : Goat's-beard. Native. Pastures and roadsides (D,E, G,Q,R,Rw). Fl.5-8(-10). Mons., Glam., Brecs., Rads., Carms., Pembs., Cards., *Monts., Mer.*, Caerns., Denbs., Flints., Angl.. 165 squares map 767.

T. hybridus L. Casual (S. Europe). Waste places. Fl.7. Glam.. 1 square : **ST**27[112].

SONCHUS L.

Sonchus asper (L.) Hill : Llaethysgallen Arw : Prickly Sow-thistle. Native. Cultivated ground (D,F,Q,R,Rg, Rw,Sc). Fl.5-7(-12). Mons., Glam., Brecs., Rads., Carms., Pembs., Cards., Monts., Mer., Caerns., Denbs., Flints., Angl.. 263 squares, all except : **SM**50,73; **SN**36,63,75, 78,86,87,91; **SO**09,20,32; **SH**51,65,72,82,92,98; **SJ**44.

S. oleraceus L. : Llaethysgallen : Smooth Sow-thistle. Native. Cultivated ground (D,E,F,Q,R,Rg,S). Fl.4-8(-10). Mons., Glam., Brecs., Rads., Carms., Pembs., Cards., Monts., Mer., Caerns., Denbs., Flints., Angl.. 266 squares, all except : **SM**62; **SN**75,76,86,87, 88,91; **SO**08; **SH**29,72,80,91,92; **SJ**03,13,44.

S. arvensis L. : Llaethysgallen yr Ŷd : Perennial Sow-thistle. Native. Cultivated ground (D,F,M,Q,R,Rg, Rw,S). Fl.7-10. Mons., Glam., Brecs., Rads., Carms., Pembs., Cards., Monts., Mer., Caerns., Denbs., Flints., Angl.. 240 squares, map 768.

LACTUCA L.

Lactuca tatarica (L.) C. A. Meyer : Russian Blue Sow-thistle. Established alien (E. Europe). Sand dunes and railways lines. *Caerns.*, Denbs.. 2 squares.
Caerns.: Llandudno, SH78, FHo, 1963[56] and GMSE, 1980[51]. **Denbs.**: Marford, SJ35, EGW, 1947[112].

L. serriola L. : Gwylaeth Bigog : Prickly Lettuce. Colonist (Europe). Roadsides (E,Q,R,Rg,Rw,S,Sc). Fl.8-10. Mons., Glam., *Brecs.*, Carms.*, Mer., Denbs.. 16 squares : **SS**48,*59,87*; **ST**06,16,17,27,38,48,58; **SN***40,42,50,90*; **SH***63*; **SJ**35.

L. sativa L. : Gwylaeth : Garden Lettuce. Casual (garden origin). Waste places. *Mons., Glam.*, Pembs.. 3 squares : **SS**19[112]; **ST**17[95],38[91].

L. saligna L. : Gwylaeth Leiaf : Least Lettuce. Established alien (England). Waste places. Glam.. 2 squares.
Glam.: Barry, ST16, EV, 1910[112]; Llandough, ST17, EV, 1917[112]; Cardiff Docks, ST17, RLS, 1924[112].

L. virosa L. : Gwylaeth Gryf-arogl : Great Lettuce. Established alien (Europe). Waste places. Fl.9. Mons., Glam., *Caerns., Denbs., Flints.*. 16 squares: **SS***68,69,78,88*; **ST***16*,17,27,38,*48*; **SH***78,87,95,97*; **SJ***07*,08,*18*.

L. pulchella (Pursh) DC. Casual (N. America). Sand dunes. Fl.7. Caerns.. 1 square : **SH**78[112].

CICERBITA Wallr.

Cicerbita macrophylla (Willd.) Wallr. subsp. **uralensis** (Rouy) P. D. Sell : Blue Sow-thistle. Established alien (Russia). Waysides (E,F,R,Rg). Fl.6-8. Mons., Glam., Rads., Carms., Cards., *Monts.*, Mer., *Caerns.*, Denbs., Flints., Angl.. 32 squares, map 769.

MYCELIS Cass.

Mycelis muralis (L.) Dumort. : Gwylaeth y Fagwyr : Wall Lettuce. Native. Walls and rocks (E,Q,R,Sc,W). Fl.6-10. Mons., Glam., Brecs., Rads., Carms., Pembs., Cards., Monts., Mer., Caerns., Denbs., Flints., *Angl.*. 32 squares, map 770.

TARAXACUM Weber

As the distribution of the microspecies has not been fully worked out in this taxonmically difficult genus, full details of v.-c. distribution and a map (where applicable) are given for each of the five Sections. Richards (1972) gives a detailed account of the microspecies as understood up to that date. Details of subsequently described species and their v.-c. distribution are given in R. J. Pankhurst, *Polyclave for* Taraxacum *in Great Britain,* version 3, 1979. Since specimens should be collected before the end of May to enable correct identification, no months of flowering have been included.
All identifications have been made or confirmed by Dr A. J. Richards or C. C. Haworth.

Sect. Spectabilia Dahlst.

Dant y Llew y Gors : Broad-leaved Marsh-dandelion. Native. Wet places, especially in upland areas (As,D,E, G,H,M,Q,R,Rg,Sc,W). Fl.3-6. Mons.*, Glam., Brecs., Rads.*, Carms., Pembs., Cards.*, Monts.*, Mer.*, Caerns., Denbs.*, Flints.*, Angl.*. 133 squares, map 771.

T. faeroense (Dahlst.) Dahlst. Native. Wet places usually in hilly areas (D,E,G,H,M,Q,R,W). Mons., Glam., *Brecs.*, Carms., *Pembs.*[205], Cards., Monts., Mer., Caerns., Denbs., *Flints.*, Angl.. 32 squares, map 772.

T. spectabile Dahlst. Native. Wet grassland (G,H,M, Q,Rg). Glam., *Brecs.*[205], Carms., Pembs., Cards., *Monts., Mer.*[206], *Caerns.*, Denbs., Flints.. 15 squares : **SS***88*; **ST**09; **SM**80; **SN**44,51,52; **SH**56,95; **SJ**01,24,25, 26,35,36,44.

T. eximium Dahlst. Native. Wet acid places (G,M). *Rads.*, Carms., Pembs.. 3 squares : **SM**80[112]; **SN**74[112]; **SO**24[104].

T. landmarkii Dahlst. Native. Streamsides and roadsides (E,G,M,R). *Brecs.*[205], Cards., Mer., Denbs., *Flints.*. 9 squares : **SN**44[112],65[112], 87[112]; **SH**52[112],53[112],85[112]; **SJ**03[112],35[245],36[245].

T. euryphyllum (Dahlst.) M. P. Christiansen Native. Roadsides and wet rocky places (E,G,Q,R). Brecs., Cards., Mer., Denbs., *Flints.*. 14 squares : **SN**65,76; **SO**11; **SH**53,74,85; **SJ**03,14,15,24,25,26,35,44.

T. maculosum A. J. Richards (*T. maculigerum* sensu A. J. Richards, non H. Lindb. fil.). Native. Wet places (G,M,Q). *Glam.*[205], Brecs., Carms., Cards., *Pembs.*[205], Mer., Denbs., *Angl.*. 9 squares : **SN**35[112],61[112]; **SO**11[112]; **SH**36[107],53[112],63[112],95[112]; **SJ**25[245],35[245].

T. praestans H. Lindb. fil. Native. Damp pathsides (E,G,R). *Pembs.*[205], Cards., *Caerns.*, Denbs., *Flints.*, *Angl.*. 6 squares : **SN**69[112]; **SH**46[209],57[107],85[112],95[112]; **SJ**26[261].

T. naevosiforme Dahlst. Native. Wet grassland, cliffs, etc. *Caerns., Denbs., Flints.*. 5 squares : **SH**66[107]; **SJ**24[245],25[245],36[245],44[245].

T. drucei Dahlst. Native. Wet places. *Caerns.*. 1 square : **SH**65[108].

T. stictophyllum Dahlst. Native. Wet places and roadsides. Denbs., *Angl.*. 3 squares : **SH**57[208],85[112]; **SJ**24[245].

T. naevosum Dahlst. Native. Grassy places. *Caerns.*. 1 square : **SH**54[107].

T. unguilobum Dahlst. Native. Wet places (D,E,G,Q, R,Sc,W). Glam., Brecs., *Rads.*, Carms., Pembs., Cards., Mer., *Caerns.*, Denbs., *Flints., Angl.*. 19 squares : **ST**09; **SM**82; **SN**69,74,76,77,*82*,84; **SO**11,*13*,14; **SH**46,52,53, 64,66,85; **SJ**16,17.

T. fulvicarpum Dahlst. Native. Damp grassland. Brecs., Cards., *Caerns.*. 4 squares : **SN**44[112],92[112]; **SO**04[107]; **SH**57[183].

T. laetifrons Dahlst. Native. Wood margins. *Brecs.*[205], *Carms., Pembs.*[205], *Flints.*. 2 squares : **SN**74[225]; **SJ**25[245].

T. nordstedtii Dahlst. (incl. *T. cambriense* A. J. Richards *p.p.*). Native. Wet places (As,D,E,F,G,H,M, Q,R,Sc). Mons., Glam., Brecs., Rads., Carms., Pembs., Cards., *Monts.*[205], Mer., Caerns., *Denbs., Flints., Angl.*. 43 squares, map 773.

T. litorale Raunk. Native. Wet meadow. Rads.. 1 square : **SN**97[112] (provisional determination).

T. adamii Claire Native. Wet meadows (D,E,G,M, Q,R,W). *Mons.*, Glam., Brecs., Carms., Pembs., Cards., Monts., *Mer.*[205], Caerns., Denbs., *Flints.*. 27 squares, map 774.

T. gelertii Raunk. Native. Woods and shaded roadsides. Cards., Mer., *Denbs.*[206]. 2 squares : **SN**58[112]; **SH**63[112].

T. britannicum Dahlst. Native. Roadside banks. *Carms.*[206], Cards., *Denbs., Flints.*. 11 squares : **SN**30,34, 35,58,64,68; **SJ**15,16,24,25,44.

T. hibernicum G. Hagl. Native. Rough grass. Cards. 1 square : **SN**34[112].

T. lancastriense A. J. Richards (*T. cambriense* A. J. Richards *p.p.*) Native. Rocky places (G,M,Q,R,Sc). *Mons.*[245], Glam., Brecs., Carms., [*Pembs.*]. 7 squares : **SS***88*[107]; **ST**07[112],18[112]; **SN**61[112],85[112]; **SO**00[112],*01*[107].

Sect. Palustria Dahlst.

T. palustre (Lyons) Symons : Dant y Llew y Waun : Narrow-leaved Marsh-dandelion. Native. Marshes. Angl., [*Mons., Glam., Brecs., Rads., Carms., Pembs., Monts., Mer., Caerns., Denbs., Flints.*]. 4 squares : **SH**46[209],48[209],57[112],58[112].

Sect. Obliqua Dahlst.

T. platyglossum Raunk. Native. Sand dunes. *Caerns.*[206], *Angl.*. 2 squares : **SH**36[209],46[209].

Sect. Erythrosperma Dahlst.

Dant y Llew Lleiaf : Lesser Dandelion. Native. Heaths, sandy places and waysides (D,E,F,G,H,M, Q,R,Rw,S,Sc). Fl.3-7. Mons., Glam., Brecs., Rads., Carms., Pembs., Cards., Monts., Mer., Caerns., Denbs., Flints., Angl.. 134 squares, map 775.

T. brachyglossum (Dahlst.) Dahlst. Native. Sandy places (D,G,Q,R,Rw,S). *Mons.*, Glam., *Brecs.*[205], Carms., *Pembs.*, Cards., Monts., Mer., Caerns., Denbs., *Flints., Angl.*. 34 squares, map 776.

T. argutum Dahlst. Native. Limestone grassland. *Denbs., Flints.*. 2 squares : **SJ**08[235],25[235].

T. lacistophyllum (Dahlst.) Raunk. Native. Sandy grassland (D,G,M,Q,R,Rw). Glam., *Brecs.*, Rads., *Carms., Pembs.*[205], *Monts.*, Mer., *Caerns.*, Denbs., Flints., *Angl.*. 23 squares : **SS**87,88; **ST***18*; **SN**61,97; **SO**11,14; **SH**36,37,48,53,67,87; **SJ**08,15,16,17,*22*,25, 27,34,36,37.

T. rubicundum (Dahlst.) Dahlst. Native. Dry sandy grassland (D,G,H,Q,S). Glam., *Brecs.*[205], Pembs., Monts., *Caerns.*, Denbs., *Flints., Angl.*. 21 squares : **SS**48,*49,58,79*; **ST***16,17,18*; **SN**10,90; **SH**36,*78*,87,88; **SJ**07,08,15,17,22,24,25,27.

T. silesiacum Dahlst. ex G. Hagl. Native. Dry calcareous grassland. Glam., *Brecs.*[205], *Monts., Caerns., Denbs., Flints., Angl.*[205]. 8 squares : **ST***18*[112]; **SH***78*[107], 87[107],88[107]; **SJ**08[245],22[245],25,[245],26[261].

T. commixtum G. Hagl. Native. Sandy heaths. *Flints.*. 1 square : **SJ**16[261].

T. arenastrum A. J. Richards Native. Grassy places. *Denbs., Angl.*. 3 squares : **SH**36[209],58[209]; **SJ**25[243].

T. disseminatum G. Hagl. Naturalized alien (Europe). Short dry grassland. *Glam.*[205].

T. glaucinum Dahlst. Denizen (Europe). Sand dunes. *Flints., Angl.*. 2 squares : **SH**36[209]; **SJ**08[245].

T. hispanicum H. Lindb. fil. Native. Dry rocky places. Brecs., Rads.. 2 squares : **SO**04[112],13[112].

T. laetum (Dahlst.) Dahlst. Native. Sandy places (D,G). *Glam.*[205], *Pembs.*[205], Cards., Mer., *Caerns., Angl.*. 4 squares : **SN**69[112]; **SH**36[107],52[112],*78*[107].

[**T. obscurans** (Dahlst. ex H. Lindb. fil.) G. Hagl. *Pembs.*]

T. dunense Van Soest Native. Sand dunes. *Glam.*[205], *Pembs.*[205], *Angl.*[205].

T. proximum (Dahlst.) Dahlst. Native. Limestone quarries. Monts., Mer., *Denbs.*[206], *Angl.*. 3 squares : **SH**36[209],53[112]; **SJ**22[112].

T. simile Raunk. Native. Rock crevices and walls (Q,Sc). Mons., *Rads., Carms.*[205], *Mer.*[205], *Flints.*. 3 squares : **ST***27*[112]; **SO**25[107]; **SJ**08[245].

T. degelii G. Hagl. Native. Rocky places. *Brecs., Rads.*[205], *Caerns.*[206], *Denbs.*[206]. 1 square : **SO**11[108].

T. pseudolacistophyllum Van Soest Native. Dry grassland and sand dunes (D,G,Q,Rg,Sc). Glam., *Brecs.*[206], *Rads.*, Carms., Cards., Denbs.. 6 squares : **SS**38[112]; **SN**24[112],61[112],69[71]; **SO**13[107]; **SH**87[112].

T. canulum G. Hagl. Naturalized alien (Europe). Limestone grassland. *Denbs.*. 1 square : **SJ**15[245].

T. proximiforme Van Soest Native. Scrub. *Flints.*. 1 square : **SJ**17[245].

T. fulvum Raunk. Native. Dry grassland (G,Q,R) Glam., Caerns., *Denbs., Flints.*. 7 squares : **SS***87*[112]; **ST***17*[112]; **SH***57*[112]; **SJ**08,[245],15[245],17[245],24[215].

T. fulviforme Dahlst. Native. Calcareous grassland (G,H,Q,R,S). Mons., Glam., Brecs., Rads., Pembs., *Monts.*, Mer., *Caerns.*, Denbs., *Flints., Angl.*. 22 squares : **SS**19,87,90; **ST**06,28; **SN***10*,69,70,97; **SO**11,14; **SH**36,52, 67,78,87,97; **SJ**15,17,22,25,27

T. oxoniense Dahlst. Native. Calcareous grassland and sand dunes (D,F,G,Q,R,S). *Mons.*, Glam., Brecs., *Rads.*, Carms., Pembs., Cards., Monts., Mer., Caerns., Denbs., Flints., *Angl.*. 49 squares, map 777.

T. glauciniforme Dahlst. Native. Calcareous grassland. Glam., Cards., *Monts., Denbs., Flints.*. 6 squares : **SS***87*[112]; **SN**58[112]; **SJ**08[245],15[245],21[107],22[245].

Sect. Taraxacum (Sect. *Vulgararia* Dahlst.)

Dant y Llew Cyffredin : Common Dandelion. Native. Grassy places (D,E,F,G,H,M,Q,R,Rg,Rw,Sc,W). Fl.2-8. Mons., Glam., Brecs., Rads., Carms., Pembs., Cards., Monts., Mer., Caerns., Denbs., *Flints., Angl.*. 220 squares, map 778.

T. obtusilobum Dahlst. ex G. Hagl. Casual (Europe). Railway bank. Glam.. 1 square : **ST**17[112].

T. cyanolepis Dahlst. Native. Roadsides. *Mons.*[205], Glam., *Pembs.*[205], *Caerns.*[205], Denbs.. 3 squares : **SS***69*[112]; **SH**76[182],97[112].

T. subcyanolepis M. P. Christiansen Native. Limestone grassland and sand dunes (D,E,G,R,Rg). Glam., Cards., *Mer.*[205], Denbs., *Flints., Angl.*. 7 squares : **SS**88[112]; **ST**07[112]; **SN**67[112]; **SH**36[209],97[112]; **SJ**24[245],26[245].

T. sellandii Dahlst. (*T. monochroum* auct., non G. Hagl.). Native. Grassy places (E,G,S). *Glam.*, Cards., Denbs., *Flints.*. 8 squares : **SN**34[112]; **SO**00[107]; **SH**88[112], 95[112],97[112]; **SJ**34[245],36[245].

T. ancistrolobum Dahlst. Native. Roadsides (E,G,R,W). Glam., Cards., *Denbs., Flints.*. 9 squares : **SS**97[112]; **ST**17[112]; **SN**14[112],54[112],65[112],66[112]; **SJ**26[245],27[261],35[245],36[245].

T. mericyclum A. J. Richards (incl. specimens formerly identified as *T. sublaciniosum* Dahlst. & H. Lindb. fil.). Colonist (Europe). Waste places. Glam., *Brecs.*[205], Cards., *Denbs., Flints.*. 7 squares : **SS***69*[112]; **SN**45[112]; **SJ**07[245], 24[245],26[261],35[245],44[245].

T. stenacrum Dahlst. Native. Roadsides. *Mons.*, Cards., *Denbs.*. 4 squares : **ST**38[91]; **SN**35[112]; **SJ**35[245],44[245].

T. procerisquameum H. Öllgaard (incl. specimens formerly identified as *T. procerum* G. Hagl.). Established alien (Europe). Roadsides. *Flints., Denbs.*. 3 squares : **SJ**16[261],35[245],36[245].

T. pannucium Dahlst. Native. Meadows (G,Rg). Glam., Cards.. 3 squares : **ST**17[112]; **SN**44[112],58[112].

[**T. linguatum** Dahlst. ex M. P. Christiansen & Wiinst. *Carms., Cards., Denbs., Flints.*]

T. pallescens Dahlst. Native. Pastures (D,G). Mons., *Glam.*[205], *Pembs.*[205], Cards.. 2 squares : **ST**28[112]; **SN**14[112].

T. alatum H. Lindb. fil. Native. Meadows (E,G,Q,R). *Glam.*[205], *Carms.*, Mer., *Caerns.*, Denbs., *Flints.*. 10 squares : **SN**40,72; **SH**23,53,76,88; **SJ**25,35,36,44.

T. lingulatum Marklund Native. Roadsides (E,G,R). *Mons.*, Glam., Cards., *Denbs., Flints.*. 10 squares : **ST**18,38; **SN**24,64; **SO**51; **SJ**24,25,36,44,45.

T. exsertum Hagendijk, Van Soest & Zevenbergen Native. Grassy places. *Mons.*, Mer., *Denbs.*, *Flints.*. 8 squares : **ST**38[91]; **SH**63[112]; **SJ**18[245],24[245],25[245],34[245],35[245], 36[245].

T. croceiflorum Dahlst. Native. Roadsides and grassy places (E,D,G,R,W). Glam., *Brecs.*[206], Cards., *Mer.*, *Caerns.*, Denbs., *Flints.*. 9 squares : **ST**18[112]; **SN**54[112], 57[112],64[112]; **SH**50[93],75[182],77[112]; **SJ**25[245],35[245].

T. undulatiflorum M. P. Christiansen Native. Roadside verges. Cards., *Denbs.*, *Flints.*. 3 squares : **SN**54[112]; **SJ**26[245],35[245].

T. macrolobum Dahlst. (incl. specimens formerly identified as *T. lacerabile* Dahlst.). Colonist (Europe). Colliery wastes. *Denbs.*. 1 square : **SJ**35[245].

T. expallidiforme Dahlst. Native. Roadsides and waste places (E,G,R). Glam., Caerns., Denbs., Flints.. 9 squares : **ST**17[112],18[112]; **SH**77[112],86[112]; **SJ**07[245],16[245], 24[245],26[245],44[245].

T. insigne E. L. Ekman ex Winnst. & K. Jessen Native. Dry grassland. Glam., *Pembs.*[205], Mer., *Denbs.*, *Flints.*[205], *Angl.*. 7 squares : **SS**87[112]; **SH**46[209],52[112]; **SJ**15[245],24[245],25[245],44[245],45[245].

T. laeticolor Dahlst. Native. Grassy roadsides. *Brecs.*[206], Cards., *Flints.*. 2 squares : **SN**25[112]; **SJ**27[245].

T. tenebricans (Dahlst.) Dahlst. Native. Roadsides and grassy places (D,E,G,R). Glam., Cards., Caerns., Denbs.. 5 squares : **ST**18[112]; **SN**34[112]; **SH**77[112],97[112]; **SJ**44[245].

T. valdedentatum Dahlst. Native. Waste places. *Monts.*, *Denbs.*. 2 squares : **SJ**21[107],44[245].

T. pannulatiforme Dahlst. Colonist (Europe). Grassy and waste places. *Denbs.*, *Flints.*. 2 squares : **SJ**25[245], 35[245].

[**T. laciniosum** Dahlst. *Mons.*, *Glam.*]

T. cherwellense A. J. Richards Native. Grassy places (E,G,R). Glam.. 1 square : **SS**97[112].

T. aequilobum Dahlst. Native. Waste places. *Glam.*[205], *Mer.*[205], *Flints.*. 1 square : **SJ**27[261].

T. ekmanii Dahlst. Native. Grassy places (E,G,M, Q,R). Glam., Brecs., Cards., *Caerns.*, *Denbs.*, *Flints.*. 10 squares : **ST**07,*17*; **SN**67,68; **SO**11; **SH**87; **SJ**16,34, 35,36.

T. porrectidens Dahlst. Native. Waste places and roadsides (E,G,M). Glam., Cards., *Caerns.*, *Flints.*. 6 squares : **ST**17[112]; **SN**78[112]; **SH**76[182]; **SJ**16[261],18[245],26[245].

T. latissimum Palmgren Colonist (Europe). Grassy places. *Denbs.*. 2 squares : **SJ**35[245],44[245].

T. copidophyllum Dahlst. Established alien (N. Europe). Churchyards. Cards.. 1 square : **SN**65[112].

T. pulchrifolium Marklund Colonist (Europe). Churchyard. Cards.. 1 square : **SN**65[112].

T. aurosulum H. Lindb. fil. Native. Roadsides. Glam., Cards., *Mer.*[205], *Denbs.*, *Flints.*. 5 squares : **ST**18[112]; **SN**58[112]; **SJ**26[245],27[245],35[245].

T. xanthostigma H. Lindb. fil. Native. Grassy places (E,G,R). *Mons.*, Cards., *Denbs.*, *Flints.*. 8 squares : **SN**24[112],34[112],54[112]; **SO**51[197]; **SJ**07[245],26[245],34[245],35[245].

[**T. dilatatum** H. Lindb. fil. *Mons.*, *Denbs.*]

T. cordatum Palmgren (incl. *T. adsimile* Dahlst.). Native. Grassy places (D,E,G,R,S). Glam., *Rads.*, *Pembs.*, Cards., *Mer.*[205], Caerns., Denbs., *Flints.*, [*Angl.*]. 20 squares : **SS**87,88; **ST**18,*27*; **SN***10*,34,45, 58,64,65,*78*; **SO**14; **SH**88,95; **SJ**16,18,25,26,34,37.

T. sagittipotens Dahlst. & R. Ohlsén ex G. Hagl. Established alien (Europe). Grassy places. Glam., Cards.. 2 squares : **ST**09[112]; **SN**58[112].

T. hemicyclum G. Hagl. Native. Laneside bank. Cards.. 1 square : **SN**68[112].

T. longisquameum H. Lindb. fil. Colonist (Europe). Grassy places (E,G,Q,R). Mons., Glam., Pembs., Cards., Monts., *Mer.*[205], *Caerns.*, Denbs., *Flints.*, *Angl.*. 12 squares : **ST**27,*49*; **SM**70; **SN**44,55,67; **SH**46,57,97; **SJ**18,37,44.

[**T. spiculatum** M. P. Christiansen (*T. spilophyllum* Dahlst.). *Caerns.*]

T. dahlstedtii H. Lindb. fil. Native. Grassy places and walls (E,G,M,Q). Mons., *Pembs.*[205], Cards., Mer.[206], *Caerns.*, *Denbs.*, *Flints.*. 8 squares : **ST**28[112]; **SN**67[112]; **SH**12[107]; **SJ**16[261],17[245],18[245],35[245],36[245].

T. reflexilobum H. Lindb. fil. Denizen (Europe). Grassy places. *Mer.*[206], *Caerns.*. 1 square : **SH**88[183].

T. ostenfeldii Dahlst. (*T. biforme* auct., non Dahlst.; *T. duplidens* H. Lindb. fil.). Colonist (N. Europe). Waste places (E,G,R). Glam., *Pembs.*[205]. 1 square : **SS**97[112].

T. melanthoides Dahlst. Native. Meadows. Denbs.. 1 square : **SJ**45[245].

T. bracteatum Dahlst. Native. Grassy places (E,G,M, Q,R,Rw,W). *Glam.*[205], Cards., *Denbs.*, *Flints.*[206]. 9 squares : **SN**57[112],58[112],65[112],67[112],68[112],78[112]; **SJ**34[245], 35[245],44[245].

T. hamatum Raunk. Native. Grassy places (E,G,H,M, Q,R,Rg,W). Mons., Glam., Pembs., Cards., Mer., *Caerns.*[205], Denbs., *Flints.*, *Angl.*. 36 squares, map 779.

T. pseudohamatum Dahlst. Native. Grassy places (D,E, G,M,Q,R). Mons., Glam., Cards., Mer., Denbs., *Flints.*. 28 squares, map 780.

T. hamatiforme Dahlst. Native. Grassy places (E,G,Q,R,Rg,W). Mons.[206], Glam., *Carms.*[205], *Pembs.*[205], Cards., Mer., Caerns., Denbs., *Flints.*. 27 squares, map 781.

T. polyhamatum H. Öllgaard Naturalized alien or casual (Europe). Roadside banks. Denbs., *Flints.*. 3 squares : **SH**87[112]; **SJ**25[245],36[245].

T. atactum Sahlin & Van Soest Native. Grassy places (E,G,H,R). Cards., Mer.. 4 squares : **SN**34[112],35[112],56[112]; **SH**93[112].

T. hamiferum Dahlst. Native. Grassy places (E,G,R). *Glam.*[206], Cards., *Denbs., Flints.*. 4 squares : **SN**46[112], 57[112]; **SJ**18[245],34[245].

T. atrovirens Dahlst. Native. Marshy roadside verges and pasture. *Mons.*, Cards., *Denbs., Flints.*. 6 squares : **SN**67[112],68[112],78[112]; **SO**51[205]; **SJ**35[245],36[245].

T. subhamatum M. P. Christiansen (including specimens formerly identified as *T. marklundii* Palmgren). Native. Grassy places (E,G,M,R). *Mons.*, Glam., Carms., *Pembs.*[205], Cards., Mer., Caerns., Denbs., *Flints., Angl.*[205]. 30 squares, map 782.

T. hamatulum Hagendijk, Van Soest & Zevenbergen Native. Grassy places (G,M,Q). *Mons.*, Cards., Denbs., *Flints.*. 9 squares : **ST**49[91]; **SN**34[112],54[112],56[112],64[112], 67[112]; **SH**95[112]; **SJ**18[245],25[245].

T. kernianum Van Soest Established alien (Europe). Waste tips. Denbs., *Flints.*. 7 squares : **SH**87[112]; **SJ**24[245],25[245],26[245],35[245],36[245],44[245].

T. boekmanii Borgvall Native. Grassy places (E,G,R) Cards., Denbs., *Flints.*. 7 squares : **SN**44[112],58[112]; **SH**87[112]; **SJ**17[112],25[245],35[245],36[245].

T. oblongatum Dahlst. Native. Grassland (E,G,M,W). *Glam.*[205], *Carms.*[205], Pembs., Cards., *Denbs., Flints.*. 12 squares : **SN**23,25,46,56,64,68; **SJ**16,24,27,34,37,44.

T. quadrans H. Öllgaard (*T. latisectum* auct., non H. Lindb. fil.). Native. Grassy places (E,G,R,Rg,W). Glam., Pembs., Cards., Denbs., Flints.. 11 squares : **SS**97; **ST**17; **SN**03,46,54,66,67,68; **SH**87,88; **SJ**16.

T. fasciatum Dahlst. Native. Roadside verges. Cards., *Denbs., Flints.*. 6 squares : **SN**57[112]; **SJ**16[261],25[245], 26[245],36[245],44[245].

T. lamprophyllum M. P. Christiansen (*T. maculatum* Jordan). Native. Roadsides and grassy places (E,G,R,W). Glam., *Pembs.*[205], Cards., *Caerns., Denbs., Flints., Angl.*. 17 squares : **ST**17,27; **SN**25,34,46,56,57,64,68;**SO**00; **SH**57,75; **SJ**16,25,26,35,44.

[**T. parvuliceps** H. Lindb. fil. *Glam., Mer.*]

T. excellens Dahlst. Native. Roadside banks. Cards.. 1 square : **SN**66[112].

T. polyodon Dahlst. Native. Grassy places. *Glam.*[205], *Pembs.*[206], Cards., *Caerns., Denbs., Flints.*. 8 squares : **SN**68[112]; **SH**64[182]; **SJ**16[261],24[245],25[245],26[245],34[245],36[245].

T. dilaceratum M. P. Christiansen Colonist (Europe). Grassy places. *Denbs.*. 1 square : **SJ**35[245].

T. ordinatum Hagendijk, Van Soest & Zevenbergen Colonist (Europe). Grassy banks. Cards., *Denbs.*. 2 squares : **SN**34[112] (provisional determination); **SJ**35[245].

T. hemipolyodon Dahlst. Native. Waste places. *Mons.*[206].

T. raunkiaeri Wiinst. (*T. duplidentifrons* auct., non Dahlst.). Native. Grassy places (D,E,G,Q,R,Rg). Glam., Carms., Cards., Mer., *Caerns., Denbs., Flints.*. 18 squares : **SS**97; **ST**17; **SN**14,24,30,44,54,56,58,68; **SH**53,63,67,76; **SJ**16,24,35,44.

T. brachylepis Marklund Colonist (Europe). Grassy places. *Denbs.*. 1 square : **SJ**16[245].

T. laciniosifrons Wiinst. (incl. specimens previously identified as *T. sinuatum* Dahlst.) Colonist (Europe). Roadsides. Cards., *Denbs.*. 2 squares : **SN**35[112]; **SJ**35[90].

T. piceatum Dahlst. Established alien (N. Europe). Grassy places. Cards., *Denbs., Flints.*. 5 squares : **SN**58[112]; **SJ**34[245],35[245],36[245],44[245].

T. crispifolium H. Lindb. fil. Native. Grassy places (E,G). *Carms.*[205], *Pembs.*[205], Cards., *Caerns.*, Denbs., *Flints.*. 6 squares : **SN**58[112]; **SH**57[107],76[182]; **SJ**15[112],25[245], 44[245].

T. tarachodum Hagendijk, Van Soest & Zevenbergen Colonist (Europe). Roadside banks. Denbs.. 1 square : **SJ**06[112].

T. trilobatum Palmgren Colonist (Europe). Grassy places. *Denbs.*. 1 square : **SJ**24[245].

T. privum Dahlst. Colonist (Europe). Grassy places (G,Q,Rg). Mons., Glam., Cards., *Denbs., Flints.*. 6 squares : **ST***17*[112],*28*[112]; **SN**65[112]; **SJ**16[261],25[245],34[245].

T. cophocentrum Dahlst. Native. Grassy places. Cards., Mer., Denbs., *Flints.*. 8 squares : **SN**34[112], 35[112],57[112]; **SH**53[112],87[112]; **SJ**16[245],24[245],35[245],44[245].

T. praeradians Dahlst. Colonist (Europe). Grassy places. Cards.. 1 square : **SN**58[112].

LAPSANA L.

Lapsana communis L.

Subsp. **communis** : Cartheig : Nipplewort. Native. Hedges (E,G,Q,R,Rg,Rw,S,Sc,W). Fl.(1)6 9. Mons., Glam., Brecs., Rads., Carms., Pembs., Cards., Monts., Mer., Caerns., Denbs., Flints., Angl.. 265 squares, all except : **SR**89; **SS**38,39,77; **ST**26,39; **SM**50,62,70,71,73,84; **SN**78,86,87; **SH**29,51.

Subsp. **intermedia** (Bieb.) Hayek (*L. intermedia* Bieb.) : Large Nipplewort. Naturalized alien (S. Europe). Limestone pastures. *Caerns., Flints.*. 2 squares.
Caerns.: Gt Ormes Head, SH78, RJP, 1977[184].
Flints.: nr Cilcain, SJ16, VG, c1970[261] and JMB, c1975[67].

CREPIS L.

Crepis paludosa (L.) Moench : Gwalchlys y Gors : Marsh Hawk's-beard. Native. Streamsides (H,M,Q,W). Fl.6-9. Glam., Brecs., *Rads., Carms.* *†, Monts., Mer., Caerns., Denbs., *Flints.*, [*Cards.*]. 66 squares, map 783.

C. mollis (Jacq.) Ascherson : Northern Hawk's-beard. Native. Mountain streamsides. Fl.8. (*Caerns.*†), *Denbs.*†. 2 squares.
(**Caerns.**: Bryncir Station, SH44, JLW, c1890[101]).
Denbs.: Llandegla, SJ15, WH, 1909[144].

C. biennis L. : Gwalchlys Garwaidd : Rough Hawk's-beard. Native. Roadsides (E,G,R). Fl.5-7. Mons., Glam., Carms.*, Pembs.*, *Caerns.*, Denbs.. 29 squares, map 784.

C. capillaris (L.) Wallr. : Gwalchlys Llyfn : Smooth Hawk's-beard. Native. Roadsides (D,E,F,G,Q,R,Rw,S). Fl.5-10. Mons., Glam., Brecs., Rads., Carms., Pembs., Cards., Monts., Mer., Caerns., Denbs., Flints., Angl.. 270 squares, all except : **SS**38,39; **ST**29; **SM**50,73; **SN**32,34,52,63,74; **SO**30; **SH**51.

C. vesicaria subsp. **haenseleri** (Boiss. ex DC.) P. D. Sell (*C. taraxacifolia* Thuill.; *C. vesicaria* subsp. *taraxacifolia* (Thuill.) Thell.) : Gwalchlys Gylfinhir : Beaked Hawk's-beard. Colonist (Europe). Roadsides (E,F,G, M,Q,R,W). Fl.5-7. Mons., Glam., Brecs., Rads., Carms., Pembs., Cards., Monts., *Mer.*, Caerns., Denbs., Flints., *Angl.*. 104 squares, map 785.

C. setosa Haller fil. : Bristly Hawk's-beard. Casual (S. Europe). Waste places. Fl.8. Glam., Denbs.. 3 squares : **ST**27[112]; **SN**80[157]; **SH**87[112].

ANDRYALA L.

Andryala integrifolia L. (incl. *A. tenuifolia* Tineo). Casual (S. Europe). Waste places. Fl.7-9. Glam.. 3 squares : **ST**16[29], 17[112],*18*[112].

PILOSELLA Hill

Pilosella peleterana (Mérat) F. W. Schultz & Schultz Bip. (*Hieracium peleteranum* Mérat). Native. Limestone scree. Fl.6-7. Monts., *Mer.*. 2 squares.
Monts.: Craig Breiddin, SJ21, WAL, 1834[108] and ARP, 1969[197]. **Mer.**: Borthwen, SH61, Ja, c1850[212]† (doubt has been expressed about the accuracy of this record).

P. officinarum F. W. Schultz & Schultz Bip. (*Hieracium pilosella* L.) : Clust y Lygoden : Mouse-ear Hawkweed. Native. Banks and pastures (D,E,G,H, M,Q,R,Rw). Fl.5-10. Mons., Glam., Brecs., Rads., Carms., Pembs., Cards., Monts., Mer., Caerns., Denbs., Flints., Angl.. 275 squares, all except : **SS**39; **SM**50,62,71,73; **SH**29,51.
Many of the determinations of the following subspecies are provisional only.

Subsp. **micradenia** (Naegeli & Peter) P. D. Sell & C. West (*P. officinarum* subsp. *concinnata* (F. J. Hanb.) P. D. Sell & C. West; *Hieracium pilosella* subsp. *micradenium* Naegeli & Peter; *H. pilosella* var. *concinnatum* F. J. Hanb.). Native. Grassy places (D,E,G,Q,R). Fl.5-10. Mons., Glam., Brecs., Rads., Pembs., Cards., Monts., *Mer., Caerns.*, Denbs.. 45 squares, map 786.

Subsp. **euronota** (Naegeli & Peter) P. D. Sell & C. West (*Hieracium pilosella* subsp. *euronotum* Naegeli & Peter). Native. Grassy places (D,E,G,M,R,Rw). Fl.5-9. Mons., Glam., Brecs., Pembs., Cards., Monts., Flints.. 20 squares : **SS**58,77,98,99; **ST**17, *18*,59; **SM**72; **SN**45,54,56,66,67,68,69,90; **SO***05*,20; **SH**90; **SJ**07.

Subsp. **officinarum** (*Hieracium pilosella* subsp. *pilosella*). Native. Grassy places (E,G,H,Q,Rw). Fl.5-9. Mons., Glam., *Brecs.*, Rads., *Pembs.*, Cards., *Monts.*, Mer., *Caerns.*, Denbs., *Flints.*. 36 squares, map 787. Under-recorded.

Subsp. **trichosoma** (Peter) P. D. Sell & C. West (*Hieracium pilosella* subsp. *trichosoma* Peter; *P. officinarum* subsp. *nigrescens* (Fries) P. D. Sell & C. West; *H. pilosella* var. *nigrescens* Fries). Native. Grassy places (D,E,G,M,Q,R,Rw,S,Sc). Fl.5-10. Mons., Glam., Brecs., Rads., Carms., Pembs., Cards., Monts., *Mer.*, Caerns., *Denbs.*, Flints.. 49 squares, map 788.

Subsp. **tricholepia** (Naegeli & Peter) P. D. Sell & C. West (*Hieracium pilosella* subsp. *tricholepium* Naegeli & Peter). Native. Grassy places (E,G,Q,R). Fl.6-7. Mons., Brecs., Cards., *Caerns.*. 8 squares. **Mons.**: Rumney, ST27, AEW, 1922[112]; Llandegfedd, ST39, AEW, 1943[112]; Kilgwrrwg, ST49, WAS, 1918[112]. **Brecs.**: Ystradfellte, SN91, HAH, 1925[112]. **Cards.**: Bryncarnedd, SN58, AOC, 1976[112]; Llangeitho, SN66, AOC, 1977[112]; Eglwys Newydd, SN77, AOC, 1977[112]. **Caerns.**: nr Pwllheli, SH43, PDS, 1952[108].

Subsp. **melanops** (Peter) P. D. Sell & C. West (*Hieracium pilosella* subsp. *melanops* Peter). Native. Grassy places (D,E,G,R,Rw). Fl.5-7. Mons., Glam., Brecs., Cards., Mer., Flints., Angl.. 11 squares : **SS**78; **ST**08,17,18,*28*; **SN**14,68; **SO***05*; **SH***57,61*; **SJ**37.

Subsp. **trichoscapa** (Naegeli & Peter) P. D. Sell & C. West (*Hieracium pilosella* subsp *trichoscapum* Naegeli & Peter). Native. Rocky river banks. Fl.6. *Brecs.*. 1 square.
Brecs.: Llangoed, SO14, AL, 1907[108].

P. praealta (Vill. ex Gochnat) F. W. Schultz & Schultz Bip. (*Hieracium praealtum* Vill. ex Gochnat). Established alien (Europe). Wall top. *Rads.*. 1 square.
Rads.: Knucklas, SO27, JRP, 1967[179].

P. caespitosa (Dumort.) P. D. Sell & C. West subsp. **colliniformis** (Peter) P. D. Sell & C. West (*Hieracium colliniforme* (Naegeli & Peter) Roffey; *H. caespitosum* subsp. *colliniforme* (Peter) P. D. Sell). Established alien (Europe). Waste places. Glam.. 1 square.
Glam.: Cardiff, ST17, EV, 1905[112].

P. aurantiaca (L.) F. W. Schultz & Schultz Bip. : Fox-and-cubs.

Subsp. **aurantiaca** (*Hieracium aurantiacum* L.). Casual (C. Europe). Waste places. Glam. 2 squares : **SS***98*[112]; **ST***18*[112].

Subsp. **carpathicola** (Naegeli & Peter) Sojók (*P. aurantiaca* subsp. *brunneocrocea* (Pugsley) P. D. Sell & C. West; *Hieracium brunneocroceum* Pugsley; *H. aurantiacum* subsp. *carpathicola* Naegeli & Peter). Established alien (Europe). Railway banks (E,G,Q, R,Rw). Fl.6-10. Mons., Glam., Brecs., Rads., Carms., Pembs., Cards., Monts., Mer., Caerns., Denbs., Flints., *Angl.**. 85 squares, map 789.

HIERACIUM L.

Most of the following records have been verified by : Mr. C. E. A. Andrews, Mr. B. A. Miles, Prof. J. N. Mills, Mr. P. D. Sell or Dr. C. West. Records in **NMW** and Pugsley's *Prodromus* not verified by one of the above are enclosed in double brackets — ((Mons.)), ((**ST**49)), and indicated by ⊙ on the maps.

Grateful thanks are due to Mr P. D. Sell and Dr. C. West for allowing the records accumulated for the *Critical Supplement to the Atlas of the British Flora* to be consulted and used, and to Mr. P. D. Sell for allowing the author to consult the extensive collections of **Hieracium** in **CGE.**

Hieracium murorum agg. : Heboglys : Hawkweed. Native (A,D,E,F,G,H,M,Q,R,Rw,W). Fl.5-10. Mons., Glam., Brecs., Rads., Carms., Pembs., Cards., Monts., Mer., Caerns., Denbs., Flints., Angl.. 237 squares, map 790.

Sect. Vulgata F. N. Williams

H. exotericum agg. Introduced (Europe). Rocky and waste places (E,Q,R,Rw). Fl.5-10. Mons., Glam., Brecs., *Rads.*, Monts., *Caerns.*, Denbs.. 25 squares, map 791.

H. exotericum Jordan ex Boreau Native. Rocky riversides. Fl.7. *Brecs.*. 1 square.
Brecs.: Abercrave, SN81, AL, 1908[108].

H. cardiophyllum Jordan ex Sudre Naturalized alien (S. Europe). Riverside. Fl.9. *Brecs.*. 1 square.
Brecs.: Llangattock, SO21, RL, 1951[108].

H. grandidens Dahlst. Established alien (C. Europe). Roadsides and waste places (E,Q,R,W). Fl.5-7. Mons., ((Glam., Carms.)), Cards., ((Angl.)). 10 squares : **ST**((*18,28,29,49*)), *59*; **SN**((31)),*77*; **SO**((*50*)),51;((**SH***67*)).

H. sublepistoides (Zahn) Druce Naturalized alien (Alps and Pyrenees). Roadsides and railway banks (E,Q, R,Rw,W). Fl.5-9. Mons., Glam., Brecs., Rads., *Monts., Mer.*, ((*Caerns.*)), Denbs.. 16 squares : **ST***18*,((*28,38*)),59; **SN**70; **SO***04*,13,20,((21,31)),50,51; **SH**71,((76)),87; **SJ**10.

H. cinderella (A. Ley) A. Ley Native or colonist. Roadsides and railway banks. Fl.6-8. Glam., Brecs., *Rads., Monts., Caerns.*[119], *Denbs.*. 11 squares : **ST***18*; **SN**80,*81*,90,*91*,96; **SO***01*,02; **SH**77; **SJ**12,*21*.

H. pellucidum Laest. Native. Limestone rocks. Fl.5-8. Mons., *Glam.*, Brecs., Carms., ((*Rads.*)). 14 squares : **ST***59*; **SN***72,80,81*,82,90,92; **SO**01,11,21,*22*,23; ((**SO***00, 50*)).

H. snowdoniense P. D. Sell & C. West Native (endemic). Mountain rocks. Fl.6-7. Caerns.. 2 squares.
Caerns.: Cwm Glas Fach, SH65, AL, 1904[108]; Cwm Idwal, SH65, PDS, 1953[108]; Nant Ffrancon, SH66, JEG, 1894[108]; Ysgolion Duon, SH66, JEG, 1892[112] and AL, 1904[112]; Cwm Perfedd, SH66, JEG, 1893[112].

H. stenstroemii (Dahlst.) K. Joh. (*H. glevense* P. D. Sell & C. West). Native. Rocky places (E,Q,R,Rw). Fl.5-8. *Mons., Glam.*, Brecs., Rads.. 15 squares : **ST***18*,49,59; **SO***00*,01,*02*,04,*11,12,13*,14,*21*,23,*50*,51.

H. discophyllum P. D. Sell & C. West Native. Base-rich rocks. Fl.7-8. *Brecs., Carms.*. 2 squares.
Brecs.: Pwll Byfre, SN81, AL, 1906[108] and BAM, 1962[108]. **Carms.**: Llyn y Fan Fach, SN82, AL, 1905[108].

H. asteridiophyllum P. D. Sell & C. West Native (endemic). Limestone rocks. Fl.6-7. Brecs., [*Mons.*]. 1 square.
Brecs.: Craig y Cilau, SO11, AL, 1900[112] and PDS, 1975[200].

H. scotostictum Hyl. Established alien (Europe). Walls and waste places. Fl.6. Mer., Caerns.. 2 squares.
Mer.: nr Dolgellau, SH61, MR, 1956[7] and PMB, 1963[7]. **Caerns.**: Bethesda, SH66, WEH, 1967[112].

H. sanguineum (A. Ley) W. R. Linton Native. Limestone rocks. Fl.5-7. Brecs., Denbs.. 3 squares.
Brecs.: Brecon Beacons, SO01, AL, 1886[106] & 1899[106]; Darren Fach, SO01, AEW, 1927[112]; Cwar yr Hendre, SO01, PDS, 1953[108] & 1975[108]; Dyffryn Crawnon, SO11, AL, 1892[106] and PDS, 1975[200]. **Denbs.**: Bryn Euryn, SH87, WEH, 1967[112].

H. pachyphylloides (Zahn) Roffey Native. Railway bank. Fl.5. Mons.. 1 square.
Mons.: nr Symonds Yat, SO51, AL, 1894[112].

H. neocoracinum Pugsley Native (endemic). Old Red Sandstone rocks. Fl.6-7. Brecs.. 1 square.
Brecs.: Craig Cerrig-gleisiad, SN92, AL, 1883[106] and BAM, 1958[108].

H. oistophyllum Pugsley Native. Rocky places. Fl.6. Denbs., Flints.. 3 squares.
Denbs.: nr Ruthin, SJ15, JMB, 1976[67]; Minera, SJ25, TE, 1972[112]. **Flints.**: Alun Gorge, SJ16, TE, 1972[112].

[**H. surrejanum** F. J. Hanb. *Mons., Caerns.*]

H. lepidulum (Stenström) Omang Established alien (Europe). Roadsides. Fl.8. *Glam..* 1 square.
Glam.: Porthkerry, ST06, FRB, 1958[7].

H. vulgatum Fries Native. Rocky places (Q,R,Rw). Fl.6-10. Mons., Glam., ((Brecs.)), Rads., *Cards.**, Monts., Mer., Caerns., Denbs., Flints., Angl.. 48 squares, map 792.

H. triviale Norrlin Native. Rocky places (E,Q,R,W). Fl.5-8. Caerns., *Denbs..* 6 squares.
Caerns.: Pentir, SH56, JMB, 1969[112]; nr Bangor, SH57, JTCWi, 1876[112] and AL, 1904[112]; Penrhyn Quarries, SH66, EFL, 1890[108]; Bethesda, SH66, AL, 1891[108] and CWe, 1953[108]; Nant Ffrancon, SH66, PDS, 1952[108]; Llanfairfechan, SH67, CWe, 1953[108]; nr Llandudno, SH87, KG, 1953[108] and JMB, 1969[112]. **Denbs.**: Nant y Glyn, SJ06, CBr, 1925[108].

H. maculatum Sm. Established alien (C. Europe). Waste places (Q,R). Fl.6. *Mons., Mer., Caerns..* 4 squares.
Mons.: Newport, ST38, TGE, 1979[91]. **Mer.**: nr Llwyngwril, SH61, PMB, 1962[108]. **Caerns.**: Bangor, SH57, PDS, 1952[108]; Cwm Idwal, SH65, JEG, 1887[107].

[**H. caesiomurorum** Lindeb. *Brecs.*]

H. orcadense W. R. Linton (*H. euprepes* F. J. Hanb., non Peter). Native. Rocky places. Fl.6-8. *Mons.,* Brecs., Carms., Caerns.. 11 squares : **SN***81*,82,*91*,92; **SO**01,*11*, 22,23; **SH***65*,*66*,76.

H. rubiginosum F. J. Hanb. Native. Rocky places. Fl.6-8. Brecs., Carms., *Caerns., Denbs..* 7 squares.
Brecs.: Craig Rhiwarth, SN81, AL, 1904[108] & 1908[112]; Dan yr Ogof, SN81, AL, 1906[112]; Pwll Byfre, SN81, AL, 1906[112] and CWe, 1959[108]; Fan Nedd, SN91, AL, 1908[106]; Cwm Tarell, SN92, AL, 1888 & 1900[106]; Craig Cerrig-gleisiad, SN92, AL, 1895[106] and MP, 1976[200]; Dyffryn Crawnon, SO11, AL, c1900[106] and MP, 1975[200]; **Carms.**: Black Mountains, SN82, AL, 1899[106] & 1905[106]. **Caerns.**; Craig Dulyn, SH76, AL, 1904[106]and PDS, 1953[108]. **Denbs.**: Eglwyseg Rocks, SJ24, AL, 1903[106].

[**H. cravoniense** (F. J. Hanb.) Roffey *Flints.*]

H. cuneifrons (W. R. Linton) Pugsley Native (endemic). Rocky places. Fl.6-7. *Mons.,* Glam., Brecs., Carms.. 7 squares.
Mons.: Tarren yr Esgob, SO23, AL, 1898[93]. **Glam.**: Llanmadog, SS49, RAG, 1955[108]; Llwydcoed, SN90, HJR, 1901[112]. **Brecs.**: Craig y Cilau, SO11, AL, 1893[106] & 1909[108]; nr Crickhowell, SO21, AL, 1897[7]; Tarren yr Esgob, SO23, AL, 1901[108] and PDS, 1953[108].
Carms.: Black Mountains, SN82, AL, c1900[108].

H. radyrense (Pugsley) P. D. Sell & C. West Native (endemic). Limestone quarry and roadsides. Fl.6-7. Glam.. 3 squares.
Glam.: Bridgend, SS97, HJR, 1916[108]; Radyr Quarry, ST17, HJR, 1910[112] and CWe, 1955[108]; Radyr, ST18, BAM, 1955[108].

H. submutabile (Zahn) Pugsley Native or colonist. Rocky places and roadsides (E,M,Q,R,Rw). Fl.6-8. *Mons., Glam.,* Brecs., Rads., Carms., Cards., *Monts., Mer.,* Caerns., Denbs.. 30 squares, map 793.

H. rectulum A. Ley Native (endemic). Rocky places (E,Q,R,W). Fl.6-7. Carms., Cards.. 8 squares.
Carms.: Clogau Mawr, SN71, AL, 1908[112] and PDS, 1975[225]. **Cards.**: Llandygwydd, SN24, AOC, 1977[112]; Mynachdy Dingle, SN46, AOC, 1978[112]; Temple Bar, SN55, AOC, 1978[107]; nr Cilcennin, SN56, AOC, 1977[112]; Betws Leucu, SN65, AOC, 1978[112]; Llangeitho, SN65, AOC, 1978[112]; nr Llanddewibrefi, SN65, AOC, 1978[112]; nr Llangybi, SN65, AOC, 1979[112]; Allt Pantygourych, SN66, AOC, 1977[112]; nr Llyn Brianne, SN75, AOC, 1979[112].

H. pulchrius (A. Ley) W. R. Linton Native (endemic). Cliff ledges. Fl.7-8. Brecs., Carms.. 5 squares.
Brecs.: Fan Nedd, SN91, AL, 1906[112]; Craig Ddu, SN92, AL, 1899[108] and BAM, 1964[108]; Craig Cerrig-gleisiad, SN92, AL, 1890[106] and BAM, 1958[108]; Brecon Beacons, SO02, AL, 1899[108]; Craig y Cilau, SO11, AL, c1890[106]; Dyffryn Crawnon, SO11, CWe, 1952[108].
Carms.: Black Mountains, SN82, AL, 1905[106].

H. acuminatum Jordan (incl. *H. lachenalii* auct., *H. strumosum* auct.). Native. Rocky places (D,E,M, Q,R,Rw,W). Fl.6-9. Mons., Glam., Brecs., Rads., Carms., Cards., Monts., Mer., Caerns., Denbs., Flints.. 131 squares, map 794.

H. cheriense Jordan ex Boreau Colonist (C. Europe). Roadsides. Fl.7. *Mons., Glam., Brecs..* 3 squares.
Mons.: Newport Docks, ST38, TGE, 1980[91]. **Glam.**: Radyr, ST18, CWe, 1958[108]. **Brecs.**: Brecon, SO02, AL, 1906[108].

H. subamplifolium (Zahn) Roffey Native. Shady rocky places. Fl.5-6. Mons., Glam., Brecs., *Rads., Carms., Mer..* 16 squares : **ST**06,49,59; **SN**61,*70*, 81,*84*,*90*,92; **SO**00,*13*,*20*,*21*,*23*,*50*; **SH**71.

H. diaphanum Fries Native. Rocky and waste places (D,G,M,Q,R,Rw). Fl.5-8. Mons., Glam., Brecs., Rads., Carms., Cards., *Monts.*, Mer., Caerns., Denbs., Flints., *Angl.*. 85 squares, map 795.

H. subminutidens (Zahn) Pugsley Native (endemic). Riverside rocks. Fl.7. Brecs., [*Angl.*]. 2 squares.
Brecs.: nr Llanwrtyd Wells, SN84, AL, 1890[106] and PDS, 1975[200]; above Abergwesyn, SN85, AL, 1897[106] & 1902[106].

H. diaphanoides Lindeb. Native. Rocky waste places (E,Q,R,Rw). Fl.5-9. Mons., Glam., Rads., Carms., *Cards.*, *Mer.*, Caerns., Denbs., Flints.. 24 squares, map 796.

Sect. Oreadea Zahn

H. schmidtii Tausch Native. Rocky places. Fl.6-7. *Mons.*, Glam., Brecs., *Rads.*, *Carms.*, Caerns., Denbs.. 12 squares : **ST***59*; **SN**61,81,*90*; **SO**04,21; **SH***64*,65, *66*,*77*,*78*,87.

H. lasiophyllum Koch Native. Basic rocks. Fl.6-8. ((*Mons.*)), Brecs., Rads., Cards., Monts., Mer., Caerns., *Denbs.*, *Flints.*. 23 squares, map 797.

H. leyi F. J. Hanb. Native. Rocky places. Fl.6-8. *Glam.*, *Brecs.*, *Monts.*, Mer., Caerns., Denbs.. 13 squares : **SN***90*,*92*; **SH***64*,65,66,67,71,((75)),76,87,95; **SJ***02*,((21)).

H. cyathis (A. Ley) W. R. Linton Native. Base-rich rocks. Fl.6-8. Brecs., *Carms.*. 4 squares.
Brecs.: Craig y Nos, SN81, AL, 1904[106] & 1908[106]; Darren Fach, SO01, AL, 1894[108] and BAM, 1958[108]; Craig y Cilau, SO11, AL, 1893[106] and MP, 1978[200]. **Carms.**: Black Mountains, SN72, AL, 1898[106].

H. pseudoleyi (Zahn) Roffey (*H. decolor* auct., non (W. R. Linton) A. Ley). Native (endemic). Limestone rocks. Fl.5-8. Caerns., *Denbs.*. 4 squares.
Caerns.: Ysgolion Duon, SH66, EFL, 1896[106] & 1900[106]; Gt Ormes Head, SH78, JEG, 1891[106]and UKD & EPB, 1964[108]; Pydew, SH87, JMB, 1969[112]. **Denbs.**: Eglwyseg Rocks, SJ24, AL, 1903[106].

H. eustomon (E. F. Linton) Roffey Native. Rocky places. Fl.6-8. Glam., Brecs., Carms., Pembs., (*Caerns.*). 11 squares : **SS**19,*48*,58; **SN***10*,*61*,*82*,90,92; **SO***02*,11; (**SH**57).

H. stenopholidium (Dahlst.) Omang Native. Cliff ledges. Fl.6-8. *Glam.*, *Brecs.*, *Carms.*, *Mer.*. 4 squares.
Glam.: Tarren y Bwllfa, SN90, HJR, 1902[109]; Craig y Llyn, SN90, BAM, 1954[108]. **Brecs.**: Craig Ddu, SN92, BAM, 1964[108], Craig Cerrig-gleisiad, SN92, CWe, 1951[108]. **Carms.**: Black Mountains, SN72, PDS, 1975[108]. **Mer.**: Cwm Croesor, Moelwyn, SH64, AL, 1903[106].

[**H. stenolepiforme** (Pugsley) P. D. Sell & C. West *Monts.*]

H. subbritannicum (A. Ley) P. D. Sell & C. West Native. Limestone rocks. Fl.5-9. *Mons.*, *Glam.*, Brecs., *Carms.*. 8 squares.
Mons.: Penmoyle Cliffs, -, AL, 1891[106]. **Glam.**: Craig y Llyn, SN90, AL, 1896[106]. **Brecs.**: Craig Ddu, SN92, AL, 1894[106] and BAM, 1964[108]; Brecon Beacons, SO01, AL, 1893[106]; Craig y Cilau, SO11, AL, 1893[106] and PDS, 1972[108]; E. Craig y Cilau, SO21, AL, 1893[106]. **Carms.**: Castell Carreg Cennen, SN61, AL, 1888[106] & 1907[106]; Clogau Mawr, SN71, BAM, 1958[108] and PDS, 1975[108]; Black Mountains, SN82, AL, 1899[106].

H. britanniciforme Pugsley Native. Limestone rocks. Fl.5-7. Caerns., Denbs., Flints.. 5 squares.
Caerns.: nr Aber, SH67, SHB, 1871[108]; Gt Ormes Head, SH78, JAWh, 1895[112] and UKD & EPB, 1964[108]; Little Ormes Head, SH88, JEG, 1893[106] and AL, 1901[106]. **Denbs.**: Colwyn Bay, SH87, JMB, 1968[112]. **Flints.**: Pant-y-mwyn, SJ16, TE, 1972[112].

H. dicella P. D. Sell & C. West Native. Limestone rocks. Fl.6-9. Brecs., Caerns.. 5 squares.
Brecs.: Dan yr Ogof, SN81, AL, 1900[106]; Craig y Cilau, SO11, AL, 1900[106] and PDS, 1975[200]; nr Crickhowell, SO21, AHGA, 1932[107]. **Caerns.**: Gt Ormes Head, SH78, SHB, 1912[108]; Bodafon, SH87, CA, 1955[7].

H. saxorum (F. J. Hanb.) P. D. Sell & C. West Native. Rocky places. Fl.6-8. Brecs., Carms., *Monts.*, *Caerns.*. 9 squares.
Brecs.: Capel Callwen, SN81, AL, 1890[106] & 1896[108]; Fan Gyhirych, SN81, AL, 1898[103]; Black Mountains, SN82, JJD, 1955[112] and PDS, 1975[108]; Fan Nedd, SN91, AL, 1906[106]; Craig Cerrig-gleisiad, SN92, CWe, 1954[108]; Brecon Beacons, SO02, AL, 1895[106]. **Carms.**: Black Mountains, SN72, PDS, 1975[118]; Black Mountains, SN82, AL, 1895[112] and JJD, 1955[112]. **Monts.**: Pistyll Rhaeadr, SJ02, An, 1891[106]; Craig Breiddin, SJ21, JRa, 1950[108]. **Caerns.**: Craig Dulyn, SH76, An, 1904[106].

H. riddelsdellii Pugsley Native (endemic). Base-rich rocks. Fl.6-7. Brecs., *Carms.*. 3 squares.
Brecs.: Craig-y-nos, SN81, AL, 1904[112] and PDS, 1975[200]; Craig Cerrig-gleisiad, SN92, AL, 1905[108] and BAM, 1958[108]. **Carms.**. Black Mountains, SN82, AL, 1906[106] and BAM, 1960[108].

H. repandulare Druce Native. Rocky places. Fl.5-7. Brecs., ((*Carms.*)), *Caerns.*. 5 squares.
Brecs.: Dan yr Ogof, SN81, AL, 1899[106]; Craig y Rhiwarth, SN81, AL, 1904[112] and MP, 1980[200]; Pwll Byfre, SN81, PDS, 1975[108]; Black Mountains, SN82, AL, 1906[106]; Craig Cerrig-gleisiad, SN92, AL, 1898[112] and PDS, 1975[200]; Glyn Tarell, SN92, PDS, 1953[108]; Dyffryn Crawnon, SO01, AL, 1904[106]. ((**Carms.**: Llyn y Fan Fach, SN82, HJR, c1900[265])). **Caerns.**: Carnedd Ddafydd, SH66, AL, 1891[106].

H. hypochoeroides Gibson Native. Limestone rocks. Fl.7-9. Brecs., Denbs.. 2 squares.
Brecs.: Dyffryn Crawnon, SO01, AL, 1892[106] and MP, 1980[200]. **Denbs.**: Eglwyseg Rocks, SJ24, AL, 1891[108] and TE, c1970[90]; Castell Dinas Bran, SJ24, AL, 1891[106] and CWe, 1912[108].

H. carneddorum Pugsley Native. Rocky places. Fl.6-8. *Rads.,* Monts., Mer., Caerns., *Denbs.*. 13 squares : **SN**54, 55,*64*,65,66,71,75,76,*77*; **SO**26; **SJ**02,03,21.

H. cillense Pugsley Native (endemic). Limestone rocks. Fl.6-7. Brecs.. 1 square.
Brecs.: Craig y Cilau, SO11, AL, 1899[106] and MP, 1980[200].

H. subrude (Arvet-Touvet) Arvet-Touvet Native. Rocky places. Fl.6-8. Cards., ((*Mer.*)), Caerns., [*Denbs.*]. 5 squares.
Cards.: Capel Ystrad-ffin, SN74, AL, 1896[108]; Devil's Bridge, SN77, JHS, 1929[112]. ((**Mer.**: Ffestiniog, SH74, HNR, pre 1930[265])). **Caerns.**: Nant Ffrancon, SH66, AL, 1894[106] and EFL & WRL, 1900[108]; Degannwy, SH77, JEW, 1947[108] and PDS, 1953[108].

H. orimeles F. J. Hanb. ex W. R. Linton Native. Cliff ledges and rocky stream sides. Fl.7-8. Carms., Mer., Caerns., *Denbs.*. 11 squares : **SN**72,82; **SH**57,65,66,71,76,77,*92*; **SJ**03,16.

H. cambricum (Baker) F. J. Hanb. Native. Limestone cliffs. Fl.6. *Glam.,* Caerns., Denbs.. 3 squares.
Glam.: nr Treorchy, SS99, HJR, 1903[108]. **Caerns.**: Gt Ormes Head, SH78, SHB, 1867[108] and UKD, 1964[108]. **Denbs.**: Cefn Fedw, SJ24, AL, 1901[112].

H. vagense (F. J. Hanb.) A. Ley Native. Limestone rocks and rocky river banks. Fl.5-7. Brecs., Rads., Caerns., Denbs.. 9 squares.
Brecs.: nr Erwood, SO04, AL, 1894[112] and MP, 1980[200]; Craig y Cilau, SO11, GCD, 1908[106]; Llangoed, SO14, AL, 1887[112] and MP, 1980[200]. **Rads.**: Boughrood, SO13, HJR, 1907[108]; nr Boughrood, SO14, PDS, 1953[108]; nr Llansteffan, SO14, CWe, 1954[108] and BAM, 1956[108]. **Caerns.**: nr Llanberis, SH65, AL, 1904[106]; Llandudno Junction, SH77, AL, 1901[106]; Penmaenmawr, SH77, CA, 1955[7]; nr Craig y Don, SH78, AOH, 1905[112]; Gloddaith, SH88, AL, 1901[108]. **Denbs.**: Colwyn Bay, SH87, JMB, 1968[112].

H. argenteum Fries Native. Basic rocks. Fl.6-8. Glam., *Brecs., Rads.,* Cards., ((*Monts.*)), Mer., Caerns., *Denbs.*. 24 squares, map 798.

H. scoticum F. J. Hanb. Native. Rocky places. Fl.6-8. Brecs., *Carms.*. 3 squares.
Brecs.: nr Penwyllt, SN81, AL, 1899[106] and WCB, 1920[112]; Black Mountains, SN82, AL, 1898[106] and PDS, 1975[200]; Cwm Hepste, SN91, AL, 1894[112] and AL, 1908[112]. **Carms.**: Black Mountains, SN82, AL, 1898[108] and PDS, 1975[210].

H. leyanum (Zahn) Roffey Native. Cliff ledges. Fl.5-8. Mons., Brecs.. 7 squares.
Mons.: Piercefield Park, ST59, AL, 1900[106]; Tarren yr Esgob, SO23, AL, 1878[112] & 1907[108]. **Brecs.**: Cwm Cadlan, SN90, MP, 1977[200]; Craig Cerrig-gleisiad, SN92, AL, 1890[106] and CWe, 1951[108]; Cwm Tarell, SN92, AL, 1888[106] and MP, 1978[200]; Dan y Graig, SO01, AL, 1894[112] and BAM, 1958[108]; Cwar-yr-hendre, SO01, PDS, 1975[108]; Cwm Cynwyn, SO02, MP, 1975[200]; Dyffryn Crawnon, SO11, AL, 1893[106] and MP, 1980[200]; Tarren yr Esgob, SO23, AL, 1898[112] and MP, 1975[200].

H. holophyllum W. R. Linton Native. Limestone rocks. Fl.5-7. Caerns., *Denbs.,* Flints.. 6 squares.
Caerns.: Gt Ormes Head, SH78, SHB, 1871[108] and CWe, 1953[108]. **Denbs.**: Denbigh, SJ06, CWe, 1947[108]; nr Four Crosses, SJ15, TE, c1970[90]; Cefn Fedw, SJ24, AL, 1891[106] & 1901[106]; nr Minera, SJ25, TE, 1978[67]. **Flints.**: Nant Alun, SJ16, TE, 1972[112].

H. caledonicum F. J. Hanb. Native. Rocky places (Q,Rw). Fl.5-9. *Brecs.,* ((*Pembs.*)), Cards., Mer., Caerns., Denbs., *Flints.*. 15 squares : ((**SS**19)); **SN**25; **SO***02,23*; **SH***54*,63,64,65,66,72,76,((*78*)); **SJ**15,16,25.

H cacuminum (A. Ley) A. Ley Native. Rocky river banks (Q,Rw). Fl.6-8. ((Mons.)), *Glam.,* Brecs., *Carms., Mer.*. 7 squares.
((**Mons.**: Wynd Cliff, ST59, WAS, 1904[112])). **Glam.**: Craig y Llyn, SN90, AL, 1890[106] and HJR, 1903[106]. **Brecs.**: Cwm Hepste, SN90, AL, 1894[112]; Craig Cerrig-gleisiad, SN92, AL, 1898[112] and PDS, 1975[200]; Pen y Fan, SN92, AEW, 1926[112] and MP, 1975[200]; Glyn Tarell, SN92, PDS, 1953[108]; Cwm Taf Fechan, SO01, AL & WAS, 1894[112]; Nant Ddu, SO01, AL, 1902[112]; Torpantau, SO01, PDS, 1953[108]; Glyn Collwng, SO02, AL, 1900[112] & 1908[112]; Brecon Beacons, SO02, AL, 1888[112]; Craig Pwllfa, SO02, AL, 1902[106]. **Carms.**: Black Mountains, SN82, AL, 1899[106]. **Mer.**: Llanuwchllyn, SH83, AL, 1891[106].

H. angustatiforme P. D. Sell & C. West Native. Rocky places. Fl.6-8. Brecs.. 3 squares.
Brecs.: Craig Cerrig-gleisiad, SN92, AL, 1899[106] and BAM, 1958[108]; Cwm Taf Fechan, SO01, AL, 1894[106] & 1908[106]; Glyn Collwng, SO02, AL, 1893[106] & 1908[106].

Sect. Cerinthoidea Koch

H. anglicum Fries Native. Rocky places. Fl.6. *Monts.*. 1 square.
Monts.: Craig Breiddin, SJ21, CWe, 1951[108].

Sect. Alpina F. N. Williams

H. holosericeum Backh. Native. Rock ledges. Fl.6-8. Caerns.. 2 squares.
Caerns.: Tryfan, SH65, JEG, 1889[112] and ER, c1960[207]; nr Llyn y Cŵn, SH65, WW, 1822[108] and ER, 1962[207]; Carnedd Ddafydd, SH66, JEG, c1890[101]; Craig yr Ysfa, SH66, ER, 1953[207].

Sect. Amplexicaulia Zahn

H. speluncarum Arvet-Touvet Established alien (Europe). Walls. *Denbs.*. 1 square.
Denbs.: Colwyn Bay, SH87, JMB, c1965[67].

Sect. Prenanthoidea Koch

H. prenanthoides Vill. Native. Limestone rocks and streamsides. Fl.7-9. *Brecs.*. 4 squares.
Brecs.: Craig Ddu, SN82, AL, 1899[106] and SN92, BAM, 1964[108]; Craig Cerrig-gleisiad, SN92, RMH, 1961[7]; nr Llanthony, SO22, AL, 1874[108]; Tarren yr Esgob, SO23, AL, 1880[107] and MP, 1975[200].

Sect. Foliosa Pugsley

H. tavense (W. R. Linton) A. Ley Native (endemic). Rocky streamsides. Fl.7. Brecs.. 1 square.
Brecs.: Upper Tawe, SN82, AL, 1899[108] and MP, 1975[200].

H. latobrigorum (Zahn) Roffey Native. Rocky places. Fl.7. Caerns.. 1 square.
Caerns.: Nant Ffrancon, SH66, AL, 1891[106] & 1906[112].

H. subcrocatum (E. F. Linton) Roffey Native. Riverside rocks. Fl.7-9. *Glam., Brecs., Rads., Carms., Cards., Monts., Mer.,* Caerns., *Denbs., Flints.*. 29 squares, map 799.

H. strictiforme (Zahn) Roffey Native. Rocky places. Fl.6-8. *Brecs.,* Rads., *Monts., Caerns.*. 4 squares.
Brecs.: Cwm Hepste, SN91, AL, 1906[108]; Cwm Mellte, SN91, WRL, 1904[108]. **Rads.**: Elan Valley, SN96, AL, 1902[112]. **Monts.**: Pistyll Rhaeadr, SJ02, AL, c1900[7]. **Caerns.**: Cwm Idwal, SH65, AL, 1887[108].

H. reticulatum (Lindeb.) Lindeb. Native. Rocky places. *Brecs., Rads.*. 2 squares.
Brecs.: Torpantau, SO01, AL, c1900[7]. **Rads.**: Elan Valley, SN96, AL, 1902[7].

[**H. subumbellatiforme** (Zahn) Roffey *Brecs.*]

Sect. Sabauda F. N. Williams

H. vagum Jordan Native. Rocky places (D,E,H,M,Q, R,Rw,W). Fl.6-10. *Mons., Glam., Brecs.,* Rads., *Carms.,* ((*Pembs.*)), Cards., Monts., Mer., Caerns., Denbs., Flints., Angl. 87 squares, map 800.

H. salticola (Sudre) P. D. Sell & C. West Native. Sand dunes and roadsides. Fl.9. Flints.. 4 squares.
Flints.: Prestatyn, SJ08, TE, 1972[112]; Efail Parcy, SJ16, TE, c1972[90]; nr Lixwm, SJ17, TE, c1972[90]; Shotton, SJ36, TE, c1970[90].

H. rigens Jordan Established alien (England). Roadsides and railway banks. Fl.7-9. Carms., ((*Flints.*)). 2 squares.
Carms.: nr Cwm-ann, SN54, JRP, 1978[112]. ((**Flints.**: nr Caerwys, SJ17, P, pre 1930[265])).

H. perpropinquum (Zahn) Druce Native. Woods and roadsides (D,E,G,M,Q,R,Rw,W). Fl.8-10. Mons., Glam., Brecs., *Rads.,* Carms., *Pembs.,* Cards., Monts., Mer., Caerns., Denbs., Flints., *Angl.*. 104 squares, map 801.

[**H. perpropinquum** × **strumosum** *Rads.*]

[**H. perpropinquum** × **subcrocatum** *Brecs.*]

Sect. Umbellata F. N. Williams

H. umbellatum L. Native. Rocky places and roadsides. Fl.7-10. Mons., Glam., Brecs., Rads., Carms., Pembs., Cards., Monts., Mer., Caerns., Denbs., *Flints., Angl.*. 117 squares, map 802.

Subsp. **umbellatum** Native. Rocky places and roadsides (D,E,G,H,M,Q,R,W). Fl.7-10. Mons., *Glam.,* Brecs., Rads., Carms., Pembs., Cards., Monts., Mer., Caerns., Denbs., [*Flints.*]. 84 squares, map 803.

Subsp. **bichlorophyllum** (Druce & Zahn) P. D. Sell & C. West (*H. bichlorophyllum* (Druce & Zahn) Pugsley). Native. Rocky places and roadsides (Ar,D,E,M,R,Rw,W). Fl.7-9. Mons., Glam., Brecs., *Rads.,* Carms., Pembs., Cards., Mer., Caerns., Denbs., *Angl.*. 46 squares, map 804.

Sect. Tridentata F. N. Williams

H. linguans (Zahn) Roffey Native (endemic). Rock ledges. Brecs., *Carms.*. 3 squares.
Brecs.: nr Capel Callwen, SN81, AL, 1898[112]; Cwm Haffes, SN81, AL, 1899[106] and BAM, 1957[108]; Upper Tywi, SN84, AL, 1906[108]. **Carms.**: Llyn y Fan Fach, SN82, AL, c1900[108].

H. sparsifolium Lindeb. Native. Rocky and grassy places. Fl.7-8. Glam., Brecs., Carms., ((Cards.)), Mer., Caerns., Denbs.. 24 squares, map 805.

H. lissolepium (Zahn) Roffey Native. Rocky places. Fl.7-8. *Brecs., Monts., Mer.,* Caerns.. 7 squares.
Brecs.: Cwm Haffes, SN81, HJR, c1900[7]. **Monts.**: nr Pontrhydgaled, SN88, AL, 1886[106]. **Mer.**: Drws-y-nant, SH82, AL, 1903[108]. **Caerns.**: Snowdon, SH65, CCB, 1847[108] and FRB, 1955[108]; Cwm Idwal, SH65, AL, 1900[112]; Nant Ffrancon, SH66, JEG, 1893[106]; nr Glanllugwy, SH66, AL, 1888[106]; nr Capel Curig, SH75, ESM, 1912[108]; nr Llandudno, SH87, JMB, 1969[112].

H. uiginskyense Pugsley Native. Rocky streamsides. Fl.6-8. Brecs., *Rads., Carms., Mer.*. 9 squares.
Brecs.: Cwm Haffes, SN81, AL, 1899[106] and MP, 1975[200]; Abercrave, SN81, AL, 1908[107]; Black Mountains, SN82, AL, 1899[106]; Cwm Hepste, SN91, AL, c1908[108] & 1918[108]; nr Llanwrtyd Wells, SN84, BAM, 1958[108]; nr Sennybridge, SN92, AL, 1890 and JNM, 1965[108]; Cwm Taf Fechan, SO01, AL, 1894[108]. **Rads.**:

Pontarelan, SN96, AL, 1889[108]; Boughrood, SO13, AL, 1907 and BAM, 1957[108]. **Carms.**: Black Mountains, SN82, BAM, 1958[108]. **Mer.**: nr Blaenau Ffestiniog, SH74, AL, 1903[106].

H. substrigosum (Zahn) Roffey Native. Rocky streamsides. Fl.6-8. *Glam.*, Brecs., *Rads.*, *Mer.*, ((*Caerns.*)). 18 squares : **SS***99*; **SN***60*,81,*82,85*,91,92,*96*; **SO**01,*02,11*; **SH**61,62,((66)),*71*,81,*83*; **SJ***03*.

H. placerophylloides Pugsley Native. Rocky and grassy places especially by streams. Fl.7-8. *Brecs.*, *Rads.*, *Mer.*, Caerns., *Denbs.*. 9 squares.
Brecs.: Cwm Haffes, SN81, JNM, 1965[7]; Torpantau, SO01, AL, 1894[106]. **Rads.**: nr Pontarelan, SN96, AL, 1889[106]. **Mer.**: Cadair Idris, SH71, AL, 1886[106]; Blaen Lliw, SH83, PMB, 1968[3]; Nant Esgairiau, SJ03, AL, 1891[106] & 1903[106]. **Caerns.**: Cwm Llefrith, SH54, JEG, 1891[112] and AL, 1904[106]; Llyn Dulyn, SH66, AL, c1900[7]; nr Capel Curig, SH75, ESM, 1912[7]. **Denbs.**: Llyn Lluncaws, SJ03, An, 1955[108].

H. nidense (F. J. Hanb.) Roffey Native (endemic). Rock ledges and streamsides. Fl.7-8. Brecs., Carms., *Mer.*. 5 squares.
Brecs.: Craig-y-nos. SN81, AL, 1906[106]; Upper Tawe, SN81, JRa, 1953[112]; Cwm Mellte, SN91, AL, 1892[112] and 1908[106]; Taf Fechan, SO01, AL, 1900[108]; Glyn Collwng, SO01, AL, 1903[112]. **Carms.**: Black Mountains, SN82, AL, 1906[112]. **Mer.**: Blaenau Ffestiniog, SH74, AL, 1903[108].

H. stewartii (F. J. Hanb.) Roffey Native. Rocky and grassy places. Fl.7. *Brecs.*, Denbs.. 3 squares.
Brecs.: Afon Tawe, SN82, RLB, 1955[7]; nr Abergwesyn, SN85, AL, 1902[106]. **Denbs.**: nr Llanrwst, SH76, JMB, 1969[112].

H. scabrisetum (Zahn) Roffey Native. Rocky places (E,M,Q,Rw). Fl.6-9. Mons., Glam., Brecs., Rads., *Carms.*, Cards., *Monts.*, Mer., Caerns., Denbs.. 44 squares, map 806.

[**H. cantianum** F. J. Hanb. *Glam.*]

H. trichocaulon (Dahlst.) K. Joh. Native. Roadsides. Fl.7-8. Glam., *Brecs.*, *Mer.*, [*Mons.*]. 4 squares.
Glam.: nr Treorchy, SS99, HJR, 1903[112]. **Brecs.**: Dyffryn Callwen, SN81, AL, 1899[108]; Taf Fechan, SO01, AL, 1901[108]. **Mer.**: nr Blaenau Ffestiniog, SH64, AL, 1903[106].

H. cambricogothicum Pugsley Colonist (Scotland). Walls. Fl.7. *Caerns.*. 1 square.
Caerns.: Llanfairfechan, SH67, CWe, 1953[108].

H. eboracense Pugsley Native. Scrub and roadsides. Fl.6-9. Mons., *Glam.*, *Brecs.*, *Carms.*, ((*Caerns.*)), Denbs.. 14 squares : **ST**39,((59)); **SN***71,90,92*; **SO***01,02,13,14*,((*22*)),*50*; **SH**((75)),95; **SJ**35.

H. calcaricola (F. J. Hanb.) Roffey Native. Grassy places. Fl.6-9. *Mons.*, *Glam.*, Brecs., *Mer.*, *Denbs.*. 14 squares : **SS**99; **ST**09,38,((*59*)),84,90,93; **SO**01,((*20,22,30,50*)); **SH***70*; **SJ**05.

MONOCOTYLEDONES

ALISMATACEAE

SAGITTARIA L.

Sagittaria sagittifolia L. : Saethlys : Arrowhead. Native. Streams (Al,Ar). Fl.7-8. Mons., Glam., *Pembs.*, (*Cards.*), *Denbs.*, *Flints.*, *Angl.*, [*Mer.*, *Caerns.*]. 20 squares, map 807.

BALDELLIA Parl.

Baldellia ranunculoides (L.) Parl. : Dŵr-lyriad Bychan : Lesser Water-plantain. Native. Marshes (Al,Ar,D,Ds,H,M,S). Fl.5-9. Mons., Glam., Brecs.*, *Carms.*, Pembs., Cards., *Monts.*, Mer., Caerns., *Denbs.*, *Flints.*, Angl.. 63 squares, map 808.

LURONIUM Rafin.

Luronium natans (L.) Rafin. : Dŵr-lyriad Nofiadwy : Floating Water-plantain. Native. Lakes (Al,Ar). Fl.4-8. *Glam.*, Brecs.*, Pembs., Cards., Monts., Mer., Caerns., Denbs., *Flints.*[230], *Angl.*, [*Carms.*]. 31 squares, map 809.

ALISMA L.

Alisma plantago-aquatica L. : Dŵr-lyriad : Water-plantain. Native. Streamsides and ponds (Al,Ar,H,M). Fl.6-9. Mons., Glam., Brecs., *Rads.*, Carms., Pembs., Cards., Monts., Mer., Caerns., Denbs., Flints., Angl.. 178 squares, map 810.

[**A. plantago-aquatica × Baldellia ranunculoides** *Angl.*]

A. lanceolatum With. : Dŵr-lyriad Culddail : Narrow-leaved Water-plantain. Native. Streamsides (Al,Ar,H,M). Fl.7-8. Mons.*, Glam.*, Brecs.*, Carms.*, Pembs.*, *Flints.*†, *Angl.*, [*Mer.*, *Caerns.*, *Angl.*]. 13 squares : **SS**19; **ST**09,28,48,49; **SN**10,20,30;**SO**00,*12*,14,24; **SJ**07.

[DAMASONIUM Miller]

[**Damasonium alisma** Miller *Mer.*]

BUTOMACEAE

BUTOMUS L.

Butomus umbellatus L. : Engraff : Flowering-rush. Native. Streams (Al,Ar,H,M). Fl.6-9. Mons., Glam., Brecs., *Pembs.*, Monts., Caerns., *Denbs.*, *Flints.*, Angl., [*Cards.*]. 36 squares, map 811.

HYDROCHARITACEAE

HYDROCHARIS L.

Hydrocharis morsus-ranae L. : Alaw Lleiaf : Frogbit. Native. Ditches (Al,Ar). Fl.7-9. Mons., Glam., Carms., *Monts.**, Denbs., *Flints.*, Angl.*, [*Caerns.*]. 14 squares : **SS**78,79; **ST**17,27,28,38,48; **SN**40; **SH**38; **SJ**21,24,35,44,45.

STRATIOTES L.

Stratiotes aloides L. : Alaw Ddiosgo : Water-Soldier. Native. Ponds. Denbs.. 2 squares.
Denbs.: Cefn Park, SJ34, LR, 1977[7]; nr Gresford, SJ35, WEJ, 1938[112] and GH, 1970[7]; Cefn Park Lake, SJ35, AGS, 1964[177] and An, 1968[7].

ELODEA Michx

Elodea canadensis Michx : Alaw Canada : Canadian Waterweed. Naturalized alien (N. America). Ponds and streams. Fl.7-9. Mons., Glam., Brecs., Rads., Carms., Pembs., Cards., Monts., Mer., Caerns., *Denbs.*, Flints., Angl.. 105 squares, map 812.

E. ernstiae St. John Naturalized alien (S. America). Canals. *Brecs.*. 1 square.
Brecs.: Llangattock, SO21, MP & RGW, 1980[245].

E. nuttallii (Planchon) St. John Naturalized alien (N. America) : Nuttall's Waterweed. Canals. *Brecs.*. 2 squares.
Brecs.: Talybont, SO12, MP, 1980[200]; Llangattock, SO21, MP, 1980[200].

LAGAROSIPHON Harvey

Lagarosiphon major (Ridley) Moss : Curly Water-thyme. Naturalized alien (S. Africa). Pools and canals. Fl.7. Mons., Glam., Rads.. 3 squares.
Mons.: Gelli-groes, ST19, AEW, 1968[112]. **Glam.:** Whitchurch Canal, ST18, ARP, 1974[112]. **Rads.:** Llanelwedd, SO05, RGW, 1980[112].

APONOGETONACEAE

APONOGETON L. fil.

Aponogeton distachyos L. fil. : Cape Pondweed. Introduced (S. Africa). Ponds. Glam., *Brecs.*, *Pembs.*. 3 squares.
Glam.: Roath Park, Cardiff, ST18, MG, 1970+[99]. **Brecs.:** Llangorse Lake, SO12, WLB, 1897[260]. **Pembs.:** Caldey Island, SS19, JEA, 1918[18].

JUNCAGINACEAE

TRIGLOCHIN L.

Triglochin maritima L. : Saethbennig Arfor : Sea Arrowgrass. Native. Salt-marshes (Ar,H,M,S). Fl.5-9. Mons., Glam., Carms., Pembs., Cards., Monts., Mer., Caerns., Denbs., *Flints.*, Angl.. 78 squares, map 813.

T. palustris L. : Saethbennig y Gors : Marsh Arrowgrass. Native. Marshes (Ar,Ds,H,M). Fl.6-9. Mons., Glam., Brecs., Rads., Carms., Pembs., Cards., Monts., Mer., Caerns., *Denbs.*, Flints., Angl.. 177 squares, map 814.

POTAMOGETONACEAE

POTAMOGETON L.

Subgen. Potamogeton

Potamogeton natans L. : Dyfrllys Llydanddail : Broad-leaved Pondweed. Native. Ponds (Al,Ar,H). Fl.5-9. Mons., Glam., Brecs., Rads., Carms., Pembs., Cards., Monts., Mer., Caerns., Denbs., Flints., Angl.. 172 squares, map 815.

P. natans × polygonifolius = P. × gessnacensis G. Fischer Spontaneous. Lakes. Fl.7-9. Caerns.. 1 square.
Caerns.: Llyn Anafon, SH66, CB, 1884[107] and RR, 1946[107].

P. polygonifolius Pourret : Dyfrllys y Gors : Bog Pondweed. Native. Moorland pools (Al,Ar,H,M). Fl.5-8. Mons., Glam., Brecs., Rads., Carms., Pembs., Cards., Monts., Mer., Caerns., Denbs., Flints., *Angl.*. 200 squares, map 816.

P. coloratus Hornem. : Dyfrllys y Migyn : Fen Pondweed. Native. Pools (Al,Ar,Ds). Fl.6-9. Glam.*, *Pembs.*, Caerns., *Denbs.*[230], Flints., Angl., [*Cards.*]. 10 squares : **SR**89,*99*; **SS**78; **SH***28*,47,48,57,58,78; **SJ**08.

P. lucens L. : Dyfrllys Disglair : Shining Pondweed. Native. Rivers and lakes (Al,Ar). Fl.6-8. Mons., Glam., Brecs., *Carms.*[230], *Monts.**, Caerns., *Denbs.*, Angl.. 14 squares : **ST**78; **SN**82; **SO**12,50,51; **SH**12,37,*47*,48; **SJ**10,20,21,*24*,33.

P. gramineus L. : Dyfrllys Amryddail : Various-leaved Pondweed. Native. Lakes (Al,Ar,H). Fl.6-7. Glam., Brecs.*, *Carms.*, Pembs., *Mer.*, *Caerns.*, *Denbs.*, *Flints.**, Angl.. 16 squares : **SS**59,*69*,78,79; **ST***17*,*27*; **SM**72,91, 92; **SO**12; **SH***28*,37,*47*,57,*93*; **SJ**07.

P. gramineus × lucens = P. × zizii Koch ex Roth : Long-leaved Pondweed. Spontaneous. Lakes. Fl.6-9. Angl.. 3 squares.
Angl.: Llyn Coron, SH36 or 37, JEG, c1890[101]; Llyn Maelog, SH37, WW, 1830[107] and EST, 1924[76]; Llyn Llwydiarth, SH57, HD, 1775[107] and JEG, 1884[112].

P. gramineus × natans = P. × sparganifolius Laest. ex Fries : Ribbon-leaved Pondweed. Spontaneous. Lakes and marshes (Al,M). Fl.6-7. Pembs., Angl.. 2 squares.
Pembs.: Trefeiddan Moor, SM72, TAWD, 1961[112]. **Angl.:** Rhos y Gad Fen, SH57, RHR, 1961[112].

P. gramineus × perfoliatus = P. × nitens Weber Spontaneous. Lakes. Fl.6. Angl.. 1 square.
Angl.: Llyn Coron, SH36, JEG, 1893[112] and E, 1919[112].

P. alpinus Balbis : Red Pondweed. Native. Lakes (Al,Ar). Fl.6-9. *Mons.**, Carms., Pembs., Mer.*, Caerns., Denbs., Angl., [*Glam.*]. 15 squares : **SN**11,62,72,82; **SO***51*; **SH***37,47*,48,*57*,66,71,*95*; **SJ**24,*34*,35.

P. alpinus × crispus = P. × olivaceus Baagöe ex G. Fischer Spontaneous. Rivers. Fl.8. Carms., *Pembs.*, Cards.. 4 squares.
Carms.: nr Cwm-cou, SN24, NH, 1978[107]; Llanfihangel-ar-arth, SN44, NH, 1978[112]; Pont-tyweli, SN44, NH, 1978[112]; **Pembs.:** nr Coedmor House, SN14, AOC, 1973[71]. **Cards.:** nr Coedmor House, SN14, AOC, 1972[112]; nr Cwm-cou, SN24, NH, 1978[112]; Newcastle Emlyn, SN34, AOC, 1979[112]; Llanfihangel-ar-arth, SN44, NH 1978[112]. Confined in Wales to the River Teifi.

P. alpinus × praelongus = P. × griffithii Ar. Benn. Spontaneous. Lakes. Fl.7-9. Caerns.. 1 square.
Caerns.: Llyn Anafon, SH66, JEG, 1882[112] and AW, 1928[107].

P. praelongus Wulfen : Long-stalked Pondweed. Native. Rivers (Al,Ar). Fl.6-7. Rads., Monts., Mer., *Denbs., Flints.,* [*Glam., Carms., Angl.*]. 8 squares.
Rads.: Llanbwchllyn, SO14, AL, 1886[107] and BS, 1964[107]; Llyn Heilyn, SO15, NYS, 1945[107] and GT, 1948[112]. **Monts.:** Llanllwchaearn, SO19, JAW, c1940[228]; Welshpool, SJ20, IAW, 1956[107]; Pool Quay, SJ21, JAW, 1940[112]; Guilsfield, SJ21, GT, 1949[107]. **Mer.:** Llynnau Cregennen, SH61, PMB, 1969[112]. **Denbs.:** Llangollen, SJ24, TR, pre 1900[146]. **Flints.:** nr St Asaph, SJ07, RHD, c1900[141].

P. perfoliatus L. : Dyfrllys Trydwll : Perfoliate Pondweed. Native. Streams (Al,Ar). Fl.7-9. Mons., Glam., Brecs., *Rads.**, Carms., *Pembs., Cards.*, Monts., Mer., Caerns., *Denbs.*, Flints., Angl.. 48 squares, map 817.

P. friesii Rupr. : Dyfrllys Gwastatgoes : Flat-stalked Pondweed. Native. Canals. Fl.7. Monts.*, *Angl.*[231], [*Glam.*]. 2 squares.
Monts.: Welshpool, SJ20, EM, 1955[109]; Guilsfield, SJ21, GT, 1949[112] and IRB, 1970[76]; nr Arddleen, SJ21, VJM, 1950+[116].

[**P. rutilus** Wolfg. *Angl.*]

P. pusillus L. : Dyfrllys Culddail : Lesser Pondweed. Native. Ditches (Al,Ar). Fl.7-9. Mons., Glam., Brecs.*, Carms.*, *Pembs.**, Cards., *Monts.**, Mer., Caerns., *Denbs.**, Flints.*, Angl.. 33 squares, map 818.

P. obtusifolius Mert. & Koch (*P. sturrockii* (Ar. Benn.) Ar. Benn.) : Dyfrllys Gwelltog : Blunt-leaved Pondweed. Native. Pools (Al, Ar). Fl.7-9. Glam., *Rads.**, Pembs.*, Cards.*, Monts., Caerns., *Denbs.**, Angl.. 18 squares : **ST**07; **SM**83; **SN**54,98; **SO**19,27; **SH**37,38,43,44,47,53,*56*,57; **SJ**20,21,*35*,53.

P. berchtoldii Fieber (*P. pusillus* auct., non L.) : Dyfrllys Eiddil : Small Pondweed. Native. Streams (Al,Ar,H). Fl.6-9. Mons., Glam., Brecs., Rads., Carms., Pembs., Cards., Monts., Mer., Caerns., Denbs., Flints., Angl.. 110 squares, map 819.

P. berchtoldii × coloratus = P. × lanceolatus Sm. : Lanceolate Pondweed. Spontaneous. Streams. Fl.6-8. Angl.. 1 square.
Angl.: Penrhosllugwy, SH48, HD, 1806[107] and RHR, 1968[208].

P. trichoides Cham. & Schlecht. : Hairlike Pondweed. Native. Ditches. (Al,Ar). Fl.7-8. Mons., Glam., Angl.. 6 squares.
Mons.: nr Whitson, ST38, BW, 1939[112] and CSc & MWa, 1974[112]; nr Llan-wern, ST38, AEW, 1941[112]; Nash, ST38, MWa, 1975[76]; Goldcliff, ST38, MWa, 1975[76]; Magor, ST48, CSc & MWa, 1975[112]; Undy, ST48, MWa, 1975[76]. **Glam.:** Kenfig Pool, SS78, HJMB, 1982[112]. **Angl.:** Llyn Coron, SH36, GCD & MSW, 1918[107] and BS, 1962[112]; Llyn Padrig, SH37, BS, 1962[112]; Cors Ddyga, SH47, HD, c1800[107].

P. compressus L. : Dyfrllys Camleswellt : Grass-wrack Pondweed. Native. Canals. Fl.7-8. Monts.. 5 squares.
Monts.: nr Newtown, SO19, JAW, 1939[112]; Glyn-cogan, SJ10, HHHa, 1935[76]; Afon Llugwy, SJ10, JAW, c1940[112]; Welshpool, SJ20, SHa, 1938[76] and IAW, 1956[76]; nr Guilsfield, SJ21, GT, 1949[76]; Pool Quay, SJ21, JAW, 1940[112]; nr Welshpool, SJ21, IAW, 1956[76] and SGH, 1970[112]; nr Burgedin Locks, SJ21, GT, 1949[184]; Walls Bridge, SJ22, BS, 1974[93].

[**P. acutifolius** Link *Angl.*]

P. crispus L. : Dyfrllys Crych : Curled Pondweed. Native. Streams (Al,Ar). Fl.6-7. Mons., Glam., Brecs., Rads., Carms., Pembs., Monts.*, Caerns., Denbs., Flints., Angl.. 98 squares, map 820.

P. crispus × friesii = P. × lintonii Fryer Spontaneous. Canals. Fl.9. Monts.. 1 square.
Monts.: Llanllwchaearn, SO19, JAW, 1939[112].

P. crispus × perfoliatus = P. × cooperi (Fryer) Fryer Spontaneous. Rivers. Fl.8-9. Pembs.. 1 square.
Pembs.: Afon Solva nr Whitchurch, SM82, CIS & NYS, 1931[76] and TAWD, 1967[112].

P. crispus × praelongus = P. × undulatus Wolfg. Spontaneous. Lakes. Fl.7. Rads., [*Angl.*]. 1 square.
Rads.: Llyn Heilyn, SO15, JAW, 1938[112] and CIS & NYS, 1945[107].

Subgen. Coleogeton Reichenb.

P. filiformis Pers. : Dyfrllys Arfor : Slender-leaved Pondweed. Native. Lakes (Al,Ar,H). Fl.7-9. Angl.. 2 squares.
Angl.: nr Llanfihangel-yn-Nhywyn, SH37, WW, 1826[76]; Cors Bodeilio, SH57, HD, 1798[112]. Recent records require confirmation[209].

P. pectinatus L. : Dyfrllys Danheddog : Fennel Pondweed. Native. Lakes and ditches (Al,Ar,H). Fl.6-9. Mons., Glam., Brecs.*, *Rads.**, Carms., Pembs.*, Cards., Monts., Mer., Caerns., Denbs., Flints., Angl.. 57 squares, map 821.

GROENLANDIA Gay

Groenlandia densa (L.) Fourr. (*Potamogeton densus* L.) : Dyfrllys Tewdws : Opposite-leaved Pondweed. Native. Streams (Al,Ar,H). Fl.6-8. Glam., *Carms.*[230], *Flints.*, *Angl.*, [*Mons.*]. 11 squares : **SS**87,96,97,*99*; **ST**06,*07*,16,*17*; **SH***36*,*47*; **SJ***07*.

RUPPIACEAE

RUPPIA L.

Ruppia cirrhosa (Petagna) Grande (*R. spiralis* L. ex Dumort.) : Spiral Tasselweed. Native. Brackish ditches. Fl.7-9. *Glam.*†, Mer., *Caerns.*, [*Cards.*]. 3 squares.
Glam.: Swansea, SS69, HDG, pre 1900[140]. **Mer.:** Dwyryd Estuary, SH63, PMB, 1971[112]. **Caerns.:** nr Afon-wen, SH43, JEG, 1897[101].

R. maritima L. : Tusw Dyfrllys : Beaked Tasselweed. Native. Brackish ditches (Al,Ar,S). Fl.7-9. Mons., Glam., *Brecs.*, Carms., *Pembs.*, Cards., Mer., Caerns., *Denbs.*, Flints., Angl.. 39 squares, map 822.

ZOSTERACEAE

ZOSTERA L.

Zostera marina L. : Gwellt y Gamlas : Eelgrass. Native. Coastal shallow seas. Fl.5-10. *Glam.*, *Carms.*[230], Pembs., *Mer.*, Caerns., Angl., [*Cards.*]. 16 squares : **ST***16*,*17*,26; **SM**70,80; **SN***69*; **SH**24,27,28,*32*,*33*,36,37, *46*, 53,*61*. Recorded from *Mer.* only as specimens washed up by the tide.

Z. angustifolia (Hornem.) Reichenb. (*Z. hornemanniana* Tutin) : Gwellt y Gamlas Culddail : Narrow-leaved Eelgrass. Native. Coastal shallow seas. Fl.6-9. Mons.*, Glam., Carms., Pembs.*, Angl.. 12 squares : **SS**58; **ST**48; **SM**80,90; **SN**00,01,31,*40*; **SH**27,28,37,39.

Z. noltii Hornem. (*Z. nana* auct.) : Dwarf Eelgrass. Native. Coastal shallow seas (Ar,S). Fl.7-9. *Glam.*, Caerns., *Angl.*, [*Carms.*]. 5 squares.
Glam.: Whitford Burrows, SS49, QONK, 1972[157]. **Caerns.:** nr Aber, SH67, JEG, 1884[112]; Deganwy, SH77, WEH, 1967[112]. **Angl.:** nr Four Mile Bridge, SH27, CCT, 1950+[209]; Holyhead, SH28, DSR, 1970+[209].

ZANNICHELLIACEAE

ZANNICHELLIA L.

Zannichellia palustris L. : Cornwlyddyn : Horned Pondweed. Native. Streams and ponds (Al,Ar,S). Fl.5-10. Mons., Glam., Brecs., *Rads.**, Carms., Pembs.*, *Cards.**, Mer.*, Caerns., Denbs., Flints., Angl.. 57 squares, map 823.

LILIACEAE

NARTHECIUM Hudson

Narthecium ossifragum (L.) Hudson : Llafn y Bladur : Bog Asphodel. Native. Bogs (Ar,H,M,Q). Fl.6-8. Mons., Glam., Brecs., Rads., Carms., Pembs., Cards., Monts., Mer., Caerns., Denbs., Flints., Angl.. 189 squares, map 824.

ASPHODELUS L.

Asphodelus fistulosus L. (incl. *A. tenuifolius* Cav.). Casual (S. Europe). Waste places. Fl.9-10. Glam., *Flints.*. 3 squares : **ST**16[112],*27*[112]; **SJ***36*[28].

HEMEROCALLIS L.

Hemerocallis lilioasphodelus L. : Day-Lily. Introduced (S.E. Asia). Roadsides. Fl.6. Carms., *Caerns.*[84]. 1 square.
Carms.: btwn Golden Grove and Cross Hands, SN61, MHS, 1960[112].

H. fulva (L.) L. : Orange Day-Lily. Naturalized alien (China). Sea cliffs. Fl.7-8. Pembs., *Angl*[119]. 1 square.
Pembs.: Dale Fort, SM80, MGe, 1960[112].

HOSTA Tratt.

Hosta sieboldiana (Loddiges) Engler : Plantain Lily. Introduced (Japan). Pondside. Fl.6. *Monts.*. 1 square.
Monts.: Gregynog Hall, SO09, DM, 1978[258].

COLCHICUM L.

Colchicum autumnale L. : Saffrwm y Gweunydd : Meadow Saffron. Native. Pastures (E,G,R,W). Fl.8-10. Mons., Glam., Brecs., *Rads.*, Pembs., Monts., *Denbs.*, [*Flints.*]. 35 squares, map 825.

LLOYDIA Salisb. ex Reichenb.

Lloydia serotina (L.) Reichenb. : Lili'r Wyddfa : Snowdon Lily. Native. Mountain cliffs. Fl.5-6. Caerns., [*Rads.*]. 2 squares.
Caerns.: Snowdon, SH65, JEG, c1897[101] and ER, 1964[207]; Glyders, SH65, CW, 1876[112] and ER, 1965[207]; Carneddau, SH66, JBa, 1865[108] and ER, 1960[207].

GAGEA Salisb.

Gagea lutea (L.) Ker-Gawler : Tylwyth y Lili : Yellow Star-of-Bethlehem. Native. Woods. Fl.5. *Glam.*, *Denbs.**. 2 squares.
Glam.: Castle Coch Woods, ST18, An, 1843[223] (doubtful record). **Denbs.:** nr Ruthin, SJ25, ECh, 1979[245].

G. bohemica (Zauschner) Schultes & Schultes fil. Native. Rock ledges. Fl.1-3. Rads.. 1 square.
Rads.: Stanner Rocks, SO25, RGW & ARP, 1975[112] and TGE, 1981[91].

FRITILLARIA L.

Fritillaria meleagris L. : Britheg : Fritillary. Naturalized alien (England). Woodland (G,M,W). Fl.4. *Mons.*, Glam., *Denbs.*. 3 squares.
Mons.: nr Chepstow, ST59, TGE, 1980[91]. **Glam.:** Llandough nr Cardiff, ST17, TGD, 1971[112]. **Denbs.:** nr Overton-on-Dee, SJ34, BB, 1944[240].

LILIUM L.

Lilium martagon L. : Llysiau Martagon : Martagon Lily. Naturalized alien (Europe). Woods and hedges (E,W). Fl.4-8. Mons., Glam., *Brecs.**, *Carms.**, *Cards.**, Monts.*, *Flints.**, *Angl.**. 11 squares : **SS**48; **ST**07,59; **SN**31,56; **SO**03,50; **SH**56; **SJ**11,21,26.

L. bulbiferum L. var. **croceum** (Chaix) Pers. (*L. croceum* Chaix) : Orange Lily. Casual (Europe). River bank. *Denbs.*. 1 square : **SH**75[144].

L. pyrenaicum Gouan : Cap y Twrc Melyn : Pyrenean Lily. Naturalized alien (Pyrenees). Woods (Ar,E,Rw,W). Fl.5-7. *Brecs.**, *Rads.**, *Carms.*, Pembs., Cards., *Monts.**, *Caerns.**, *Denbs.**. 35 squares, map 826.

ORNITHOGALUM L.

Subgen. Beryllis (Salisb.) Baker

Ornithogalum pyrenaicum L. : Seren Fethlehem Hir : Spiked Star-of-Bethlehem. Casual (S. England). Hedges (E,G,W). Fl.7. *Glam.*, Pembs.. 2 squares : **SS**19[112]; **ST**17[95].

Subgen. Ornithogalum

O. umbellatum L. : Seren Bethlehem : Star-of-Bethlehem. Naturalized alien (E. England). Pastures (D,E,G,M,Q,R,Rg). Fl.5-6. Mons., Glam., *Brecs.*, *Carms.**, Pembs., *Cards.**, Monts., Caerns., Denbs., *Flints.*, *Angl.*. 51 squares, map 827.

Subgen. Myogalum (Link) Baker

O. nutans L. : Drooping Star-of-Bethlehem. Casual (Europe). Grassy places (G,W). Fl.4. Glam.*, *Denbs.*. 2 squares : **SS**87[112]; **SJ***35*[198].

SCILLA L.

Subgen. Scilla

Scilla bifolia L. Casual (Europe). Waste places. Fl.3. Angl.. 1 square : **SH**57[112].

Subgen. Petranthe (Salisb.) Chouard

S. verna Hudson : Seren y Gwanwyn : Spring Squill. Native. Grassy cliff-slopes and heaths (D,G,H,Sc). Fl.4-5. Glam., *Carms.**, Pembs., Cards., Mer., Caerns., *Denbs.*, *Flints.*, Angl.. 64 squares, map 828.

Subgen. Prospero (Salisb.) Chouard

S. autumnalis L. : Seren yr Hydref : Autumn Squill. Native. Grassy cliff-slopes. Glam.†. 2 squares.
Glam.: nr Marcross, SS96, JS, c1880[112]; (Barry, ST16, RLS, c1923[22]).

HYACINTHOIDES Medicus

Hyacinthoides non-scripta (L.) Chouard ex Rothm. (*Endymion non-scriptus* (L.) Garcke) : Clychau'r Gog : Bluebell. Native. Woods (Ar,D,E,G,H,M,Q,R,Rw,W). Fl.4-6. Mons., Glam., Brecs., Rads., Carms., Pembs., Cards., Monts., Mer., Caerns., Denbs., Flints., Angl.. 277 squares, all except : **SR**89; **SS**39; **SM**50; **SH**51,98.

H. hispanica (Miller) Rothm. (*Endymion hispanicus* (Miller) Chouard) : Spanish Bluebell. Naturalized alien (S. Europe). Woods (G,R,Rg,Rw,W). Fl.4-5. Glam., *Brecs.**, Rads.*, Carms., *Pembs.*, *Cards.**, *Mer.**, *Caerns.*, *Denbs.**, *Angl.*. 23 squares, map 829.

MUSCARI Miller

Subgen. Leopoldia (Parl.) Zahar.

Muscari comosum (L.) Miller : Tassel Hyacinth. Naturalized alien (Mediterranean region). Grassland and dunes (D,F,G,R). Fl.5-7. Glam., Carms., Pembs., (*Angl.*). 5 squares.
Glam.: The Leys nr Aberthaw, ST06, RLS, 1937[112]; Barry, ST16, AEW, 1923[112] and DM, 1974[93]. **Carms.:** Pembrey, SN40, GWH, 1935[112] and OW, 1978[184]. **Pembs.:** Tenby, SS19, FLR, 1944[112]. (**Angl.:** Cymyran nr Caergeiliog, SH27, MPl, 1976[157]).

Subgen. Botryanthus (Kunth) Zahar.

M. neglectum Guss. ex Ten. (*M. atlanticum* Boiss. & Reuter) : Grape Hyacinth. Casual. (S. England). Roadsides and waste places. *Glam.*, *Angl.*[119], [*Pembs.*]. 2 squares : **SS**87[95]; **ST**17[95].

ALLIUM L.

Allium schoenoprasum L. : Seifys : Chives. Native. Rocky banks (G,M,Q,R,Sc). Fl.5-8. (Mons., *Glam.*), Brecs., Rads., Carms., Pembs., (*Cards.*, Caerns.), *Denbs.*. 15 squares : **SM**72,93; **SN**(31,*42*,45),61,(62); **SO**(*00*),04,05,13,14,(51; **SH***57*); **SJ***15*.

A. roseum L. subsp. **bulbiferum** (DC.) E. F. Warb. : Rosy Garlic. Established alien (Mediterranean region). Roadsides (G,R). Fl.5-6. Mons., Glam., Cards.. 4 squares.
Mons.: Llanofer Fawr, SO30, HJF, 1951[112]. **Glam.:** Marcross, SS96, KR, 1920[112] and RGE & AEW, 1973[112]; Barry, ST16, RLS, 1938[112]. **Cards.:** Henfynyw, SN46, AOC, 1977[112].

A. triquetrum L. : Three-cornered Leek. Naturalized alien (W. Mediterranean). Hedgerows and waste places (E,Q,R,W). Fl.4-6. Glam., Carms.*, Pembs., Cards.*, Caerns.*, Denbs.*, Angl.*. 18 squares : **SR**99; **SS**09,19,48, 58,87; **SM**62,72; **SN**10,40,*42*,58; **SH**33,34,46,57,77,87.

A. ursinum L. : Craf y Geifr : Ramsons. Native. Woods (E,M,Q,W). Fl.4-6. Mons., Glam., Brecs., Rads., Carms., Pembs., Cards., Monts., Mer., Caerns., Denbs., *Flints., Angl.*. 188 squares, map 830.

A. oleraceum L. : Garlleg Rhesog y Maes : Field Garlic. Native. Pastures (D,E,G,R). Fl.7-8. Mons., Glam., Carms., *Pembs., Denbs.**, *Flints.*. 13 squares : **SS***19*,49,87; **ST**06,*17*,18,*48*,*58*,*59*; **SN**20,*31*; **SJ**06,*08*.

A. carinatum L. : Garlleg Fynyddig : Keeled Garlic. Established alien (Europe). Roadsides (D,E,R). Fl.9. Glam., Monts., Mer., *Denbs.*. 7 squares.
Glam.: Kenfig Burrows, SS78, DPr, 1962[112]; Llanishen, ST18, EV, 1944[112]. **Monts.:** Penegoes, SH70, IAW, 1959[112] and WCo, 1979[60]; nr Castell Caereinion, SJ10, RPT, 1974[116]. **Mer.:** nr Bala, SH93, PMB, 1970+[3]. **Denbs.:** Colwyn Bay, SH87, JMB, 1965+[67]; Bryn Euryn, SH88, JMB, 1968[67].

A. sativum L. : Garlleg : Garlic. Established alien (C. Asia). Sea cliffs. Fl.10. *Pembs., Denbs.*. 2 squares.
Pembs.: Wiseman's Bridge nr Tenby, SN10, TJW, c1923[22]. **Denbs.:** Penmaen Head nr Colwyn Bay, SH87, JMB, 1968[180].

A. ampeloprasum L. : Garlleg Mawr Pengrwn : Wild Leek. Denizen (Mediterranean region). Rocky sea coasts (Q,Sc). Fl.6-8. Glam., *Pembs.*, (Caerns.*). 4 squares.
Glam.: Porthkerry, ST06, HJR, c1906[140]; Flat Holm, ST26, JG, 1837[108] and RGE, 1980[95]. **Pembs.:** Wiseman's Bridge nr Tenby, SN10, TJW, c1923[23]. (**Caerns.:** Bangor, SH57, JTCW, 1876[112]).

A scorodoprasum L. : Craf y Nadroedd : Sand Leek. Casual (England). Waste places. Cards., [*Carms., Mer.*]. 1 square : **SN**68[112].

A. vineale L. : Garlleg Gwyllt : Wild Onion. Native. Pastures (D,E,G,M,Q,R,Rw). Fl.5-7. Mons., Glam., Brecs., Rads., Carms., Pembs., Cards., Monts., Mer., Caerns., Denbs., *Flints.*, Angl.. 94 squares, map 831.

IPHEION Rafin.

Ipheion uniflorum (R. C. Graham) Rafin. (*Brodiaea uniflora* (Lindley) Engler; *Milla uniflora* R. C. Graham) : Spring Starflower. Casual (S. America). Waste places. Fl.4. Glam.. 1 square : SS87[112].

CONVALLARIA L.

Convallaria majalis L. : Clych Enid : Lily-of-the-valley. Native. Woods (E,G,Q,R,Rw,W). Fl.5-6. Mons., Glam., Brecs., Rads., Carms.*, (*Cards.**), Caerns., *Denbs., Flints.*, (*Angl.**). 35 squares, map 832.

POLYGONATUM Miller

Polygonatum multiflorum (L.) All. : Sêl Solomon : Solomon's-seal. Native. Woods (D,E,R,Rw,W). Fl.4-8. Mons., Glam., Brecs., Rads., Carms., Pembs.*, (Cards.), *Monts.*, (Mer., Caerns.), *Denbs.*, (*Angl.*). 62 squares, map 833.

P. multiflorum × odoratum = P. × hybridum Brügger Naturalized alien (garden origin). Woodland. Fl.7. Cards.. 2 squares.
Cards.: nr Llanfair, SN44, JRP, 1978[112]; Llanafan, SN67, AOC, 1978[71]. It is possible that many records of **P. multiflorum** refer to this hybrid.

P. odoratum (Miller) Druce : Angular Solomon's-seal. Native. Wooded cliffs (D,Q,Sc,W). Fl.5-6. Mons., Brecs., *Pembs.,* (*Monts.*, Mer., *Denbs.**, Angl.*). 7 squares.
Mons.: Wynd Cliff, ST59, WAS, 1914[112]. **Brecs.:** Craig y Cilau, SO11, AL, 1904[112] and MP, 1970+[200]. **Pembs.:** Penally Burrows, SS19, HT, 1867[107]; Black-rock nr Tenby, SS19, BL, 1931[79]. (**Monts.:** Llanidloes, SN98, DH, c1870[163]. **Mer.:** Mochras, SH52, RDr, c1901[112] and PMB, 1972[3]. **Denbs.:** Gresford, SJ35, HDr, 1910[83]. **Angl.:** Beaumaris, SH67, RDr, 1901[112] (there is some doubt as to whether this plant was actually collected at Beaumaris)).

PARIS L.

Paris quadrifolia L. : Cwlwm Cariad : Herb-Paris. Native. Woods (M,Sc,W). Fl.4-7. Mons., Glam., Brecs., Rads., Carms., *Cards.**[71], *Monts.,* Caerns., Denbs., Flints., *Angl.*. 69 squares. map 834.

ASPARAGUS L.

Asparagus officinalis L. : Merllys : Wild Asparagus.
Subsp. **officinalis** Naturalized alien (Europe). Sea coast and waste places (D,R,Rw,S). Fl.6-8. *Mons.**, Glam., Carms., Monts., Mer., Flints., [*Cards.*]. 15 squares : **SS***48*,*58*,69,*78*,*87*; **ST***16*,18,*38*; **SN**30,40,*69*; **SO**39,51; **SJ**08,36.

Subsp. **prostratus** (Dumort.) Corb. (*A. prostratus* Dumort.). Native. Rocky slopes near the sea (D,Sc). Fl.5-6. Glam., *Carms.,* Pembs., *Flints.*, Angl., [*Mons., Cards.*]. 14 squares : **SS**19,*38*,48,*49*,58,*59*,*79*,87; **ST***06*; **SN***40*; **SH***37*,*46*; **SJ***08*,*36*.

RUSCUS L.

Ruscus aculeatus L. : Celynnen Fair : Butcher's-broom. Native. Copses (Ds,E,Sc,W). Fl.(2-)5-7. Mons., Glam., (*Brecs.**, *Rads.**), Carms., Pembs., (*Cards.,* Monts., Mer.), Caerns., Denbs., Flints., (Angl.*). 60 squares, map 835.

AGAVACEAE

YUCCA L.

Yucca gloriosa L. : Spanish-dagger. Casual (N. America). Waste places. *Glam.*[84], *Carms.*. 1 square : **SN**30[201].

AMARYLLIDACEAE

LEUCOJUM L.

[**Leucojum vernum** L. *Glam.*]

L. aestivum L. : Eiriaidd : Summer Snowflake. Naturalized alien (S. England). Damp woods (G,W). Fl.3-5. Mons.*, Glam., *Carms.**, Pembs., *Cards.*, Mer.*. 15 squares : **SS**58,68,88; **ST***06*,28; **SM**82; **SN**00,03,57, 58,62,*65*; **SO**51; **SH**60,*71*.

GALANTHUS L.

Galanthus nivalis L. : Eirlys : Snowdrop. Native but frequently introduced. Woods (E,G,M,W). Fl.1-4. Mons., Glam., Brecs., *Rads.**, Carms., Pembs., *Cards.*, (Monts., *Mer.*, Caerns.), *Denbs.*, Flints., (Angl.). 122 squares, map 836. It has not proved possible to differentiate between native and introduced populations on the map but it may be assumed that many are introduced.

NARCISSUS L.

Narcissus tazetta L. : Bunch-flowered Narcissus. Casual (Mediterranean region). Waste places. Fl.5. Glam.. 1 square : **ST***17*[112].

N. poeticus L. (*N. majalis* Curt.) : Gylfinog Farddol : Pheasant's-eye Narcissus. Naturalized alien (S. France). Pastures and woods. Fl.4-5. Mons., Glam., Pembs., Monts., *Denbs., Angl.*. 23 squares, map 837.

N. poeticus × pseudonarcissus = N. × incomparabilis Miller *s.s.* : Nonesuch Daffodil. Naturalized alien (garden origin). Pastures (D,G). Fl.4. Glam.. 5 squares. **Glam.:** nr Swansea, SS69, JAW, pre 1950[7]; Kenfig Burrows, SS78, EMT, 1943[112]; Margam, SS88, DL, c1840[81]; nr Pen-coed, SS98, JS, c1876[223]; nr Pontcanna, ST17, EV, 1933[112].

N. poeticus × tazetta = N. × medioluteus Miller (*N. × biflorus* Curtis) : Gylfinog Dauflodeuog : Primrose-peerless. Naturalized alien (garden origin). Pastures (D,E,G,Rg,W). Fl.5. Mons., Glam., Pembs., *Cards.*, Caerns., *Angl.*. 17 squares : **SR**99; **SS***48*,87,88; **ST***06,16,17*,28; **SM**72; **SN**00,*46*; **SO***30,40*,51; **SH***36*,38,57.

N. pseudonarcissus L.

Subsp. **pseudonarcissus** (incl. *N. telamonias* Link) : Cenhinen Bedr : Wild Daffodil. Native but frequently introduced. Woods and hedgebanks (E,G, R,W). Fl.3-5. Mons., Glam., *Brecs., Rads.*, Carms., (*Pembs.*, Cards.), Monts., (Mer.), Caerns., Denbs., Flints., *Angl.*. 96 squares, map 838. Possibly over-recorded as a native in the east.

Subsp. **major** (Curtis) Baker (*N. hispanicus* Gouan) : Spanish Daffodil. Naturalized alien (S. Europe). Pastures. Fl.4. Mons.*, Glam., Carms., Monts., *Denbs.*. 20 squares, map 839.

N. obvallaris Salisb. : Tenby Daffodil. Naturalized alien (garden origin?). Pastures (G,M,W). Fl.3-4. Carms., Pembs., [*Cards.*]. 16 squares : **SR**99; **SS**19; **SM**80,81; **SN**02,03,10,11,12,13,20,21,22,23,24,42.

DIOSCOREACEAE

TAMUS L.

Tamus communis L. : Gwinwydden Ddu : Black Bryony. Native. Hedges (E,G,M,Q,W). Fl.5-8. Mons., Glam., Brecs., Rads., Carms., Pembs., Cards., Monts., Mer., Caerns., Denbs., Flints., *Angl.*. 196 squares, map 840.

IRIDACEAE

SISYRINCHIUM L.

Sisyrinchium montanum E. L. Greene (*S. bermudiana* sensu Coste, non L.; *S. angustifolium* Miller) : Glaswellt Llygadlas : Blue-eyed-grass. Naturalized alien (E. N. America). Sand dunes and waste places (D,G,Q,R, Rg,Rw). Fl.6-7. *Mons.**, Glam., Carms., Pembs.*, *Cards.**, *Monts.**, Caerns.*. 16 squares : **SR**99; **SS**19, 78,*87*,88,97; **ST**17,*18*,39; **SN**10,30,69; **SO**30; **SH**53, 55; **SJ**22.

S. californicum (Ker-Gawler) Aiton fil. : Yellow-eyed-grass. Naturalized alien (California). Cliffs and woods (Ar,Sc,W). Fl.6. *Mons., Pembs.*. 2 squares. **Mons.**: Llwynycelyn, ST49, TGE, c1970[91]. **Pembs.**: Skomer Island, SM70, ARPa, 1973[183].

HERMODACTYLUS Miller

Hermodactylus tuberosus (L.) Miller : Snake's-head Iris. Established alien (S. Europe). Disturbed ground. Fl.4. Pembs..
Pembs.: Manorbier, SS09, DRe, 1981[92] and SBE, 1982[112].

IRIS L.

Subgen. Iris

Iris foetidissima L. : Iris Ddrewllyd : Stinking Iris. Native. Woods (D,E,G,R,W). Fl.5-7. Mons., Glam., (Brecs.*), *Rads.**, Carms., Pembs., (*Cards.**), *Monts., Caerns., Denbs., Flints.**, Angl., [*Mer.*]. 67 squares, map 841.

I. pseudacorus L. : Iris Felen : Yellow Iris. Native. Marshes (Al,Ar,Ds,H,M,W). Fl.5-7. Mons., Glam., *Brecs.*, Rads., Carms., Pembs., *Cards.*, Monts., *Mer.*, Caerns., Denbs., Flints., Angl.. 231 squares, map 842.

I. germanica L. : Purple Flag. Established alien (Mediterranean region). Grassy places. *Glam.*†. 1 square.
Glam.: Flat Holm, ST26, THT & JS, 1890[70].

Subgen. Xiphion (Miller) Spach

I. latifolia (Miller) Voss (*I. xiphioides* Ehrh.). Casual (Europe). Waste places. *Glam.*. 1 square : **SS***69*[81].

CROCUS L.

Crocus vernus (L.) Hill (*C. purpureus* Weston) : Saffrwm y Gwanwyn : Spring Crocus. Naturalized alien (Italy). Pastures (G,W). Fl.3-4. Glam., *Carms., Denbs.*. 5 squares.
Glam.: Fairyhill Wood, SS49, JAW, 1926[112]; Laleston, SS87, EV, 1932[91]. **Carms.**: Llansteffan, SN31, BMO, 1869[113]; nr Carmarthen, SN42, An, pre 1930[224]. **Denbs.**: Clocaenog Forest, SJ05, AGS, 1979[197].

C. biflorus Miller : Scotch Crocus. Casual (Europe). Pastures. *Glam.*. 1 square : **SS***48*[22].

C. sativus L. : Saffyr Meddygol : Saffron Crocus. Casual (garden origin). Field. *Mons.*. 1 square : **ST**48[226].

TRITONIA Ker-Gawler

Tritonia aurea Pappe ex Hooker × **pottsii** (Baker) Baker = **T.** × **crocosmiflora** (Lemoine) Nicholson (*Crocosmia* × *crocosmiflora* (Lemoine) N. E. Br.) : Montbretia. Naturalized alien (garden origin). Hedgerows (E,M,R,Rw,Sc,W). Fl.8-9. *Mons.*, Glam., *Brecs.*, Rads., Carms., *Pembs.*, Cards., Monts., *Mer.*, Caerns., *Denbs., Angl.*. 101 squares, map 843.

CURTONUS N. E. Br.

Curtonus paniculatus (Klatt) N. E. Br. : Aunt-eliza. Established alien (S. Africa). Roadsides. Brecs..
Brecs.: nr Alltmawr, SO04, RGE, 1981[112].

JUNCACEAE

JUNCUS L.

Subgen. Juncus

Juncus maritimus Lam. : Brwynen Arfor : Sea Rush. Native. Salt-marshes (Ds,M,S). Fl.5-9. Mons., Glam., Carms., Pembs., Cards., Monts., Mer., Caerns., Denbs., Flints., Angl.. 63 squares, map 844.

J. acutus L. : Llymfrwynen : Sharp Rush. Native. Dune-slacks (D,Ds,Sm). Fl.6-9. Glam., Carms., Pembs., Mer., Caerns., [*Flints.*]. 19 squares : **SS**19,49,58,69,78, 79,*87*; **SN**10,20,30,40; **SH**12,32,50,51,52,53,61,63.

Subgen. Genuini Buchenau

J. filiformis L. : Thread Rush. Native. Lake margins. Fl.10. Caerns.. 1 square.
Caerns.: nr Dolgarrog, SH76, THB, 1980[112].

J. inflexus L. : Brwynen Galed : Hard Rush. Native. Wet places (Ar,G,H,M,R,Sc,W). Fl.6-10. Mons., Glam., Brecs., Rads., Carms., Pembs., Cards., *Monts.** Mer., Caerns., Denbs., Flints., Angl.. 165 squares, map 845.

J. effusus L. : Brwynen Babwyr : Soft Rush. Native. Wet places (Al,Ar,E,G,H,M,Q,R,W). Fl.5-10. Mons., Glam., Brecs., Rads., Carms., Pembs., Cards., Monts., Mer., Caerns., Denbs., *Flints.*, Angl.. 276 squares, all except : **SR**89; **SS**39; **SM**50,73; **SH**51,98.

J. effusus × **inflexus** =**J.** × **diffusus** Hoppe Spontaneous. Wet places (G,M). Fl.7-9. Mons., Glam., Brecs., *Pembs., Caerns.*, Angl.. 10 squares **SS**19,49,*69*, 88; **ST***07,27*; **SO**02; **SH**23,47,48.

J. conglomeratus L. (*J. subuliflorus* Drejer) : Brwynen Babwyr Bellennaidd : Compact Rush. Native. Marshy meadows (Al,G,H,M,Q,S). Fl.6-9 Mons., Glam., Brecs., *Rads.*, Carms., Pembs., Cards., *Monts.*, Mer., Caerns., Denbs., *Flints.*, Angl.. 226 squares, map 846.

J. conglomeratus ×**effusus** = **J.** × **kern-reichgeltii** Jansen & Wachter ex Van Ooststr. Spontaneous. Wet places in uplands. *Glam., Caerns.*, [*Pembs.*]. 2 squares.
Glam.: Barry, ST16, An, no date[123]. **Caerns.**: nr Llyn Idwal, SH65, ADQA, 1950+[238]; Cwm Glas, SH65, ADQA, 1950+[238]; Llyn Llydaw, SH65, ADQA, 1950+[238].

[**J. conglomeratus** × **inflexus** = **J.** × **ruhmeri** Ascherson & Graebner *Glam.*]

Subgen. Pseudotenageia V. Krecz. & Gontsch.

J. squarrosus L. : Brwynen Droellgorun : Heath Rush. Native. Moorlands (G,H) Fl.5-9. Mons., Glam., Brecs., Rads., Carms., Pembs., Cards., Monts., Mer., Caerns., Denbs., *Flints., Angl.*. 181 squares, map 847.

J. compressus Jacq. : Brwynen Talgron : Round-fruited Rush. Native. Marshes. Fl.6-8. *Mons.**, Brecs., Caerns., *Flints.*, Angl., [*Glam., Mer.*]. 11 squares : **ST**48; **SO***02*,12,41; **SH**36,*45,46*,47,67,*77*; **SJ***08*.

J. gerardi Loisel. : Brwynen Gerard : Saltmarsh Rush. Native. Salt-marshes (Ar,H,M,Q,R,S). Fl.6-8. Mons., Glam., Carms., Pembs., Cards., Monts., Mer., Caerns., Denbs., Flints., Angl.. 84 squares, map 848.

J. tenuis Willd. : Slender Rush. Naturalized alien (America). Roadsides (F,G,M,Q,R,Rw,W). Fl.6-10. Mons.*, Glam., Brecs., Rads., Carms., Pembs., Cards., Monts., Mer., Caerns., Flints.*, *Angl.*. 117 squares, map 849.

Subgen. Poiophylli Buchenau

J. bufonius agg. : Brwynen y Llyffant : Toad Rush. Native. Moist waysides (Ar,Ds,G,H,M,Q,R,Rw,S,W). Fl.6-9. Mons., Glam., Brecs., Rads., Carms., Pembs., Cards., Monts., Mer., Caerns., Denbs., Flints., Angl.. 268 squares, all except : **SS**38,39; **ST**26; **SM**73,81,94; **SN**21,36,66,77,87; **SO**08; **SH**98; **SJ**37.

J. foliosus Desf. Native. Wet places (H,G,M). Fl.6-9. Glam., Brecs., Pembs., Monts., Mer., *Caerns.*, Flints., Angl.. 36 squares, map 850.

J. bufonius L. *s.s.* (incl. *J. minutulus* V. Krecz. & Gontsch.). Native. Wet places (Al,Ar,Ds,G,H,M, Q,R,Rw,S,W). Fl.6-9. Mons., Glam., Brecs., Rads., Carms., Pembs., Cards., Monts., Mer., Caerns., Denbs., Flints., Angl.. 146 squares, map 851.

J. **ambiguus** Guss. (*J. ranarius* Song. & Perr.). Native. Coastal mud-flats (G,M,S). Fl.6-8(-10). Glam., Pembs., Mer., Caerns., *Flints.,* Angl.. 19 squares : **SS**10,49,*78*; **ST**06,*17*; **SN***10*; **SH**28,32,34, 36,37,46,50,52,53,*57*,61,78; **SJ**26.

Subgen Juncinella V. Krecz. & Gontsch.

J. capitatus Weigel : Dwarf Rush. Native. Moist sandy places. Fl.6. Angl.. 2 squares.
Angl.: Newborough Warren, SH36, An, 1937[36]; nr Rhosneigr, SH37, BK, 1917[17] and GCD, 1925[112].

Subgen. Septati Buchenau

J. subnodulosus Schrank : Brwynen Glymog a Blodau Blaendwn : Blunt-flowered Rush. Native. Marshes (Al, D,Ds,G,H,M,R,S,W). Fl.6-8. Mons.*, Glam., *Rads.,* Carms., Pembs., *Cards., Monts.,* Mer., Caerns., Denbs., Flints., Angl.. 49 squares, map 852.

J. bulbosus agg. : Brwynen Fwlbaidd : Bulbous Rush. Native. Wet heaths and bogs (Al,Ar,G,H,M,W). Fl.6-9. Mons., Glam., Brecs., *Rads.,* Carms., Pembs., Cards., Monts., Mer., Caerns., Denbs., Flints., Angl.. 211 squares, map 853.

J. bulbosus L. *s.s.* Native. Wet heaths (Al,Ar,H,M). Fl.5-9. Mons., Glam., Brecs., *Rads.,* Carms., Pembs., Cards., Monts., Mer., Caerns., Denbs., Angl.. 54 squares, map 854.

J. kochii F. W. Schultz Native. Wet heaths and bogs (Al,Ar,G,H,M,Q). Fl.6-9. Mons., Glam., Brecs., Rads.*, Carms., Pembs., Cards., Monts.*, Mer., Caerns., Denbs.*, *Flints., Angl.**. 142 squares, map 855.

J. acutiflorus Ehrh. ex Hoffm. : Brwynen Flodfain : Sharp-flowered Rush. Native. Marshes (Ar,D,Ds,G,H, M,Q,W). Fl.6-10. Mons., Glam., Brecs., Rads., Carms., Pembs., Cards., Monts., Mer., Caerns., Denbs., *Flints.,* Angl.. 237 squares, map 856.

J. acutiflorus × articulatus = J. × surrejanus Druce Spontaneous. Marshy ground (G,M). Fl.6-10. *Rads.,* Pembs., Cards., Monts., Mer., Caerns., *Denbs.,* Angl.. 24 squares, map 857.

J. articulatus L. : Brwynen Gymalog : Jointed Rush. Native. Marshes (Al,Ar,D,Ds,G,H,M,R,W). Fl.6-10. Mons., Glam., Brecs., Rads., Carms., Pembs., Cards., Monts., Mer., Caerns., Denbs., Flints., Angl.. 264 squares, all except : **SS**38,39; **ST**26,49,58; **SM**50,71,73; **SN**00,24,43; **SO**40; **SH**29,51,80,98; **SJ**34,45.

Subgen. Alpini Buchenau

J. triglumis L. : Brwynen Tri-flodeuog : Three-flowered Rush. Native. Boggy places. Fl.7-8. Caerns., [*Brecs., Cards.*]. 2 squares.
Caerns.: Cwm Idwal, SH65, JHS, 1925[112] and ER, 1964[207]; Ysgolion Duon, SH66 JEG, 1892[112] and ER, 1960[207]; Carnedd Llywelyn, SH66, AW, 1928[112] and ER, 1964[207]; nr Llyn Melynllyn, SH66, ER, 1963[207].

LUZULA DC.

Subgen. Luzula

Luzula campestris (L.) DC. : Coedfrwynen y Maes : Field Wood-rush. Native. Pastures (D,Ds,E,G,H,M,Q, R,W). Fl.3-6. Mons., Glam., Brecs., Rads., Carms., Pembs., Cards., Monts., Mer., Caerns., Denbs., Flints., Angl.. 267 squares, all except : **SR**89; **SS**39; **SM**50,90, 94; **SN**24,25,32,43,59; **SO**30,40,41; **SH**51,98.

L. multiflora (Retz.) Lej. : Coedfrwynen Luosben : Heath Wood-rush. Native. Heaths (G,H,M,Q,R,W). Fl.5-8. Mons., Glam., Brecs., Rads., Carms., Pembs., Cards., Monts., Mer., Caerns., Denbs., Flints., Angl.. 221 squares, map 858.

[**L. pallescens** Swartz *Rads.*]

[**L. spicata** (L.) DC. *Caerns.*]

Subgen. Anthelaea (Griseb.) Buchenau

L. sylvatica (Hudson) Gaudin : Coedfrwynen Fawr : Great Wood-rush. Native. Woods (M,Q,W). Fl.3-8. Mons., Glam., Brecs., Rads., Carms., Pembs., Cards., Monts., Mer., Caerns., Denbs., Flints., *Angl.*. 190 squares, map 859.

L. luzuloides (Lam.) Dandy & Wilmott : White Wood-rush. Naturalized alien (Europe). Woods. *Mons.*. 1 square.
Mons.: btwn Wynd Cliff and Tintern, ST59, MCo, 1922[226].

Subgen. Pterodes (Griseb.) Buchenau

L. pilosa (L.) Willd. : Coedfrwynen Flewog : Hairy Wood-rush. Native. Woods (E,M,Rw,W). Fl.3-6(-8). Mons., Glam., Brecs., Rads., Carms., Pembs., Cards., Monts., Mer., Caerns., Denbs., Flints., Angl.. 185 squares, map 860.

L. forsteri (Sm.) DC. : Coedfrwynen Gulddail : Southern Wood-rush. Native. Woods (E,R,W). Fl.4-6. Mons., Brecs.*, Rads.*, *Cards.*[230], [*Glam.*]. 10 squares : **ST***49*,59; **SO**05,13,14,21,22,41,50,51.

L. forsteri × pilosa = L. × borreri Bromf. ex Bab. Spontaneous. Woods. Fl.5-7. Mons., Brecs.. 3 squares.
Mons.: Wynd Cliff, ST59, AL, 1879[107]; Lady Mill nr Mounton, ST59, WAS, 1894[112]; Barbadoes Hill, SO50, WAS, c1900[216]. **Brecs.**: Grwyne Ffechan, SO22, MP, 1973[112]. There are specimens from *Caerns.* in **NMW** & **ABS** that closely resemble this hybrid, but one of the parents, **L. forsteri,** has yet to be recorded from that vice-county.

COMMELINACEAE

COMMELINA L.

Commelina diffusa Burm. fil. (*C. nudiflora* auct., non L.) : Day-flower. Casual (Tropics). Waste places. Fl.9. Mons.. 1 square : **ST**29[112].

GRAMINEAE

SASA Makino & Shibata

Sasa palmata (Burbidge) E. G. Camus (*S. senanensis* auct. brit.) : Bamboo. Naturalized alien (Japan). Roadsides and woods. Fl.9. Glam., Carms., Cards.. 5 squares.
Glam.: Merthyr Mawr, SS87, RGE, 1982[112]; Margam, SS88, RGE, 1982[112]; Castell Coch, ST18, MG, 1972[99]. **Carms.**: Edwinsford, SN63, AOC, 1965[112]. **Cards.**: Llanbadarn Fawr, SN68, AOC, 1980[112].

ARUNDINARIA Michx

Arundinaria japonica Siebold & Zucc. ex Steudel : Bamboo. Naturalized alien (Japan). Woods. Glam., Brecs., *Carms.,* Cards.. 8 squares.
Glam.: Margam, SS88, RGE, 1982[112]; nr Garn-swllt, SN61, RGE, 1981[112]. **Brecs.**: nr Erwood, SO04, RGE, 1982[112]. **Carms.**: Middleton Hall, SN51, RDP, 1980[201]. **Cards.**: Henllan Bridge, SN34, AOC, 1980[61]; Trawsgoed, SN67, AOC, 1978[112]; Llanbadarn Fawr, SN68, AOC, 1980[61]; Lodge Park, SN69, AOC, 1978[112].

FESTUCA L.

Festuca altissima All. : Peisgwellt y Gwigoedd : Wood Fescue. Native. Woods (Ar,M,Q,W). Fl.5-7. Mons., Glam., Brecs., Cards., Monts., Mer.. 19 squares : **ST**59; **SN**70,*75*,77,79,80,81,89; **SO***02*,22,*50*,51; **SH**61, 71,72,74,82,*92*,*93*.

F. gigantea (L.) Vill. : Peisgwellt Mawr : Giant Fescue. Native. Woods and hedgebanks (E,G,Sc,W). Fl.6-9. Mons., Glam., Brecs., Rads., Carms., Pembs., Cards., Monts., Mer., Caerns., Denbs., Flints., *Angl.*. 191 squares, map 861.

F. pratensis Hudson : Peisgwellt y Waun : Meadow Fescue. Native but frequently introduced. Pastures (Ar,E,G,M,R,Rw). Fl.6-10. Mons., Glam., Brecs., Rads., Carms., Pembs., Cards., Monts., *Mer.*, Caerns., *Denbs., Flints.,* Angl.. 141 squares, map 862.

F. arundinacea Schreber : Peisgwellt Tal : Tall Fescue. Native but frequently introduced. Wood borders and upper regions of salt-marshes (E,F,G,M,Q,R,S). Fl.5-8. Mons., Glam., Brecs., Rads., Carms., Pembs., Cards., Monts., Mer., Caerns., Denbs., *Flints.,* Angl.. 126 squares, map 863.

F. heterophylla Lam. : Grandmother's hair. Introduced (Europe). Roadside. Fl.6. *Mons.*. 1 square.
Mons.: nr Monmouth, SO51, RL, 1943[107].

F. nigrescens Lam. (*F. rubra* subsp. *commutata* Gaudin; *F. rubra* var. *fallax* (Thuill.) Hackel) : Chewings Fescue. Colonist (S. England). Grassy roadsides (F,G,Q,R). Fl.5-6. Mons., Glam., Brecs., [*Mer.*] 5 squares.
Mons.: Newport, ST38, TGE, 1978[112]. **Glam.**: Pyle, SS88, EV, c1930[112]; Llandough nr Cardiff, ST17, AEW, 1920[112]; Cardiff, ST17, AEW, 1972[112]; Castell Coch, ST18, EV, 1932[112]; Caerphilly Common, ST18, RGE & AEW, 1974[117]. **Brecs.**: nr Merthyr Tydfil, SO00, EV, 1932[112].

F. rubra L. : Peisgwellt Coch : Red Fescue. Native. Pastures (D,E,F,G,H,M,Q,R,S,Sc,W). Fl.5-7. Mons., Glam., Brecs., Rads., Carms., Pembs., Cards., Monts., Mer., Caerns., *Denbs., Flints.,* Angl.. 275 squares, all except : **SN**21,32,43; **SO**32,41; **SH**51; **SJ**10.

The following subspecies have been recorded from Wales but their distribution is very imperfectly known.

Subsp. **rubra** Native. Pastures (E,F,G,H,Q,R,Rw, Sc). Fl.6-7. Mons., Glam., Brecs., Rads., Carms., Pembs., Cards., Monts., Mer., Caerns., *Denbs., Flints.,* Angl.. Distribution probably much the same as for the species and not mapped separately.

Subsp. **litoralis** (G. F. W. Meyer) Auquier Native. Tidal banks of rivers (Ar,G,M,S). Fl.6-7. Mons., *Glam., Carms.*. 7 squares.
Mons.: Rumney, ST27, AEW, 1932[112]; Newport, ST38, TGE, 1978[112]; Collister Pill, ST48, TGE, 1980[91]; Chepstow, ST59, TGE, 1978[112]. **Glam.**: Kenfig Burrows, SS78, TGE, 1978[184]; Sully Island, ST16, TGE, 1979[60]. **Carms.**: Tywyn Burrows, SN30, RDP, 1981[201]

Subsp. **arenaria** (Osbeck) Syme (*F. rubra* var. *arenaria* Fries). Native. Sand dunes. Fl.6. Glam., Pembs., *Cards., Angl.*. 5 squares.
Glam.: Crymlyn Burrows, SS79, JAW, 1939[112]. **Pembs.**: Tenby Burrows, SS19, BS, 1963[112]. **Cards.**: Tan-y-bwlch Beach, SN57, JHS, c1940[211]. **Angl.**: nr Rhosneigr, SH37, RHR, 1970+[209]; Newborough Warren, SH46, AHLH, c1970[156].

Subsp. **pruinosa** (Hackel) Piper (*F. rubra* var. *pruinosa* (Hackel) Howarth). Native. Grassy places (G,R,Rw,S,Sc). Fl.5-7. Mons., Glam., Pembs., Angl.. 12 squares: **SR**99; **SS**96; **ST**17; **SM**80,82,83; **SN**04; **SO**51; **SH**28,36,37,46.

F. juncifolia St-Amans : Rush-leaved Fescue. Native. Dunes. Fl.6-7. Glam., Denbs.. 4 squares.
Glam.: Gower, SS48, CAS & RC, 1970+[242]; Oxwich, SS58, HCE, 1959[104]; Three Cliffs Bay, SS58, QONK, 1978[157]; Kenfig Burrows, SS78, PLT, c1973[217]. **Denbs.**: Kinmel Bay, SH98, JMB, 1976[112].

F. diffusa Dumort. (*F. megastachys* Hegetschw.; *F. rubra* var. *grandiflora* Hackel). Casual (Europe). Waste places. Fl.5. Mons., *Caerns.*. 3 squares : **ST**27[112],38[112]; **SH**23[74].

F. tenuifolia Sibth. (*F. ovina* subsp. *tenuifolia* (Sibth.) Peterm.) : Fine leaved Sheep's-fescue. Native. Grassy places (D,E,G,H,R). Fl.6-8. Mons., Glam., *Brecs.**, Rads.*, *Carms.*[85], *Pembs.*[85], Cards., *Monts.*[85], *Mer.*[85], Caerns.[85], *Denbs.**, *Angl.*[85]. 14 squares : **SS**58,79,99; **ST**07,27,48,58; **SN**68; **SO**05,11,25,51; **SH**85; **SJ**05. Although recorded from all but one of the Welsh counties, there are very few localized records.

F. ovina L. : Peisgwellt y Defaid : Sheep's-fescue. Native. Heaths (D,Ds,G,H,M,Q,R,Rw,Sc,W). Fl.5-7. Mons., Glam., Brecs., Rads., *Carms.*, Pembs., *Cards.*, Monts., Mer., Caerns., *Denbs.*, *Flints.*, *Angl.*. 266 squares, all except : **SS**39; **ST**27,39; **SM**50; **SN**00,20, 21,22,36; **SO**30,32; **SH**51,90,98; **SJ**10,45.

F. guestfalica Boenn. ex Reichenb. (*F. longifolia* auct., non Thuill.). Casual (Europe). Waste places. Fl.6. Mons.. 1 square : **ST**28[112].

F. arvernensis Auquier, Kerguélen & Markgr. -Dannenb. (*F. glauca* Lam., non Vill.). Casual (Europe). Heaths. *Angl.*. 1 square : **SH**36[30].

F. vivipara (L.) Sm. : Peisgwellt Bywhiliog : Viviparous Fescue. Native. Mountains. Fl.6-8. Brecs., Cards.*, *Monts.**, Mer., Caerns., *Denbs.**, *(Angl.*)*. 29 squares, map 864.

FESTUCA × LOLIUM = × FESTULOLIUM Ascherson & Graebner

Festuca gigantea × **Lolium perenne** = × **Festulolium brinkmanii** (A. Braun) Ascherson & Graebner Spontaneous. *Pembs.*, *Mer.*. 2 squares.
Pembs.: Haverfordwest, SM91, An, 1972[49]. **Mer.**: nr Pennal, SN79, PMB, 1973[3].

F. pratensis × **L. multiflorum** = × **F. braunii** (K. Richter) A. Camus Spontaneous. Wetland. *Denbs.*. 1 square.
Denbs.: nr Denbigh, SJ06, JMB, 1978[59].

F. pratensis × **L. perenne** = × **F. loliaceum** (Hudson) P. Fourn. : Hybrid Fescue. Spontaneous. Meadows (G,R). Fl.6-7. *Mons.*, Glam., Brecs., *Rads.*, *Carms.*, *Pembs.*, *Cards.*, Monts., Angl.. 28 squares, map 865.

FESTUCA × VULPIA = × FESTULPIA Melderis ex Stace & R. Cotton

Festuca juncifolia × **Vulpia fasciculata** = × **Festulpia melderisii** Stace & R. Cotton Spontaneous. Sand dunes. *Glam.*. 1 square.
Glam.: Kenfig Burrows, SS78, PLT, 1973[217].

F. rubra × **V. fasciculata** = × **F. hubbardii** Stace & R. Cotton Spontaneous. Sand dunes Fl.6-7. Glam., Carms.[217], Mer., Caerns., Angl.. 6 squares.
Glam.: Llangennith Burrows, SS49, CAS & RC, c1970[242]; Kenfig Burrows, SS78, QONK, 1979[157]; Jersey Marine, SS79, PLT, 1969[217]. **Mer.**: Morfa Harlech, SH53, PMB, 1958[4]. **Caerns.**: Llandudno, SH78, WEH, 1968[112]. **Angl.**: Newborough Warren, SH46, PMB, 1958[112].

F. rubra × **V. myuros** Spontaneous. Roadside and railway. Fl.7. Mer.. 1 square.
Mer.: Arthog, SH61, PMB, 1957[112] & 1970[3].

LOLIUM L.

Lolium perenne L. : Rhygwellt Lluosflwydd : Perennial Rye-grass. Native. Pastures (D,E,G,M,R,W). Fl.5-11. Mons., Glam., Brecs., Rads., Carms., Pembs., Cards., Monts., *Mer.*, Caerns., Denbs., Flints., Angl.. 277 squares, all except : **SS**39; **SM**50,71,73; **SH**51.

L. multiflorum Lam. (*L. perenne* subsp. *multiflorum* (Lam.) Husnot; *L. perenne* var. *aristatum* Schumacher) : Rhygwellt Eidalaidd : Italian Rye-grass. Naturalized alien (S. Europe). Pastures (E,G,R,Rg). Fl.6-10(-12). Mons., Glam., *Brecs.*, Rads., Carms., Pembs., Cards., Monts., Mer., Caerns., Denbs., Flints., *Angl.*. 155 squares, map 866.

L. multiflorum × **perenne** = **L.** × **hybridum** Hausskn. Spontaneous or introduced. Grassy places and roadsides. *Glam.*[217], *Mer.*[4], *Angl.*. 2 squares.
Angl.: Hirdrefaig, SH47, RHR, 1978[208]; nr Pentraeth, SH57, RHR, 1968[208].

L. multiflorum × **rigidum** = **L.** × **hubbardii** Jansen & Wachter Spontaneous. Rubbish tip. Mons.. 1 square : **ST**38[112].

L. multiflorum × **temulentum** Spontaneous. Rubbish tip. Mons.. 1 square : **ST**38[112].

L. rigidum Gaudin Casual (Mediterranean region). Waste places. Fl.7. Mons., *Glam.*. 3 squares : **SS***78*[140]; **ST**38[112]; **SO***00*[107].

L. temulentum L. : Efrau : Darnel. Established alien or casual (Europe). Waste places (G,R,Rg). Fl.5-8. Mons., Glam., Carms.*, *Cards.*, Caerns., *Denbs.*, *Flints.*, Angl.. 17 squares : **SS***78*; **ST**17,*27*,38,*49*; **SN**31,*40*,58, *90*; **SH***28*,*57*,*67*,*78*; **SJ***08*,12,*25*,*26*.

VULPIA C. C. Gmelin

Vulpia geniculata (L.) Link (*Festuca geniculata* (L.) Willd.). Casual (S. Europe). Waste places. Fl.7. *Glam.*. 1 square : **ST***27*[25].

V. ligustica (All.) Link (*Festuca ligustica* (All.) Bertol.). Casual (Europe). Waste places. Fl.5. Glam.. 1 square : **ST***27*[112].

V. fasciculata (Forskål) Samp. (*V. membranacea* auct., non (L.) Dumort.) : Peisgwellt Uncib : Dune Fescue. Native. Sand dunes (D,G,S). Fl.5-7. Glam., (*Brecs.*), Carms., Pembs., Cards., Mer.*, Caerns., Angl.. 36 squares, map 867.

V. bromoides (L.) S. F. Gray : Peisgwellt Anhiliog : Squirreltail Fescue. Native. Dry sandy places (D,G,H, Q,R,Rw,W). Fl.5-7. Mons., Glam., Brecs., Rads., Carms., Pembs., Cards., Monts., Mer., Caerns., Denbs., *Flints.*, Angl.. 187 squares, map 868.

V. myuros (L.) C. C. Gmelin : Peisgwellt y Fagwyr : Rat's-tail Fescue. Native but frequently introduced, especially in the north. Waysides (Q,R,Rw). Fl.5-7. Mons., Glam., *Brecs.*, Rads., Carms., Pembs., Cards.*, *Monts.*, Mer., Caerns., Denbs., Flints.*, *Angl.*. 100 squares, map 869.

[**V. ciliata** Dumort. subsp. **ambigua** (Le Gall) Stace & Auquier (*V. ambigua* (Le Gall) More; *Festuca ambigua* Le Gall). *Glam.*]

DESMAZERIA Dumort.

Desmazeria marina (L.) Druce (*Catapodium marinum* (L.) C. E. Hubbard) : Corwenithwellt y Morfa : Sea Fern-grass. Native. Sandy shores (Q,R,Rw,S,Sc). Fl.6-8. Glam., Carms., Pembs., Cards., Mer., Caerns., *Denbs.**, Flints., Angl.. 74 squares, map 870.

D. marina × rigida Spontaneous. Railway ballast. Fl.5. *Mer.*†. 1 square.
Mer.: Aberdyfi Station, SN69, PMB, 1960[4].

D. rigida (L.) Tutin (*Catapodium rigidum* (L.). C. E. Hubbard) : Gweunwellt Anhyblyg : Fern-grass. Native. Walls (Ar,D,M,Q,R,Rw). Fl.5-9. Mons., Glam., Brecs.*, Rads.*, Carms., Pembs., Cards., Monts.*, Mer., Caerns., Denbs., Flints., *Angl.*. 109 squares, map 871.

POA L.

Poa annua L. : Gweunwellt Unflwydd : Annual Meadow-grass. Native. Waysides (D,E,F,G,Q,R, Rg,Rw,S,Sc,W). Fl.4-10. Mons., Glam., Brecs., Rads., Carms., Pembs., Cards., *Monts.*, Mer., Caerns., Denbs., Flints., Angl.. 279 squares, all except : **SR**89; **SS**39; **SH**51.

P. trivialis L. : Gweunwellt Llederw : Rough Meadow-grass. Native. Meadows (Ar,E,G,M,R,Rw,W). Fl.5-8. Mons., Glam., Brecs., Rads., Carms., Pembs., Cards., *Monts.*, Mer., Caerns., Denbs., *Flints.*, Angl.. 271 squares, all except : **SS**39,77; **SM**50,84; **SN**22,25, 36,99; **SO**26,41; **SJ**00.

P. pratensis agg. Native. Pastures (D,E,G,Q,R,Rw,S). Fl.5-8. Mons., Glam., Brecs., Rads., Carms., Pembs., Cards., Monts., Mer., Caerns., Denbs., *Flints.*, Angl.. 271 squares, all except : **SS**39; **SM**50,71,73,90; **SN**32, 51,63,88,98; **SH**80.

P. subcaerulea Sm. (*P. pratensis* subsp. *subcaerulea* (Sm.) Tutin; *P. pratensis* subsp. *irrigata* (Lindm.) Lindb. fil.) : Spreading Meadow-grass. Native. Grassland (D,Ds,G,M,Q,R,Rw). Fl.5-8. Mons., Glam., Brecs., Carms.*, Pembs., Cards.*, *Monts.**, Mer., Caerns.*, Denbs.*, *Flints.**, Angl.. 105 squares, map 872.

P. pratensis L. *s.s.* (*P. pratensis* subsp. *pratensis*) : Gweunwellt Llyfn : Smooth Meadow-grass. Native. Pastures (D,G,M,Q,R,Rw). Fl.5-7. Mons., Glam., Brecs., Rads., Carms., Pembs., Cards., Monts., Mer., Caerns., Denbs., *Flints.*, Angl.. 131 squares, map 873. Very much under-recorded as the segregate.

P. angustifolia L. (*P. pratensis* subsp. *angustifolia* (L.) Gaudin) : Narrow-leaved Meadow-grass. Native but frequently introduced. Rough grassland (D,G,M, Q,R,Rw). Fl.5-6. Mons.*, Glam., *Brecs.**, *Rads.**, *Carms.**, *Pembs.**, Cards.*, *Monts.**, Mer.*, *Caerns.**, *Denbs.**, *Flints.**, *Angl.**. 53 squares, map 874.

P. chaixii Vill. : Broad-leaved Meadow-grass. Established alien (Europe). Plantations (E,G,Rg,W). Fl.6-7. *Mons.**, Glam., *Brecs.**, *Cards.**, Monts.*, Caerns.*, *Denbs.**. 9 squares.
Mons.: Pen-clawdd, SO40, NYS, 1946[45]. **Glam.**: Aberdare, SO00, HJR, 1905[112]. **Brecs.**: Llangamarch, SN94, MP, 1976[200]; Llysdinam, SO05, FMS, 1975[200]. **Cards.**: Llanfihangel Ystrad, SN55, AOC, 1978[112]. **Monts.**: Llangurig, SN97, JAW, 1953[112] and JH, 1970+[116]. **Caerns.**: nr Ro-wen, SH77, AW, 1936[7]. **Denbs.**: Bodnant Gardens, SH77, JMB, 1970+[67]; nr Chirk, SJ23, VG, 1963[93].

P. compressa L. : Gweunwellt Cywasg : Flattened Meadow-grass. Colonist (England). Walls (E,Q,R, Rg,Rw). Fl.7-9. Mons., Glam., Brecs., *Rads.*, Carms., Pembs., Cards.*, Monts.*, Mer., *Caerns.**, *Denbs.**, *Flints.**, Angl.. 77 squares, map 875.

P. palustris L. : Swamp Meadow-grass. Naturalized alien (England). Marshy ground (M,R,Rg). Fl.7-9. *Mons.*, Glam., Mer., Caerns., *Denbs.*. 7 squares.
Mons.: Newport, ST38, DM, 1954[53]. **Glam.**: Aberthaw, ST06, ARP, 1979[112]; Barry, ST16, GCD, 1925[24]; Cardiff Docks, ST17, HJR, 1904[24]; Splott, Cardiff, ST27, AEW, 1926[112]. **Mer.**: nr Minffordd, SH53, DIH & GWe, 1961[4] and PMB, 1970[112]. **Caerns.**: nr Portmadoc, SH53, APC, 1961[112]. **Denbs.**: Trefnant, SJ07, JMB, 1978[184].

P. glauca Vahl : Glaucous Meadow-grass. Native. Mountain rock-ledges. Fl.6-8. Caerns.. 3 squares.
Caerns.: Snowdon, SH65, HD, c1800[112] and ER, 1956[207]; Cwm Idwal, SH65, ER, 1954[207]; Ysgolion Duon, SH66, JEG, 1892[112] and ER, 1952[207]; Carnedd Ddafydd, SH66, AL, 1891[112]; Moel Siabod, SH75, ER, 1952[207].

P. balfourii Parnell : Balfour's Meadow-grass. Native. Mountain rock-ledges. Fl.6-9. Caerns.. 2 squares.
Caerns.: Snowdon, SH65, CCB, 1847[108] and ER, 1956[207]; Cwm Idwal, SH65, ER, 1950[207]; Cwm Cywion, SH65, ER, 1956[207]; Cwm Gafr, SH66, JEG, 1888[112]; Carnedd Ddafydd, SH66, JEG, c1890[101]; Ysgolion Duon, SH66, ER, 1959[207]; Carnedd Llywelyn, SH66, ER, 1959[207]. Doubt has been expressed over some of these records.

P. nemoralis L. : Gweunwellt y Coed : Wood Meadow-grass. Native. Woods (E,F,M,Q,R,W). Fl.5-8(-10). Mons., Glam., Brecs., Rads., Carms., *Pembs.*, Cards., Monts., Mer., Caerns., Denbs., *Flints.*, Angl.. 151 squares, map 876.

P. bulbosa L. var. **vivipara** Koeler : Bulbous Meadow-grass. Native. Shingle beaches. Fl.3-6. Glam.. 2 squares. **Glam.**: Coldknap nr Barry, ST06 and ST16, HJR, 1907[112] and ARP & AEW, 1974[95].

P. alpina L. : Alpine Meadow-grass. Native. Mountain rocks. Fl.6-8. Mer., Caerns.. 3 squares.
Mer.: Aran Benllyn and Aran Fawddwy, SH82, FS, 1871[113]. **Caerns.**: Snowdon, SH65, JSH, 1826[112] and ER, 1959[207]; Carnedd Ddafydd, SH66, JEG, c1890[101]; Carnedd Llywelyn, SH66, JLW, c1890[101] and ER, 1955[209].

PUCCINELLIA Parl.

Puccinellia distans (L.) Parl. : Gweunwellt Gwrthblygedigaidd : Reflexed Saltmarsh-grass. Native. Saltmarshes (Q,S). Fl.5-9. Mons., Glam., Carms.*, Cards., Caerns., *Denbs., Flints.,* Angl.*. 24 squares, map 877.

P. fasciculata (Torrey) E. P. Bicknell : Borrer's Saltmarsh-grass. Native. Sandy mud-flats. Fl.6-7. Glam., *Carms.**, [*Mons.*]. 2 squares.
Glam.: Cardiff, ST27, HJR, 1905[112]. **Carms.**: Llansteffan, SN31, JFH & PCH, c1950[161].

P. maritima (Hudson) Parl. : Gweunwellt Arfor : Common Saltmarsh-grass. Native. Salt-marshes (D,M,S). Fl.5-7(-9). Mons., Glam., Carms., Pembs., *Cards.,* Monts., Mer., Caerns., Denbs., *Flints.,* Angl.. 75 squares, map 878.

P. rupestris (With.) Fernald & Weatherby : Stiff Saltmarsh-grass. Native. Mud-flats (Ar,M,R,S). Fl.5-9. Mons., Glam., Pembs.*, (*Caerns.**), [*Mer., Denbs.*]. 11 squares : **SS***38*; **ST***17,27*,28,38,*59*; **SM***72*,82,83,91; (**SH**12).

DACTYLIS L.

Dactylis glomerata L. : Troed y Ceiliog : Cock's-foot. Native. Pastures (D,E,G,M,Q,R,Rw,Sc,W). Fl.5-10. Mons., Glam., Brecs., Rads., Carms., Pembs., Cards., Monts., Mer., Caerns., Denbs., Flints., Angl.. 278 squares, all except : **SS**39; **SM**50; **SN**59; **SH**51.

BECKMANNIA Host

Beckmannia eruciformis (L.) Host Casual (Europe). Waste places (R,Rw). Fl.7-8. Mons., Glam.. 2 squares : **ST***27*[112]; **SO***51*[112].

B. syzigachne (Steudel) Fernald Casual (Asia and N. America). Waste places. *Glam.*. 2 squares : **ST**16[51],*27*[51].

CYNOSURUS L.

Cynosurus cristatus L. : Rhonwellt y Ci : Crested Dog's-tail. Native. Pastures (D,E,G,M,R,Rw,W). Fl.5-8. Mons., Glam., Brecs., Rads., Carms., Pembs., Cards., Monts., Mer., Caerns., Denbs., *Flints.,* Angl.. 278 squares, all except : **SS**39; **SM**50; **SH**51,98.

C. echinatus L. : Rough Dog's-tail. Casual (S. Europe). Waste places (R,Rw). Fl.5-8. *Mons.**, Glam., Monts., Caerns., *Angl.*. 10 squares : **ST**08,16,*17,27*; **SO**30; **SH**37,57,*67,77*; **SJ**22.

C. elegans Desf. Casual (Mediterranean region). Waste places. Fl.5. Glam.. 1 square : **ST***27*[112].

LAMARCKIA Moench

Lamarckia aurea (L.) Moench Casual (Mediterranean region). Waste places. *Mons., Glam.*. 2 squares : **ST***27*[26],38[245].

CATABROSA Beauv.

Catabrosa aquatica (L.) Beauv. : Brigwellt Dyfrdrig : Whorl-grass. Native. Pond margins (Al,Ar,G,H,M,Sc). Fl.5-8. Mons., Glam., Brecs., Rads.*, *Carms., Pembs.,* Caerns., Denbs., Flints., Angl., [*Cards., Mer.*]. 32 squares, map 879.

APERA Adanson

Apera spica-venti (L.) Beauv. : Maeswellt Sidanaidd : Loose Silky-bent. Casual (Europe). Waste places (R,Rg). Fl.6-7(-10). Glam., Brecs., *Denbs.*[85], *Angl.*[104]. 6 squares : **SS***69*[140],*78*[107]; **ST***17*[112],*27*[112]; **SO***00*[107],*02*[112].

A. interrupta (L.) Beauv. : Dense Silky-bent. Casual (S. Europe). Waste places. Fl.7-8. Glam.. 2 squares : **ST***17*[112]; **SO***00*[112].

MIBORA Adanson

Mibora minima (L.) Desv. : Eiddil-welltyn Cynnar : Early Sand-grass. Native. Sandy places. Fl.4-5. Glam.*, Angl.. 4 squares.
Glam.: Whitford Burrows, SS49, EAGD & GTG, 1964[112] and TGE, 1980[91]. **Angl.**: Llyn Coron, SH36, JTCW, 1876[112]; Aberffraw, SH36, GCD, 1914[112] and RHR, 1970+[209]; nr Rhosneigr, SH37, RHR, 1970+[209]; nr Llyn Maelog, SH37, CPH, 1905[112]; Maelog Sands, SH37, JEG, 1915[112] and An, 1970+[5]; nr Valley Airfield, SH37, RHR, 1970+[209]; Bodorgan, SH37, JSi, 1876[112]; Malltraeth, SH46, RHR, 1970[112].

BRIZA L.

Briza media L. : Crydwellt : Quaking-grass. Native. Pastures (D,E,G,H,M,Q,R,Rw,Sc,W). Fl.6-8. Mons., Glam., Brecs., Rads., Carms., *Pembs., Cards., Monts.,* Mer., Caerns., Denbs., Flints., Angl.. 190 squares, map 880.

B. maxima L. : Crydwellt Mwyaf : Great Quaking-grass. Casual but occasionally becoming established (Mediterranean region). Waste places (E,G,Q,R,Rg). Fl.5-9. Mons., Glam., Pembs., Cards., Caerns., *Denbs., Angl.*. 14 squares : **SS**87; **ST**07,*18,27*,38; **SN**10,58,68; **SH**23,32,33,37,57; **SJ**34.

B. minor L. : Crydwellt Bychan : Lesser Quaking-grass. Casual (S. Europe). Waste places. *Mons., Glam.,* [*Flints.*]. 2 squares : **ST***27*[25],59[91].

SESLERIA Scop.

Sesleria albicans Kit. ex Schultes (*S. caerulea* (L.) Ard. subsp. *calcarea* (Čelak.) Hegi) : Corswelltyn Rhuddlas : Blue Moor-grass. Native. Limestone pastures. *Brecs.*†, Flints.. 2 squares.
Brecs.: nr Hirwaun, SN90, HJR, c1900[139]. **Flints.**: nr Cilcain, SJ16, JPS, 1956[7].

MELICA L.

Melica nutans L. : Meligwellt Gogwydd : Mountain Melick. Native. Rocky limestone woods (Q,R,Rw,W). Fl.5-7. Mons., Glam., Brecs., Rads., *Monts.*, Mer., Denbs.*. 24 squares, map 881. There is an unconfirmed record from *Caerns.*: **SH**75.

M. uniflora Retz. : Meligwellt : Wood Melick. Native. Woods (E,M,Q,R,W). Fl.5-7. Mons., Glam., Brecs., Rads., Carms., *Pembs.*, Cards., *Monts.*, Mer., Caerns., Denbs., Flints., *Angl.*. 171 squares, map 882.

GLYCERIA R. Br.

Glyceria maxima (Hartman) Holmberg : Perwellt : Reed Sweet-grass. Native. Ditches (Al,Ar,H,M). Fl.7-8. Mons., Glam., Brecs.*, Carms., *Pembs.*, *Cards.*, Monts., *Caerns.*, *Denbs.*, *Flints.*, *Angl.*, [*Mer.*]. 60 squares, map 883.

G. declinata Bréb. : Perwellt Llwydlas : Small Sweet-grass. Native. Muddy pondsides (Al,Ar,G,H,M,W). Fl.5-9. Mons., Glam., Brecs., Rads., Carms., *Pembs.*, Cards., Monts.*, Mer., Caerns., *Denbs.*, Flints.*, Angl.*. 188 squares, map 884.

G. fluitans (L.) R. Br. : Glaswellt y Dŵr : Floating Sweet-grass. Native. Streamsides (Al,Ar,G,H,M,W). Fl.5-10. Mons., Glam., Brecs., Rads., Carms., Pembs., Cards., Monts., Mer., Caerns., Denbs., Flints., Angl.. 258 squares, all except : **SR**89; **SS**19,38,39,68,77; **ST**26; **SM**50,71,73,84,94; **SN**00,04,11,20,21,25,36; **SO**28,50; **SH**68,80,98.

G. fluitans × plicata = G. × pedicellata Townsend : Hybrid Sweet-grass. Spontaneous. Streamsides (Al,Ar,M). Fl.5-8. Mons., Glam., Brecs., *Rads.*[123], *Pembs.*, Cards., *Monts.*, Mer., *Caerns.*, *Denbs.*, *Flints.*, *Angl.*. 54 squares, map 885.

G. plicata (Fries) Fries : Perwellt Plygedig : Plicate Sweet-grass. Native. Marshes (Al,Ar,G,H,M). Fl.5-8. Mons., Glam., Brecs., Rads., Carms., *Pembs.*, *Cards.*, *Monts.*, Caerns., *Denbs.*, *Flints.*, Angl., [*Mer.*]. 114 squares, map 886.

BROMUS L.

Bromus diandrus Roth (*B. gussonii* Parl.; *B. maximus* auct.; *B. rigens* auct.; *Anisantha diandra* (Roth) Tutin; *A. gussonii* (Parl.) Nevski) : Great Brome. Established alien or casual (Europe). Waste places and dunes (D,R, Rw). Fl.5-8. Mons., Glam., *Carms.*, *Pembs.*, *Monts.*, Mer., Caerns., *Flints.*. 18 squares : **SS**48,69,78,87; **ST**06,16,17,18, 27,38,59; **SN**10,51,69; **SO**19; **SH**53,77; **SJ**08.

B. rigidus Roth (*Anisantha rigida* (Roth) Hyl.). Casual (Channel Isles). Waste places. Fl.7. *Glam.*. 3 squares : **SS**78[140],87[109]; **ST**17[140].

B. sterilis L. (*Anisantha sterilis* (L.) Nevski) : Bromwellt Hysb : Sterile Brome. Native. Roadsides and waste places (D,E,F,G,H,M,Q,Rg,Rw,S). Fl.5-8. Mons., Glam., Brecs., Rads., Carms., Pembs., Cards., Monts., Mer., Caerns., Denbs., *Flints.*, *Angl.*. 212 squares, map 887.

B. tectorum L. (*Anisantha tectorum* (L.) Nevski) : Drooping Brome. Casual (Europe). Waste places. Fl.6-9. Mons., Glam.. 7 squares : **SS**78[112]; **ST**16[112],17[112],27[112],38[184],59[112]; **SN**90[107].

B. madritensis L. (*Anisantha madritensis* (L.) Nevski) : Bromwellt Dwysedig : Compact Brome. Colonist (S. England). Waste places (Q,R,Rg). Fl.5-7. Mons., Glam., *Carms.*, Pembs., Cards., *Flints.*. 17 squares : **SS**09,19,59,69,78,87; **ST**06,16,17,18,27,48,59; **SM**90; **SN**00,58; **SJ**08.

B. rubens L. (*Anisantha rubens* (L.) Nevski) : Foxtail Chess. Casual (S. Europe). Waste places. Fl.5-6. Glam.. 2 squares : **ST**17[112],27[112].

B. inermis Leysser (*Zerna inermis* (Leysser) Lindman) : Hungarian Brome. Casual (Europe). Sandy ground. Fl.7. Mons., *Pembs.*, Mer., *Caerns.*. 6 squares : **ST**38[91],59[112]; **SM**72[92]; **SH**23[74],50[5],53[112].

B. ramosus Hudson (*Zerna ramosa* (Hudson) Lindman) : Bromwellt Blewog : Hairy-brome. Native. Hedges (E,G,Q,W). Fl.6-10. Mons., Glam., Brecs., Rads., Carms., Pembs., Cards., Monts., Mer., Caerns., Denbs., Flints., Angl.. 150 squares, map 888.

B. benekenii (Lange) Trimen (*Zerna benekenii* (Lange) Lindman; *B. ramosus* var *benekenii* (Lange) Druce) : Lesser Hairy-brome. Native. Woods and hedges. Fl.6-7. Mons., *Brecs.**, *Rads.**. 5 squares.
Mons.: Runston, ST49, WAS, c1920[226]; Piercefield Park, ST59, AL, 1878[107] and AEW, c1930[216]; Wynd Cliff, ST59, AL, 1876[107] and TGE, 1972[91]; nr Tintern, SO50, SHB, 1905[107]. **Brecs.**: nr Callwen, SN81, AL, 1899[106]. **Rads.**: Aberedw, SO04, AL, 1886[107].

B. erectus Hudson (*Zerna erecta* (Hudson) S. F. Gray) : Bromwellt Syth : Upright Brome. Native. Dry limestone fields (E,G,Q,R,Rw,W). Fl.5-7(-11). Mons., Glam., Brecs., Carms.*, Pembs., (Mer.), *Caerns.*, Denbs., Flints.*, *Angl.*. 44 squares, map 889.

B. arvensis L. : Field Brome. Casual (Europe). Waste places (G,R). Fl.6-8. Glam., *Carms.*, *Cards.*, Mer., *Denbs.*, *Flints.*. 13 squares : **SS**69,78,87; **ST**17,18,27; **SN**58,73,80; **SO**00; **SH**53,87; **SJ**16.

B. pseudosecalinus P. M. Sm. Established alien (England). Upland pasture. Fl.7. *Brecs.*, Monts.. 3 squares.
Brecs.: Hay-on-Wye, SO24, MP, 1980[200]. **Monts.**: Carno, SN99, IAW, 1956[107]; Betws, Lower Garth, SJ09, EM, 1955[109].

B secalinus L. : Bromwellt Ller : Rye Brome. Casual or established alien (Europe). Waste places (F,G,R,Rg). Fl.6-9. Mons., Glam., Brecs.*, *Rads.**, Carms., Cards., *Monts.**, *Mer.**, Caerns., *Flints.*, *Angl.*, [*Pembs.*, *Denbs.*]. 27 squares, map 890.

B. commutatus Schrader : Pawrwellt Mwyaf y Maes : Meadow Brome. Native. Meadows (F,G,R). Fl.5-7. Mons., Glam., *Brecs.**, *Rads.*[230], Carms., *Pembs.*, *Cards.*, *Monts.*, *Mer.*, Caerns., Denbs., *Flints.**. 31 squares, map 891.

B. racemosus L. : Bromwellt Llyfn : Smooth Brome. Native. Meadows (E,F,G,R,Rw,W). Fl.6-8. Mons., Glam., Carms.*, *Pembs.**, Cards., (Mer.), *Caerns.*, *Denbs.*[230], *Angl.*, [*Rads.*, *Monts.*]. 34 squares, map 892.

B. hordeaceus L. Native. Grassy places (D,G,R,Rw, S,Sc). Fl.6-7. Mons., Glam., Brecs., Rads., Carms., Pembs., Cards., Monts., Mer., Caerns., Denbs., Flints., Angl.. 265 squares, all except : **SS**39; **SM**50; **SN**35,59, 63,88,89; **SO**09; **SH**29,65,80,81,91,92; **SJ**00,11,13.

Subsp. **hordeaceus** (*B. mollis* L.) : Bromwellt Maswy : Soft-brome. Native. Meadows (D,E,G,H). Fl.5-10. Mons., Glam., Brecs., Rads., Carms., Pembs., Cards., Monts., Mer., Caerns., Denbs., Flints., Angl.. 217 squares, map 893.

Subsp. **ferronii** (Mabille) P. M. Sm. (*B. ferronii* Mabille) : Least Soft-brome. Native. Sea cliffs (D,Q,Sc). Fl.5-6. Glam.*, Pembs.*, Cards., Caerns.*, Angl.. 28 squares, map 894. Some of the identifications of this species are provisional only.

Subsp. **thominii** (Hard.) Maire & Weiller Native. Sand dunes (D,G). Fl.5-8. Glam., *Carms.**, *Cards.*, Mer.*, Angl., [*Mons.*, *Brecs.*, *Rads.*, *Pembs.*, *Monts.*, *Caerns.*, *Flints.*]. 7 squares.
Glam.: Kenfig Burrows, SS78, TGE, 1980[112]; Crymlyn Burrows, SS79, WRL, 1888[112]. **Carms.**: Burry Port, SN40, PMS, 1965[238]. **Cards.**: Gwbert, SN14, PMB, 1954[112]; Ynys-las, SN69, PMS, 1963[238]. **Mer.**: Morfa Dyffryn, SH52, PMB, 1956[112]. **Angl.**: Aberffraw Common, SH37, JEG, 1892[112] and RHR, 1961[112]. The identifications of specimens in **NMW** are provisional only.

B. hordeaceus × lepidus = B. × pseudothominii P. M. Sm. (*B. thominii* sensu Tutin, non Hard.) : Lesser Soft-brome. Spontaneous. Roadsides (D,E,F,G, M,Q,R,Rw). Fl.5-9. Mons., Glam., Brecs., Rads., *Carms.*, Pembs., Cards., *Monts.*, Mer., Caerns., Flints., Angl.. 69 squares, map 895.

B. lepidus Holmberg : Pawrwellt Gweirglawdd : Slender Soft-brome. Naturalized alien (Europe). Meadows (D,F,G,R,Rw). Fl.5-8. Mons., Glam., Brecs., *Rads.**, Pembs., Cards.*, *Monts.**, Mer., Caerns.*, *Denbs.**, *Angl.**. 49 squares, map 896.

B. lanceolatus Roth (*B. macrostachys* Desf.). Casual (Europe). Waste places. Fl.6. Mons., Glam.. 2 squares : **ST**27[112],38[112].

B. japonicus Thunb. Casual (Europe). Waste places. Fl.6. Glam.. 1 square : **ST***27*[112].

B. squarrosus L. Casual (Europe). Waste places. Fl.6. Glam.. 2 squares : **SS***79*[112]; **ST***27*[112].

B. briziformis Fischer & C. A. Meyer Casual (Europe). Waste places. *Glam.*. 1 square : **SN***90*[107].

B. willdenowii Kunth (*B. unioloides* Kunth; *Ceratochloa unioloides* (Willd.). Beauv.) : Rescue Brome. Casual (S. America). Waste places. Fl.6-10. Mons., Glam., *Flints.*. 12 squares : **SS***78*; **ST***16*,*17*,27,38,59; **SN***50*; **SO***00*,*40*; **SH***26*, *27*,36.

B. carinatus Hooker & Arnott (*Ceratochloa carinata* (Hooker & Arnott) Tutin) : California Brome. Casual (N. America). Waste places. Fl.7-10. Glam., Denbs.. 2 squares : **ST**16; **SJ**23.

BRACHYPODIUM Beauv.

Brachypodium sylvaticum (Hudson) Beauv. : Breichwellt y Coed : False Brome. Native. Hedges (E,G,M,Q,R,Rw,Sc,W). Fl.6-10. Mons., Glam., Brecs., Rads., Carms., Pembs., Cards., Monts., Mer., Caerns., Denbs., Flints., Angl.. 242 squares, map 897.

B. pinnatum (L.) Beauv. : Breichwellt y Tŵr : Torgrass. Native. Limestone pastures (G,H,R,Rw,W). Fl.5-8. Mons., Glam., (Carms.*), Pembs., (Cards.*), [*Caerns.*]. 13 squares : **SR**99; **SS**96; **ST**06,16,48,59; (**SN**14,30,40,54); **SO***00*,25,51.

LEYMUS Hochst.

Leymus arenarius (L.) Hochst. (*Elymus arenarius* L.) : Amdowellt : Lyme-grass. Native. Sea shores (D,S). Fl.5-9. Glam., Carms.*, Pembs., *Cards.*†, Mer., Caerns., Denbs., Flints., *Angl.*. 34 squares, map 898.

ELYMUS L.

Elymus caninus (L.) L. (*Agropyron caninum* (L.) Beauv.) : Marchwellt y Coed : Bearded Couch. Native. Wood borders (E,G,M,Q,R,Rw,W). Fl.6-9. Mons., Glam., Brecs., Rads., Carms., Pembs., Cards., Monts., Mer., Caerns., Denbs., Flints., Angl.. 127 squares, map 899.

E. pycnanthus (Godron) Melderis (*Agropyron pycnanthum* (Godron) Gren. & Godron; *A. pungens* auct., non (Pers.) Roemer & Schultes) : Marchwellt Arfor : Sea Couch. Native. River estuaries and salt-marshes (D,M,R,S,Sc). Fl.6-9. Mons., Glam., Carms.*, Pembs., *Cards.*, Mer., Caerns., *Denbs.**, *Flints.*, *Angl.*. 47 squares, map 900.

E. pycnanthus × repens (*Agropyron × oliveri* Druce). Spontaneous. Estuaries. Fl.7-8. *Glam.*, Mer., [*Angl.*]. 2 squares.
Glam.: Port Talbot, SS78, HJR, 1904[107]. **Mer.**: Dysynni Broadwater, SH50, PMB, 1954[112].

E. repens (L.) Gould (*Agropyron repens* (L.) Beauv.) : Marchwellt : Common Couch. Native. Cultivated ground (D,E,F,G,Q,R,Rg,Rw,S). Fl.6-10. Mons., Glam., Brecs., Rads., Carms., Pembs., Cards., *Monts.,* Mer., Caerns., Denbs., *Flints.,* Angl.. 236 squares, map 901.

E. farctus (Viv.) Runemark ex Melderis subsp. **boreali-atlanticus** (Simonet & Guinochet) Melderis (*Agropyron junceiforme* (Á. & D. Löve) Á. & D. Löve) : Marchwellt Tywyn : Sand Couch. Native. Sand dunes (D,G, S,Sc). Fl.6-7. Glam., Carms., Pembs., Cards., *Monts.,* Mer., *Caerns.,* Denbs., Flints., Angl.. 69 squares, map 902.

E. farctus × pycnanthus (*Agropyron acutum* auct.; *A.* × *obtusiusculum* Lange). Spontaneous. Sand dunes (D, Rw,S). Fl.6-9. *Glam., Pembs.,* Mer., *Caerns., Angl..* 12 squares : **SS**48,49,58,68,69,79; **ST**06; **SN**10; **SH**37,53,61,*67*.

E. farctus × repens (*Agropyron* × *laxum* (Fries) Tutin). Spontaneous. Shore line (D,G,R,S). Fl.7-8. Glam., Mer., Caerns., *Angl..* 13 squares : **SS***78,87*; **ST**06; **SH**33,46,48,51,52,53,57,58,61,*67*.

AEGILOPS L.

Aegilops ventricosa Tausch (*Triticum ventricosum* (Tausch) Cesati, Passer. & Gibelli). Casual (Europe). Waste places. Fl.6. Glam.. 1 square : **ST***27*[112].

A. cylindrica Host (*Triticum cylindricum* (Host) Cesati, Passer. & Gibelli). Casual (Europe). Rubbish tip. Glam.. 1 square : **ST**17[112].

A. triuncialis L. (*Triticum triunciale* (L.) Raspail). Casual (Europe). Waste places. Fl.6. *Glam..* 2 squares : **ST***17*[218],*27*[107]

A. geniculata Roth (*A. ovata* L. p.p.; *Triticum ovatum* (L.) Gren. & Godron.). Casual (Europe). Waste places. Fl.6-7. Glam.. 2 squares : **SS***78*[107]; **ST***27*[112].

A. neglecta Req. ex Bertol. (*A. triaristata* Willd.; *Triticum triaristatum* Gren. & Godron). Casual (Europe). Cultivated ground. Fl.8. Glam.. 1 square : **ST***27*[112].

TRITICUM L.

Triticum durum Desf. Casual (cultivated origin). Rubbish tip. Mons.. 1 square : **ST**38[112].

T. turgidum L. : Rivet Wheat. Casual (cultivated origin). Waste places. *Carms..* 1 square : **SN**30[178].

T. spelta L. Casual (cultivated origin). Waste places. Fl.6. Glam.. 1 square : **ST***27*[112].

T. aestivum L. (*T. vulgare* Vill.) : Gwenith : Bread Wheat. Casual (cultivated origin). Waste places. Fl.7-9. Mons., Glam., Rads., Mer., Flints.. 14 squares : **SS**69,78,79; **ST**06,09, 16,17,27,28; **SN**80,90; **SO**36; **SH**61; **SJ**36.

SECALE L.

Secale cereale L. : Rhyg : Rye. Casual (cultivated origin). Waste places (F,R). Fl.8-10. Glam., Carms.. 7 squares : **SS***78*[140]; **ST***16*[112],*17*[112],18[112],*27*[112]; **SN**30[112]; **SO***00*[140].

HORDEUM L.

Hordeum distichon L. : Two-rowed Barley. Casual (cultivated origin). Waste places. Fl.6-10. Mons., Glam., Carms., *Caerns.,* Flints.. 16 squares : **SS***69*,98; **ST**06,09,17, 18,27,38,59; **SN**31,51,*90*; **SH***12*,*77*; **SJ**18,36.

H. vulgare L. : Llys y Bara : Six-rowed Barley. Casual (cultivated origin). Waste places. Fl.6. Glam., *Carms., Angl..* 10 squares : **SS**69,*78*,*79*; **ST**06,17; **SN**30,42,80; **SO***00*; **SH**68.

H. murinum L. : Heiddwellt y Mur : Wall Barley. Native. Waste places (D,E,Q,R,Rw). Fl.6-10. Mons., Glam., Brecs., *Rads.**, Carms., Pembs., Cards., Monts., Mer., Caerns., Denbs., Flints., Angl.. 100 squares, map 903.

H. marinum Hudson : Heiddwellt y Morfa : Sea Barley. Native. Banks by the sea (M,R,S). Fl.6-8. Mons., Glam., *Carms.**, [*Mer., Caerns.*]. 11 squares : **SS***69,79*; **ST**06*17*,27,28,*48*,58,59; **SN**31,40. Records from *Mer.* and *Caerns.* require confirmation.

H. hystrix Roth (*H. gussoneanum* Parl.) : Mediterranean Barley. Casual (S. Europe). Waste places. Fl.6. Glam.. 1 square : **ST***27*[112].

H. secalinum Schreber : Heiddwellt y Maes : Meadow Barley. Native. Meadows (Ar,G,M,R,S). Fl.6-7. Mons., Glam., *Brecs.**, Carms., Pembs.*, (Cards.*), *Monts.**, Caerns., Denbs., *Flints., Angl..* 48 squares, map 904.

H. jubatum L. : Foxtail Barley. Casual (N. America). Waste places. Fl.8-9. Mons., Glam., Pembs., *Caerns..* 5 squares : **ST***17*[112],*27*[112]; **SN**92[112]; **SO**11[112]; **SH**12[74].

HORDELYMUS (Jessen) C. O. Harz

Hordelymus europaeus (L.) C. O. Harz : Heiddwellt y Coed : Wood Barley. Native. Woods. Fl.6-10. Mons., Monts.*. 4 squares.
Mons.: Slade Wood, ST48, TGE, 1960+[91]; Wynd Cliff, ST59, WHPu, 1847[112]; Redding's Inclosure, SO51, SGC, 1934[112]; Lady Park Wood, SO51, SGC, 1941[112]. **Monts.**: Abermule, SO19, EM, 1970[112].

AVENA L.

Avena strigosa Schreber : Blewgeirch : Bristle Oat. Casual (Europe). Cornfields (F,R). Fl.6-7. Mons., Glam., *Brecs.,* Pembs., *Cards., Mer., Caerns.,* Angl.. 23 squares, map 905.

A. fatua L. : Ceirchwellt Gwyllt y Gwanwyn : Wild-oat. Established alien (Europe). Cultivated ground (E,F, R,Rw). Fl.7-9. Mons., Glam., Brecs., *Carms., Pembs.**, *Cards., Monts.**, *Mer.**, *Caerns., Flints.**, *Angl.**. 44 squares, map 906.

A. sativa L. : Ceirch : Cultivated Oat. Casual (cultivated origin). Waste places and roadsides (F,G,R). Fl.6-9. Mons., Glam., *Brecs., Rads.,* Carms., Monts., Caerns., Denbs., *Flints., Angl..* 38 squares, map 907.

A. sterilis L.
Subsp. **sterilis** (incl. *A. orientalis* Schreber) : Animated Oat. Casual (Europe). Waste places. Fl.6-10. Mons., Glam., *Angl..* 4 squares : **ST**16[112],17[29],*27*[112]; **SH***37*[21].

Subsp. **ludoviciana** (Durieu) Nyman (*A. ludoviciana* Durieu) : Ceirchwellt Gwylltyr Hydref : Winter Wild-oat. Casual (Europe). Waste places. Glam., *Flints.*. 3 squares : **ST***16*[26]; **SN**61[112]; **SJ**08[261].

AVENULA (Dumort.) Dumort.

Subgen. Pubavenastrum (Vierh.) J. Holub

Avenula pubescens (Hudson) Dumort. (*Helictotrichon pubescens* (Hudson) Pilger) : Ceirchwellt Blewog : Downy Oat-grass. Native. Pastures (D,E,G,H,Q,R,Rw,S,Sc). Fl.5-7. Mons., Glam., Brecs., *Rads.**, Carms., Pembs., Cards., *Monts.**, Mer., Caerns., *Denbs.*, Flints., Angl.. 102 squares, map 908.

Subgen. Avenula

A. pratensis (L.) Dumort. (*Helictotrichon pratense* (L.) Besser) : Ceirchwellt Culddail : Meadow Oat-grass. Native. Limestone pastures (G,Sc). Fl.6-7. Mons., Glam., *Brecs.*, *Carms.**[161], *Pembs.*, *Monts.**, Caerns., *Denbs.*, *Flints.**, Angl.. 28 squares, map 909.

ARRHENATHERUM Beauv.

Arrhenatherum elatius (L.) Beauv. ex J. & C. Presl
Subsp. **elatius** : Ceirchwellt Tal : False Oat-grass. Native. Hedgerows (D,E,G,M,Q,R,S,Sc,W). Fl.6-8. Mons., Glam., Brecs., Rads., Carms., Pembs., Cards., Monts., Mer., Caerns., Denbs., Flints., *Angl.*. 276 squares, all except : **SS**39; **SM**50,73,84; **SN**98; **SO**32.

Subsp. **bulbosum** (Willd.) Schübler & Martens (*A. bulbosum* (Willd.) C. Presl; *A. tuberosum* (Gilib.) F. W. Schultz; *A. elatius* var. *tuberosum* Ascherson) : Onion Couch. Native. Hedgerows (E,F,G,R). Fl.7-8. *Mons.*, Glam., *Brecs.*, *Carms.*, Pembs., *Cards.*, Monts., Mer., Caerns., *Angl.*[209]. 18 squares : **ST***16*,18,28,*58*,59; **SM**70,*72*; **SN**21,36,*58*,61,90,*96*; **SO**11,19; **SH***45*,61,71. Certainly much more common than records indicate.

GAUDINIA Beauv.

Gaudinia fragilis (L.) Beauv. : French Oat. Casual (S. Europe). Waste places (G,R,S). Fl.6-9. Glam., *Carms.*[241]. 4 squares : **ST**06[241],*16*[112],*17*[24],*27*[112].

KOELERIA Pers.

Koeleria macrantha (Ledeb.) Schultes (*K. cristata* (L.) Pers.; *K. albescens* auct., p.p.) : Cribwellt : Crested Hair-grass. Native. Dry pastures and sea cliffs (D,G,R,S,Sc). Fl.5-7. Mons., Glam., *Carms.*, Pembs., Cards., Caerns., *Denbs.*, Flints., Angl., [*Mer.*]. 72 squares, map 910.

K. glauca (Schrader) DC. (*K. albescens* auct., p.p.; *K. arenaria* Dumort.) Native. Sandy places. Fl.5-7. Specimens resembling this species have been recorded from Glam., Pembs., Cards., Caerns., Angl., but further research is needed before its detailed distribution can be worked out.

LOPHOCHLOA Reichenb.

Lophochloa cristata (L.) Hyl. (*Koeleria phleoides* (Vill.) Pers.) Casual (Europe). Waste places. Glam.. 3 squares : **ST***16*[23],*17*[112],*27*[112].

TRISETUM Pers.

Trisetum flavescens (L.) Beauv. : Ceirchwellt Melyn : Yellow Oat-grass. Native. Pastures (D,E,G,M,R,Rw,S,W). Fl.5-7(-11). Mons., Glam., Brecs., Rads., Carms., *Pembs.*, Cards., Monts., Mer., Caerns., *Denbs.*, Flints., Angl.. 139 squares, map 911.

LAGURUS L.

Lagurus ovatus L. : Cwt Ysgyfarnog : Hare's-tail. Casual (Channel Isles). Waste places (Q,R,Rg). Fl.6. *Mons.*, Glam., *Mer.*. 7 squares : **SS***69*[26],*87*[22]; **ST**16[112],*17*[140],*27*[112],38[184]; **SH***61*[126].

DESCHAMPSIA Beauv.

Deschampsia cespitosa (L.) Beauv.
Subsp. **cespitosa** : Brigwellt Cydynnog : Tufted Hair-grass. Native. Wet grassland (E,G,H,M,Q,R,Rw,W). Fl.6-9. Mons., Glam., Brecs., Rads., Carms., Pembs., Cards., Monts., Mer., Caerns., Denbs., *Flints.*, Angl.. 261 squares, all except : **SR**89; **SS**19,38,39,77,96; **ST**26; **SM**50,71,73,84,94; **SN**04,15,20,25,36; **SH**51,78,97,98.

Subsp. **alpina** (L.) Tzvelev (*D. alpina* (L.) Roemer & Schultes) : Alpine Hair-grass. Native. High mountain rocks. Fl.6-9. Caerns.. 2 squares. **Caerns.**: Clogwyn Du'r Arddu, SH55, JEG, c1890[101]; Snowdon, SH65, JEG, c1890[101] and ER, 1950+[207]; Glyders, SH65, JEG, 1889[104] and ER, 1950+[207]; Cwm Cywion, SH65, BS, 1961[112].

D. setacea (Hudson) Hackel : Bog Hair-grass. Native. Wet peaty ground. Fl.7-8. *Pembs.*†, Caerns.. 3 squares. **Pembs.**: Dowrog Pool, SM72, AGM, c1884[129]. **Caerns.**: Mynydd Cilan, SH22, APC, 1974[112] and APC & PMB, 1980[3]; Cwmistir Heath, SH23, APC & PMS, 1956[74].

D. flexuosa (L.) Trin. : Brigwellt Main : Wavy Hair-grass. Native. Heaths (D,G,H,M,Q,R,Rw,W). Fl.5-9. Mons., Glam., Brecs., Rads., Carms., Pembs., Cards., Monts., Mer., Caerns., Denbs., *Flints.*, Angl.. 217 squares, map 912.

AIRA L.

Aira praecox L. : Brigwellt y Gwanwyn : Early Hair-grass. Native. Heaths (D,E,G,H,Q,R,Sc). Fl.5-7. Mons., Glam., Brecs., Rads., Carms., Pembs., Cards., Monts., Mer., Caerns., Denbs., *Flints.*, *Angl.*. 234 squares, map 913.

A. caryophyllea L. : Brigwellt Arian : Silver Hair-grass. Native. Heaths (D,E,G,H,Q,R,Rw,W). Fl.5-8. Mons., Glam., Brecs., Rads., Carms., Pembs., Cards., Monts., Mer., Caerns., Denbs., *Flints.*, Angl.. 214 squares, map 914.

ANTHOXANTHUM L.

Anthoxanthum odoratum L. : Chwyth yr Wŷdd : Sweet Vernal-grass. Native. Meadows (Ar,D,E,F,G,H, M,Q,R,Rw,Sc,W). Fl.4-9. Mons., Glam., Brecs., Rads., Carms., Pembs., Cards., Monts., Mer., Caerns., Denbs., *Flints., Angl.*. 276 squares, all except : **SR**89; **SS**39; **SM**50; **SO**30; **SH**51,98.

A. aristatum Boiss. (*A. puelii* Lecoq & Lamotte) : Annual Vernal-grass. Casual (Europe). Waste places (R,Rw). Glam.. 2 squares : **ST***16*[26]; **SO***00*[112].

HOLCUS L.

Holcus lanatus L. : Maswellt : Yorkshire-fog. Native. Meadows (Al,Ds,E,G,H,M,Q,R,Rw,S,W). Fl.6-9. Mons., Glam., Brecs., Rads., Carms., Pembs., Cards., Monts., Mer., Caerns., Denbs., Flints., Angl.. 282 squares.

H. lanatus × **mollis** = **H.** × **hybridus** K. Wein. Spontaneous. Woodland edges. *Cards., Caerns.*. 2 squares.
Cards.: Cwm Rhyddnant, SN77, CPC & KJ, c1960[56].
Caerns.: Porth Neigwl, SH22, PMB & APC, 1977[74],

H. mollis L. : Maswellt Rhedegog : Creeping Soft-grass. Native. Woods (E,G,H,R,S,W). Fl.6-9. Mons., Glam., Brecs., *Rads.,* Carms., Pembs., Cards., Monts., Mer., Caerns., Denbs., *Flints., Angl.*. 243 squares, map 915.

CORYNEPHORUS Beauv.

Corynephorus canescens (L.) Beauv. : Grey Hair-grass. Naturalized alien (S. England). Sandy fields (D,G,S). Fl.7. Glam.. 2 squares.
Glam.: Aberafan, SS78, HJR, 1904[112] and EV, 1938[112]; Cardiff Docks, ST17, JS, c1876[218].

AGROSTIS L.

Agrostis canina L. : Maeswellt y Cŵn : Brown Bent. Native. Wet grassland and heaths (D,Ds,F,G,H,M,Q, R,S,W). Fl.(1-)6-9. *Mons.,* Glam., Brecs., Rads., *Carms.,* Pembs., Cards., *Monts.,* Mer., Caerns., *Denbs.,* Flints., Angl.. 229 squares, map 916.

A. vinealis Schreber (*A. canina* L. subsp. *montana* (Hartman) Hartman; *A. canina* var. *arida* Schlect.) : Heath Bent. Native. Dry heaths (H,M). Fl.6-8. Mons., Glam., Brecs., Pembs., Mer., Caerns., *Angl.*. 23 squares, map 917. Under-recorded due to confusion with the preceding species.

A. curtisii Kerguélen (*A. setacea* Curtis, non Vill.) : Maeswellt Gwrychog : Bristle Bent. Native. Heaths (G,H,R,Sc). Fl.6-7(-9). Glam.. 12 squares : **SS**48,49,58, 59,68,69,87,88,97,98; **ST**07,*18*.

A. scabra Willd. Established alien (N. America). Railway ballast. Mons.. 1 square.
Mons.: Newport Docks, ST38, TGE, & AG, 1980[245].

A. capillaris L. (*A. tenuis* Sibth.) : Maeswellt Cyffredin : Common Bent. Native. Pastures (D,E,F,G,H,M,Q,R, Rw,S,W). Fl.6-10. Mons., Glam., Brecs., Rads., Carms., Pembs., Cards., Monts., Mer., Caerns., Denbs., *Flints.,* Angl.. 270 squares, all except : **SR**89; **SS**39; **SM**50; **SO**30,41; **SH**51,98; **SJ**10,33,34,44,45.

A. capillaris × **gigantea** = **A.** × **bjoërkmannii** Widén Spontaneous. Grassland. Fl.9. Caerns.. 1 square.
Caerns.: nr Bangor, SH57, RHR, 1979[112].

A. capillaris × **stolonifera** = **A.** × **murbeckii** Fouillade ex P. Fourn. Spontaneous. Grassland. Fl.7. *Cards.,* Mer., *Caerns.,* Flints., Angl.. 11 squares : **SN**69; **SH**32,33,46,48,49,52,53,57,61; **SJ**18.

A. gigantea Roth : Maeswellt Mawr : Black Bent. Native or colonist. Cultivated ground (Ar,Ds,E,F, G,M,R,Rg,Rw,W). Fl.7-9. Mons., Glam., Brecs.*, Rads.*, Carms., Pembs.*, *Cards., Monts.*,* (Mer.*), Caerns., Denbs., *Flints.*,* Angl.. 69 squares, map 918.

A. stolonifera L. : Maeswellt Gwyn y Maes : Creeping Bent. Native. Meadows (Ar,D,Ds,E,G,H,M,Q,R,Rg,Rw, S,Sc,W). Fl.6-8. Mons., Glam., Brecs., *Rads.,* Carms., Pembs., Cards., Monts., Mer., Caerns., Denbs., Flints., Angl.. 270 squares, all except : **SS**39; **SM**50,84; **SN**15, 25,35,36,43,45,88; **SO**32; **SH**82.

A. avenacea J. F. Gmelin (*Calamagrostis filiformis* (G. Forster) Pilger ex Ostenf.; *Deyeuxia retrofracta* Kunth). Casual (Australasia). Waste places. Fl.8. Glam.. 2 squares : **ST***16*[24],*17*[112].

AGROSTIS × POLYPOGON = × AGROPOGON P. Fourn.

Agrostis stolonifera × **Polypogon monspeliensis** = × **Agropogon littoralis** (Sm.) C. E. Hubbard : Perennial Beard-grass. Spontaneous. Muddy pond margins. Fl.7-10. Glam.. 1 square.
Glam.: Cardiff Docks, ST27, JS, c1876[218] and EV, 1929[112] (habitat now destroyed).

GASTRIDIUM Beauv.

Gastridium ventricosum (Gouan) Schinz & Thell. : Nit-grass. Native. Limestone cliffs (E,G,R). Fl.6-8. (Mons.), Glam., (*Pembs.*, Mer.*, Caerns.**), [*Denbs., Flints.*]. 13 squares : **SS**48,58,*96,97*; **ST**06, (*16,17,27, 38,59*; **SN***695*; **SO**40; **SH***67*). The *Pembs.* record is based on an unlocalized specimen in the herbarium of Bristol Museum dated c1804.

POLYPOGON Desf.

Polypogon monspeliensis (L.) Desf. : Annual Beard-grass. Established alien (S. England). Waste places (M,R, Rg). Fl.6-9. Mons.*, Glam.. 4 squares.
Mons.: Newport Docks, ST38, TGE & CT, 1973[112] and TGE, 1980[91]. **Glam.**: Barry Docks, ST16, AEW, 1923[112]; Penarth Ferry, ST17, JHS, 1902[112]; Cardiff, ST17, HJR, 1907[112] and RLS, 1938[112]; Splott, Cardiff, ST27, AEW, 1926[112] & 1937[112].

P. maritimus Willd. Casual (Europe). Waste places. Fl.9. Glam., Caerns.. 2 squares : **ST***27*[112]; **SH**67[112].

P. viridis (Gouan) Breistr. (*P. semiverticillatus* (Forskål) Hyl.; *Agrostis semiverticillata* (Forskål) C. Chr.) : Water Bent. Established alien (S. Europe). Waste places. Fl.8-10. Mons., Glam.. 4 squares.
Mons.: Newport, ST38, TGE, 1980[112]. **Glam.**: Barry Docks, ST16, AEW, 1924[112] and TGE & AG, 1981[91]; East Moors, Cardiff, ST27, JS, 1876[112] and GCD, c1924[23].

AMMOPHILA Host

Ammophila arenaria (L.) Link : Moresg : Marram.
Subsp. **arenaria** Native. Sand dunes (Ar,D,Ds,R,S). Fl.6-9. (Mons.*), Glam., Carms., Pembs., Cards., Mer., Caerns., Denbs., *Flints.,* Angl., [*Brecs.*]. 64 squares, map 919.

Subsp. **breviligulata** (Fernald) Maire & Weiller (*A. breviligulata* Fernald). Introduced (N. America). Sand dunes. *Angl.*. 1 square.
Angl.: Newborough Warren, SH46, AHLH, 1950+[156] and RHR, 1970+[209].

CALAMAGROSTIS Adanson

Calamagrostis canescens (Weber) Roth : Mawnwellt Blewog : Purple Small-reed. Native. Fens (M,R,Rw). Fl.7. Glam., (Mer.), *Denbs.*. 4 squares.
Glam.: Crymlyn Fen, SS69, AJL, 1977[112]; (Llanishen, ST18, JPC, 1973[112] & 1978[197]. **Mer.**: Tywyn, SN59, KS, 1969[112] & 1972[112]). **Denbs.**: Llyn Bedydd, SJ43, JMB, 1978[184].

C. epigejos (L.) Roth : Mawnwellt : Wood Small-reed. Native. Bushy places (E,G,M,Q,Rw,Sc,W). Fl.7-10. Mons., Glam., Brecs.*, Carms., Pembs.*, Cards.*, *Monts.**, Mer., Caerns., *Flints.**, Angl.. 50 squares, map 920.

PHLEUM L.

Phleum pratense L. Mons., Glam., Brecs., Rads., Carms., Pembs., Cards., Monts., Mer., Caerns., Denbs., Flints., Angl.. 244 squares, map 921.

Subsp. **pratense** : Rhonwellt : Timothy. Native but frequently introduced. Meadows (G,M,R). Fl.5-10. Mons., Glam., Brecs., Rads., Carms., *Pembs.,* Cards., Monts., Mer., Caerns., Denbs., Flints., Angl.. 182 squares, map 922.

Subsp. **bertolonii** (DC.) Bornm. (*P. bertolonii* DC.; *P. nodosum* L.) : Rhonwellt Penfain : Smaller Cat's-tail. Native but frequently introduced. Pastures (D,G,R,S). Fl.6-9. Mons., Glam., *Brecs.**, *Rads.**, Carms., Pembs.*, Cards.*, *Monts.**, Mer., Caerns., Denbs., *Flints.**, Angl.. 152 squares, map 923.

P. phleoides (L.) Karsten : Purple-stem Cat's-tail. Casual (Europe). Waste places. *Glam.*. 1 square : **ST***17*[218].

P. arenarium L. : Rhonwellt y Tywyn : Sand Cat's-tail. Native. Sand dunes (D,Ds,R). Fl.5-9. Glam., Carms., Pembs., Cards., Mer., Caerns., *Denbs.,* Flints., Angl.. 48 squares, map 924.

ALOPECURUS L.

Alopecurus pratensis L. : Cynffonwellt y Maes : Meadow Foxtail. Native. Meadows (E,F,G,M,Q,R,Rw). Fl.4-8. Mons., Glam., Brecs., Rads., Carms., Pembs., Cards., Monts., Mer., Caerns., Denbs., *Flints., Angl.*. 200 squares, map 925.

A. geniculatus L. : Cynffonwellt Elinog : Marsh Foxtail. Native. Marshes (Al,Ar,D,G,H,M). Fl.5-9. Mons., Glam., Brecs., Rads., Carms., Pembs., Cards., Monts., Mer., Caerns., Denbs., Flints., *Angl.*. 232 squares, map 926.

A. aequalis Sobol. : Orange Foxtail. Native. Margins of pools (Al,Ar,M). Fl.6-9. Mons., Brecs.*, Monts.*, Mer.*, *Denbs.,* [*Glam.*]. 8 squares.
Mons.: Llangattock-nigh-Usk, SO30, TGE, 1980[91]; Penpergwm, SO30, RWR, 1929[112] and DCO, 1982[112]; nr Abergavenny, SO31, TGE, 1974[112]. **Brecs.**: Tal-y-bont Reservoir, SO01, MP, 1973[112]; Llys-wen, SO13, MP, 1968[112]. **Monts.**: Welshpool, SJ20, MH, 1970+[93]. Llansanffraidd Deuddwr, SJ21, AMP, 1976[112]. **Mer.**: Mochras, SH52, PMB, 1961[112]. **Denbs.**: Gresford, SJ35, JEB, c1835[144] and An, 1973[7].

A. aequalis × **geniculatus** = **A.** × **haussknechtianus** Ascherson & Graebner Spontaneous. Marshy edge of pool. Fl.7. *Mer.*. 1 square.
Mer.: Mochras, SH52, PMB, 1972[3].

A. bulbosus Gouan : Rhonwellt Cadno Oddfynog : Bulbous Foxtail. Native. Salt-marshes (M,S). Fl.5-6. Mons., Glam., Carms., *Pembs.*[230]. 11 squares : **SS***59*; **ST**06,*16,17,27*,38,39,48,58,59; **SN**31.

A. myosuroides Hudson : Cynffonwellt Ddu : Black-grass. Established alien (S. England). Cultivated ground (E,F,R,Rg). Fl.5-10. Mons., Glam., Brecs.*, Carms., *Cards., Monts., Caerns., Denbs., Flints., Angl.*. 36 squares, map 927.

A. rendlei Eig (*A. utriculatus* auct., non Solander). Casual (Europe). Waste places. Fl.6-8. Glam.. 2 squares : **ST***16*[24],*27*[112].

PARAPHOLIS C. E. Hubbard

Parapholis incurva (L.) C. E. Hubbard : Corwelltyn Camaidd : Curved Hard-grass. Native. Shingle. Fl.6-7. Glam., *Pembs.,* Mer.*, [*Angl.*]. 10 squares : **SR**89,99; **SS***69*,87; **ST**06,(*16,17,27*); **SM**70; **SH**53.

P. strigosa (Dumort.) C. E. Hubbard : Corwellt y Morfa : Hard-grass. Native. Salt-marshes (M,R,S). Fl.6-9. Mons., Glam., Carms., Pembs., Cards., Monts.*, Mer., Caerns., *Denbs.,* Flints., Angl.. 53 squares, map 928.

HAINARDIA W. Greuter

Hainardia cylindrica (Willd.) W. Greuter (*Lepturus cylindricus* (Willd.) Trin.; *Monerma cylindrica* (Willd.) Cosson & Durieu). Casual (Mediterranean region). Waste places. Fl.7. Glam.. 1 square : **ST***16*[112].

PHALARIS L.

Phalaris arundinacea L. : Gwyran : Reed Canary-grass. Native. Marshes (Al,Ar,H,M,Rw,W). Fl.6-9(-11). Mons., Glam., Brecs., Rads., Carms., Pembs., *Cards.*, Monts., Mer., Caerns., Denbs., Flints., Angl.. 241 squares, map 929.

P. coerulescens Desf. (*P. aquatica* auct., non L.). Casual (Europe). Waste places. Fl.10. Glam., Mer.. 2 squares : **ST27**[112]; **SN69**[27].

P. canariensis L. : Pefrwellt Amaethol : Canary-grass. Established alien (N. Africa). Waste places (E,R,Rw). Fl.6-12. Mons., Glam., Brecs., Rads., Carms., Pembs., Cards., *Monts.*, Mer., Caerns., *Denbs., Flints., Angl.*. 77 squares, map 930.

P. minor Retz. : Lesser Canary-grass. Casual (Mediterranean region). Waste places. Fl.5-6(-10). Glam.. 4 squares : **SS***78*[140]; **ST***16*[112],17[112],*27*[112].

P. paradoxa L. : Awned Canary-grass. Casual (Mediterranean region). Waste places. Fl.6-8. Mons., Glam., *Caerns.*. 4 squares : **ST***17*[112],*27*[112],28[112]; **SH***56*[18].

P. angusta Nees ex Trin. Casual (N. America). Waste places. *Glam.*. 1 square : **ST***27*[26].

MILIUM L.

Milium effusum L. : Miledwellt : Wood Millet. Native. Woods (Q,W). Fl.5-6. Mons., Glam., Brecs., *Rads.*, Carms., Pembs., *Cards.*, Monts., Mer., *Caerns.*, Denbs., Flints., *Angl.*. 96 squares, map 931.

PIPTATHERUM Beauv.

Piptatherum miliaceum (L.) Cosson (*Oryzopsis miliacea* (L.) Bentham & Hooker ex Ascherson & Graebner; *P. multiflorum* (Cav.) Beauv.). Casual (Europe). Waste places. Fl.7-9(-11). Mons., Glam.. 3 squares : **ST***16*[112],*17*[112],38[112].

ARUNDO L.

Arundo donax L. : Giant Reed. Casual (Mediterranean region). Waste places. *Glam.*. 1 square : **SS69**[221].

PHRAGMITES Adanson

Phragmites australis (Cav.) Trin. ex Steudel (*P. communis* Trin.) : Corsen : Common Reed. Native. Swamps (Ar,H,M,R,W) Fl.6-10. Mons., Glam., Brecs., *Rads.*, Carms., Pembs., Cards., Monts., Mer., Caerns., Denbs., Flints., Angl.. 155 squares, map 932.

CORTADERIA Stapf

Cortaderia selloana (Schultes & Schultes fil.) Ascherson & Graebner : Pampas-grass. Established alien (S. America). Fixed dunes. *Carms.*. 1 square.
Carms.: Laugharne Burrows, SN20/30, IMV, 1972[181].

DANTHONIA DC.

Danthonia decumbens (L.) DC. (*Sieglingia decumbens* (L.) Bernh.) : Glaswellt y Rhos : Heath-grass. Native. Pastures (D,G,H,R). Fl.6-9. Mons., Glam., Brecs., Rads., Carms., Pembs., Cards., Monts., Mer., Caerns., Denbs., *Flints.*, Angl.. 244 squares, map 933.

MOLINIA Schrank

Molinia caerulea (L.) Moench

Subsp. **caerulea** : Glaswellt y Bwla : Purple Moor-grass. Native. Moors (G,H,M,R,W). Fl.6-9. Mons., Glam., Brecs., Rads., Carms., Pembs., Cards., Monts., Mer., Caerns., Denbs., *Flints.*, Angl.. 237 squares, map 934.

Subsp. **arundinacea** (Schrank) H. Paul (*M. littoralis* Host). Introduced (Europe). Grassland. Fl.8. *Rads.*. 1 square.
Rads.: Llandrindod Wells, SO06, WCB, 1916[109].

NARDUS L.

Nardus stricta L. : Cawnen Ddu : Mat-grass. Native. Heaths and pastures (G,H,M,Q). Fl.3-6(-8). Mons., Glam., Brecs., Rads., Carms., Pembs., Cards., Monts., Mer., Caerns., Denbs., *Flints., Angl.*. 215 squares, map 935.

ERAGROSTIS N. M. Wolf

Eragrostis pectinacea (Michx) Nees Casual (N. America). Waste places. Fl.8. Glam.. 1 square : **ST17**[112].

E. cilianensis (All.) F. T. Hubbard (*E. megastachya* (Koeler) Link). Casual (Europe). Waste places. Fl.7-9. Glam.. 2 squares : **ST17**[112],*27*[112].

E. minor Host (*E. poaeoides* Beauv.; *E. eragrostis* (L.) Druce). Casual (Europe). Waste places. Fl.7. Glam.. 2 squares : **SS***78*[140]; **ST27**[112].

E. chloromeles Steudel Casual (S. Africa). Waste places. Glam.. 1 square : **ST27**[112].

SPOROBOLUS R.Br.

Sporobolus pyramidatus (Lam.) A. S. Hitchc. (*S. argutus* (Nees) Kunth). Casual (S. Africa). Waste places. *Glam.*. 1 square : **ST***17*[23].

MUHLENBERGIA Schreber

Muhlenbergia distichophylla (Presl) Kunth Casual (America). Waste places. Fl.7. *Glam.*. 1 square : **SN***90*[107].

CRYPSIS Aiton

Crypsis aculeata (L.) Aiton (*Pallasia aculeata* (L.) Druce). Casual (Europe). Waste places. *Glam.*. 1 square : **ST***17*[140].

C. schoenoides (L.) Lam. (*Heleochloa schoenoides* (L.) Host). Casual (Europe). Waste places. *Glam.*. 1 square : **ST***17*[140].

CYNODON L. C. M. Richard

Cynodon dactylon (L.) Pers. : Bermuda-grass. Established alien (S. England). Sandy waste places (R,S). Fl.7-10. Mons., Glam., *Caerns.*. 8 squares.
Mons.: Aber-carn, ST29, FP, 1959[112]. **Glam.**: West Cross, SS68, CH & HH, 1974[117] and QONK, 1980[157]; Swansea, SS69, JWLo, 1934[112]; Llansamlet, SS69, SJH & WHW, 1944[112]; Blackpill, SS69, CH & HH, 1974[112]; Kenfig Burrows, SS78, TGE, 1962[91]; Cardiff Docks, ST17 & 27, JS, c1876[218] and ST27, RGE, 1973[112]; nr Pontardawe, SN60, EMP, 1958[112]. **Caerns.**: Portmadoc, SH53, GCD, 1917[16].

SPARTINA Schreber

[**Spartina maritima** (Curtis) Fernald *Glam., Cards.*]

S. alterniflora Loisel. × **maritima** = **S.** × **townsendii** H. & J. Groves Naturalized alien (S. England). Salt-marshes. Fl.10. Mons., *Mer.*. 3 squares.
Mons.: Newport, ST38, TGE, 1978[112]; Sudbrook, ST58, TGE, 1978[112]; Blackrock, ST58, TGE, 1978[91]. **Mer.**: Barmouth, SH61, PMB, 1955[3] & 1971[3].

S. anglica C. E. Hubbard (*S.* × *townsendii* auct., non H. & J. Groves) : Cordwellt : Common Cord-grass. Naturalized alien (S. England). Mud-flats. Fl.7-11. Mons., Glam., Carms., Pembs., Cards., Monts.*, Mer., Caerns.*, Denbs.*, Flints., *Angl.**. 48 squares, map 936.

PANICUM L.

Panicum miliaceum L.: Common Millet. Casual (Asia). Waste places (E,F,R). Fl.8-10. Mons., Glam., Carms., Pembs., Cards., *Mer.*, Angl.. 15 squares : **SS**69,79,97; **ST**17,18,27,28,38,59; **SM**72; **SN**40,58; **SO***00*; **SH***52*,57.

P. capillare L. : Old-witch-grass. Casual (N. America). Waste places. Fl.9. Glam.. 6 squares : **ST***16*[112],*17*[22],18[24],27[112]; **SN***90*[140]; **SO***00*[140].

P. maximum Jacq. Casual (Africa). Waste places. *Glam.*. 1 square : **ST***27*[24].

ECHINOCHLOA Beauv.

Echinochloa crus-galli (L.) Beauv. (*Panicum crus-galli* L.) : Cibogwellt Rhydd : Cockspur. Casual (Tropics and warm temperate regions). Waste places (R,Rg). Fl.6-9(-11). Mons., Glam., Carms., Pembs., *Monts.*. 16 squares : **SS**69,*78*,79,88; **ST***16*,17,18,27,38,59; **SM**80; **SN**31,40; **SO***00*,51; **SJ**21.

E. utilis Ohwi & Yabunó (*E. frumentacea* auct., non Link). Casual (garden origin). Rubbish tips. Fl.10. Mons.. 2 squares : **ST**38[112],59[112].

DIGITARIA Haller

Digitaria sanguinalis (L.) Scop. : Hairy Finger-grass. Casual (Mediterranean region). Waste and cultivated ground (F,R,Rg). Fl.9-10. Mons., Glam., Carms., Cards., Mer.. 7 squares : **ST**08[112],17[112],18[112],27[112]; **SN**31[161], 68[112]; **SH**50[112].

D. ciliaris (Retz.) Koeler Casual (Tropics & Subtropics). Rubbish tip. *Mons.*. 1 square : **ST**38[245].

D. ischaemum (Schreber) Muhl. : Smooth Finger-grass. Casual (S. England). Waste places. *Mons.*, Glam.. 2 squares : **ST***17*[112],*38*[226].

PASPALUM L.

Paspalum dilatatum Poiret Casual (S. America). Waste places. Fl.10. Glam.. 1 square : **ST***17*[112].

SETARIA Beauv.

Setaria pumila (Poiret) Schultes (*S. glauca* auct., non (L.) Beauv.; *S. lutescens* F. T. Hubbard) : Yellow Bristle-grass. Casual (S. Europe). Waste places. Fl.8-10. *Mons.*, Glam., Carms., Mer., *Caerns.*. 9 squares : **ST***16*[24],*17*[112],*18*[24],*27*[112],38[112]; **SN**61[112]; **SO***00*[140]; **SH***53*[16],*93*[126].

S. geniculata (Lam.) Beauv. Casual (S. America). Waste places. Fl.11. *Mons.*, Glam.. 2 squares : **ST**17[112],38[245].

S. verticillata (L.) Beauv. : Cibogwellt Troellog : Rough Bristle-grass. Casual (Europe). Waste places. Fl.9. *Mons.*, Glam., Cards., *Mer.*. 6 squares : **ST**16[107],*17*[112],27[112],38[245]; **SN***68*[112]; **SH***93*[126].

S. viridis (L.) Beauv. : Cibogwellt Gwyrddlas : Green Bristle-grass. Casual (Europe). Waste places (R,Rg). Fl.7-10. Mons., Glam., *Carms.*, Cards., *Mer.*, *Caerns.*, Denbs., *Flints.*, *Angl.*. 18 squares : **SS***69*,*78*,*79*,*97*; **ST***16*,17,18,27,38,59; **SN***42*,58,*90*; **SO***00*,51; **SH***93*; **SJ***18*,35.

S. italica (L.) Beauv. : Foxtail Bristle-grass. Casual (garden origin). Waste places. Fl.9-10. Mons., Glam., Brecs., Carms.. 8 squares : **ST**17[112],18[112],27[112],38[112],59[91]; **SN**31[161],40[112],90[112].

SORGHUM Moench

Sorghum halepense (L.) Pers. Casual (Mediterranean region). Waste places. Fl.9-10. Mons., Glam., Cards.. 4 squares : **ST**17[112],27[112],38[112]; **SN**67[112].

S. bicolor (L.) Moench (*S. cernuum* (Ard.) Host; *S. vulgare* Pers.; *S. saccharatum* (L.) Pers.). Casual (Africa). Waste places. Fl.9-10. *Mons.*, Glam.. 2 squares : **ST**17[112],38[184].

ZEA L.

Zea mays L. : Maize. Casual (Tropical America). Waste places. Fl.9. *Mons.*, Glam.. 6 squares : **SS**79[140]; **ST***16*[24],*17*[112], 18[112],38[91]; **SO***00*[140].

PALMAE

PHOENIX L.

Phoenix dactylifera L. : Date Palm. Casual (N. Africa). Waste places. Glam., *Caerns.*. 2 squares : **ST***27*[112]; **SH***33*[18].

ARACEAE

ACORUS L.

Acorus calamus L. : Gellesgen Beraroglaidd : Sweet-flag. Naturalized alien (S. Asia). Streamsides (Al,Ar,M). Fl.6. Glam., *Brecs.**, *Rads.**, *Carms.**, *Pembs.**, Monts., Denbs.*. 16 squares : **SR**99; **SS**69,*79*,88; **ST**07,17,27; **SN***50*,95; **SO**12,19,37; **SJ**10,20,21,35.

LYSICHITON Schott

Lysichiton americanus Hultén & St. John : Yellow Skunk-cabbage. Naturalized alien (N. America). Lakesides and ditches. Fl.3-7. Glam., Cards., *Mer.*, *Caerns.*, *Denbs.*. 7 squares.
Glam.: Margam, SS88, MG, 1974[99] and RGE, 1982[112]; Llanishen, ST18, RAC, 1969[112]. **Cards.**: Cwm Einon, SN69, ABP & RL, 1979[112]. **Mer.**: nr Harlech, SH53, OHB & HHan, 1967[3] and BMH, 1970[3]. **Caerns.**: nr Rhydyclafdy, SH33, APC, 1981[74]. **Denbs.**: nr Gresford, SJ35, JMB, c1965[67]; nr Maenan, SH76, MCh, 1981[3].

L. camtschatcense (L.) Schott : White Skunk-cabbage. Introduced (Japan). Lakesides. Glam.. 1 square.
Glam.: Margam, SS88, DMC, 1956[7].

ZANTEDESCHIA Sprengel

Zantedeschia aethiopica (L.) Sprengel : Arum-lily. Introduced (S. Africa). Waste places. Glam.. 1 square.
Glam.: Blackpill, SS69, An, 1960[248].

ARUM L.

Arum italicum Miller

Subsp. **italicum** : Italian Lords-and-Ladies. Naturalized alien (S. Europe). Roadsides (E,R,Rg, Rw,W). Fl.2-5. *Brecs.*, Carms., Pembs., *Mer.*, Caerns.. 14 squares : **SR**99; **SS**19; **SM**72,90; **SN**10,11,20,30,31; **SO**11; **SH**32,53,57,71.

Subsp. **neglectum** (Townsend) Prime (*A. neglectum* (Townsend) Ridley). Native. Limestone cliff in woodland. Fl.6. Glam., (*Mer.*). 2 squares.
Glam.: Dinas Powys, ST17, EV, 1930[112] and AEW, 1953[112]. (**Mer.**: nr Minffordd, SH53, An, 1976[7]).

A. italicum × maculatum Spontaneous. Woods. Fl.3. Glam.. 1 square.
Glam.: nr Dinas Powys, ST17, AEW, 1955[112].

A. maculatum L. : Pidyn y Gog : Lords-and-Ladies. Native. Hedges (E,G,M,Q,R,Rg,Rw,S,Sc,W). Fl.4-5. Mons., Glam., Brecs., Rads., Carms., Pembs., Cards., Monts., Mer., Caerns., Denbs., Flints., *Angl.*. 221 squares, map 937.

LEMNACEAE

WOLFFIA Horkel ex Schleiden

Wolffia arrhiza (L.) Horkel ex Wimmer : Rootless Duckweed. Native. Ponds. *Glam.*†. 2 squares.
Glam.: Cardiff, ST17, AHT, 1899[223]; Cardiff Docks, ST27, JS, c1876[218].

LEMNA L.

Subgen. Staurogeton Reichenb.

Lemna trisulca L. : Llinad Fiddew : Ivy-leaved Duckweed. Native. Ditches (Al,Ar). Fl.5-7. Mons., Glam., Brecs., *Rads.*, *Carms.*, *Pembs.*[120], *Monts.*, *Caerns.*[221], Denbs., Flints., *Angl.*. 53 squares, map 938.

Subgen. Lemna

L. gibba L. : Llinad y Dŵr Crythog : Fat Duckweed. Native. Ditches (Al,Ar). Mons., Glam., Brecs.*, *Carms.*, *Monts.**, *Denbs.*, Flints., *Angl.**, [*Caerns.*]. 20 squares, map 939.

L. minor L. : Llinad : Common Duckweed. Native. Ponds and streams (Al,Ar,M). Fl.6-7. Mons., Glam., Brecs., Rads., Carms., Pembs., Cards., Monts., Mer., Caerns., Denbs., Flints., Angl.. 208 squares, map 940.

L. minuscula Herter Naturalized alien (N. America). Lakes. *Flints.*. 1 square.
Flints.: Llyn Helyg, SJ17, AOC, RWD & GW, 1981[71].

SPIRODELA Schleiden

Spirodela polyrhiza (L.) Schleiden (*Lemna polyrhiza* L.) : Llinad Mawr : Greater Duckweed. Native. Streams and ponds. Mons., Glam., Brecs., *Carms.*[161], *Monts.**, *Denbs.*, *Flints.*, [*Pembs.*, *Caerns.*, *Angl.*]. 24 squares, map 941.

SPARGANIACEAE

SPARGANIUM L.

Subgen. Sparganium

Sparganium erectum L. (*S. ramosum* Hudson) : Cleddlys : Branched Bur-reed. Native. Rivers and ditches (Al, Ar,H,M). Fl.6-9. Mons., Glam., Brecs., *Rads.*, Carms., Pembs., Cards., Monts., Mer., Caerns., Denbs., Flints., Angl.. 205 squares, map 942.

Subsp. **erectum** (*S. erectum* var. *erectum*). Native. Rivers and ditches (Al,Ar,M). Fl.6-8. Mons., Glam., Brecs., *Rads.*, Carms., Pembs., *Cards.*, Monts.*, Mer., *Caerns.*, *Denbs.*, *Flints.*, Angl.. 87 squares, map 943. Under-recorded.

Subsp. **oocarpum** (Čelak.) Domin (*S. erecta* var. *oocarpum* (Čelak.) Hayek). Native. Swampy bogs. Fl.7. Pembs., Angl.*. 3 squares.
Pembs.: Marloes Mere, SM70, TAWD, 1971[112]; Cwm Dewi nr Dinas, SN03, TAWD, 1960+[79]. **Angl.**: Llyn Coron, SH36, BS, 1962[112].

Subsp. **microcarpum** (Neuman) Domin (*S. erectum* var. *microcarpum* (Neuman) Hayek; *S. ramosum* var. *microcarpum* Neuman). Native. Streamsides (Al,Ar). Fl.7-8. Mons.*, Glam., *Rads.*, Carms.*, *Pembs.**, Caerns., Denbs.*, *Flints.**. 13 squares : **SR**99; **SS***69*; **ST***07,17*,28; **SN**30; **SO***06*,42; **SH**32,43,*56*; **SJ**17,54.

Subsp. **neglectum** (Beeby) Schinz & Thell. (*S. neglectum* Beeby; *S. erectum* var. *neglectum* (Beeby) Fiori & Paol.). Native. Rivers and ditches (Al,Ar,M). Fl.7-9. Mons., Glam., *Brecs.**, Carms., *Pembs.*, *Cards.*, Monts., Mer., Caerns., Denbs., Flints., Angl.*. 75 squares, map 944. Under-recorded.

Subgen. Xanthosparganium Holmberg

S. emersum Rehmann (*S. simplex* Hudson) : Cleddlys Digainc : Unbranched Bur-reed. Native. Streams and ditches (Al,Ar,H,M). Fl.7-9. Mons., Glam., Brecs., *Rads.,* Carms., Pembs., *Cards., Monts.,* Mer., Caerns., Denbs., Flints., *Angl..* 117 squares, map 945.

S. angustifolium Michx : Cleddlys Culddail : Floating Bur-reed. Native. Lakes (Al,Ar). Fl.7-8. Glam., Brecs.*, Carms., *Pembs.*[232], Cards., Monts.*, Mer., Caerns., *Denbs.**, *Angl.*[231]. 40 squares, map 946.

S. minimum Wallr. : Cleddlys Bach : Least Bur-reed. Native. Lakes (Al,Ar,H,M). Fl.7-10. Glam., Brecs.*, Carms., Cards.*, Mer., *Caerns., Denbs., Flints.,* Angl., [*Mons.*]. 27 squares, map 947. Some of these records may refer to **S. angustifolia** with which this species is frequently confused.

TYPHACEAE

TYPHA L.

Typha angustifolia L. : Cynffon-y-Gath Gulddail : Lesser Bulrush. Native. Swamps (Al,Ar,H,M). Fl.6-9. (*Mons.**), Glam., *Brecs.,* Rads.*, *Carms.**, *Pembs.,* Mer., Caerns., Denbs., Flints., Angl.. 30 squares, map 948.

T. latifolia L. : Cynffon-y-Gath : Bulrush. Native. Swamps (Al,Ar,H,M). Fl.7-10. Mons., Glam., *Brecs.,* Rads., Carms., Pembs., Cards., Monts., Mer., *Caerns., Denbs.,* Flints., Angl.. 156 squares, map 949.

CYPERACEAE

SCIRPUS L.

Scirpus sylvaticus L. : Clwbfrwynen y Coed : Wood Club-rush. Native. Marshes (Ar,H,M,W). Fl.6-8. Mons., Glam., Brecs., Rads., Carms., Pembs., Cards.*, Monts., *Caerns.,* Denbs., *Flints..* 55 squares, map 950.

S. maritimus L. : Clwbfrwynen Arfor : Sea Club-rush. Native. Salt-marshes (Al,Ar,M,Q,S). Fl.6-9. Mons., Glam., Carms., Pembs., Cards., Monts., Mer., Caerns., Denbs., Flints., Angl.. 80 squares, map 951.

S. lacustris L.

Subsp. **lacustris** (*Schoenoplectus lacustris* (L.) Palla) : Llafrwynen : Common Club-rush. Native. Streams (Al,Ar,R). Fl.5-9. (Mons.*), Glam., *Brecs.**, Carms., *Pembs.**, *Cards., Monts.**, Mer., Caerns., Denbs., *Flints.,* Angl.. 52 squares, map 952.

Subsp. **tabernaemontani** (C. C. Gmelin) Syme (*Schoenoplectus tabernaemontani* (C. C. Gmelin) Palla) : Llafrwynen Arfor : Grey Club-rush. Native. Streams especially near the sea (Al,Ar,D,G,H,M,S). Fl.6-10. Mons., Glam., Brecs., Carms., Pembs., Cards., Monts., Mer., Caerns., Denbs., Flints., Angl.. 80 squares, map 953.

S. holoschoenus L. (*Holoschoenus vulgaris* Link) : Clwbfrwynen Sypynog : Round-headed Club-rush. Naturalized alien (England). Sea coasts (M,R,S). Fl.5-8(-10). Mons.*, Glam.. 4 squares.

Mons.: Newport Docks, ST38, CT & TGE, 1973[112] and TGE, 1980[91]. **Glam.**: Neath, SS79, EMT, 1937[112]; Jersey Marine, SS79, TRL, 1959[104]; Barry Docks, ST16, RLS, 1926[112] and GCD, 1930[112]; East Moors, Cardiff, ST27, JS, 1879[218].

S. setaceus L. (*Isolepis setacea* (L.) R. Br.) : Clwbfrwynen Fach : Bristle Club-rush. Native. Wet gravelly places (Al,Ar,D,G,H,M,Q,Sc,W). Fl.6-9. Mons., Glam., Brecs., Rads., Carms., Pembs., Cards., Monts., Mer., Caerns., *Denbs., Flints., Angl..* 200 squares, map 954.

S. cernuus Vahl (*Isolepis cernua* (Vahl) Roemer & Schultes : Clwbfrwynen Gwychog Eiddilaidd : Slender Club-rush. Native. Wet sandy ground, especially near the coast (Al,Ar,D,Ds,G,M,R). Fl.6-9. Glam., Carms., Pembs., Cards., Monts., Mer., Caerns., *Flints.*[230], Angl.. 56 squares, map 955.

S. fluitans L. (*Eleogiton fluitans* (L.) Link) : Clwbfrwynen Nawf : Floating Club-rush. Native. Pools (Al,Ar,Ds,H,M). Fl.5-10. Mons.*, Glam., Brecs., Rads., Carms., Pembs., Cards., *Monts.**, Mer., Caerns., Denbs., *Flints.,* Angl.. 87 squares map 956.

S. hudsonianus (Michx) Fernald (*Trichophorum alpinum* (L.) Pers.) There is an unconfirmed record of this species from **Caerns.**: Cwm Idwal, SH65, LWF, c1955; 'no specimen was gathered and seeds collected at the time failed to germinate'[7].

S. cespitosus L. subsp. **germanicus** (Palla) Broddeson (*Trichophorum cespitosum* subsp. *germanicum* (Palla) Hegi) : Clwbfrwynen y Mawn : Deergrass. Native. Moorlands (H,M,Q). Fl.5-8. Mons., Glam., Brecs., Rads., Carms., Pembs., Cards., *Monts.,* Mer., Caerns., *Denbs., Flints., Angl..* 151 squares, map 957.

BLYSMUS Panzer

Blysmus compressus (L.) Panzer ex Link (*Scirpus compressus* (L.) Pers., non Moench) : Flat-sedge. Native. Marshes. Fl.6-7. *Caerns.*[230]†, *Flints.*†, [*Mer.*]. 1 square. **Flints.**: nr Prestatyn, SJ08, B, c1800[141].

B. rufus (Hudson) Link (*Scirpus rufus* (Hudson) Schrader) : Corsfrwynen Rudd : Saltmarsh Flat-sedge. Native. Salt-marshes (M,S). Fl.6-9. Glam., Pembs., Cards.*, Mer., Caerns., *Denbs.*[196], *Flints.,* Angl.. 12 squares : **SS***49*; **SN**03,69; **SH**27,36,45,46,52,53,58,61; **SJ**37.

ERIOPHORUM L.

Eriophorum angustifolium Honckeny : Plu'r Gweunydd : Common Cottongrass. Native. Bogs (Al,Ar,Ds,G,H,M,W). Fl.4-9. Mons., Glam., Brecs., Rads., Carms., Pembs., Cards., Monts., Mer., Caerns., Denbs., Flints., Angl.. 209 squares, map 958.

E. latifolium Hoppe : Plu'r Gweunydd Llydanddail : Broad-leaved Cottongrass. Native. Bogs (H,M). Fl.5-8. Mons., Glam., Brecs., *Rads., Carms., Pembs.**, Cards.*, Caerns.*, Denbs., *Flints.*, Angl.. 35 squares, map 959.

E. gracile Koch ex Roth : Plu'r Gweunydd Eiddil : Slender Cottongrass. Native. Acid fens. Fl.6. Glam., Caerns.. 3 squares.
Glam.: Crymlyn Fen, SS69, AJL, 1977[112]. **Caerns.**: nr Cwmistir, SH23, APC & PMB, 1980[3]; Cors Geirch, SH33, RHR & AMS, 1973[112] and AVJ, 1975[112].

E. vaginatum L. : Plu'r Gweunydd Unben : Hare's-tail Cottongrass. Native. Bogs (Al,H,M). Fl.3-7. Mons., Glam., Brecs., *Rads.*, Carms., Pembs., Cards., *Monts.*, Mer., Caerns., Denbs., *Flints., Angl.*[232]. 143 squares, map 960.

ELEOCHARIS R. Br.

Eleocharis quinqueflora (F. X. Hartmann) O. Schwarz (*Scirpus pauciflorus* Lightf.) : Sbigfrwynen Goch : Few-flowered Spike-rush. Native. Marshes (Ar,D,Ds,G,H,M,Q,S). Fl.6-8. Glam., Brecs., Rads., Carms., Pembs., Cards., *Monts.*, Mer., *Caerns., Denbs.*, Flints., Angl.. 75 squares, map 961.

E. parvula (Roemer & Schultes) Link ex Bluff, Nees & Schauer : Sbigfrwynen Morafon : Dwarf Spike-rush. Native. Sandy sea-flats. Fl.6-8(-9). Mer., Caerns.. 6 squares.
Mer.: Minffordd, SH53, DAJ, 1921[112] and PMB, 1956[112]; Traeth Bach, SH53, PMB, 1953[112]; Traeth Mawr, SH53, WEH, 1968[112]; Penmaenpool, SH61, PMB, 1968[112] & 1979[3]; Dwyryd Estuary, SH63, PMB, 1972[3] and OBH & HHan, 1975[3]; Maentwrog, SH64, PMB, 1971[3]. **Caerns.**: Aber-soch, SH32, JEG, 1890[112] and PWR, 1955[114]; Pwllheli, SH33, GCD, 1919[112] and JHS, 1933[112].

E. acicularis (L.) Roemer & Schultes : Sbigfrwynen Leiaf : Needle Spike-rush. Native. Pool margins. Fl.7-8. Brecs., *Rads., Cards.*†, *Monts.*, Mer., Caerns., Flints., Angl., [*Glam.*]. 27 squares, map 962.

E. palustris (L.) Roemer & Schultes subsp. **vulgaris** Walters : Sbigfrwynen y Gors : Common Spike-rush. Native. Pool margins (Al,Ar,Ds,G,H,M). Fl.5-8. Mons., Glam., Brecs., Rads., Carms., Pembs., Cards., Monts.*, Mer., Caerns., Denbs., Flints., Angl.. 211 squares, map 963.

E. uniglumis (Link) Schultes : Sbigfrwynen Un Plisgyn : Slender Spike-rush. Native. Marshes (Al,Ar, Ds,M,S). Fl.5-8. Mons.*, Glam., Carms.*, Pembs., Cards., Monts., Mer., Caerns., *Denbs.**, *Flints.*, Angl.. 43 squares, map 964.

E. multicaulis (Sm.) Desv. : Sbigfrwynen Gadeiriog : Many-stalked Spike-rush. Native. Marshes (Al,Ar,D,G, H,M). Fl.6-8. Glam., Brecs., *Rads.**, Carms., Pembs., Cards., *Monts.**, Mer., Caerns., Denbs., Angl., [*Mons., Flints.*]. 76 squares, map 965.

CYPERUS L.

Cyperus longus L. : Ysnoden Fair : Galingale. Native. Marshes (Al,Ar,M,R). Fl.7-11. (Mons.*, Glam., Brecs.*), Pembs., Caerns., *Flints.*. 13 squares : **SR**99; **SS**10,(88, 99; **ST**17,18,38); **SM***72*; **SN**10; (**SO**14); **SH**52,*77*; **SJ***36*.

C. eragrostis Lam. (*C. declinatus* Moench). Established alien (N. America). Waste places. Fl.7. Mons., Glam.†. 2 squares.
Mons.: Newport Docks, ST38, CT & TGE, 1973[112].
Glam.: Penarth Ferry, ST17, JS, c1876[112].

CLADIUM Browne

Cladium mariscus (L.) Pohl : Llemfrwynen : Great Fen-sedge. Native. Pool margins (Al,H,M). Fl.5-8. Glam., Brecs.*, Rads.*, *Pembs., Cards.*[230], Caerns., Angl., [*Flints.*]. 18 squares : **SS***19*,69,78; **ST**07,28; **SM**93; **SN**92; **SO**14; **SH**23,27,32,33,39,47,48,57,58,75.

RHYNCHOSPORA Vahl

Rhynchospora alba (L.) Vahl : Corsfrwynen Wen : White Beak-sedge. Native. Wet moorlands (Ar,H,M). Fl.5-9. Mons., Glam., Brecs.*, Rads.*, Carms., Pembs.*, Cards., Monts., Mer., Caerns., *Denbs., Flints., Angl.*[230]. 56 squares, map 966.

R. fusca (L.) Aiton fil. : Brown Beak-sedge. Native. Bogs. Fl.6-7. *Glam.*†, Cards., Mer.*, [*Brecs.*]. 4 squares.
Glam.: Crymlyn Fen, SS69, EF, 1805[223]. **Cards.**: Tregaron Bog, SN66, HAH, 1936[112] and PMB & EH, 1959[238]; Cors Fochno, SN69, AL, 1885[106] and TF, 1977[96]. **Mer.**: Arthog, SH61, PMB, 1963[112] & 1980[3].

SCHOENUS L.

Schoenus nigricans L. : Corsfrwynen Ddu : Black Bog-rush. Native. Marshes (Ds,H,M,Sc). Fl.(2-)6-8. Glam., Pembs., Cards., Monts., Caerns., *Denbs.*, Angl., [*Mer.*]. 27 squares, map 967.

CAREX L.

Subgen. Vignea (Beauv. ex Lestib.) Kük.

Carex paniculata L. : Hesgen Rafunog Fwyaf : Greater Tussock-sedge. Native. Marshes (Ar,G,H,M, Rw,W). Fl.5-8. Mons., Glam., Brecs., Rads., Carms., Pembs., Cards., Monts.*, Mer., Caerns., Denbs., Flints.*, Angl.. 184 squares, map 968.

C. paniculata × remota = C. × boenninghausiana Weihe Spontaneous. Moors. *Glam.*. 1 square.
Glam.: nr Cwm Park, SS99, CSN, c1900[32].

C. appropinquata Schumacher : Hesgen Manedafedd : Fibrous Tussock-sedge. Native. Acid fen. Fl.6-7. Pembs.. 1 square.
Pembs.: Dowrog Common, SM72, TAWD, 1963[112] and RWD, 1978[197]; nr Caerfarchell, SM72, SBE, LF & DRS, 1981[92].

C. diandra Schrank : Hesgen Rafunog Leiaf : Lesser Tussock-sedge. Native. Marshes (Al,H,M). Fl.6-8. Glam., Rads., Pembs., *Caerns.**, *Flints.**, Angl.. 14 squares : **SS***69*; **ST***09*; **SM**80,83; **SO***06*,14; **SH**23,33,37, 47,48,57,58; **SJ**16.

C. otrubae Podp. : Hesgen Dywysennog Fwyaf : False Fox-sedge. Native. Ditch and pond sides and salt-marshes (Al,Ar,D,E,G,H,M,R,S). Fl.5-10. Mons., Glam., *Brecs.*, Carms., Pembs., Cards., Monts., Mer., Caerns., Denbs., Flints., *Angl.*. 121 squares, map 969.

C. otrubae × **remota** = **C. pseudaxillaris** K. Richter Spontaneous. Roadsides. Fl.6-9. Glam., Pembs., *Mer.*, *Caerns.*, *Flints.*. 7 squares.
Glam.: nr Lavernock, ST16, EV, c1900[107]; Cogan, ST17, EV, 1905[112]. **Pembs.**: Tenby, SN10, FLR, 1946[112]; nr Cardigan, SN14, RWD, 1978[112]; Cilgerran, SN14, TAWD, 1976[112]. **Mer.**: nr Barmouth, SH61, PMB, 1961[4]. **Caerns.**: Glyder Fawr, SH65, RBa, 1886[107]. **Flints.**: Caerwys, SJ17, N, c1910[141].

C. spicata Hudson (*C. contigua* Hoppe) : Hesgen Dywysennog : Spiked Sedge. Native. Hedges (Ar,E,M, Q,R,Rw,W). Fl.5-8. Mons., Glam., *Brecs.*, Rads., Carms.*, *Pembs.**, Monts.*, Mer., *Caerns.*, *Denbs.*, Flints., Angl., [*Cards.*]. 37 squares, map 970. Only records confirmed by RWD, AOC, or ACJ are mapped.

C. muricata L.

Subsp. **muricata** Native. Limestone scree. Fl.5-7. Denbs.. 1 square.
Denbs.: nr Minera, SJ25, JEBo, 1840[44] and AN & RWD, 1980[112].

Subsp. **lamprocarpa** Čelak. (*C. muricata* subsp. *pairaei* (F. W. Schultz) Čelak.; *C. pairaei* F. W. Schultz) : Hesgen Lleiaf Dywysennog : Prickly Sedge. Native. Roadsides and hedgebanks (Ar,E,G, M,R,Rw,Sc). Fl.5-9. Mons., Glam., Brecs., Rads., Carms., Pembs., Cards., Monts., Mer., Caerns., Denbs., Flints., Angl.. 58 squares, map 971. Only records confirmed by RWD, AOC or ACJ are mapped.

C. divulsa Stokes

Subsp. **divulsa** : Hesgen Llwydlas : Grey Sedge. Native. Hedgebanks (E,G,R,Rw). Fl.5-8. Mons., Glam., Brecs., *Rads.*, Pembs., *Monts.*, Denbs., *Flints.*, [*Carms.*, *Cards.*, *Mer.*, *Caerns.*, *Angl.*]. 20 squares : **SS***58*; **ST**16,*17*,*18*,48,59; **SN**00; **SO**12,21,*24*,30,40, 41,42,51; **SJ**06,07,16,*20*,*24*. Only records confirmed by RWD, AOC or ACJ are listed.

Subsp. **leersii** (Kneucker) W. Koch (*C. muricata* subsp. *leersii* (Kneucker) Ascherson & Graebner; *C. polyphylla* Kar. & Kir.). Native. Hedgebanks (E,G,Sc). Fl.5-8. Mons., Glam., *Monts.**, Caerns., *Denbs.**, *Flints.**, [*Pembs.*, *Angl.*]. 11 squares : **SS**68; **ST**26, *49*,58; **SO**19; **SH**77,88; **SJ**06,08,15,16. Only records confirmed by RWD, AOC or ACJ are listed.

C. divulsa × **otrubae** Spontaneous. Woods. Fl.7. Mons.. 1 square.
Mons.: Portskewett, ST58, ESM & WAS, 1902[112].

[**C. divulsa** × **remota** = **C.** × **emmae** H. Gross *Brecs.*]

C. arenaria L. : Hesgen Arfor : Sand Sedge. Native. Sand dunes (D,Ds,R,S,Sc). Fl.5-9. Mons.*, Glam., (*Rads.**), Carms., Pembs., Cards., Mer., Caerns., *Denbs.*, *Flints.*, Angl.. 84 squares, map 972.

C. disticha Hudson : Hesgen Lygliw Benblydd : Broom Sedge. Native. Marshes (D,G,H,M). Fl.5-7(-10). Mons.*, Glam., Brecs., Carms., Pembs., *Caerns.*, Denbs., *Flints.*, Angl., [*Cards.*, *Monts.*, *Mer.*]. 33 squares, map 973.

C. divisa Hudson : Hesgen Blodeilog : Divided Sedge. Native. Salt-marshes (D,M,S). Fl.5-7. Mons.*, Glam., Carms., *Denbs.*[196], Flints., [*Mer.*]. 6 squares.
Mons.: Newport Docks, ST38, TGE & CT, 1973[112] and TGE, 1980[91]. **Glam.**: Blackpill Burrows, SS69, JAW, 1941[112]; nr Cardiff, ST17, EV, 1937[112]. **Carms.**: Llanelli, SS59, M, c1850[161]. **Flints.**: Rhiw Marsh, SJ08, JEBo, 1840[109]; Acre-fair, SJ08, HS, 1962[112]; nr Saltney, SJ36, Th, c1900[144].

C. remota L. : Hesgen Anghyfagos : Remote Sedge. Native. Damp woods (E,H,M,Q,R,Rw,W). Fl.5-8. Mons., Glam., Brecs., Rads., Carms., Pembs., Cards., Monts., Mer., Caerns., Denbs., *Flints.*, Angl.. 216 squares, map 974.

C. ovalis Good. : Hesgen Hirgylchaidd : Oval Sedge. Native. Pastures (E,G,H,M,Q,R,W). Fl.5-7. Mons., Glam., Brecs., Rads., Carms., Pembs., Cards., Monts., Mer., *Caerns.*, Denbs., *Flints.*, Angl.. 234 squares, map 975.

C. echinata Murray : Hesgen Seraidd : Star Sedge. Native. Boggy places (Ar,G,H,M,R,Rw,Sc,W). Fl.6-9. Mons., Glam., Brecs., Rads., Carms., Pembs., Cards., Monts., Mer., Caerns., Denbs., *Flints.*, Angl.. 223 squares, map 976.

C. dioica L. : Hesgen Ysgar : Dioecious Sedge. Native. Bogs (G,H,M). Fl.5-8. Glam., Brecs., Carms.*, Pembs., Cards.*, Monts., Mer., Caerns., Denbs., *Flints.*[230], Angl.. 60 squares, map 977.

C. dioica × **echinata** = **C.** × **gaudiniana** Guthnick Spontaneous. Moorland flush. *Denbs.*. 1 square.
Denbs.: Mynydd Hiraethog, SH95, IRB, 1970[242].

C. elongata L. : Elongated Sedge. Native. Canal banks. Fl.6. Denbs.. 2 squares.
Denbs.: nr Chirk, SJ23, RLB, 1938[7] and AGS, 1977[112]; Bettisfield, SJ43, MJW, 1979[67].

C. curta Good. : Hesgen Benwen : White Sedge. Native. Bogs (A,H,M,R,W). Fl.5-7. Mons., Glam., Brecs., Rads., Carms., Pembs.*, Cards., *Monts.*, Mer., Caerns., Denbs., *Flints.*, *Angl.*. 126 squares, map 978.

C. curta × paniculata = C. × ludibunda Gay Spontaneous. Bogs. Fl.7. *Brecs.*. 1 square.
Brecs.: nr Coelbren, SN81, AL, 1899[106].

Subgen. Carex

C. hirta L. : Hesgen Flewog : Hairy Sedge. Native. Grassy roadsides (D,Ds,E,G,H,M,Q,R,Sc). Fl.5-7. Mons., Glam., Brecs., Rads., Carms., Pembs., Cards., Monts., Mer., Caerns., *Denbs., Flints.*, Angl.. 181 squares, map 979.

C. lasiocarpa Ehrh. : Hesgen Feindwf : Slender Sedge. Native. Lake-sides (Al,H,M). Fl.6-9. *Brecs.**, Pembs.*, Cards.*, Monts.*, Mer., *Caerns.*, Denbs.*, Angl.. 23 squares, map 980.

C. acutiformis Ehrh. : Hesgen Ganolig-dywysennog : Lesser Pond-sedge. Native. Streamsides (Al,Ar,G,H,M). Fl.5-7. Mons., Glam., Brecs., Rads., Carms.*, Pembs., Cards., Monts.*, *Mer.*, Caerns., Denbs., Flints., Angl.. 95 squares, map 981.

C. acutiformis × riparia Spontaneous. Marshes. *Glam.*. 2 squares.
Glam.: Peterston-super-Ely, ST07, HJR, c1907[140]; Fairwater, Cardiff, ST17, HJR, c1908[142]. Both records require confirmation.

C. riparia Curtis : Hesgen Braff-dywysennog : Greater Pond-sedge. Native. Marshes (Al,Ar,M,W). Fl.5-8. Mons., Glam., Brecs., *Rads.**, Carms., Pembs., Cards., Monts., Mer.*, Caerns., *Denbs.*, Flints., Angl.. 83 squares, map 982.

C. pseudocyperus L. : Hesgen Hopysaidd : Cyperus Sedge. Native. Ditches (Al,Ar,H,M). Fl.5-8. Mons., Glam., Brecs.*, *Carms.**[161], *Monts.**, Mer.*, Denbs., Flints., Angl., [*Caerns.*]. 30 squares, map 983.

C. rostrata Stokes : Hesgen Chwysigennaidd Ylfinfain : Bottle Sedge. Native. Marshes (Al,Ar,G,H,M,W). Fl.5-8. Mons., Glam., Brecs., Rads., Carms., Pembs., Cards., Monts., Mer., Caerns., Denbs., Flints., Angl.. 154 squares, map 984.

C. rostrata × vesicaria = C. × involuta (Bab.) Syme Spontaneous. Lakesides. Fl.6-7. Glam., Brecs., *Mer.*, Denbs.. 4 squares.
Glam.: Mynydd y Glew, ST07, RM, 1923[112] and GCD & EV, 1927[26]. **Brecs.**: Llangorse Lake, SO12, BS, 1962[112]. **Mer.**: Bala Lake, SH83, PMB, 1979[245]. **Denbs.**: Llyn Fawnog, SH87, GAH, 1906[112].

C. vesicaria L. : Hesgen Chwysigennaidd Ber-dywysennog : Bladder-sedge. Native. Marshes (Al,Ar,Ds,G,H,M). Fl.5-9. Mons., Glam., Brecs., Rads.*, Carms., Pembs.*, Cards., *Monts.**, Mer., Caerns., Denbs., Flints.*, Angl.. 69 squares, map 985.

C. pendula Hudson : Hesgen Ddibynnaidd Fwyaf : Pendulous Sedge. Native. Wet woods (Al,Ar,M,R,Rg,W). Fl.4-7. Mons., Glam., *Brecs.*, Carms., Pembs., *Cards., Monts.**, Mer., Caerns., Denbs., *Flints.*, Angl.. 84 squares, map 986.

C. sylvatica Hudson : Hesgen Ddibynnaidd y Goedwig : Wood-sedge. Native. Woods (E,G,M,R,W). Fl.5-7. Mons., Glam., Brecs., Rads., Carms., Pembs., Cards., *Monts.*, Mer., Caerns., Denbs., Flints., Angl.. 167 squares, map 987.

C. capillaris L. : Hesgen Llechwedd y Mynydd : Hair Sedge. Native. Wet mountain slopes. Fl.6-7. Caerns.. 2 squares.
Caerns.: Snowdon, SH65, REH, 1962[112]; Creigiau Gleision, SH76, EPE, 1946[112] and ER, 1965[207].

C. strigosa Hudson : Hesgen Ysbigog Denau : Thin-spiked Wood-sedge. Native. Woods (M,Q,R,W). Fl.5-8. Mons., Glam., Brecs.*, Carms.*, Pembs.*, Monts.*, Caerns.*, Denbs., Flints.*. 22 squares, map 988.

C. flacca Schreber : Hesgen Oleulas Wyrgamddail : Glaucous Sedge. Native. Pastures (Al,D,Ds,E,G,H,M,Q,R,Sc,W). Fl.4-7. Mons., Glam., Brecs., Rads., Carms., Pembs., Cards., Monts., Mer., Caerns., Denbs., Flints., Angl.. 258 squares, all except : **SS**39,99; **SM**50,73; **SN**12,21,22,36,59,63,86,87,88,98; **SO**07,08,42; **SH**51,80,91,98; **SJ**02,04,45.

C. panicea L. : Hesgen Benigen Ddail : Carnation Sedge. Native. Marshes (G,H,M,Q,R,W). Fl.5-7. Mons., Glam., Brecs., Rads., Carms., Pembs., Cards., Monts., Mer., Caerns., Denbs., *Flints.*, Angl.. 247 squares, map 989.

C. depauperata Curtis ex With. : Starved Wood-sedge. Native. Old quarry. Fl.6. Angl.. 1 square.
Angl.: Holyhead, SH28, AGH & WH, 1936[233] and WEH, 1967[112].

C. laevigata Sm. : Hesgyn Ylfinog Lefn : Smooth-stalked Sedge. Native. Wet woods (Al,Ar,E,G,H,M,R,W). Fl.5-7. Mons., Glam., Brecs., Rads., Carms., Pembs., Cards., Monts.*, Mer., Caerns., Denbs., Flints., *Angl.*. 127 squares, map 990.

C. binervis Sm. : Hesgen Ddeulasnod : Green-ribbed Sedge. Native. Moors (G,H,M,Q,R,Sc,W). Fl.5-8. Mons., Glam., Brecs., *Rads.*, Carms., Pembs., Cards., *Monts.*, Mer., Caerns., Denbs., Flints., Angl.. 183 squares, map 991.

C. binervis × laevigata = C. × deserta Merino Spontaneous. Damp roadside. *Caerns.*. 1 square.
Caerns.: Llyn Du nr Tremadoc, SH54, AOC, 1961[71].

C. binervis × punctata Spontaneous. Wet rock face. Fl.6-7. Mer.†. 1 square.
Mer.: Barmouth, SH61, PMB, 1955[112] & 1957[107].

C. distans L. : Hesgen Anghysbell : Distant Sedge. Native. Marshes near the sea (Ar,G,M,S,Sc). Fl.5-8. Mons., Glam., Carms., Pembs., Cards., Mer., Caerns., *Denbs.*, Flints., Angl.. 63 squares, map 992.

C. distans × extensa = C. × tornabenii Chiov. Spontaneous. Salt-marshes. Fl.6-7. Mer.. 2 squares.
Mer.: Mochras, SH52, GM, 1948[112] and PMB, 1956[112]; Morfa Harlech, SH53, PMB, 1959[112].

C. punctata Gaudin : Hesgen Pigiad : Dotted Sedge. Native. Marshes near the sea (D,E,G,M,Q,Sc). Fl.6-8. Pembs., Cards.*, Mer., Caerns., *Angl.*. 15 squares : **SS**19; **SM**72,80; **SN**10,14,59,69; **SH**22,32,33,53,*57*,61, 63,64.

C. extensa Good. : Hesgen Hiriain : Long-bracted Sedge. Native. Salt-marshes (D,Ds,M,Sm). Fl.6-8. Mons., Glam., Carms., Pembs., Cards., Mer., Caerns., Denbs., Flints., Angl.. 55 squares, map 993.

C. hostiana DC. : Hesgen Dywyll-felen : Tawny Sedge. Native. Marshes (Al,Ar,Ds,G,H,M). Fl.5-8(-10). Mons., Glam., Brecs., Rads., Carms., Pembs., Cards., *Monts.,* Mer., Caerns., Denbs., *Flints.**, Angl.. 142 squares, map 994.

C. hostiana × **lepidocarpa** = **C.** × **fulva** Good. Spontaneous. Fens. Fl.6-8. Mons., Pembs., *Caerns.,* Angl.. 6 squares.
Mons.: nr Pant y Spig, SO22, AEW, 1949[112]. **Pembs.**: Castlemartin, SR89, WL, 1981[112]. **Caerns.**: Cors Geirch, SH33, AOC, 1961[71]. **Angl.**: Cors y Farl, SH47, RHR, 1970+[209]; Cors Goch, SH48, An, 1937[36] and RHR, 1970+[209]; Cors Erddreiniog, SH48, RHR, 1961[112]; Rhos y Gad, SH57, PMB, 1958[112]; Cors Bodeilio, SH57, RHR, 1970+[209].

[**C. hostiana** × **serotina** = **C.** × **appeliana** Zahn *Glam., Caerns.*]

C. lepidocarpa Tausch. : Hesgen Felen Paladr Hir : Long-stalked Yellow-sedge.

Subsp. **lepidocarpa** Native. Marshes (Al,Ar,D,Ds,G, H,M). Fl.5-9. Mons., Glam., Brecs., Rads., *Carms.,* Pembs., *Mer.,* Caerns., *Denbs., Flints.,* Angl., [*Cards., Monts.*]. 42 squares, map 995. Only records accepted by RWD, AOC, or ACJ are mapped.

Subsp. **scotica** E. W. Davies Native. Mountain flushes. Fl.8-9. *Caerns..* 1 square.
Caerns.: Cwm Glas, SH65, SMW, 1950[7] and ER, 1955[207]; Clogwyn Du'r Arddu, SH65, ER, 1956[207]. Recent research has indicated that this may not be worth maintaining as a separate subspecies.

C. demissa Hornem. : Hesgen Felen : Common Yellow-sedge. Native. Marshes (Ar,E,H,M,Q,R). Fl.5-10. Mons., Glam., Brecs., Rads., Carms., Pembs., Cards., Monts., Mer., Caerns., Denbs., *Flints.,* Angl.. 225 squares, map 996.

C. demissa × **hostiana** Spontaneous. Bogs. Fl.7-8. Brecs., *Carms., Cards.,* Mer., Caerns., *Denbs., Angl..* 28 squares, map 997.

C. demissa × **laevigata** Spontaneous. *Mer..* 1 square.
Mer.: nr Egryn Abbey, SH52, PMB, 1970[217].

C. demissa × **lepidocarpa** subsp. **scotica** Spontaneous. Mountain rocks. *Caerns..* 1 square.
Caerns.: Cwm Glas, SH65, EDa, c1950[235].

C. serotina Mérat

Subsp. **serotina** : Hesgyn Oeder : Small-fruited Yellow-sedge. Native. Wet sandy places (D,Ds,G, H,M,S). Fl.6-9. Glam., Carms., *Pembs.,* Cards., Mer., Caerns., *Denbs.**, *Flints.,* Angl., [*Mons., Monts.*]. 20 squares, map 998. Only records confirmed by RWD, AOC or ACJ are mapped.

Subsp. **pulchella** (Lönnr.) Van Ooststr. (*C. scandinavica* E. W. Davies). Native. Sea-flats. Fl.7-9. *Pembs., Mer., Caerns., Flints..* 6 squares.
Pembs.: St David's, SM72, WRL, 1900[110]. **Mer.**: Aberdyfi, SN59, MAL, 1867[107]; nr Tywyn, SH50, AL, 1875[107]; Llanbedr, SH52, WGT, 1911[110]; Mochras, SH52, AOC, 1960[71]; Harlech, SH53, CB, 1880[107]. **Caerns.**: Portmadoc, SH53, AL, 1878[107]. **Flints.**: Llyn Helyg, SJ17, AAD & JAWh, 1906[110]. Recent research has indicated that this may not be worth maintaining as a separate subspecies.

C. pallescens L. : Hesgen Welwlas : Pale Sedge. Native. Meadows (E,G,H,M,Q,R,Rw,W). Fl.5-9. Mons., Glam., Brecs., Rads., Carms., *Pembs.**, Cards., *Monts.**, Mer., Caerns., *Denbs.,* Flints., *Angl..* 123 squares, map 999.

C. digitata L. : Fingered Sedge. Native. Woods on limestone (E,R,Rw,W). Fl.4-6. Mons.. 2 squares.
Mons.: nr Tintern, ST59, CB, 1875[112] and AEW, 1925[112]; Black Cliff, ST59, AEW, 1951[112] and TGE, 1980[91]; Wynd Cliff, ST59, RWD, 1977[197] and TGE, 1980[91]; nr Monmouth, SO51, SGC, 1925[112] & 1946[112].

C. caryophyllea Latourr. : Hesgen Gynnar : Spring-sedge. Native. Pastures (Al,D,E,G,M,Q,R,Rw,Sc,W). Fl.4-6. Mons., Glam., Brecs., Rads., Carms., Pembs., Cards., Monts., Mer., Caerns., *Denbs., Flints.,* Angl.. 182 squares, map 1000.

C. montana L. : Hesgen Mynydd-dir : Soft-leaved Sedge. Native. Heaths (G,H). Fl.4-7. Mons., Glam., Brecs., *Carms.**, Denbs.. 16 squares : **SS**87,97; **ST**59; **SN**71,72,81,82,85,90,91,92; **SO**00,01,21,51; **SJ***24*.

C. pilulifera L. : Hesgen Bengron : Pill Sedge. Native. Heaths (G,H,Q,R,W). Fl.4-7. Mons., Glam., Brecs., Rads., Carms., Pembs., Cards., *Monts.,* Mer., Caerns., Denbs., *Flints.,* Angl.. 153 squares, map 1001.

C. limosa L. : Hesgen Eurwerdd : Bog-sedge. Native. Bogs (Al,H). Fl.6. Glam., Cards., Monts.*, Mer.*, Caerns., Angl.. 20 squares : **SS**69; **SN**66,67,75; **SH**23,28, 32,33,37,38,39,44,48,54,58,65,66,73,75,91.

C. magellanica Lam. subsp. **irrigua** (Wahlenb.) Hiitonen (*C. paupercula* Michx) : Tall Bog-sedge. Native. Bogs (H,M). Fl.6. Monts.*, Mer.*, *Denbs..* 4 squares.
Monts.: Bwlch y Fedwen, SH91, FAC, 1978[112]; E. Llyn Coch-hwyad, SH91, RGW, 1978[112]. **Mer.**: Trawsfynydd, SH73, MWe & PMB, 1963[112]; nr

Llanuwchllyn, SH82, PMB, 1971[3]; Arennig, SH83, PMB, 1974[3]. **Denbs.**: Vicarage Moss, SJ43, JEB, c1835[176].

C. atrata L. : Hesgen Ddu : Black Alpine-sedge. Native. Mountain rocks. Fl.6-7(-9). Caerns.. 1 square.
Caerns.: Snowdon, SH65, JEG, 1890[112] and ER, 1965[207]; Glyders, SH65, DAR, 1952[7] and ER, 1965[207].

C. aquatilis Wahlenb. : Hesgen y Dŵr Mynyddog : Water Sedge. Native. Stream margins (Al,Ar,H,M). Fl.(5-)7-9. Carms.*, Pembs., Cards., Mer.*, *Denbs.**, [*Brecs.*]. 13 squares : **SN**14,24,34,44,54,63,65,66,76; **SH**83,84,85,93.

C. bigelowii Torrey ex Schweinitz : Stiff Sedge. Native. Mountain screes. Fl.4-8. Mer., Caerns.. 5 squares.
Mer.: Craig Las, SH61, ME, 1965[155]; Cadair Idris, SH71, AL, 1875[104] and PMB & KS, 1966[3]; Aran Fawddwy, SH82, JHS, 1931[4] and AOC, 1962[112]; Aran Benllyn, SH82, AOC, 1962[4]. **Caerns.**: Snowdon, SH65, JHL, 1875[112] and JMB, 1965[112]; Glyders, SH65, JHS, 1925[112] and ER, 1965[207]; Carnedd Llywelyn, SH66, ER, 1962[207] and JMB, 1971[112]; Ysgolion Duon, SH66, ER, 1955[207]; Foel Grach, SH66, ER, 1963[207].

C. elata All. : Hesgen Oleulas Sythddail : Tufted-sedge. Native. Marshes. Fl.5-7. Glam., Brecs.*, Pembs.*, *Monts.**, Caerns., Denbs., *Flints.**, Angl.. 20 squares, map 1002.

C. elata × nigra = C. × turfosa Fries Spontaneous. Marshes and fens. Fl.6-7. *Glam., Caerns.,* Angl.. 4 squares.
Glam.: Flemingston Moors, ST07, HJR, 1904[107]. **Caerns.**: Cors Geirch, SH33, PMB, 1979[245]. **Angl.**: Cors y Farl, SH47, RHR, 1959[112] and PMB, 1980[112]; Cors Erddreiniog, SH48, RHR, 1970+[209].

C. nigra (L.) Reichard : Swp-hesgen y Fawnog : Common Sedge. Native. Marshes (Al,Ar,D,Ds,G,H,M,Q,W). Fl.4-8. Mons., Glam., Brecs., Rads., Carms., Pembs., Cards., Monts., Mer., Caerns., Denbs., Flints., Angl.. 246 squares, map 1003.

C. acuta L. : Hesgen Eiddil Dywysennog : Slender Tufted-sedge. Native. Streamsides (Al,Ar,G,M). Fl.5-7. Mons., Glam., Brecs., Rads., Carms., Cards., *Monts.**, Mer., Denbs., Angl., [*Caerns.*]. 30 squares, map 1004.

C. acuta × acutiformis = C. × subgracilis Druce Spontaneous. Riversides. Fl.8. *Caerns.*. 1 square.
Caerns.: Afon Soch, SH22, APC, 1979[245].

C. acuta × elata = C. × prolixa Fries Spontaneous. Marshy places. Fl.5-6. *Glam.*, Brecs.. 2 squares.
Glam.: Crymlyn Fen, SS69, HJR, 1904[107]. **Brecs.**: Llangorse Lake, SO12, BS, 1964[112].

C. acuta × nigra = C. × elytroides Fries Spontaneous. Marshy places. Fl.5. *Glam.*, Brecs., *Carms., Mer., Caerns.*, Angl.. 7 squares.
Glam.: Crymlyn Fen, SS69, HJR, c1904[140]. **Brecs.**: Llangorse Lake, SO12, BS, 1964[112]. **Carms.**: Llanelli, SS59, JM, c1850[143]. **Mer.**: Bala Lake, SH83, PMB, 1979[245]. **Caerns.**: Llyn Padarn, SH56, CES, 1916[149]. **Angl.**: nr Trearddur, SH27, JEG, c1889[101]; nr Rhosneigr,SH37, JEG, c1889[101].

Subgen. Primocarex Kük.

C. pauciflora Lightf. : Few-flowered Sedge. Native. Bogs. Fl.8. Caerns.. 2 squares.
Caerns.: Moel Hebog, SH54, FCR, c1895[101]; Migneint, SH74, DR, 1976[112].

C. pulicaris L. : Chwein-hesgen : Flea Sedge. Native. Bogs (Ar,Ds,G,H,M,Q,W). Fl.5-8. Mons., Glam., Brecs., Rads., Carms., Pembs., Cards., Monts., Mer., Caerns., *Denbs., Flints.,* Angl.. 185 squares, map 1005.

ORCHIDACEAE

[CYPRIPEDIUM L.]

[**Cypripedium calceolus** L. *Caerns.*]

EPIPACTIS Zinn

Epipactis palustris (L.) Crantz : Caldrist y Gors : Marsh Helleborine. Native. Marshes and dune slacks (D,Ds,G,H,M,R). Fl.6-8. Mons., Glam., *Brecs., Rads.,* Carms., Pembs., *Cards.*, Monts.,* Mer., Caerns., Denbs., Flints., Angl.. 52 squares, map 1006.

E. helleborine (L.) Crantz : Caldrist Lydanddail : Broad-leaved Helleborine. Native. Woods and roadsides (D,Ds,E,G,Q,R,W). Fl.6-8. Mons., Glam., Brecs., Rads., Carms., Pembs., Cards., Monts., Mer., *Caerns.*, Denbs., Flints., Angl.. 146 squares, map 1007.

E. leptochila (Godfery) Godfery : Narrow-lipped Helleborine. Native. Woods. Fl.7. *Mons.,* [*Flints., Angl.*]. 1 square.
Mons.: Wynd Cliff, ST59, CES, 1920[237].

E. dunensis (T. & T. A. Stephenson) Godfery : Dune Helleborine. Native. Dune slacks. Fl.7-8. Angl.. 1 square.
Angl.: Newborough Warren, SH46, GCD, 1919[113] and RHR, 1978[208].

[**E. purpurata** Sm. *Cards., Caerns.*]

E. phyllanthes G. E. Sm. (incl. *E. cambrensis* C. Thomas) : Green-flowered Helleborine. Native. Dune slacks and woods. Fl.6-9. Glam., Mer.*, *Flints.*. 6 squares.
Glam.: Oxwich, SS58, QONK, 1975[157]; Margam Burrows, SS78, CTh, 1941[233]; Kenfig Burrows, SS78, HWP, 1935[107] and DPY, 1949[107]. **Mer.**: Morfa Dyffryn, SH52, PMB, 1959[112] & 1964[177]. **Flints.**: Pant-y-mwyn, SJ16, EPAJ, 1948[107]; Llyn Helyg, SJ17, JAWh, 1906[112]; Rhyd-y-mwyn, SJ26, THE, 1954[7].

E. atrorubens (Hoffm.) Besser : Caldrist Rhuddgoch: Dark-red Helleborine. Native. Rocky ground (E,Q,Sc). Fl.6-8. *Brecs.,* Caerns., *Denbs.**, Flints., [*Mons.*]. 9 squares.
Brecs.: Builth Wells, SO04, CBC, 1864[107]. **Caerns.**: Gt Ormes Head, SH78, RD, 1861[112] and RHR, 1977[208]; nr Gloddaith, SH88, JEG, c1890[101] and RHR, 1977[208]. **Denbs.**: Llysfaen, SH87, AAD, c1910[144] and AMS, 1958[7]; nr Llangollen, SJ24, EMT, c1910[146]. **Flints.**: Talargoch Hill, SJ07, RB, 1885[107]; nr Dyserth, SJ07, JAWh, 1909[112] and AMS, 1956[7]; Meliden, SJ08, JRC, 1952[112]; The Leete, SJ16, AAD, c1908[143]; Moel Findeg, SJ26, AAD, c1908[143].

[**E. atrorubens** × **helleborine** = **E.** × **schmalhousenii** K. Richter *Caerns., Denbs.*]

CEPHALANTHERA L. C. M. Richard

[**Cephalanthera damasonium** (Miller) Druce *Mer.*]

C. longifolia (L.) Fritsch : Narrow-leaved Helleborine. Native. Woods (M,Rw,W). Fl.5-7. Mons., *Cards.**, *Monts.**, Mer., Caerns., *Angl.**. 8 squares.
Mons.: Chepstow, ST59, WAS, 1879[112]; Wynd Cliff, ST59, EV, 1914[112]; btwn Chepstow and Tintern, ST59, TGE, 1970[226] & 1980[91]. **Cards.**: Eglwys-fach, SN69, IMC & WCo, 1956[69]. **Monts.**: Kerry, SO18, EHW, 1978[184]. **Mer.**: Aberdyfi, SN69, JHS, 1928[112] and PMB, 1978[3]; Arthog, SH61, An, 1920[19] and An, 1978[3]; Harlech, SH63, DAJ, 1905[112] and OHB & HHan, 1971[3] & 1981[3]. **Caerns.**: Llanberis, SH56, JEB, c1830[112] and WSL, RHR & PMB, 1979[3]. **Angl.**: Newborough Forest, SH36, PE, 1981[208].

NEOTTIA Ludwig

Neottia nidus-avis (L.) L. C. M. Richard : Tegeirian Nyth Aderyn : Bird's-nest Orchid. Native. Woods. Fl.4-7. Mons., Glam., *Brecs., Rads., Carms.,* Pembs., Cards., *Monts.**, *Mer., Caerns., Denbs.,* Flints.. 48 squares, map 1008.

LISTERA R. Br.

Listera ovata (L.) R.Br. : Ceineirian : Common Twayblade. Native. Woods (D,Ds,E,G,M,Q,Rw,W). Fl.5-7. Mons., Glam., Brecs., Rads., Carms., Pembs., Cards., *Monts.,* Mer., Caerns., Denbs., Flints., Angl.. 170 squares, map 1009.

L. cordata (L.) R.Br. : Ceineirian Bach : Lesser Twayblade. Native. Woods and heaths. Fl.6-7. Brecs.*, *Monts., Mer.,* Caerns., *Denbs., Flints..* 32 squares, map 1010.

SPIRANTHES L. C. M. Richard

Spiranthes spiralis (L.) Chevall. : Ceineirian Troellog: Autumn Lady's-tresses. Native. Calcareous pastures (D,Ds,G,Q,R,Rg). Fl.8-9. Mons., Glam., Carms., Pembs., Cards., *Monts.,* Mer., Caerns., Denbs., Flints., Angl.. 89 squares, map 1011.

HERMINIUM Guett.

Herminium monorchis (L.) R.Br. : Musk Orchid. Native. Sand dunes. Fl.7. Glam., [*Mons.*]. 2 squares.
Glam.: Kenfig Burrows, SS78, AMa, 1958[93] and JWL, 1968[93]; Beaupre, ST07, Tr, c1880[223].

PLATANTHERA L. C. M. Richard

Platanthera bifolia (L.) L. C. M. Richard : Tegeirian Llydanwyrdd Bach : Lesser Butterfly-orchid. Native. Damp heathy places (E,G,H,M,W). Fl.6-7. Mons., *Glam.,* Brecs., Rads., Carms., Pembs., Cards., *Monts.**, Mer., Caerns., *Denbs.,* Flints., Angl.. 94 squares, map 1012.

P. chlorantha (Custer) Reichenb. : Tegeirian Llydanwyrdd : Greater Butterfly-orchid. Native. Woods and grassland (E,F,G,M,Rw,W). Fl.5-7. Mons., Glam., Brecs., Rads., Carms., Pembs., *Cards.,* Monts., Mer., Caerns., *Denbs.,* Flints., [*Angl.*]. 106 squares, map 1013.

GYMNADENIA R.Br.

Gymnadenia conopsea (L.) R.Br. : Tegeirian Pêr : Fragrant-orchid.

Subsp. **conopsea** Native. Bogs (D,Ds,G,H,M,Q,Rw,Sc). Fl.6-7. Mons., Glam., Brecs., Rads., Carms., *Pembs.,* Cards., *Monts.,* Mer., Caerns., Denbs., Flints., Angl.. 81 squares, map 1014.

Subsp. **densiflora** (Wahlenb.) G. Camus, Bergen & A. Camus (*G. conopsea* var. *densiflora* (Wahlenb.) Lindley). Native. Sand dunes. Fl.7. Caerns., Angl.. 2 squares.
Caerns.: Cors Edern, SH23, PMB, 1980[74]. **Angl.**: Newborough Warren, SH36, MJG, 1919[100] and RPB, 1966[7].

PSEUDORCHIS Séguier

Pseudorchis albida (L.) Á. & D. Löve (*Leucorchis albida* (L.) E. H. F. Meyer) : Tegeirian Broga Gwyn : Small-white Orchid. Native. Pastures (G,Rw). Fl.6-7. Mons., Glam., *Brecs., Carms.,* Cards., *Monts.**, Mer., Caerns., *Flints..* 25 squares, map 1015.

COELOGLOSSUM Hartman

Coeloglossum viride (L.) Hartman : Paladr Blodeuwyrdd : Frog Orchid. Native. Calcareous pastures (D,G,H,Q,R,Rw,W). Fl.6-8. Mons., Glam., *Brecs., Rads., Carms., Pembs.**, *Cards., Monts.,* Mer., Caerns., Denbs., Flints., Angl.. 49 squares, map 1016.

COELOGLOSSUM × DACTYLORHIZA = × DACTYLOGLOSSUM P. F. Hunt & Summerhayes

Coeloglossum viride × **Dactylorhiza fuchsii** = × **Dactyloglossum mixtum** (Ascherson & Graebner) P. F. Hunt Spontaneous. Grassland. *Flints..* 1 square.
Flints.: Maeshafn, SJ26, EH, 1953[52].

DACTYLORHIZA Necker ex Nevski

Dactylorhiza incarnata (L.) Soó (*Orchis strictifolia* Opiz; *Dactylorchis incarnata* (L.) Vermeulen) : Tegeirian y Gors : Early Marsh-orchid. Native. Marshes (Al,D,Ds,G,H,M,R). Fl.6-7. Mons., Glam., *Brecs.**, Rads., Carms., Pembs., Cards., Monts.*, Mer., Caerns., Denbs.*, Flints., Angl.. 98 squares, map 1017.

Subsp. **incarnata** (*Dactylorchis incarnata* subsp. *incarnata*). Native. Marshes (D,Ds,G,H,M). Fl.5-7. Mons., Glam., Brecs., Rads.*, Carms., Pembs., Cards., *Monts.**, Mer., Caerns., *Denbs.**, Flints., Angl.. 60 squares, map 1018.

Subsp. **coccinea** (Pugsley) Soó (*Orchis strictifolia* subsp. *coccinea* (Pugsley) Clapham; *Dactylorchis incarnata* subsp. *coccinea* (Pugsley) (H.-Harrison). Native. Dune slacks (D,Ds,M,R). Fl.5-7. Glam., Carms., *Cards., Mer., Caerns.,* Flints., Angl.. 18 squares : **SS**49,58,78,87; **SN**20,30,31,40,69; **SH**36,37,43, 46,52,53,78; **SJ**36,37.

Subsp. **pulchella** (Druce) Soó (*Orchis strictifolia* var. *pulchella* (Druce) Clapham; *Dactylorchis incarnata* subsp. *pulchella* (Druce) H.-Harrison). Bogs (Ds,G, H,M). Fl.6-7. *Glam., Brecs., Rads., Pembs.,* Cards., *Mer., Caerns., Denbs.,* Angl.. 25 squares, map 1019.

Subsp. **ochroleuca** (Boll) P. F. Hunt & Summerhayes (*Dactylorchis incarnata* subsp. *ochroleuca* (Boll) H.-Harrison). Native. Wet places. *Carms.*. 1 square.
Carms.: nr Kidwelly, SN40, TSt & TAS, c1919[148].

D. incarnata × maculata = D. × claudiopolitana (Soó) Soó Spontaneous. Marshes and bogs. Fl.6. *Pembs., Mer., Angl.,* [*Glam.*]. 7 squares. Some of these records were made as *Orchis latifolia × maculata*, indicated by*.
Pembs.: Penally Marsh, SS19, RFT, 1900[134]. **Mer.**: nr Cross Foxes, SH71, PMB, 1956[4]. **Angl.**: Bodorgan, SH36, GCD, c1899[135]; Newborough Warren, SH46, EM, 1937[36]*; Cors Goch, SH48, JFGC, 1937[36]*; Cors Erddreiniog, SH48, RHR, 1971[208]; Cors Bodeilio, SH57, JFGC & EM, 1937[36]*; Llaniestyn Common, SH57, RHR, 1971[208]; Red Wharf Bay, SH58, EM, 1937[36]*.

D. incarnata × majalis subsp. **praetermissa** (*D. × wintonii* (A. Camus) P. F. Hunt). Spontaneous. Dunes. Glam., *Brecs., Carms., Pembs.*[217]. 4 squares.
Glam.: Kenfig Burrows, SS78, DA & AA, 1952[112]. **Brecs.**: nr Cwmwysg, SN82, MP, 1980[200]. **Carms.**: Pendine, SN20, BDH, 1978[61]; Kidwelly, SN40, TSt & TAS, c1919[148].

D. incarnata × majalis subsp. **purpurella** (*D. × latirella* (P.M. Hall) Soó). Spontaneous. Dune slacks and marshes. Fl.6. *Brecs., Cards., Mer., Angl.*. 8 squares.
Brecs.: Penderyn, SN90, MP, 1977[200]. **Cards.**: nr Aberystwyth, SN- -, TSt & TAS, 1919[148]; btwn Borth and Moel Ynys, SN69, PMH, 1935[107]. **Mer.**: Morfa Dyffryn, SH52, PMB, 1958[4] & 1968[3]. **Angl.**: Rhoscolyn Marsh, SH27, JDG, 1935[107]; Newborough Warren, SH46, EM, 1937[36] and RHR, 1958[208]; nr Llanbedr-goch, SH48, EM, 1937[36]; Cors Bodeilio, SH57, JFGC, 1937[36]; Red Wharf Bay, SH58, EM, 1937[36] and RHR, 1960[208].

D. incarnata × traunsteineri = D. × lehmanii (Klinge) Soó Spontaneous. Fens. Fl.6-7. *Caerns.,* Angl.. 3 squares.
Caerns.: Cors Geirch, SH33, RHR & WSL, c1957[174]; Rhydyclafdy, SH33, RHR & WSL, 1957[54]. **Angl.**: Cors Erddreiniog, SH48, RHR, 1958[112]; Cors Bodeilio, SH57, RHR, 1956[112]; Rhos y Gad, SH57, RHR & NB, 1977[184].

D. majalis (Reichenb.) P. F. Hunt & Summerhayes (*Dactylorchis majalis* (Reichenb.) Vermeulen)

Subsp. **cambrensis** (R. H. Roberts) R. H. Roberts : Tegeirian y Gors Cymreig : Welsh Marsh-orchid. Native. Marshes. Fl.6-7. Cards., *Mer.,* Caerns., Angl.. 4 squares.
Cards.: Ynys-las, SN69, RHR, 1958[112]. **Mer.**: nr Minffordd, SH53, RHR & THB, 1979[245]; nr Pontcroesor, SH54, DTE & RHR, 1979[245]. **Caerns.**: Glaslyn Valley, SH54, RHR, 1962[112]. **Angl.**: Newborough Warren, SH46, RHR, 1959[112]; Malltraeth, SH46, RA, 1979[208].

Subsp. **cambrensis** × subsp. **purpurella.** Spontaneous. Marshes. *Angl.*†. 1 square.
Angl.: Cefni Estuary, SH46, RHR, 1960+[208].

Subsp. **purpurella** (T. & T. A. Stephenson) D. Moresby Moore & Soó (*D. purpurella* (T. & T. A. Stephenson) Soó; *Dactylorchis purpurella* (T. & T. A. Stephenson) Vermeulen; *Orchis purpurella* T. & T. A. Stephenson) : Tegeirian y Migin : Northern Marsh-orchid. Native. Marshes (D,G,M,Rw). Fl.6-7. Brecs.*, *Rads.**, Carms., Pembs., Cards., Monts., Mer., Caerns., Denbs., Flints., Angl., [*Glam.*]. 84 squares, map 1020.

[Subsp. **purpurella** × **traunsteineri** *Caerns.*]

Subsp. **praetermissa** (Druce) D. Moresby Moore & Soó (*D. praetermissa* (Druce) Soó; *Dactylorchis praetermissa* (Druce) Vermeulen; *Orchis praetermissa* Druce) : Tegeirian y Gors Deheulog : Southern Marsh-orchid. Native. Marshes (D,Ds,G,H,M,O,Rw). Fl.5-7. Mons., Glam., Brecs., *Rads.,* Carms., Pembs., Cards., *Monts.**, Mer., *Caerns., Denbs., Flints.**, [*Angl.*]. 103 squares, map 1021.

Subsp. **praetermissa** × subsp. **purpurella** (*D. × insignis* (T. & T. A. Stephenson) Soó). Spontaneous. Dune slacks (D,Ds,G). Cards., *Mer.*. 3 squares.
Cards.: nr Aberystwyth, SN68, MJG, pre 1930[100] and TSt, c1940[151]; Ynys-las, SN69, JHS, 1942[112]. **Mer.**: Morfa Harlech, SH53, PMB, 1960+[4] and RHR, c1965[238].

D. traunsteineri (Sauter) Soó (*Dactylorchis traunsteineri* (Sauter) Vermeulen; *Orchis traunsteineri* Sauter) : Narrow-leaved Marsh-orchid. Native. Fens. Fl.5-7. Caerns., Angl.*, [*Glam., Mer.*]. 5 squares.
Caerns.: Cors Edern, SH23, RHR & SDW, 1976[184]; Cors Hirdre, SH23, PMB & APC, 1980[74]; Cors Geirch, SH33, WSL, 1956[112] and APC & PMB, 1980[74]. **Angl.**: Cors Bodeilio, SH47 and SH57, RHR, 1956[112]; Gwenfro Uchaf, SH47 and SH57, RHR, 1959[112]; Cors Erddreiniog, SH48, RHR, 1957[172] and WEH, 1966[112]; Nant Isaf nr Bryn-teg, SH48, RHR, 1960[208]; Rhos y Gad, SH57, RHR, 1958[112].

D. maculata (L.) Soó (incl. *D. maculata* subsp. *ericetorum* (E. F. Linton) P. F. Hunt & Summerhayes; *Dactylorchis maculata* subsp. *ericetorum* (E. F. Linton) Vermeulen; *Orchis ericetorum* (E. F. Linton) E. S. Marshall) : Tegeirian Brych : Heath Spotted-orchid. Native. Wet heaths (D,G,H,M,Q,R,W). Fl.5-7. Mons., Glam., Brecs., Rads., Carms., Pembs., Cards., Monts., Mer., Caerns., *Denbs., Flints.*, Angl.. 202 squares, map 1022.

D. maculata × majalis subsp. **praetermissa** (*D.* × *hallii* (Druce) Soó). Spontaneous. Woods and sand dunes (D,H,M,W). Fl.6. Mons., Glam., Carms., Pembs., *Cards., Mer.*, [*Angl.*]. 15 squares : **SS***48*,49, 59,*78*; **SM**72,80; **SN**10,*32,40*,56,*73*; **SO***00,21*,50; **SH**53.

D. maculata × majalis subsp. **purpurella** (*D.* × *formosa* (T. & T. A. Stephenson) Soó). Spontaneous. Wet meadows. Fl.6-7. *Brecs., Carms.*, Cards., Monts., Mer., Caerns., Denbs., *Flints., Angl.*, [*Glam.*]. 21 squares, map 1023.

D. maculata × traunsteineri = D. × jenensis (Brand) Soó Spontaneous. Base-rich fens. *Caerns.*, Angl.. 4 squares.
Caerns.: Cors Edern, SH23, RHR & SDW, 1977[208]; Cors Hirdre, SH23, PMB & APC, 1980[74]; Cors Geirch, SH33, RHR, GR & AMS, 1973[182]. **Angl.**: Cors Erddreiniog, SH48, RHR, 1957-58[174]; Rhos y Gad, SH57, RHR, 1961[112].

D. fuschii (Druce) Soó (*Dactylorchis fuchsii* (Druce) Vermeulen; *Orchis fuchsii* Druce) : Tegeirian Brych Cyffredin : Common Spotted-orchid. Native. Wet pastures (D,Ds,E,F,G,H,M,Q,R,Rw,W). Fl.6-7. Mons., Glam., Brecs., Rads., Carms., Pembs., Cards., *Monts.**, Mer., Caerns., Denbs., Flints., Angl.. 195 squares, map 1024.

D. fuchsii × incarnata = D. × kernerorum (Soó) Soó Spontaneous. Marshes and dune slacks. *Mer., Flints., Angl.*. 5 squares.
Mer.: Morfa Harlech, SH53, PMB, 1962[4]. **Flints.**: Point of Air, SJ18, TE, 1970+[90]. **Angl.**: Cefni Estuary, SH46, RHR, 1960[208]; Cors Bodeilio, SH57, EM, c1937[36]; Rhos y Gad, SH57, RHR, 1970+[209]; nr Llanddona, SH57, RHR, 1970+[209]; Llanfihangel Dinsylwy, SH58, EM, 1937[36].

D. fuchsii × maculata = D. × transiens (Druce) Soó Spontaneous. Marshes and dune slacks. Fl.6. *Mer., Denbs., Angl.*. 3 squares.
Mer.: Morfa Harlech, SH53, PMB & RHR, 1964[178]. **Denbs.**: nr Glyndyfrdwy, SJ14, VG, 1958[7]. **Angl.**: Llyn Cadarn, SH48, EM, 1937[36].

D. fuchsii × majalis subsp. **cambrensis** Spontaneous. Base-rich marshes. *Angl.*. 1 square.
Angl.: Cefni Estuary, SH46, RHR, 1959-1965[208]†; Malltraeth, SH46, RHR, 1978[208].

D. fuchsii × majalis subsp. **praetermissa** (*D.* × *grandis* (Druce) P. F. Hunt). Spontaneous. Marshes and dune slacks (D,Ds,M). Fl.6-7. Mons., Glam., Brecs., *Carms.*, Pembs., Mer., *Denbs., Flints.*, [*Angl.*]. 14 squares : **SS**78; **SM**70; **SN**10,30.90; **SO**00,*01*,10,22; **SH**53; **SJ**26,*35*,37,43.

D. fuschsii × majalis subsp. **purpurella** (*D.* × *venusta* (T. & T. A. Stephenson) Soó). Spontaneous. Marshes. Fl.6-7. *Glam., Brecs., Mer., Caerns., Denbs.*, Flints., Angl.. 16 squares : **SN**81,84,90; **SH**23,28,47,53,57, 58,71,87; **SJ**08,17,25,35,43.

D. fuchsii × traunsteineri = D. × kellerana P. F Hunt Spontaneous. Fens. Fl.6. *Caerns., Angl.*. 3 squares.
Caerns.: Cors Edern, SH23, RHR & SDW, 1976[184]. **Angl.**: Cors Erddreiniog, SH48, RHR, 1957[174] and 1964[180]; Rhos y Gad, SH57, RHR, 1970[180].

DACTYLORHIZA × GYMNADENIA = × DACTYLOGYMNADENIA Soó

Dactylorhiza fuchsii × Gymnadenia conopsea = × Dactylogymnadenia cookei (H.-Harrison) Soó Spontaneous. *Brecs., Flints.*. 2 squares.
Brecs.: Cwm Cadlan, SN90, MP, 1973[200]. **Flints.**: nr Ysgeifiog, SJ17, GW, 1970+[261].

D. maculata × G. conopsea = × D. legrandiana (Camus) Soó Spontaneous. Grassland. Fl.6-7. Cards., Mer., *Caerns., Angl.*. 6 squares.
Cards.: Tregaron, SN66, TAS, 1915[16]; nr Devil's Bridge, SN77, MJG & TSt, 1919[100] and PMH, 1935[107]; Glasbwll, SN79, JHS, 1938[112]. **Mer.**: Brithdir, SH71, MR, 1928[112] and DCL, 1970[3]; nr Cross Foxes, SH71, JHS, 1939[112] and MR, 1948[4]. **Caerns.**: nr Bangor, SH57, WHP, 1888[135]. **Angl.**: Red Wharf Bay, SH58, EM, 1937[36].

ORCHIS L.

Orchis morio L. : Tegeirian y Waun : Green-winged Orchid. Native. Calcareous pastures (D,G,M). Fl.4-6. Mons., Glam., Brecs., Rads., Carms., Pembs., *Cards., Monts.*, Mer., Caerns., *Denbs., Flints.*, Angl.. 87 squares, map 1025.

O. ustulata L. : Cordegeirian : Burnt Orchid. Native. Limestone grassland. Fl.5. Caerns.. 1 square.

Caerns.: Pabo nr Llandudno, SH87, DEV, 1953[112] (doubt has been expressed as to whether this plant was actually collected in Wales).

O. mascula (L.) L. : Tegeirian Coch : Early-purple Orchid. Native. Pastures and woods (E,G,Q,R,Sc,W). Fl.4-5. Mons., Glam., Brecs., Rads., Carms., Pembs., Cards., Monts., Mer., Caerns., Denbs., Flints., Angl.. 177 squares, map 1026.

O. mascula × morio = O. × morioides Brand Spontaneous. Limestone grassland. *Angl.*†. 1 square.
Angl.: Cors Goch, SH48, RHR, 1959[173].

ANACAMPTIS L. C. M. Richard

Anacamptis pyramidalis (L.) L. C. M. Richard : Tegeirian Bera : Pyramidal Orchid. Native. Calcareous pastures (D,Ds,E,G,Q,R). Fl.6-7. Mons., Glam., Carms., Pembs., Cards., Monts., Mer., Caerns., Denbs., Flints., Angl., [*Rads.*]. 60 squares, map 1027.

OPHRYS L.

Ophrys insectifera L. : Tegeirian Pryfyn : Fly Orchid. Native. Wood borders and fens (H,M,Rw,W). Fl.5-7. Mons., Denbs., Angl., [*Glam.*]. 8 squares.
Mons.: Common Coed, ST48, DWH, pre 1930[226]; Shirenewton, ST49, EJL, 1869[226]; Chepstow, ST59, WAS, 1883[112]; Wynd Cliff ST59, WAS, c1890[226]; nr Monmouth, SO51, SGC, 1926[112]. **Denbs.**: N. Denbs., — , WBR, c1900[144]; Llanddulas, SH97, MDa & JMG, 1938[112]. **Angl.**: Cors Bodeilio, SH47, JTCWi, 1877[112] and JEG, 1882[112]; Cors Bodeilio, SH47 and SH48, RHR, 1970+[208]; Penrhosllugwy SH48, JEG, 1891[112]; Cors Erddreiniog, SH48, RHR, 1958[112] & 1977[208]; Rhos y Gad, SH57, RHR, 1958-1978[208].

O. sphegodes Miller : Tegeirian Copyn : Early Spider-orchid. Native. Pastures. Fl.5. Denbs.†. 1 square.
Denbs.: btwn Llandrillo and Colwyn Bay, SH87, SHB, 1891[112].

O. apifera Hudson : Tegeirian Gwenynen : Bee Orchid. Native. Calcareous pastures (D,Ds,G,Q,R,Rw). Fl.5-8. Mons., Glam., Carms., Pembs., *Cards., Monts.* *, Mer., Caerns., Denbs., Flints., *Angl.*. 62 squares, map 1028.

LIPARIS L. C. M. Richard

Liparis loeselii (L.) L. C. M. Richard : Tegeirian y Fign : Fen Orchid. Native. Dune slacks. Fl.6-9. Glam., Carms.. 7 squares.
Glam.: Whitford Burrows, SS49, AEW, 1956[227] and QONK, 1977[157]; Oxwich Burrows, SS58, QONK, 1972[157]; Margam Burrows, SS78, EV, 1927[224]; Kenfig Burrows, SS78, PMH, 1935[107] and AEW & RGE, 1977[95]; Baglan Burrows, SS79, RLS, 1926[112]†; Crymlyn Burrows, SS79, QONK, 1975[157]. **Carms.**: Pendine Burrows, SN20, JFT, 1938[112] and RGE, 1982[93]; Tywyn Burrows, SN30, RBr, 1898[1] and IMV & TAWD, 1971[59]; Pembrey Burrows, SN40, HHK, 1906[107].

HAMMARBYA O. Kuntze

Hammarbya paludosa (L.) O. Kuntze : Gefell-lys y Gors : Bog Orchid. Native. Bogs (H,M). Fl.6-9. Brecs., *Carms.* *, Pembs., *Cards.* *, Mer., Caerns.. 17 squares : **SN**03,13,71,77,87,96; **SO***02*; **SH**44,*55*,*61*,64,*65*,*66*,70,71, 75,81.

Distribution maps

on all maps

○ pre 1930

● 1930 onwards

\+ introduced

on maps for *Hieracium* only

⊙ record not verified by specialist

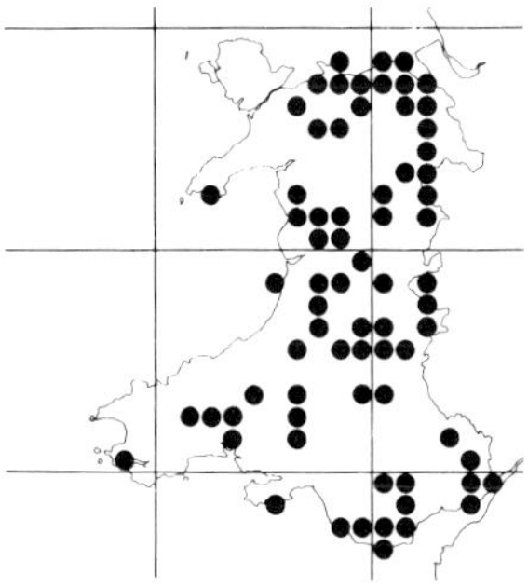

1 Larix decidua

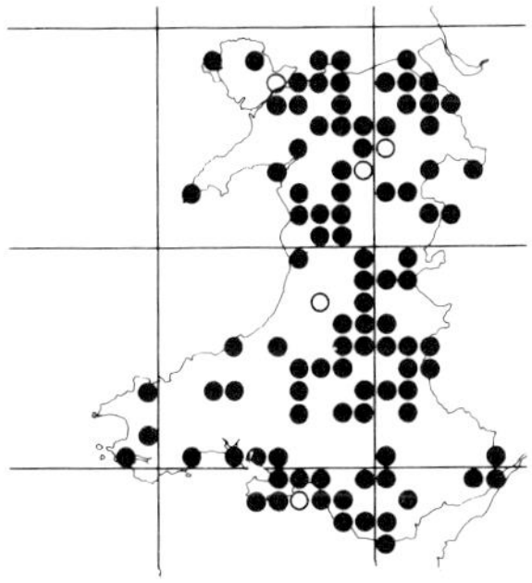

2 Pinus sylvestris

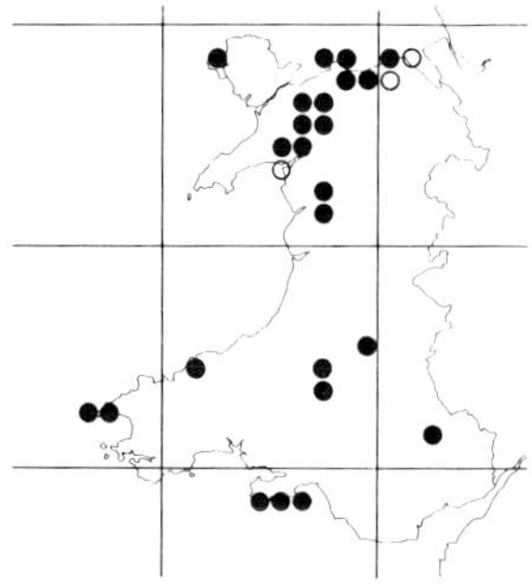

3 Juniperus communis

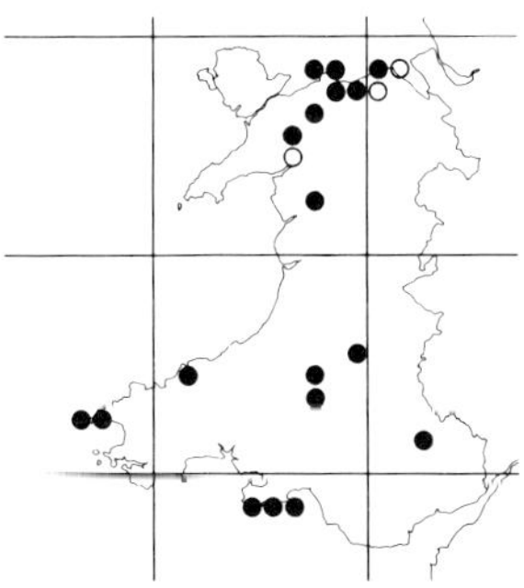

4 Juniperus communis
subsp. communis

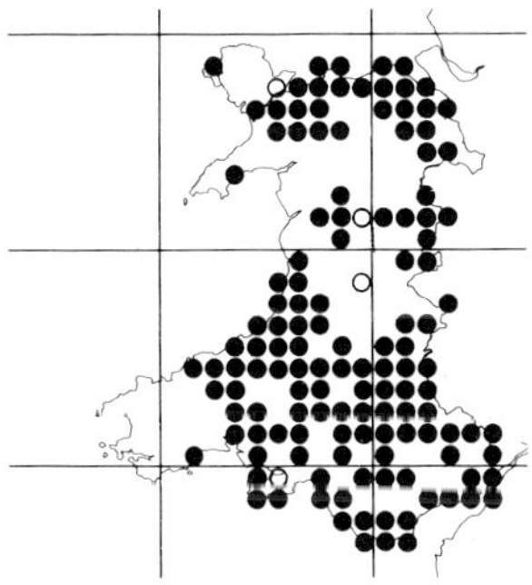

5 Taxus baccata

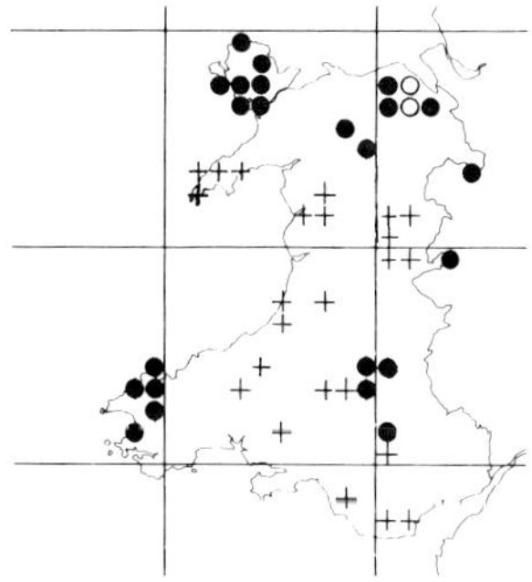

6 Salix pentandra

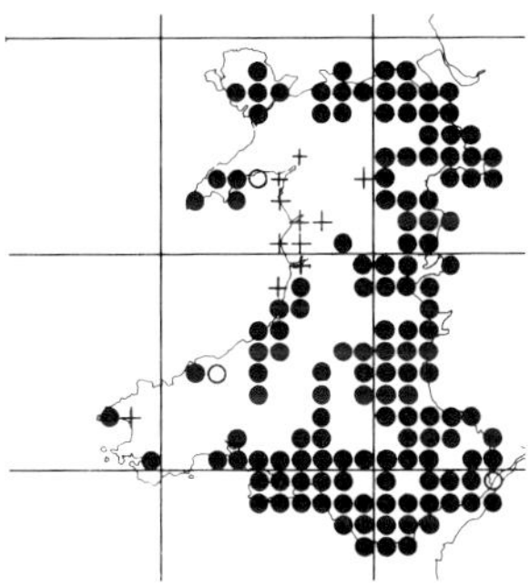

7 Salix fragilis

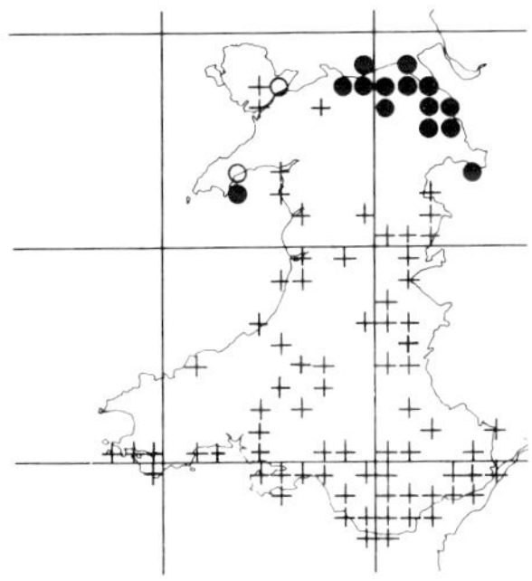

8 Salix alba
subsp. alba

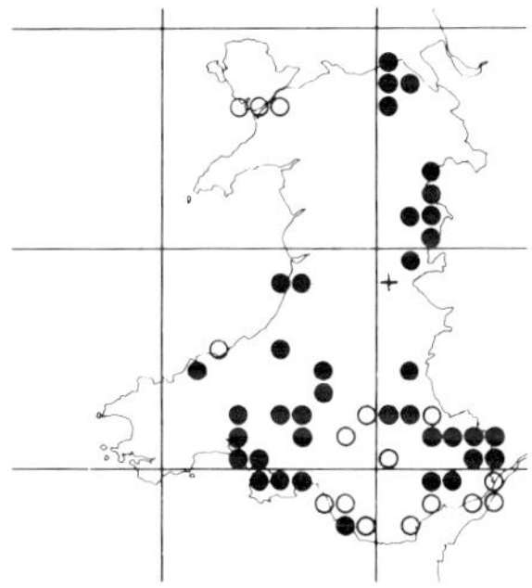

9 Salix triandra

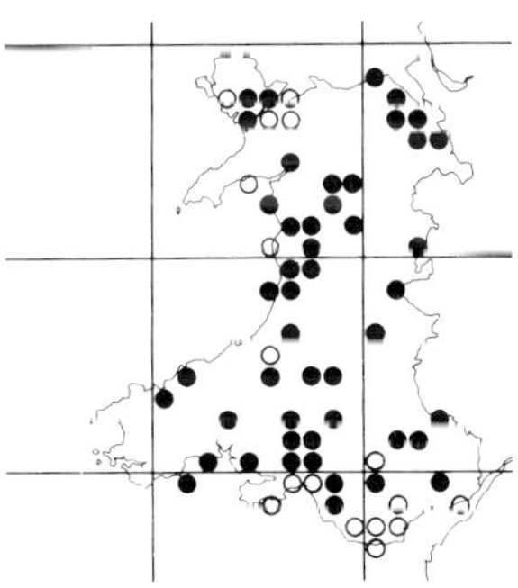

10 Salix cinerea × viminalis
= S. × smithiana

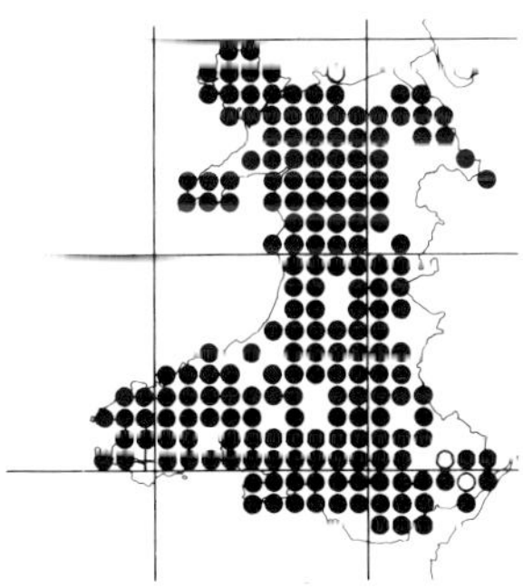

11 Salix aurita

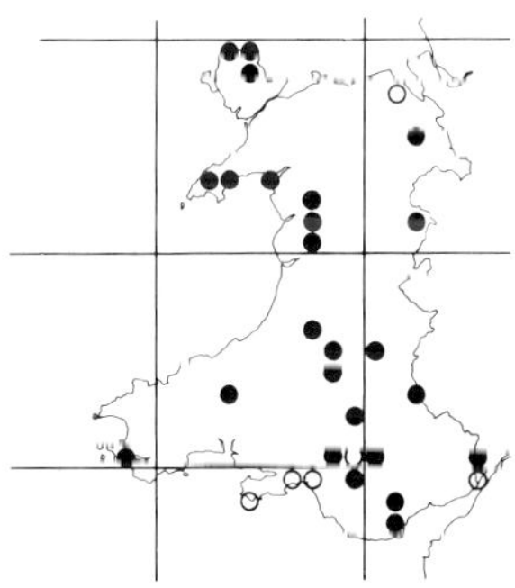

12 Salix aurita × cinerea
= S. × multinervis

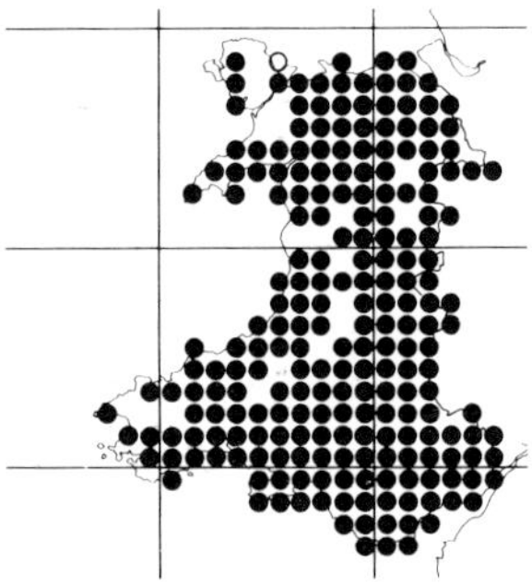

13 Salix caprea

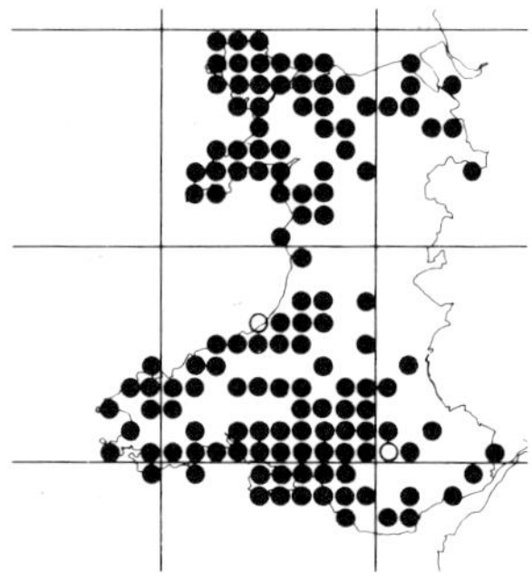

14 Salix repens agg.

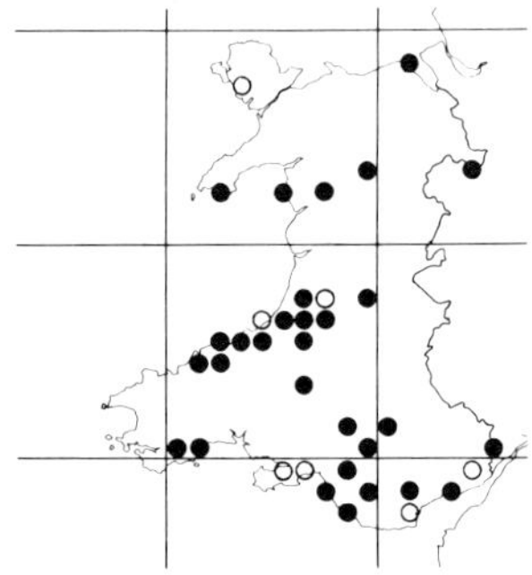

15 Salix repens s.s.

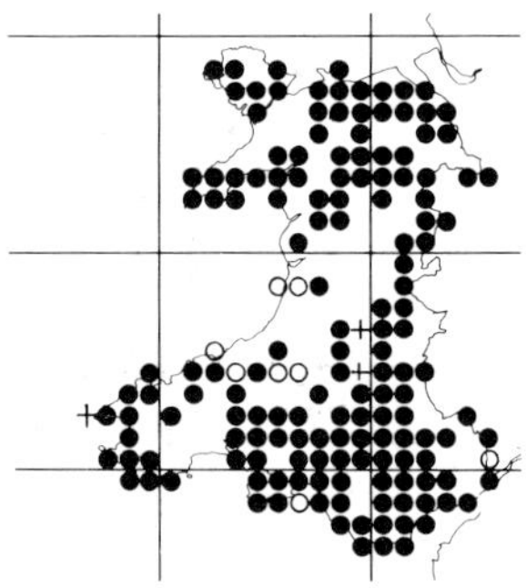

16 Salix viminalis

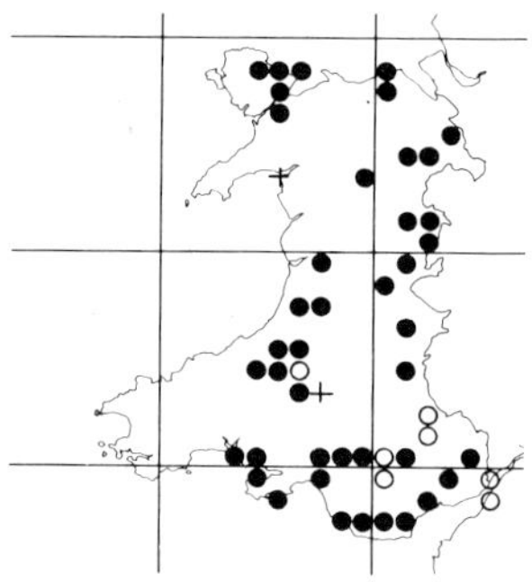

17 Salix purpurea

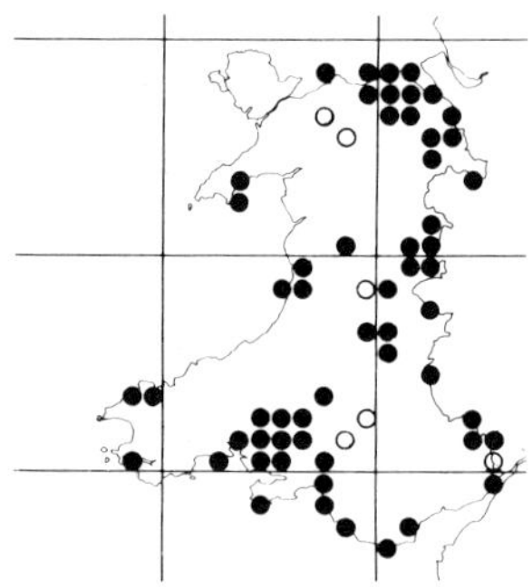

18 Populus alba

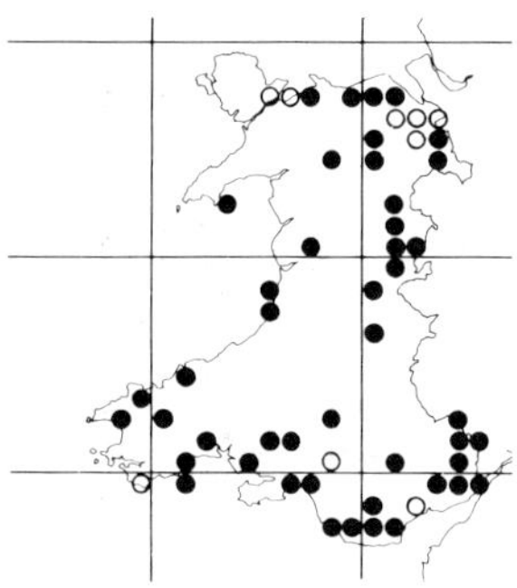

19 Populus alba × tremula
= P. × canescens

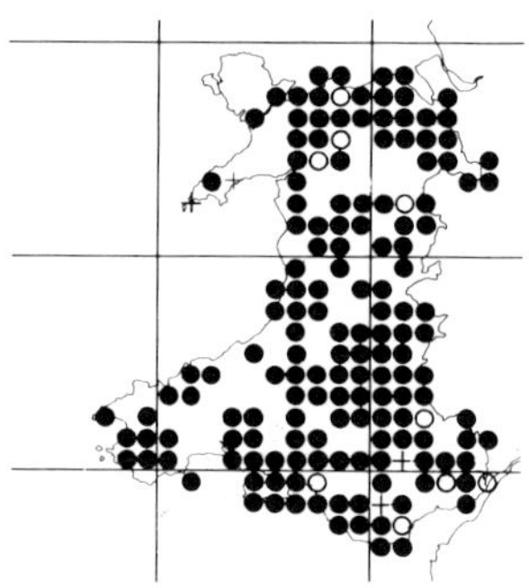

20 Populus tremula

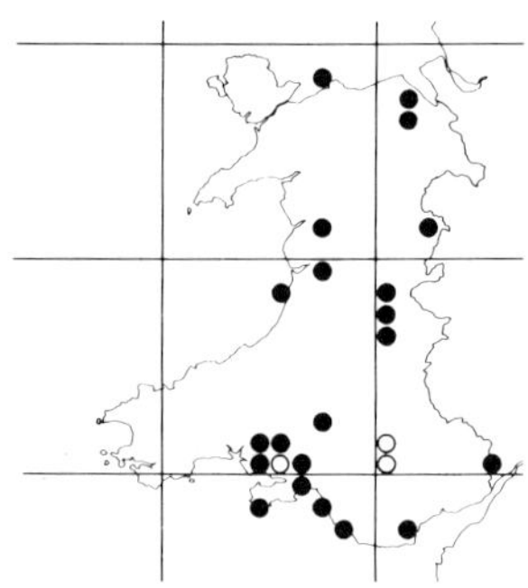

21 Populus gileadensis

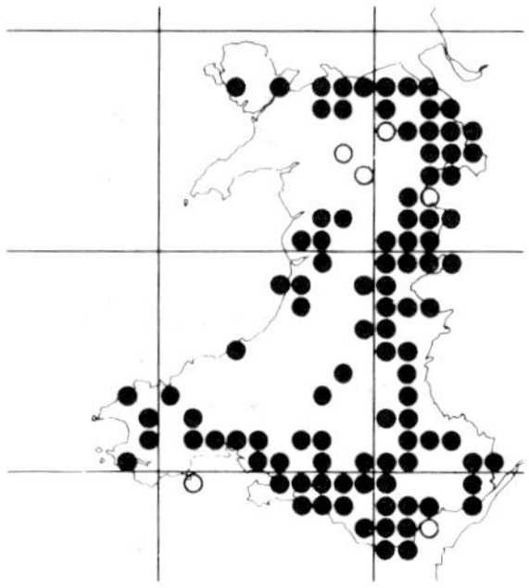

22 Populus nigra agg.

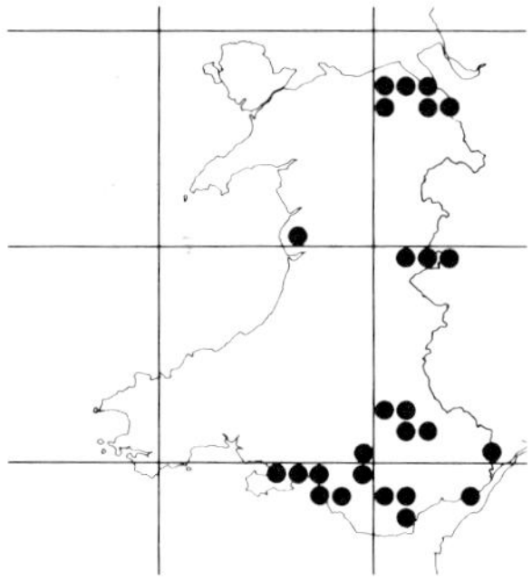

23 Populus nigra s.s.

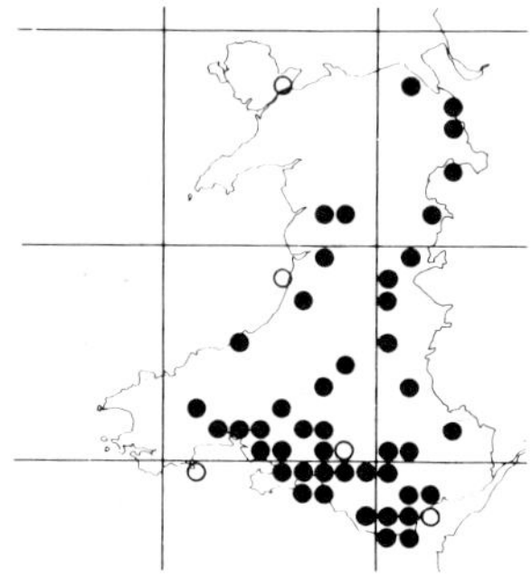

24 Populus deltoides × nigra
= P. × canadensis

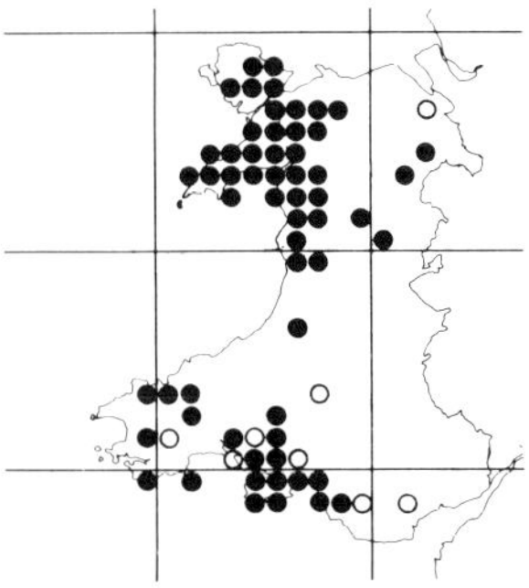

25 Myrica gale

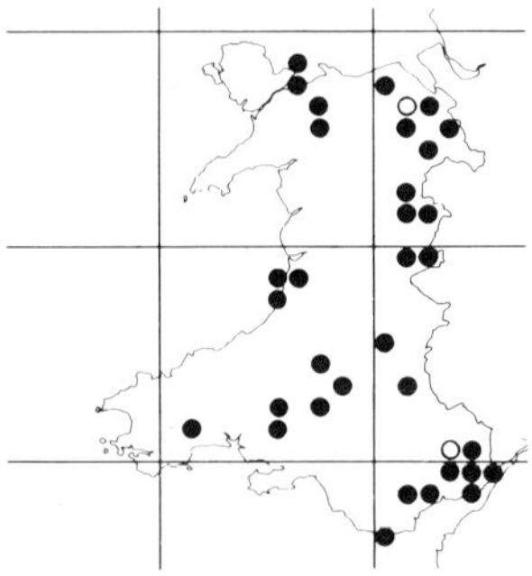

26 Juglans regia

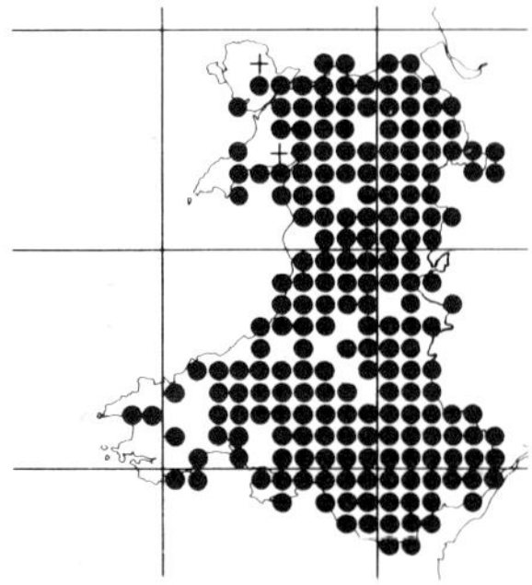

27 Betula pendula

28 Betula pubescens
subsp. pubescens

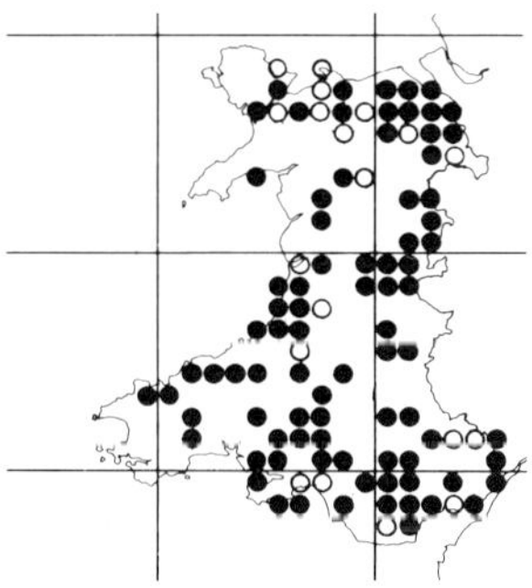

29 Carpinus betulus

30 Fagus sylvatica

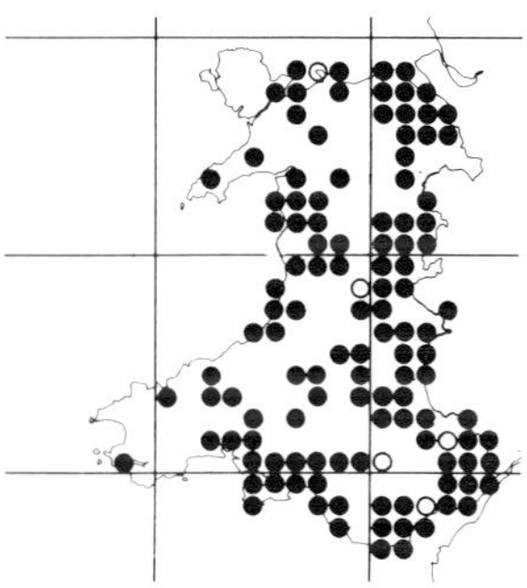

31 Castanea sativa

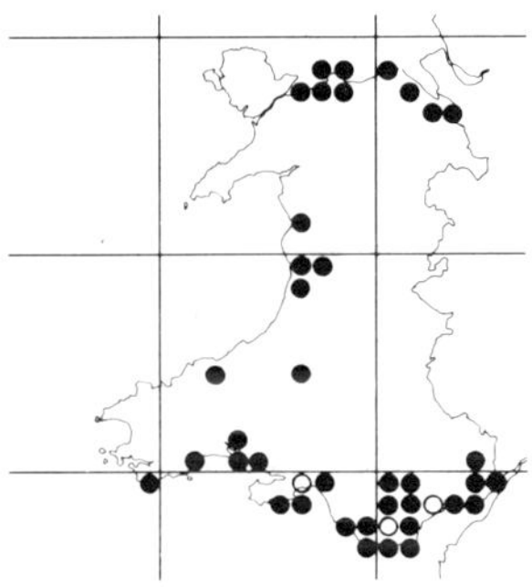

32 Quercus ilex

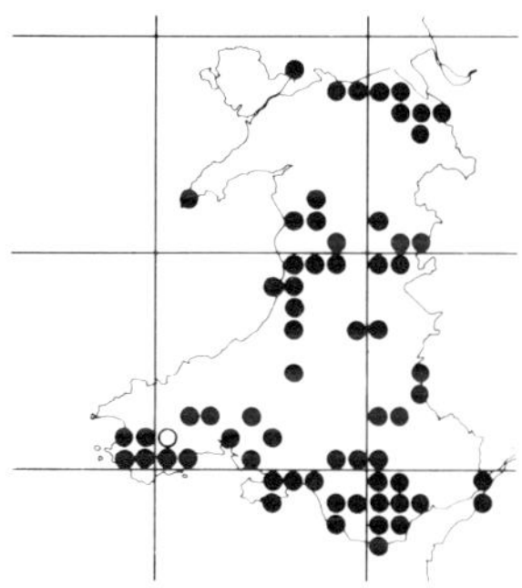

33 Quercus cerris

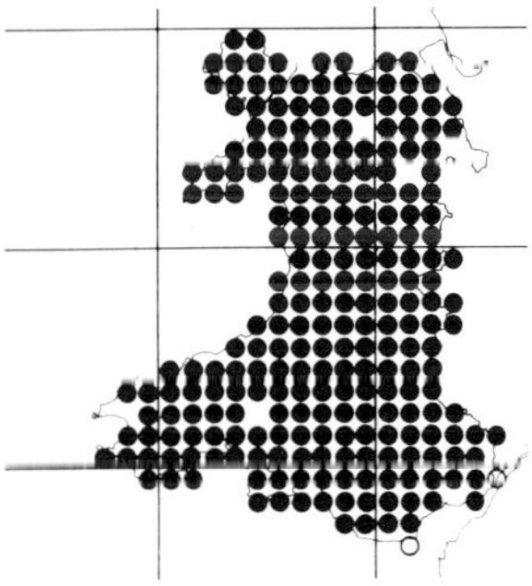

34 Quercus petraea

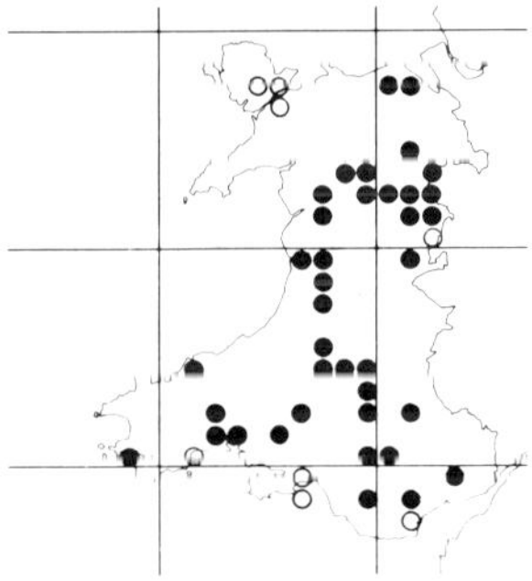

35 Quercus petraea × robur
= Q. × rosacea

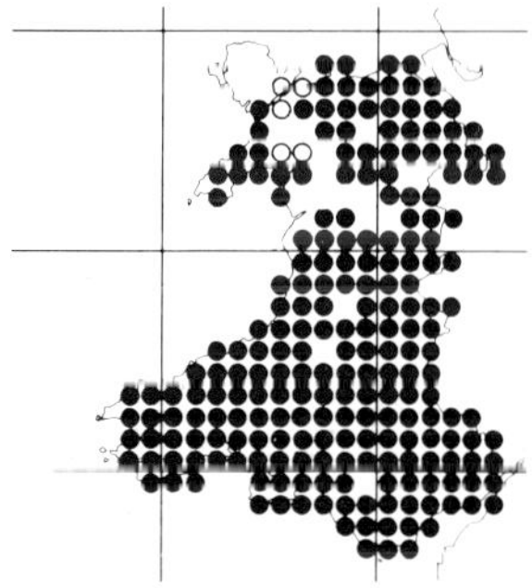

36 Quercus robur

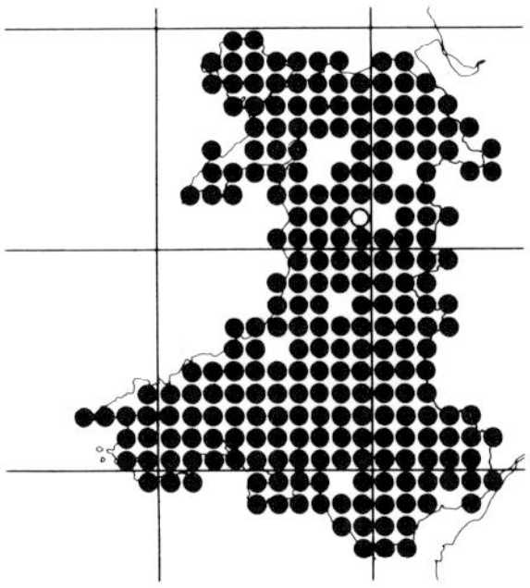

37 Ulmus glabra

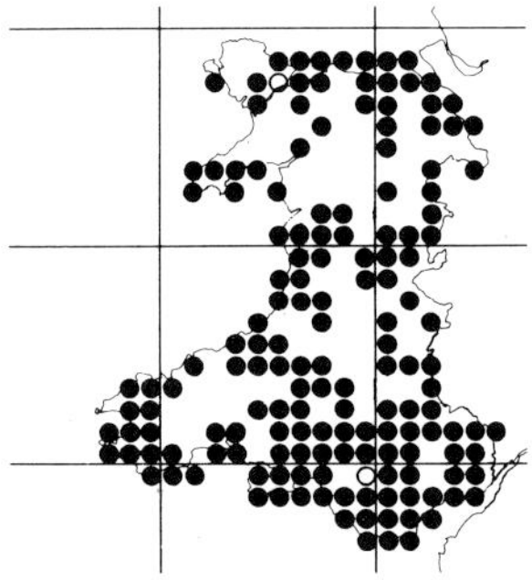

38 Ulmus procera

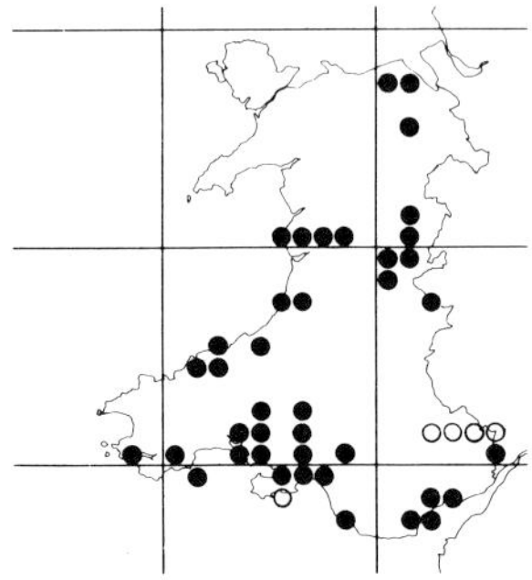

39 Ulmus carpinifolia × glabra × plotti = U. × hollandica

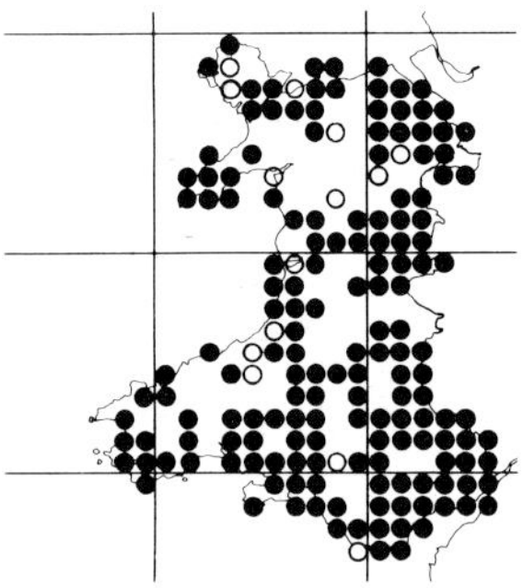

40 Humulus lupulus

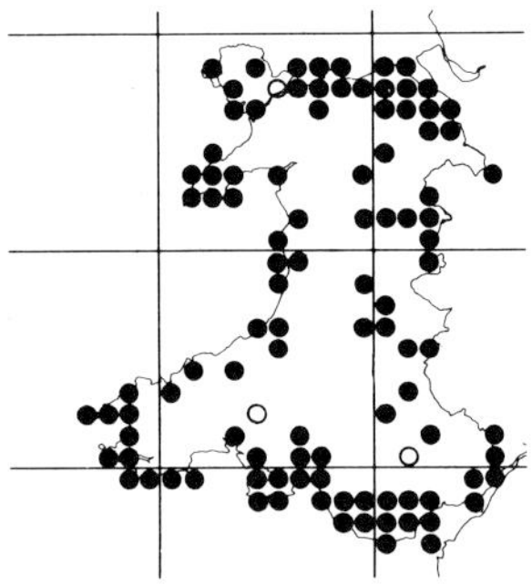

41 Urtica urens

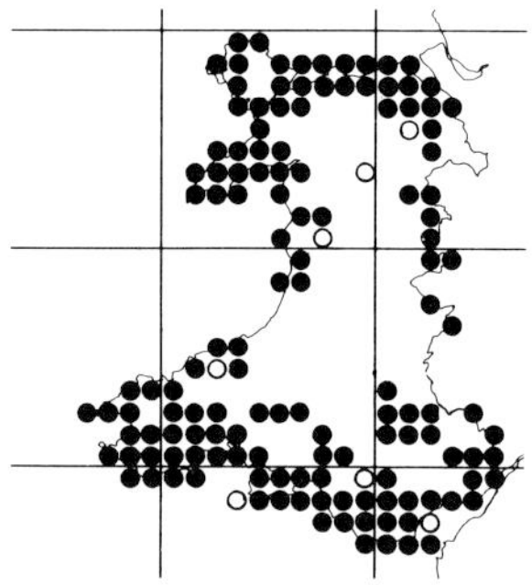

42 Parietaria judaica

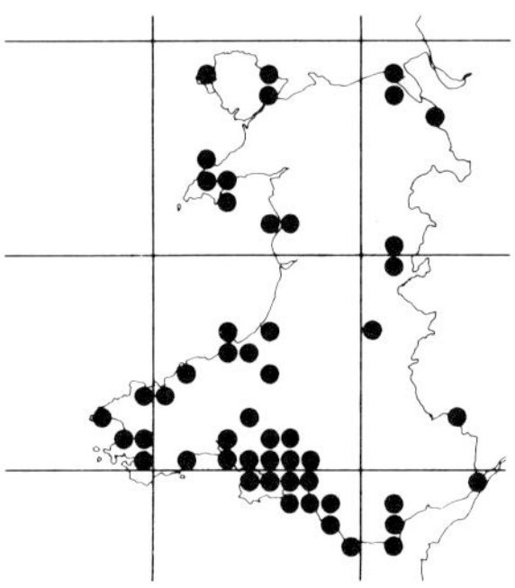

43 Soleirolia soleirolii

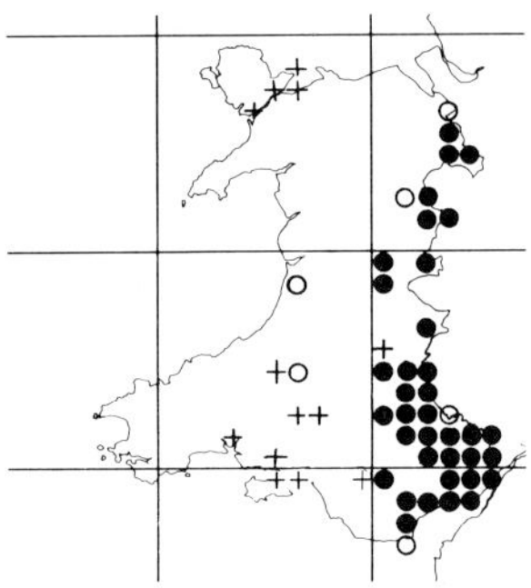

44 Viscum album

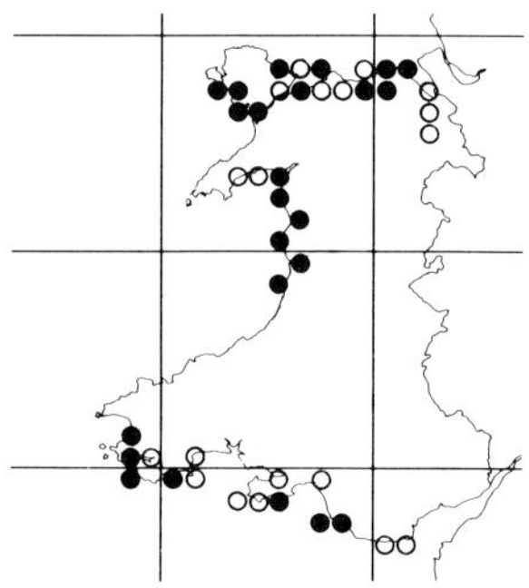

45 Polygonum oxyspermum

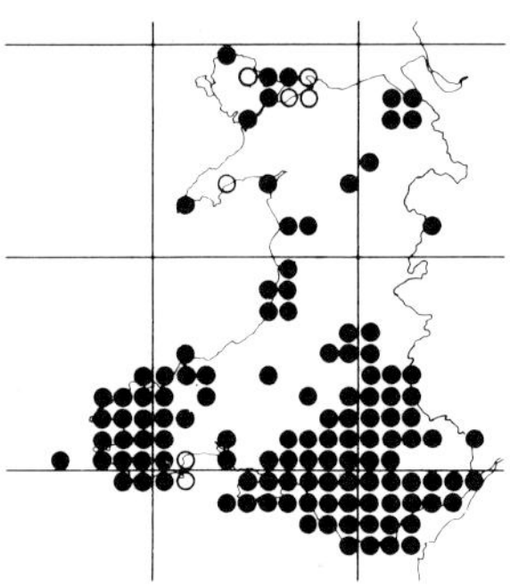

46 Polygonum aviculare s.s.

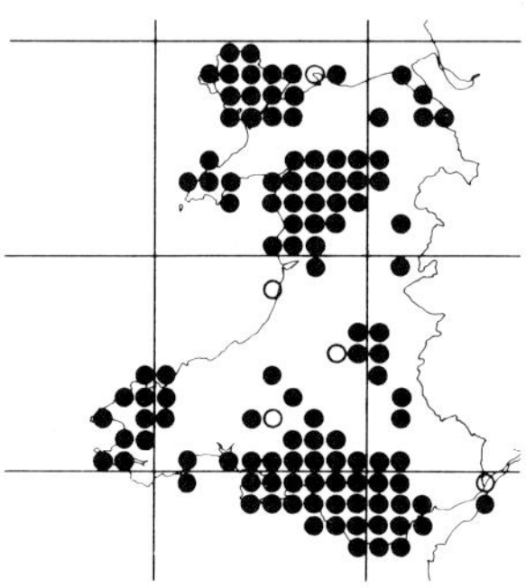

47 Polygonum arenastrum

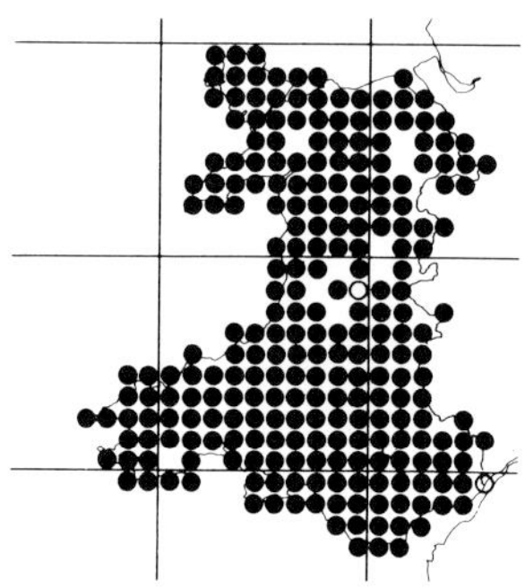

48 Polygonum hydropiper

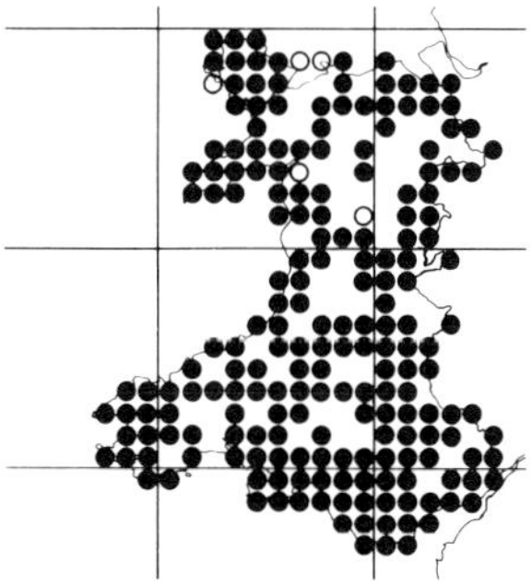

49 Polygonum lapathifolium

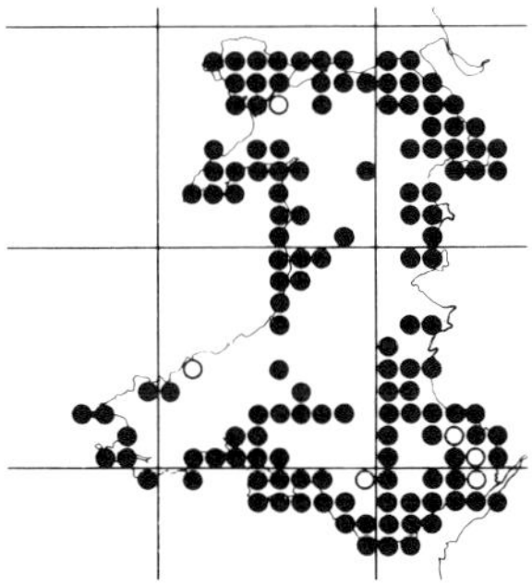

50 Polygonum amphibium

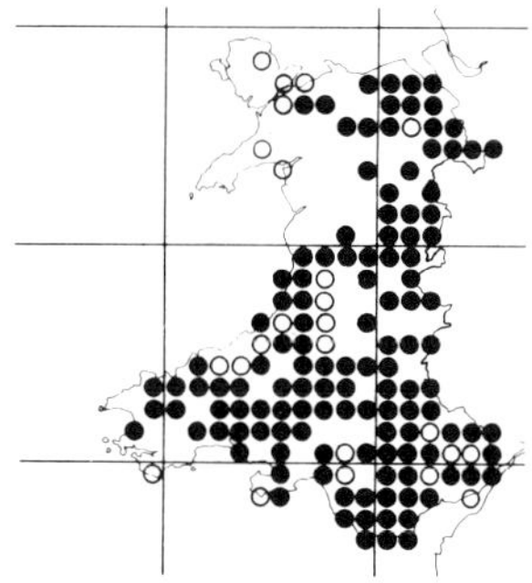

51 Polygonum bistorta

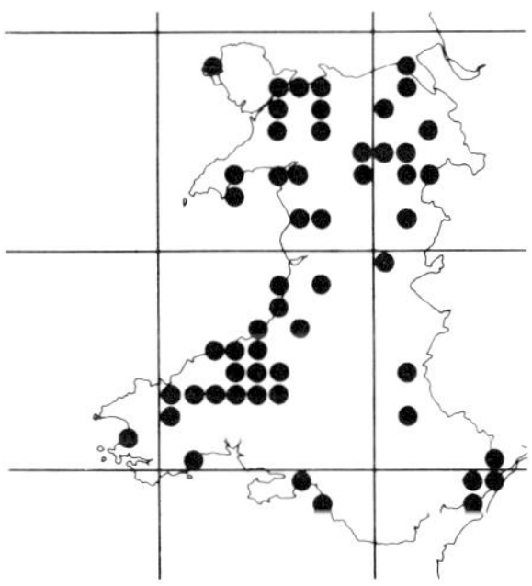

52 Polygonum polystachyum

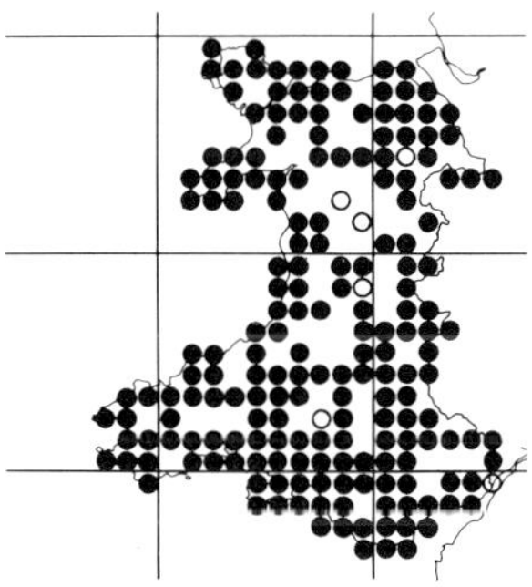

53 Fallopia convolvulus

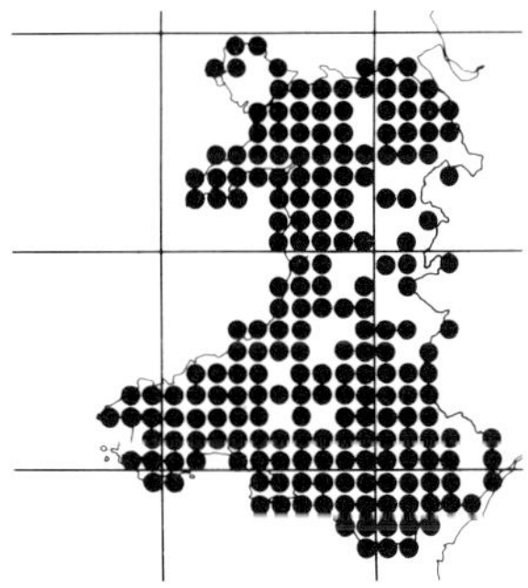

54 Reynoutria japonica

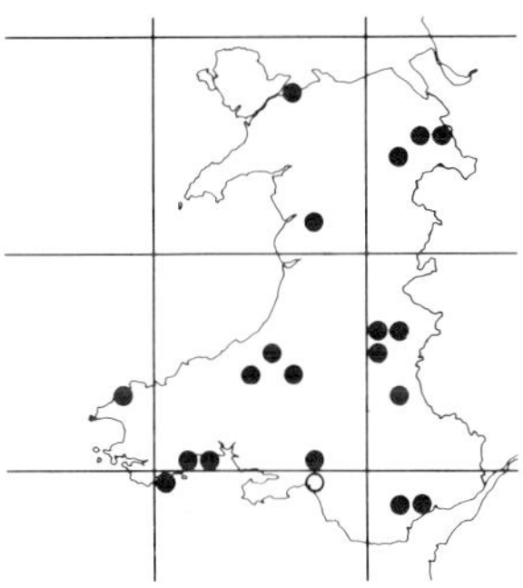

55 Reynoutria sachalinensis

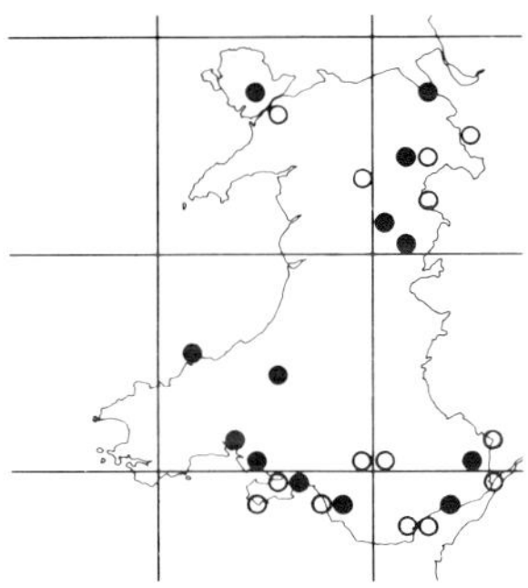

56 Fagopyrum esculentum

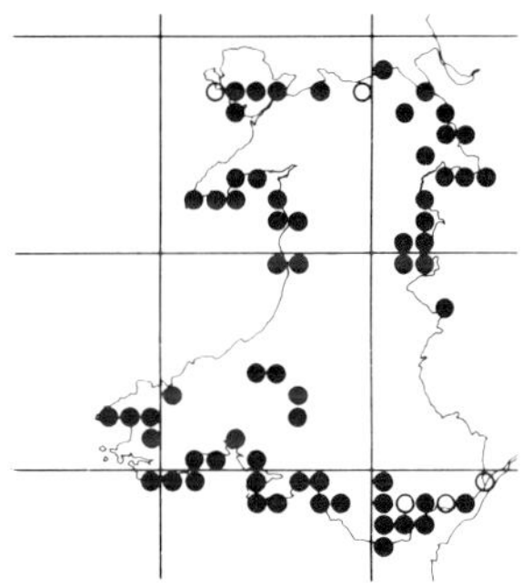

57 Rumex hydrolapathum

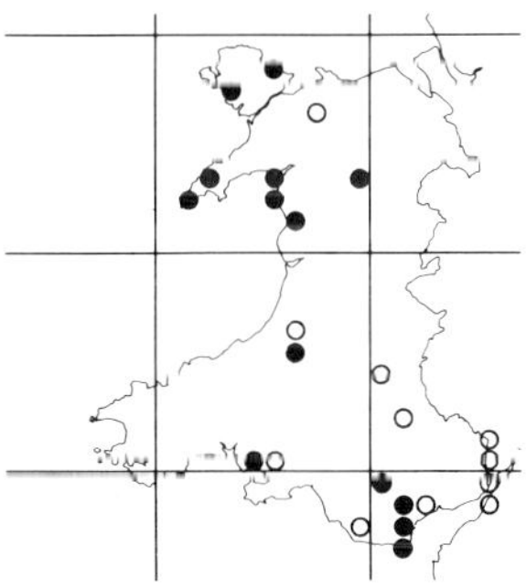

58 Rumex crispus × obtusifolius
= R. × pratensis

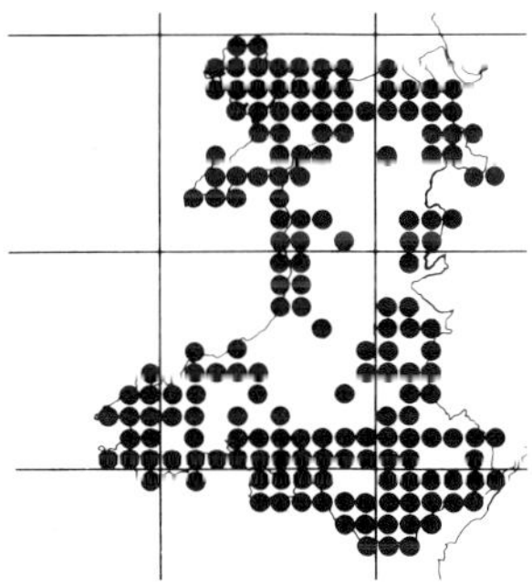

59 Rumex conglomeratus

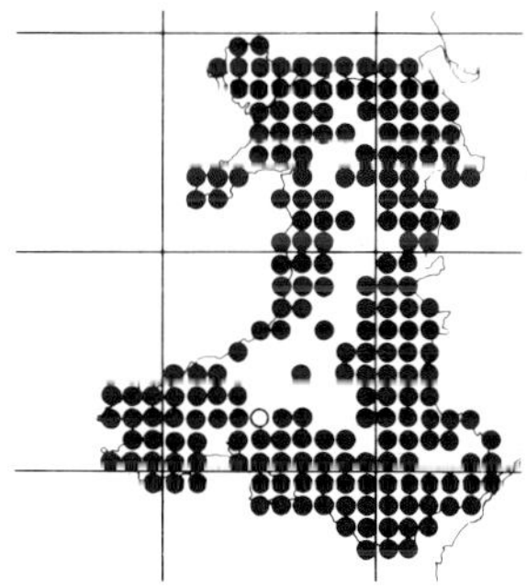

60 Rumex sanguineus

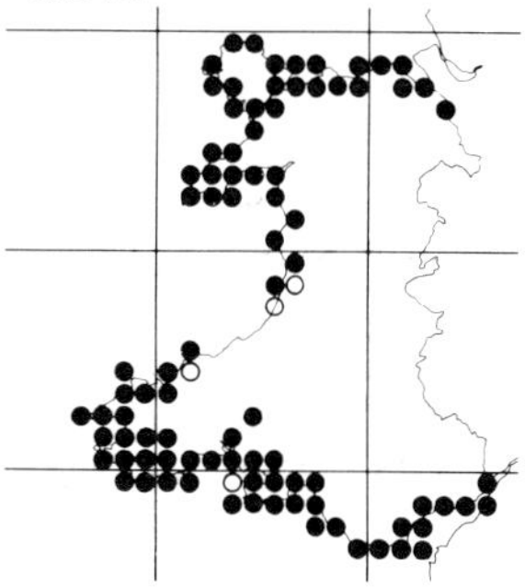
61 Beta vulgaris subsp. maritima

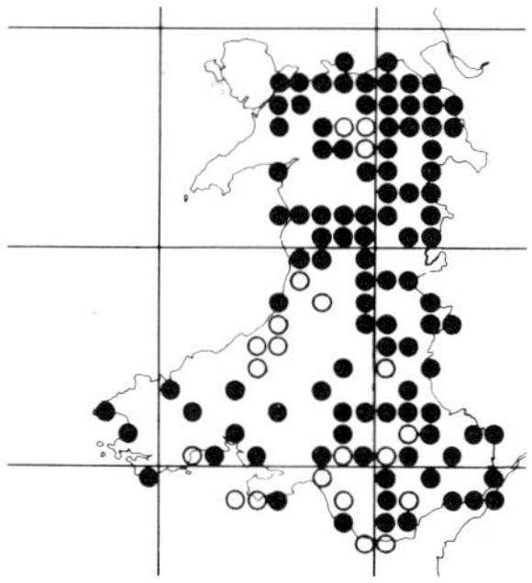
62 Chenopodium bonus-henricus

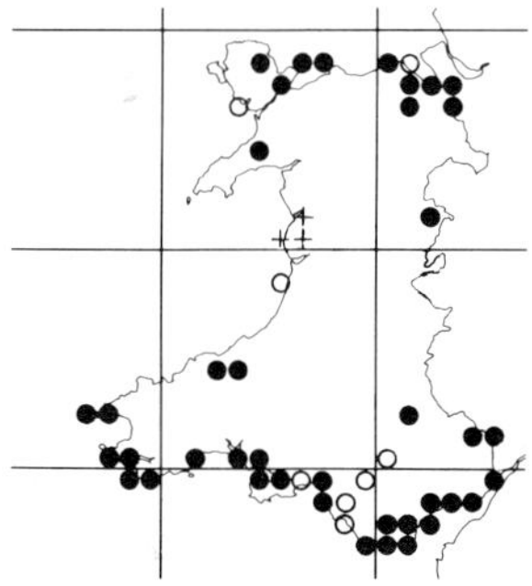
63 Chenopodium rubrum

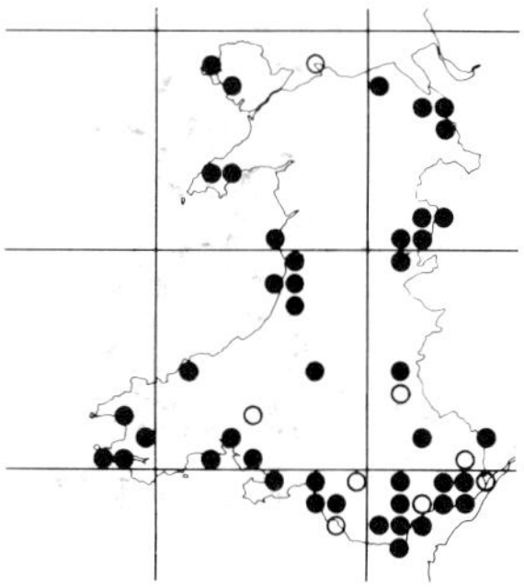
64 Chenopodium polyspermum

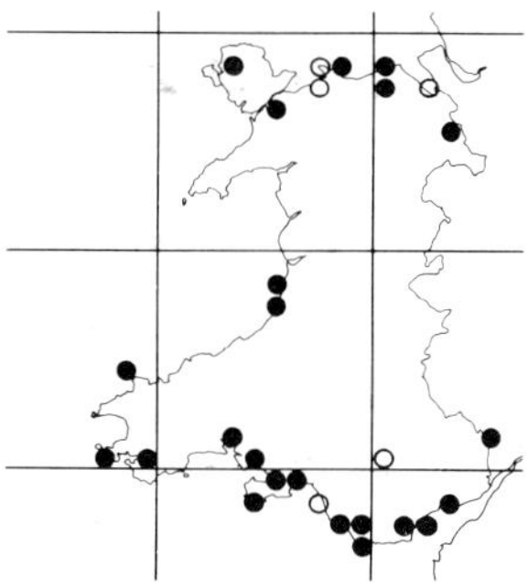
65 Chenopodium murale

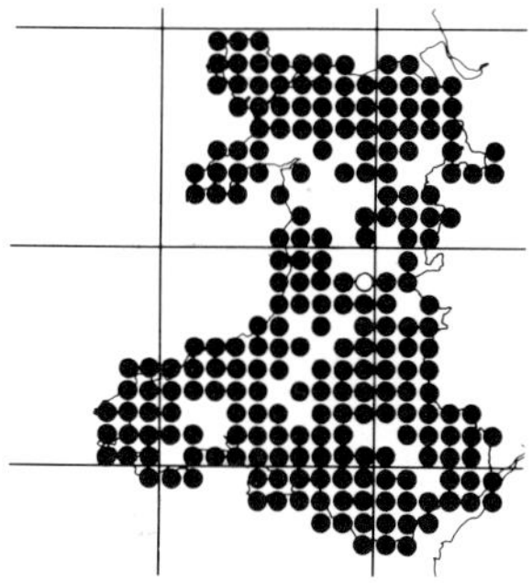
66 Chenopodium album subsp. album

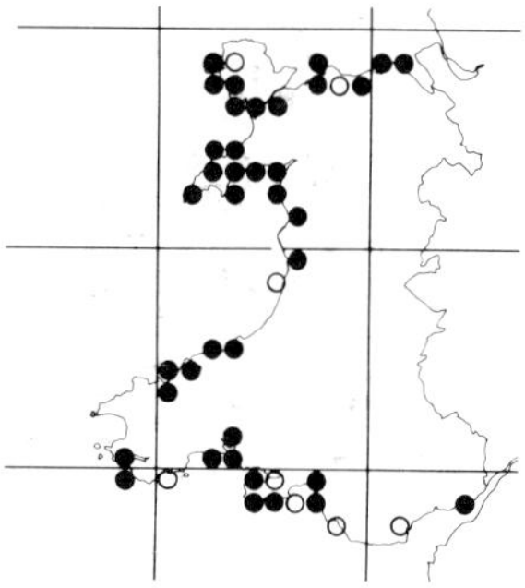
67 Atriplex laciniata

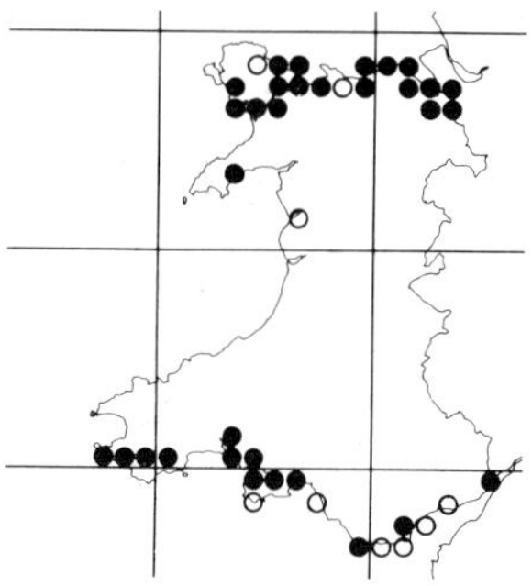
68 Atriplex littoralis

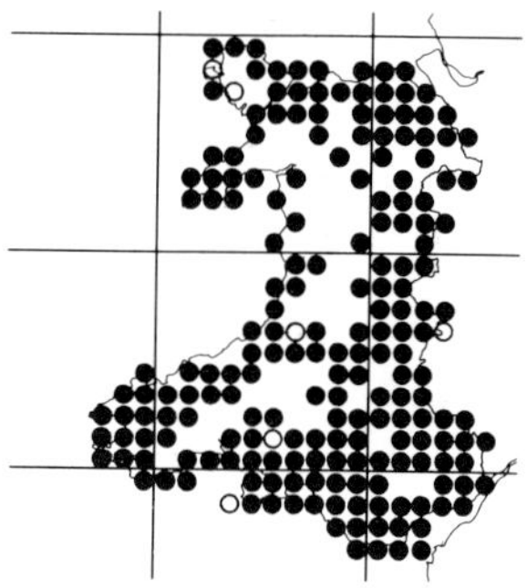
69 Atriplex patula

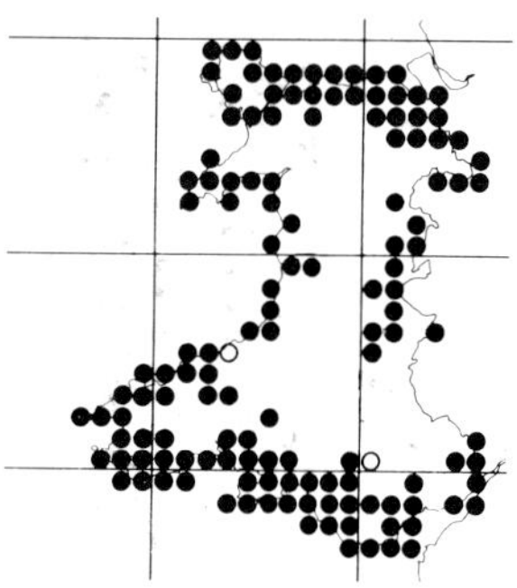
70 Atriplex prostrata

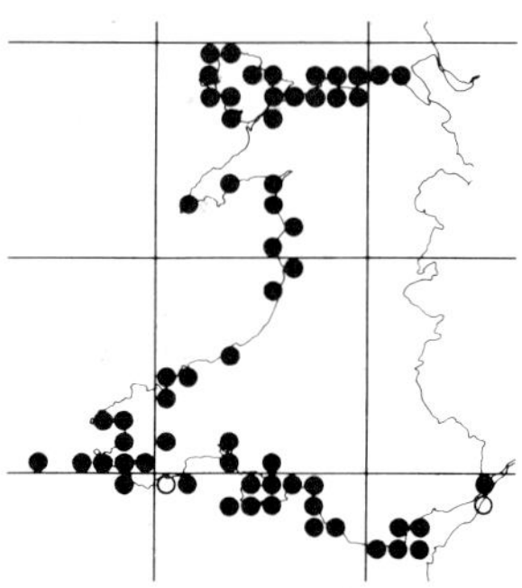
71 Atriplex glabriuscula

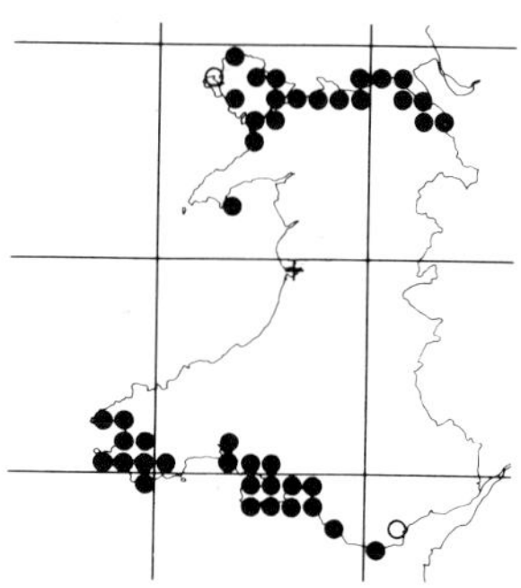
72 Halimione portulacoides

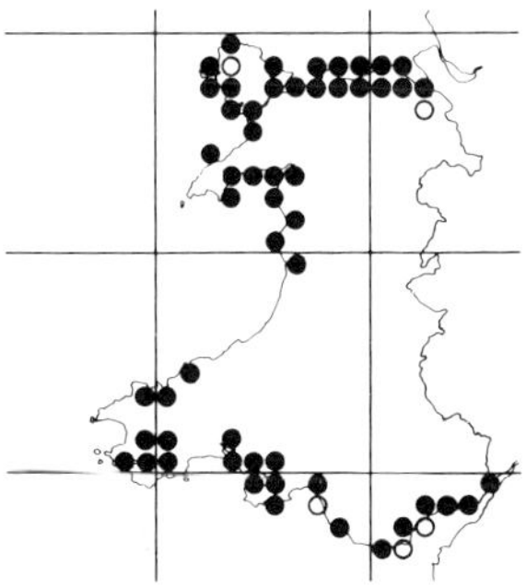
73 Salicornia agg.

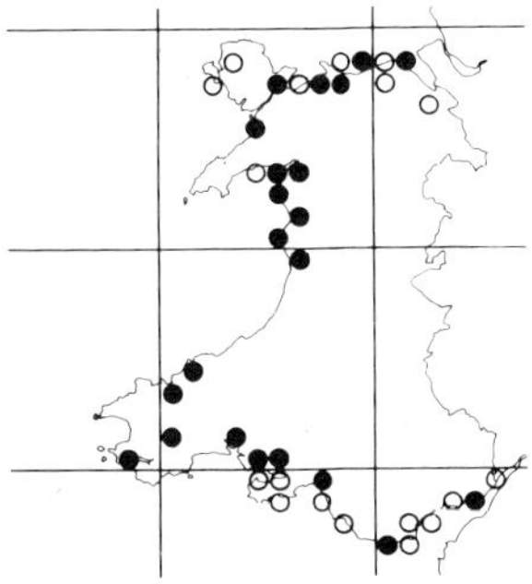
74 Salicornia europaea

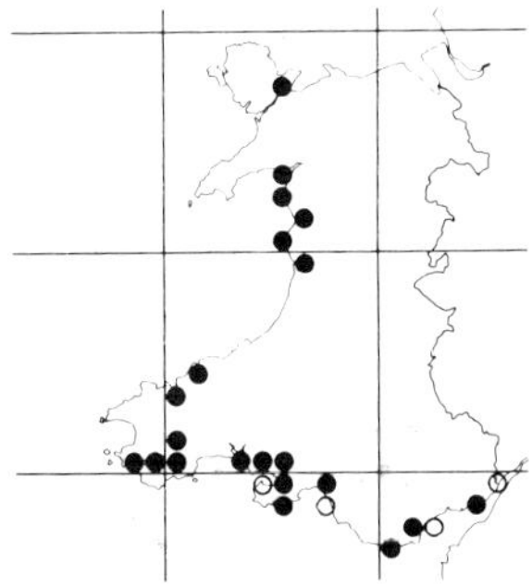
75 Salicornia ramosissima

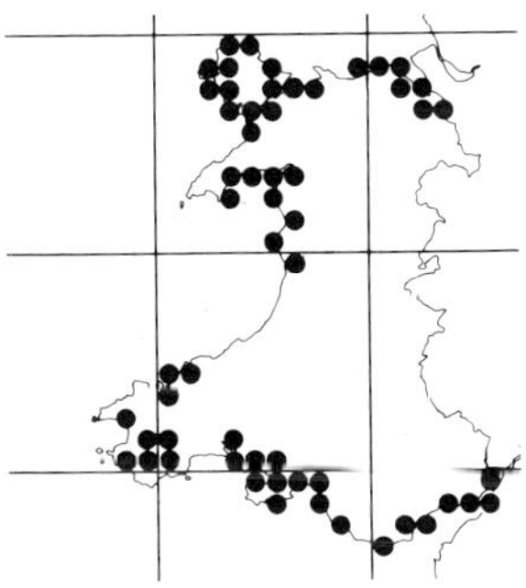
76 Suaeda maritima

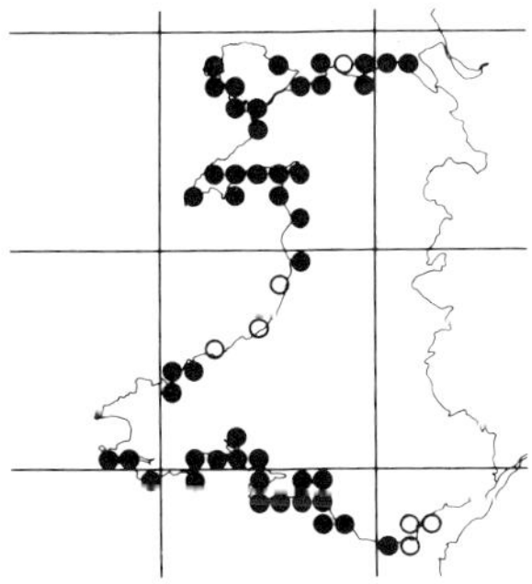
77 Salsola kali
subsp. kali

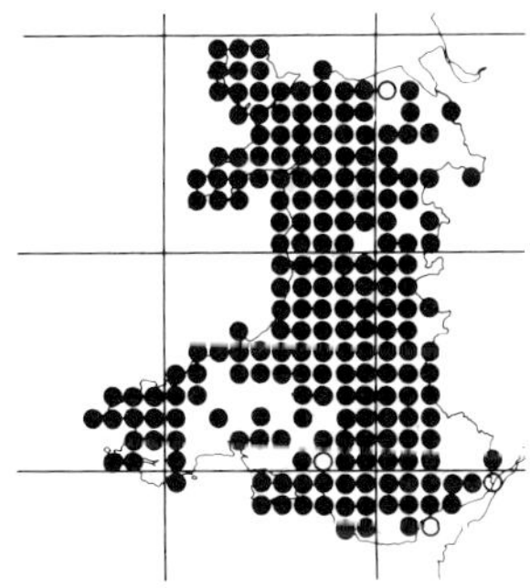
78 Montia fontana

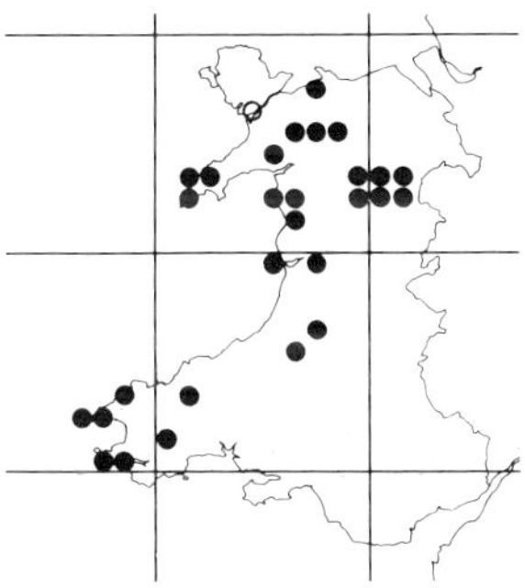
79 Montia fontana
subsp. fontana

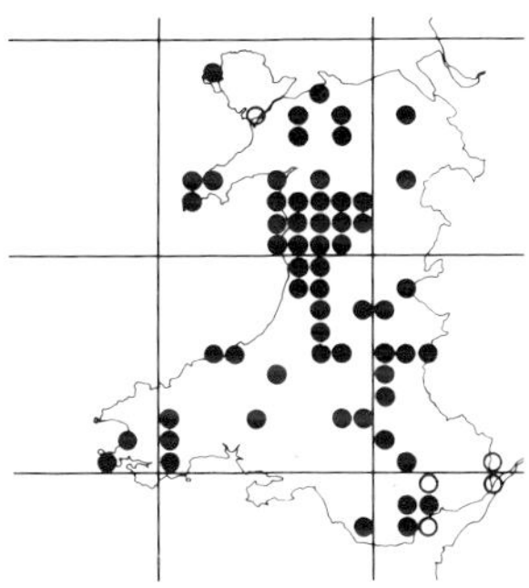
80 Montia fontana
subsp. variabilis

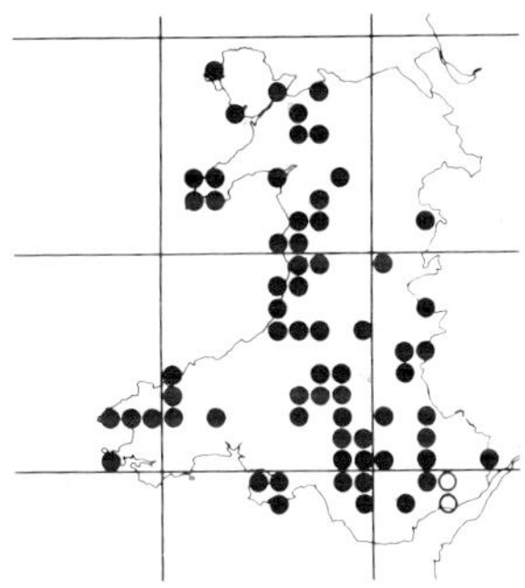
81 Montia fontana
subsp. amporitana

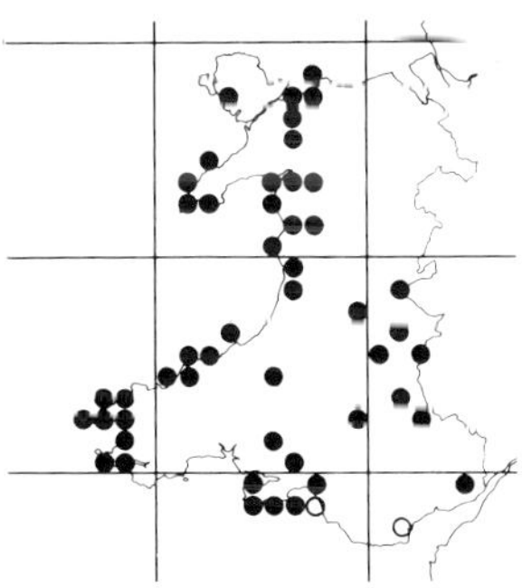
82 Montia fontana
subsp. chondrosperma

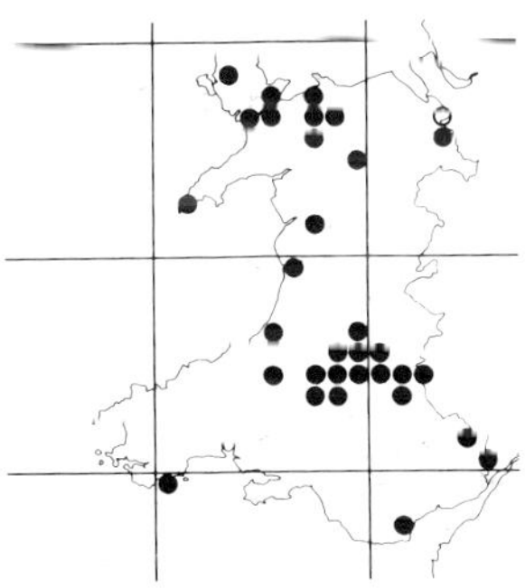
83 Montia sibirica

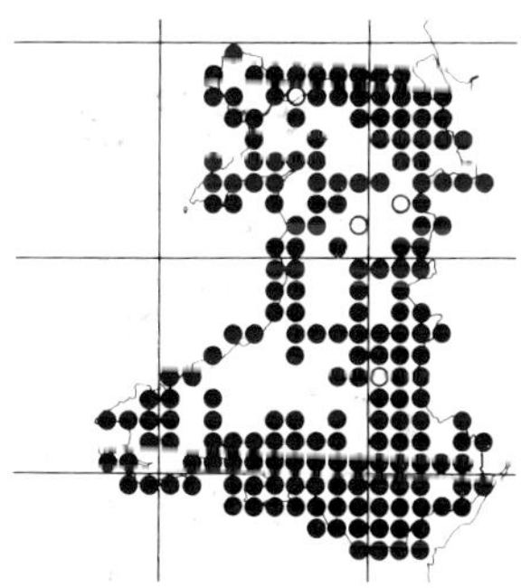
84 Arenaria serpyllifolia

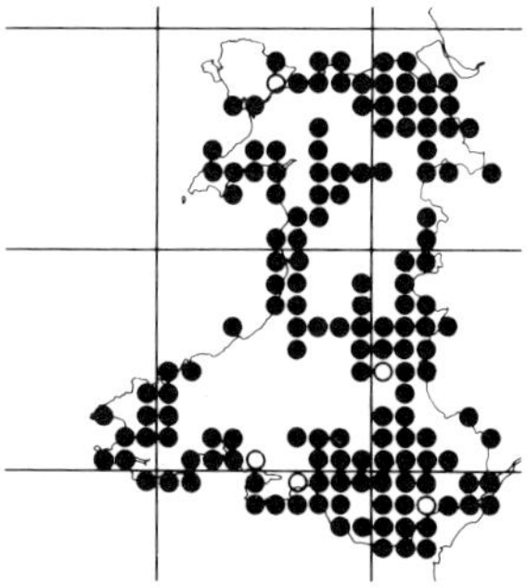

85 Arenaria serpyllifolia subsp. serpyllifolia

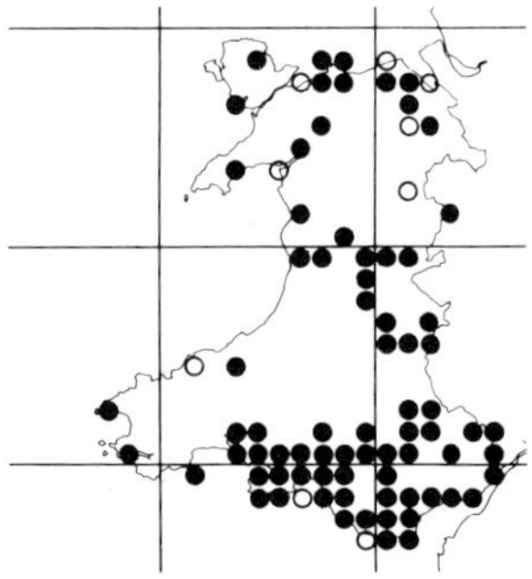

86 Arenaria serpyllifolia subsp. leptoclados

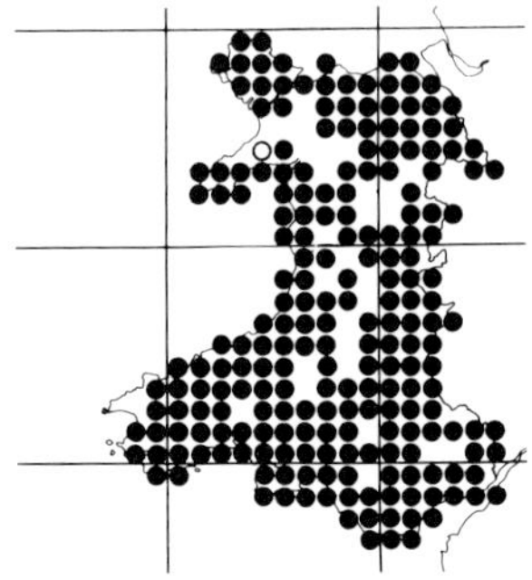

87 Moehringia trinervia

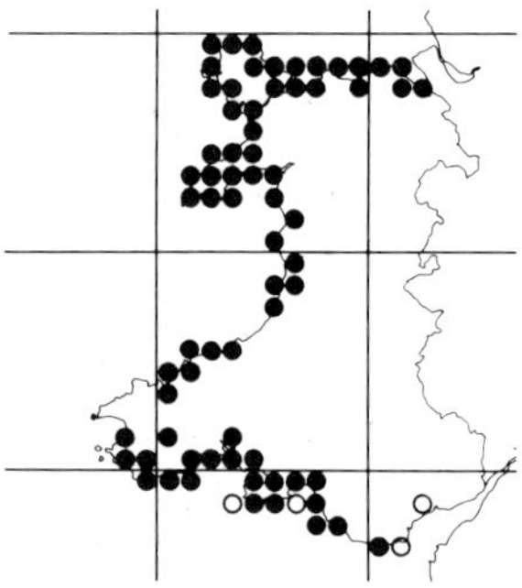

88 Honkenya peploides

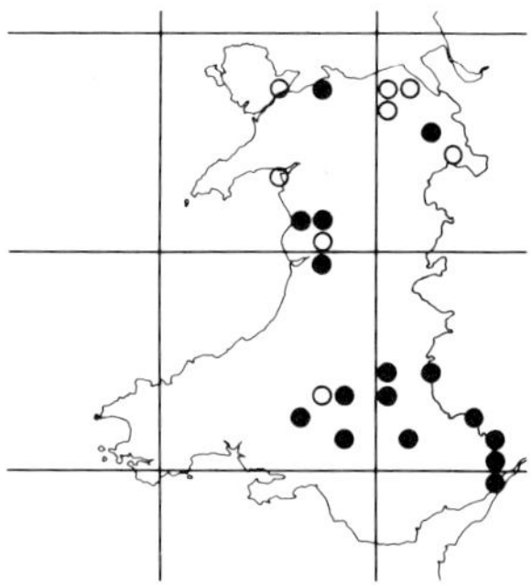

89 Stellaria nemorum

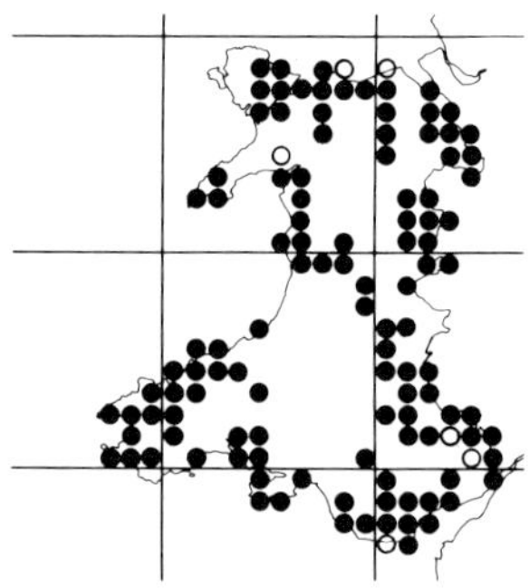

90 Stellaria neglecta

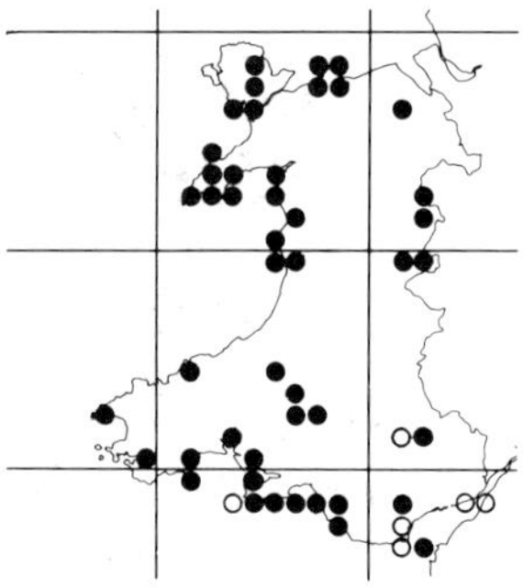

91 Stellaria pallida

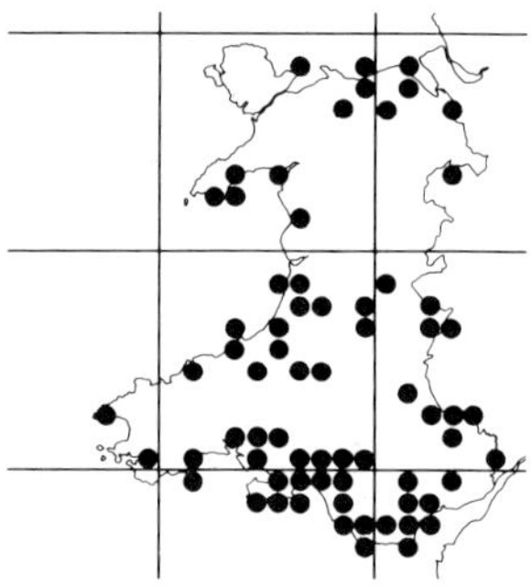

92 Cerastium tomentosum

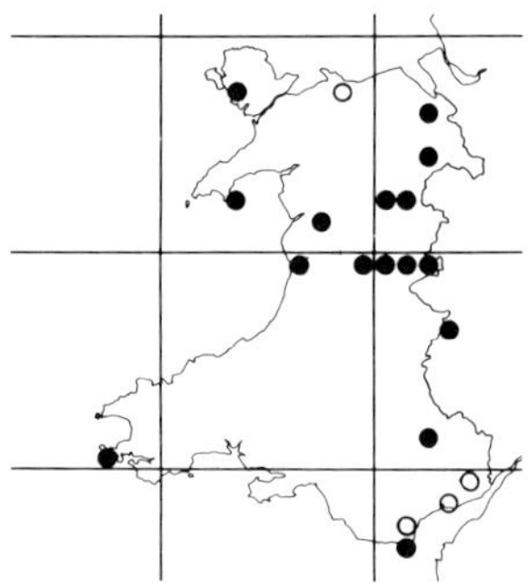

93 Cerastium arvense

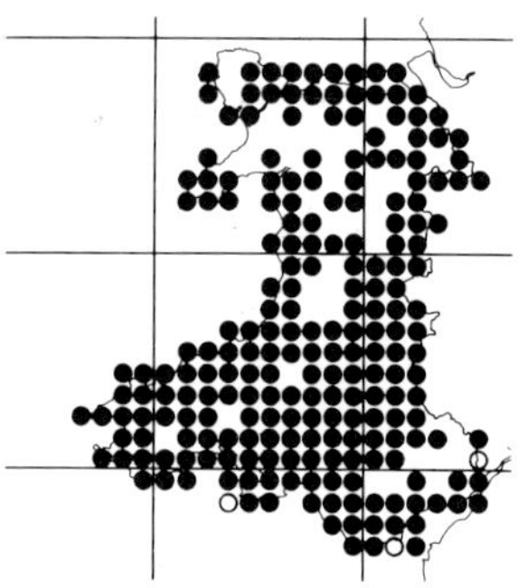

94 Cerastium glomeratum

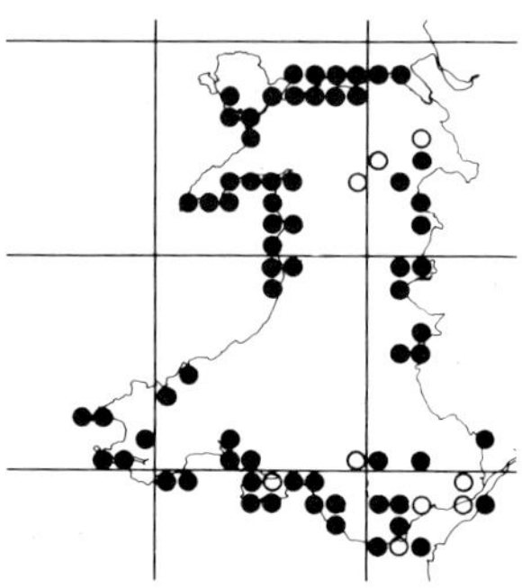

95 Cerastium semidecandrum

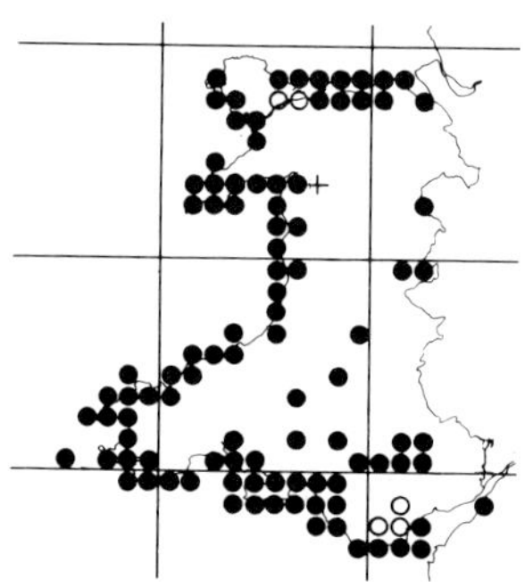

96 Cerastium diffusum

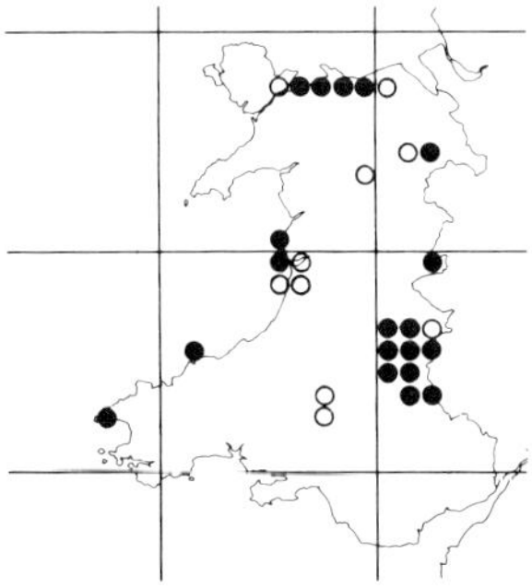

97 Moenchia erecta

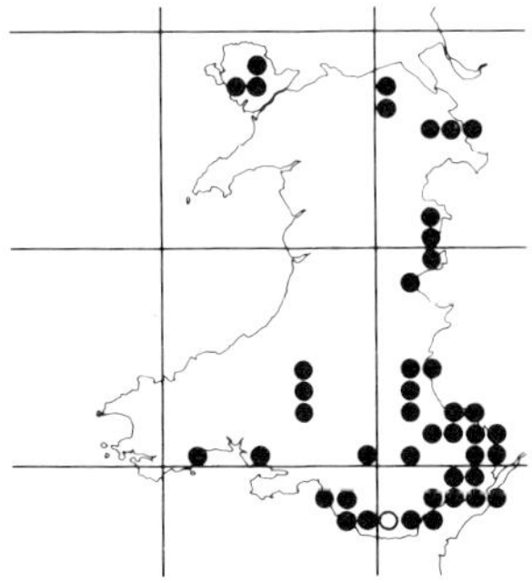

98 Myosoton aquaticum

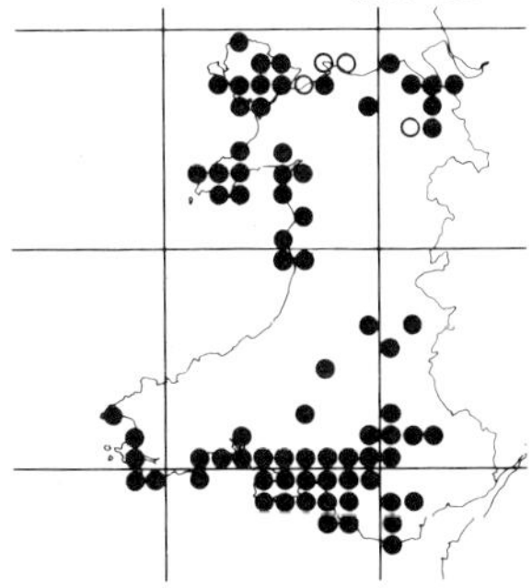

99 Sagina nodosa

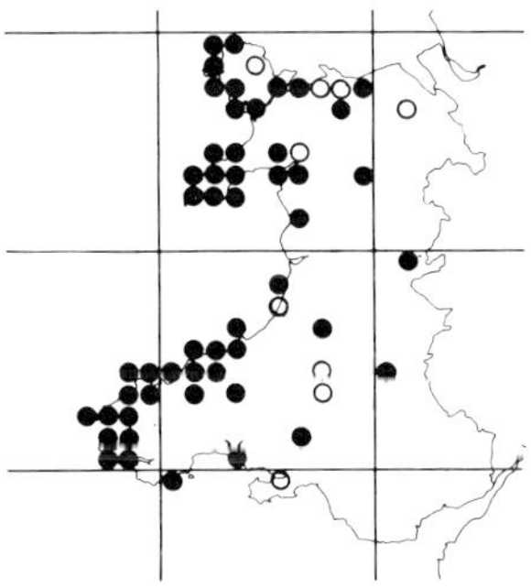

100 Sagina subulata

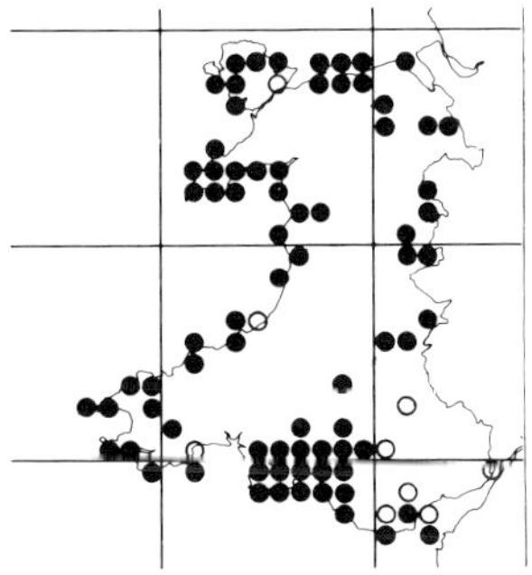

101 Sagina apetala subsp. apetala

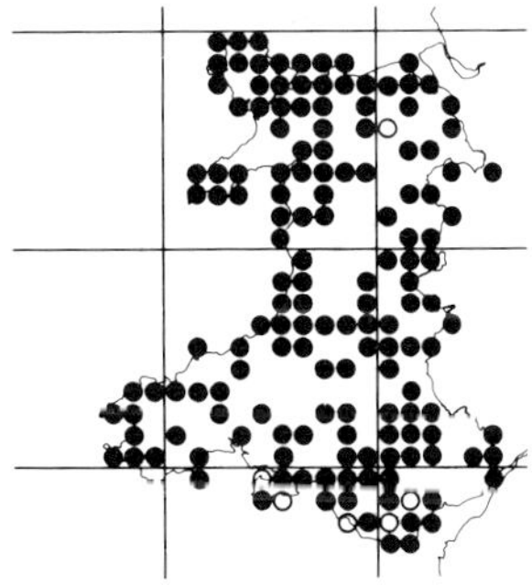

102 Sagina apetala subsp. erecta

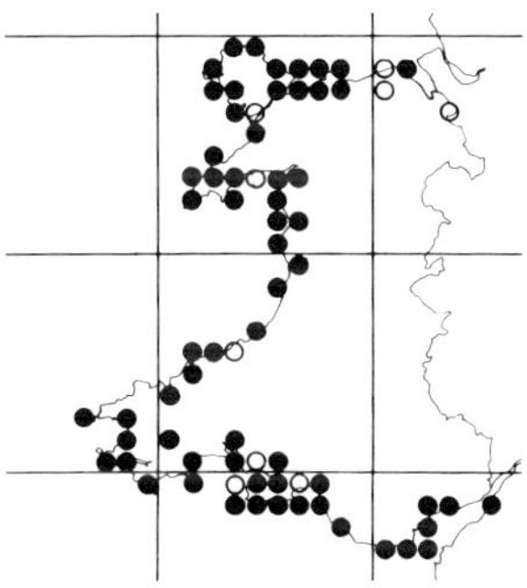

103 Sagina maritima

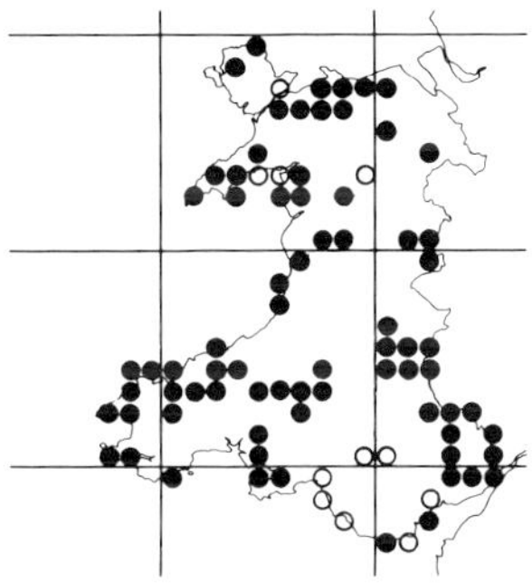

104 Scleranthus annuus

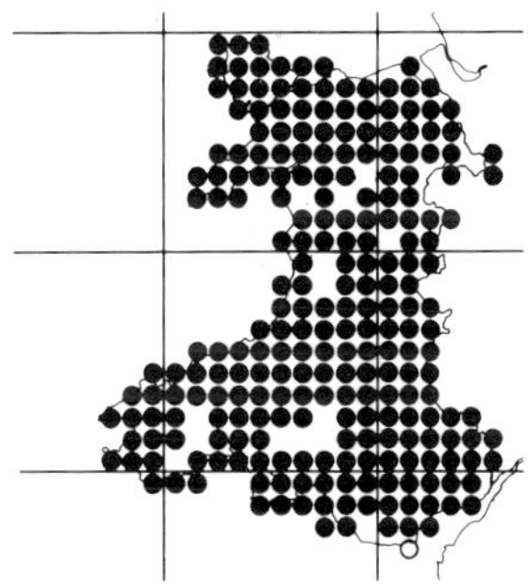

105 Spergula arvensis

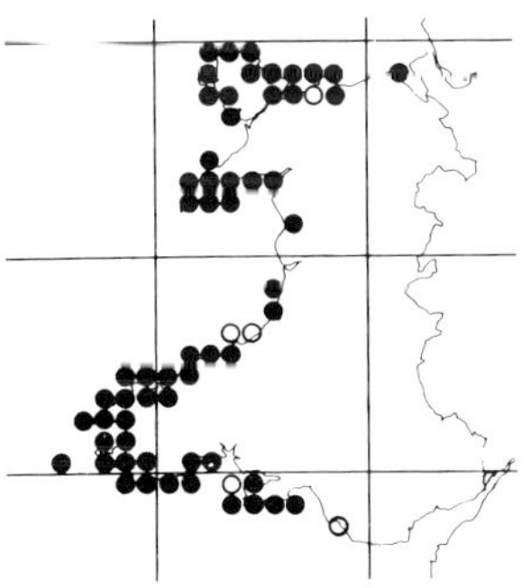

106 Spergularia rupicola

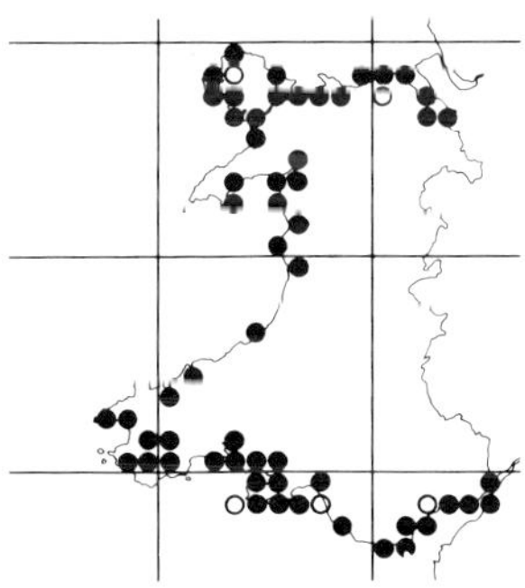

107 Spergularia media

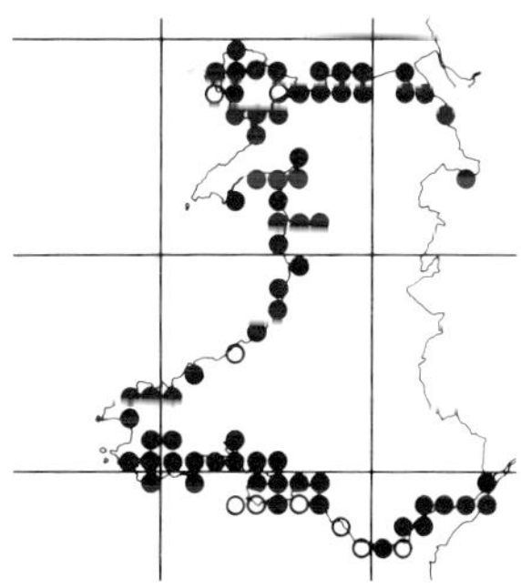

108 Spergularia marina

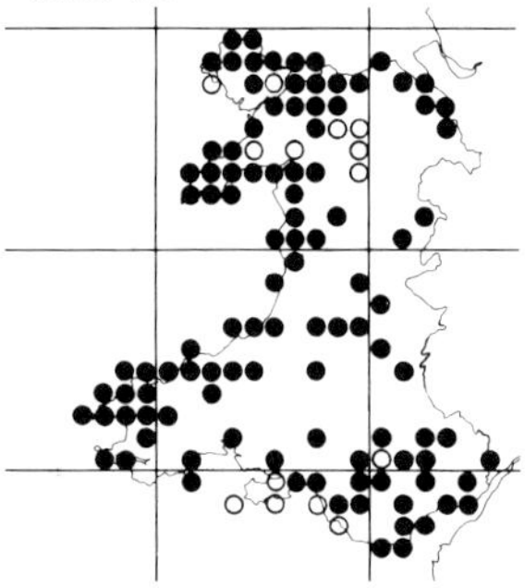

109 Spergularia rubra

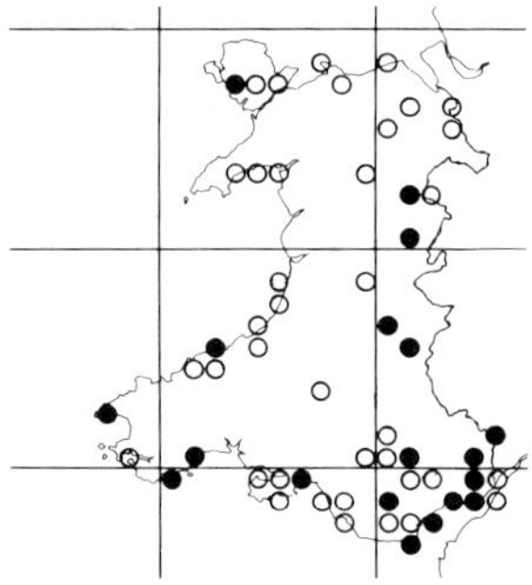

110 Agrostemma githago

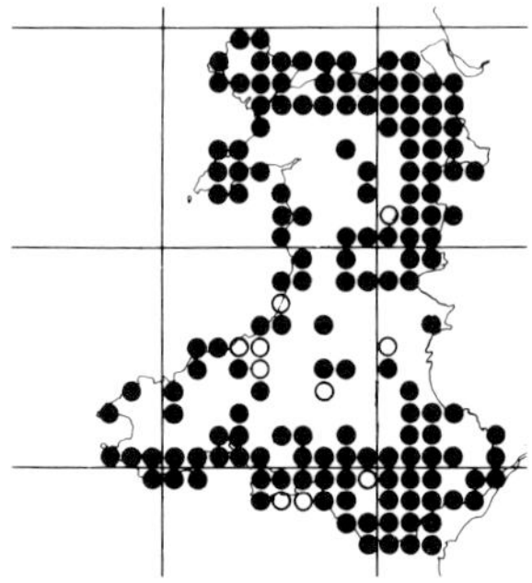

111 Silene vulgaris
subsp. vulgaris

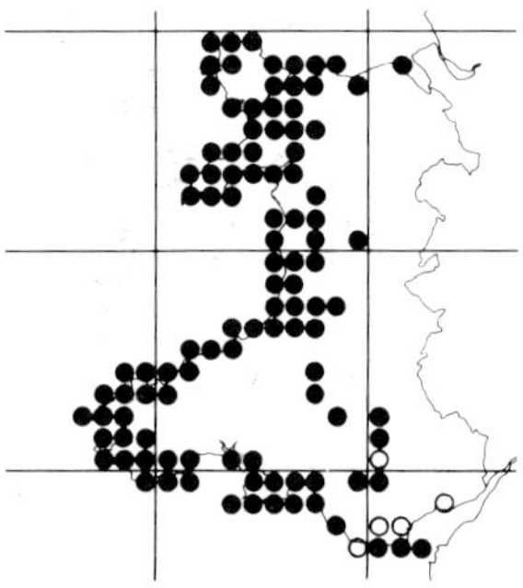

112 Silene vulgaris
subsp. maritima

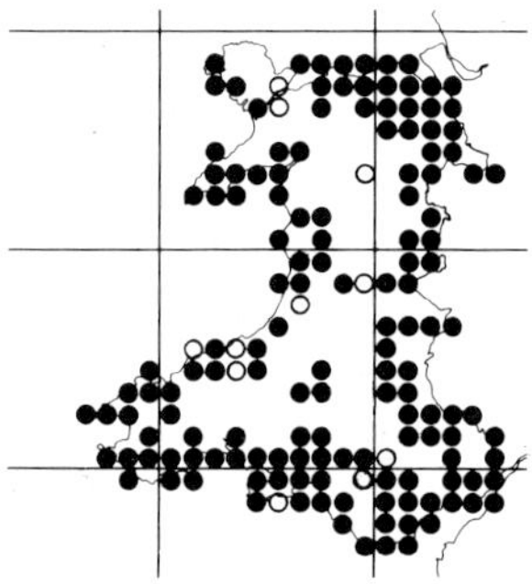

113 Silene pratensis

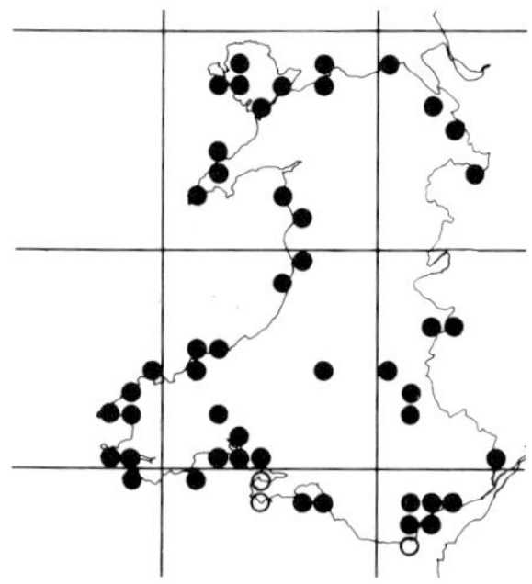

114 Silene dioica × pratensis

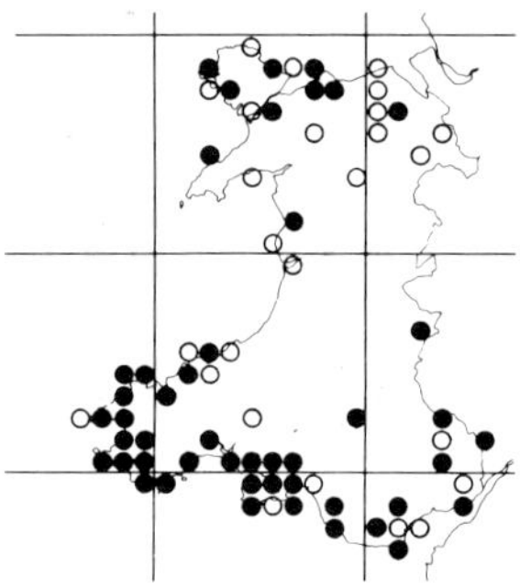

115 Silene gallica

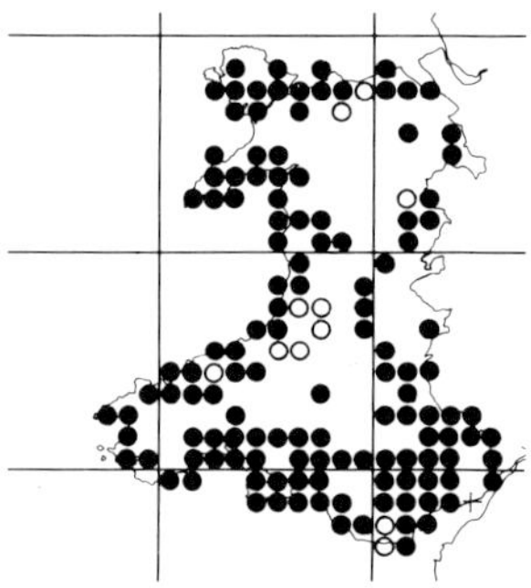

116 Saponaria officinalis

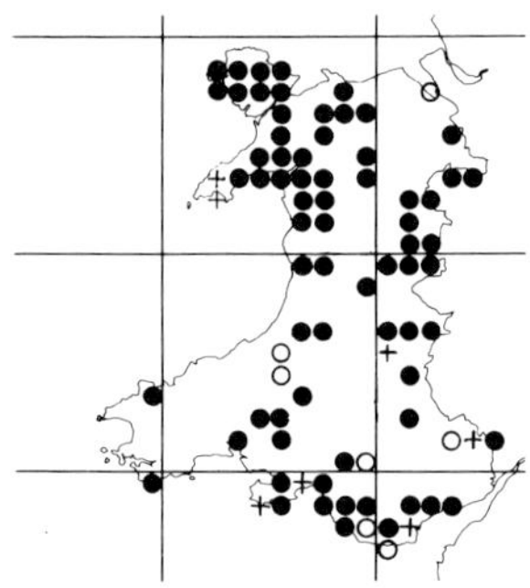

117 Nymphaea alba

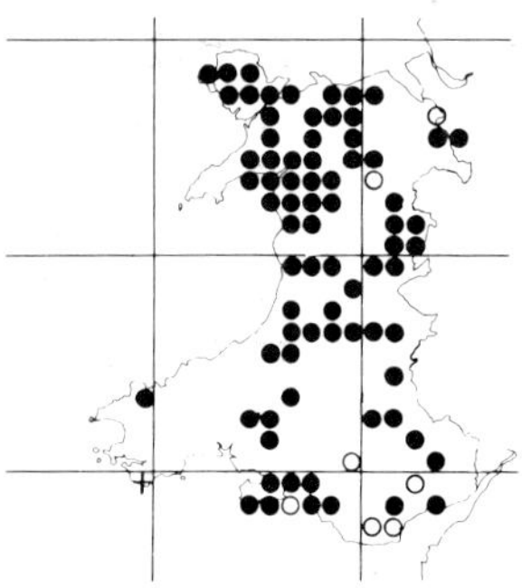

118 Nuphar lutea

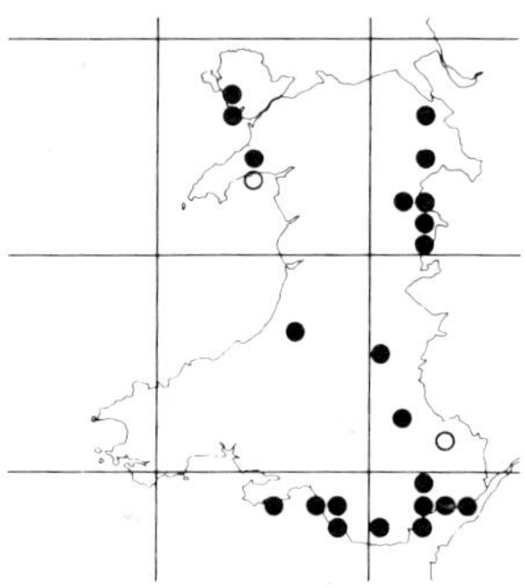

119 Ceratophyllum demersum

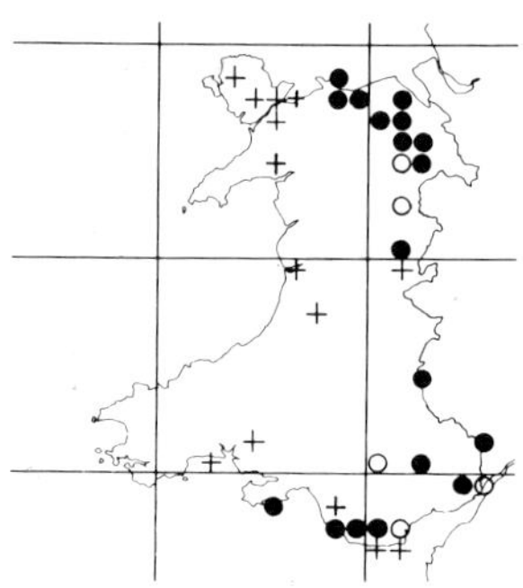

120 Helleborus foetidus

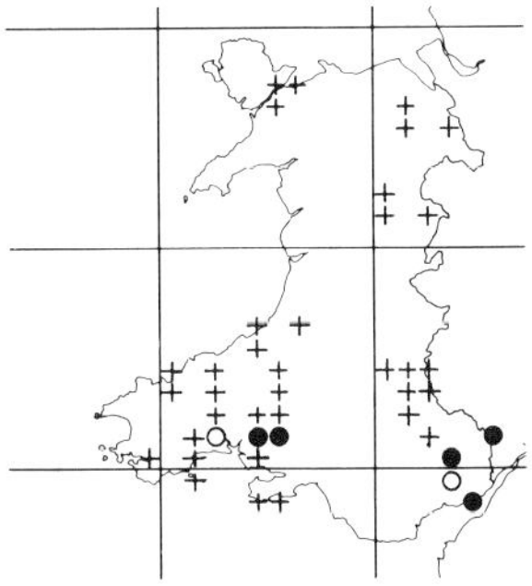

121 Helleborus viridis

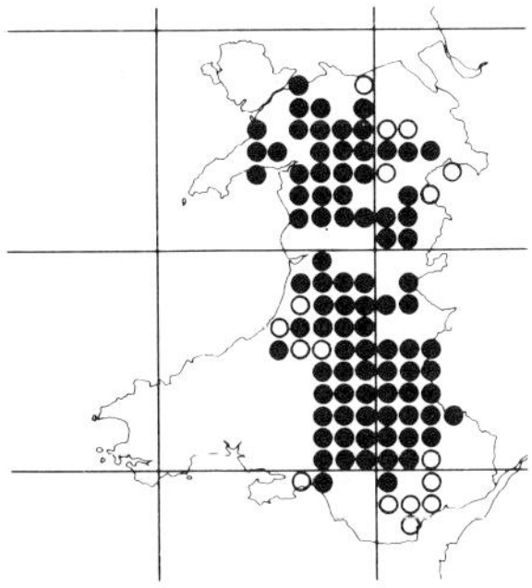

122 Trollius europaeus

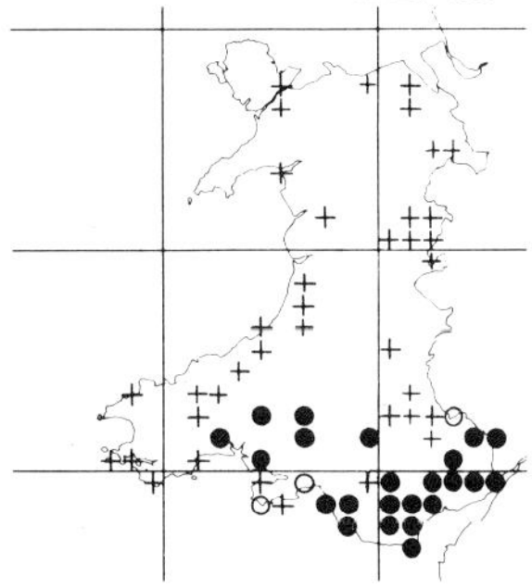

123 Aconitum napellus

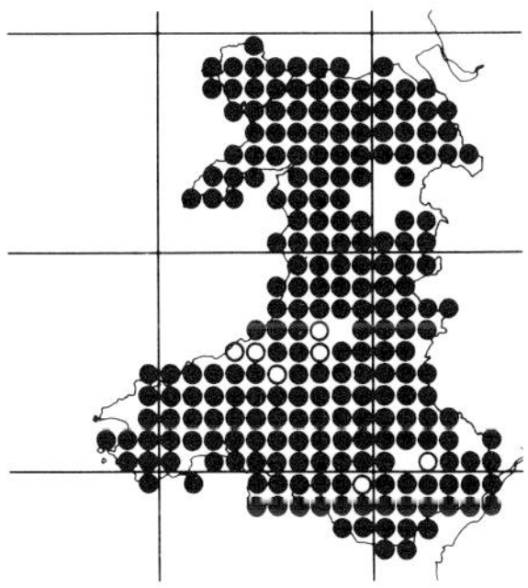

124 Anemone nemorosa

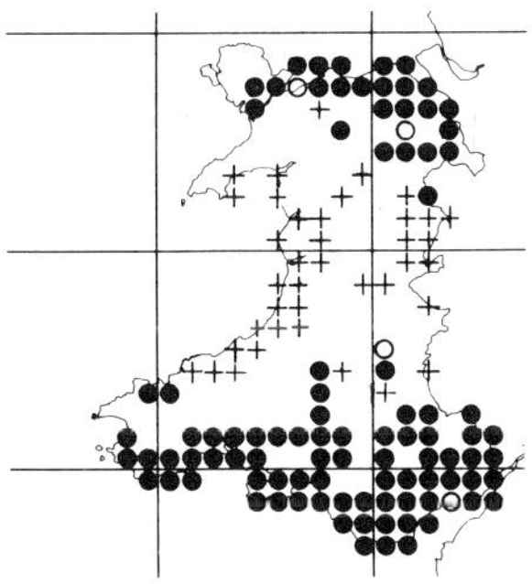

125 Clematis vitalba

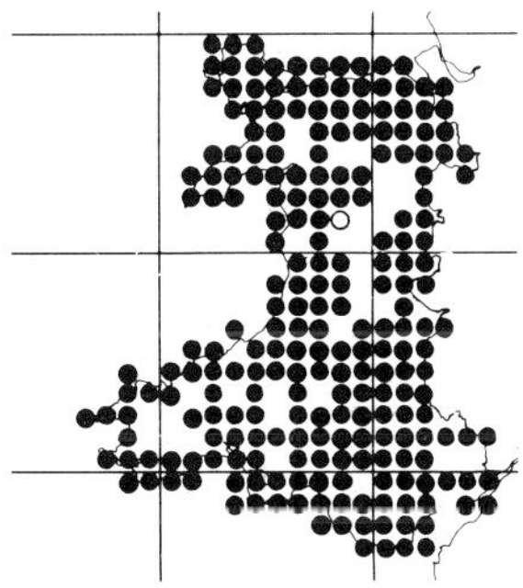

126 Ranunculus bulbosus

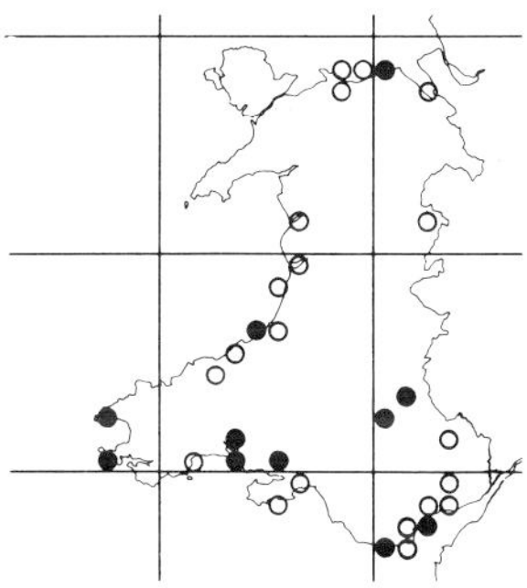

127 Ranunculus sardous

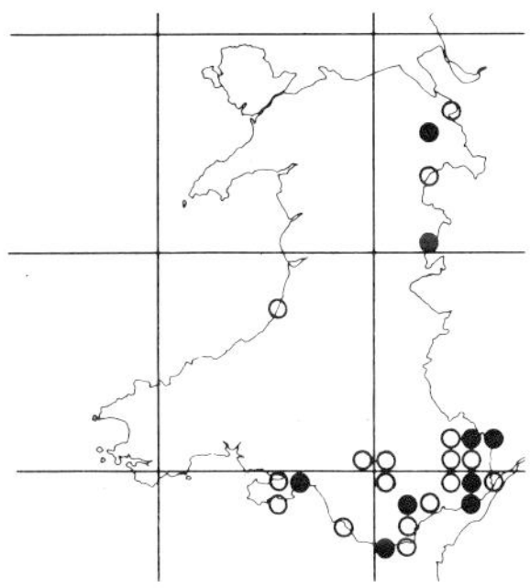

128 Ranunculus arvensis

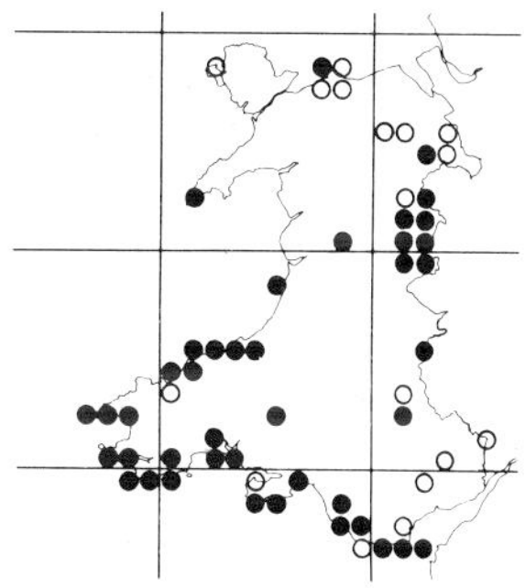

129 Ranunculus parviflorus

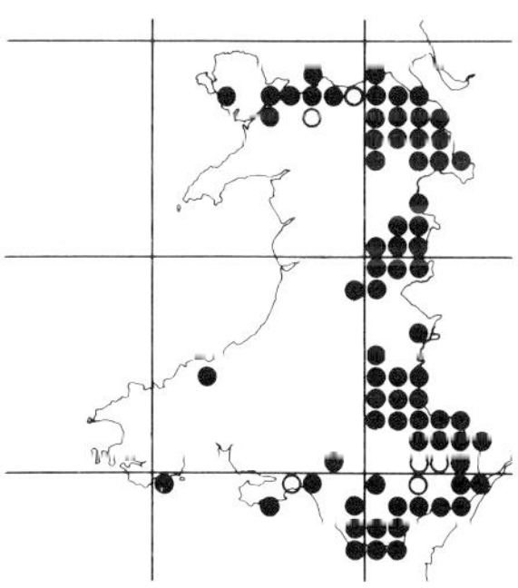

130 Ranunculus auricomus

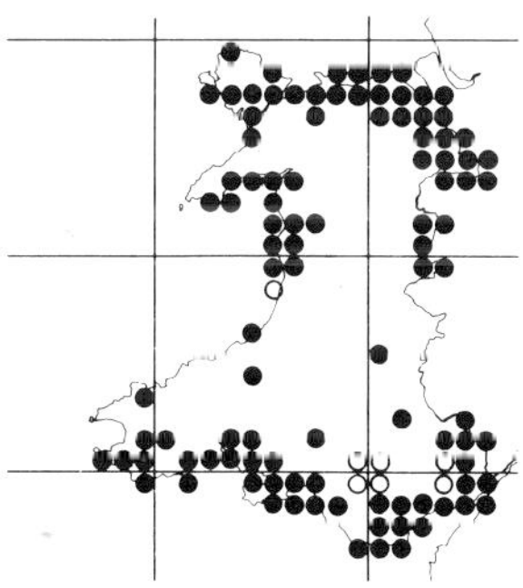

131 Ranunculus sceleratus

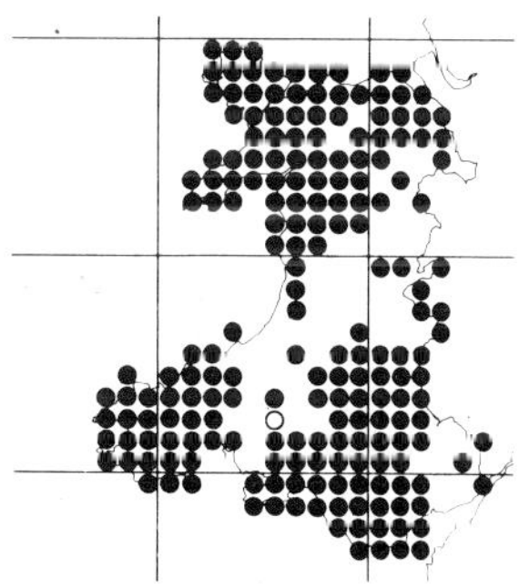

132 Ranunculus ficaria
subsp. ficaria

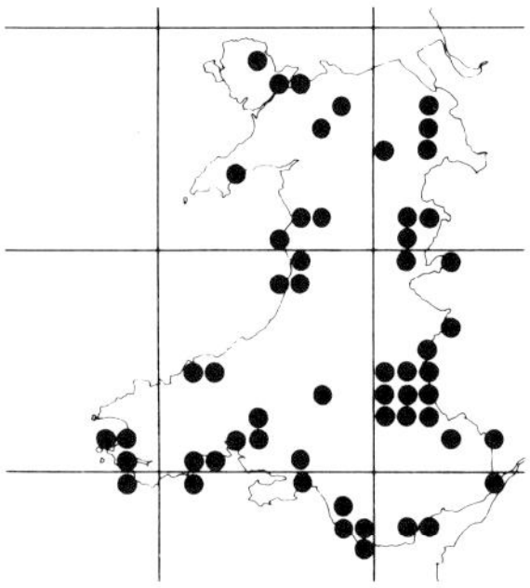

133 Ranunculus ficaria subsp. bulbifer

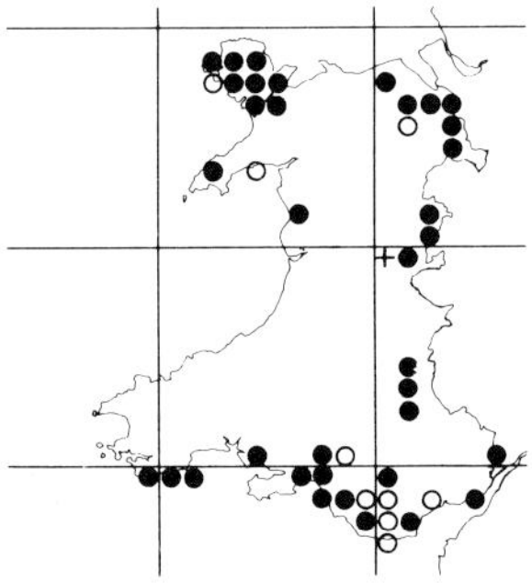

134 Ranunculus lingua

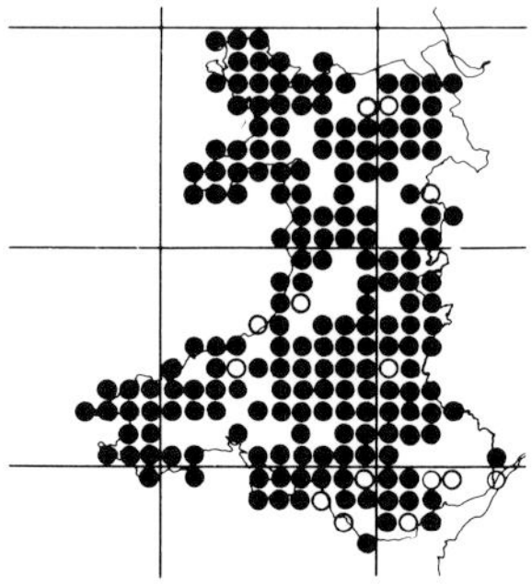

135 Ranunculus hederaceus

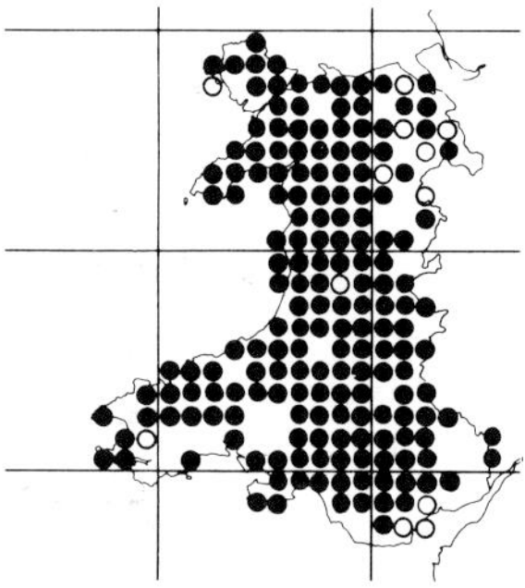

136 Ranunculus omiophyllus

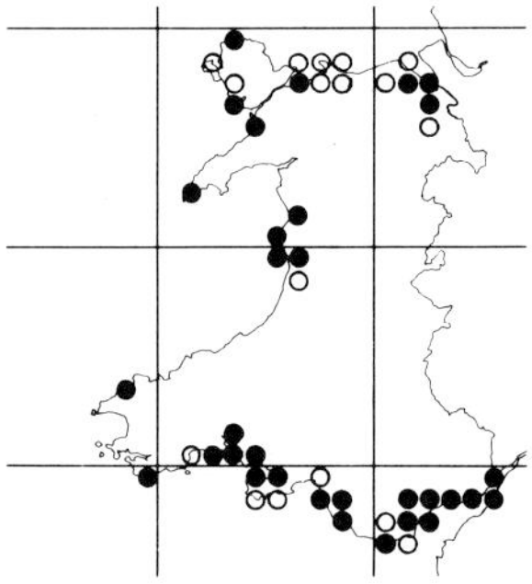

137 Ranunculus baudotii

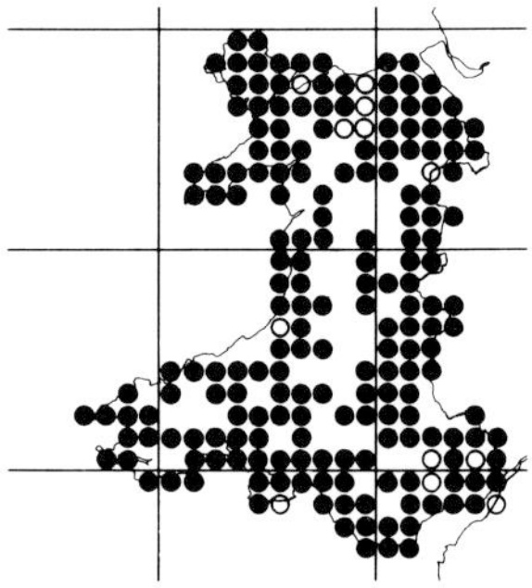

138 Ranunculus aquatilis agg.

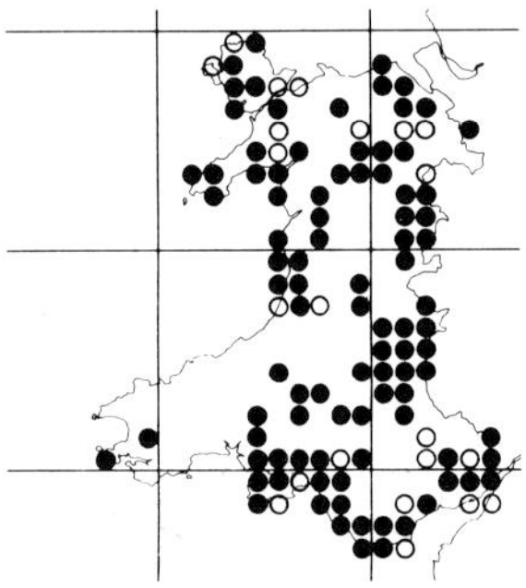

139 Ranunculus peltatus

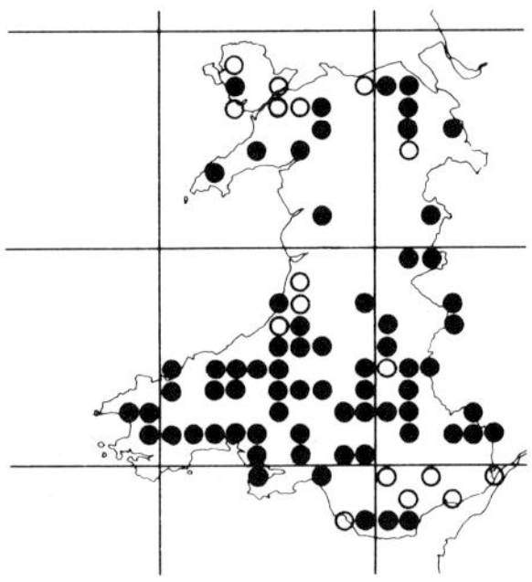

140 Ranunculus penicillatus

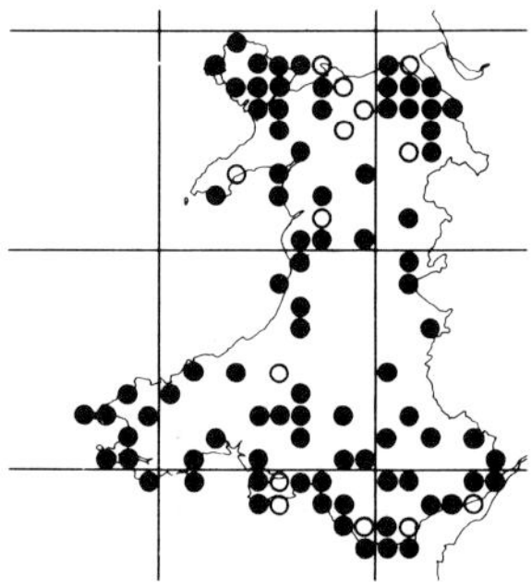

141 Ranunculus aquatilis s.s.

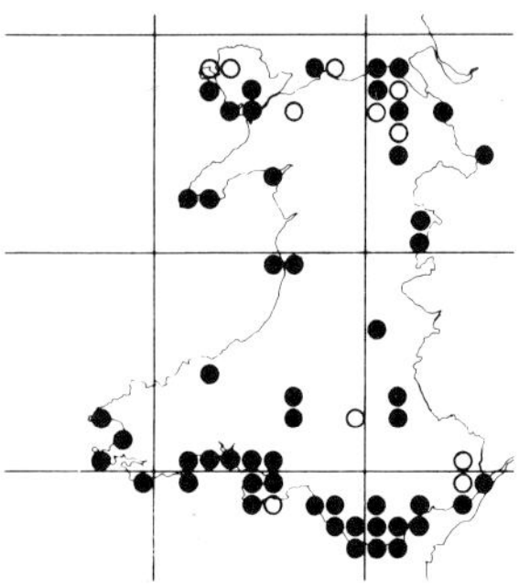

142 Ranunculus trichophyllus

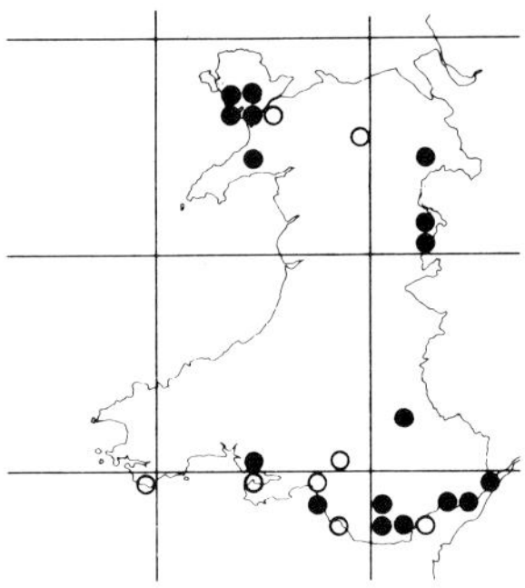

143 Ranunculus circinatus

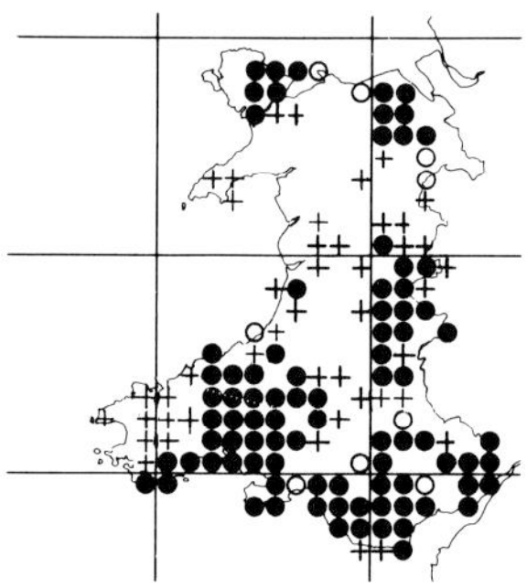

144 Aquilegia vulgaris

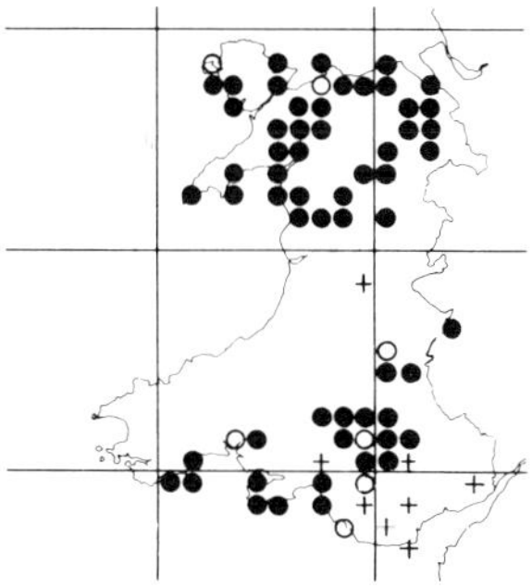
145 Thalictrum minus

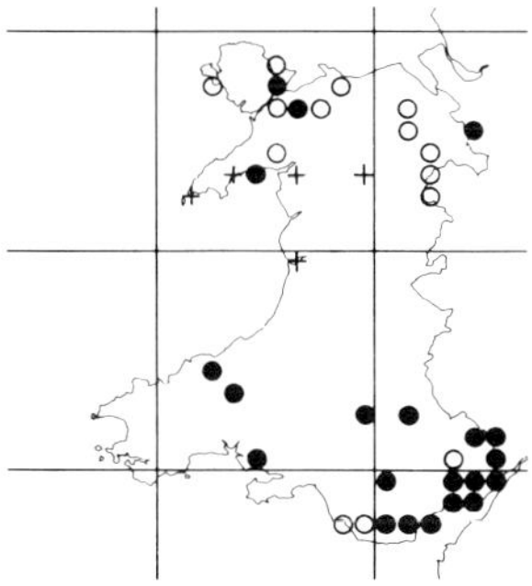
146 Thalictrum flavum

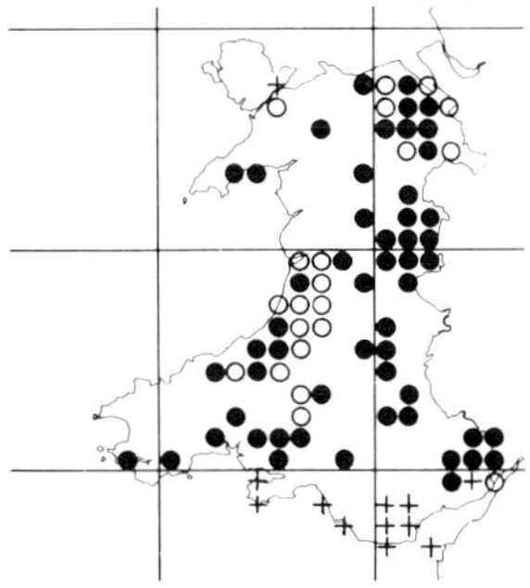
147 Berberis vulgaris

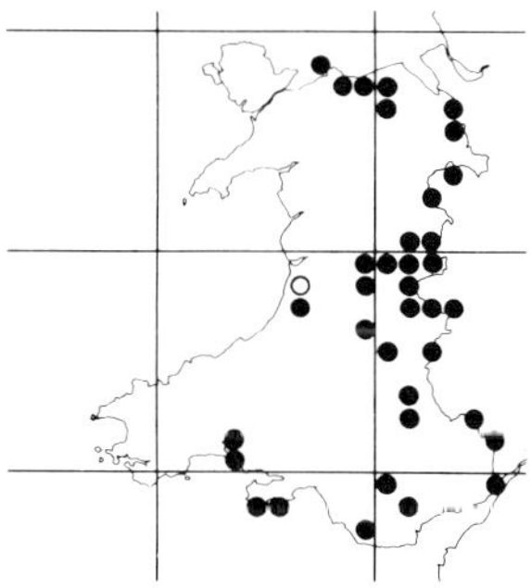
148 Mahonia aquifolium

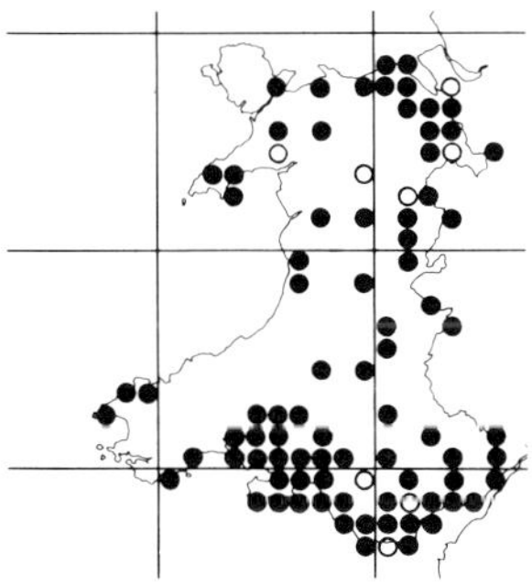
149 Papaver somniferum

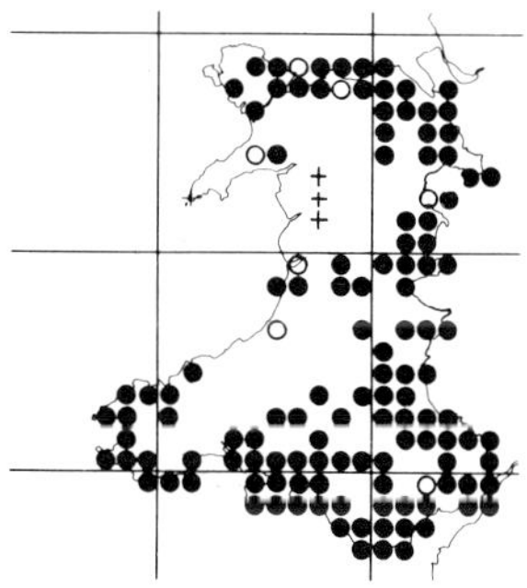
150 Papaver rhoeas

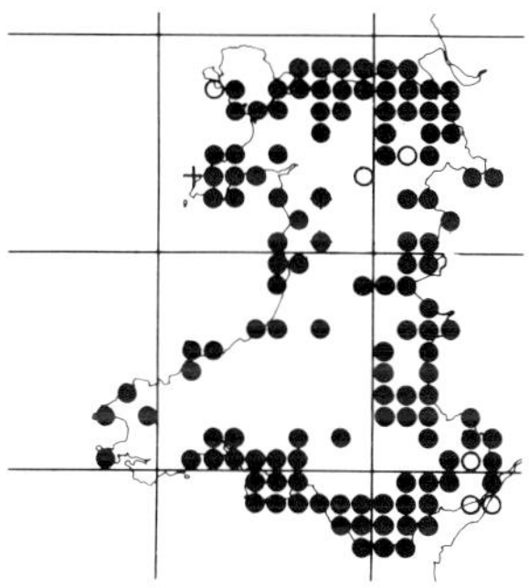
151 Papaver dubium agg.

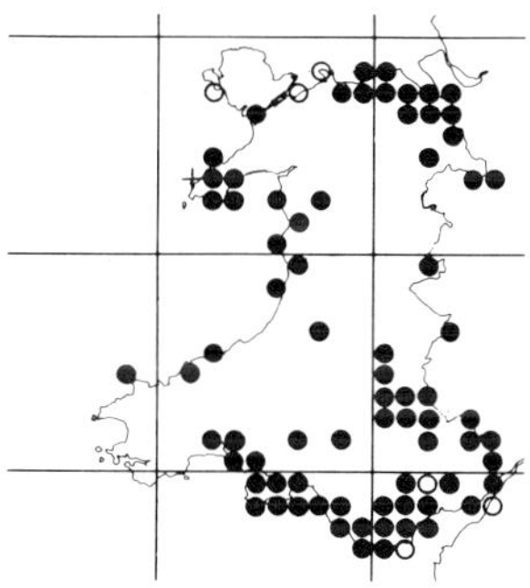
152 Papaver dubium s.s.

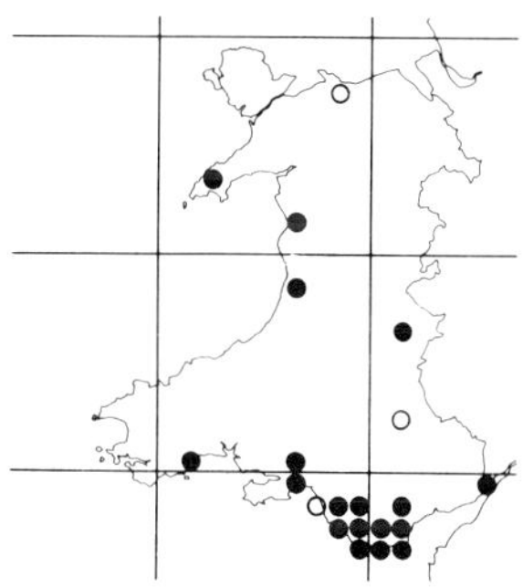
153 Papaver lecoqii

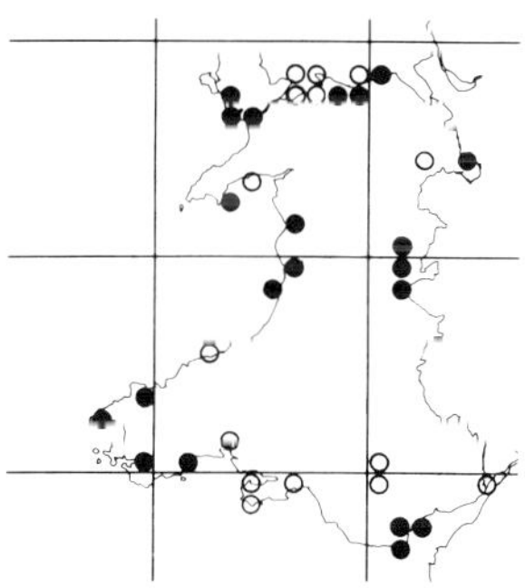
154 Papaver argemone

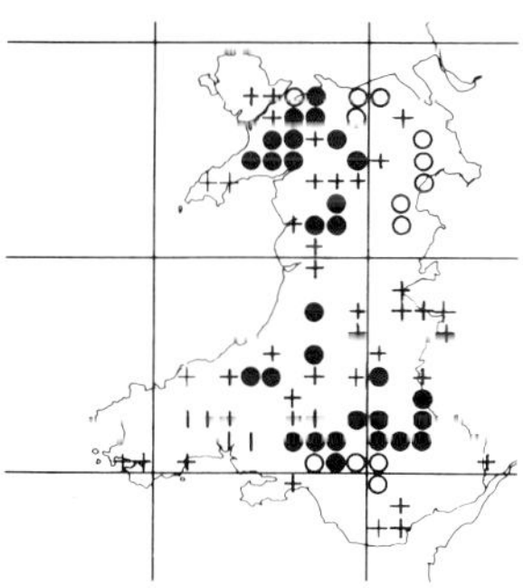
155 Meconopsis cambrica

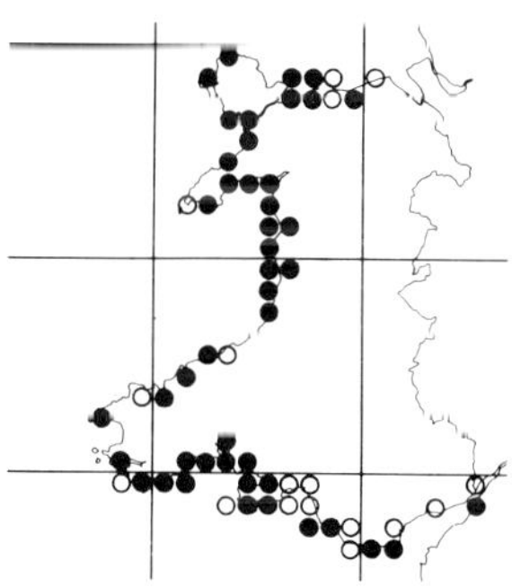
156 Glaucium flavum

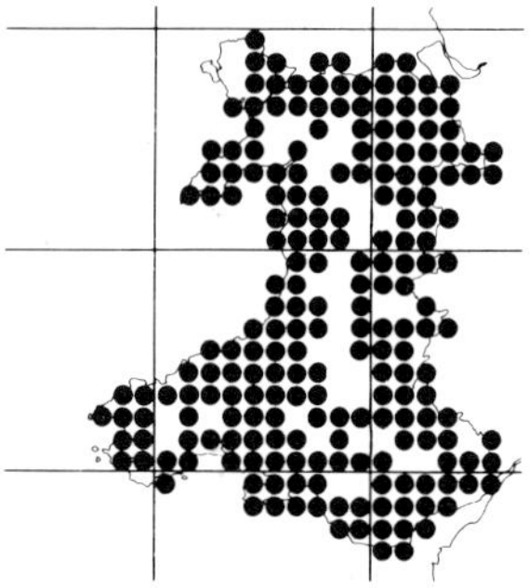

157 Chelidonium majus

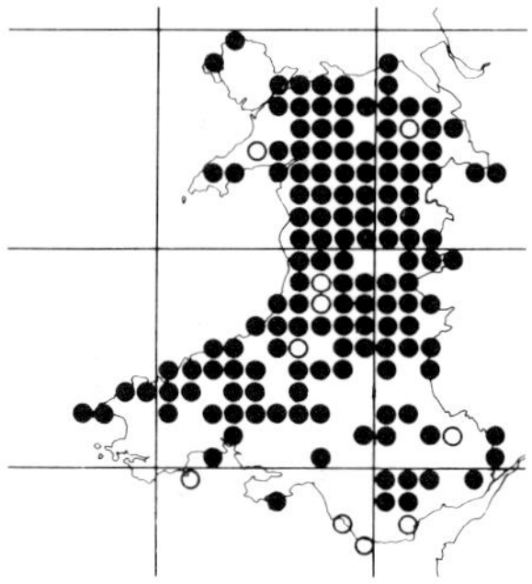

158 Corydalis claviculata

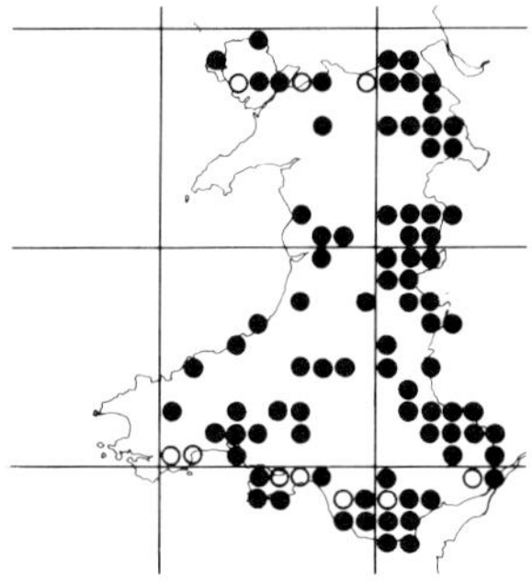

159 Corydalis lutea

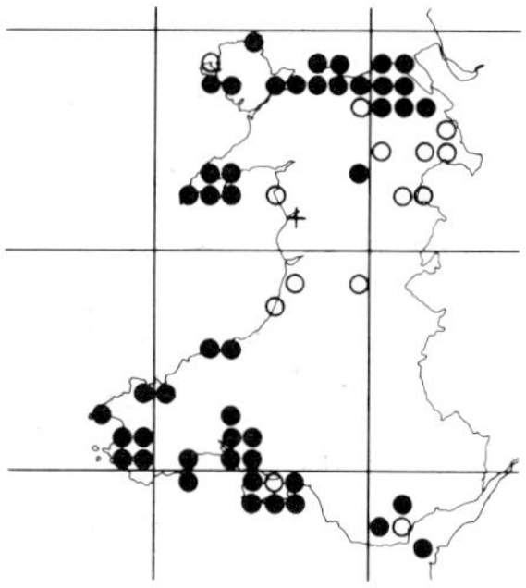

160 Fumaria capreolata

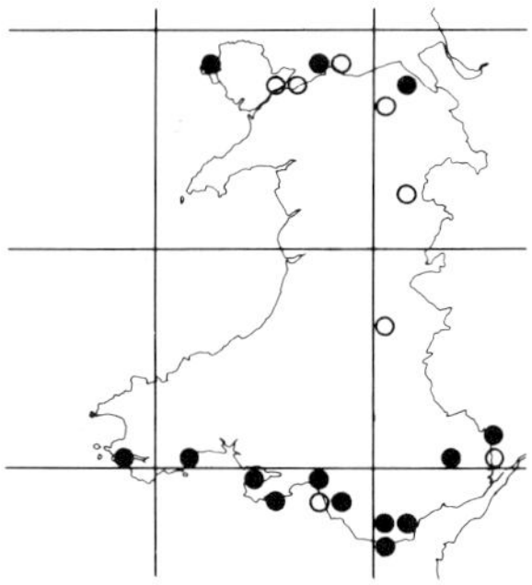

161 Fumaria purpurea

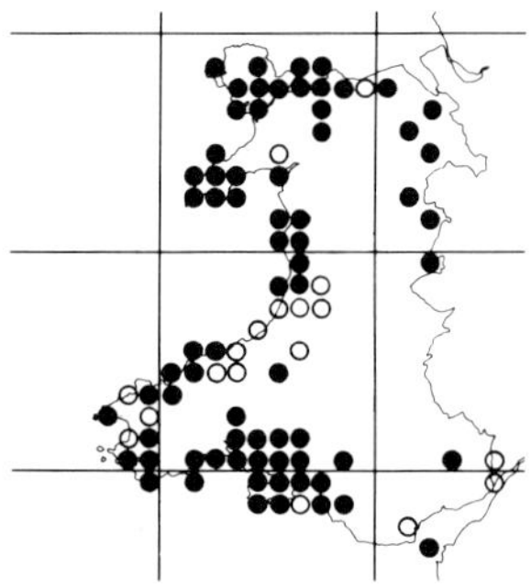

162 Fumaria bastardii

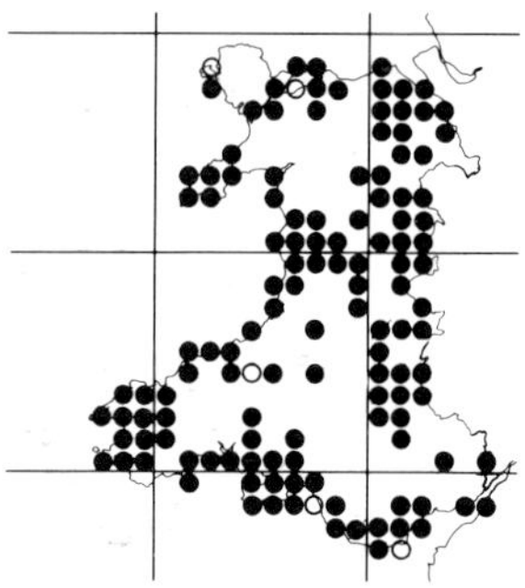

163 Fumaria muralis subsp. boraei

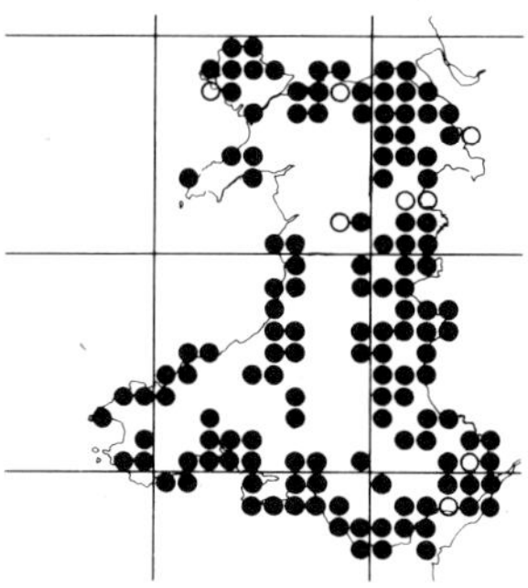

164 Fumaria officinalis

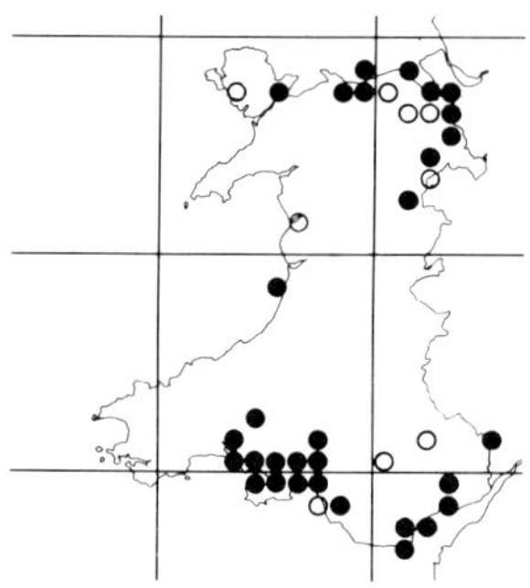

165 Sisymbrium altissimum

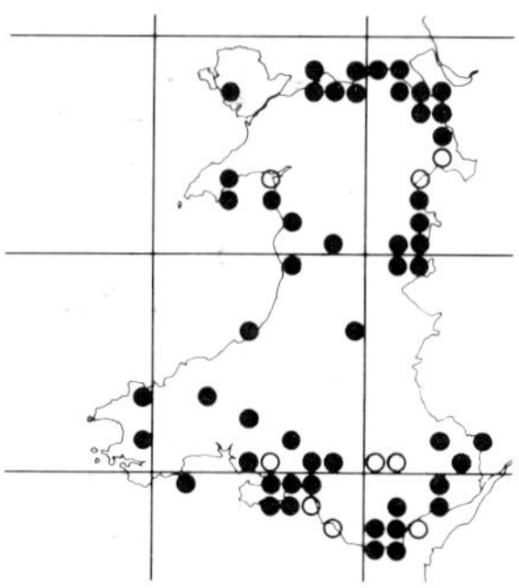

166 Sisymbrium orientale

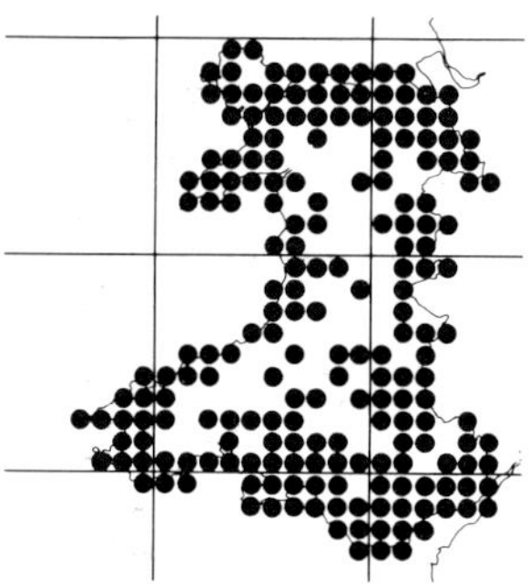

167 Sisymbrium officinale

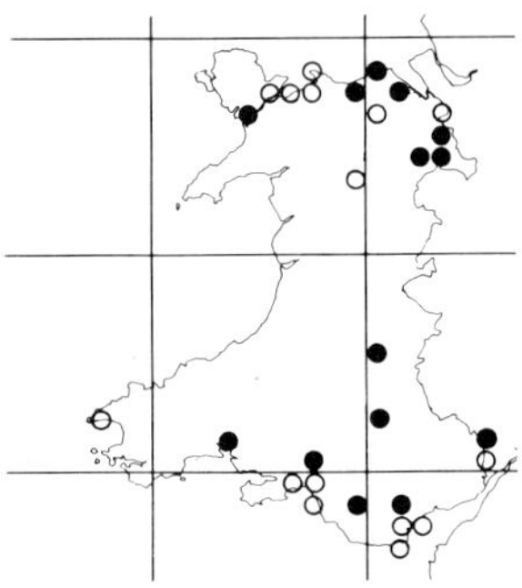

168 Descurainia sophia

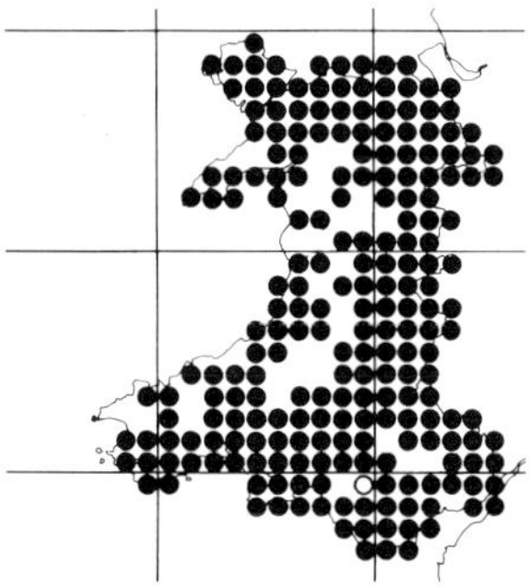

169 Alliaria petiolata

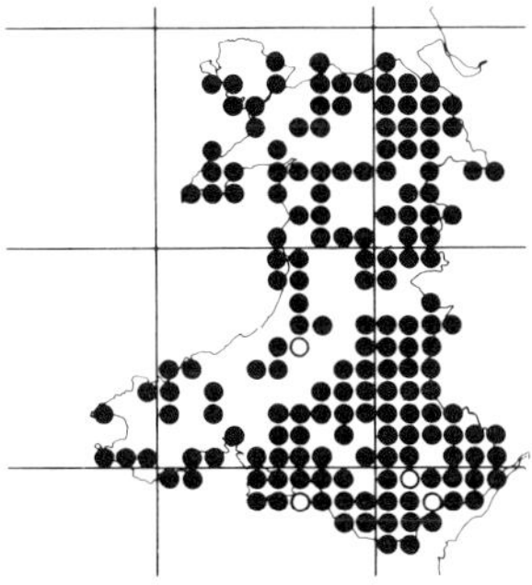

170 Arabidopsis thaliana

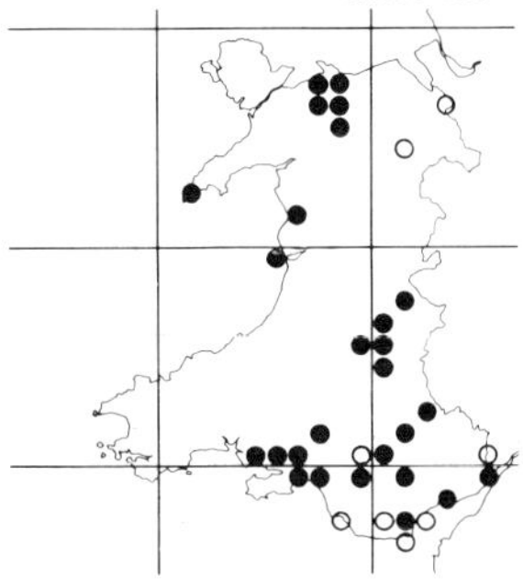

171 Erysimum cheiranthoides

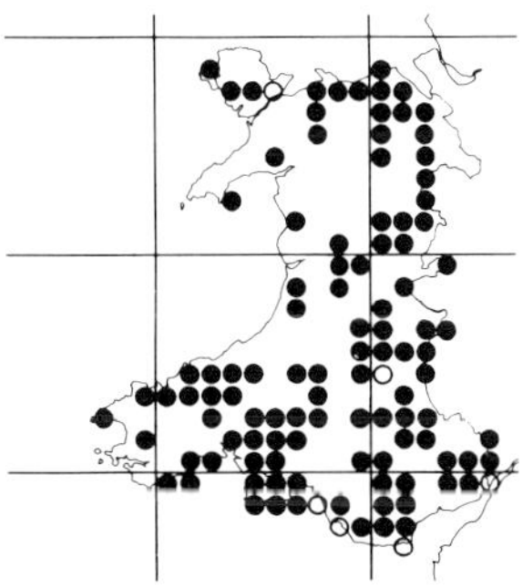

172 Hesperis matronalis

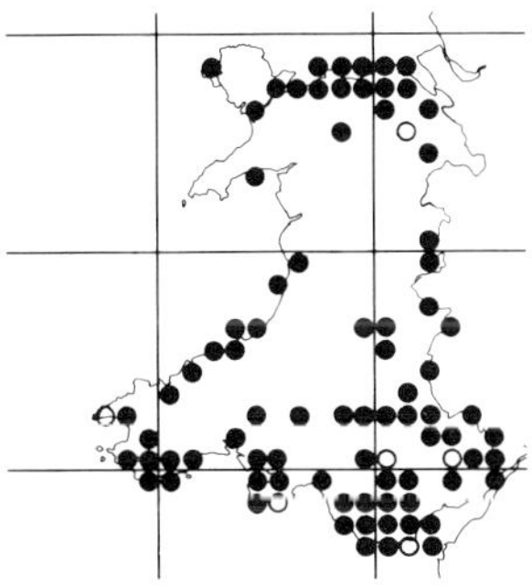

173 Cheiranthus cheiri

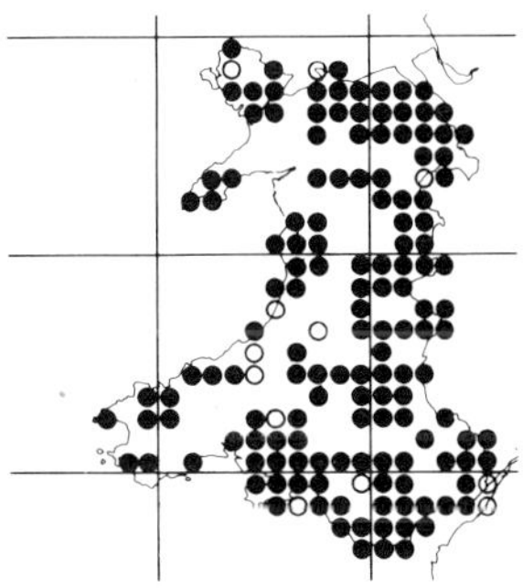

174 Barbarea vulgaris

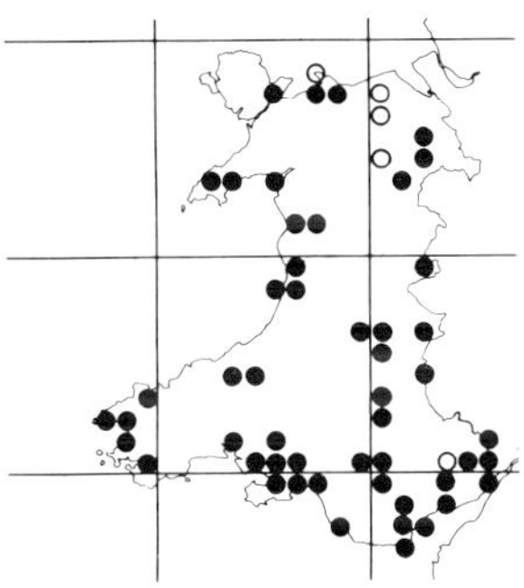

175 Barbarea verna

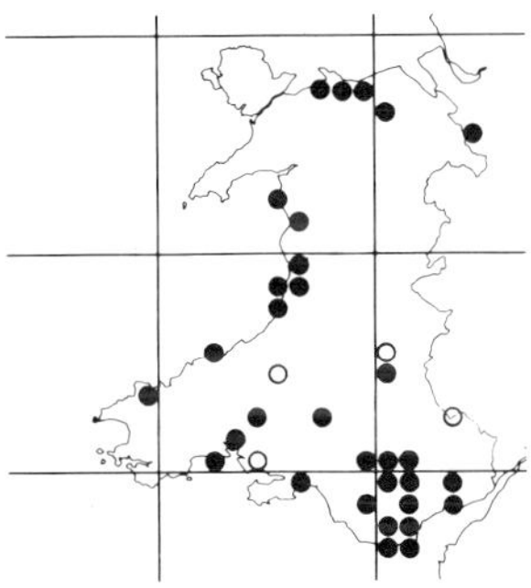

176 Barbarea intermedia

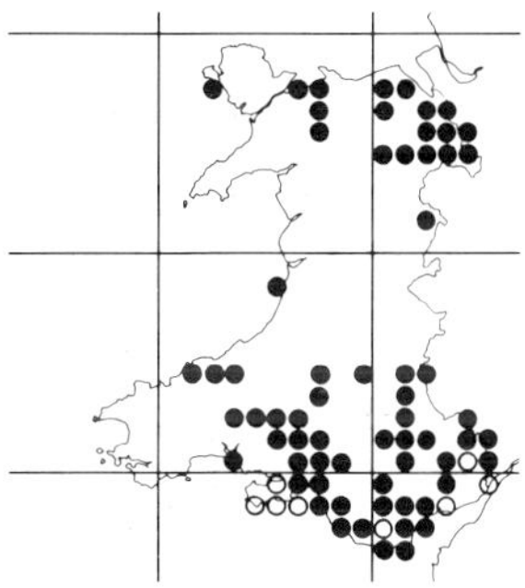

177 Rorippa sylvestris

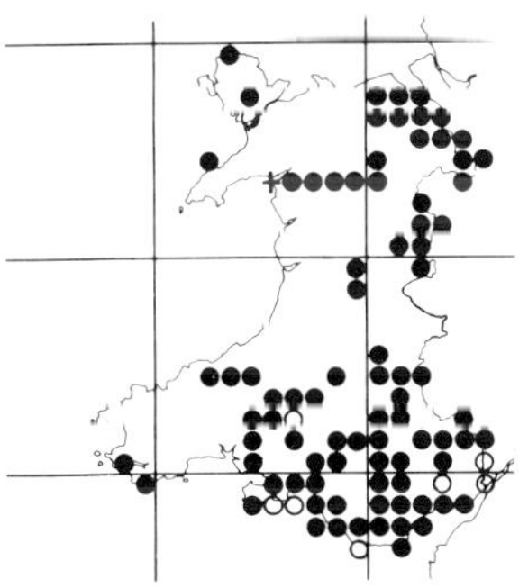

178 Rorippa palustris

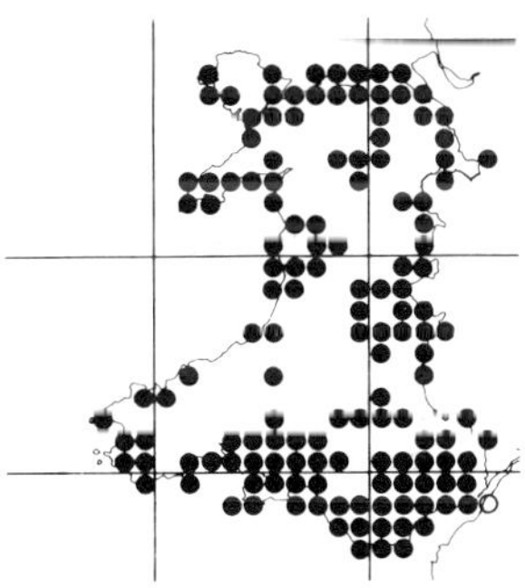

179 Armoracia rusticana

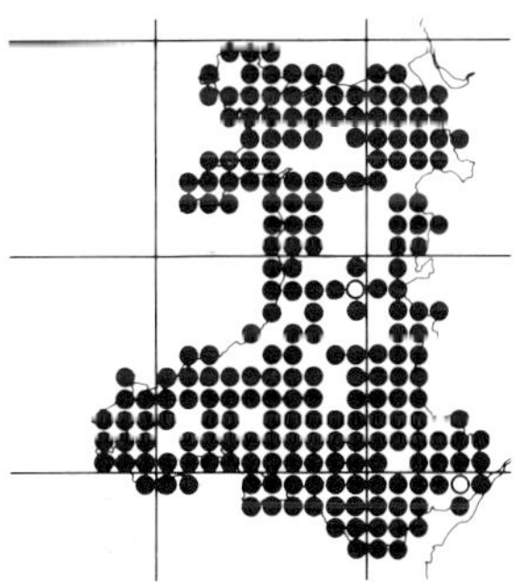

180 Nasturtium officinale agg.

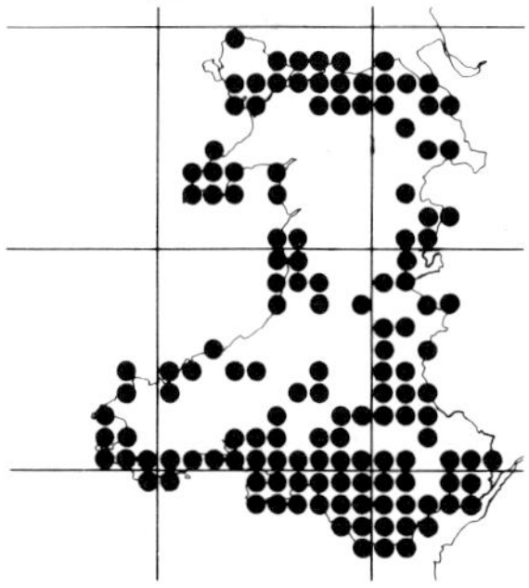

181 Nasturtium officinale s.s.

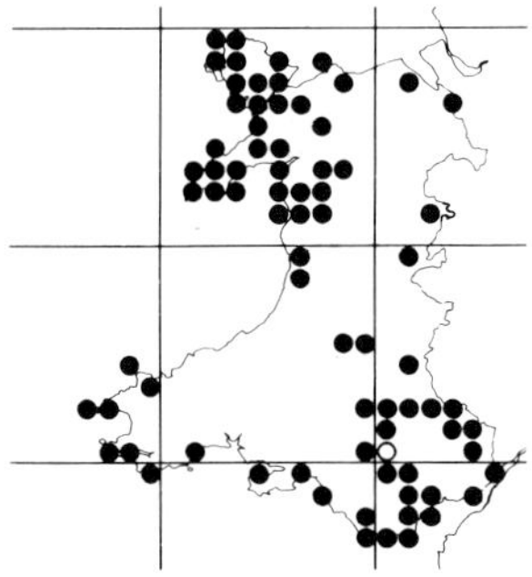

182 Nasturtium microphyllum

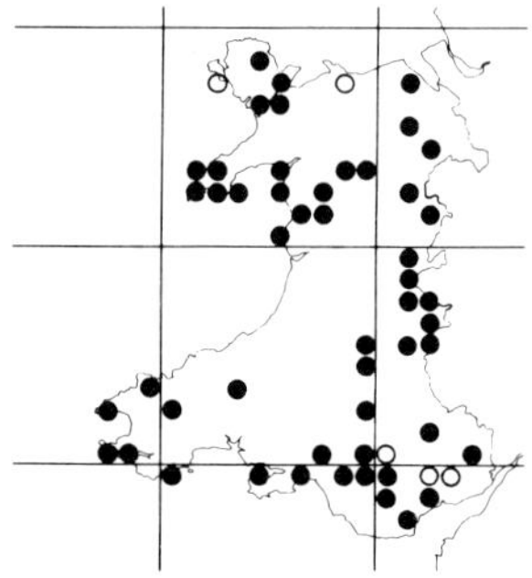

183 Nasturtium microphyllum × officinale

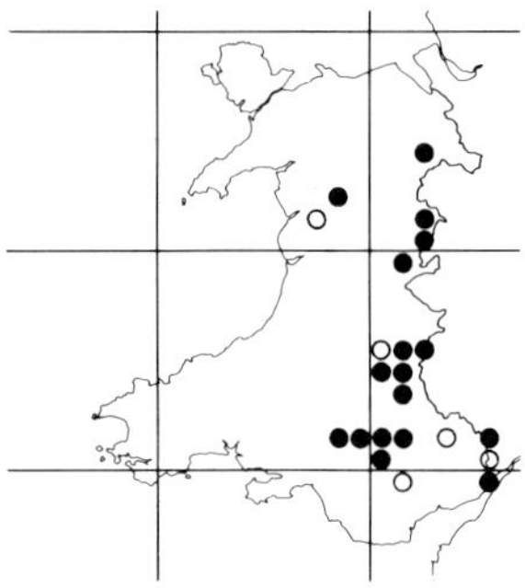

184 Cardamine impatiens

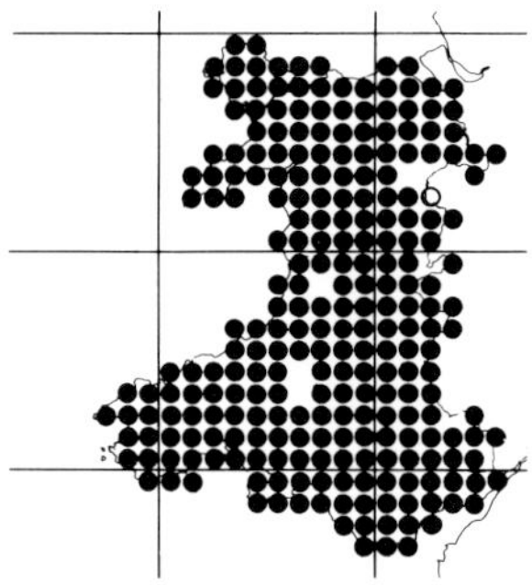

185 Cardamine flexuosa

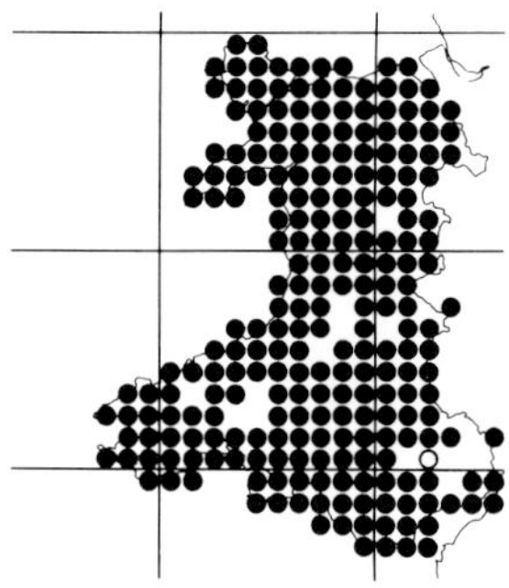

186 Cardamine hirsuta

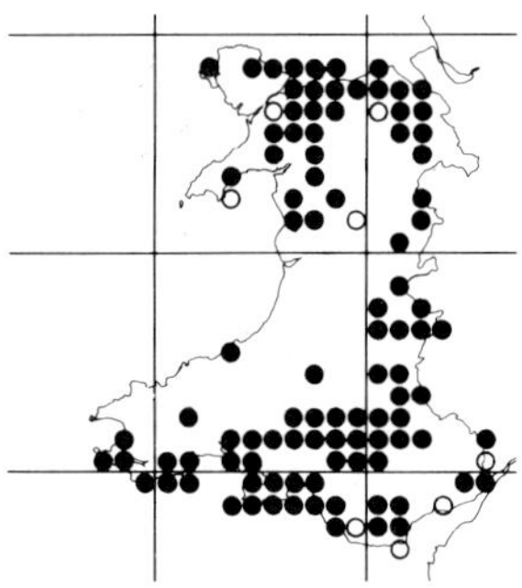

187 Arabis hirsuta

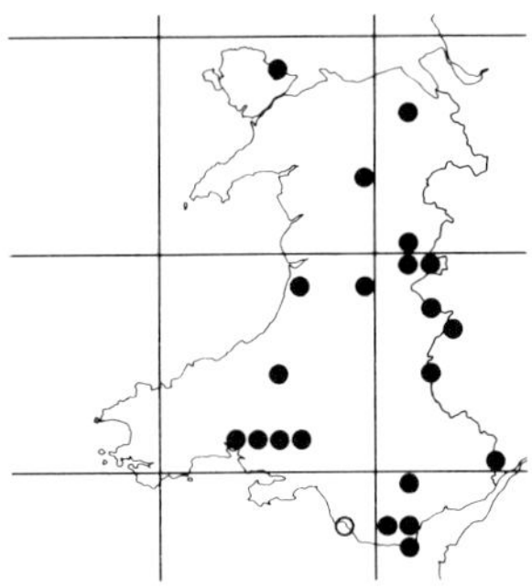

188 Arabis caucasica

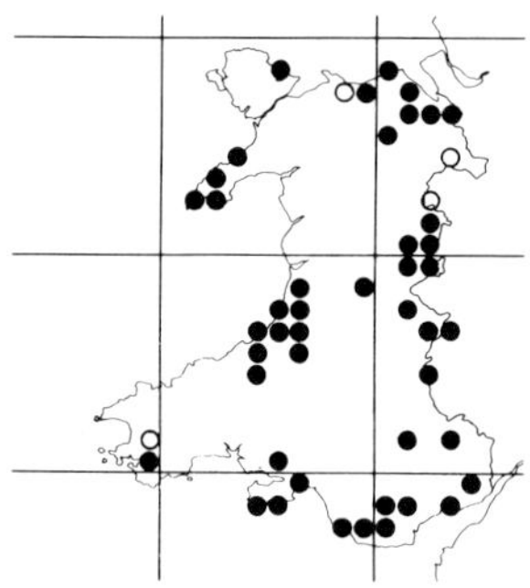

189 Lunaria annua

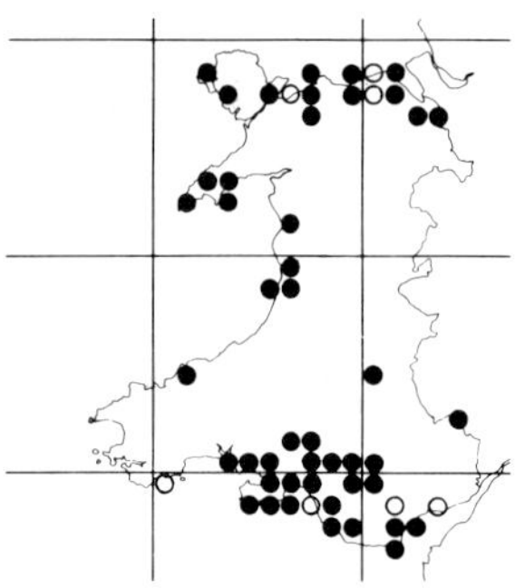

190 Lobularia maritima

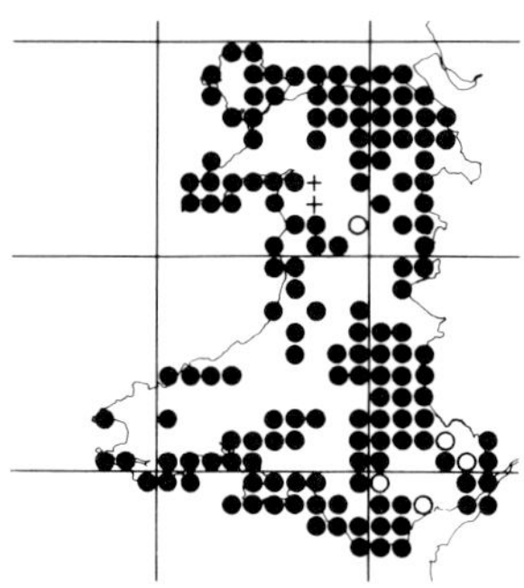

191 Erophila verna

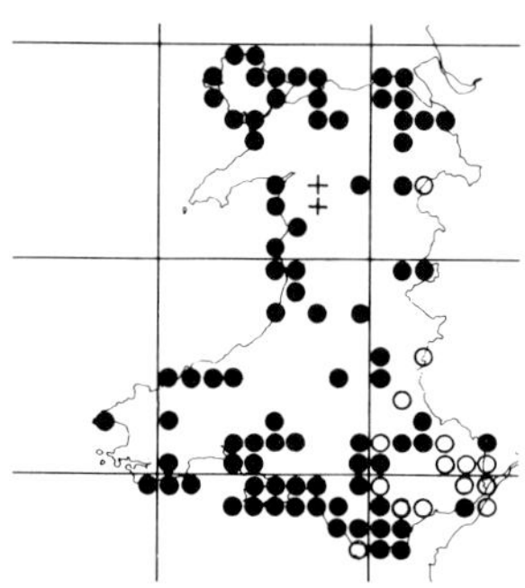

192 Erophila verna subsp. verna

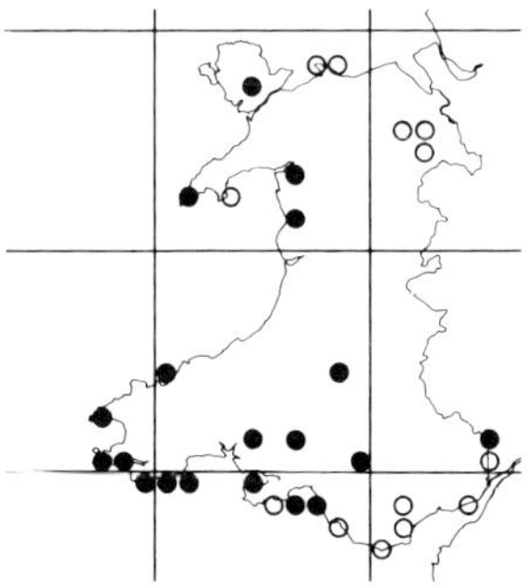

193 Erophila verna subsp. spathulata

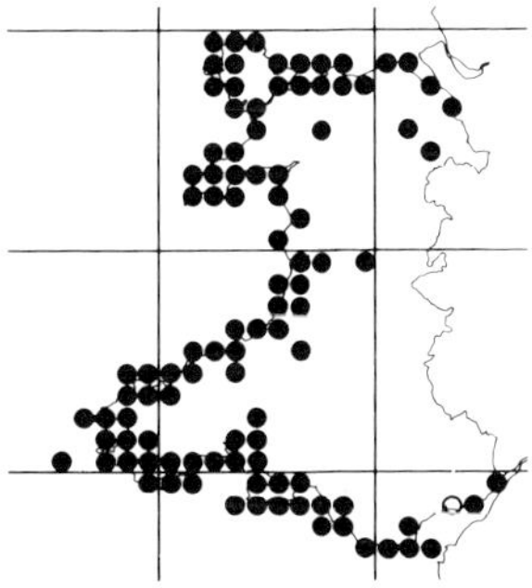

194 Cochlearia danica

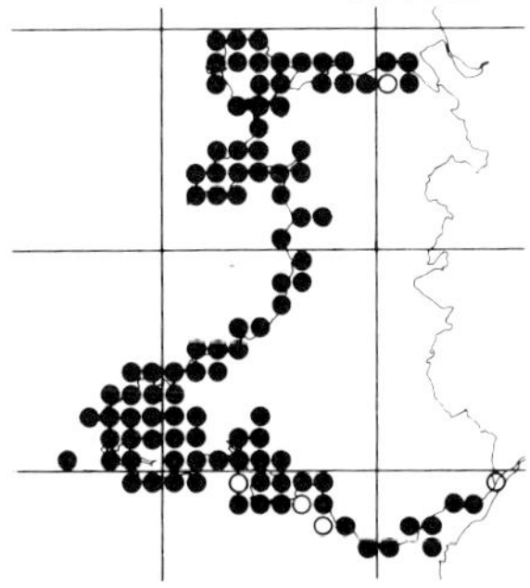

195 Cochlearia officinalis

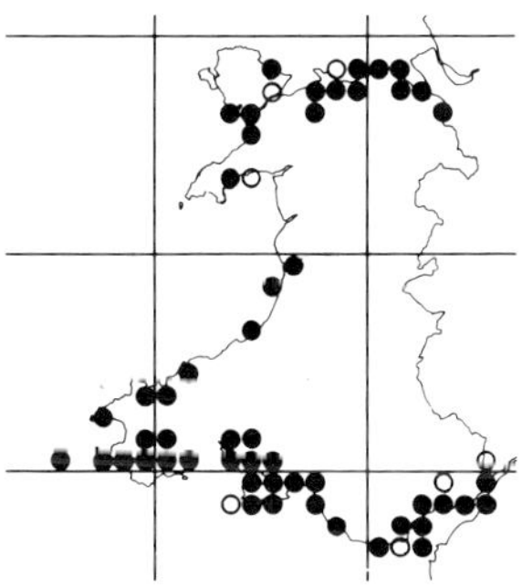

196 Cochlearia anglica

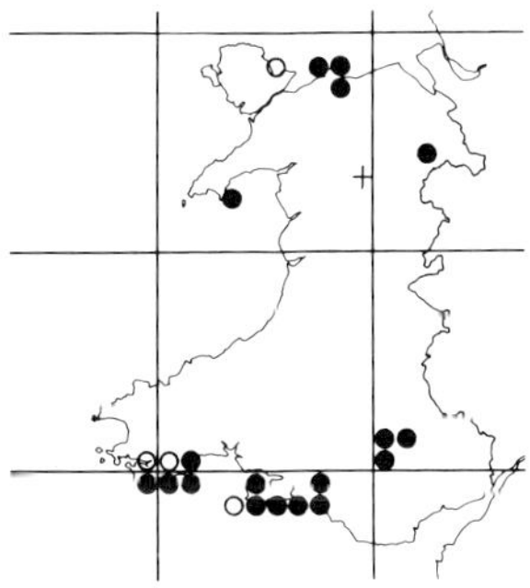

197 Hornungia petraea

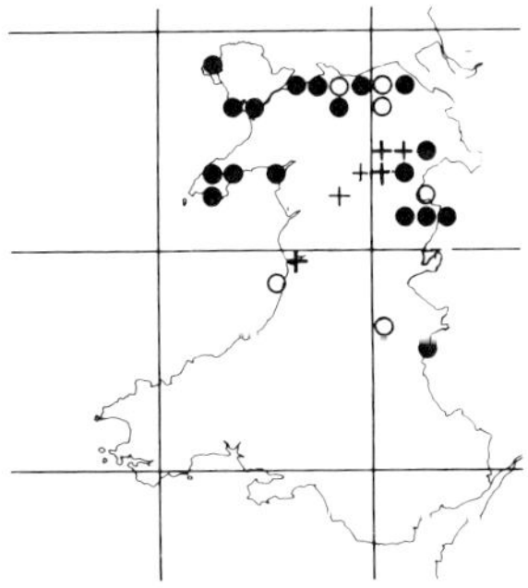

198 Teesdalia nudicaulis

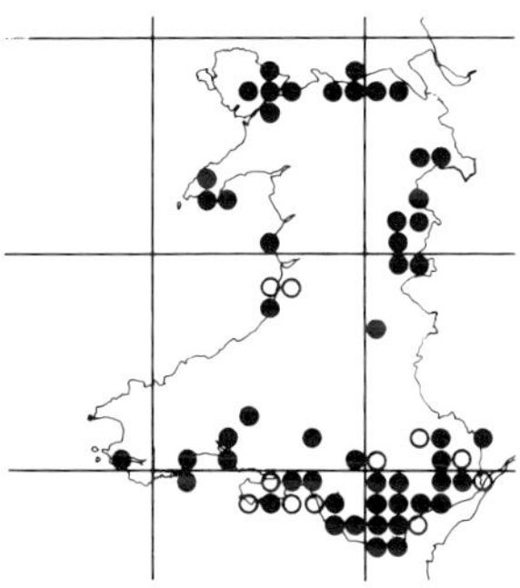

199 Thlaspi arvense

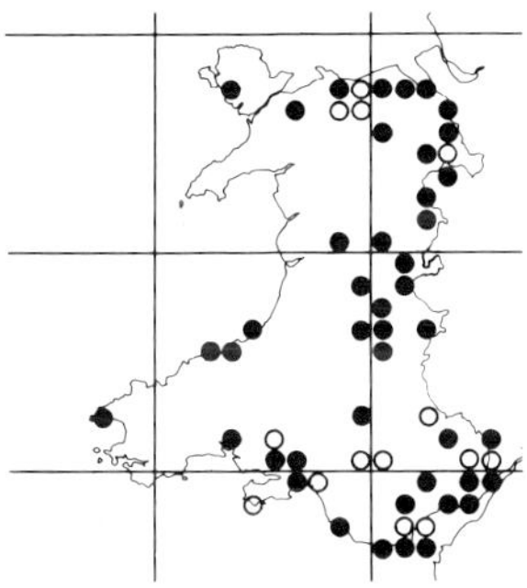

200 Lepidium campestre

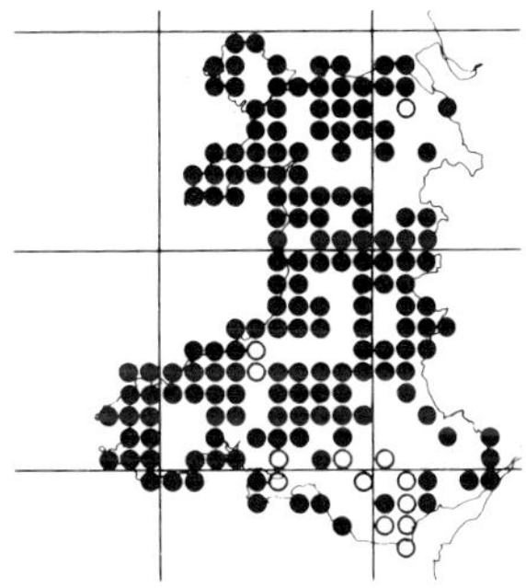

201 Lepidium heterophyllum

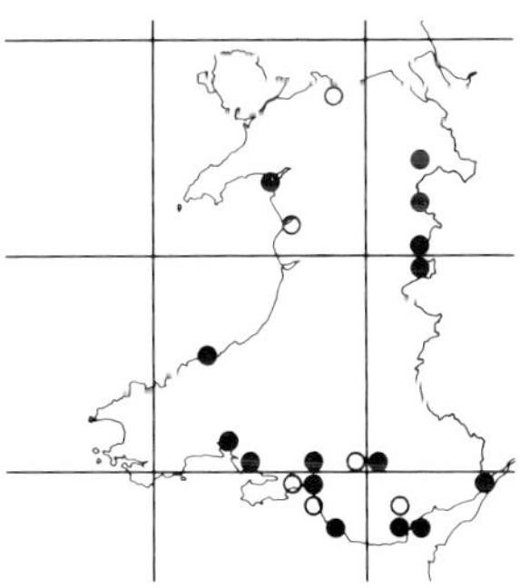

202 Lepidium sativum

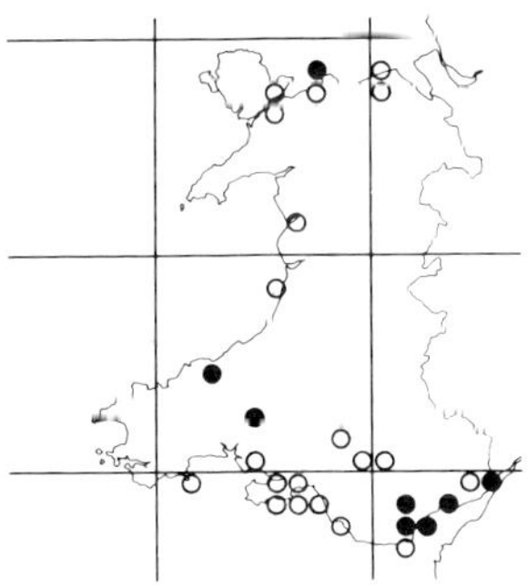

203 Lepidium ruderale

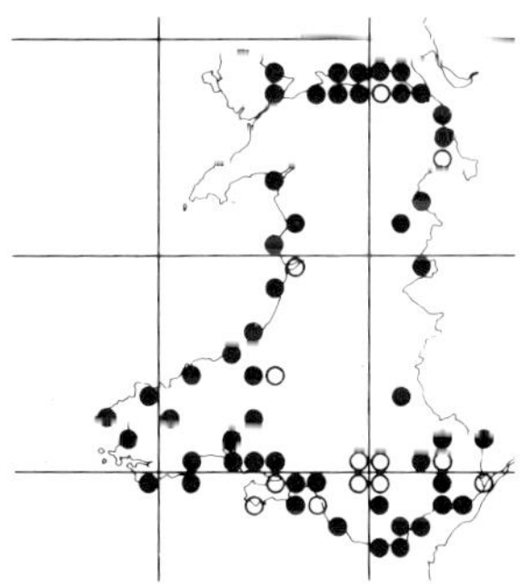

204 Cardaria draba

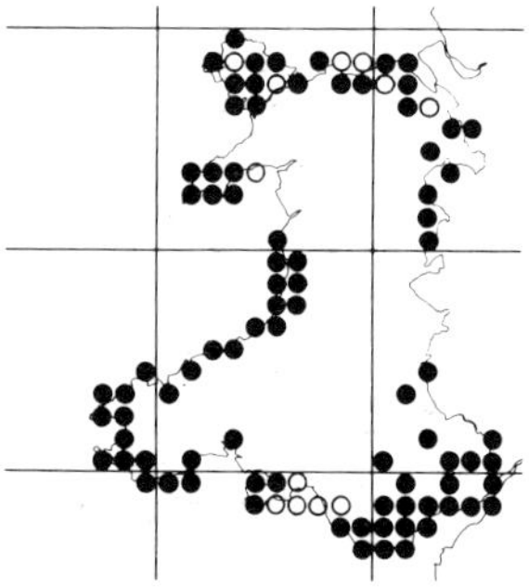

205 Coronopus squamatus

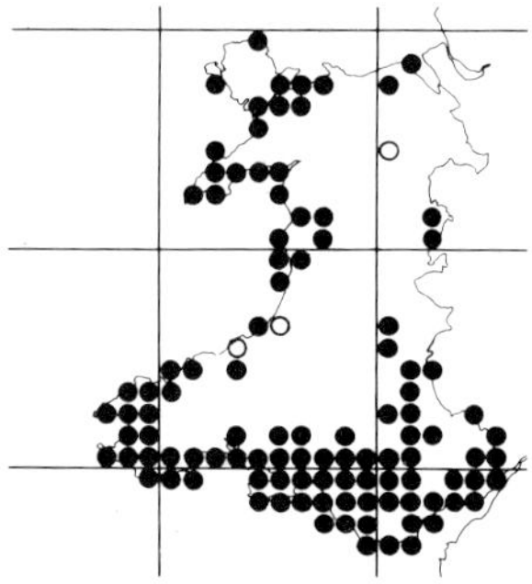

206 Coronopus didymus

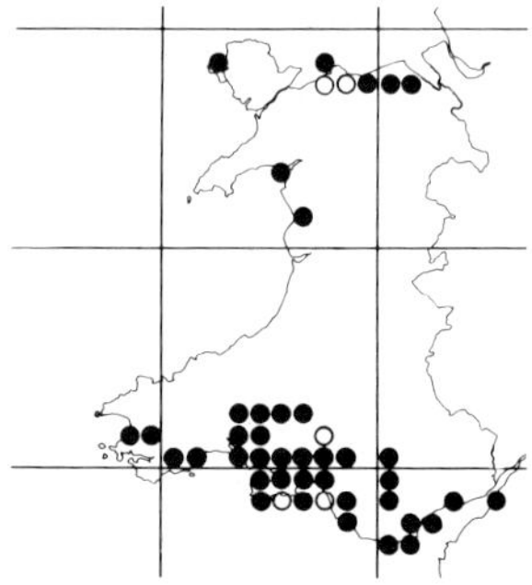

207 Diplotaxis tenuifolia

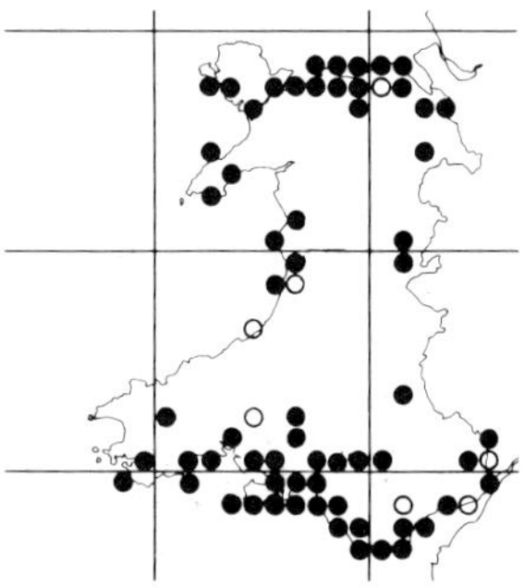

208 Diplotaxis muralis

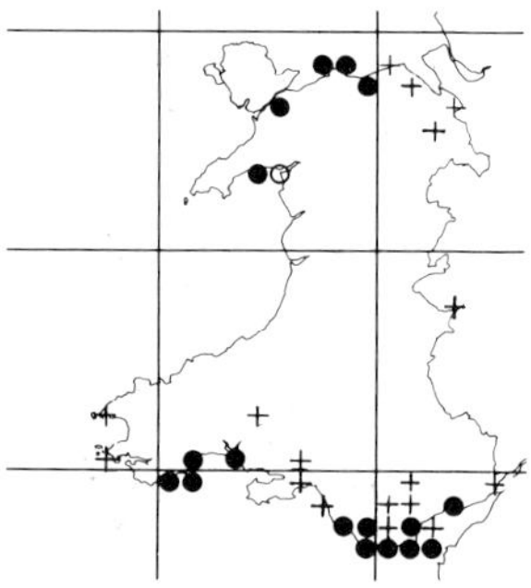

209 Brassica oleracea

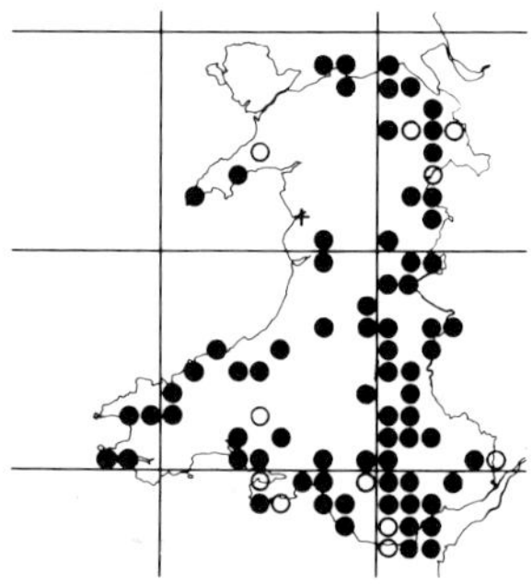

210 Brassica napus

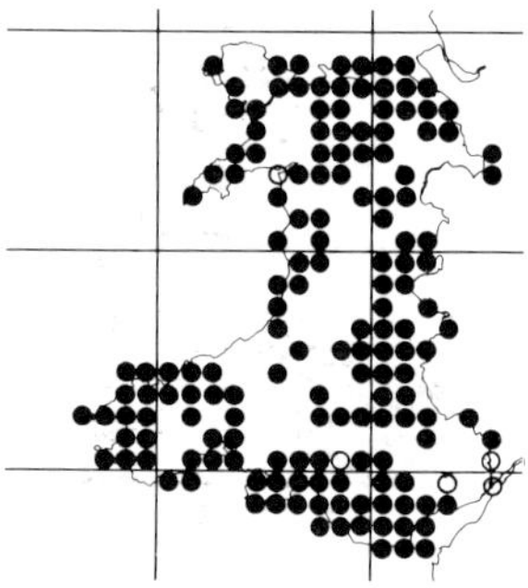

211 Brassica rapa

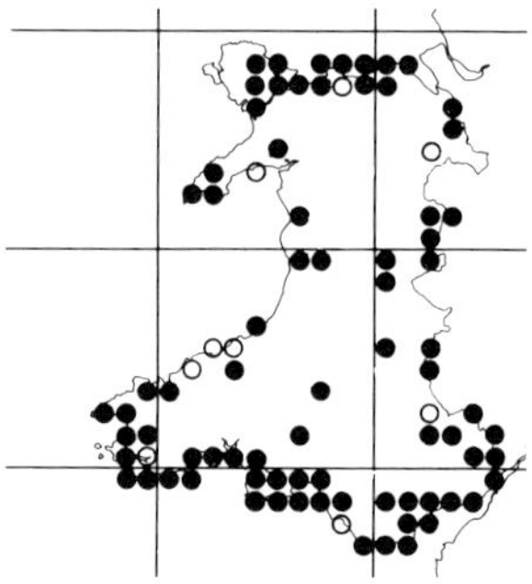

212 Brassica nigra

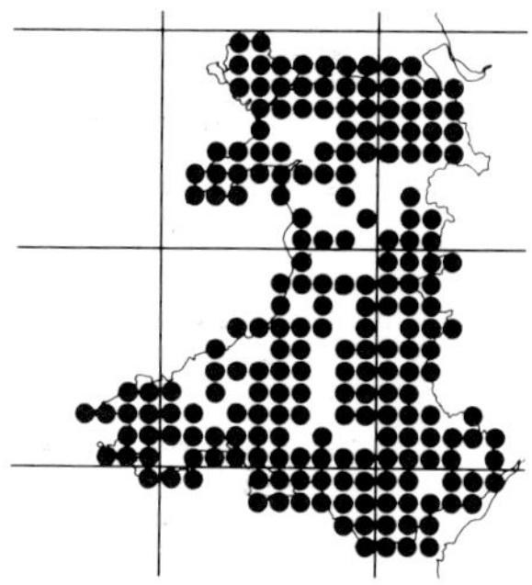

213 Sinapis arvensis

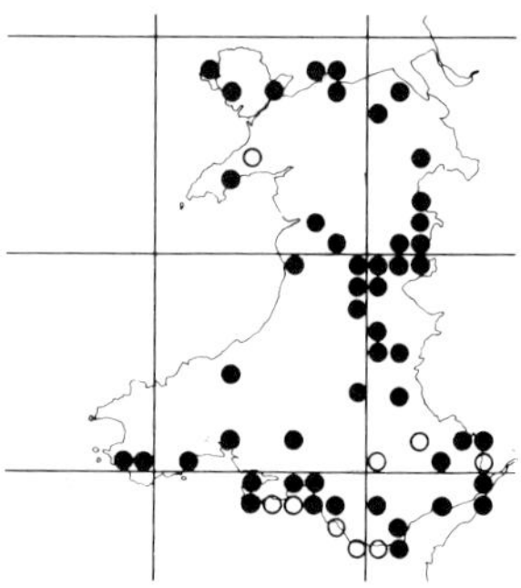

214 Sinapis alba
subsp. alba

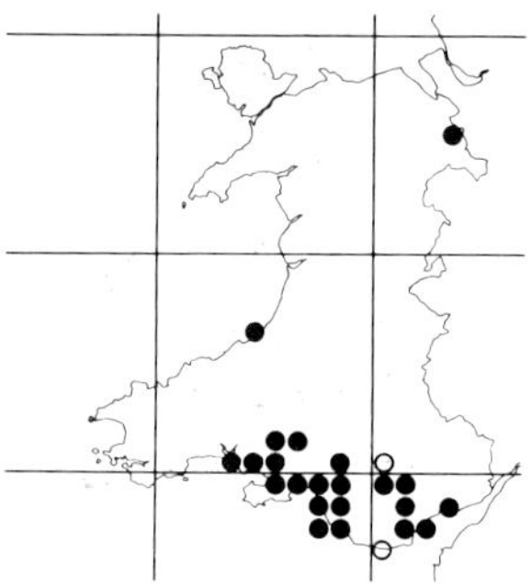

215 Hutera cheiranthos

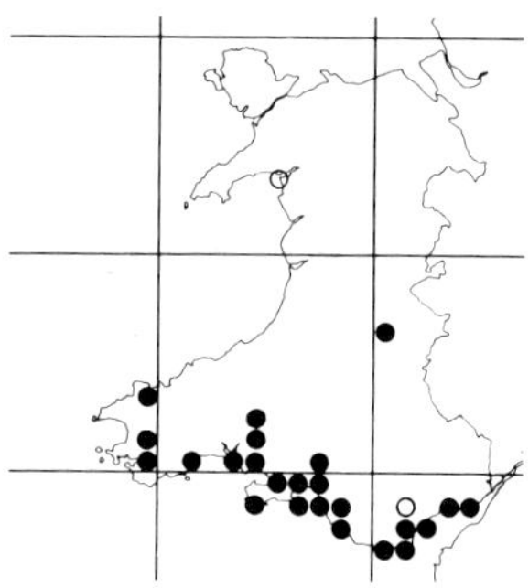

216 Hirschfeldia incana

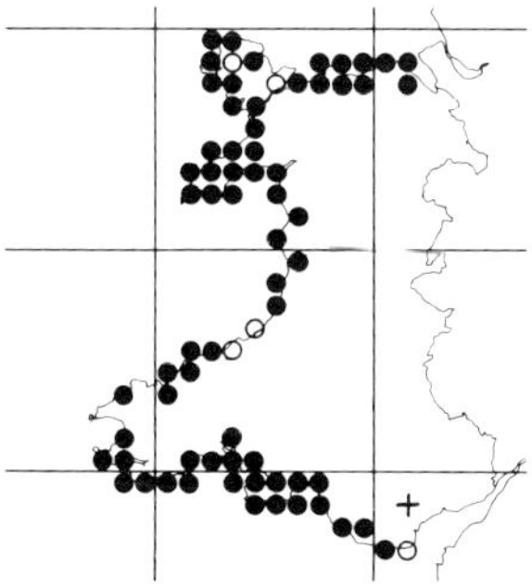

217 Cakile maritima

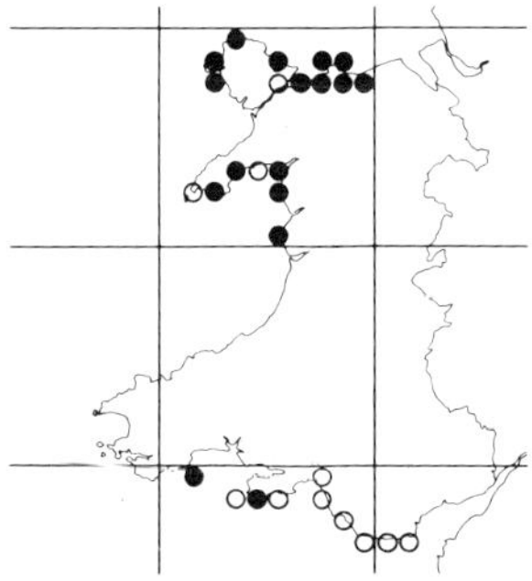

218 Crambe maritima

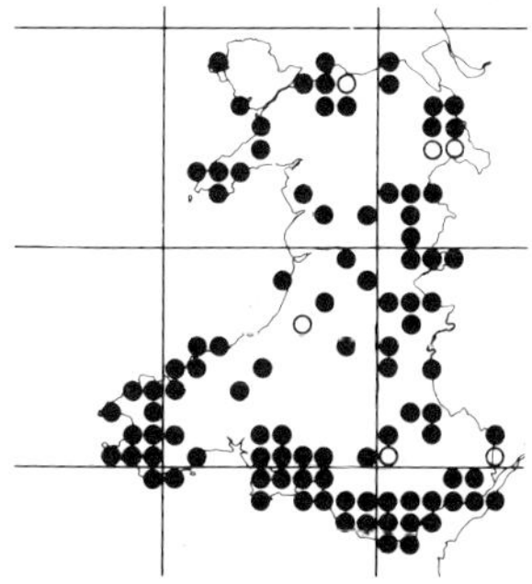

219 Raphanus raphanistrum subsp. raphanistrum

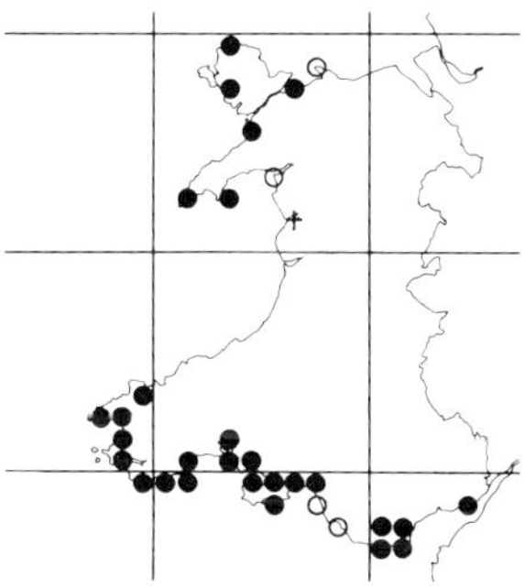

220 Raphanus raphanistrum subsp. maritimus

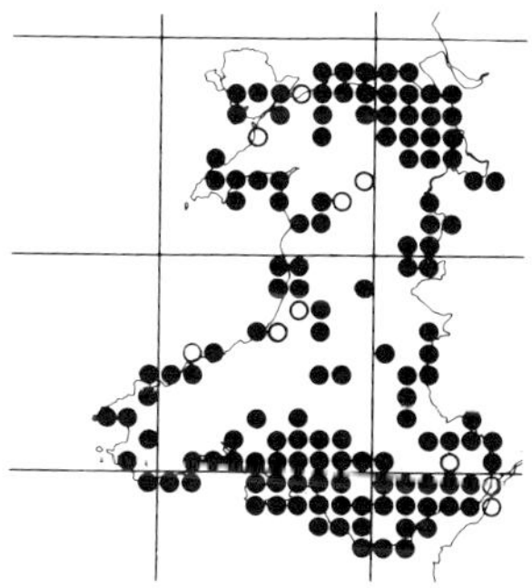

221 Reseda luteola

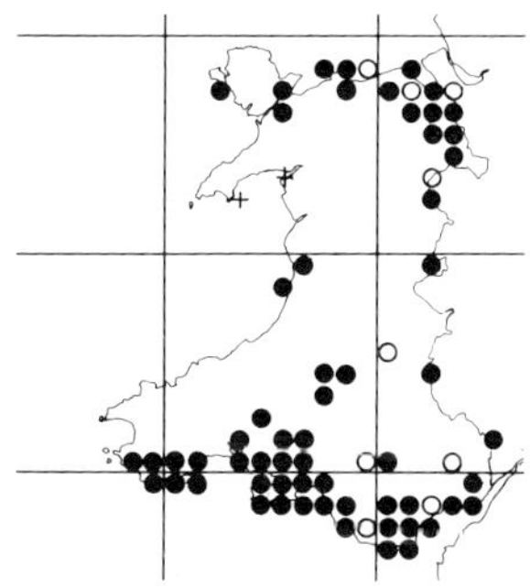

222 Reseda lutea

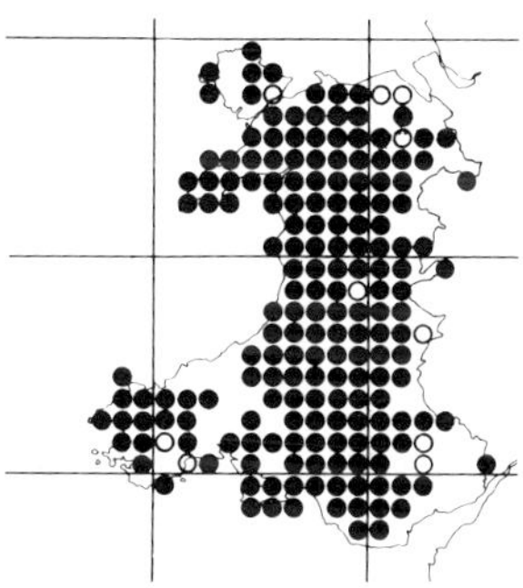

223 Drosera rotundifolia

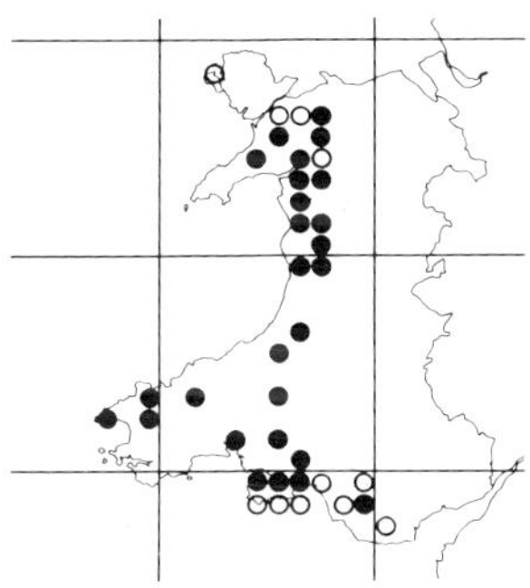

224 Drosera intermedia

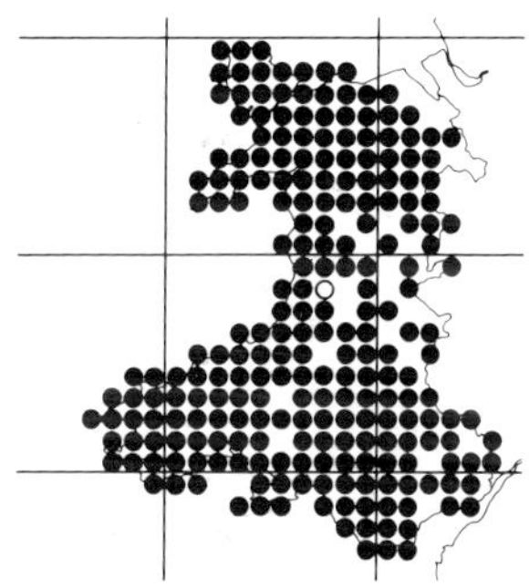

225 Umbilicus rupestris

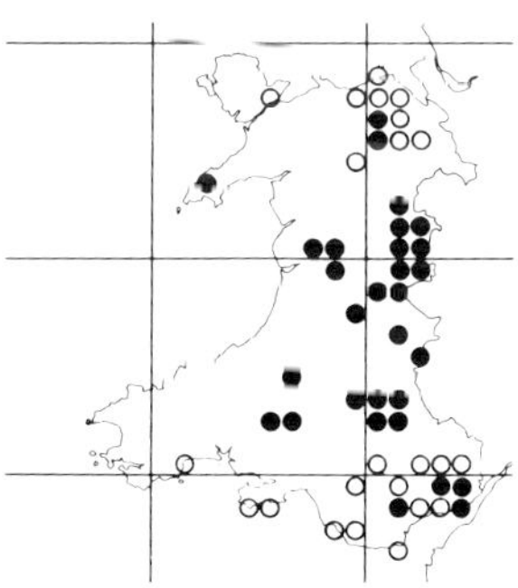

226 Sempervivum tectorum

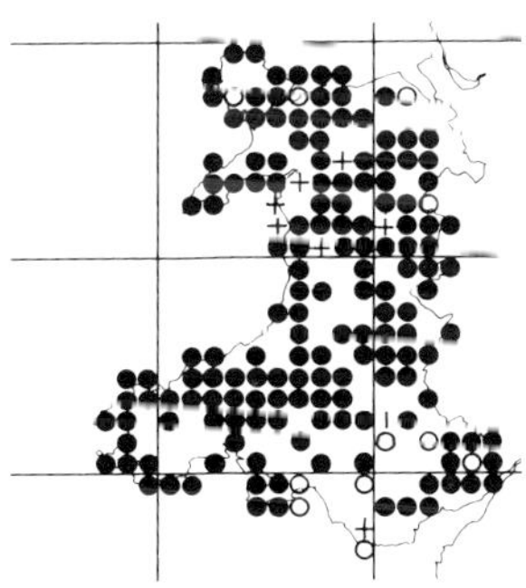

227 Sedum telephium

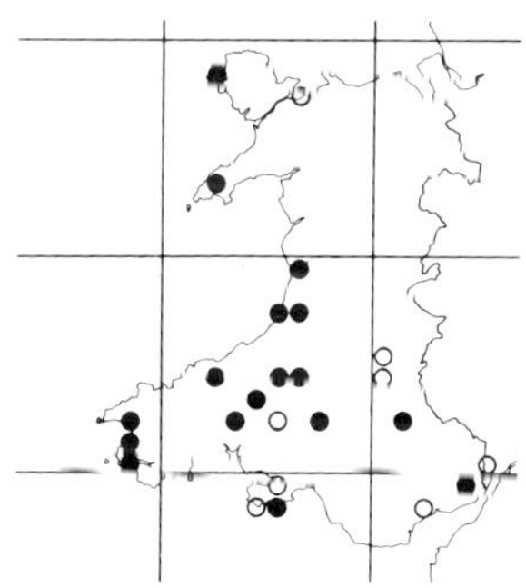

228 Sedum telephium subsp. telephium

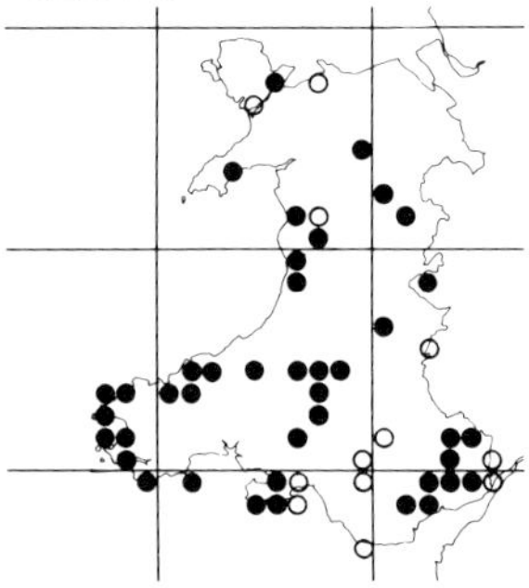

229 Sedum telephium subsp. fabaria

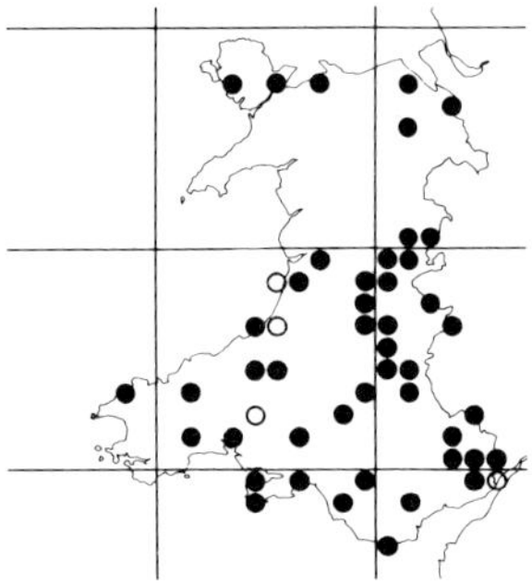

230 Sedum spurium

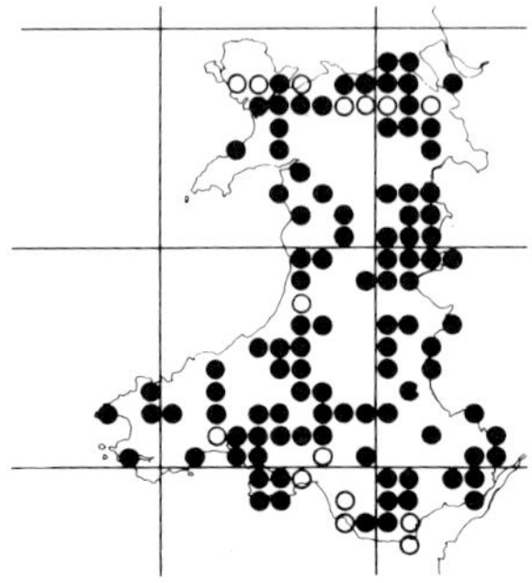

231 Sedum reflexum

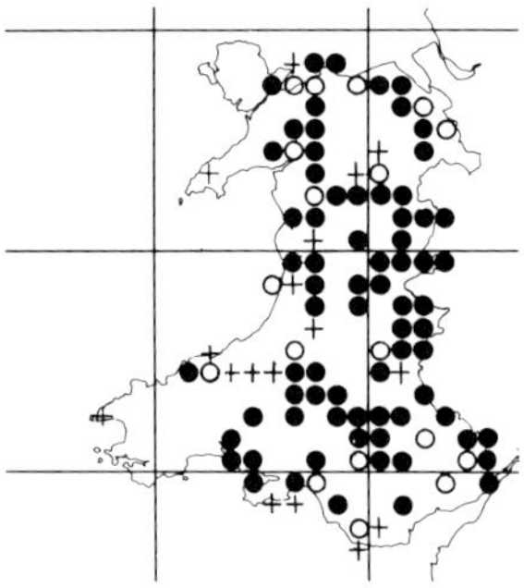

232 Sedum forsteranum

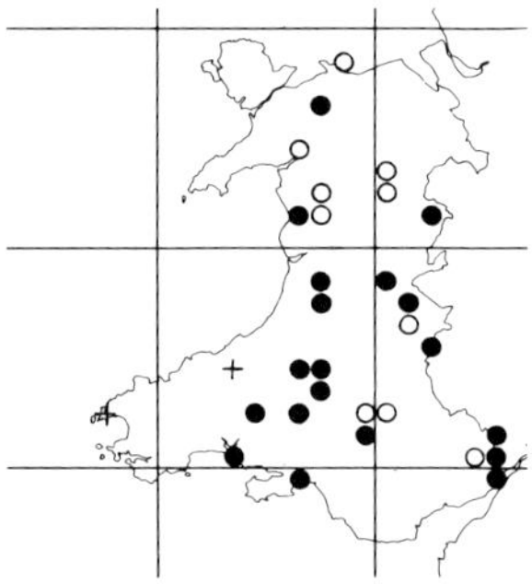

233 Sedum forsteranum subsp. forsteranum

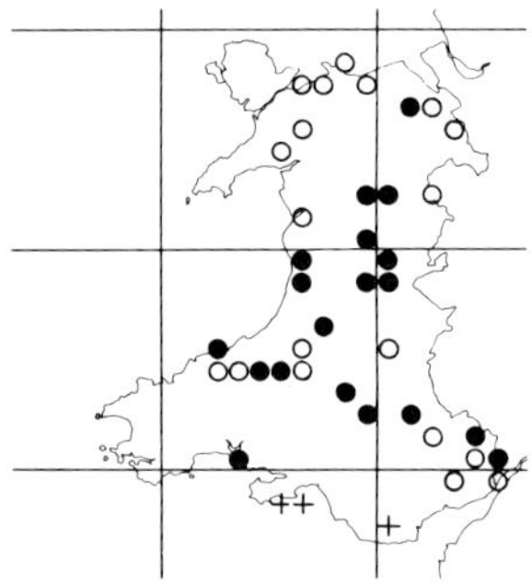

234 Sedum forsteranum subsp. elegans

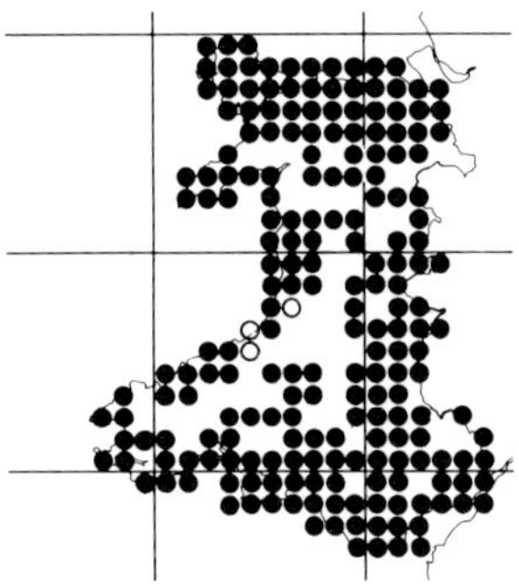

235 Sedum acre

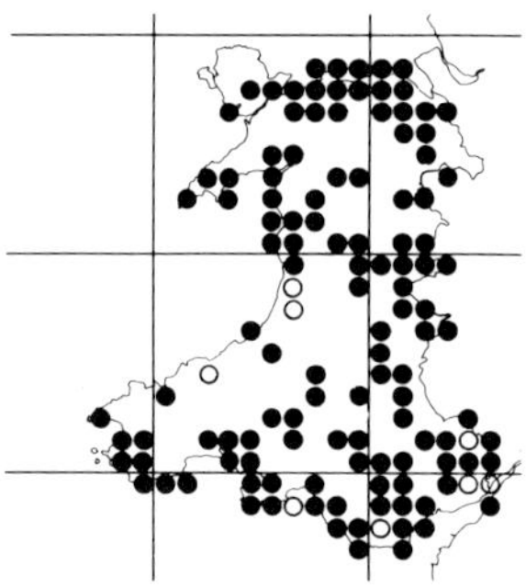

236 Sedum album subsp. album

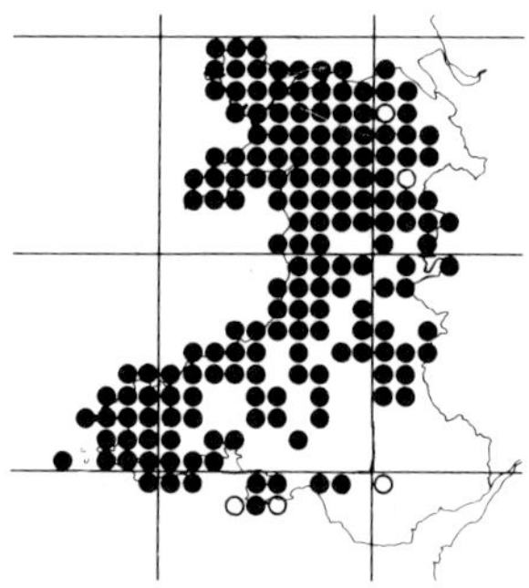

237 Sedum anglicum

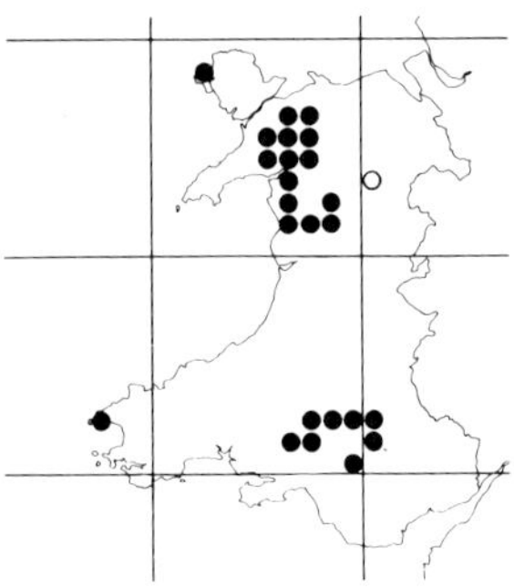

238 Rhodiola rosea

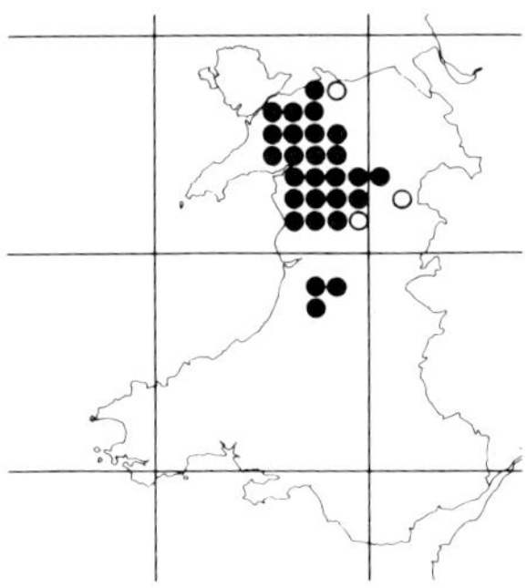

239 Saxifraga stellaris

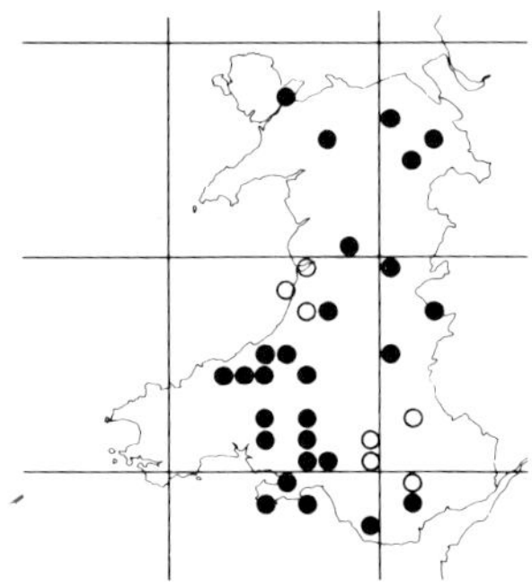

240 Saxifraga spathularis × umbrosa = S. × urbium

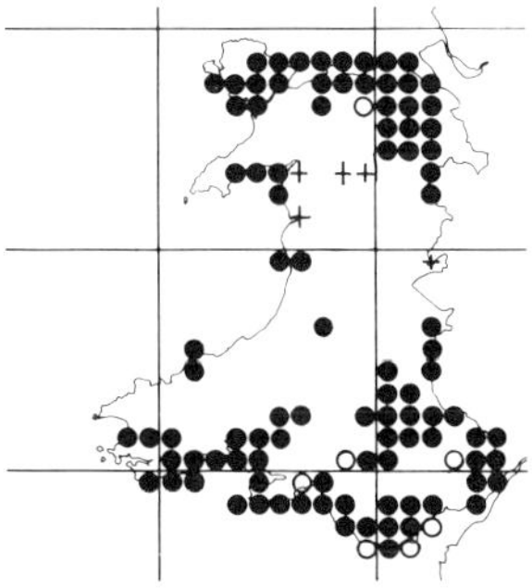

241 Saxifraga tridactylites

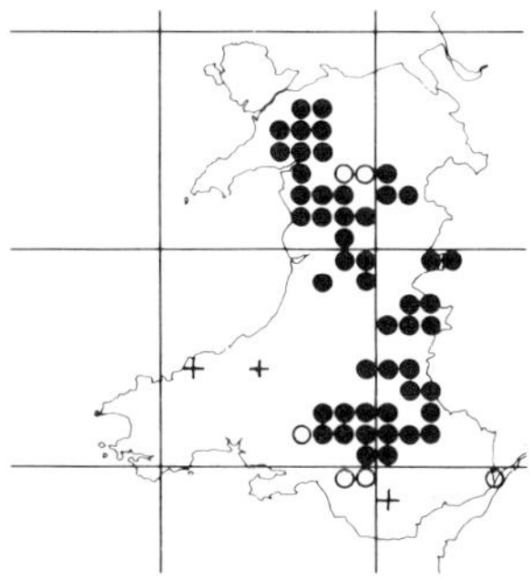

242 Saxifraga hypnoides

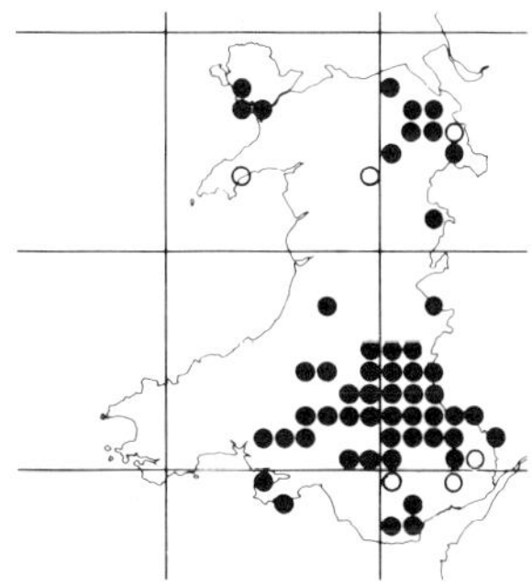

243 Saxifraga granulata

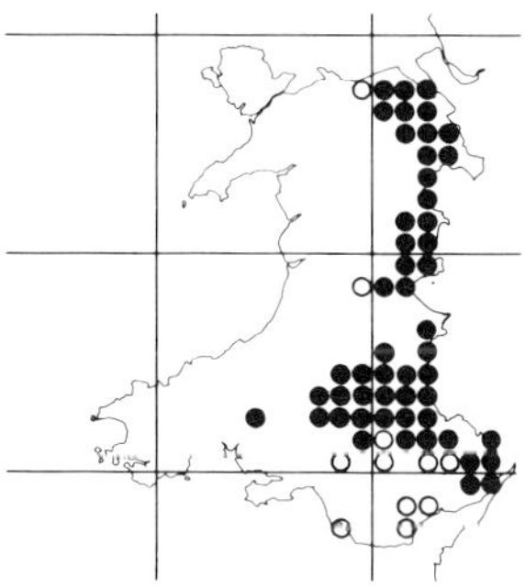

244 Chrysosplenium alternifolium

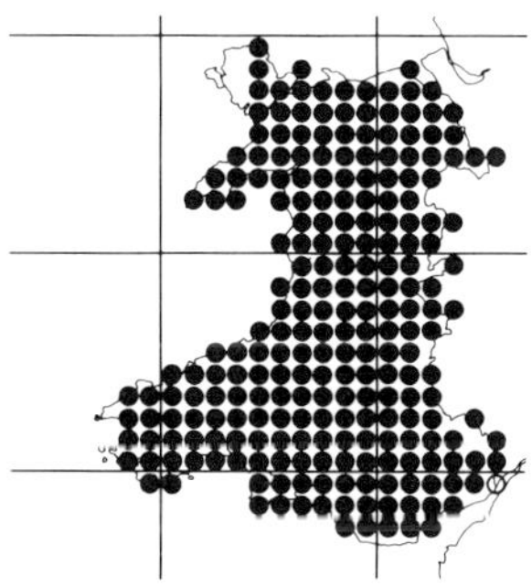

245 Chrysosplenium oppositifolium

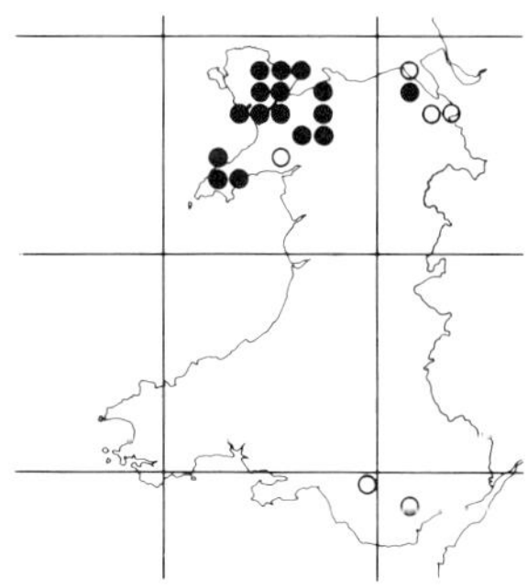

246 Parnassia palustris

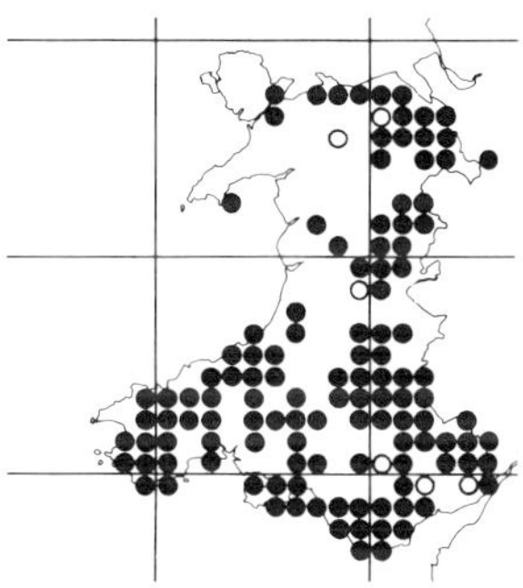

247 Ribes rubrum

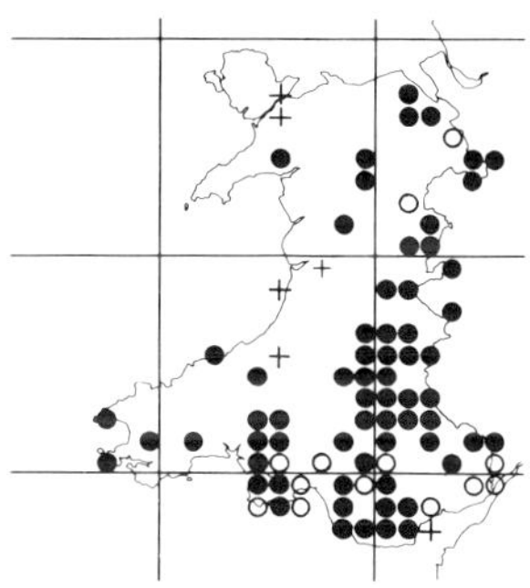

248 Ribes nigrum

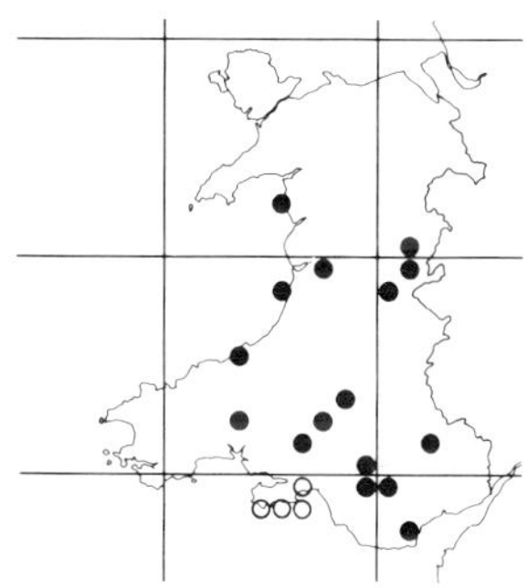

249 Ribes sanguineum

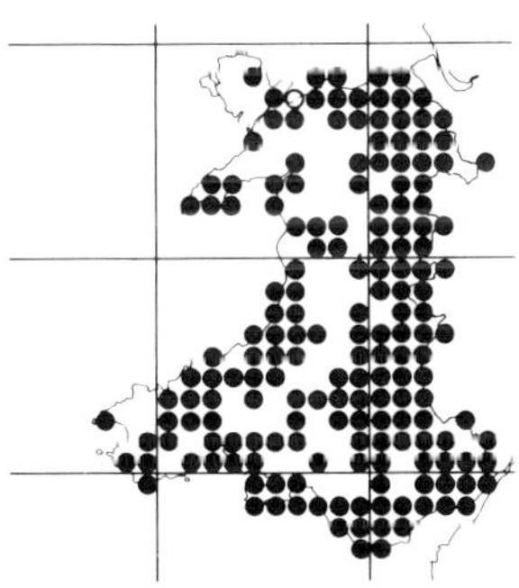

250 Ribes uva-crispa

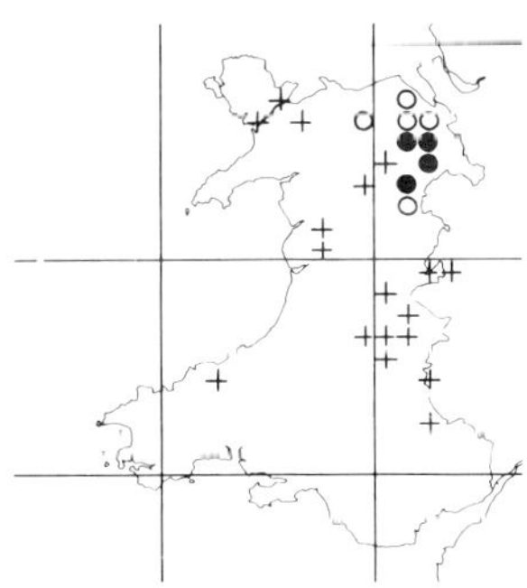

251 Ribes alpinum

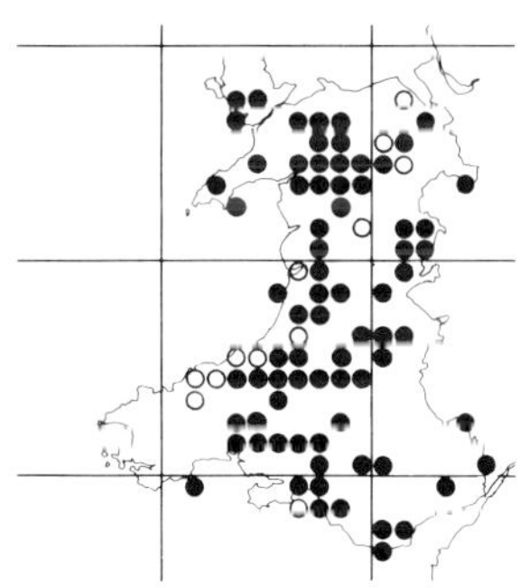

252 Spiraea salicifolia agg.

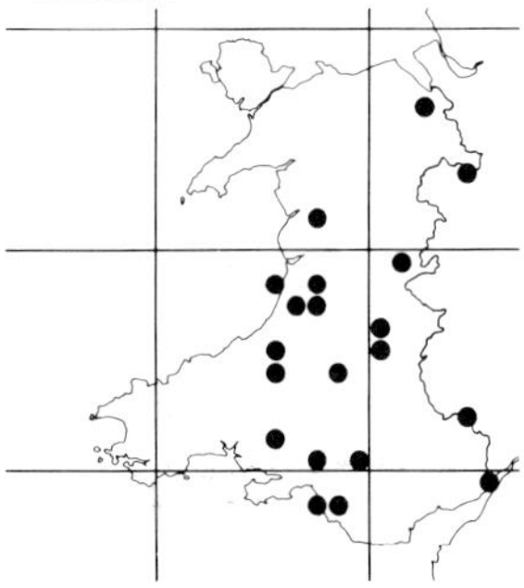

253 Spiraea douglasii × salicifolia = S. × billardii

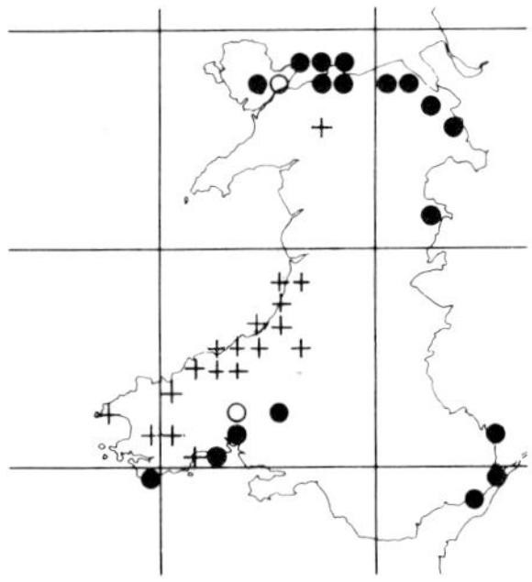

254 Filipendula vulgaris

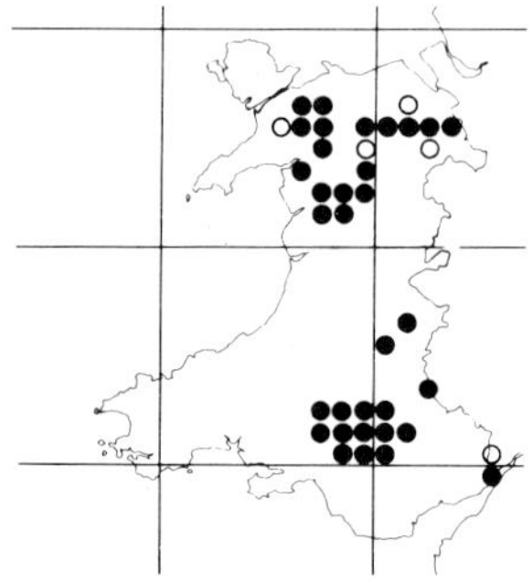

255 Rubus saxatilis

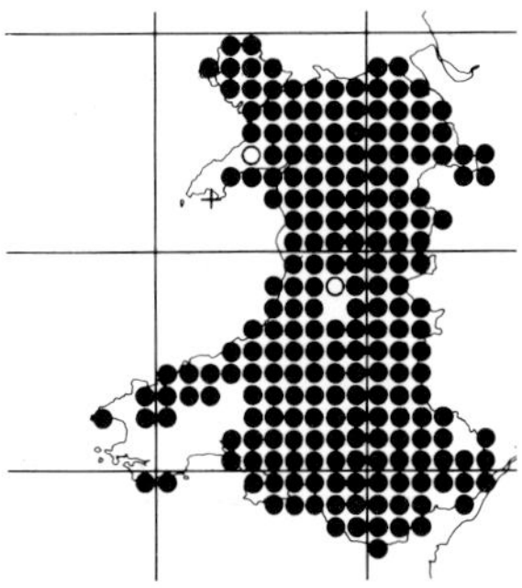

256 Rubus idaeus

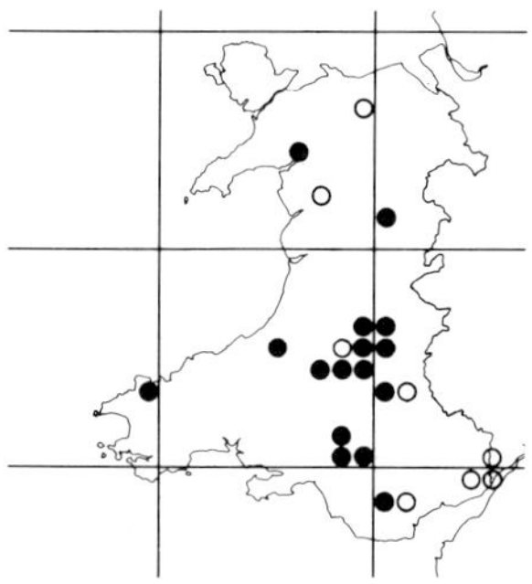

257 Rubus nessensis

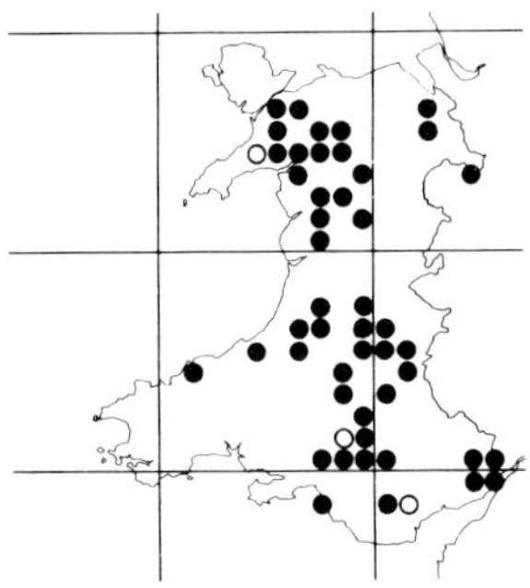

258 Rubus scissus

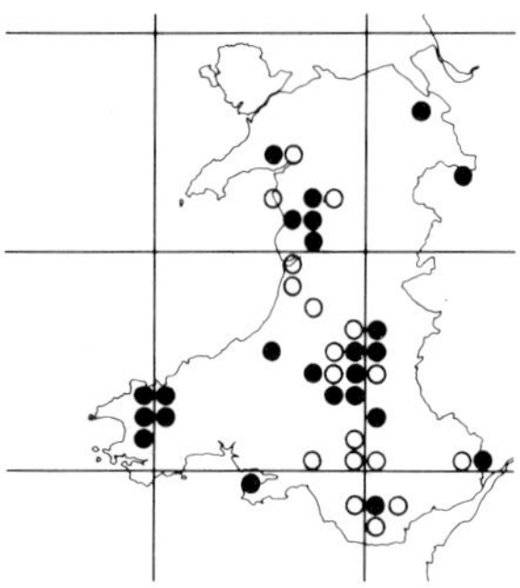

259 Rubus bertramii

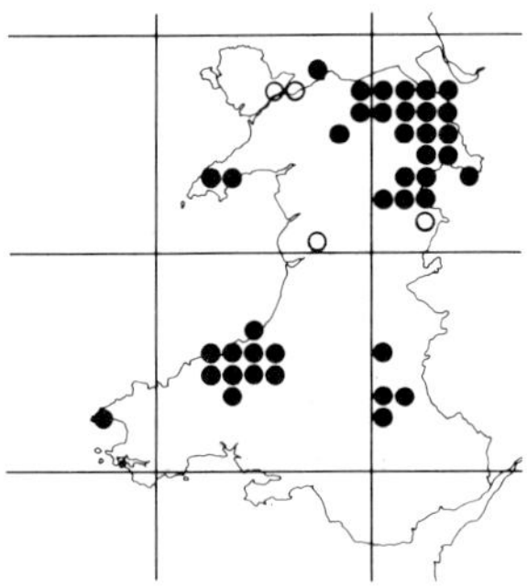

260 Rubus tuberculatus

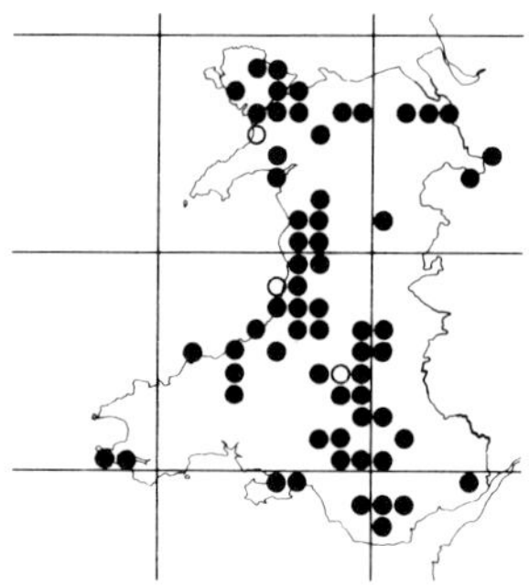

261 Rubus nemoralis

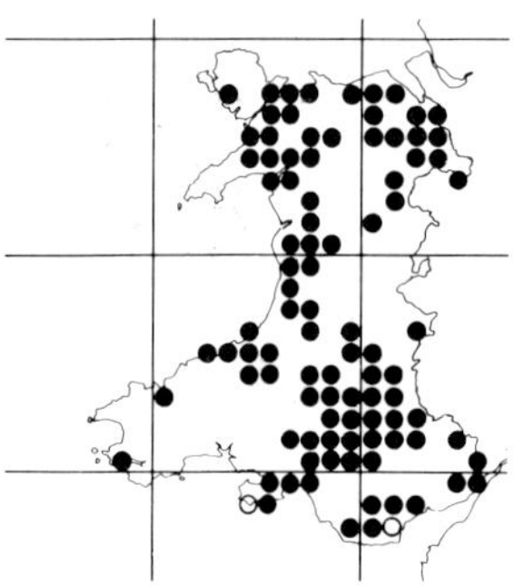

262 Rubus lindleianus

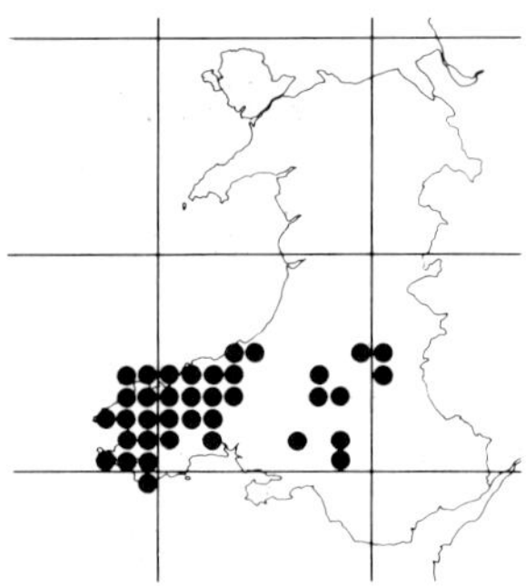

263 Rubus perdigitatus

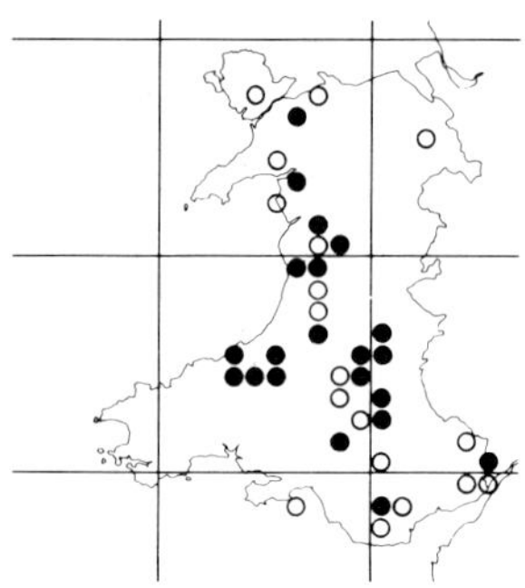

264 Rubus pyramidalis

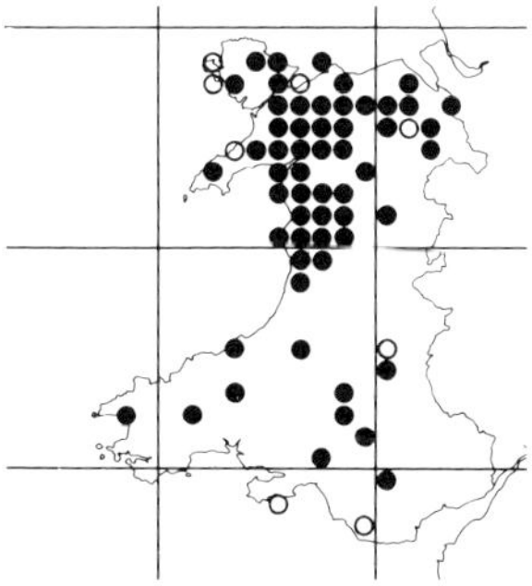

265 Rubus incurvatus

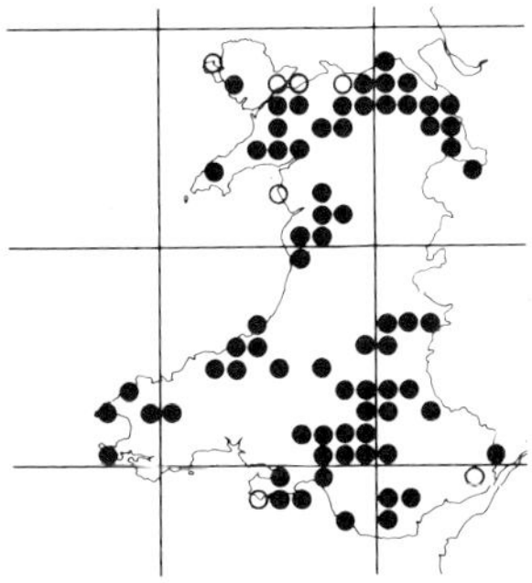

266 Rubus polyanthemus

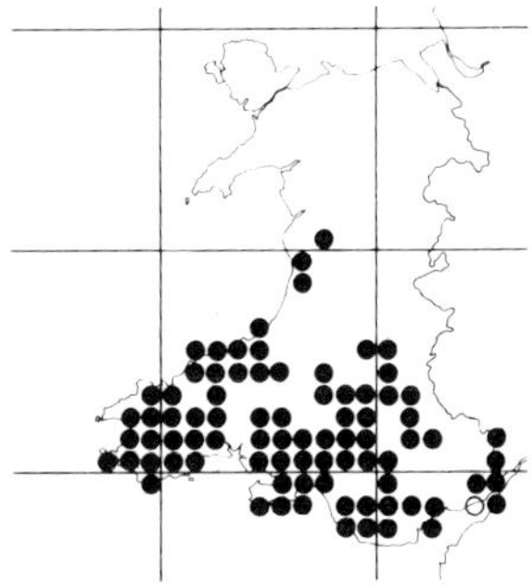

267 Rubus rubritinctus

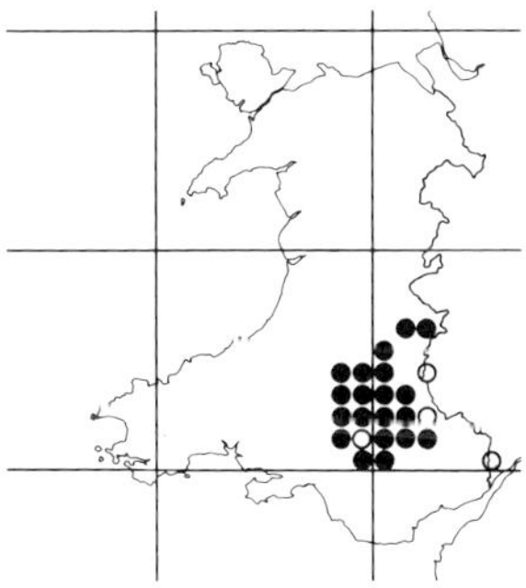

268 Rubus acclivitatum

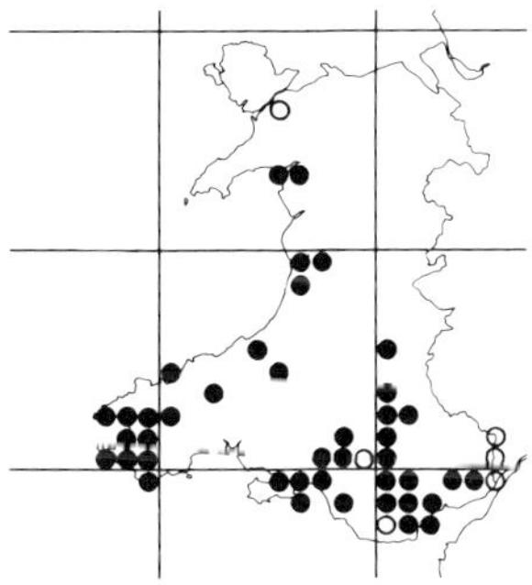

269 Rubus prolongatus

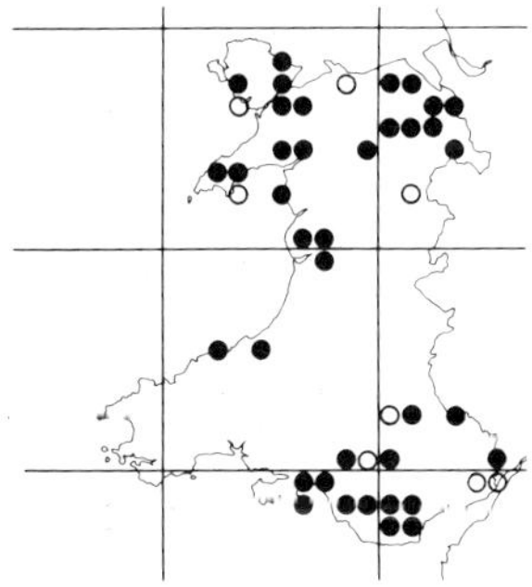

270 Rubus cardiophyllus

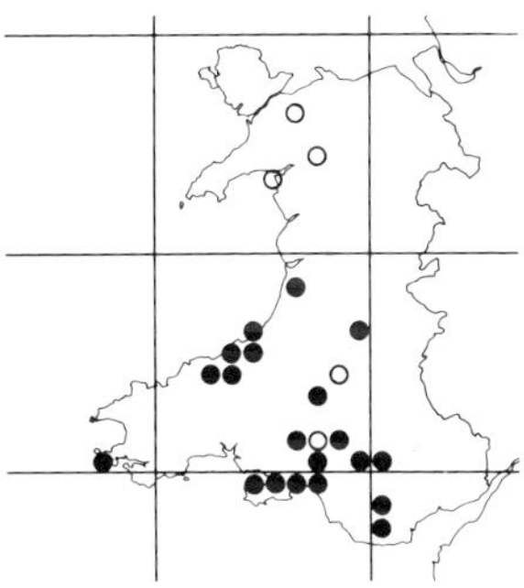

271 Rubus dumnoniensis

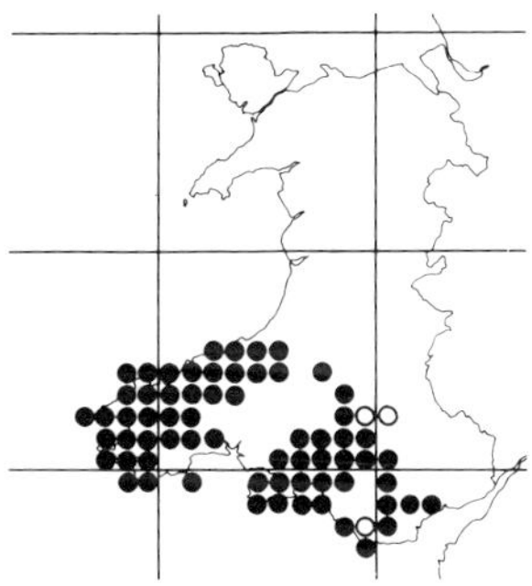

272 Rubus altiarcuatus

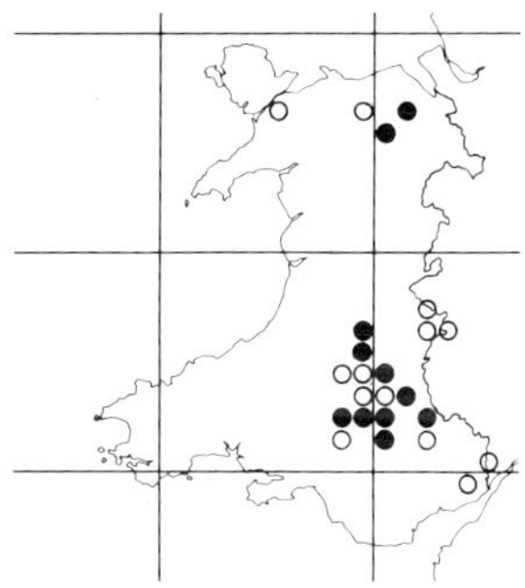

273 Rubus lindebergii

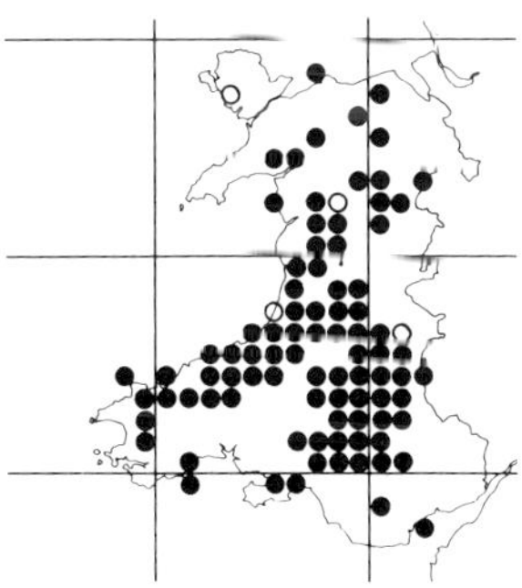

274 Rubus silurum

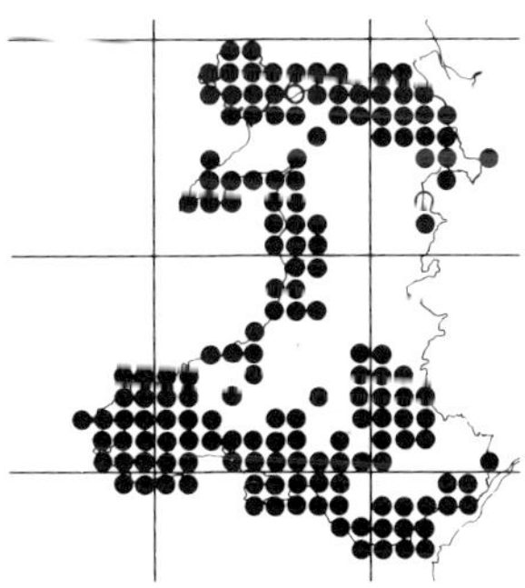

275 Rubus ulmifolius

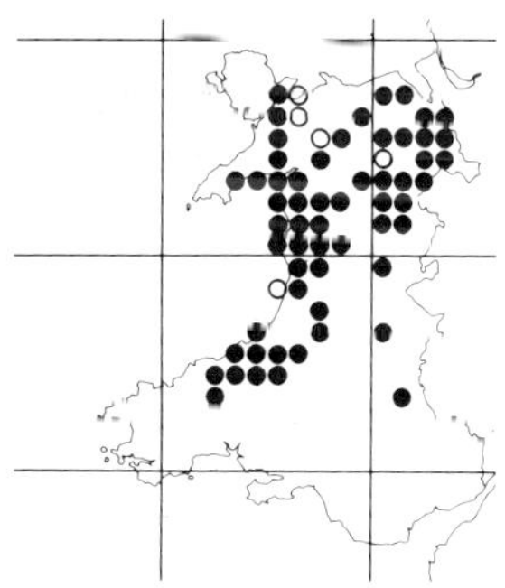

276 Rubus bartonii

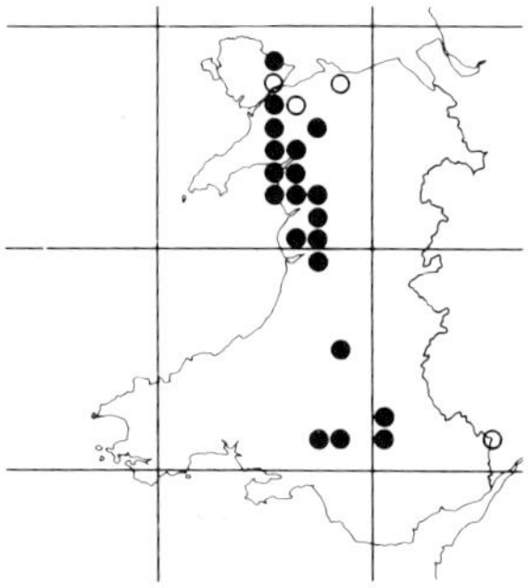

277 Rubus lanaticaulis

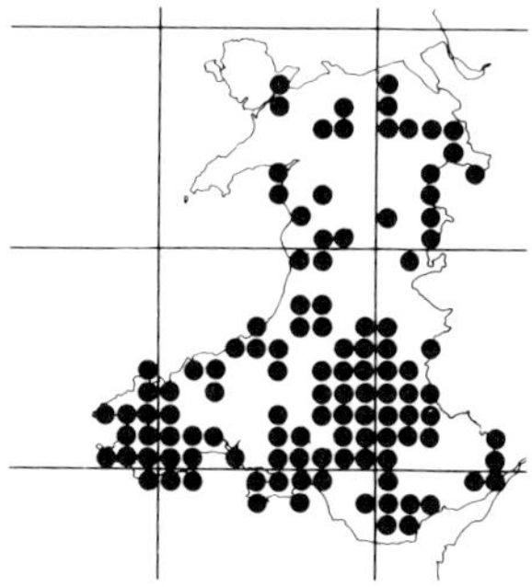

278 Rubus vestitus

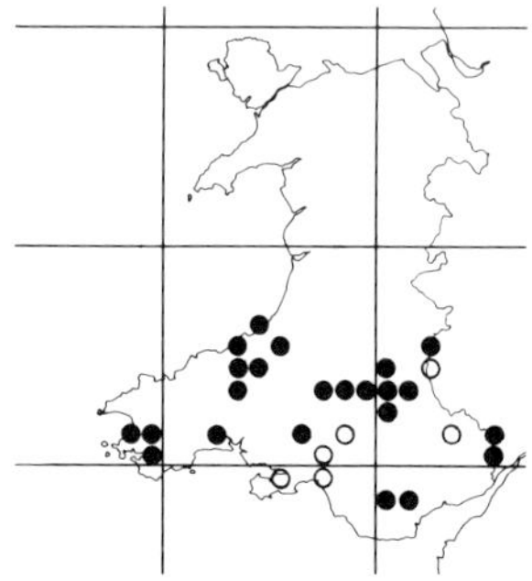

279 Rubus longus

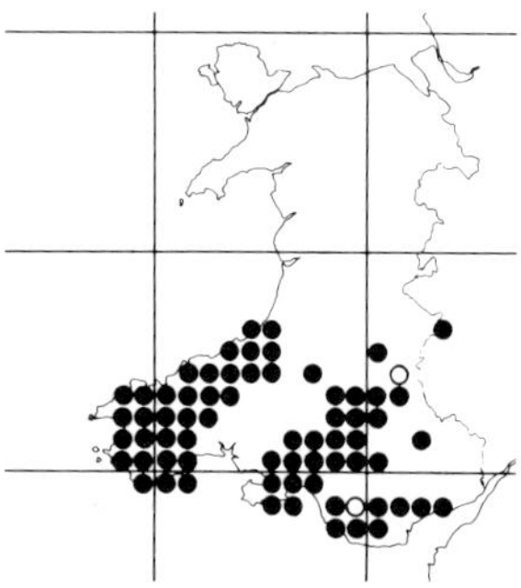

280 Rubus leyanus

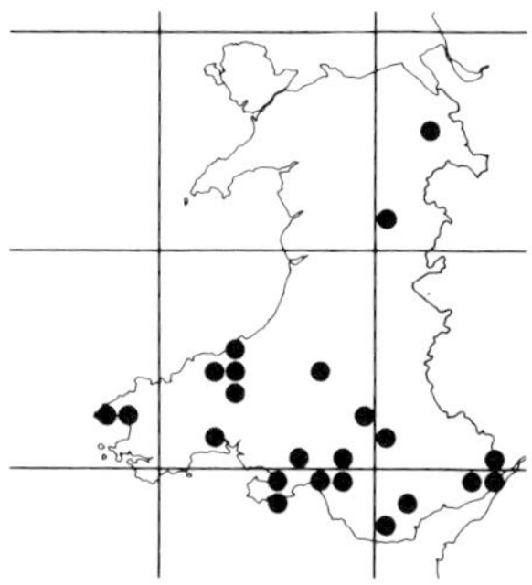

281 Rubus dentatifolius

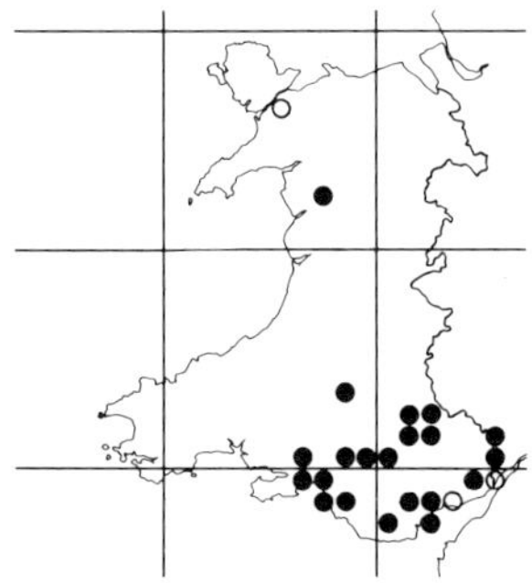

282 Rubus rufescens

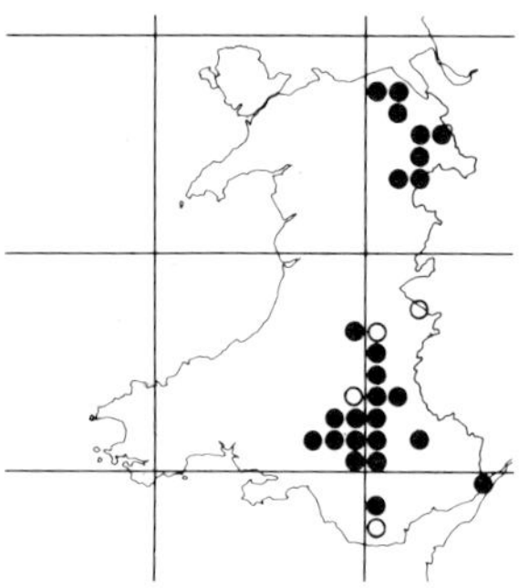

283 Rubus raduloides

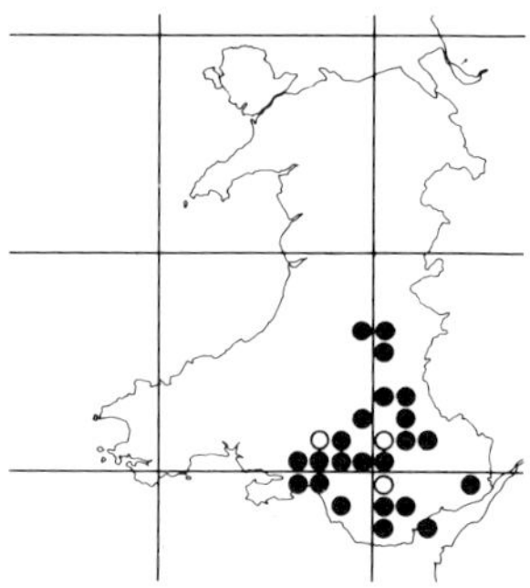

284 Rubus moylei

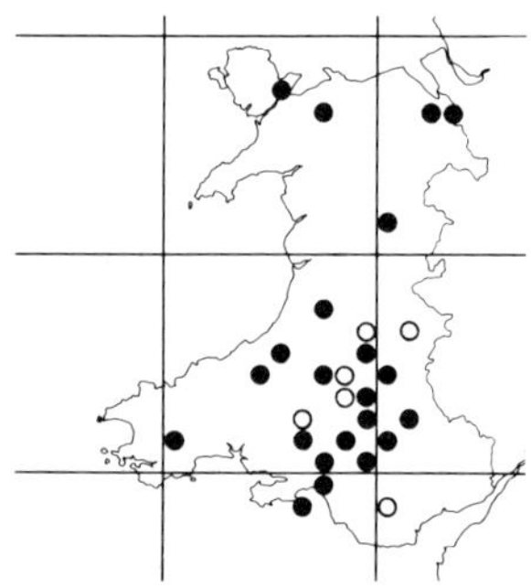

285 Rubus hylocharis

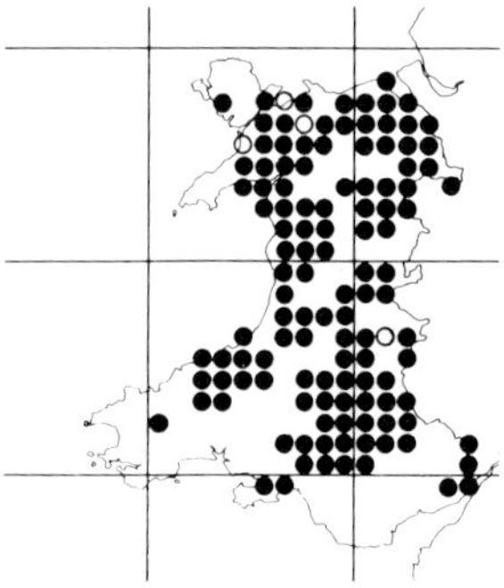

286 Rubus dasyphyllus

287 Rubus caesius

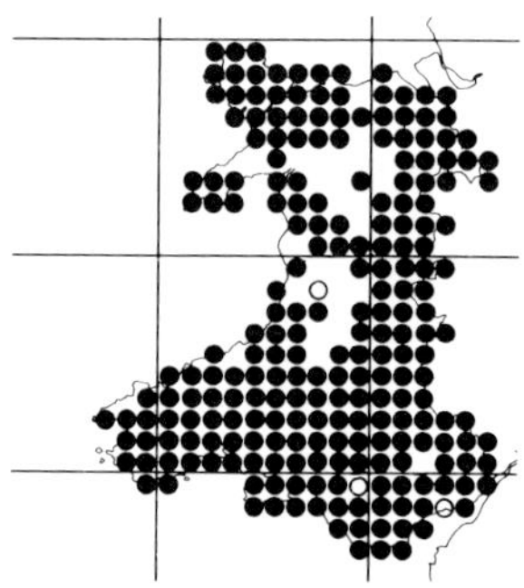

288 Rosa arvensis

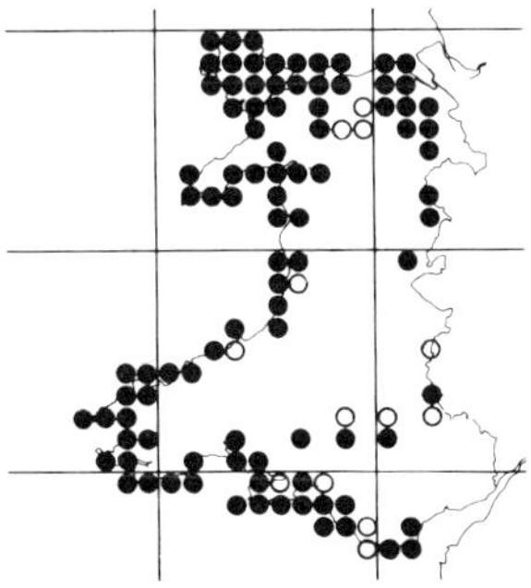

289 Rosa pimpinellifolia

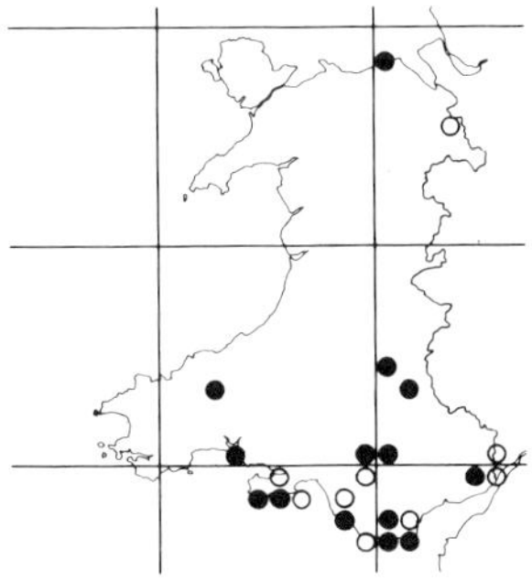

290 Rosa stylosa

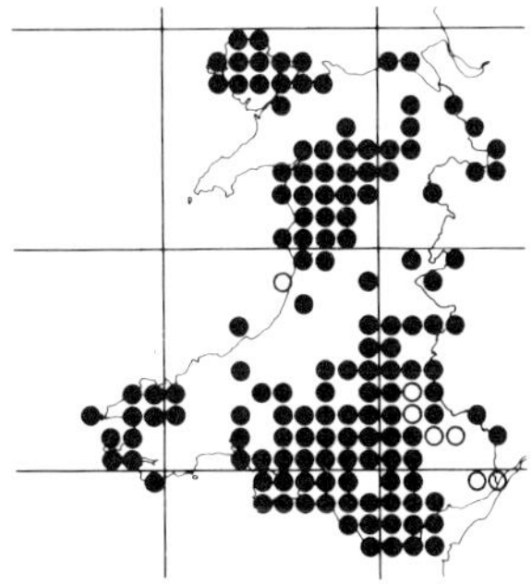

291 Rosa canina s.s.

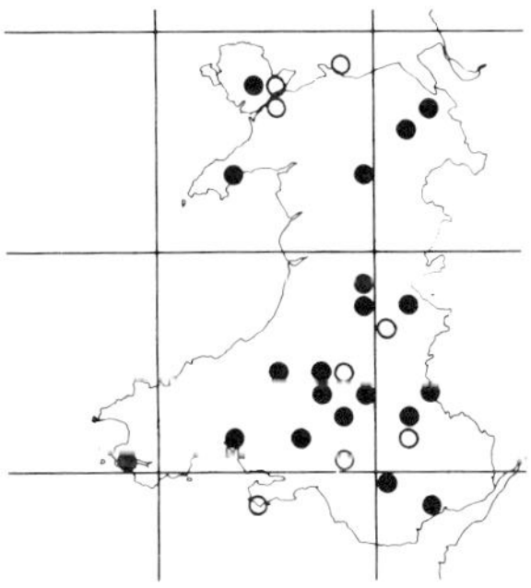

292 Rosa afzeliana

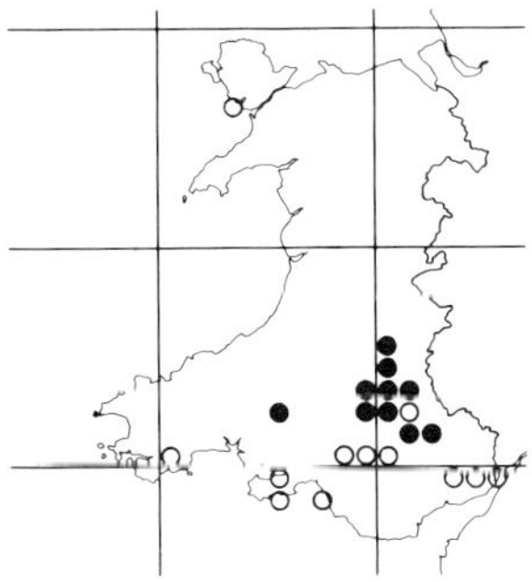

293 Rosa obtusifolia

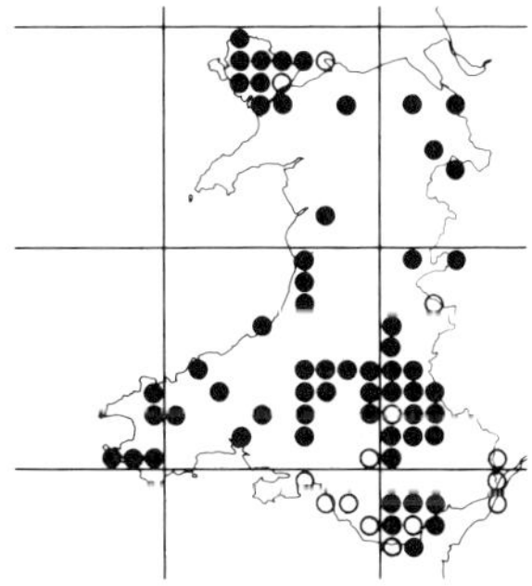

294 Rosa dumetorum

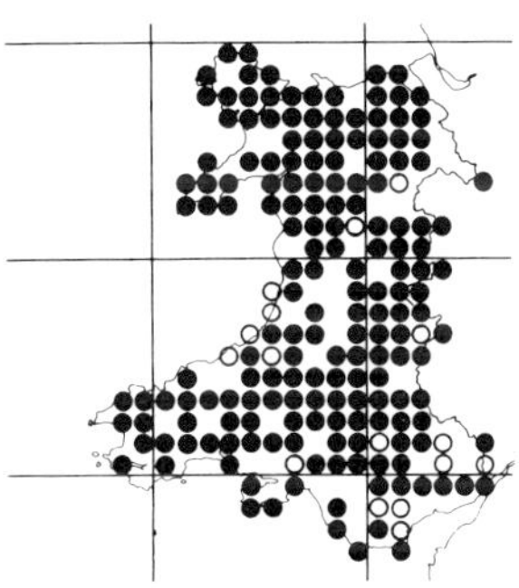

295 Rosa tomentosa agg.

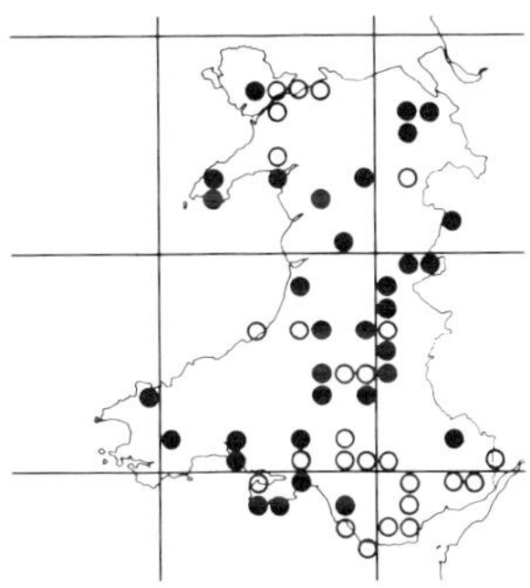

296 Rosa tomentosa s.s.

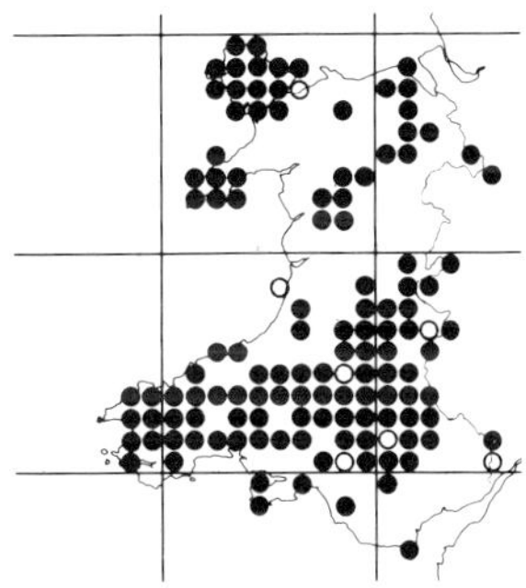

297 Rosa sherardii

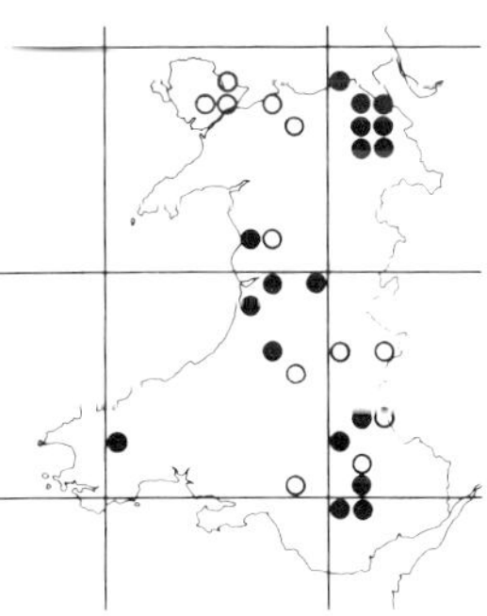

298 Rosa villosa

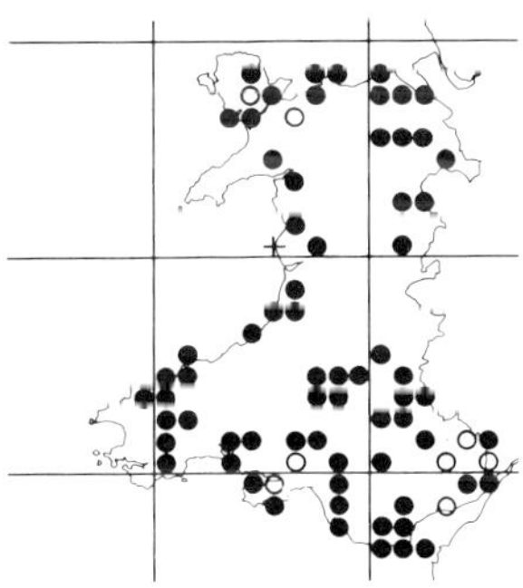

299 Rosa rubiginosa agg.

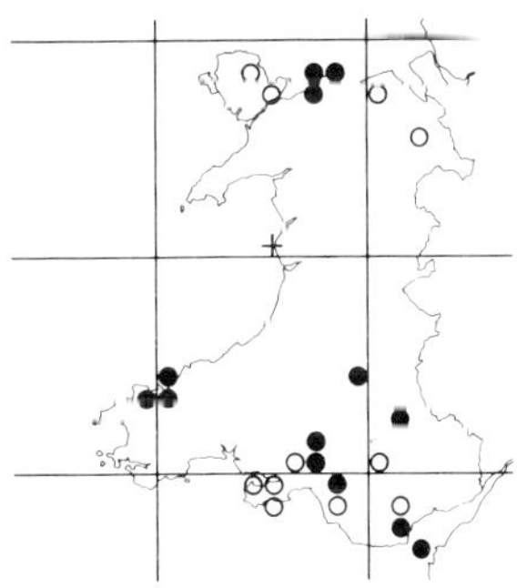

300 Rosa rubiginosa s.s.

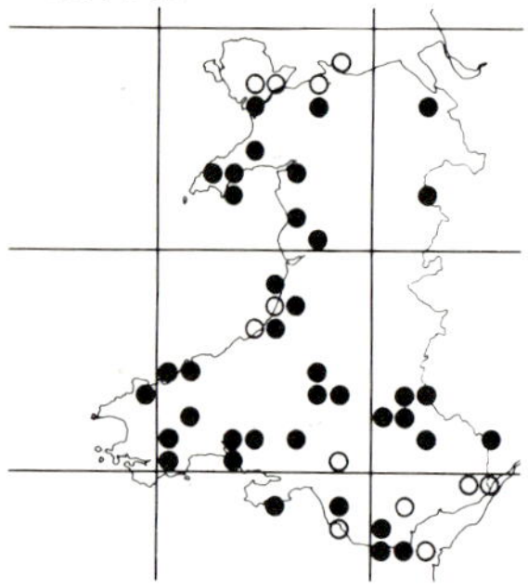

301 Rosa micrantha

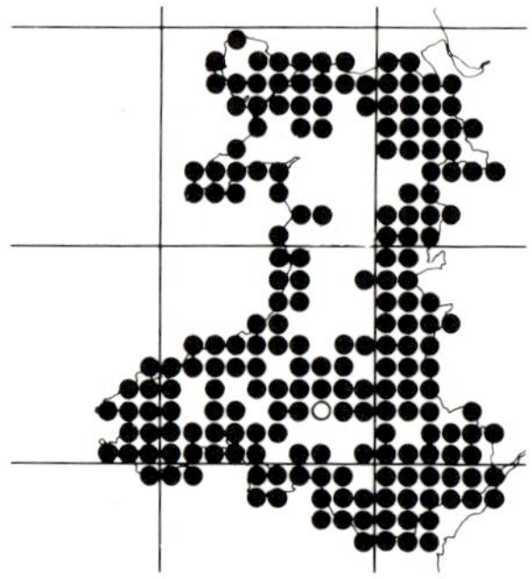

302 Agrimonia eupatoria

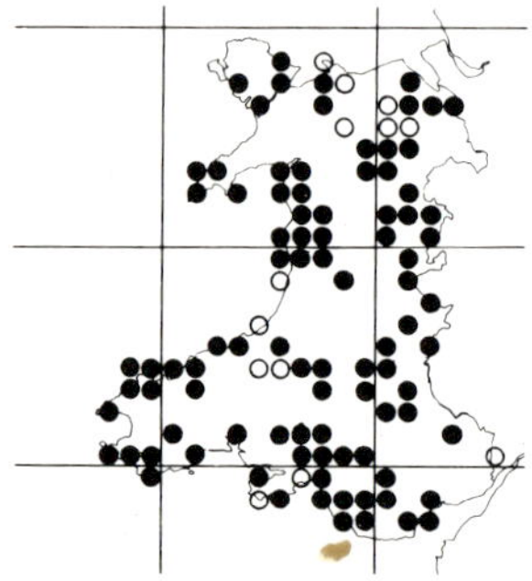

303 Agrimonia procera

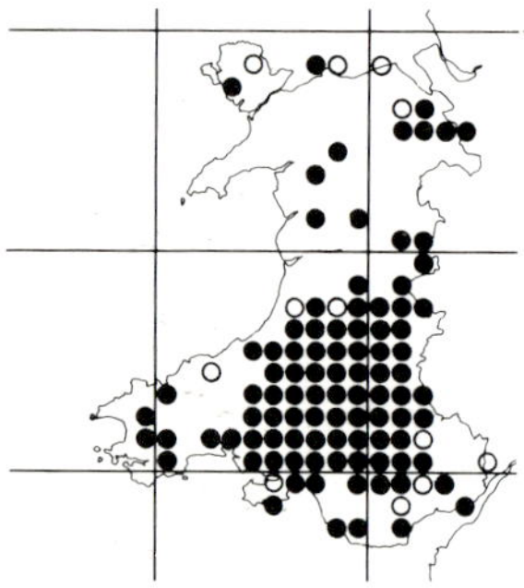

304 Sanguisorba officinalis

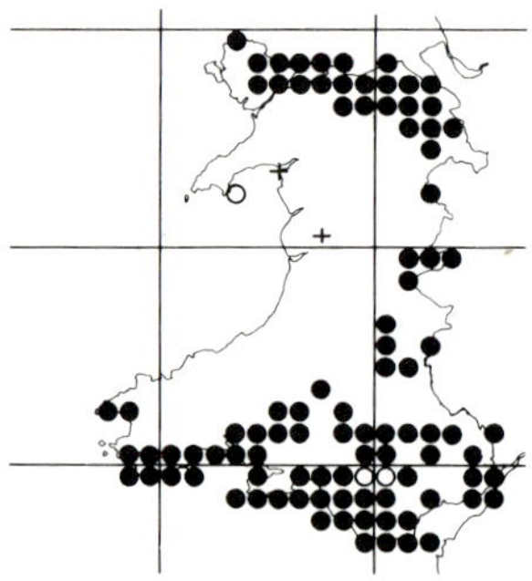

305 Sanguisorba minor subsp. minor

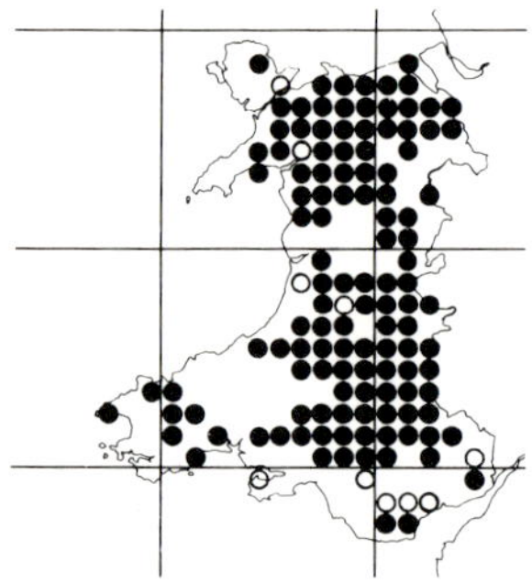

306 Geum rivale

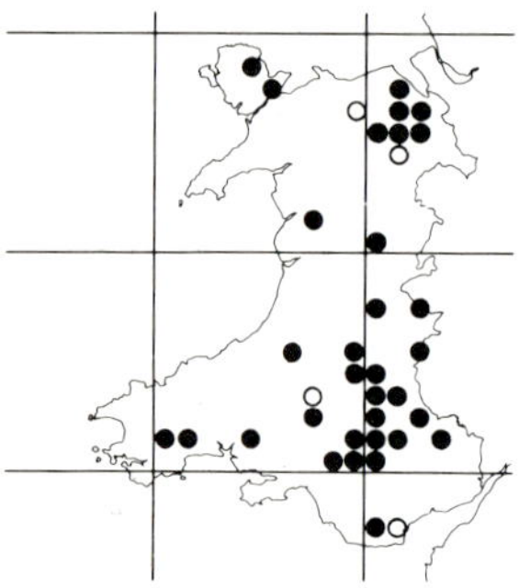

307 Geum rivale × urbanum = G. × intermedium

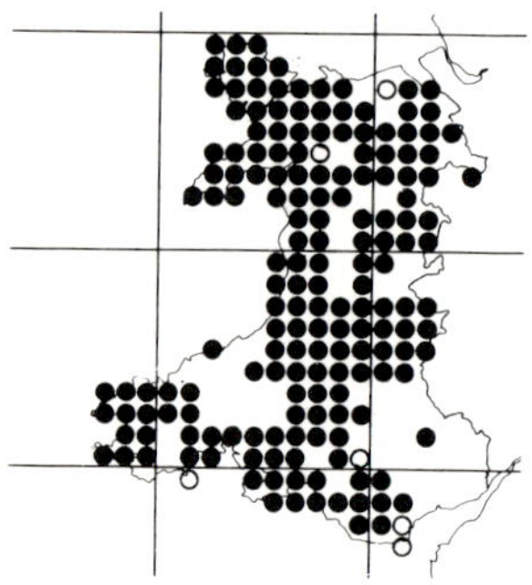

308 Potentilla palustris

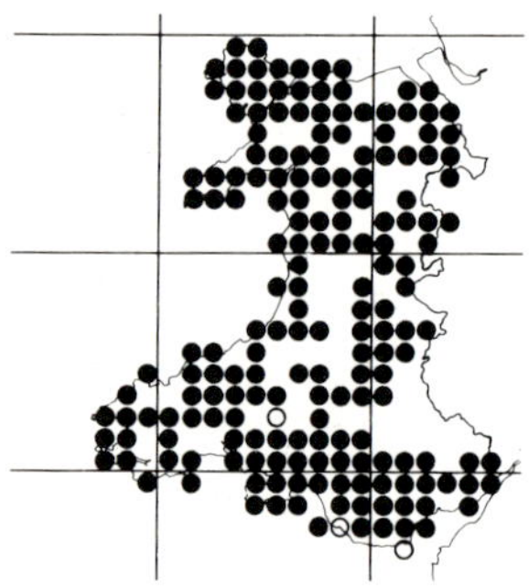

309 Potentilla anglica

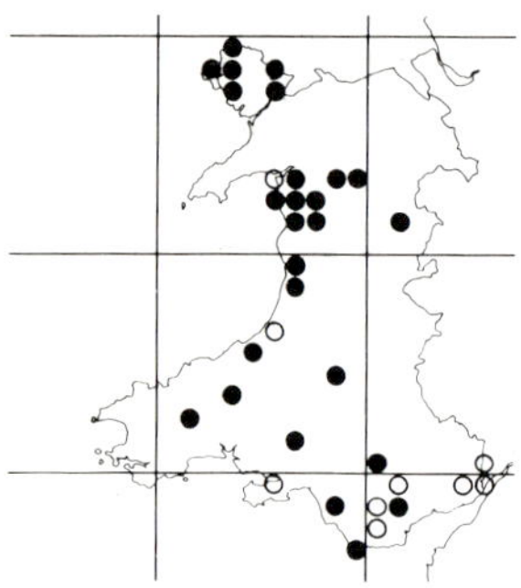

310 Potentilla anglica × erecta = P. × suberecta

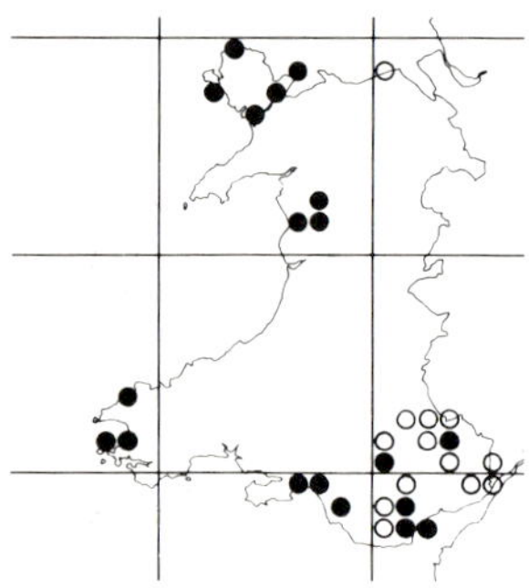

311 Potentilla anglica × reptans = P. × mixta

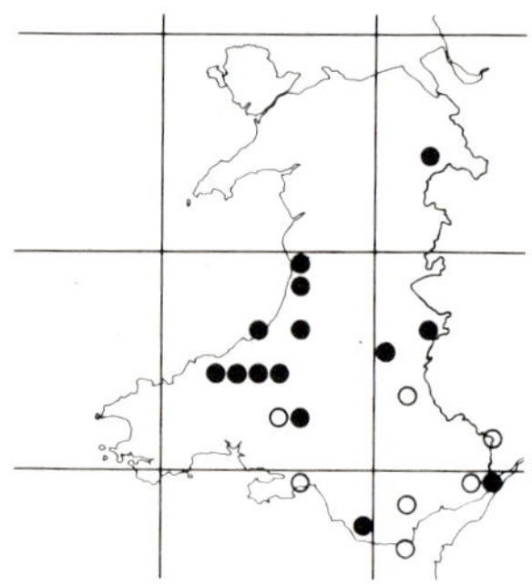

312 Fragaria moschata

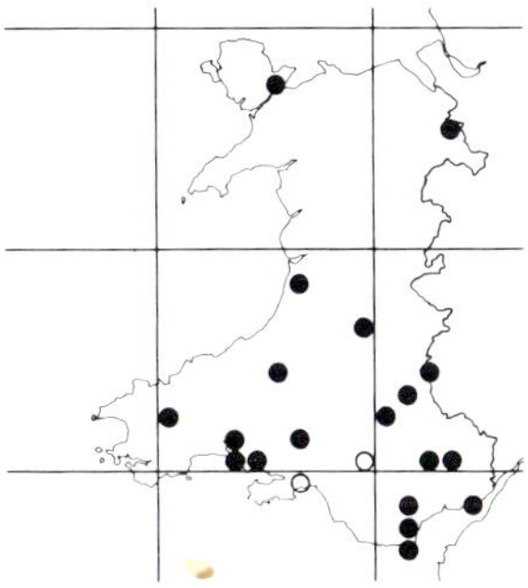

313 Fragaria chiloensis
× virginiana
= F. × ananassa

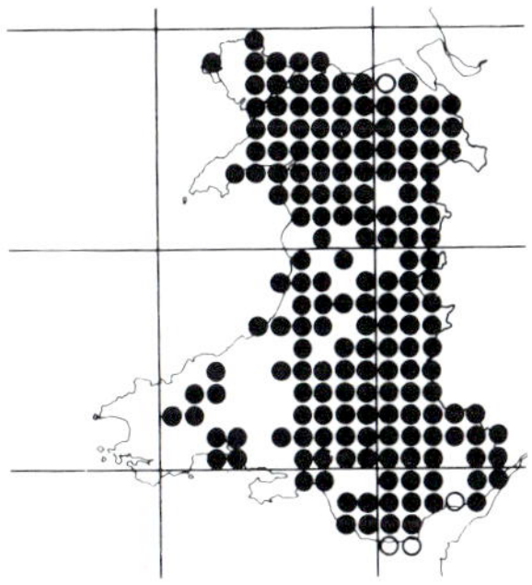

314 Alchemilla vulgaris agg.

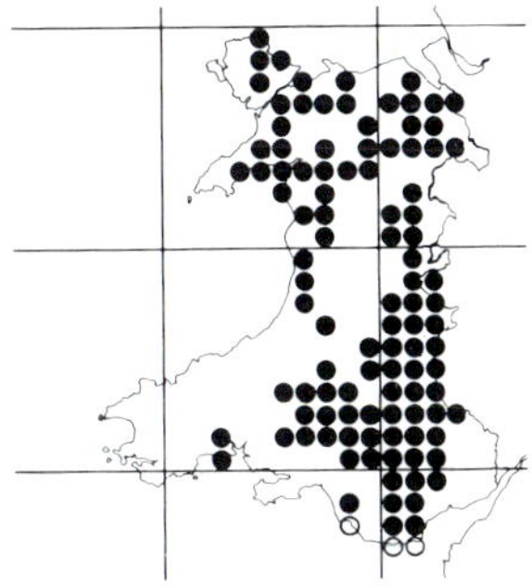

315 Alchemilla xanthochlora

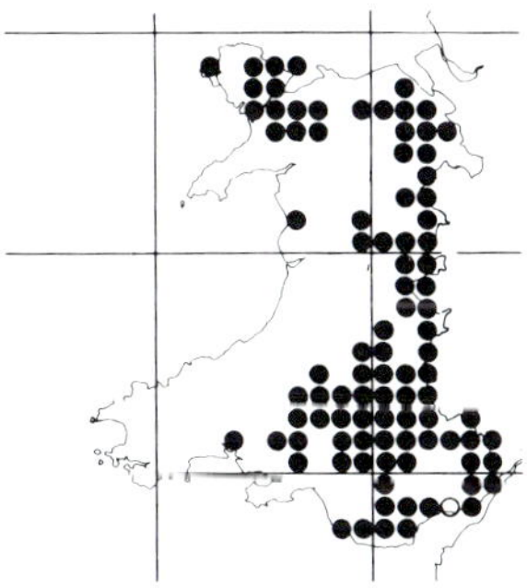

316 Alchemilla filicaulis
subsp. vestita

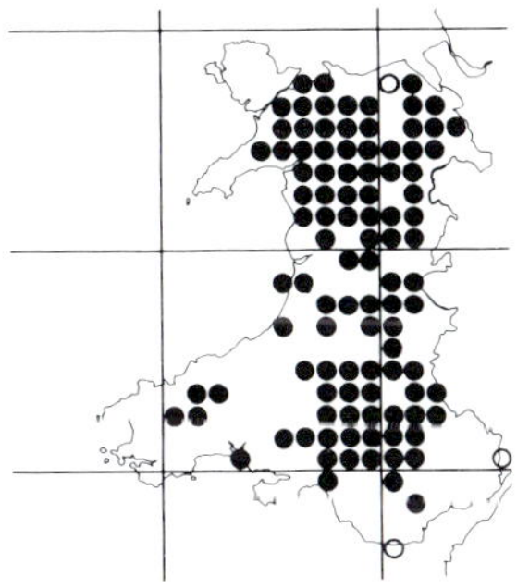

317 Alchemilla glabra

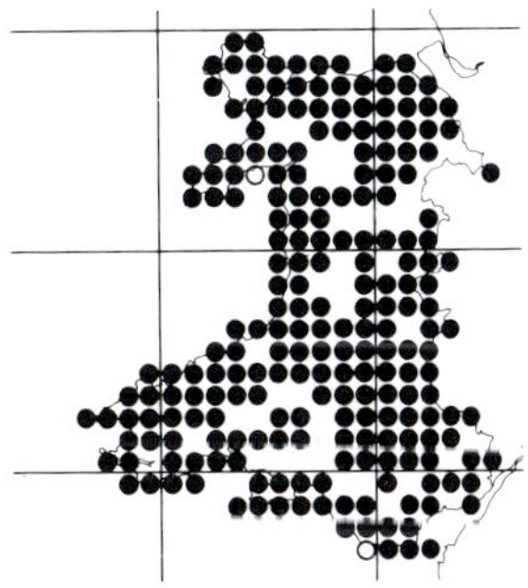

318 Aphanes arvensis agg.

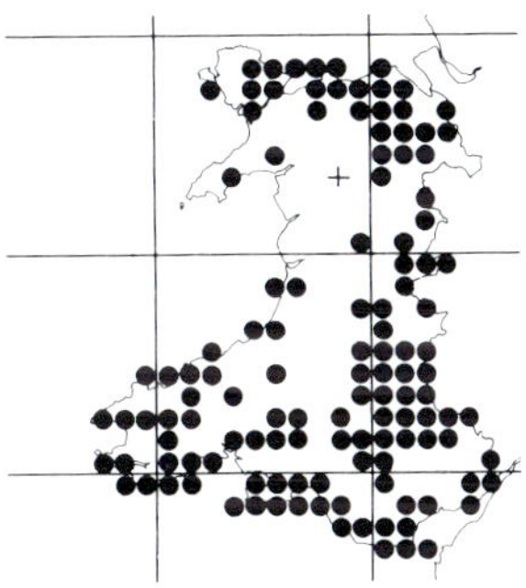

319 Aphanes arvensis s.s.

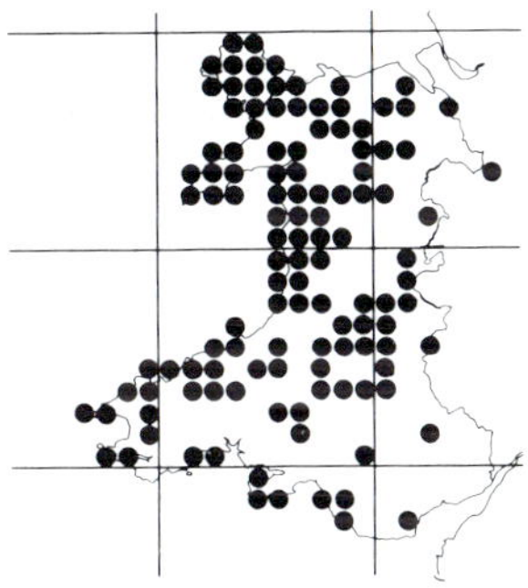

320 Aphanes microcarpa

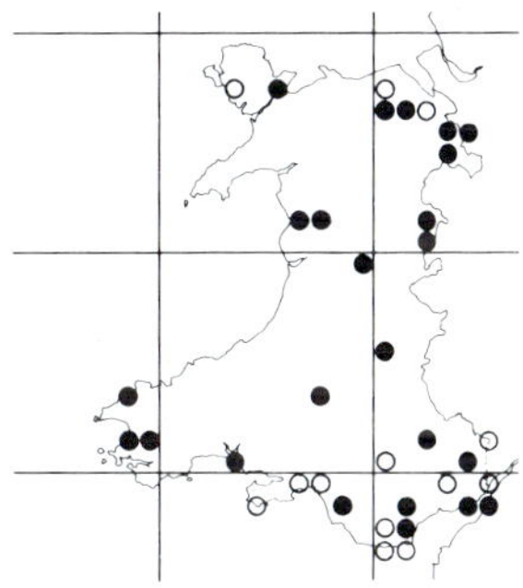

321 Pyrus pyraster

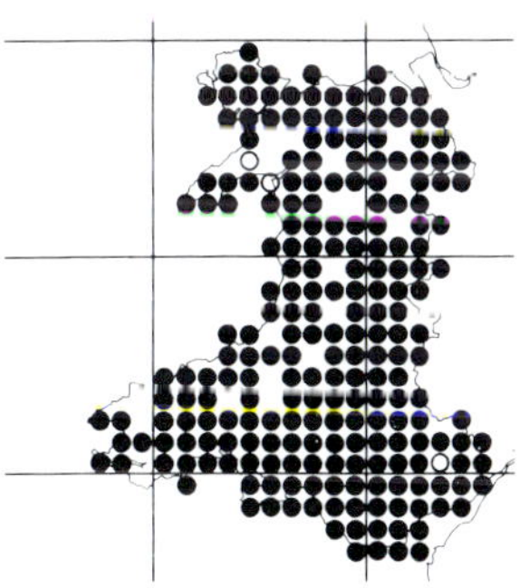

322 Malus sylvestris agg.

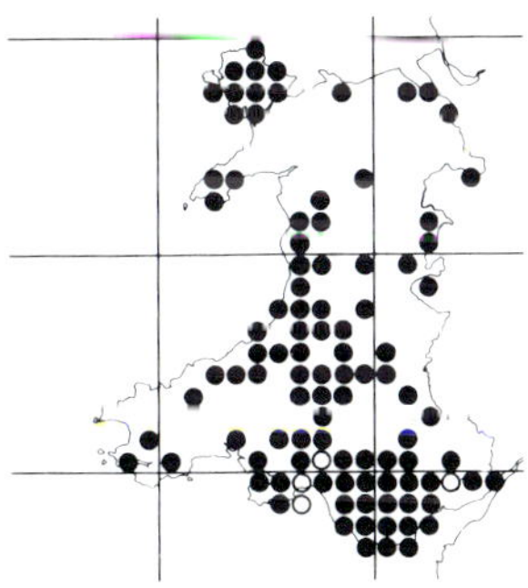

323 Malus sylvestris s.s.

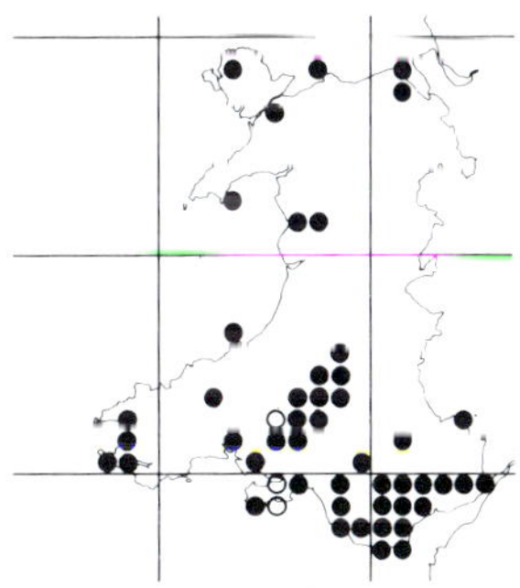

324 Malus domestica

325 Sorbus aucuparia

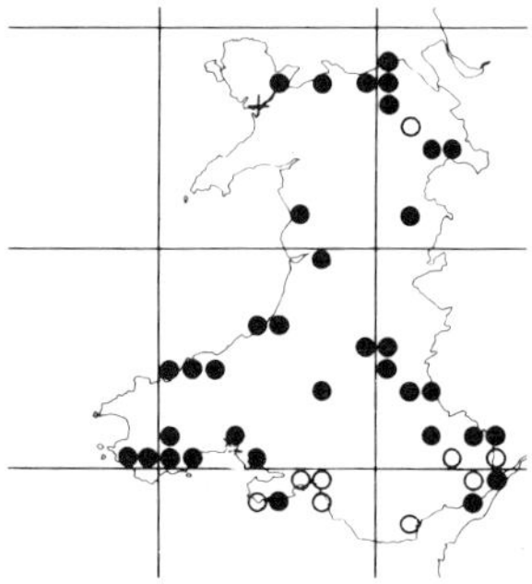

326 Sorbus torminalis

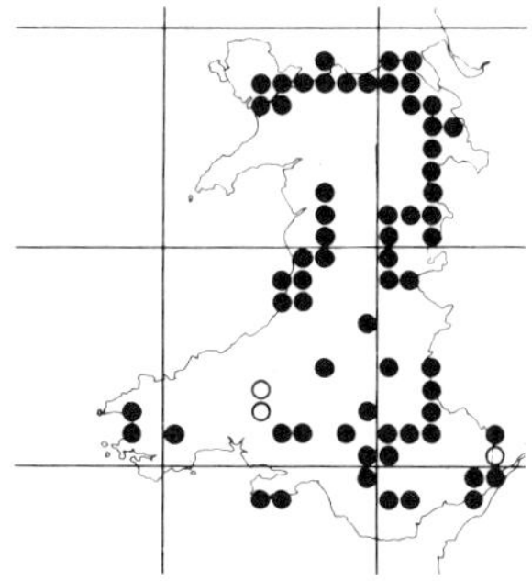

327 Sorbus aria agg.

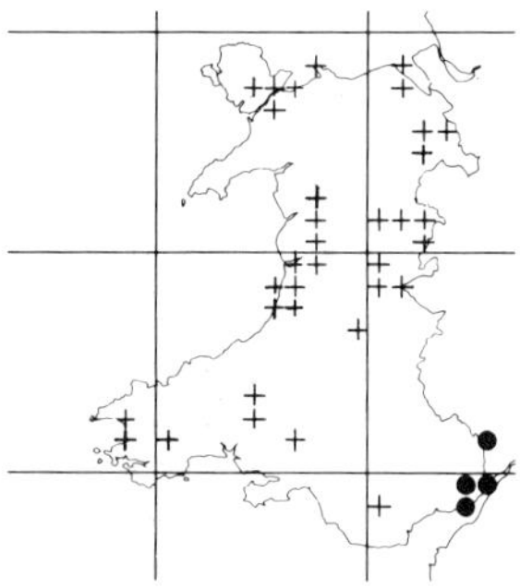

328 Sorbus aria s.s.

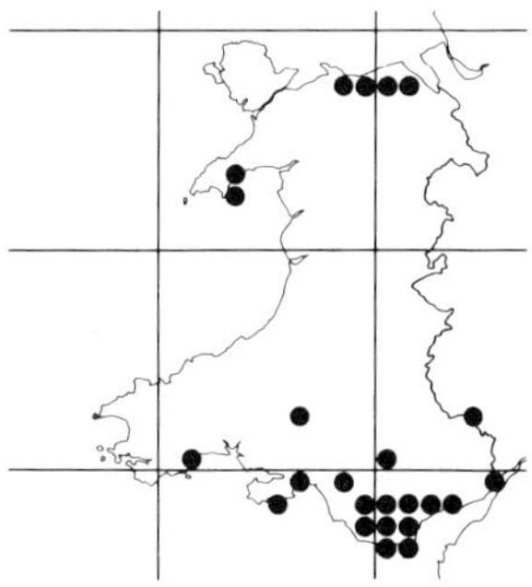

329 Cotoneaster horizontalis

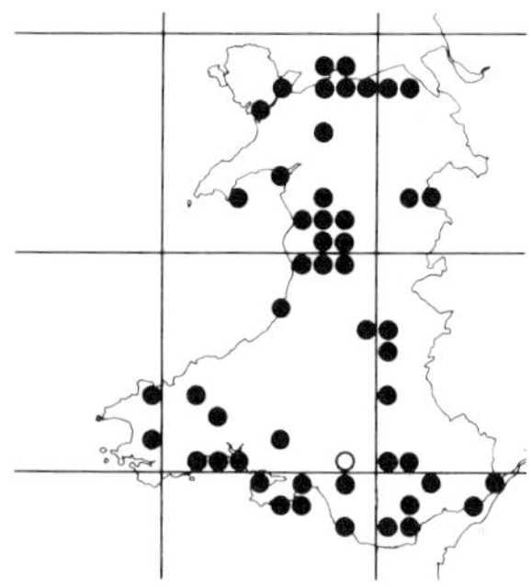

330 Cotoneaster simonsii

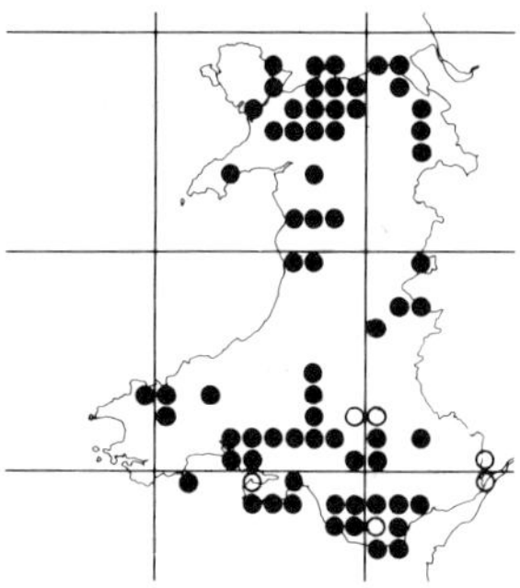

331 Cotoneaster microphyllus

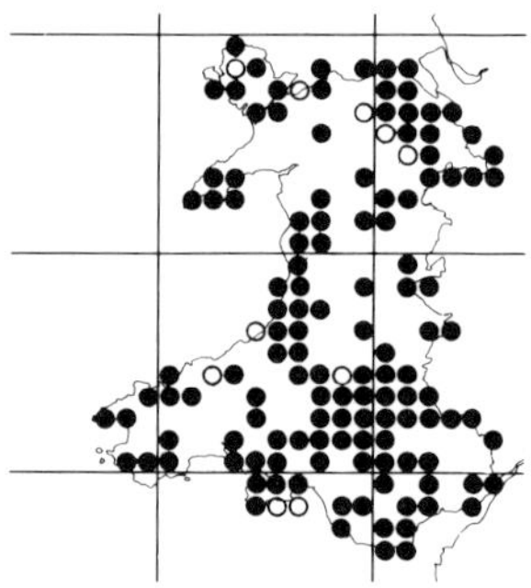

332 Prunus domestica

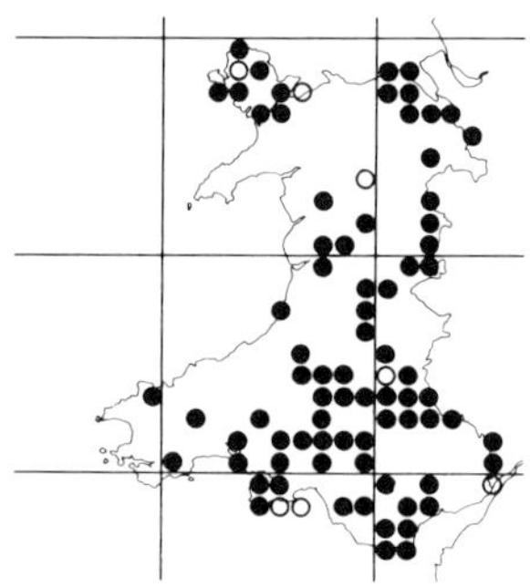

333 Prunus domestica subsp. domestica

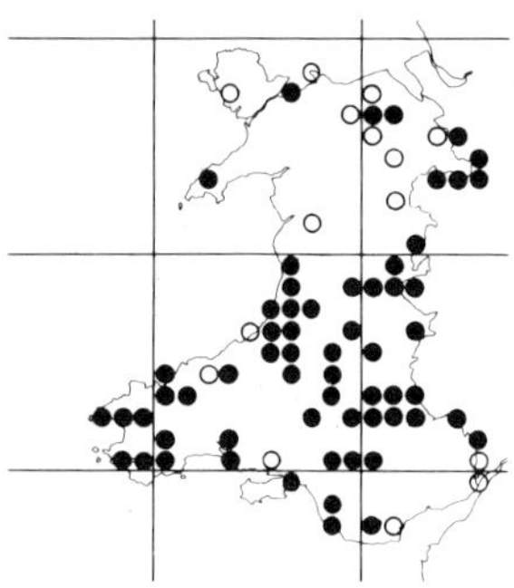

334 Prunus domestica subsp. insititia

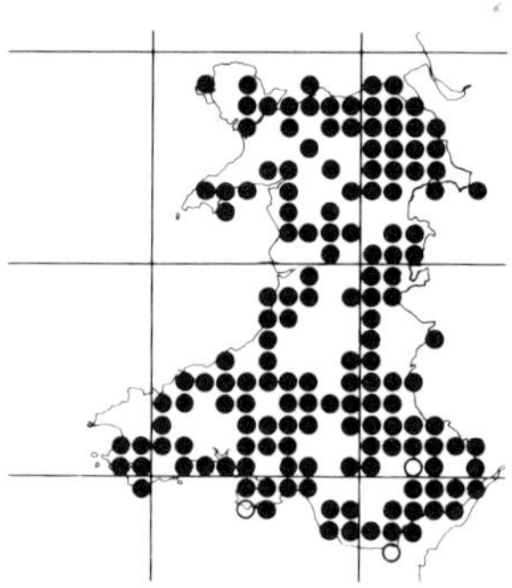

335 Prunus avium

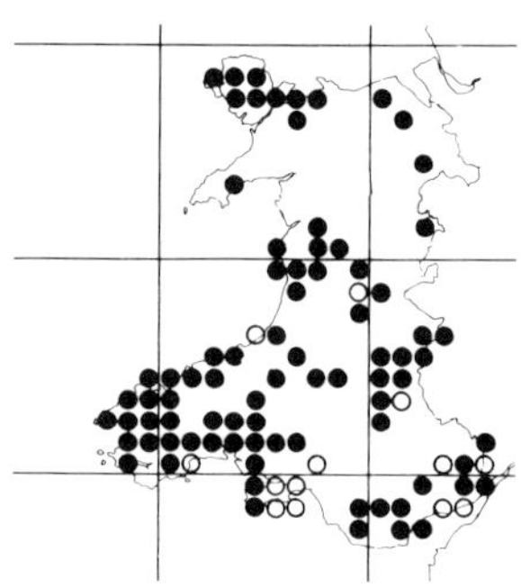

336 Prunus cerasus

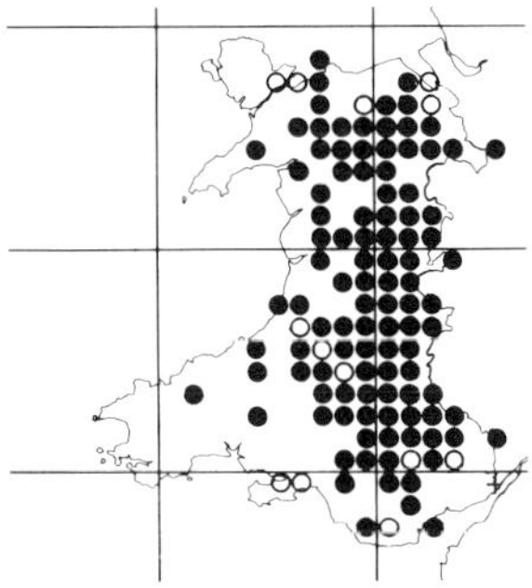

337 Prunus padus

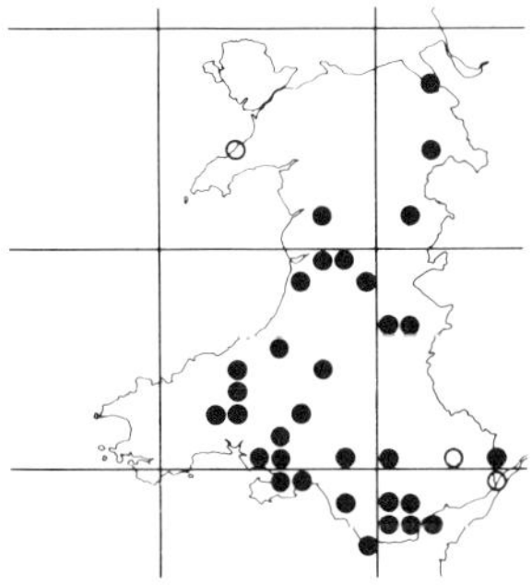

338 Prunus laurocerasus

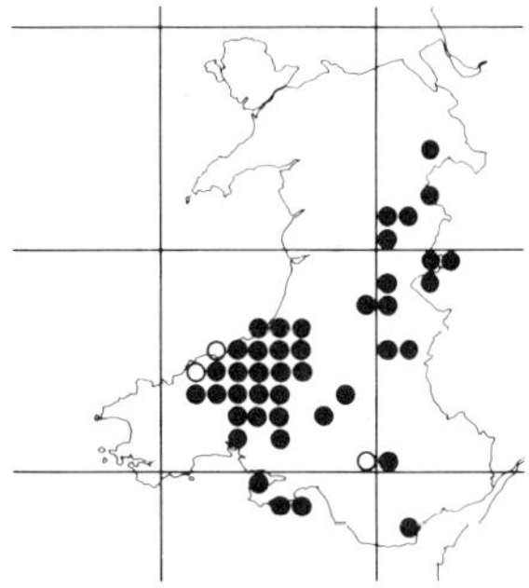

339 Laburnum anagyroides

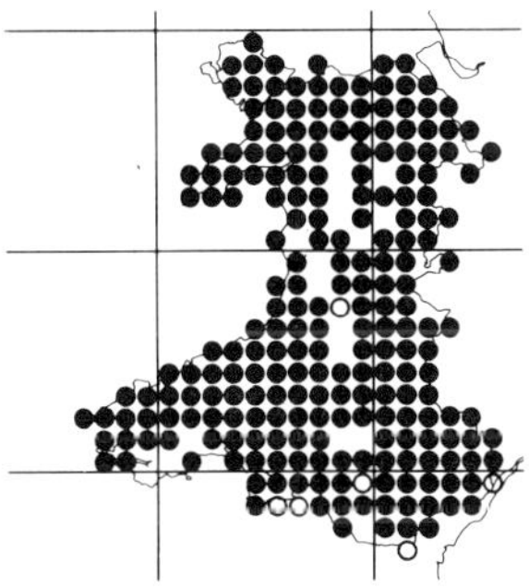

340 Cytisus scoparius

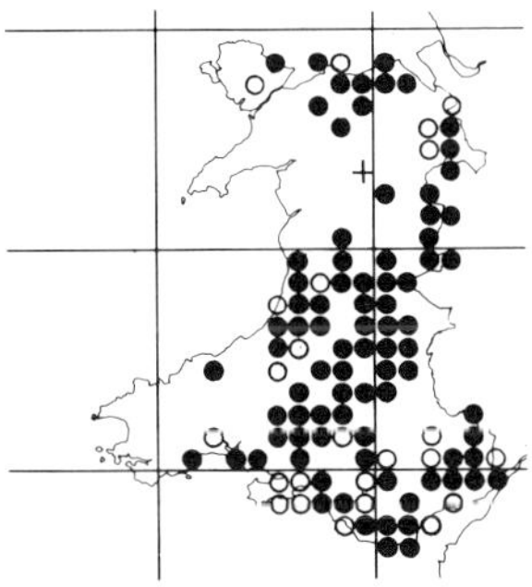

341 Genista tinctoria subsp. tinctoria

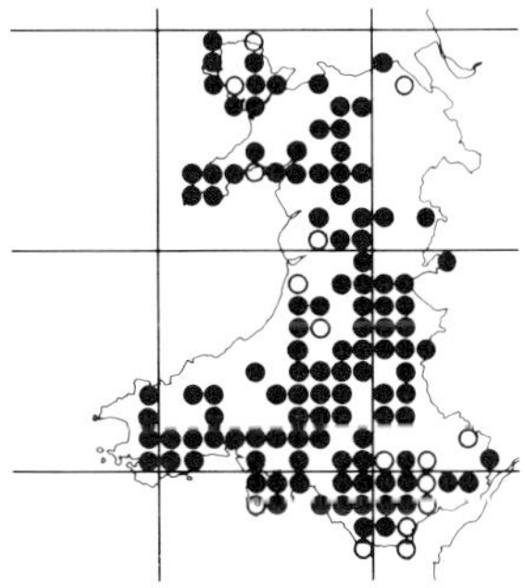

342 Genista anglica

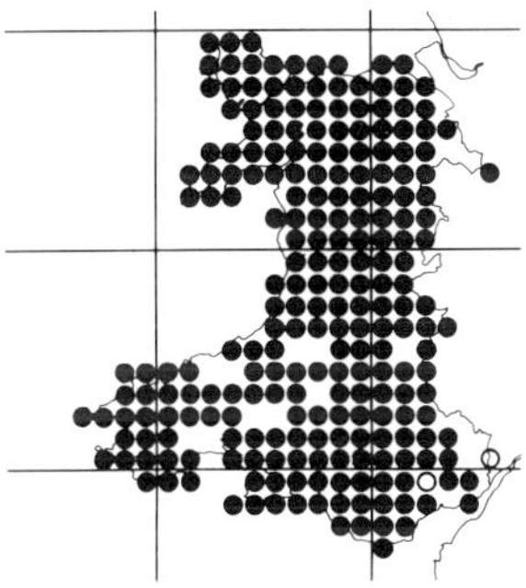

343 Ulex gallii

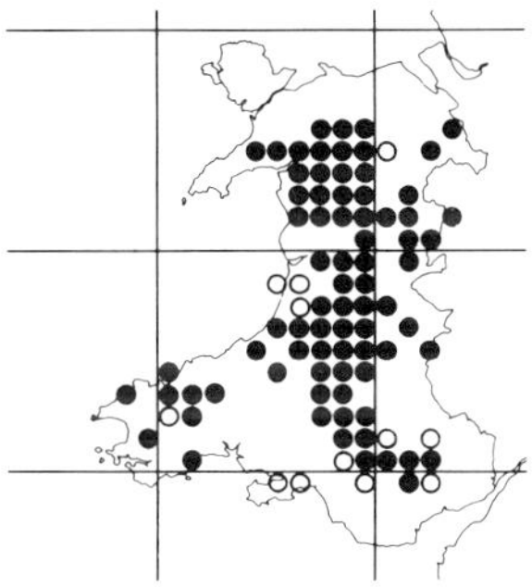

344 Vicia orobus

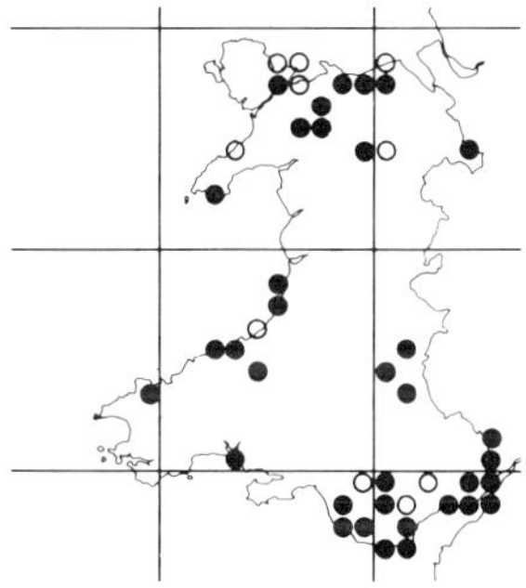

345 Vicia sylvatica

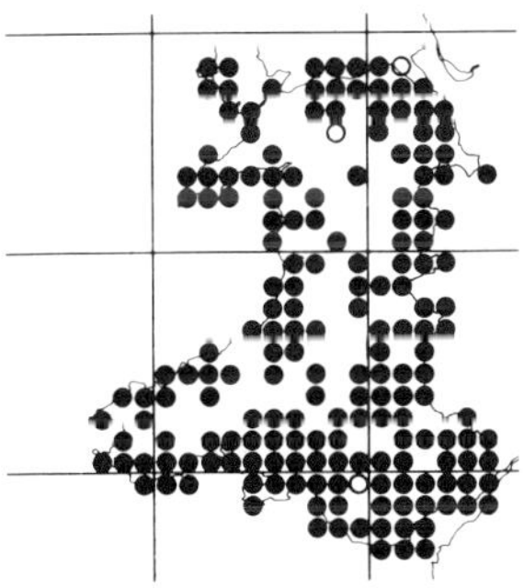

346 Vicia hirsuta

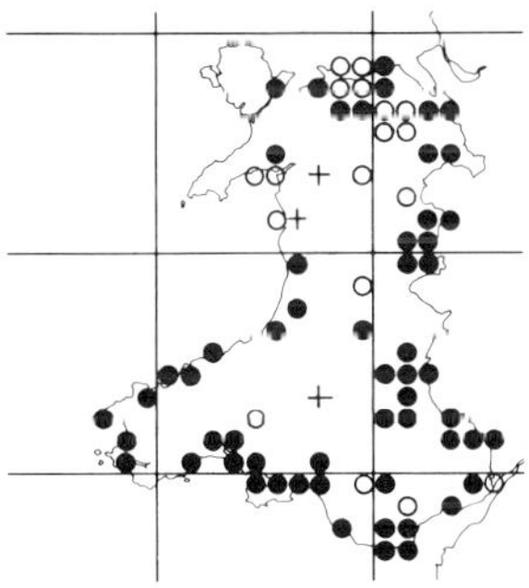

347 Vicia tetrasperma

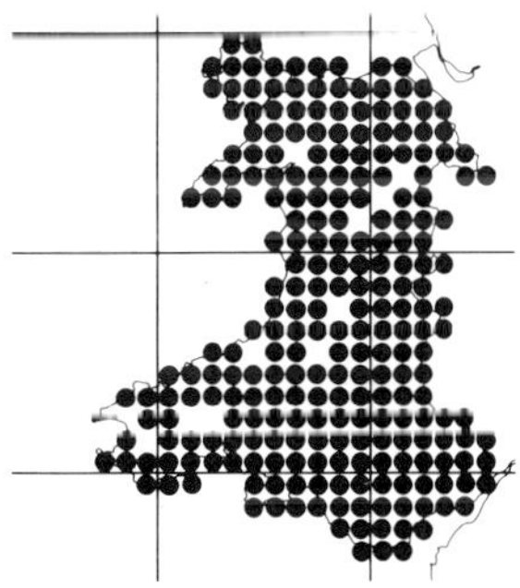

348 Vicia sepium

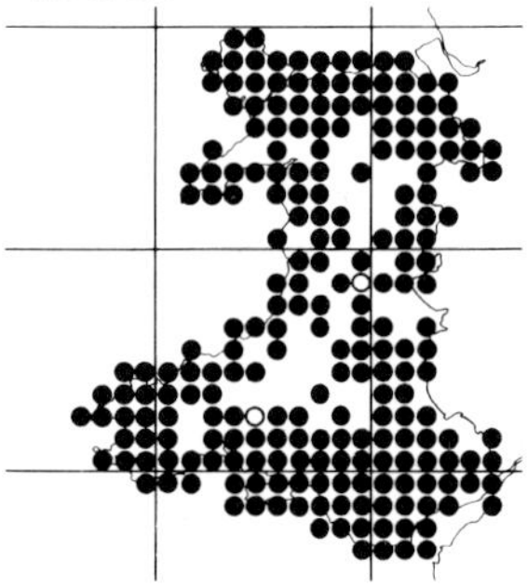

349 Vicia sativa subsp. nigra

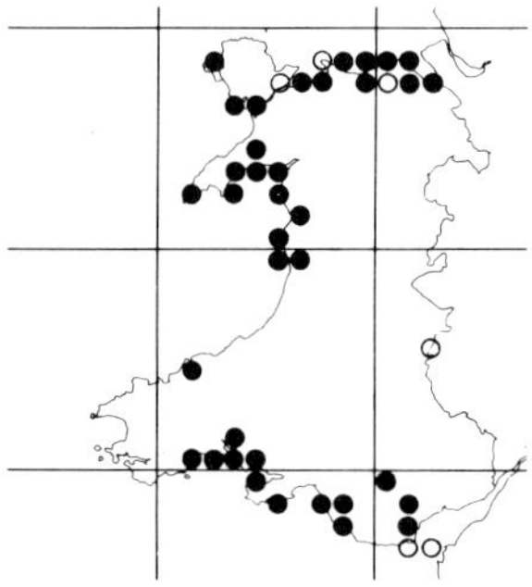

350 Vicia lathyroides

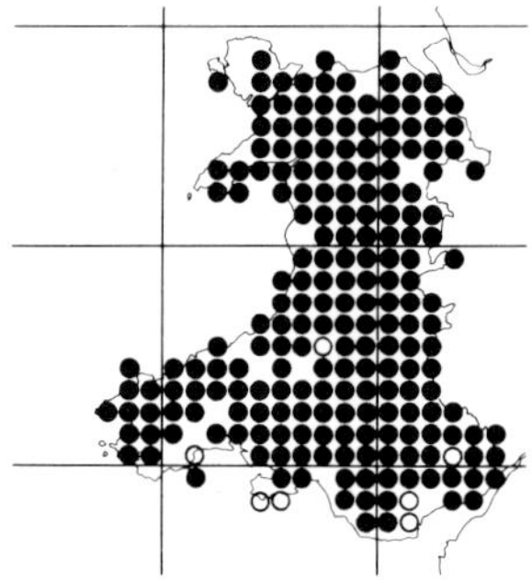

351 Lathyrus montanus

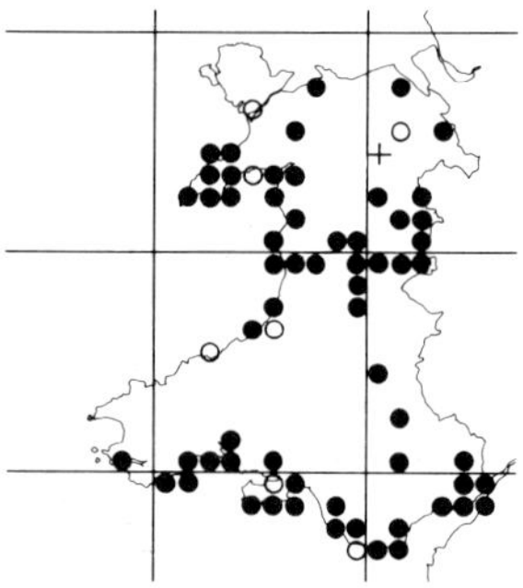

352 Lathyrus sylvestris

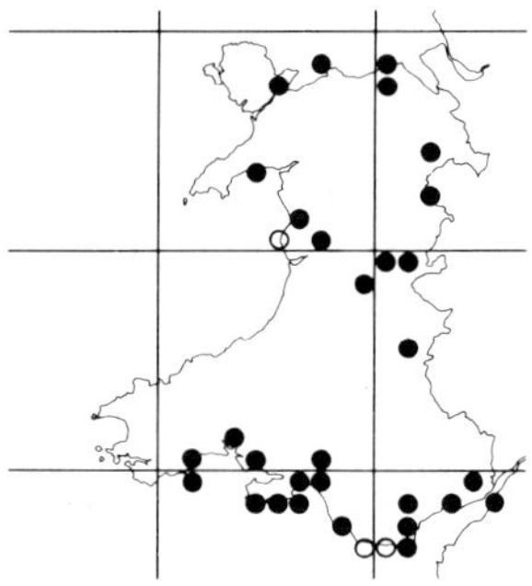

353 Lathyrus latifolius

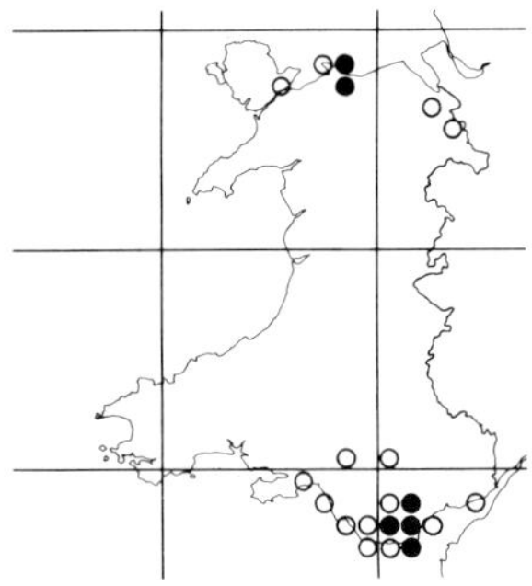

354 Lathyrus aphaca

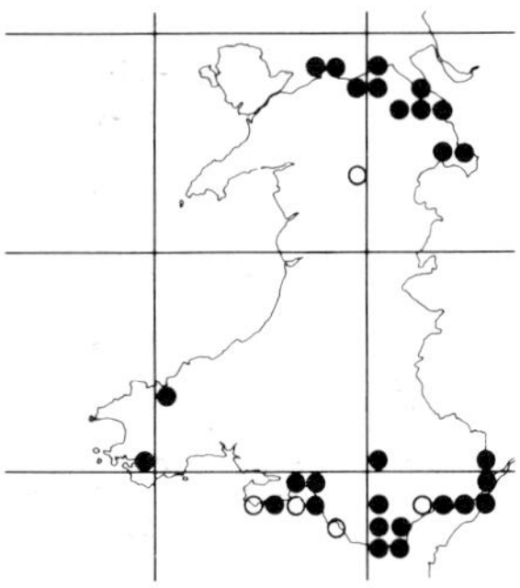

355 Ononis spinosa

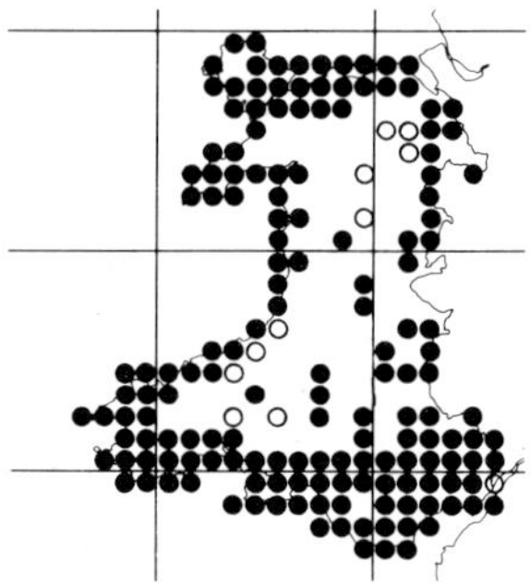

356 Ononis repens

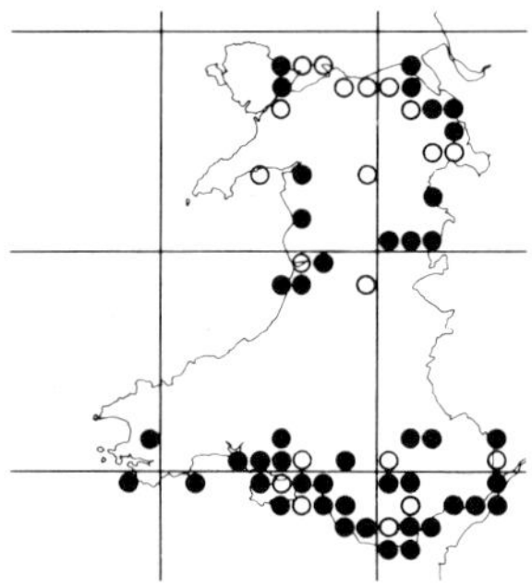

357 Melilotus altissima

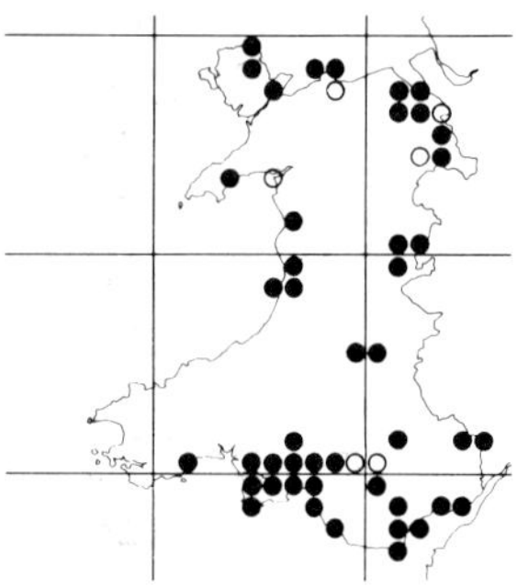

358 Melilotus alba

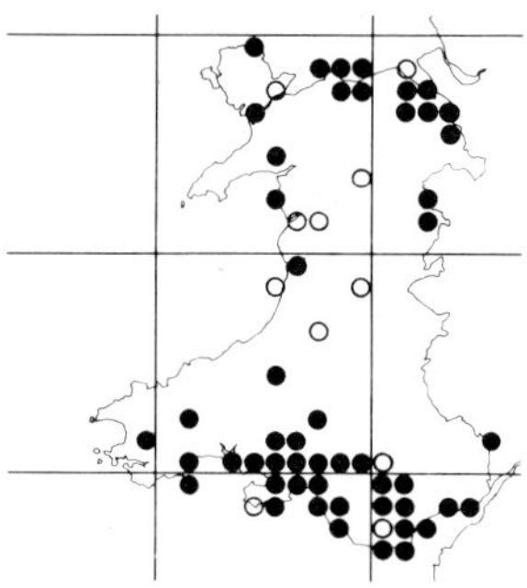

359 Melilotus officinalis

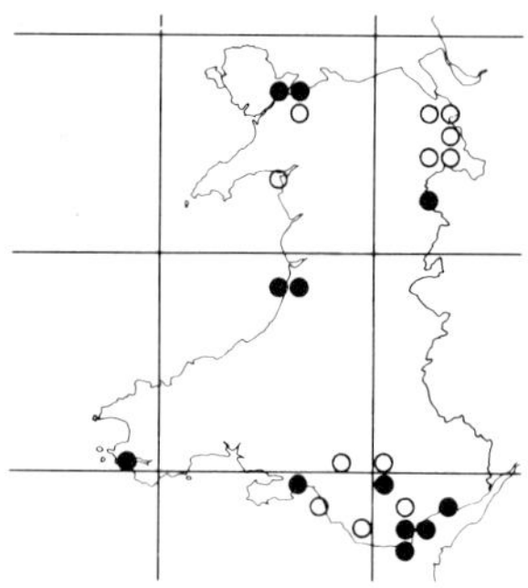

360 Melilotus indica

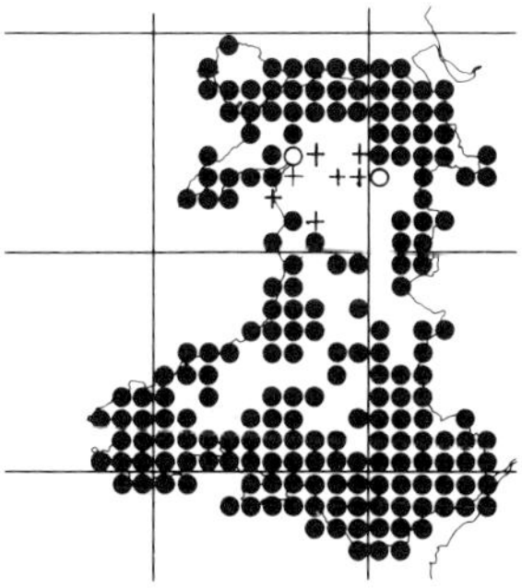

361 Medicago lupulina

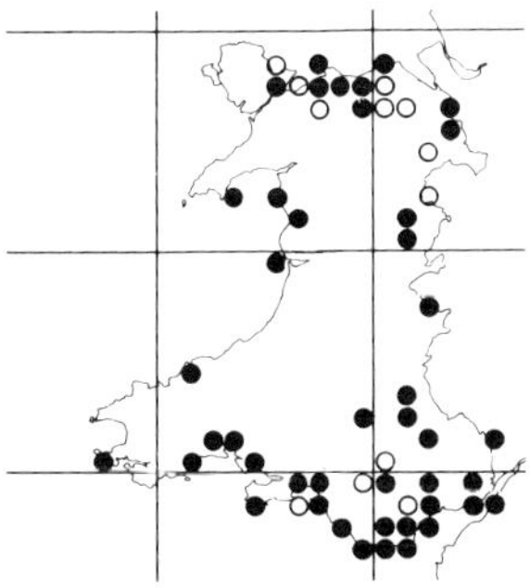

362 Medicago sativa subsp. sativa

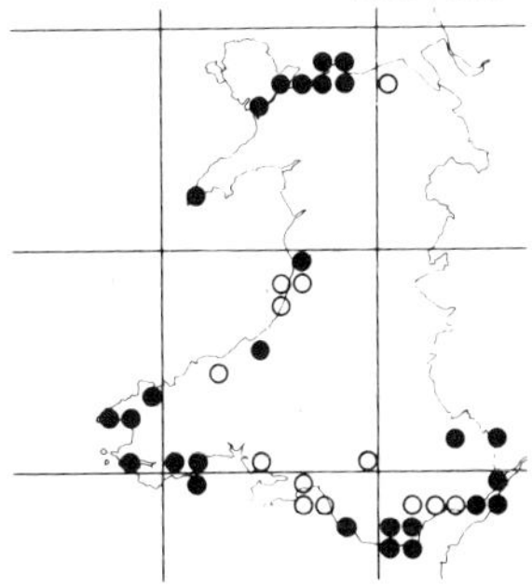

363 Medicago arabica

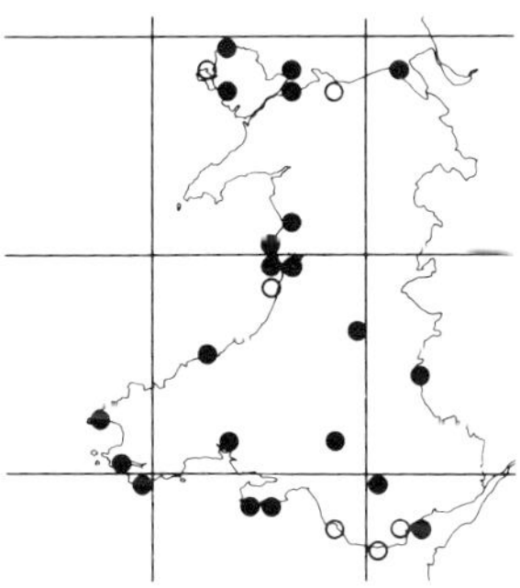

364 Trifolium ornithopodioides

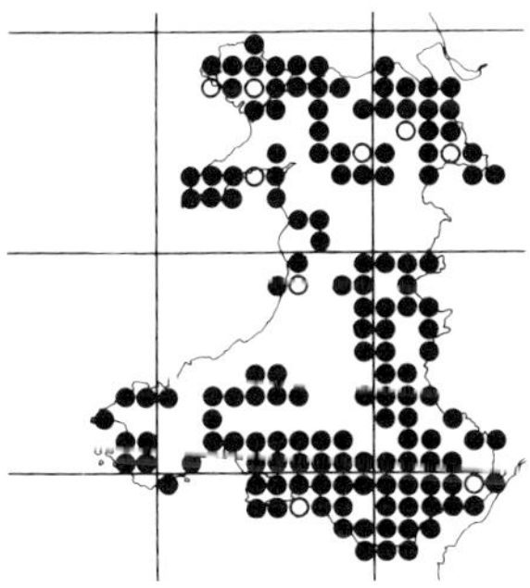

365 Trifolium hybridum subsp. hybridum

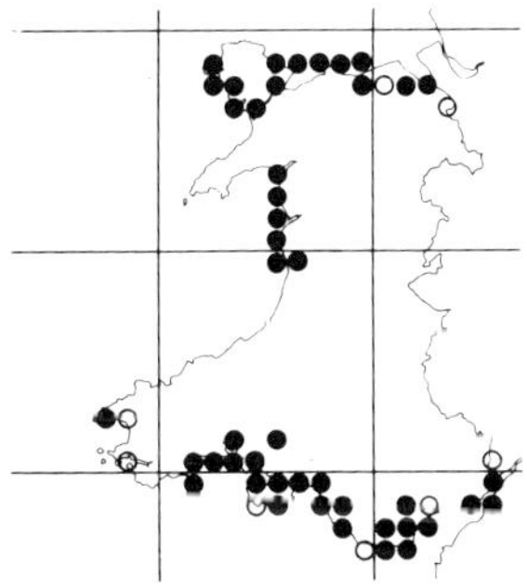

366 Trifolium fragiferum

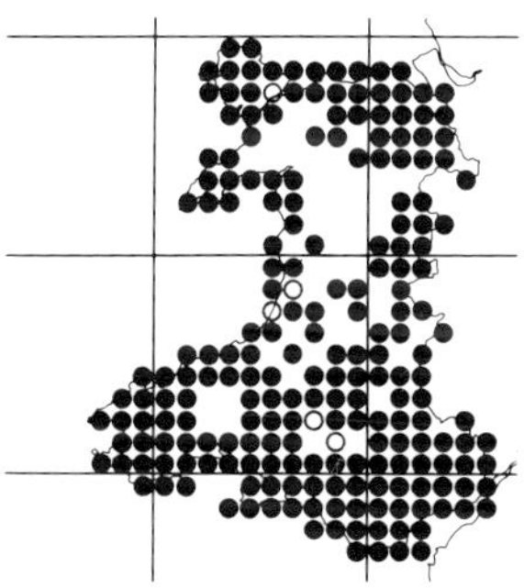

367 Trifolium campestre

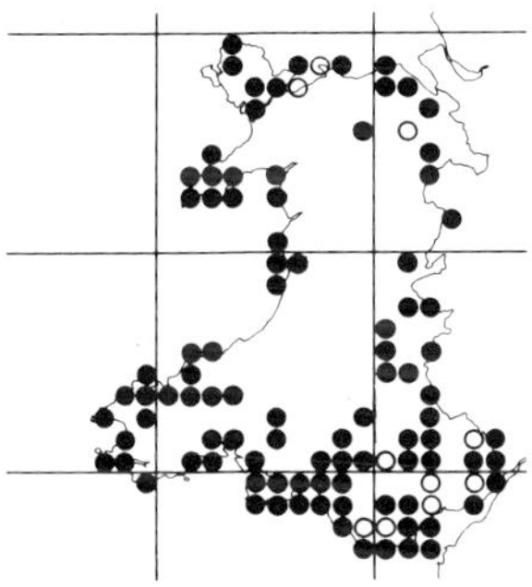

368 Trifolium micranthum

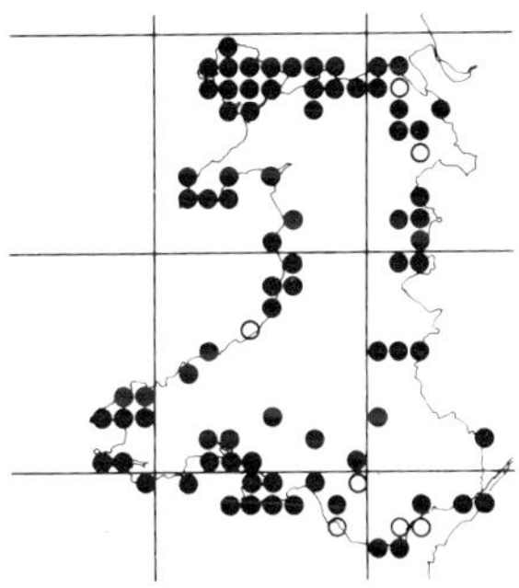

369 Trifolium striatum

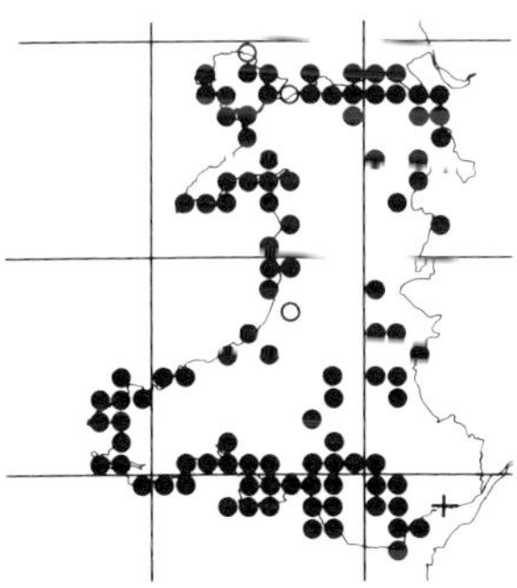

370 Trifolium arvense

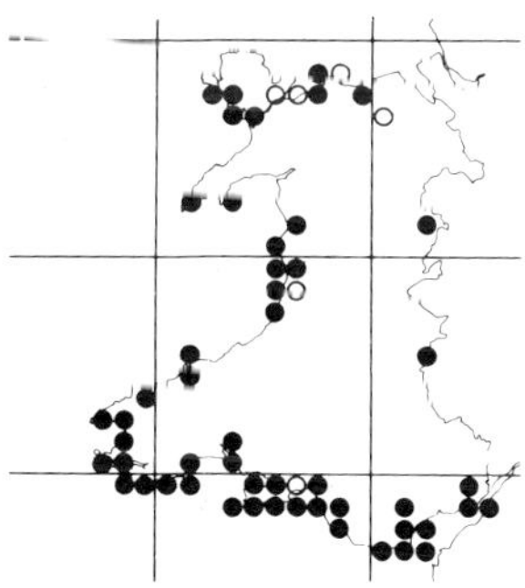

371 Trifolium scabrum

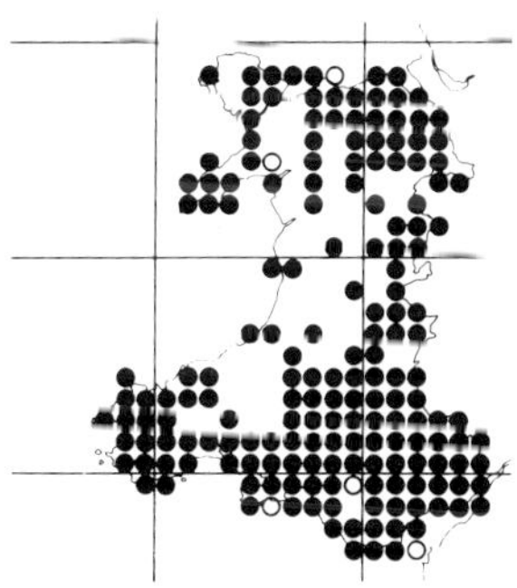

372 Trifolium medium

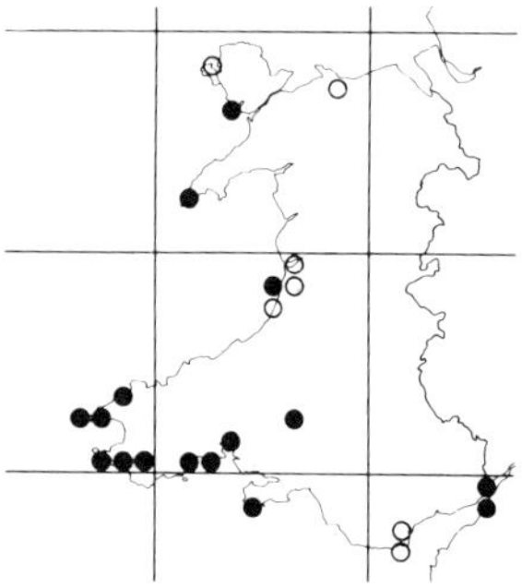

373 Trifolium subterraneum

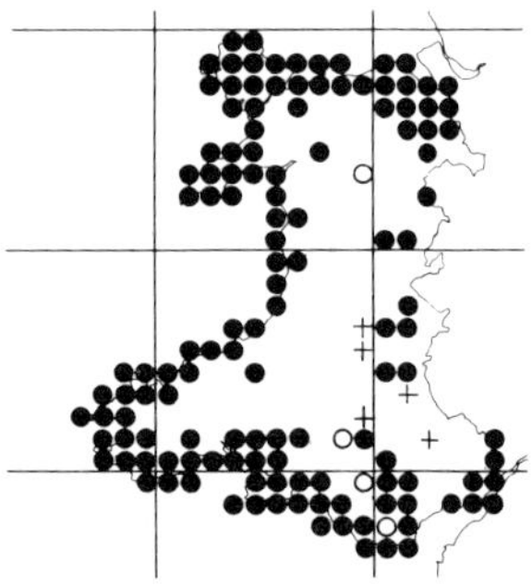

374 Anthyllis vulneraria subsp. vulneraria

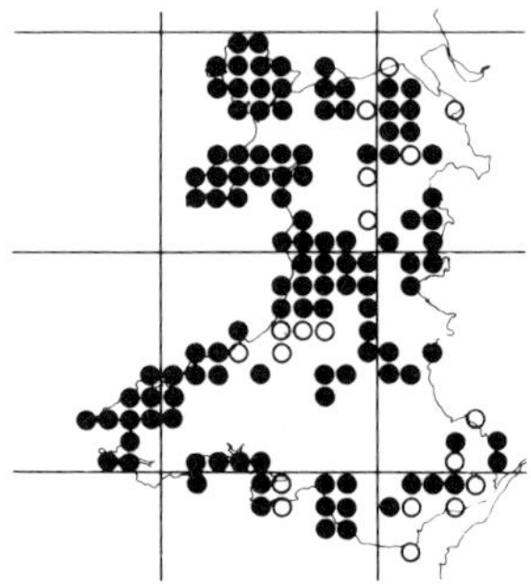

375 Ornithopus perpusillus

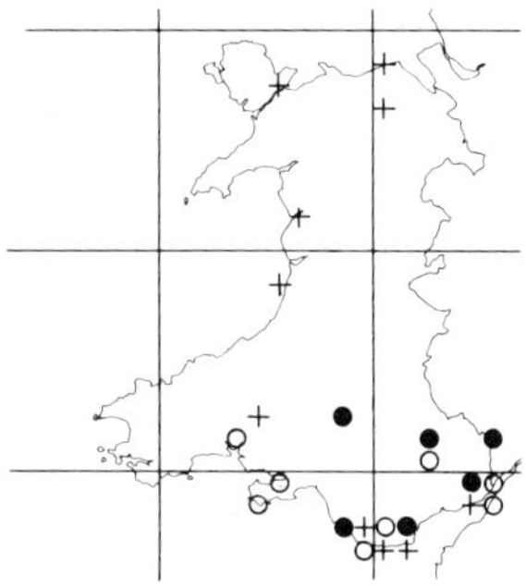

376 Onobrychis viciifolia

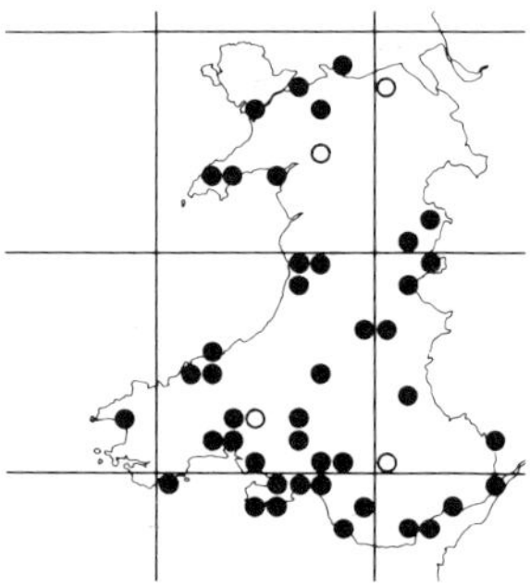

377 Oxalis corniculata

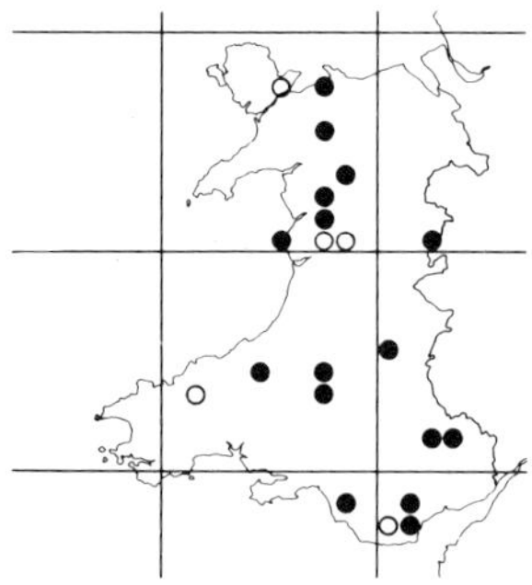

378 Oxalis europaea

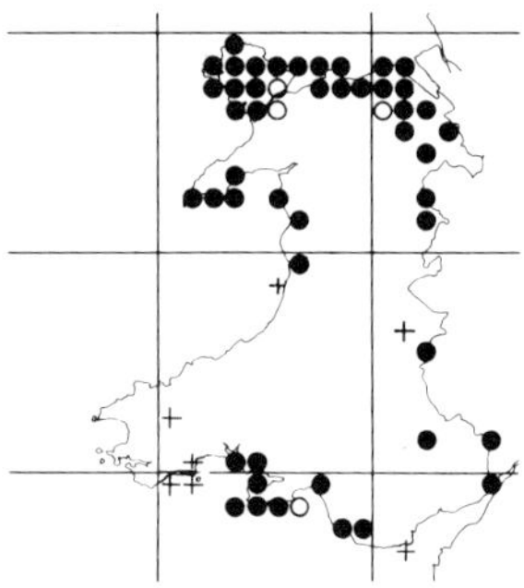

379 Geranium sanguineum

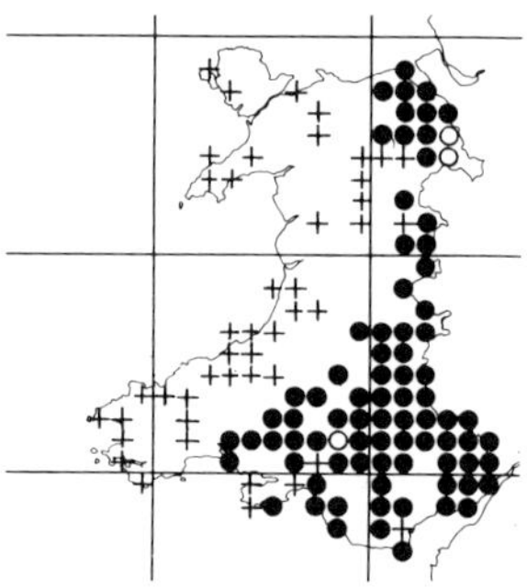

380 Geranium pratense

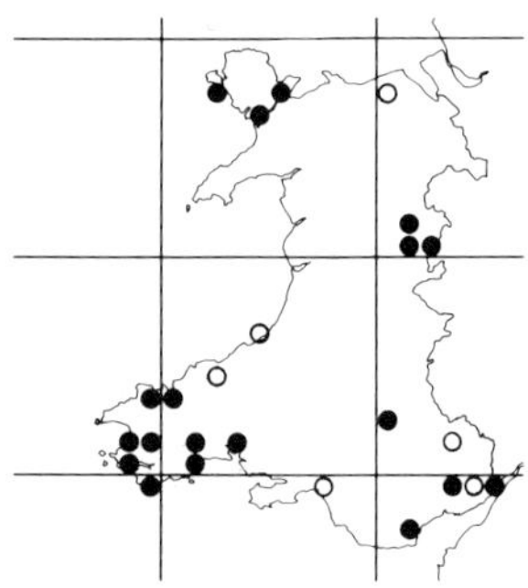

381 Geranium versicolor

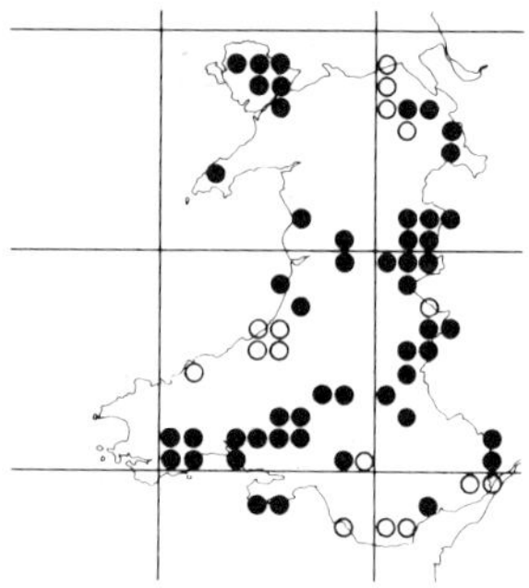

382 Geranium phaeum

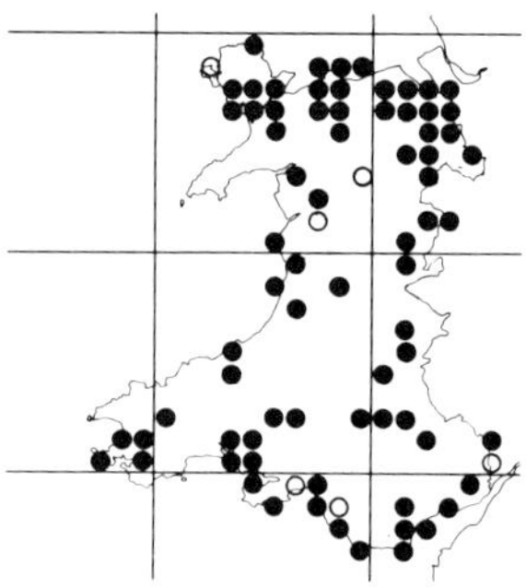

383 Geranium pyrenaicum

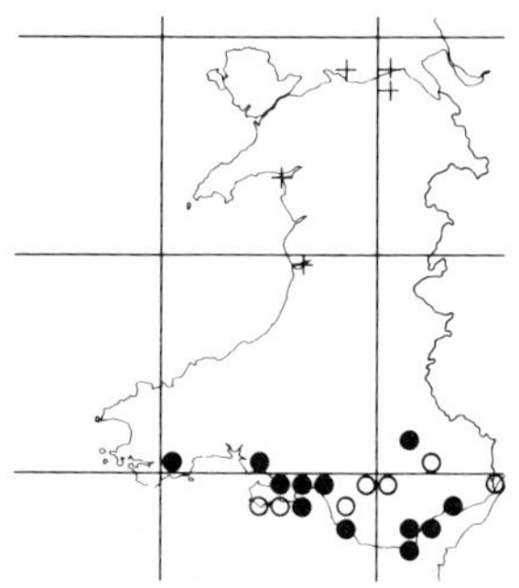

384 Geranium rotundifolium

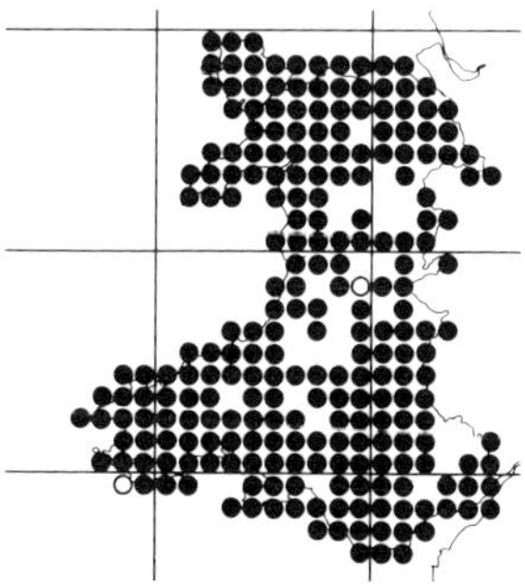

385 Geranium molle

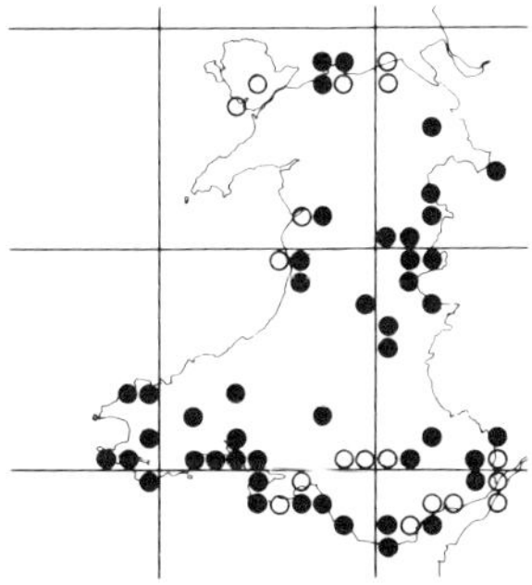

386 Geranium pusillum

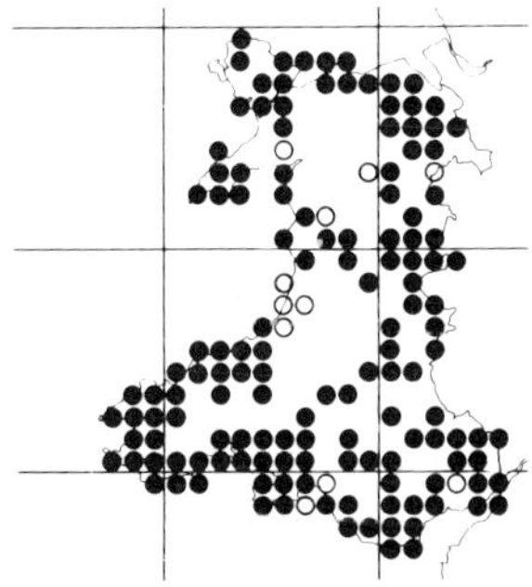

387 Geranium columbinum

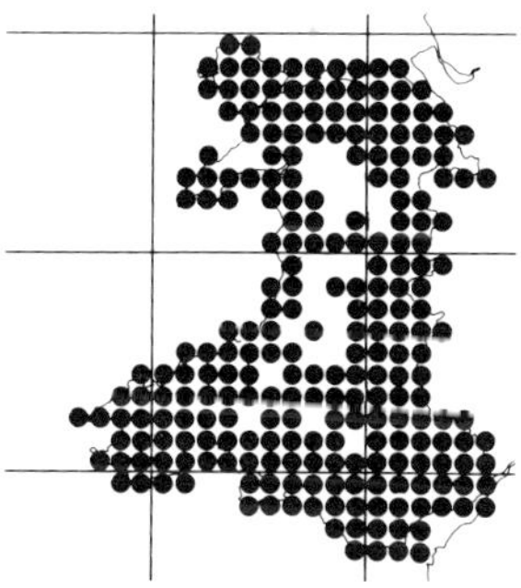

388 Geranium dissectum

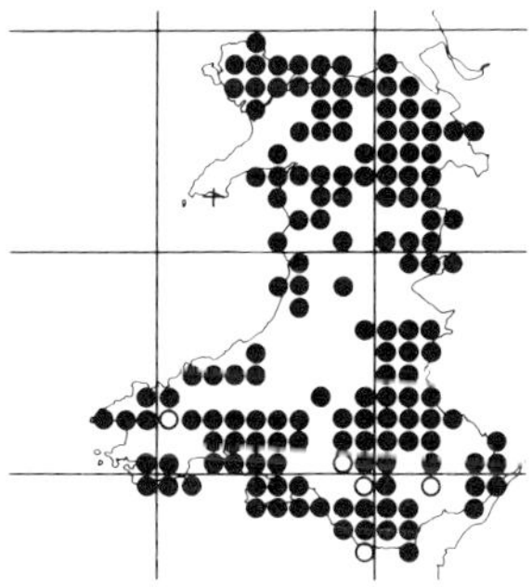

389 Geranium lucidum

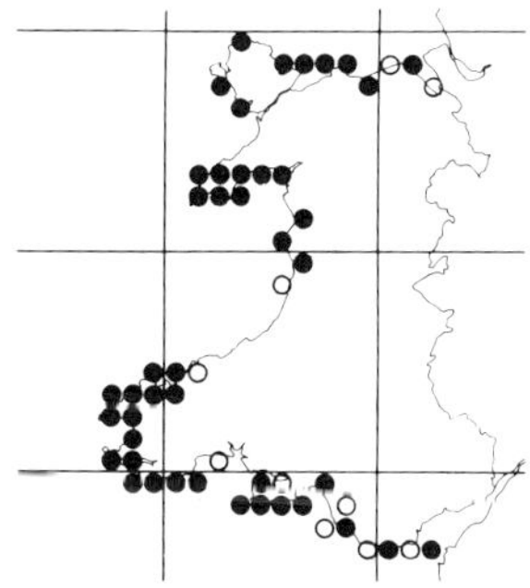

390 Erodium maritimum

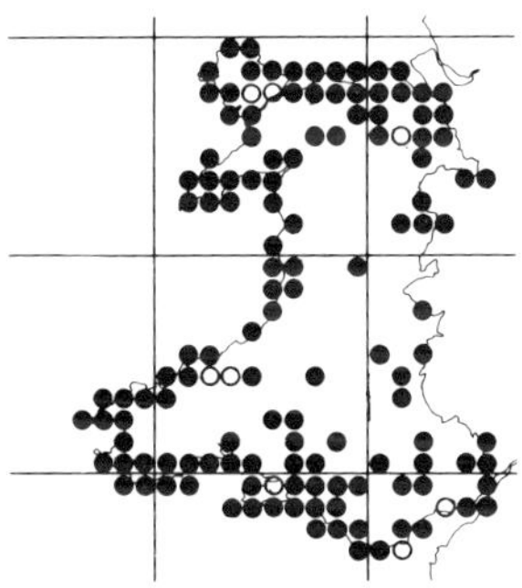

391 Erodium cicutarium

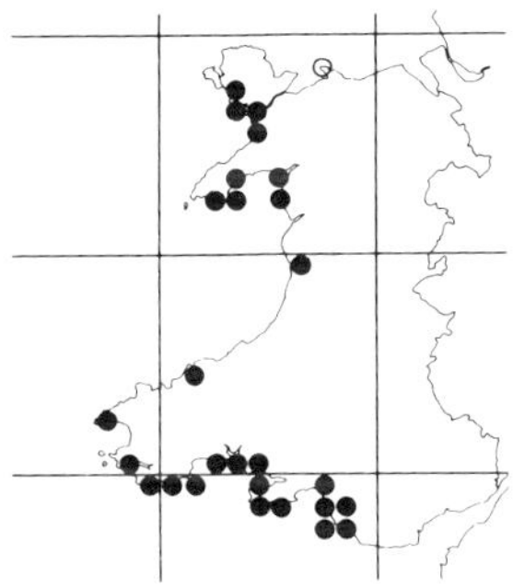

392 Erodium glutinosum

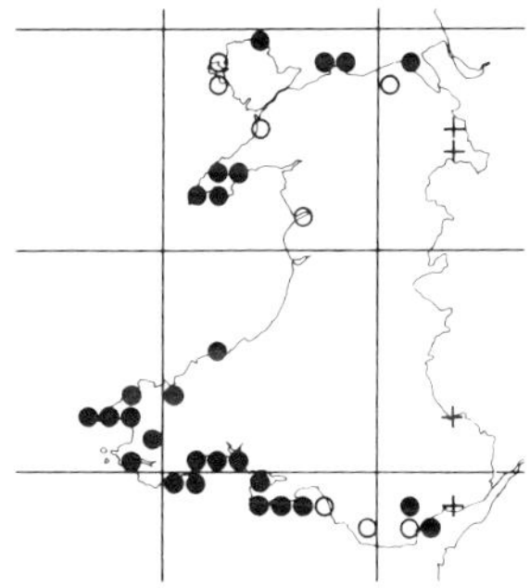

393 Erodium moschatum

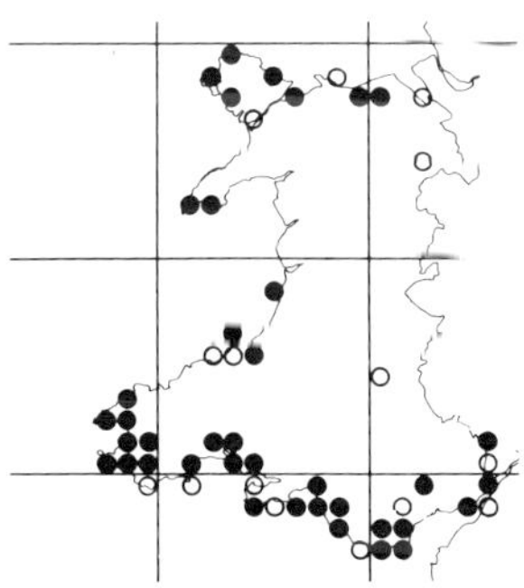

394 Linum bienne

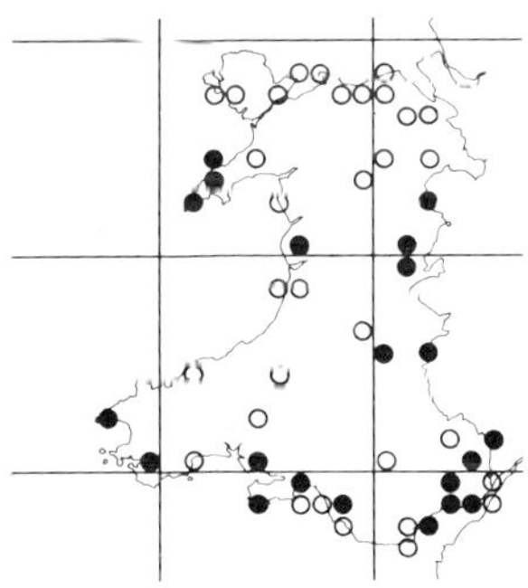

395 Linum usitatissimum

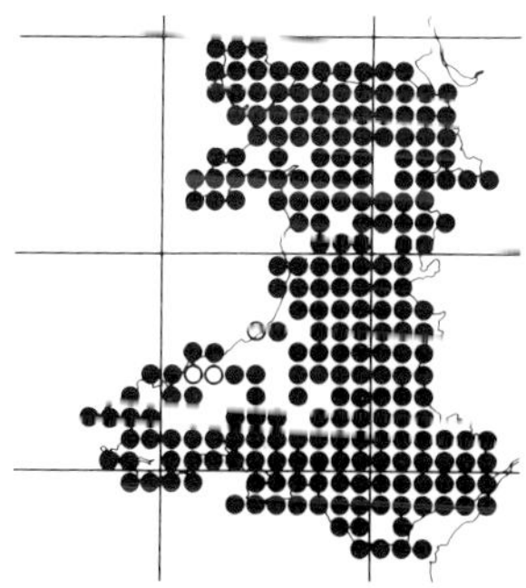

396 Linum catharticum

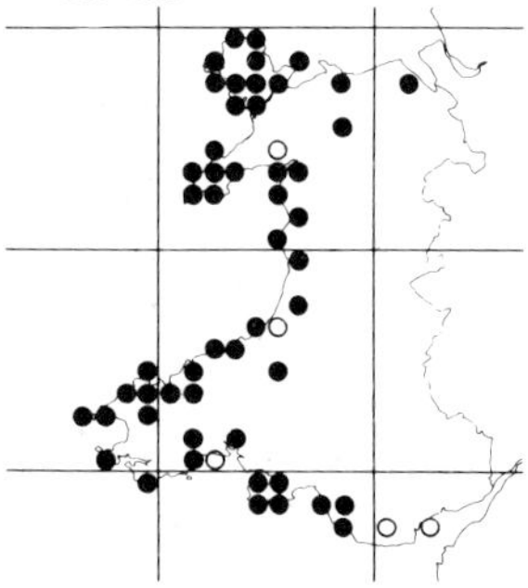

397 Radiola linoides

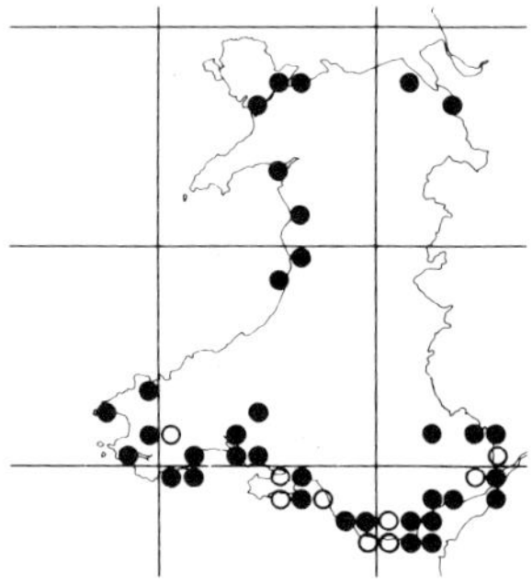

398 Mercurialis annua

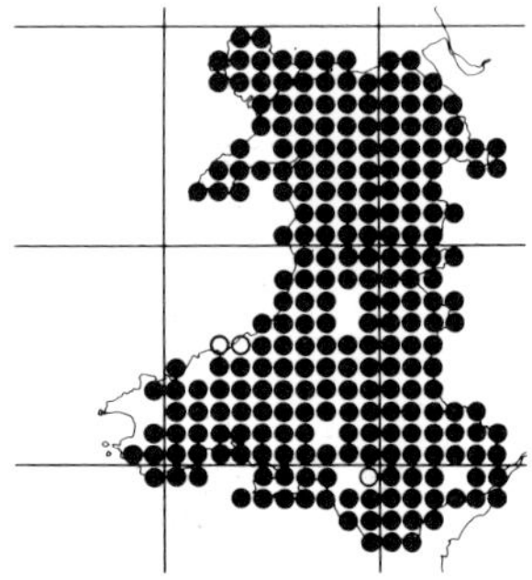

399 Mercurialis perennis

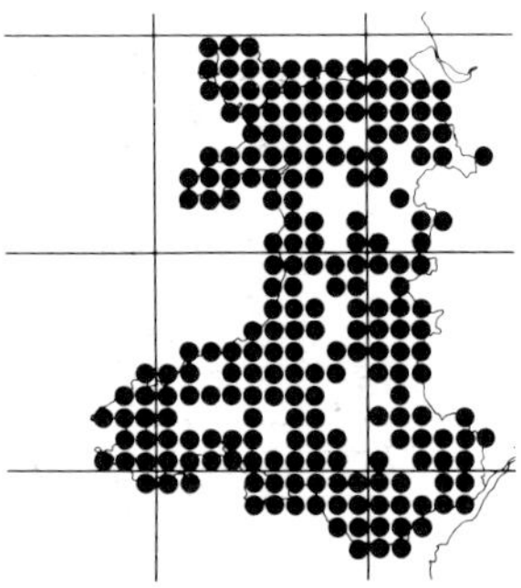

400 Euphorbia helioscopia

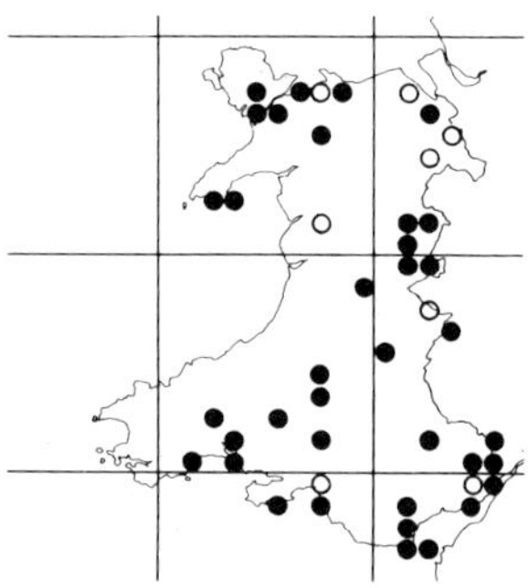

401 Euphorbia lathyris

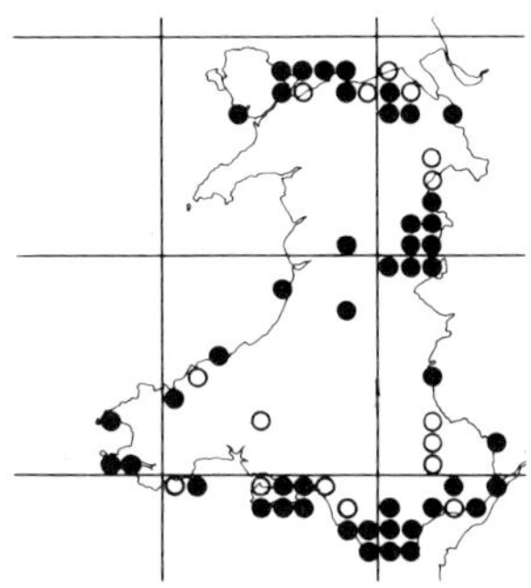

402 Euphorbia exigua

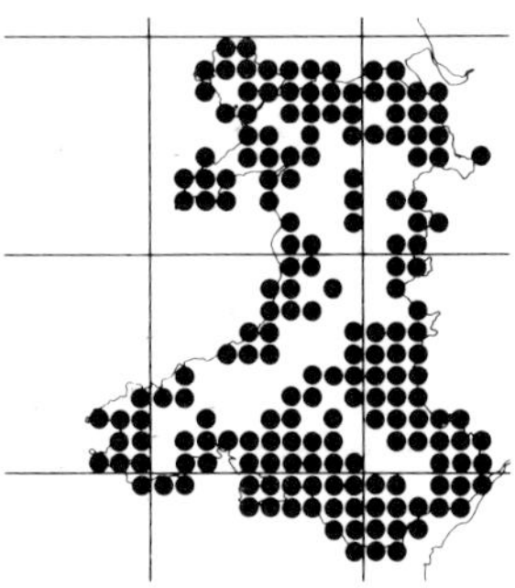

403 Euphorbia peplus

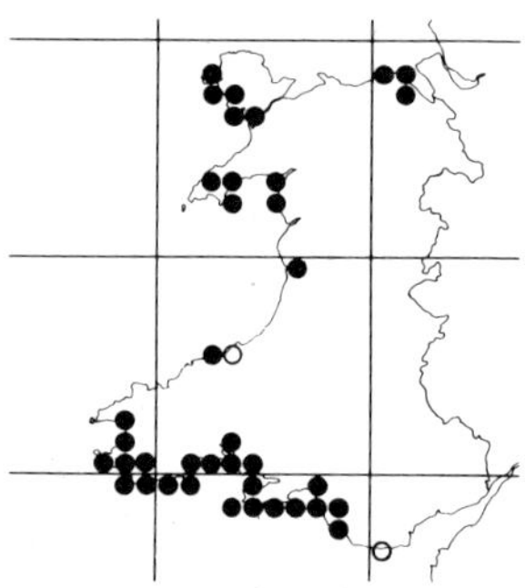

404 Euphorbia portlandica

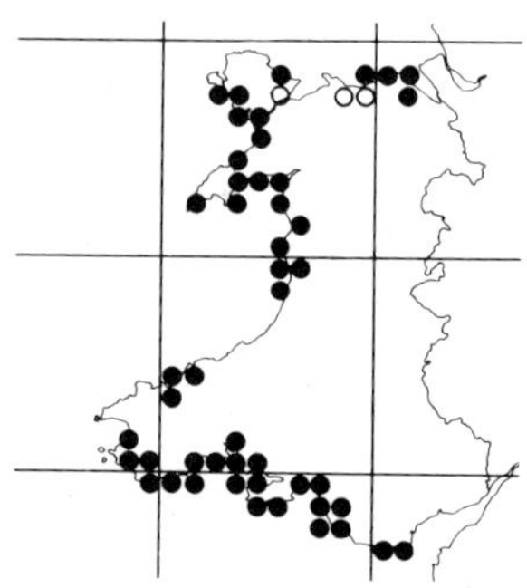

405 Euphorbia paralias

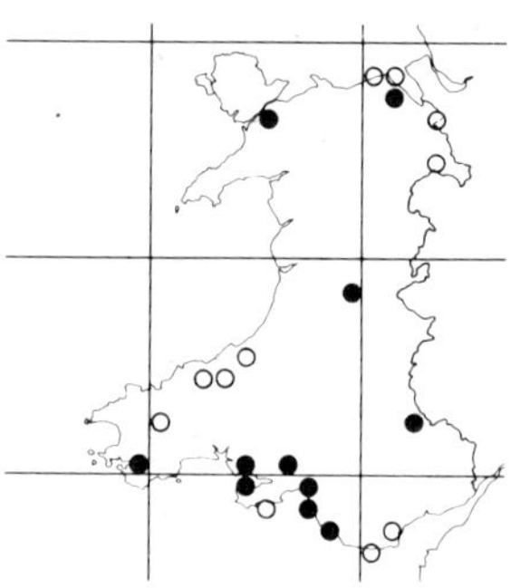

406 Euphorbia cyparissias

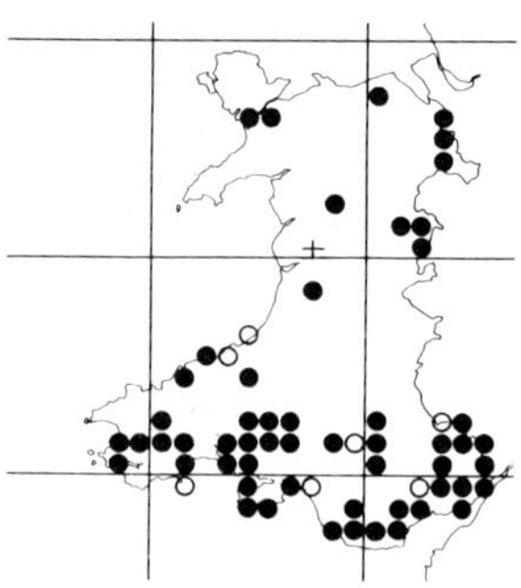

407 Euphorbia amygdaloides

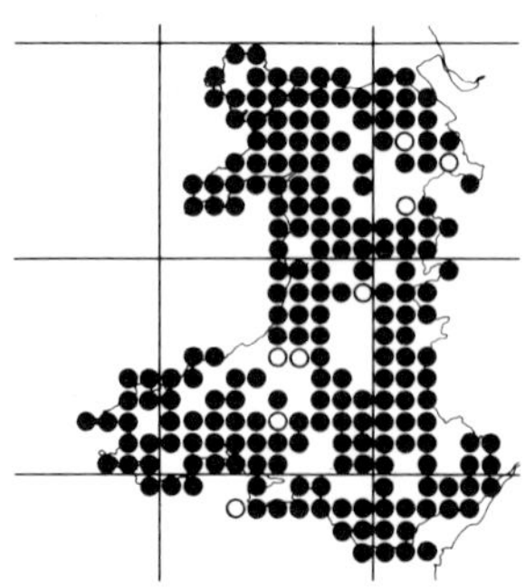

408 Polygala vulgaris

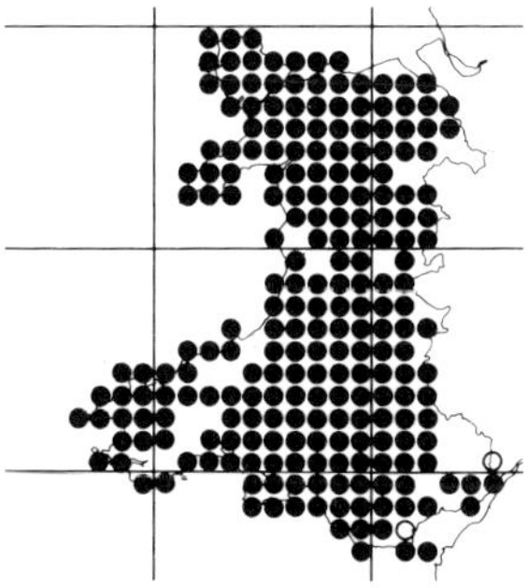

409 Polygala serpyllifolia

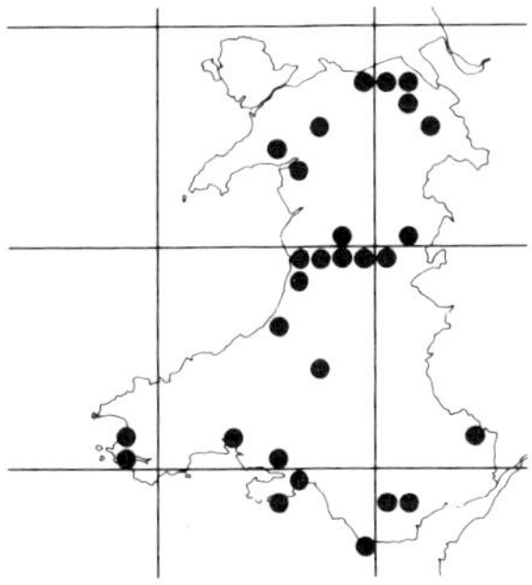

410 Acer platanoides

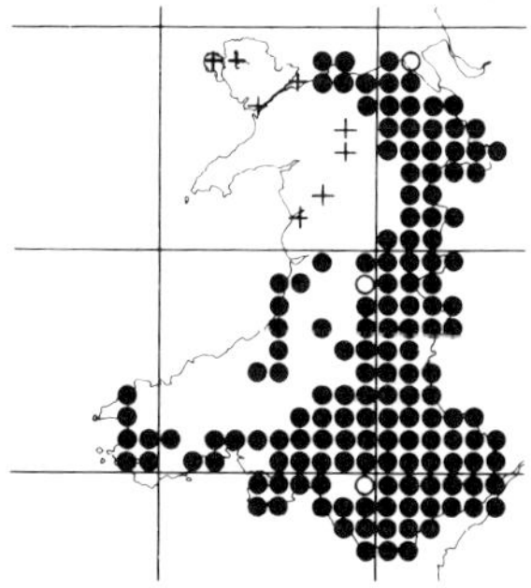

411 Acer campestre

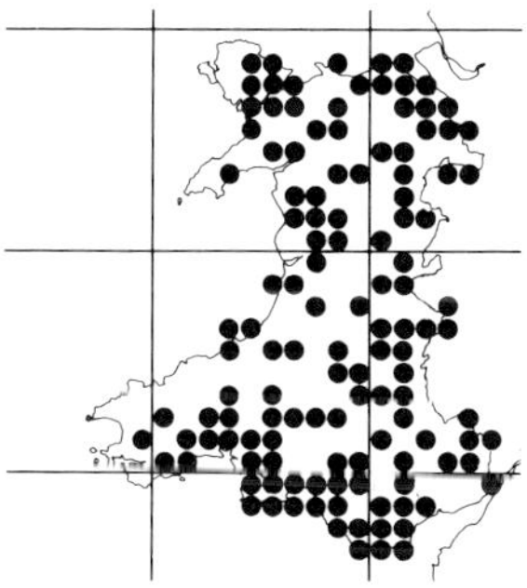

412 Aesculus hippocastanum

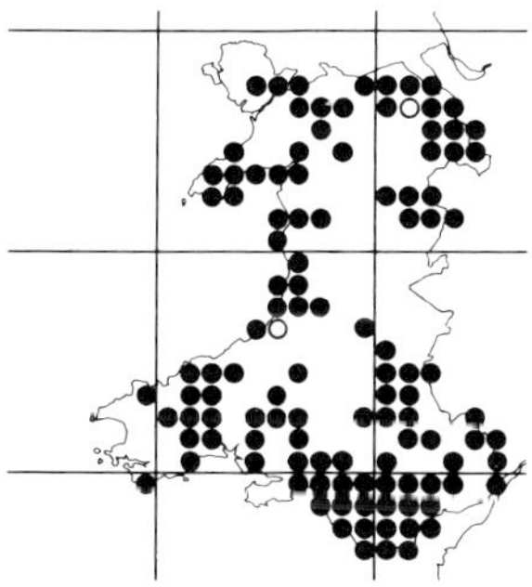

413 Impatiens glandulifera

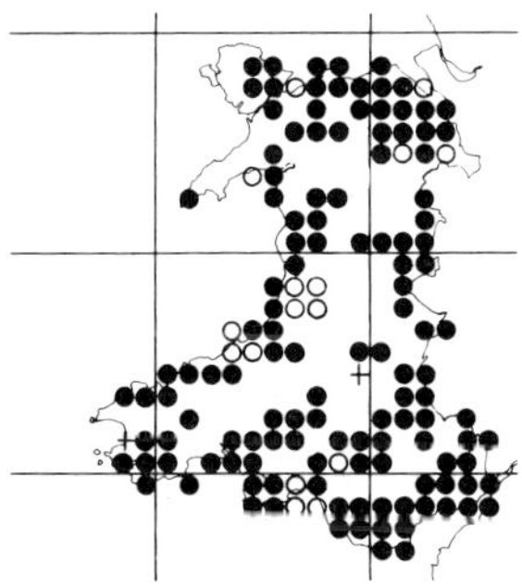

414 Euonymus europaeus

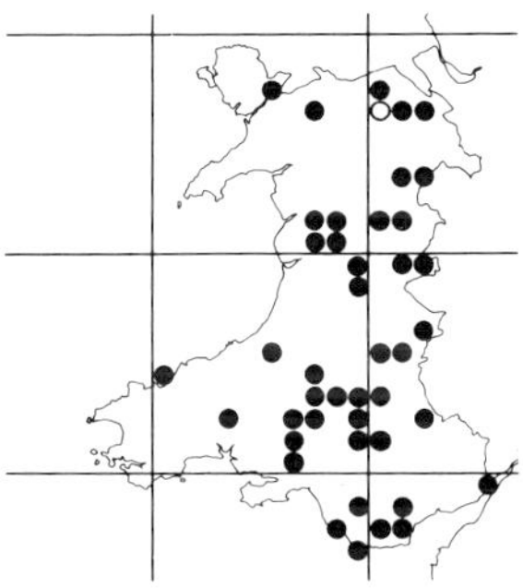

415 Buxus sempervirens

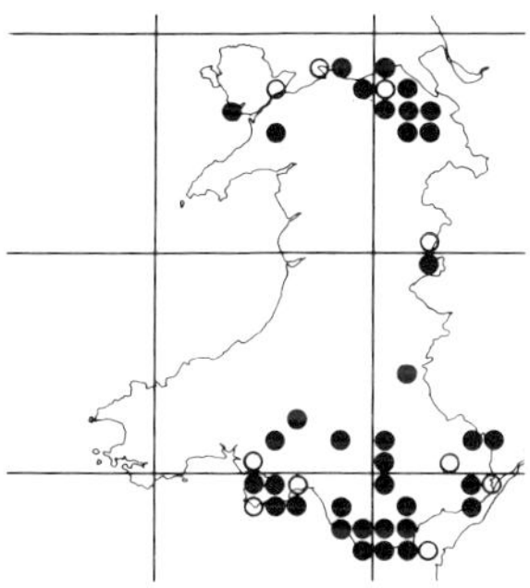

416 Rhamnus catharticus

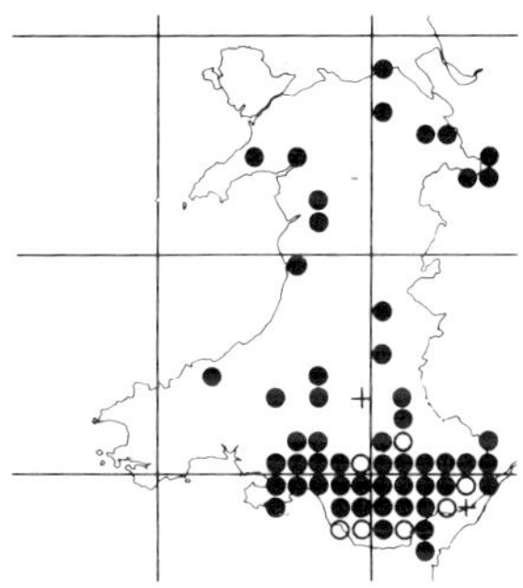

417 Frangula alnus

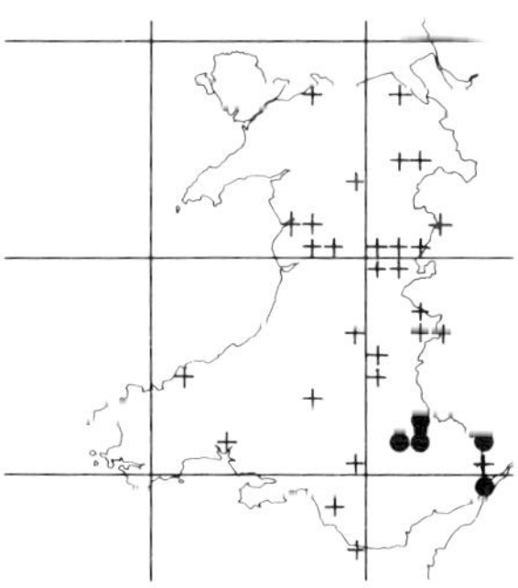

418 Tilia platyphyllos

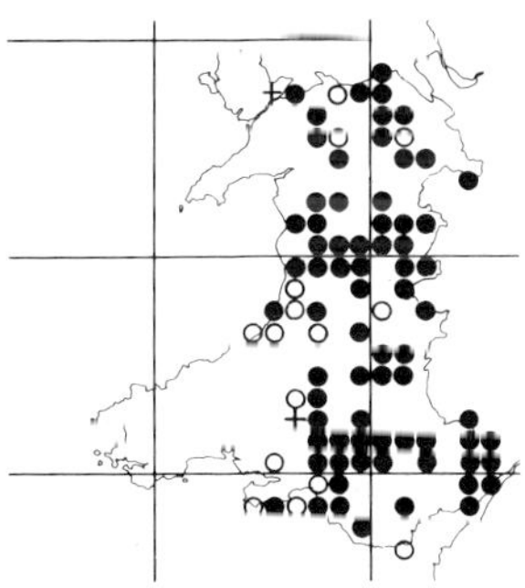

419 Tilia cordata

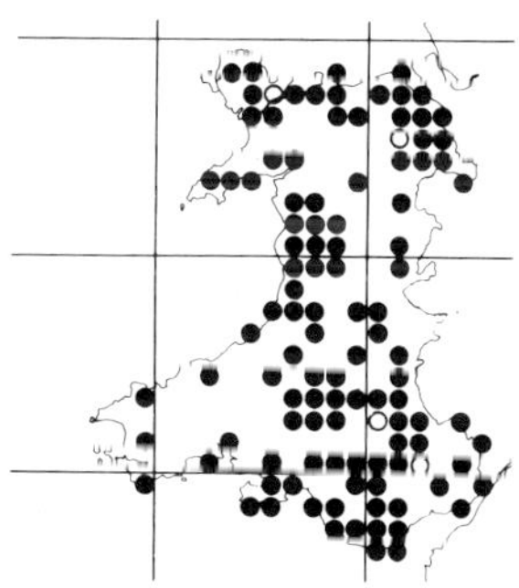

420 Tilia cordata × platyphyllos
= T. × vulgaris

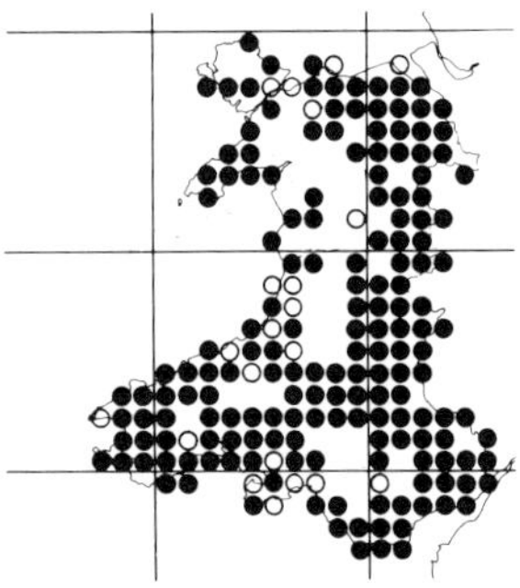
421 Malva moschata

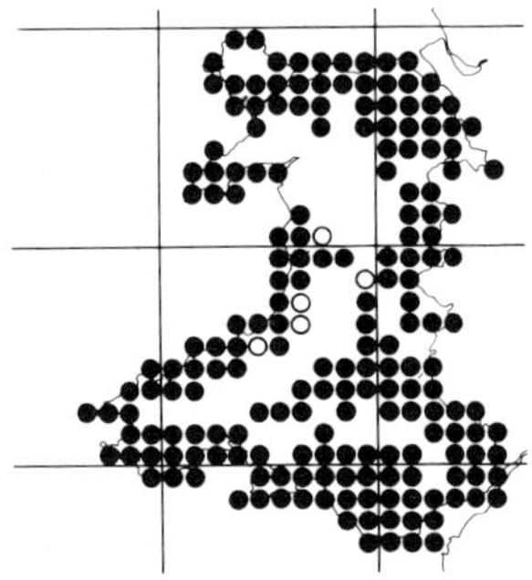
422 Malva sylvestris

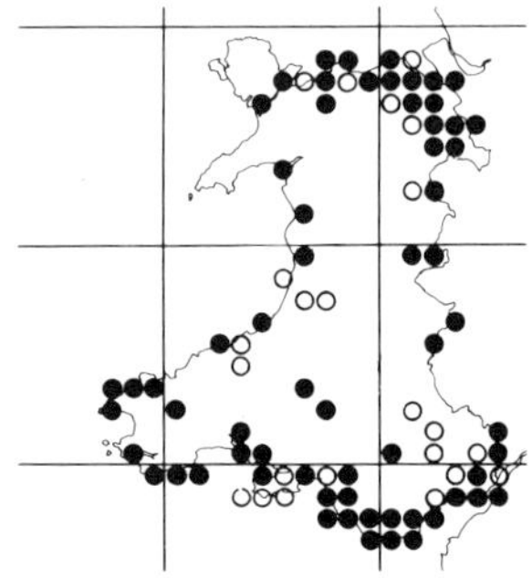
423 Malva neglecta

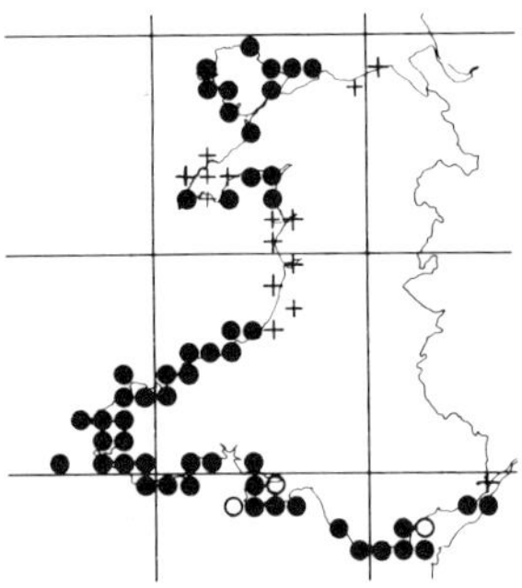
424 Lavatera arborea

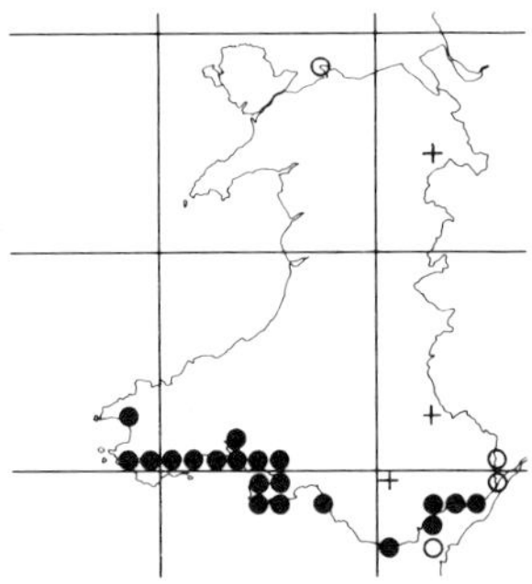
425 Althaea officinalis

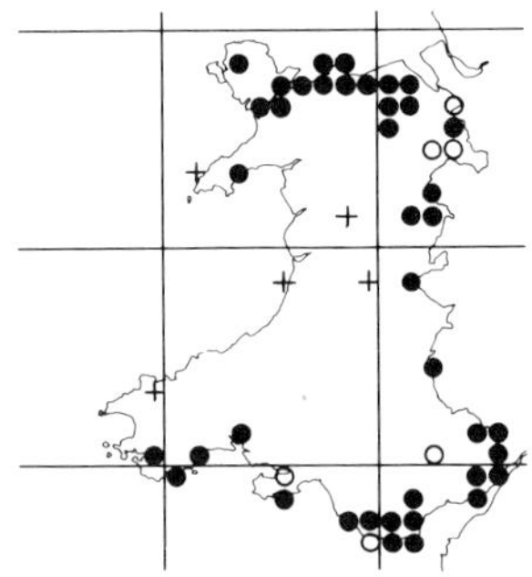
426 Daphne laureola

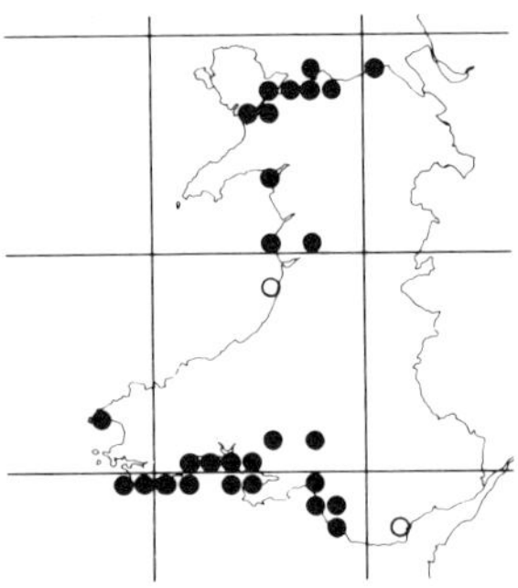
427 Hippophae rhamnoides

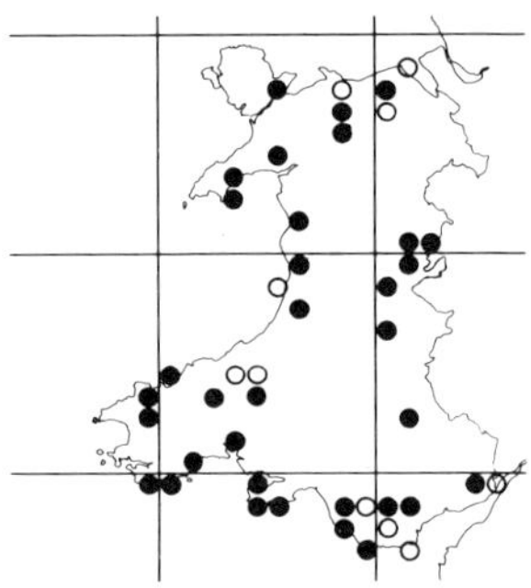
428 Hypericum calycinum

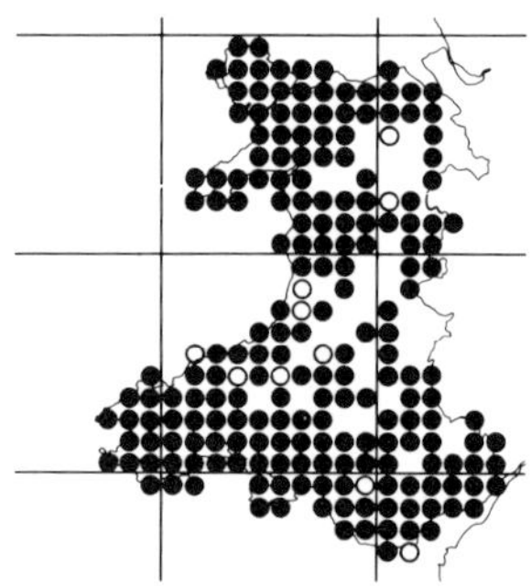
429 Hypericum androsaemum

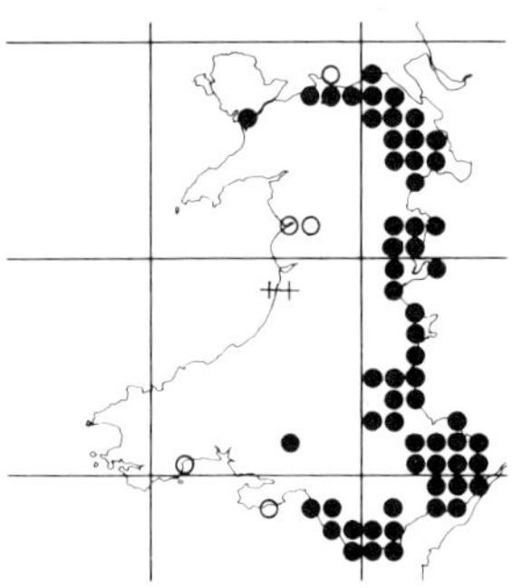
430 Hypericum hirsutum

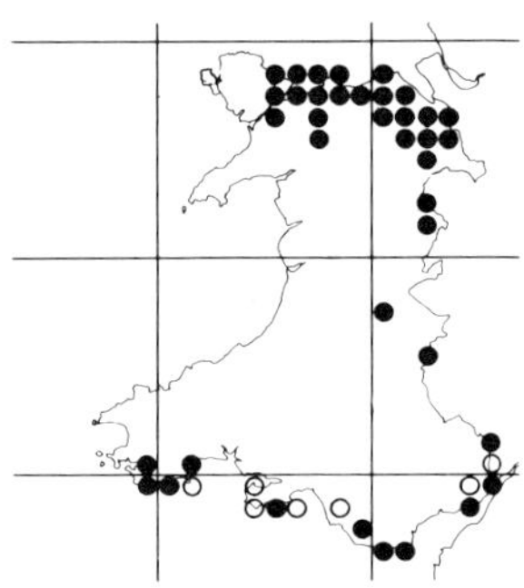
431 Hypericum montanum

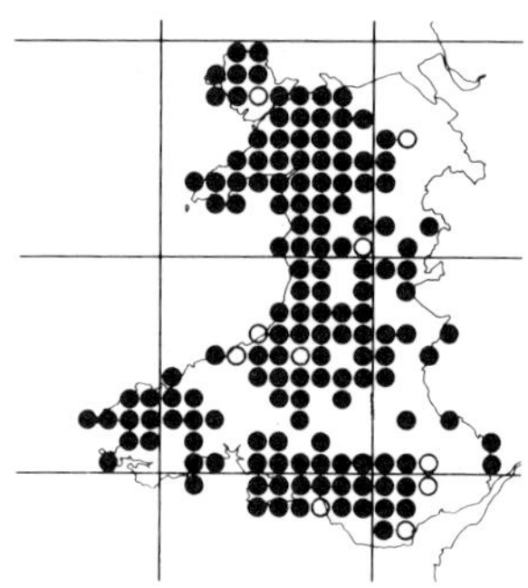
432 Hypericum elodes

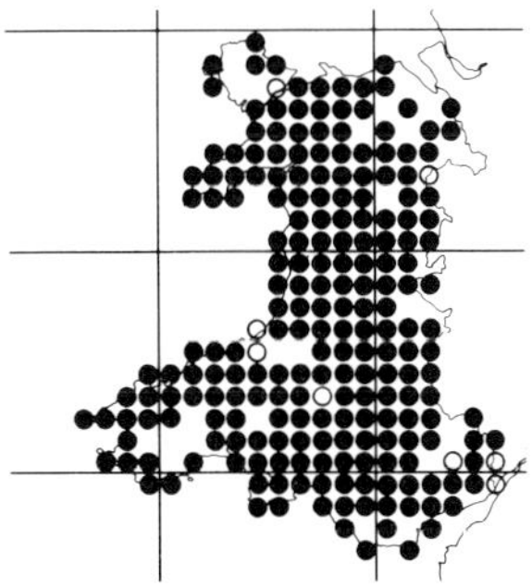

433 Hypericum humifusum

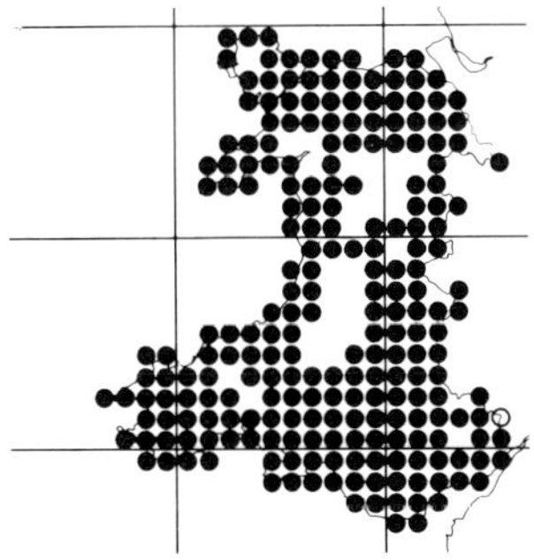

434 Hypericum tetrapterum

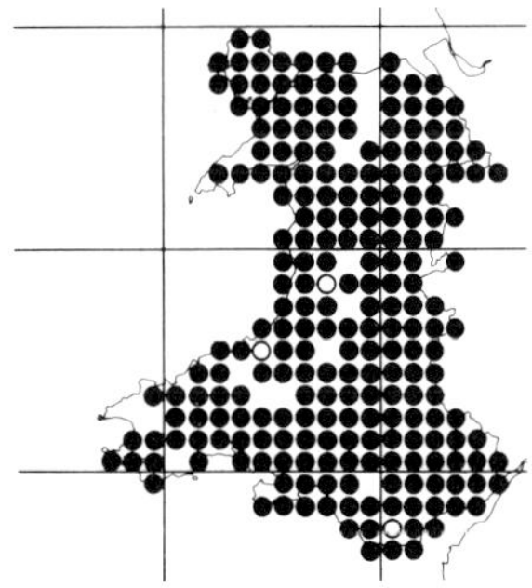

435 Hypericum maculatum

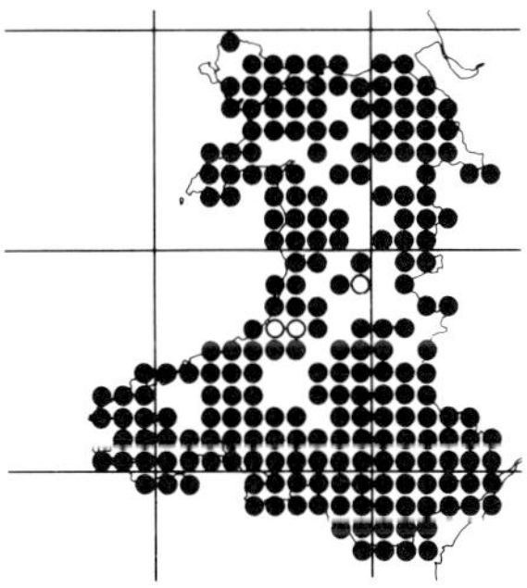

436 Hypericum perforatum

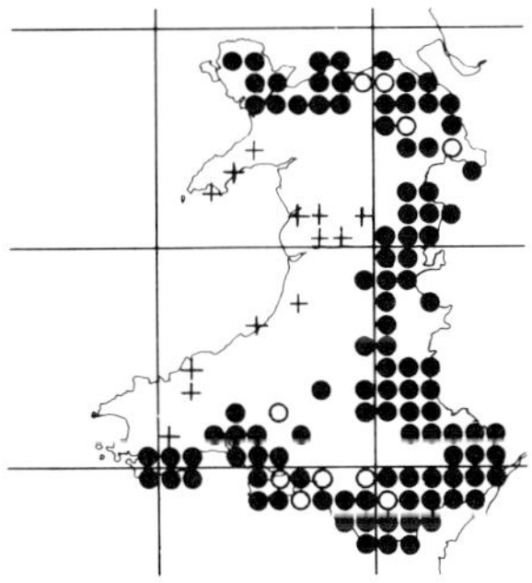

437 Viola odorata

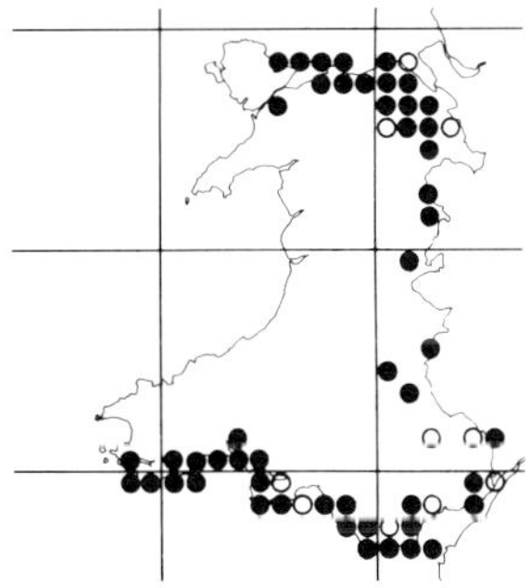

438 Viola hirta subsp. hirta

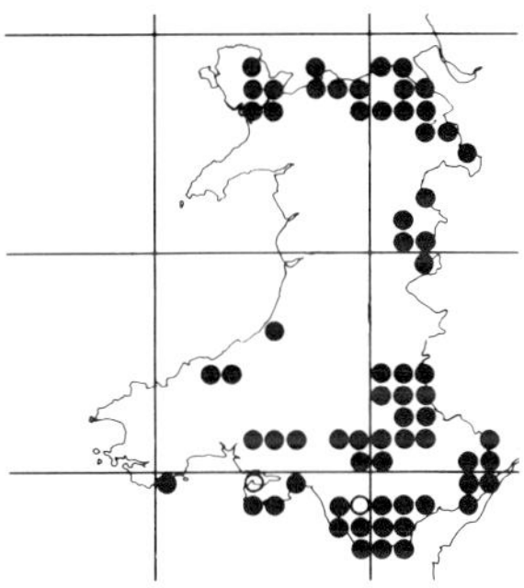

439 Viola reichenbachiana

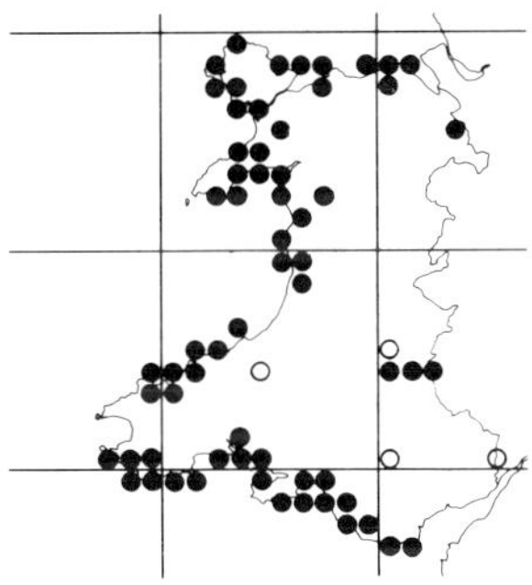

440 Viola canina

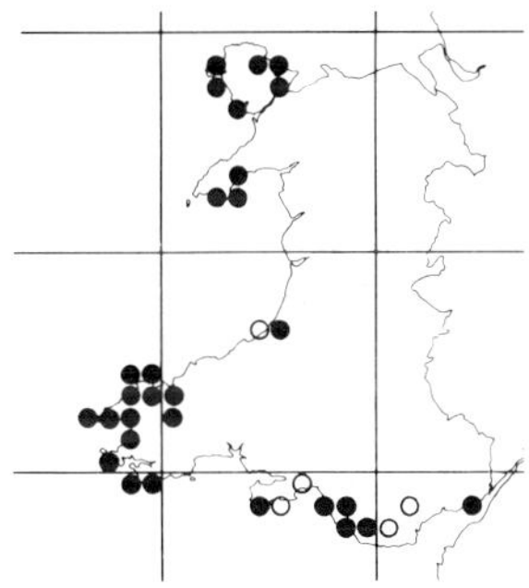

441 Viola lactea

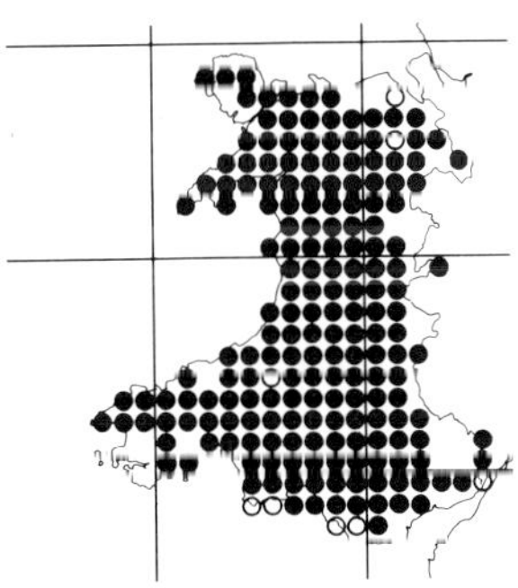

442 Viola palustris subsp. palustris

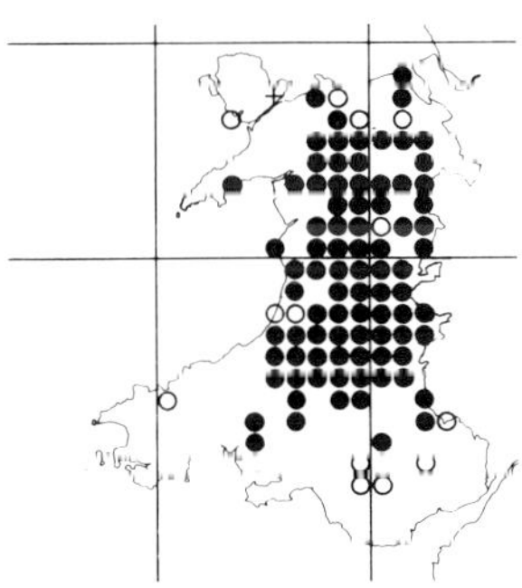

443 Viola lutea

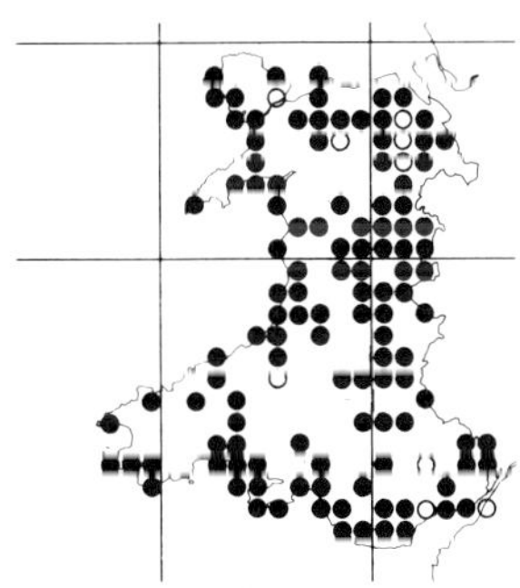

444 Viola tricolor subsp. tricolor

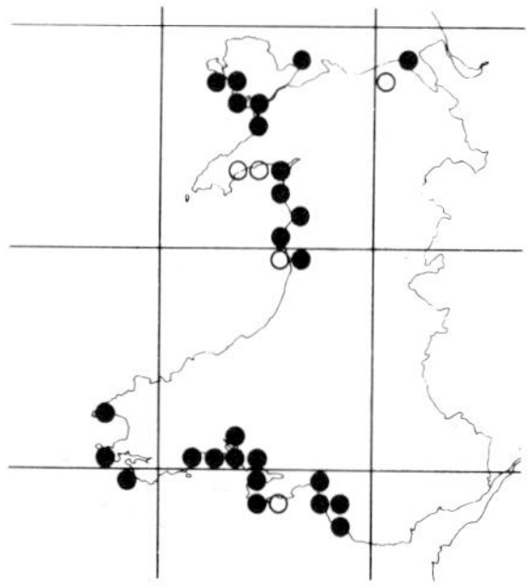

445 Viola tricolor subsp. curtisii

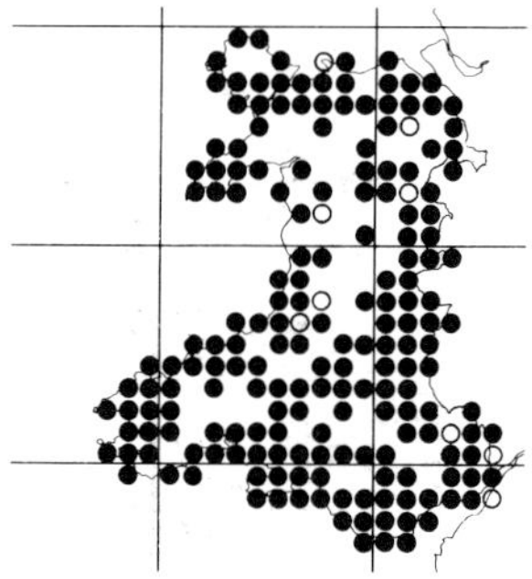

446 Viola arvensis

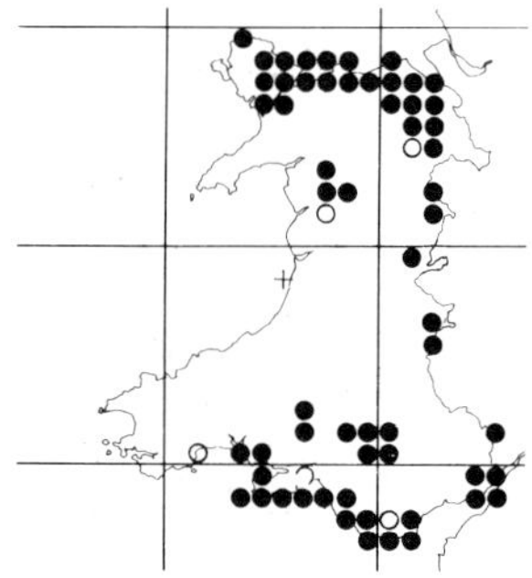

447 Helianthemum nummularium subsp. nummularium

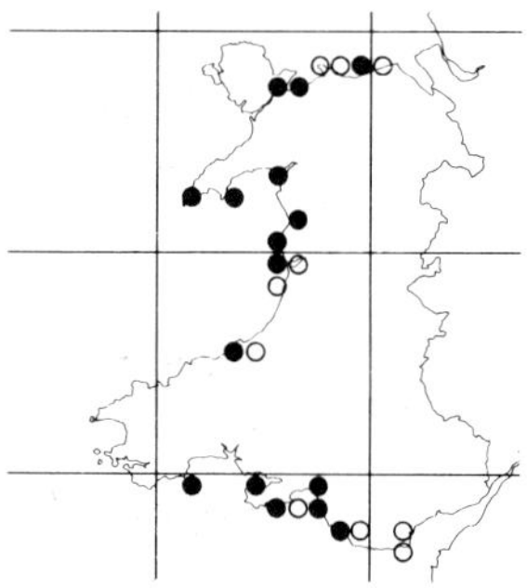

448 Tamarix gallica

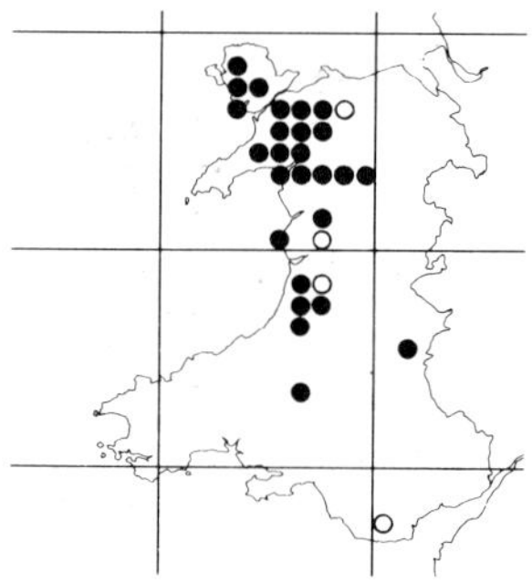

449 Elatine hexandra

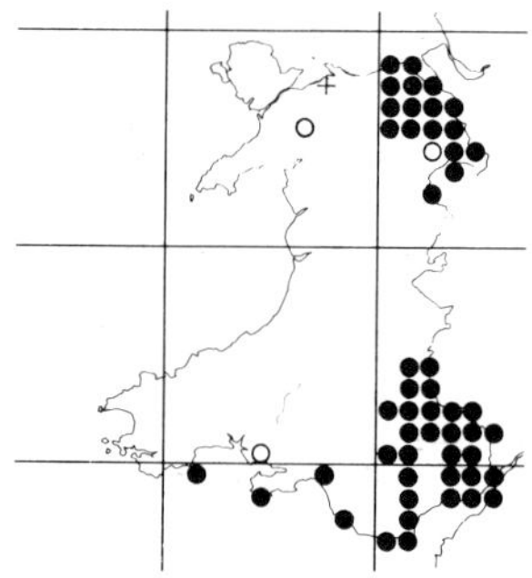

450 Bryonia cretica

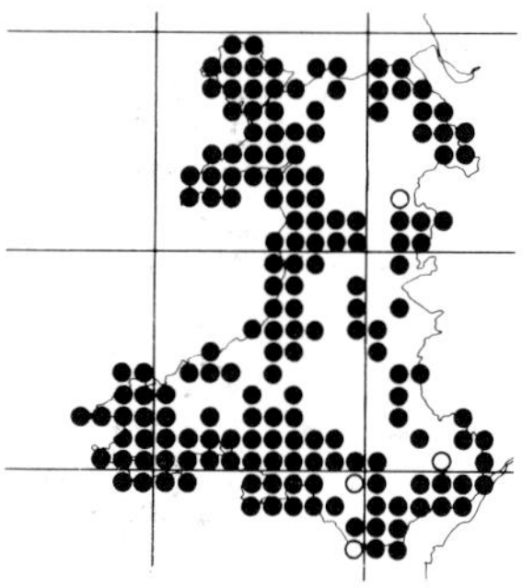

451 Lythrum salicaria

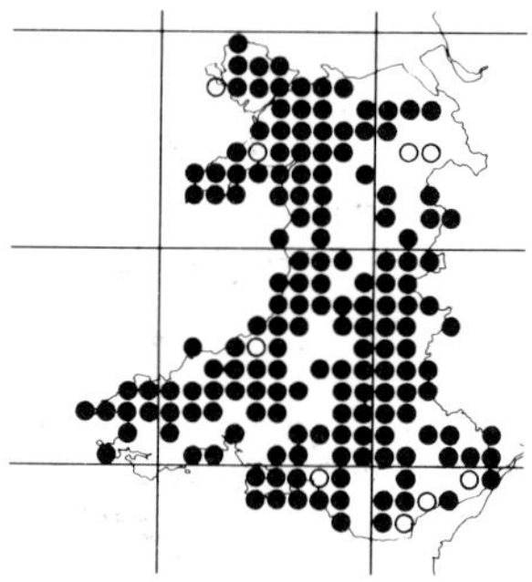

452 Lythrum portula

453 Circaea lutetiana

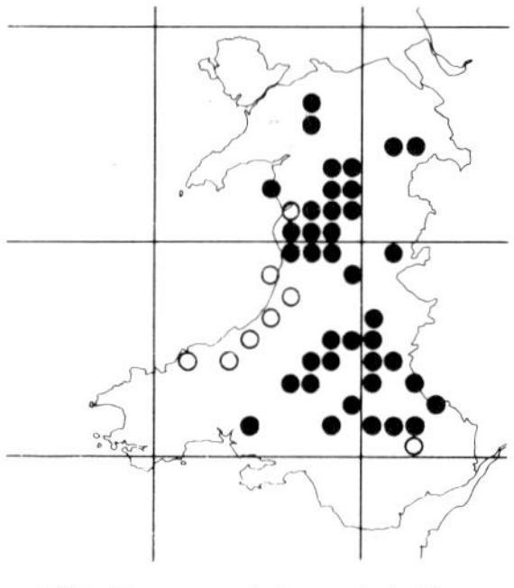

454 Circaea alpina × lutetiana = C. × intermedia

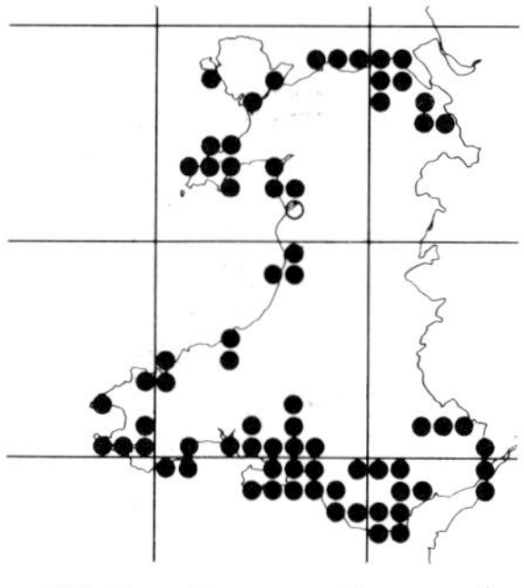

455 Oenothera erythrosepala

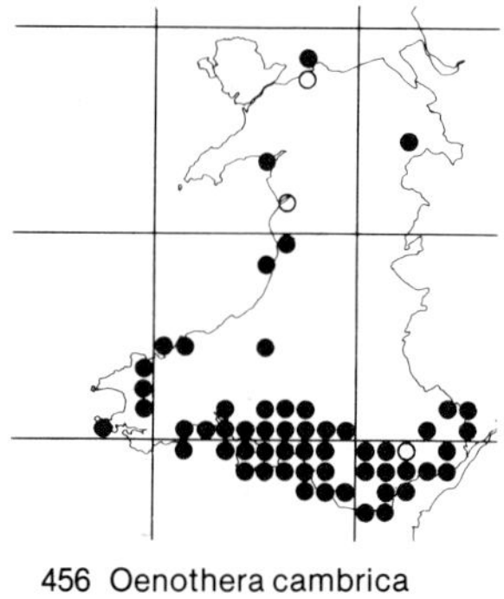

456 Oenothera cambrica

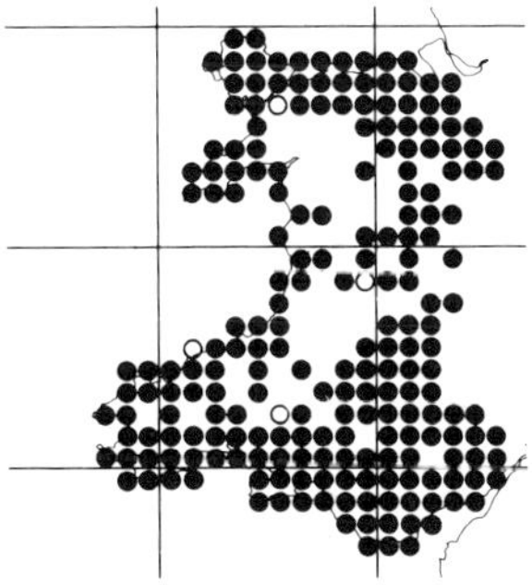

457 Epilobium hirsutum

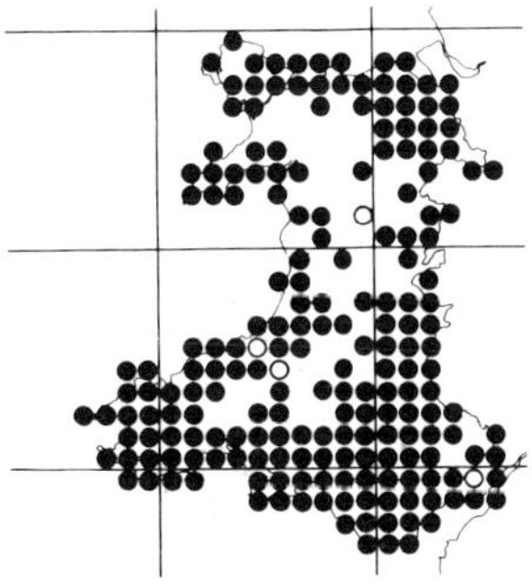

458 Epilobium parviflorum

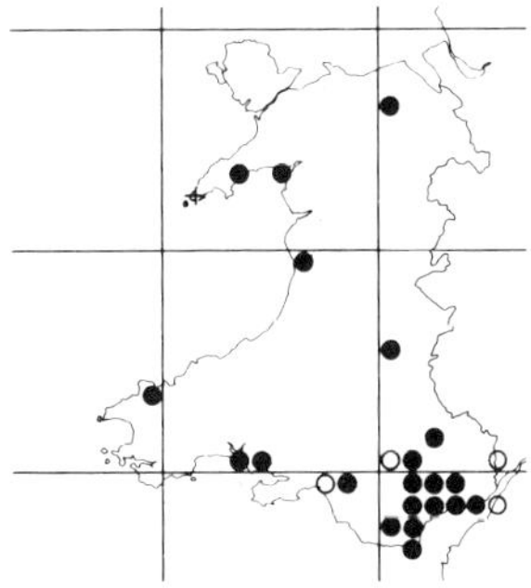

459 Epilobium lanceolatum

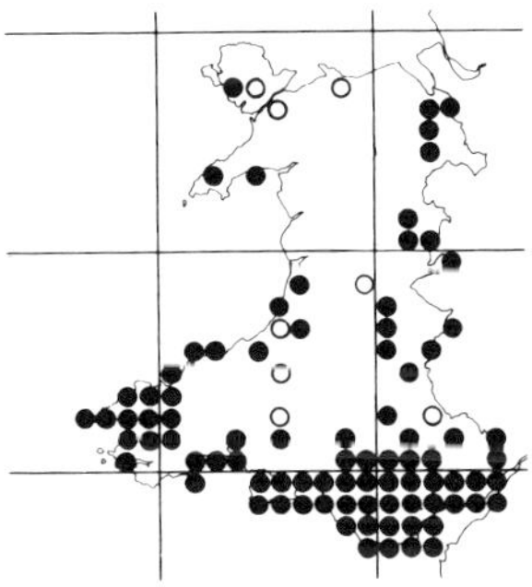

460 Epilobium tetragonum subsp. tetragonum

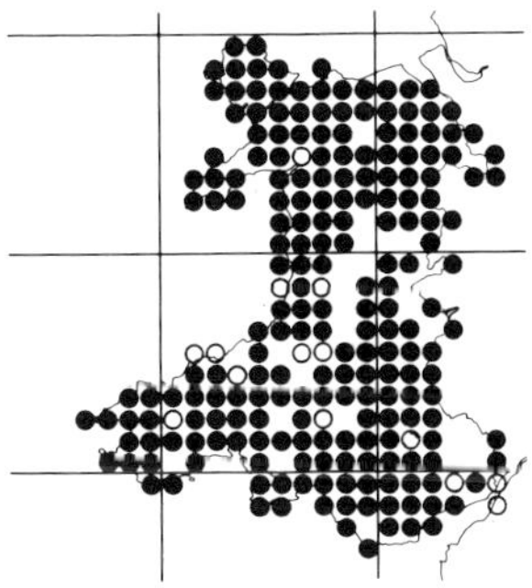

461 Epilobium obscurum

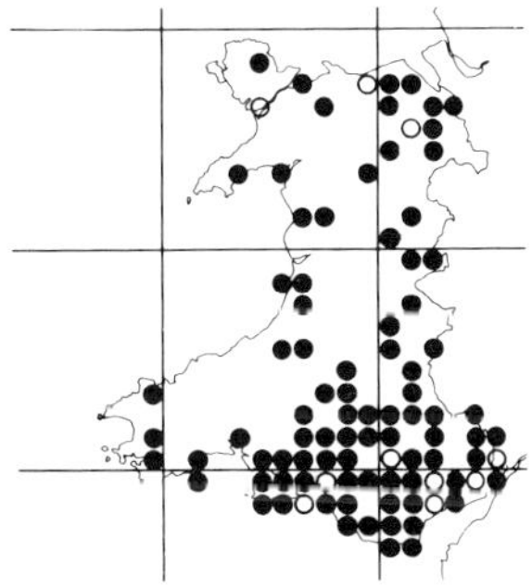

462 Epilobium roseum

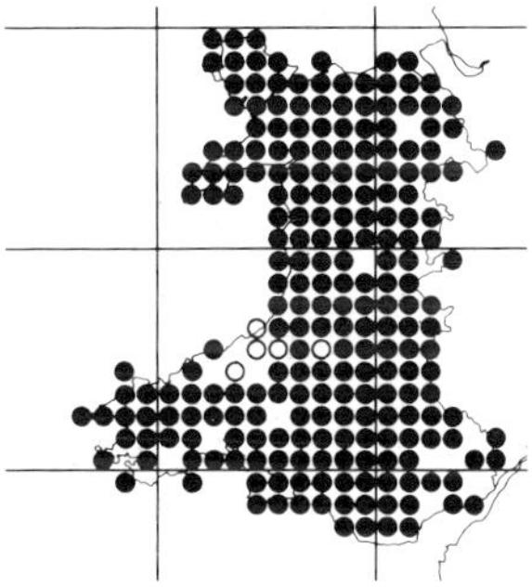

463 Epilobium palustre

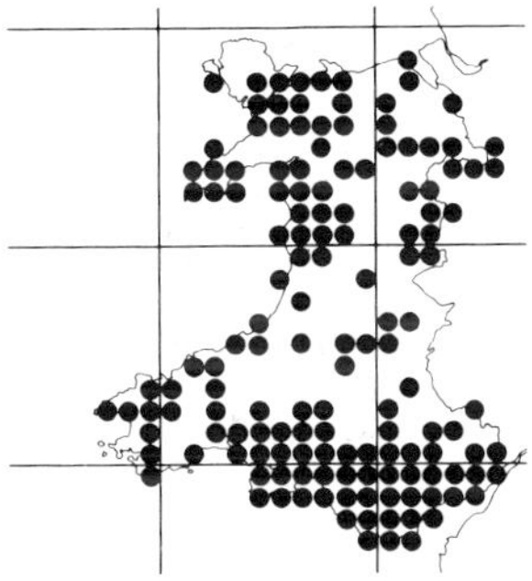

464 Epilobium ciliatum

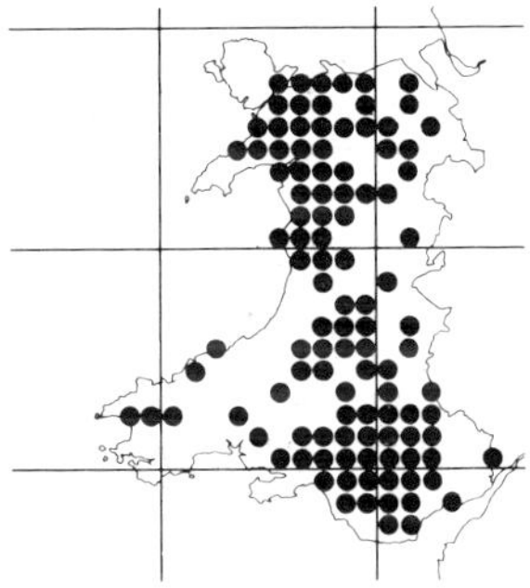

465 Epilobium brunnescens

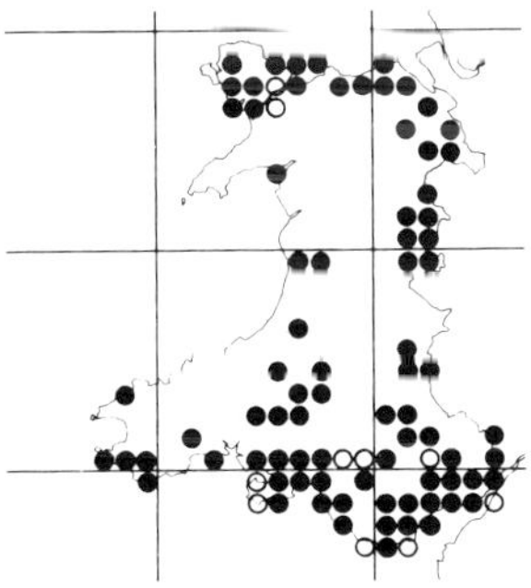

466 Myriophyllum spicatum

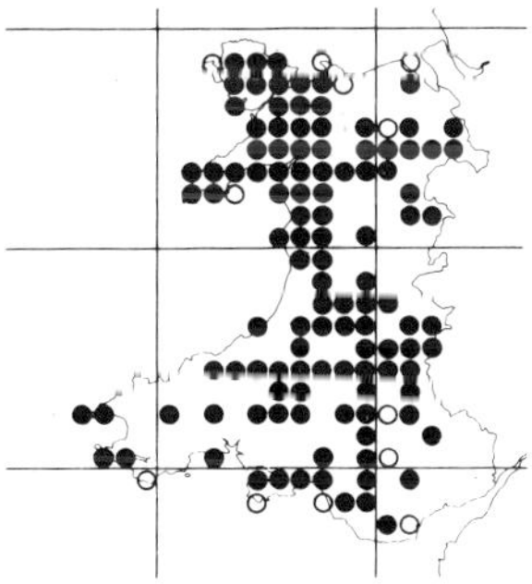

467 Myriophyllum alterniflorum

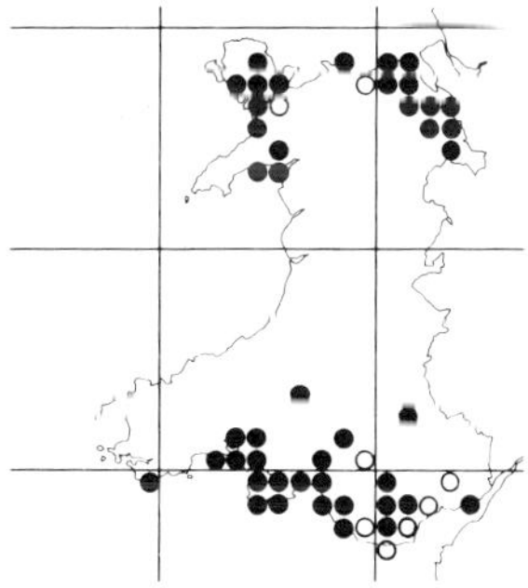

468 Hippuris vulgaris

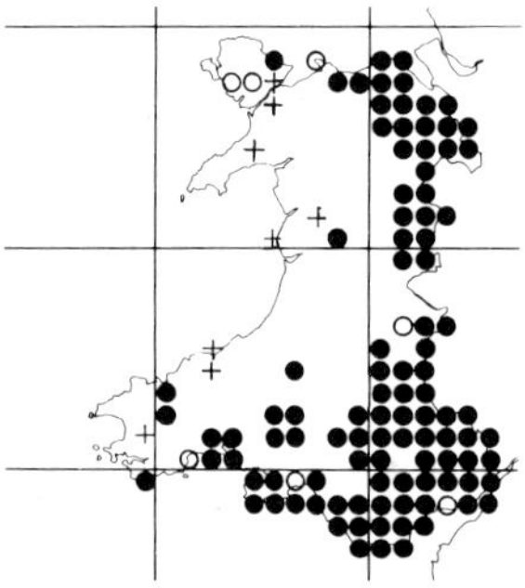

469 Cornus sanguinea

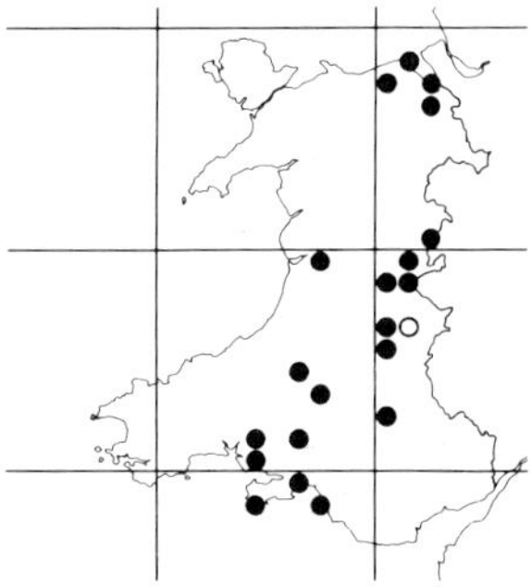

470 Cornus sericea

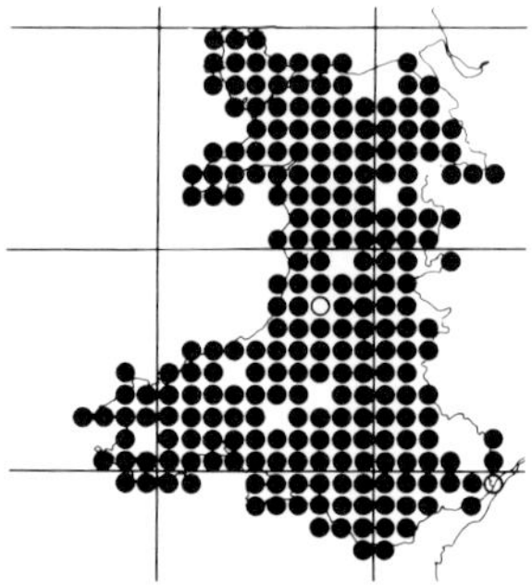

471 Hydrocotyle vulgaris

472 Sanicula europaea

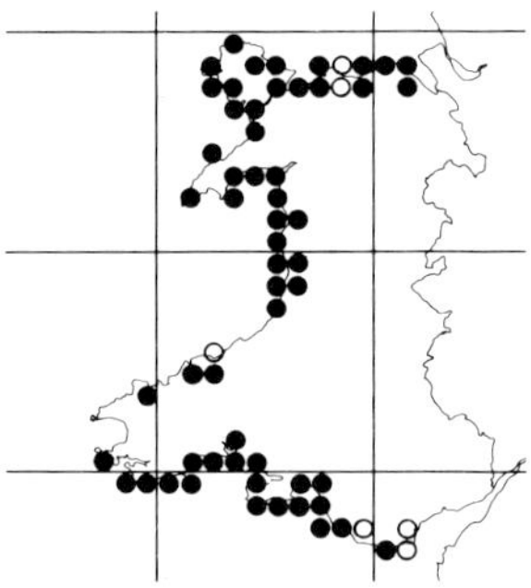

473 Eryngium maritimum

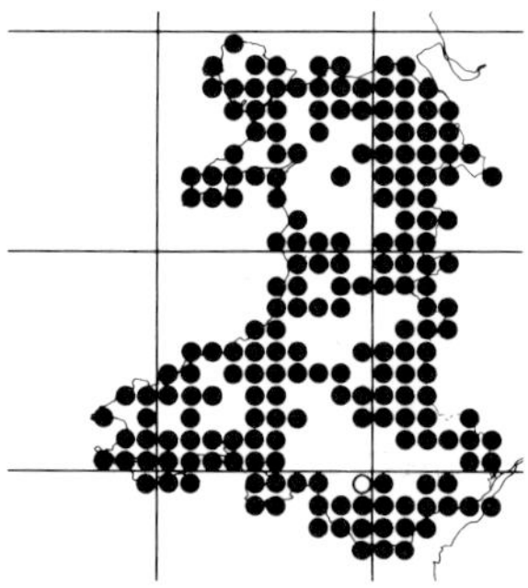

474 Chaerophyllum temulentum

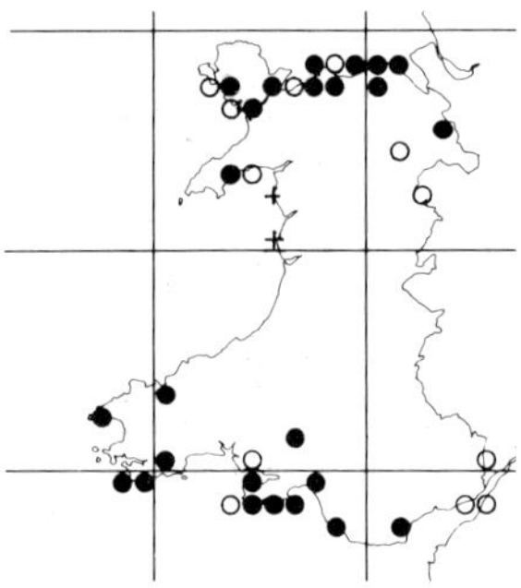

475 Anthriscus caucalis

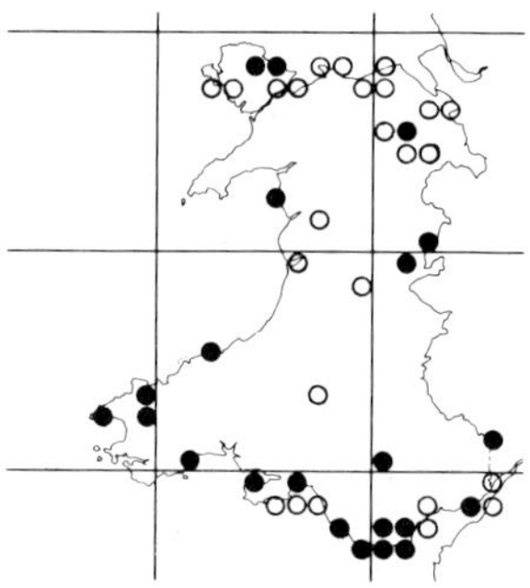

476 Scandix pecten-veneris

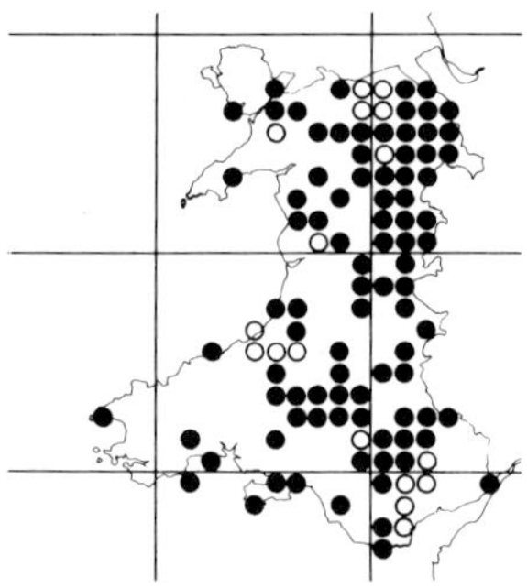

477 Myrrhis odorata

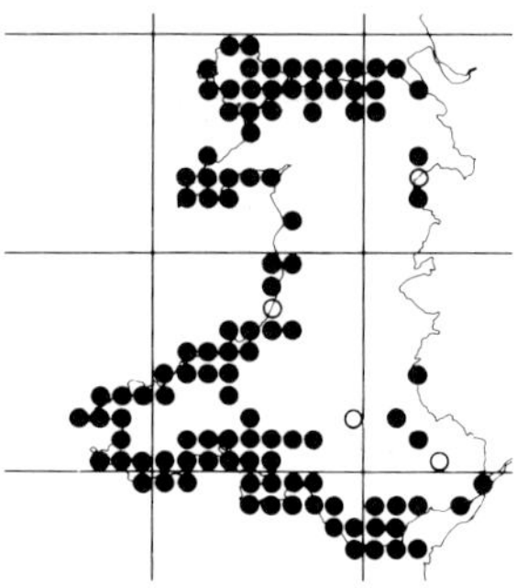

478 Smyrnium olusatrum

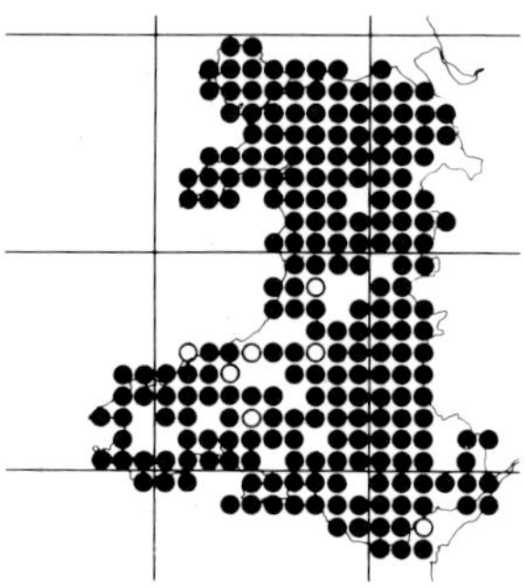

479 Pimpinella saxifraga

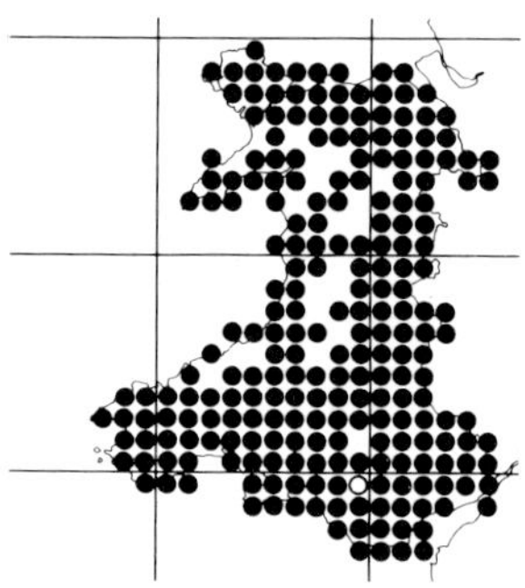

480 Aegopodium podagraria

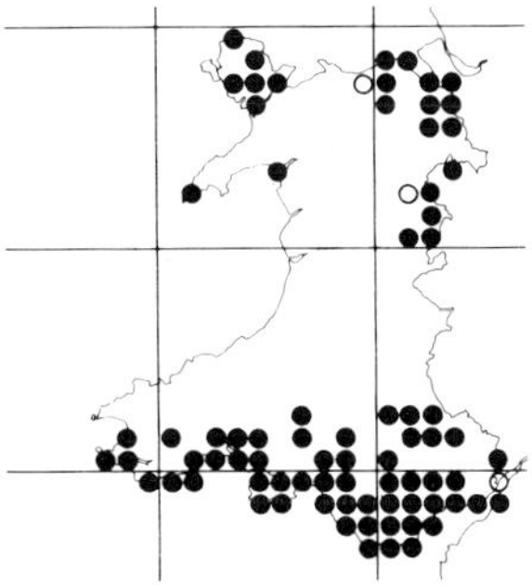

481 Berula erecta

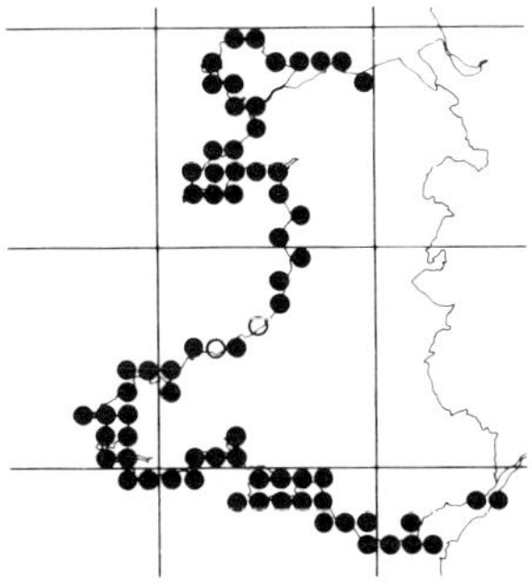

482 Crithmum maritimum

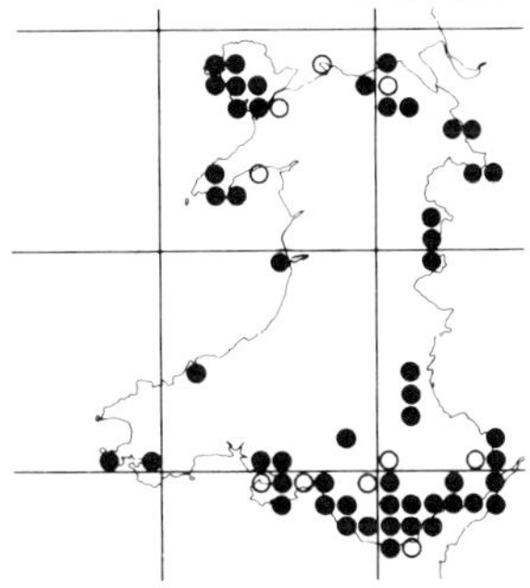

483 Oenanthe fistulosa

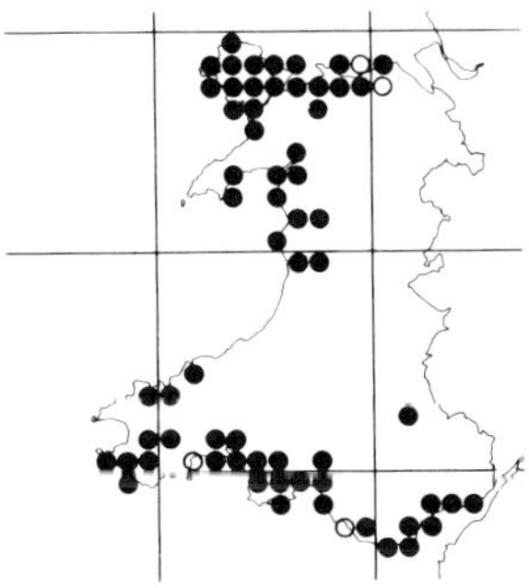

484 Oenanthe lachenalii

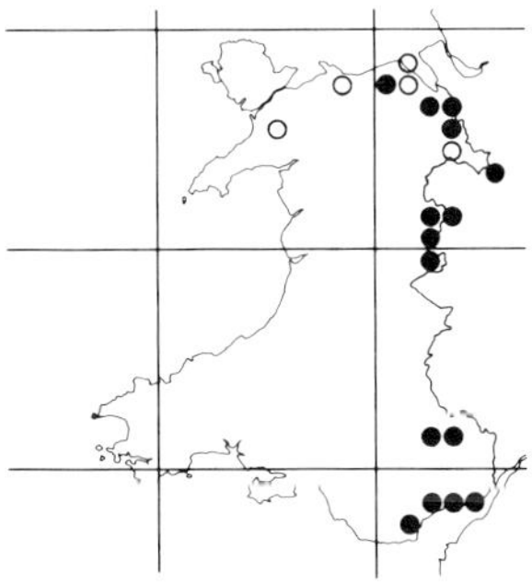

485 Oenanthe aquatica

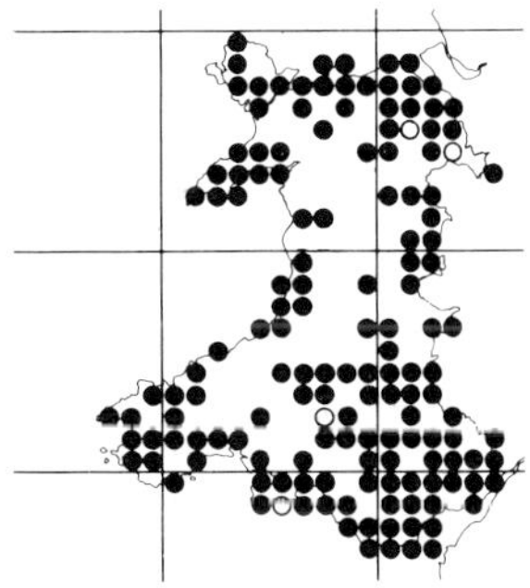

486 Aethusa cynapium

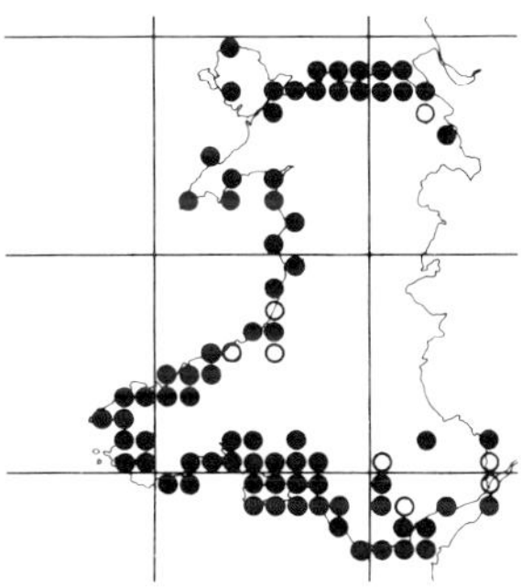

487 Foeniculum vulgare

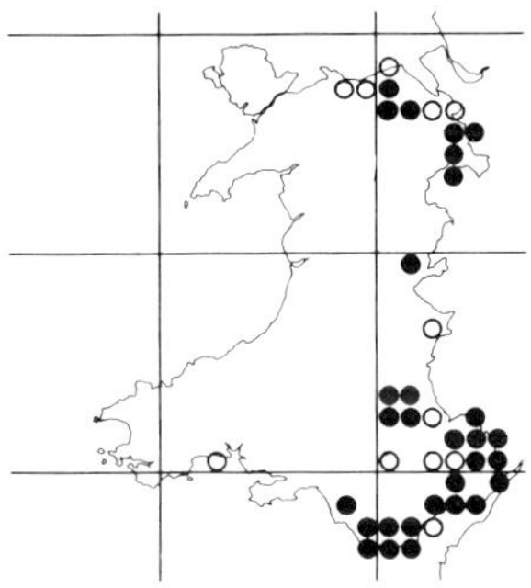

488 Silaum silaus

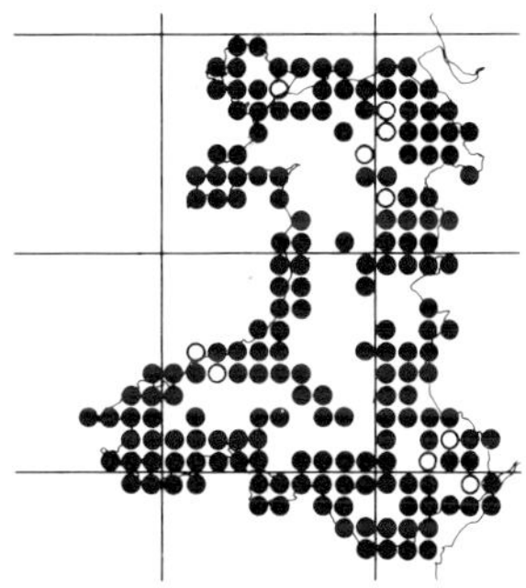

489 Conium maculatum

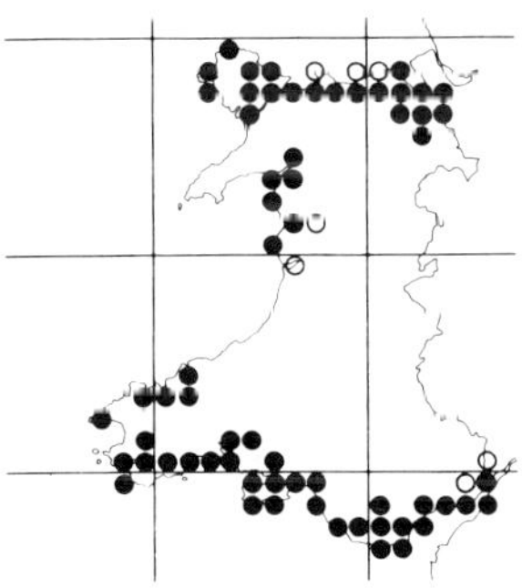

490 Apium graveolens

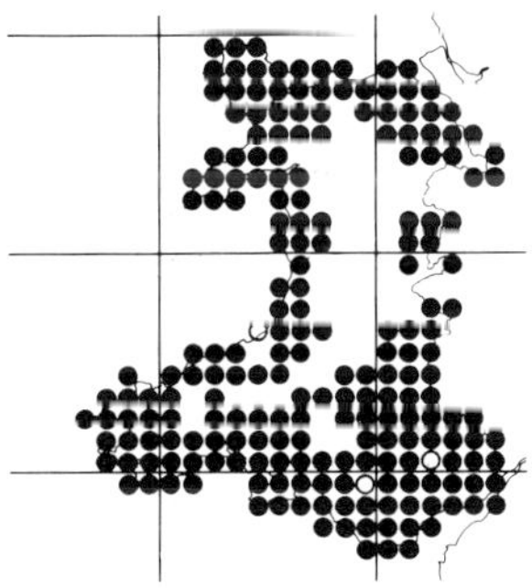

491 Apium nodiflorum

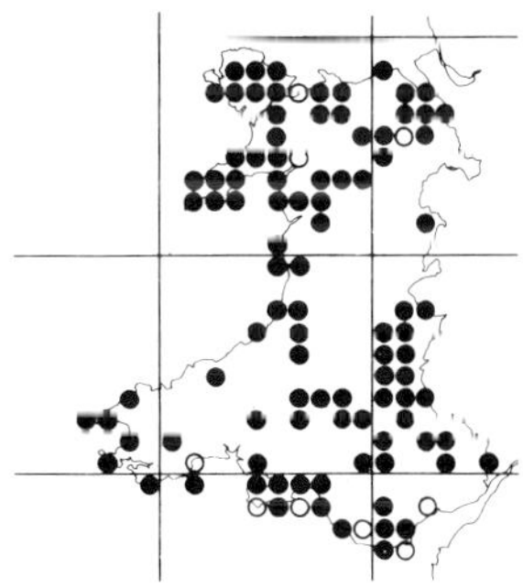

492 Apium inundatum

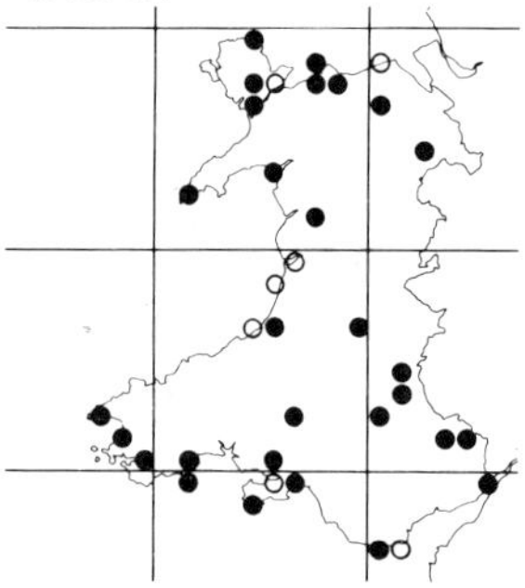

493 Petroselinum crispum

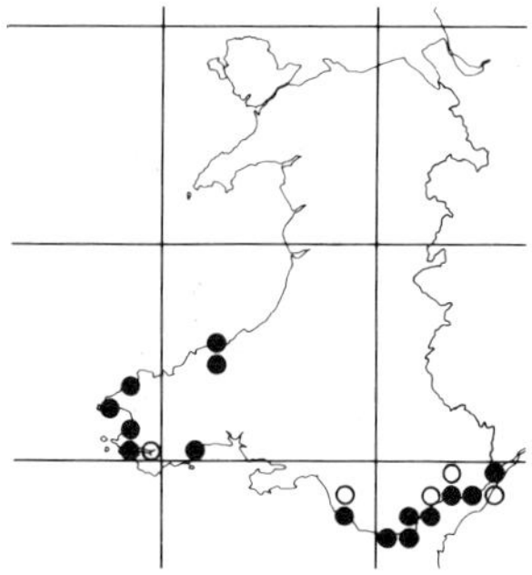

494 Petroselinum segetum

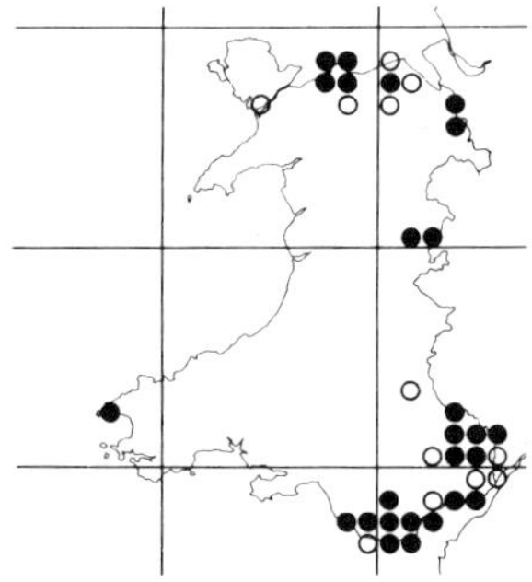

495 Sison amomum

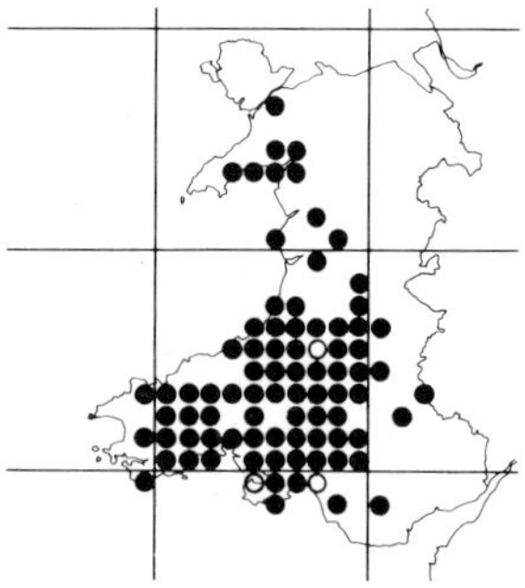

496 Carum verticillatum

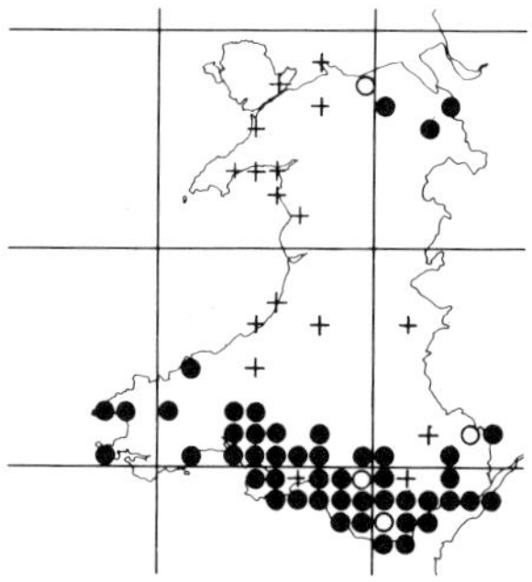

497 Pastinaca sativa

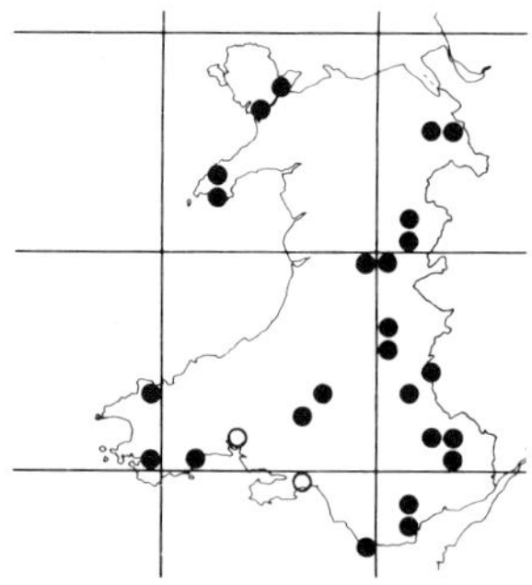

498 Heracleum mantegazzianum

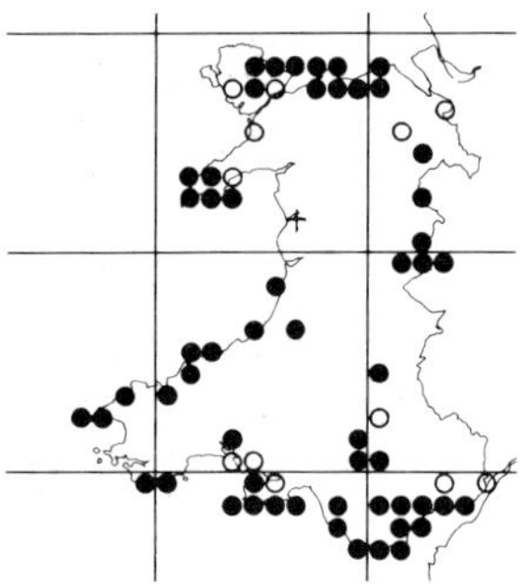

499 Torilis nodosa

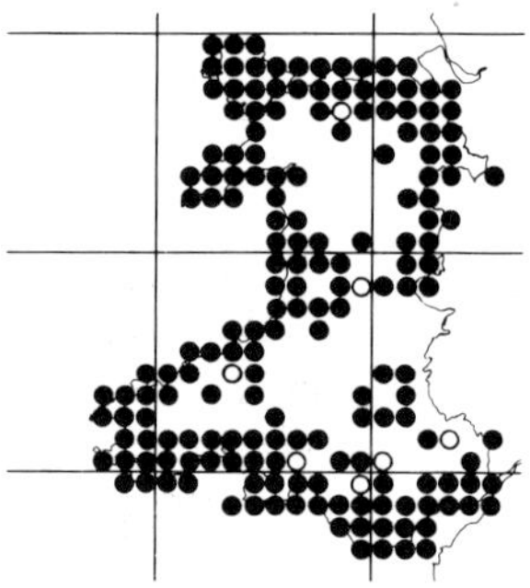

500 Daucus carota subsp. carota

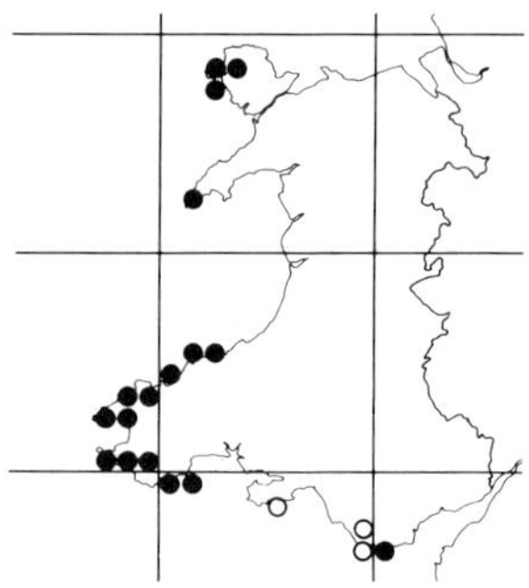

501 Daucus carota subsp. gummifer

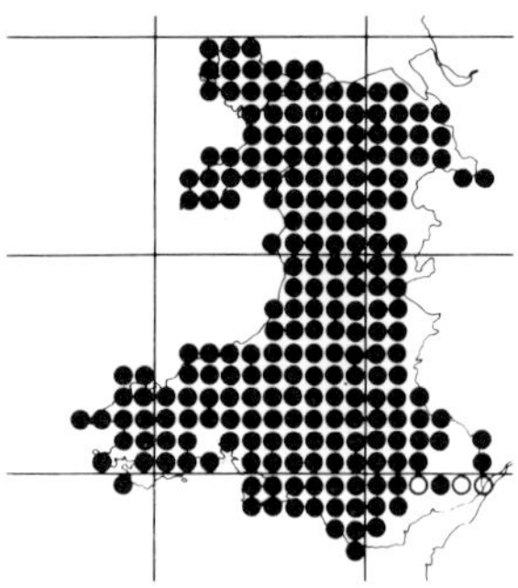

502 Erica tetralix

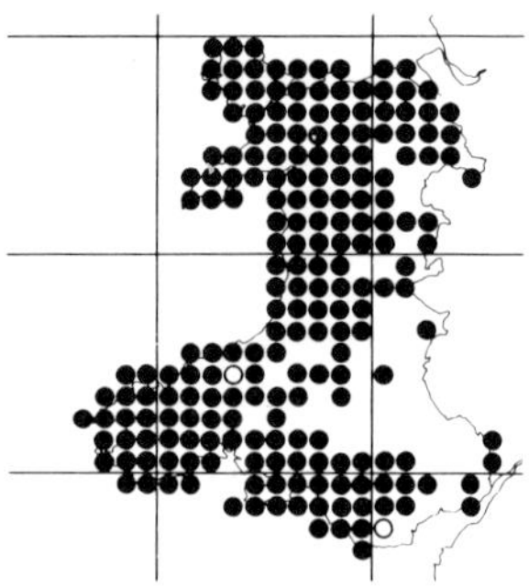

503 Erica cinerea

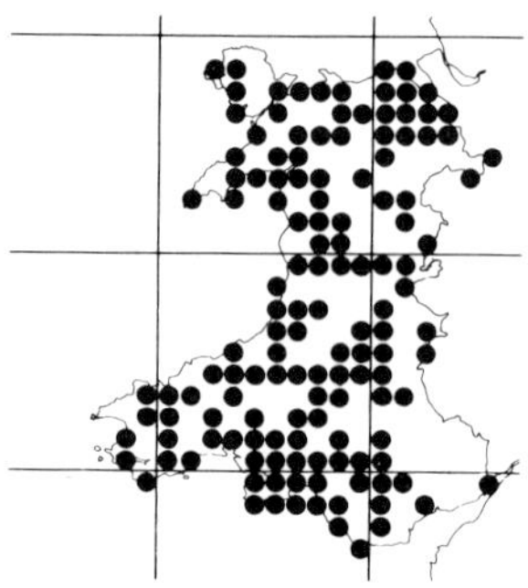

504 Rhododendron ponticum

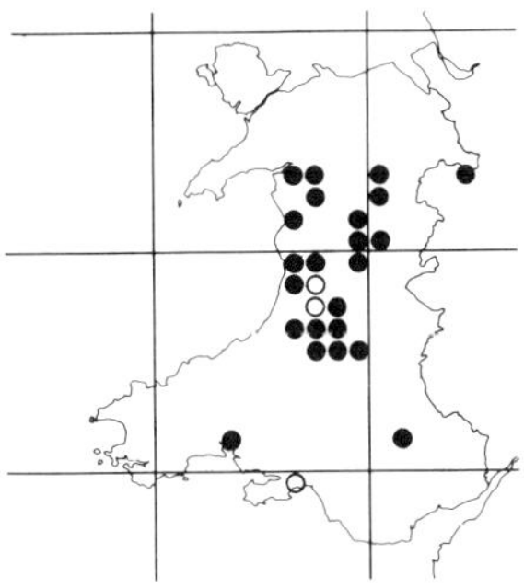
505 Andromeda polifolia

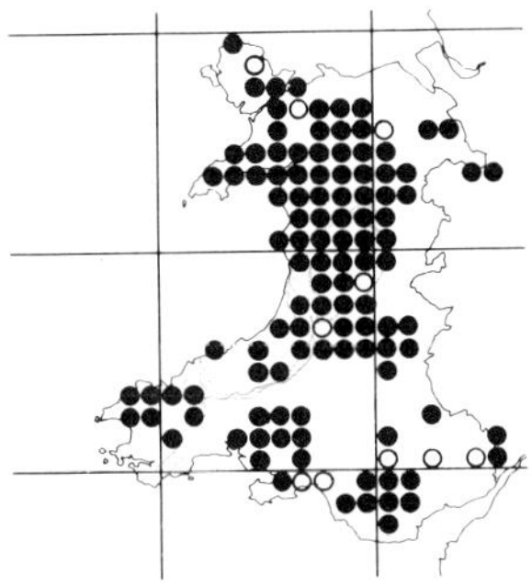
506 Vaccinium oxycoccus

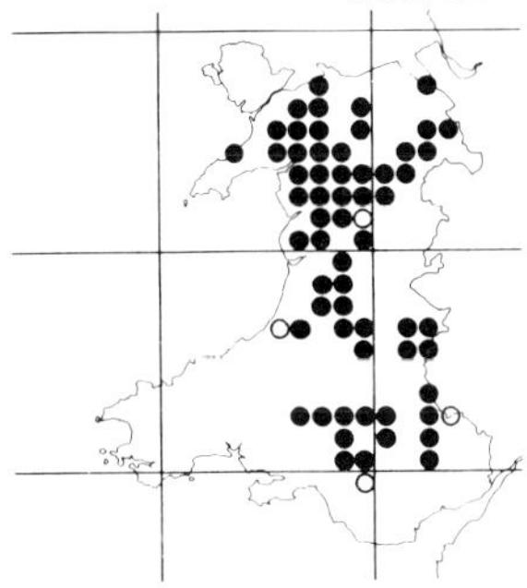
507 Vaccinium vitis-idaea

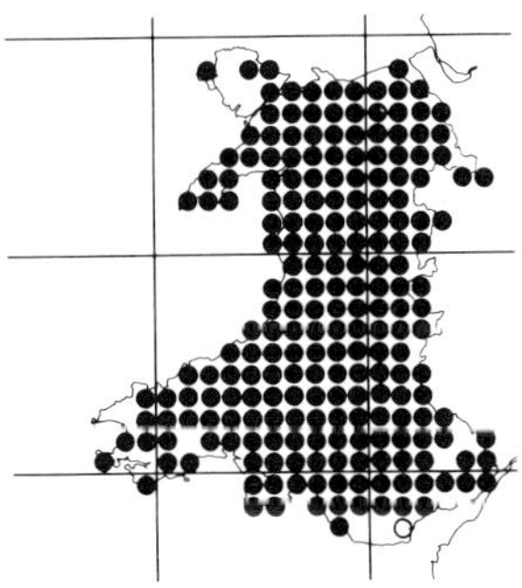
508 Vaccinium myrtillus

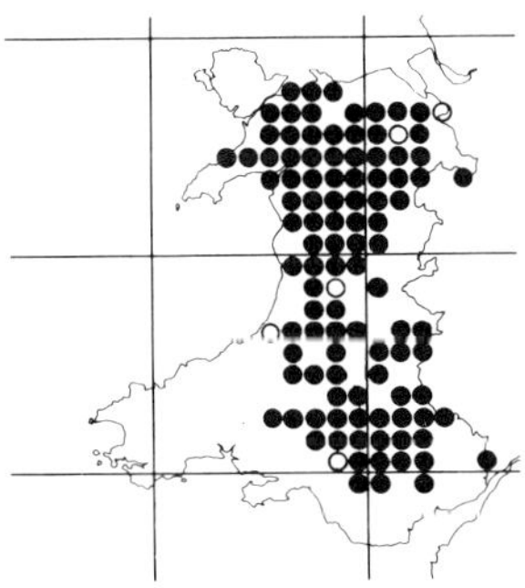
509 Empetrum nigrum subsp. nigrum

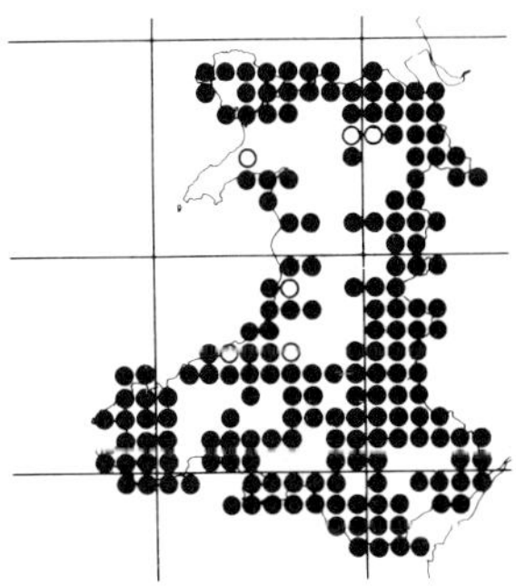
510 Primula veris

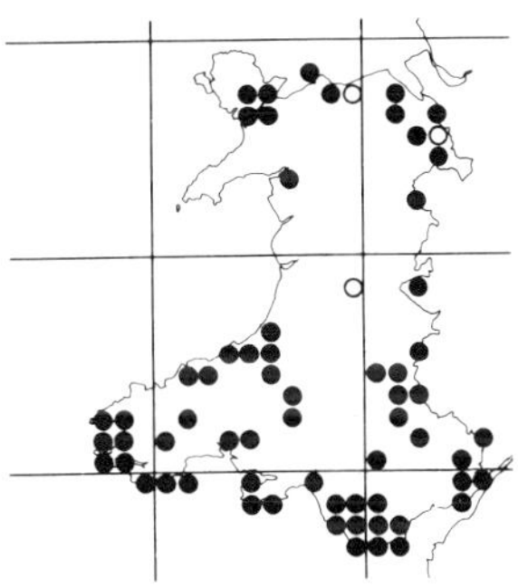
511 Primula veris × vulgaris = P. × tommasinii

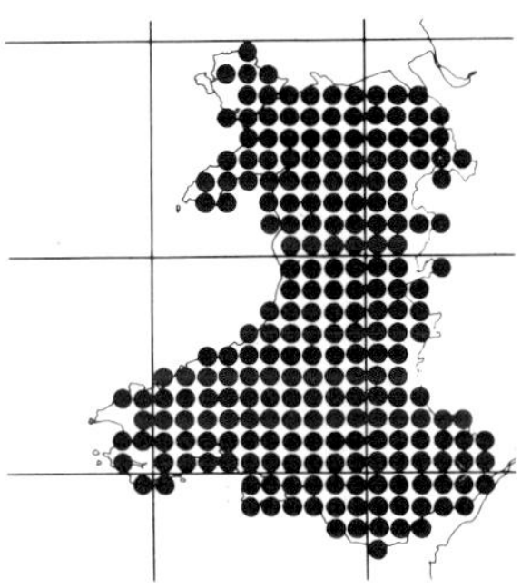
512 Lysimachia nemorum

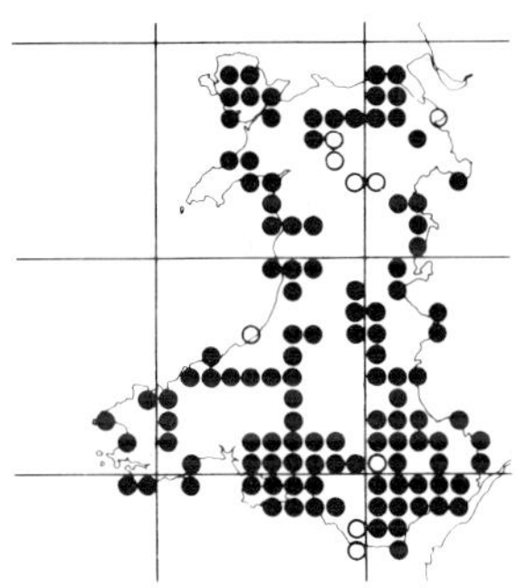
513 Lysimachia vulgaris

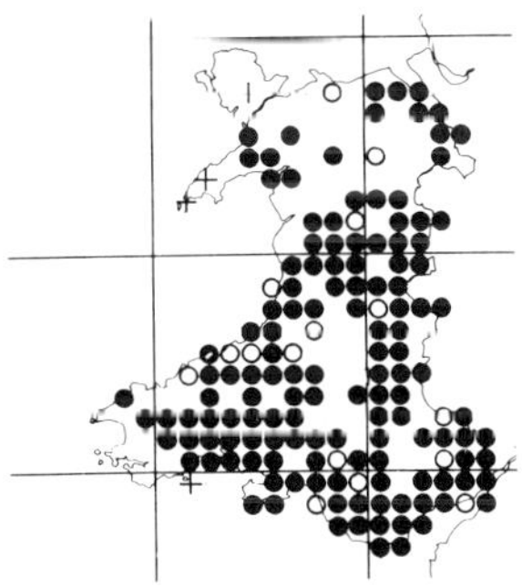
514 Lysimachia nummularia

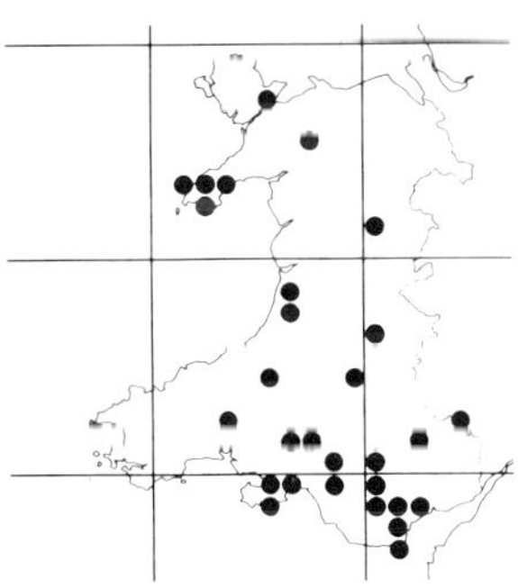
515 Lysimachia punctata

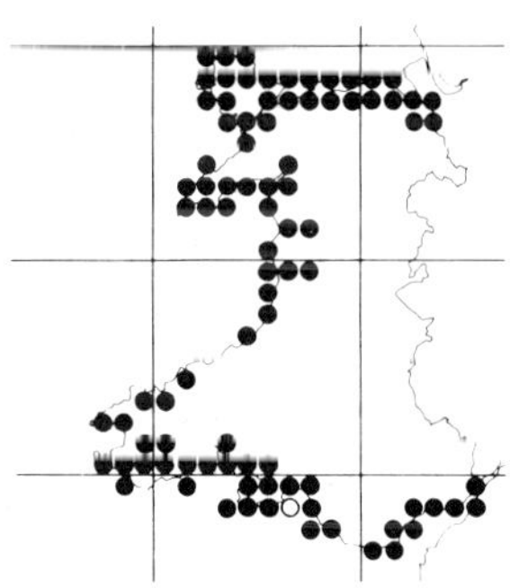
516 Glaux maritima

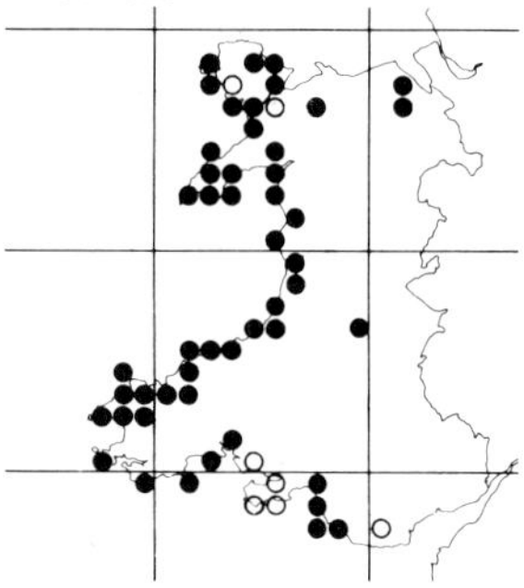

517 Anagallis minima

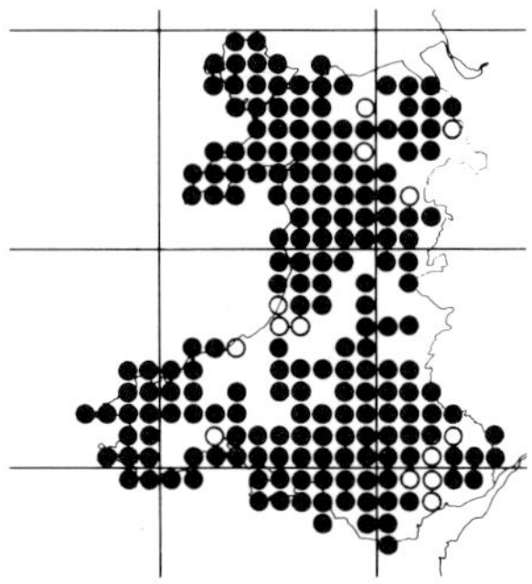

518 Anagallis tenella

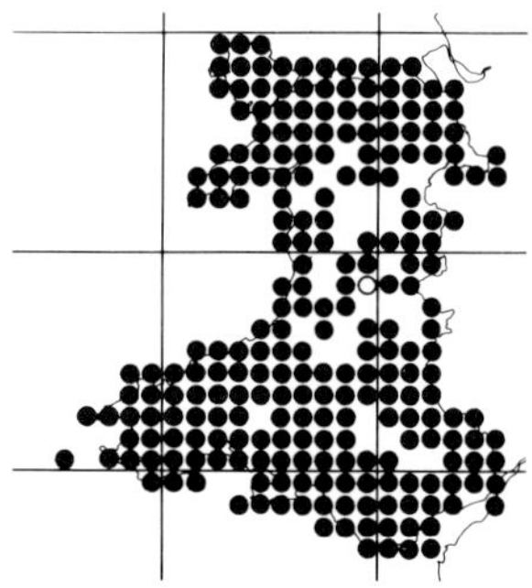

519 Anagallis arvensis

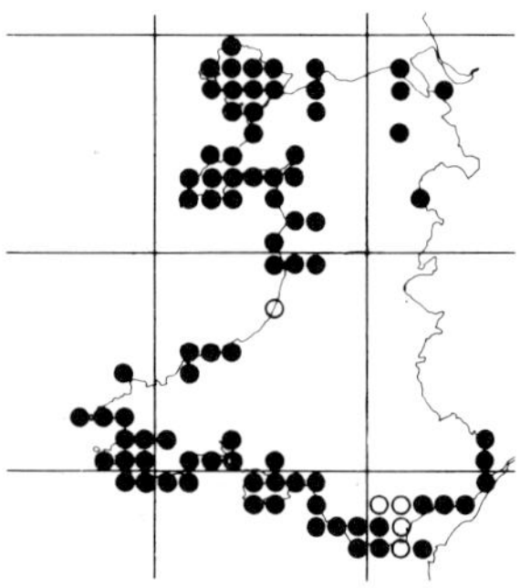

520 Samolus valerandi

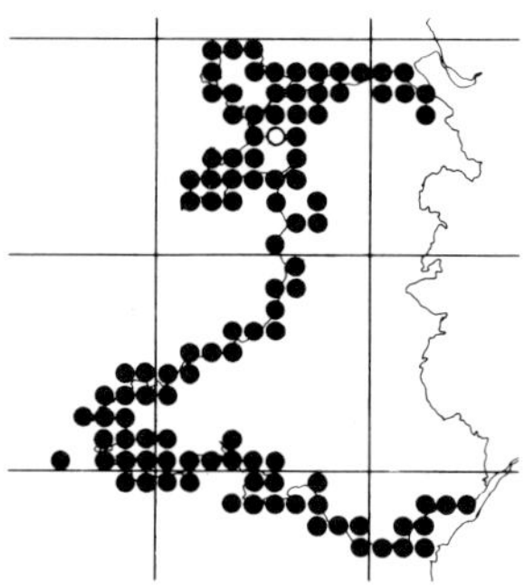

521 Armeria maritima

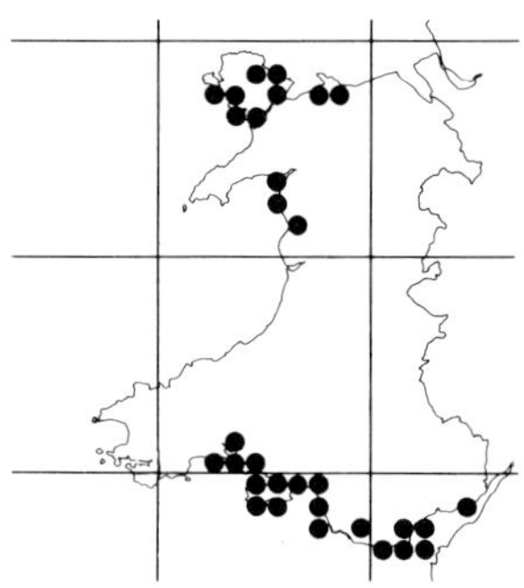

522 Limonium vulgare

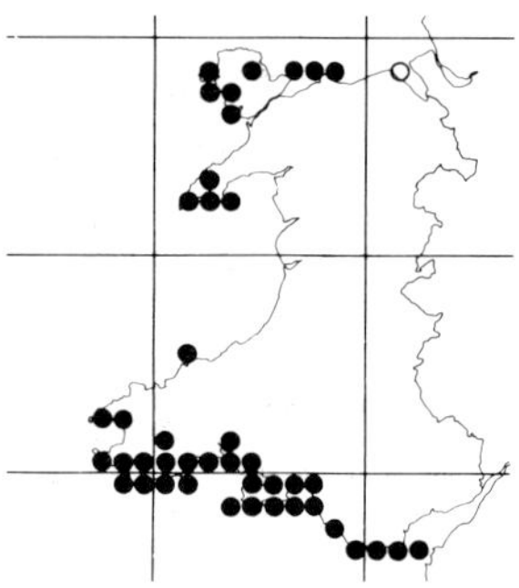

523 Limonium binervosum

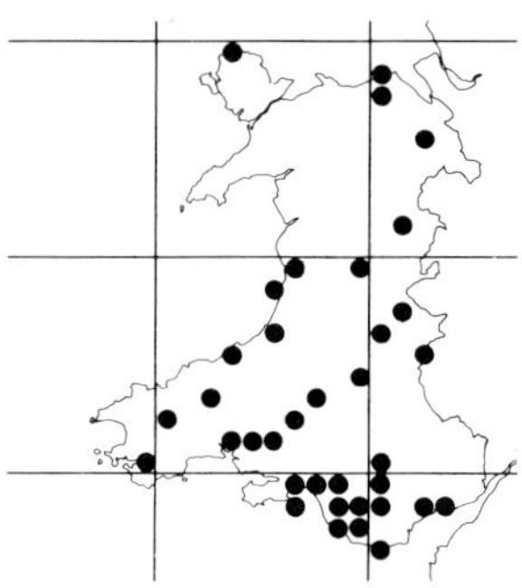

524 Syringa vulgaris

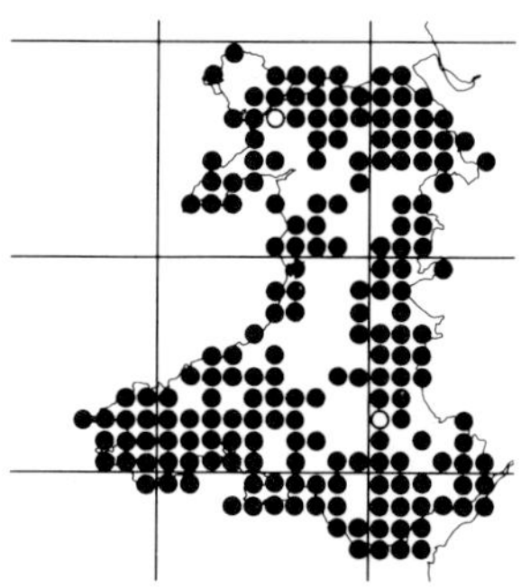

525 Ligustrum vulgare

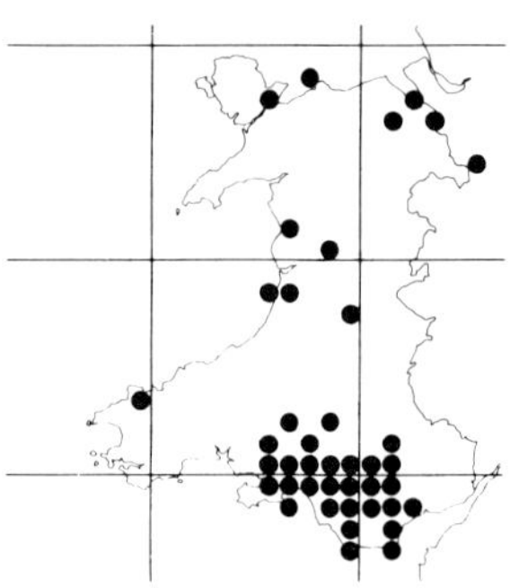

526 Ligustrum ovalifolium

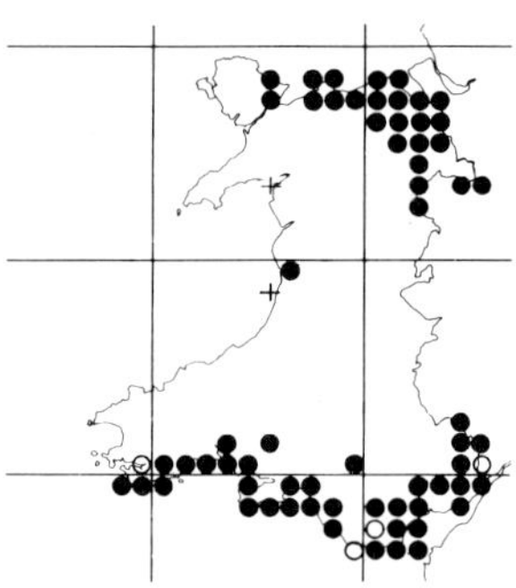

527 Blackstonia perfoliata

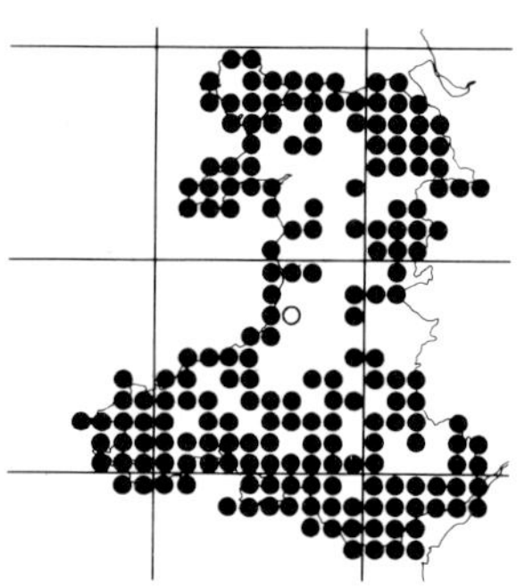

528 Centaurium erythraea

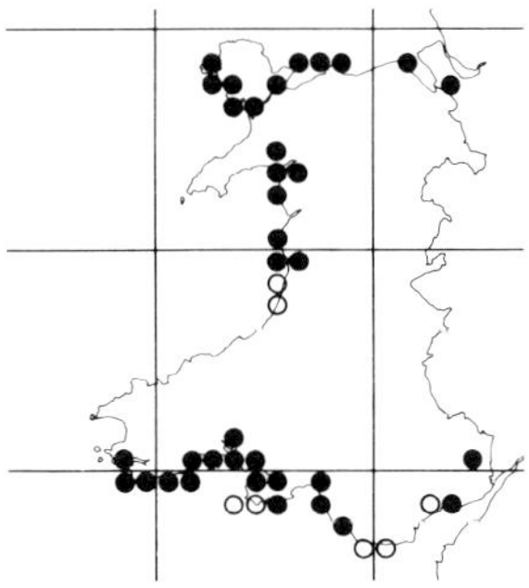

529 Centaurium pulchellum

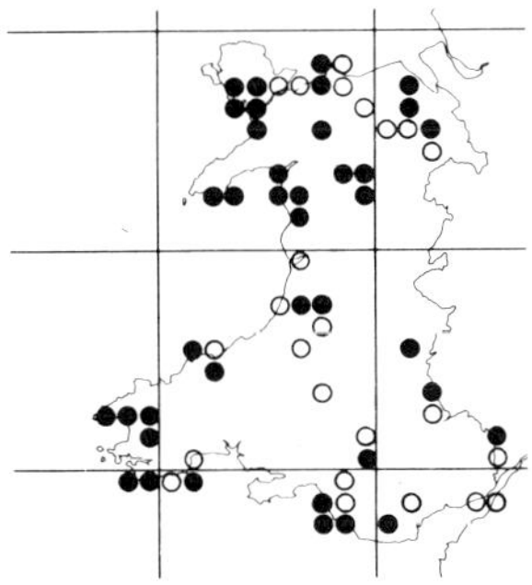

530 Gentianella campestris

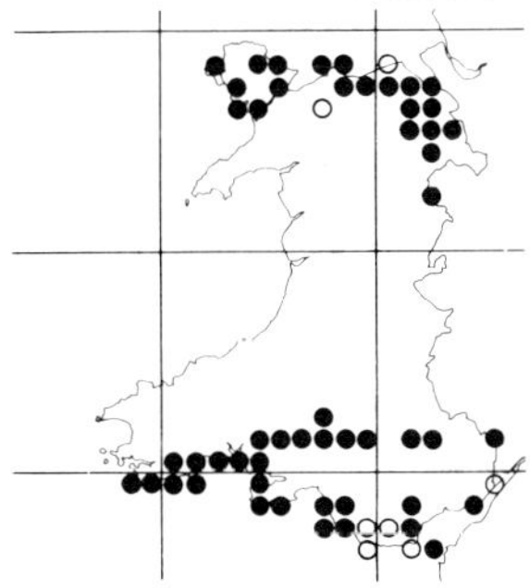

531 Gentianella amarella

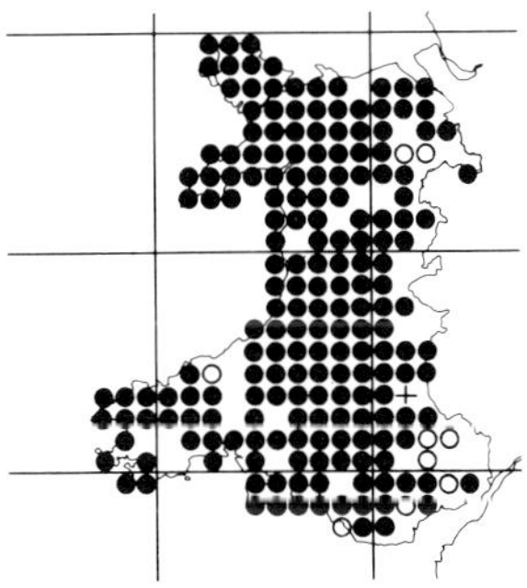

532 Menyanthes trifoliata

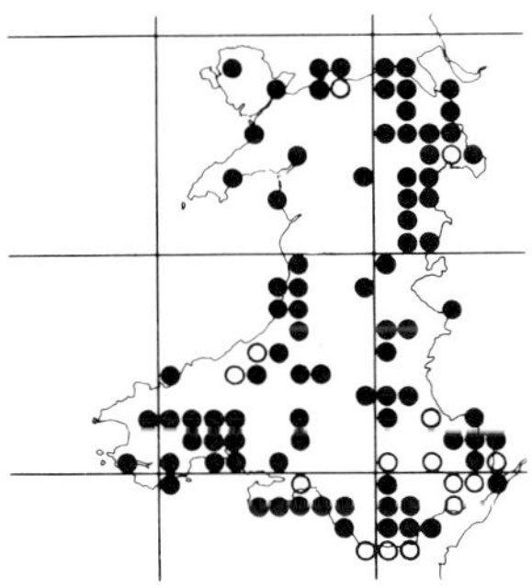

533 Vinca minor

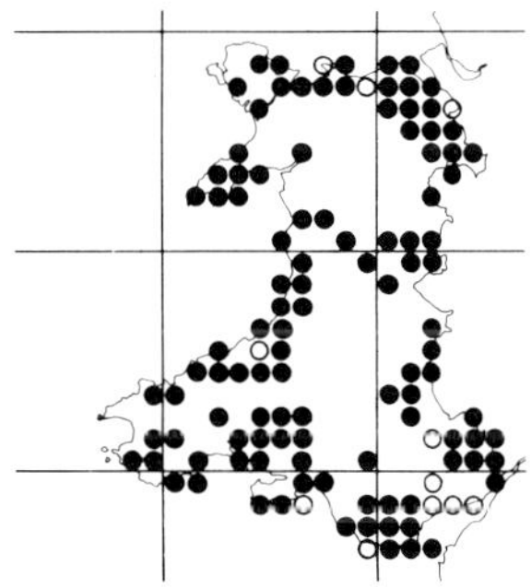

534 Vinca major

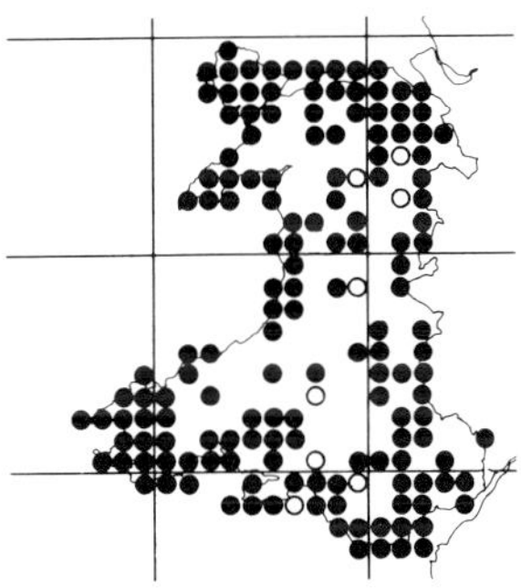

535 Sherardia arvensis

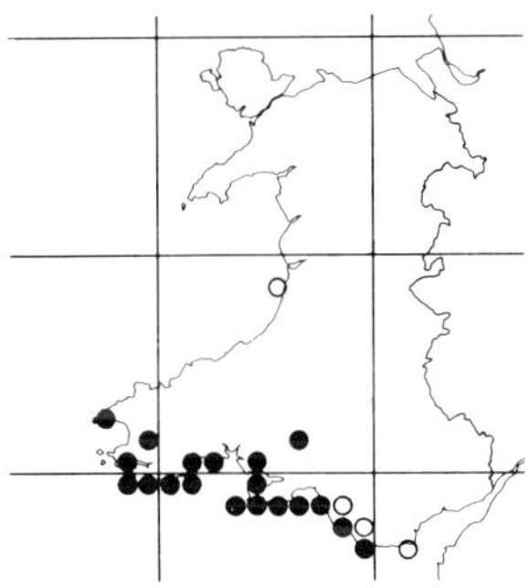

536 Asperula cynanchica agg.

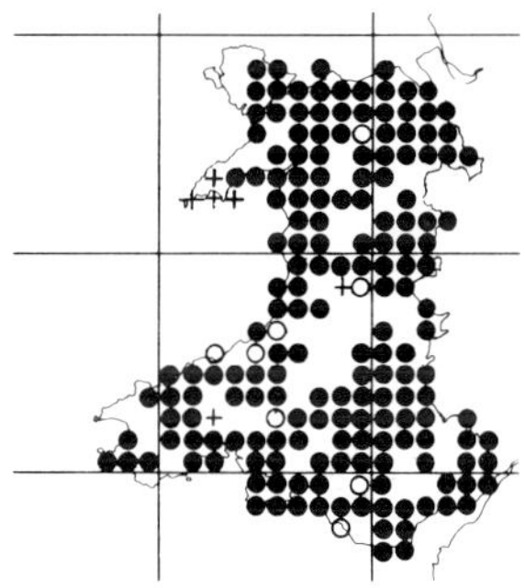

537 Galium odoratum

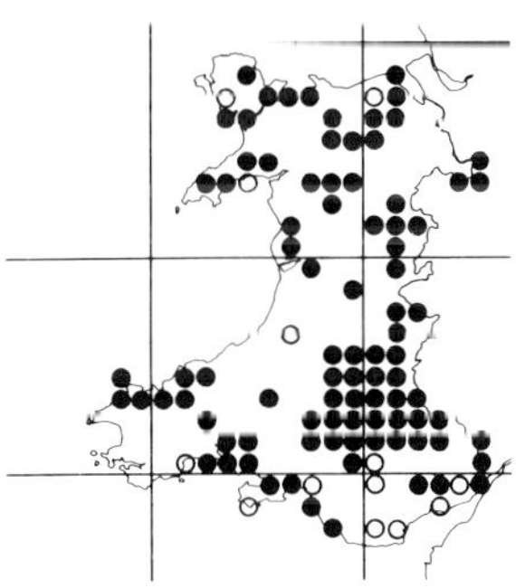

538 Galium uliginosum

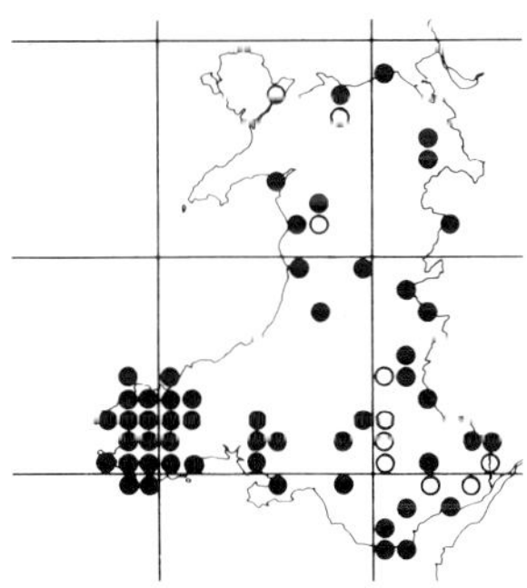

539 Galium palustre s.s.

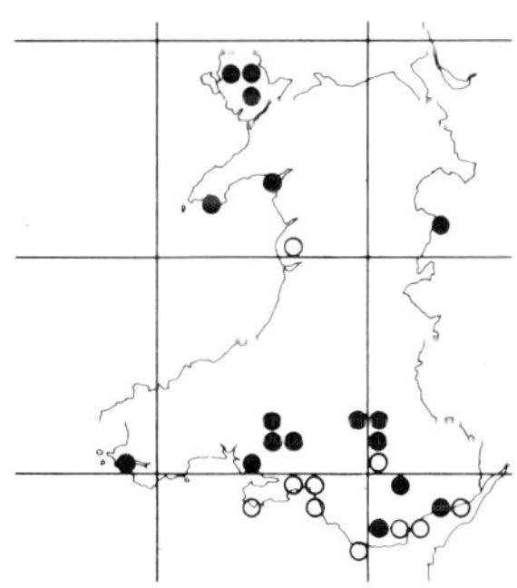

540 Galium elongatum

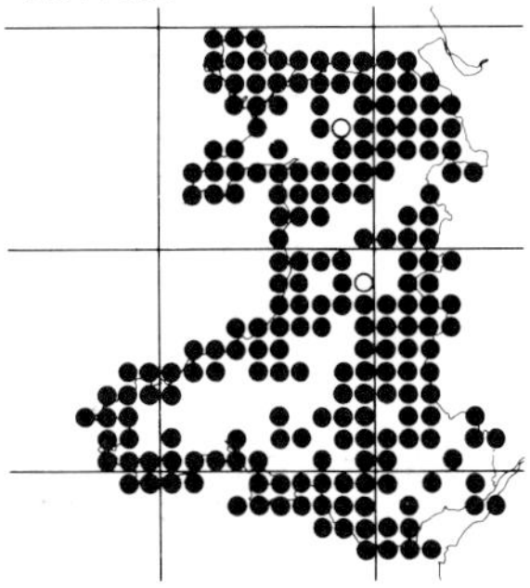

541 Galium verum

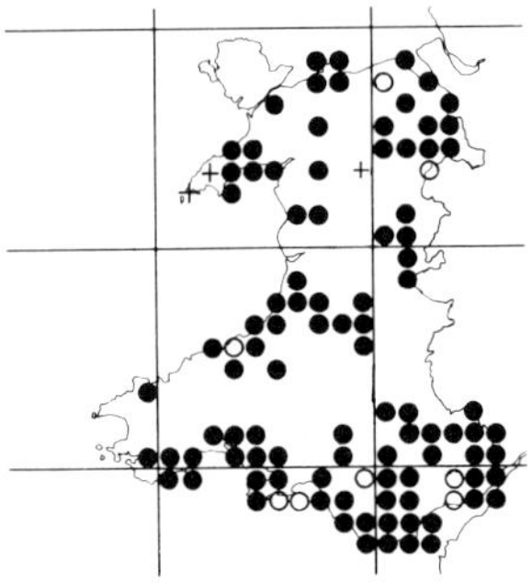

542 Galium mollugo agg.

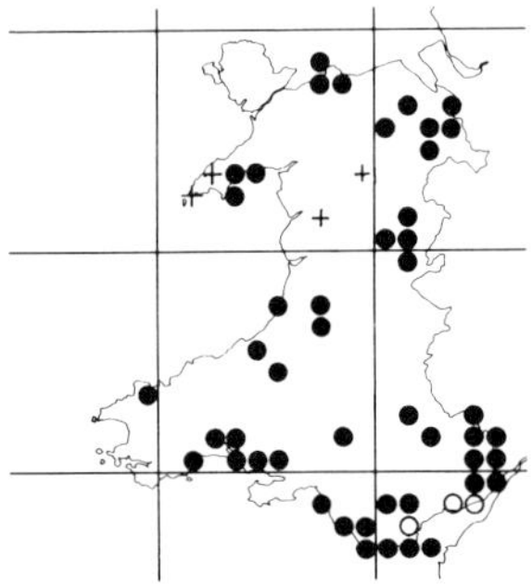

543 Galium mollugo s.s.

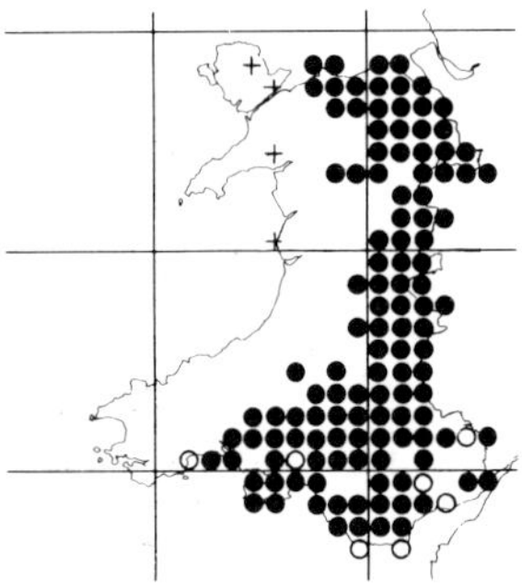

544 Cruciata laevipes

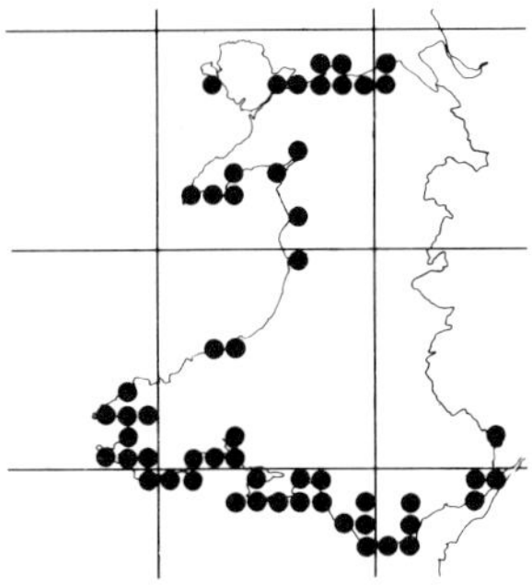

545 Rubia peregrina

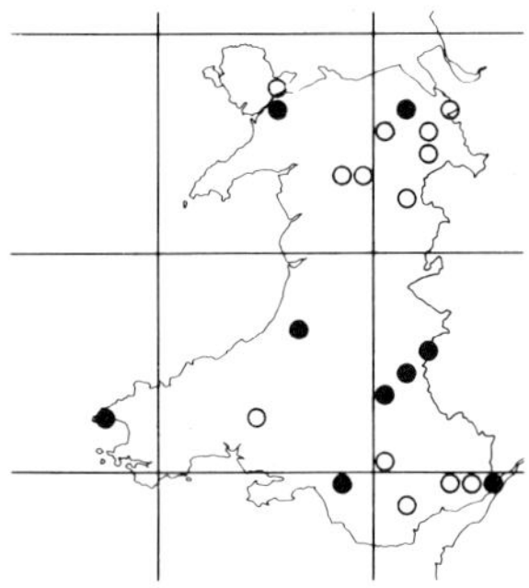

546 Polemonium caeruleum

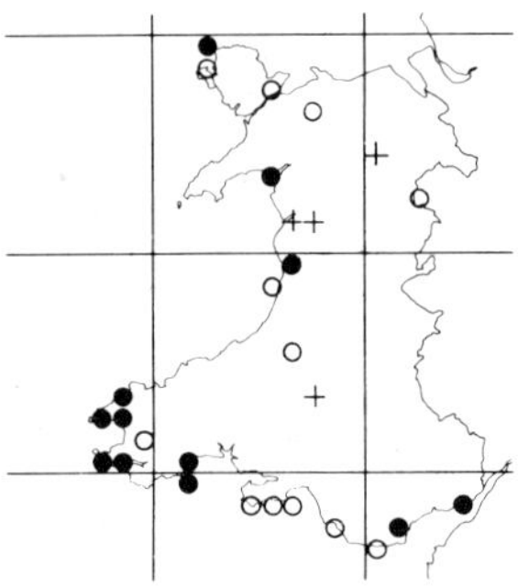

547 Cuscuta epithymum

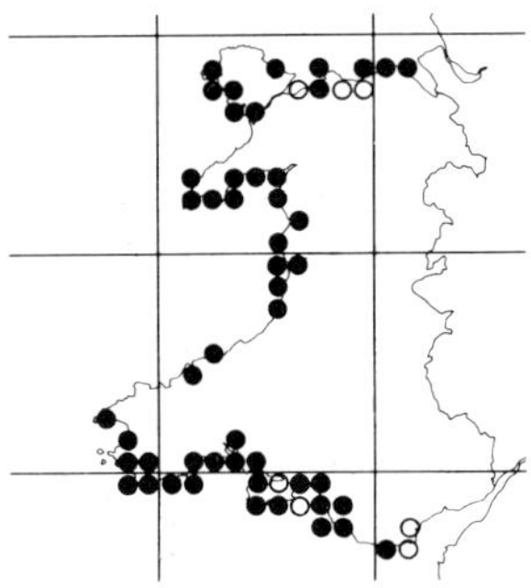

548 Calystegia soldanella

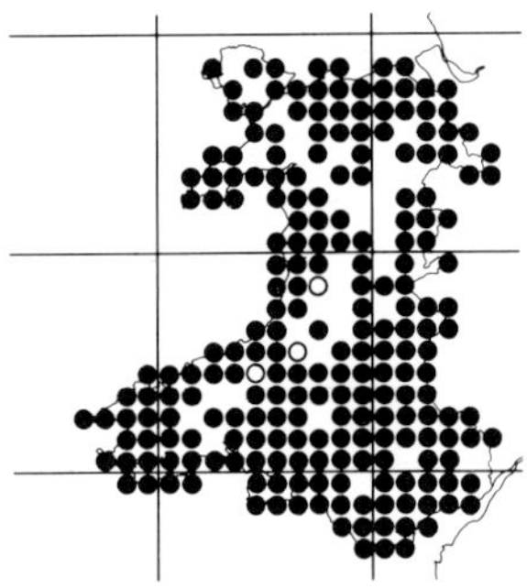

549 Calystegia sepium subsp. sepium

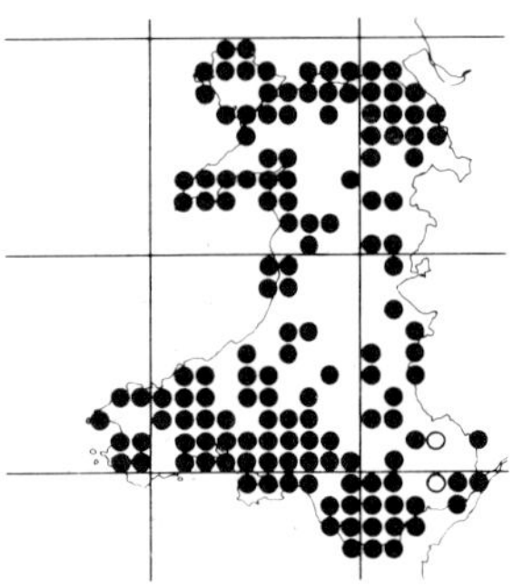

550 Calystegia silvatica

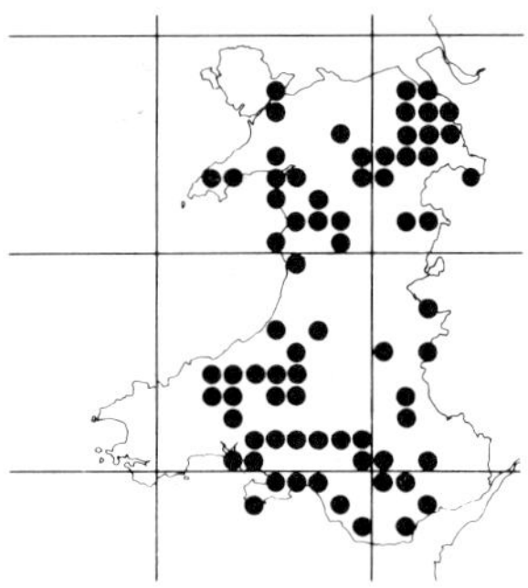

551 Calystegia pulchra

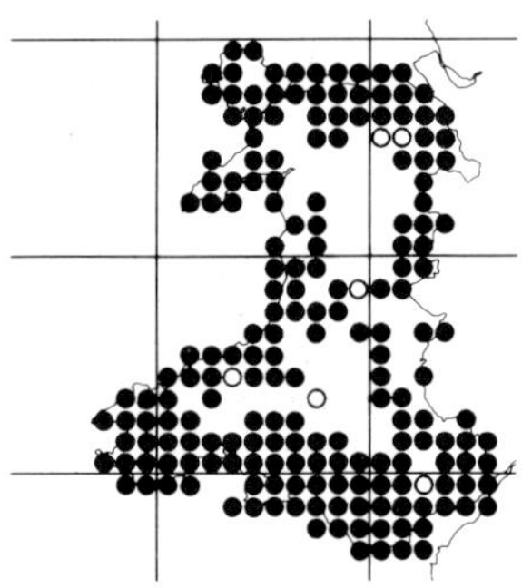

552 Convolvulus arvensis

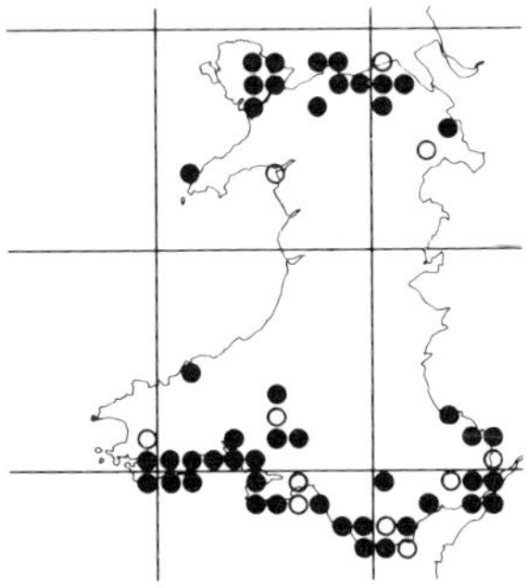

553 Lithospermum officinale

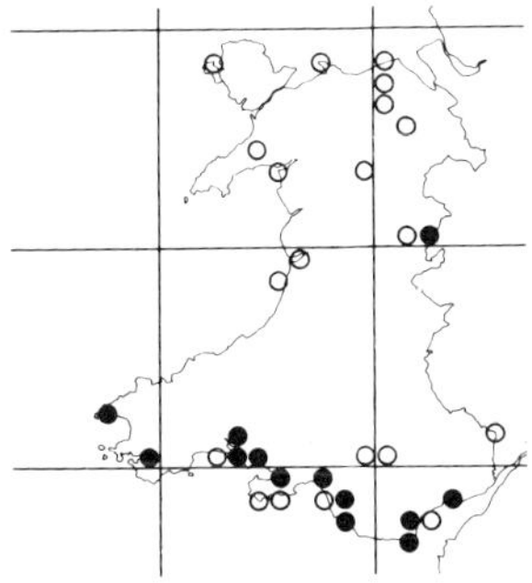

554 Buglossoides arvensis

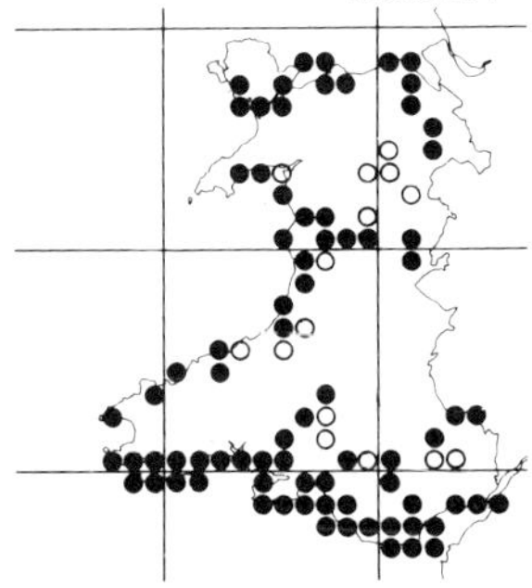

555 Echium vulgare

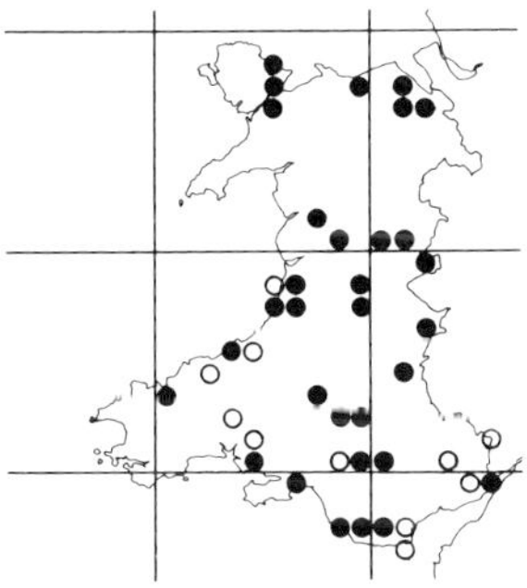

556 Pulmonaria officinalis

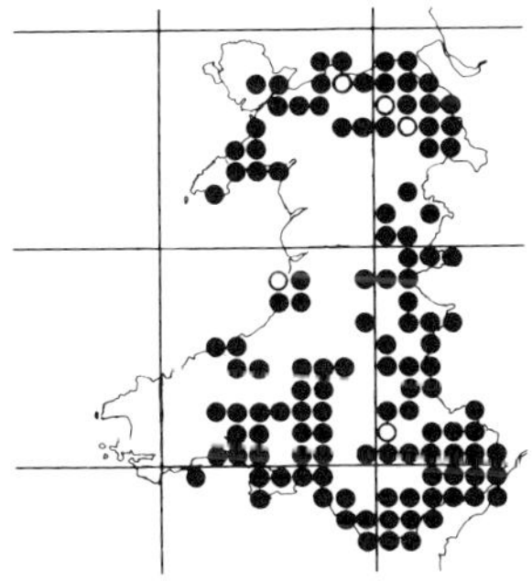

557 Symphytum officinale

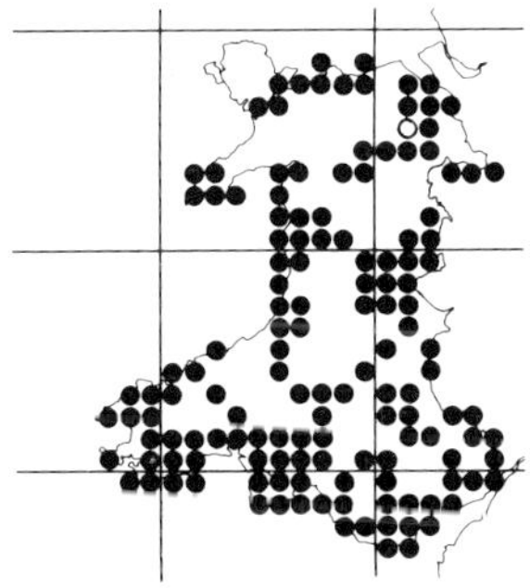

558 Symphytum asperum × officinale = S. × uplandicum

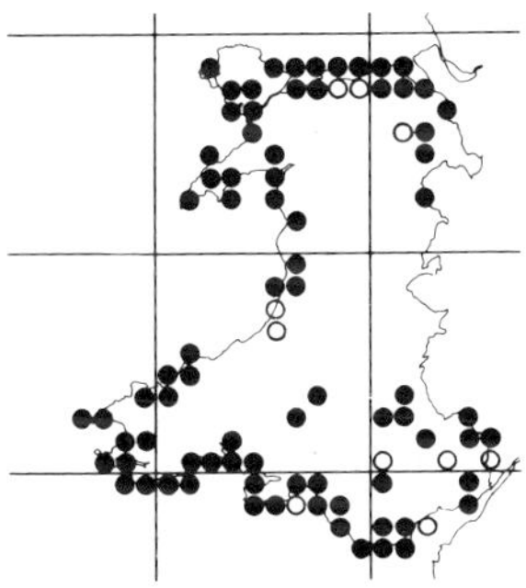

559 Anchusa arvensis

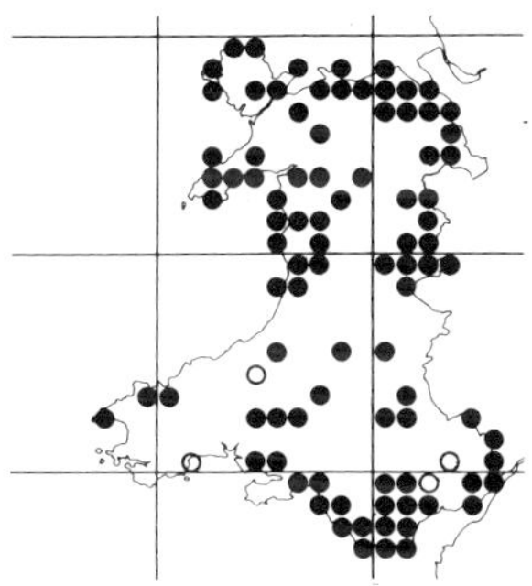

560 Pentaglottis sempervirens

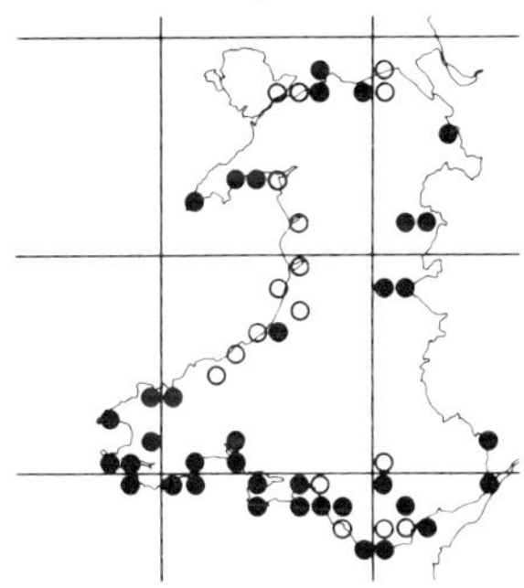

561 Borago officinalis

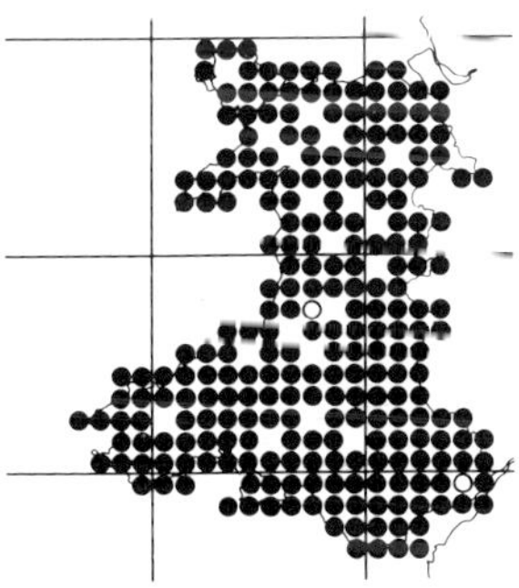

562 Myosotis arvensis

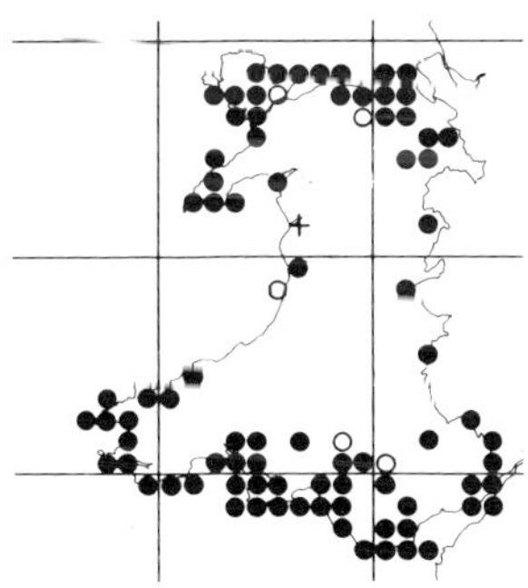

563 Myosotis ramosissima

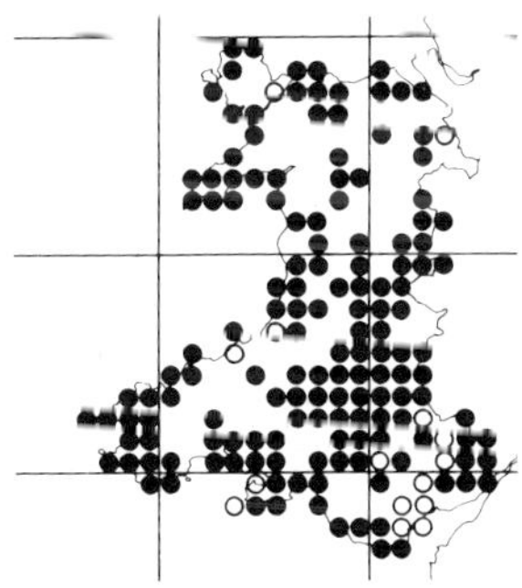

564 Myosotis discolor

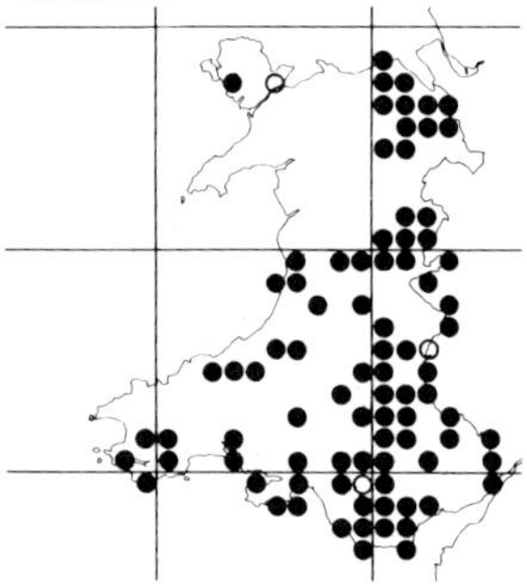

565 Myosotis sylvatica

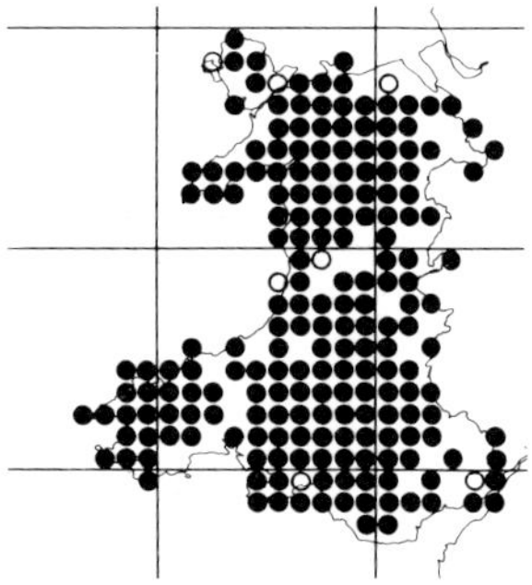

566 Myosotis secunda

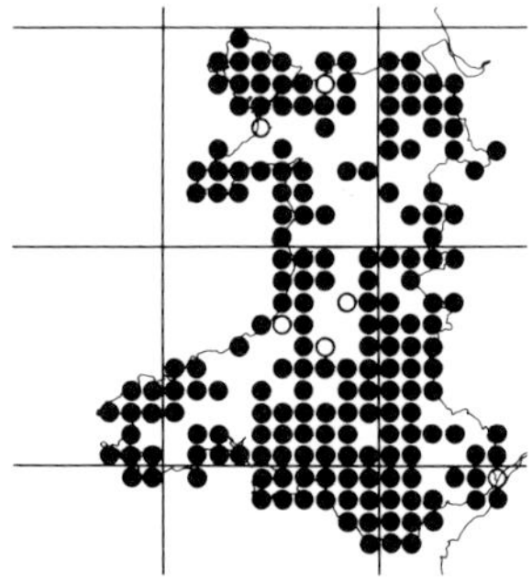

567 Myosotis laxa
subsp. caespitosa

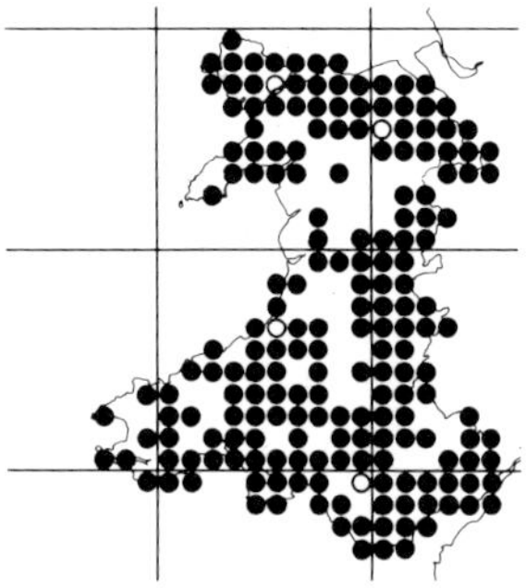

568 Myosotis scorpioides

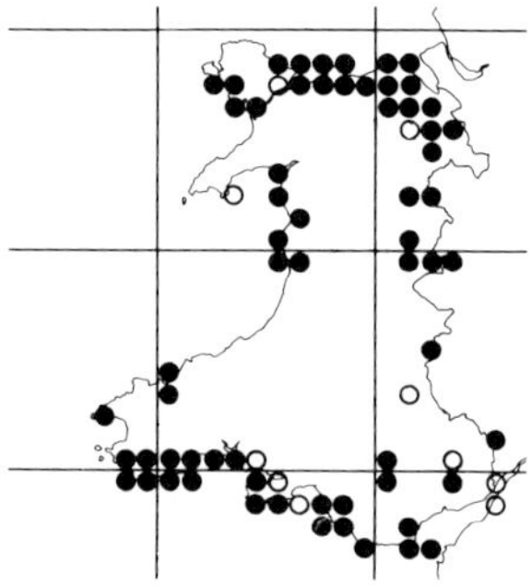

569 Cynoglossum officinale

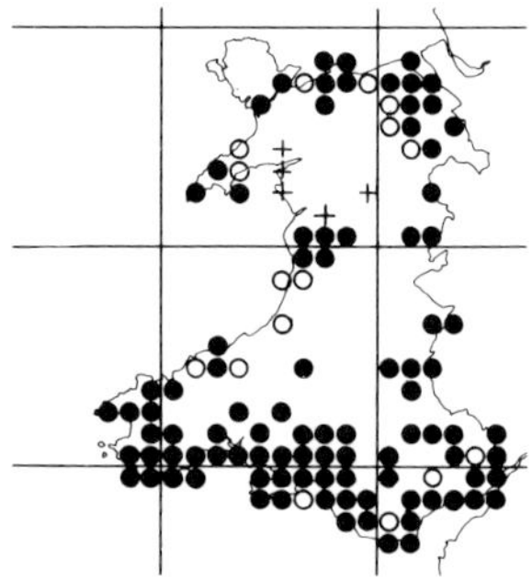

570 Verbena officinalis

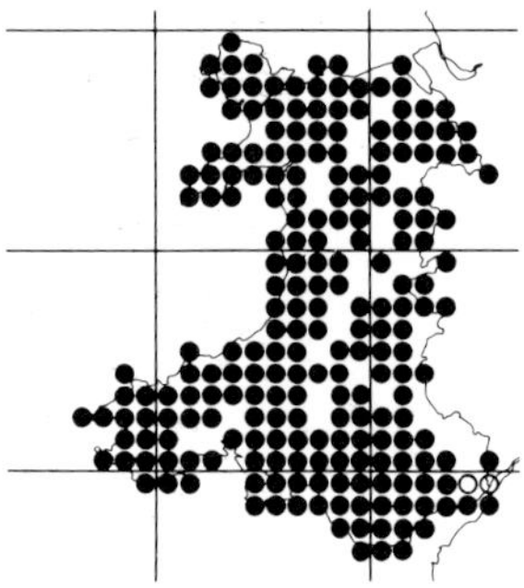

571 Callitriche stagnalis agg.

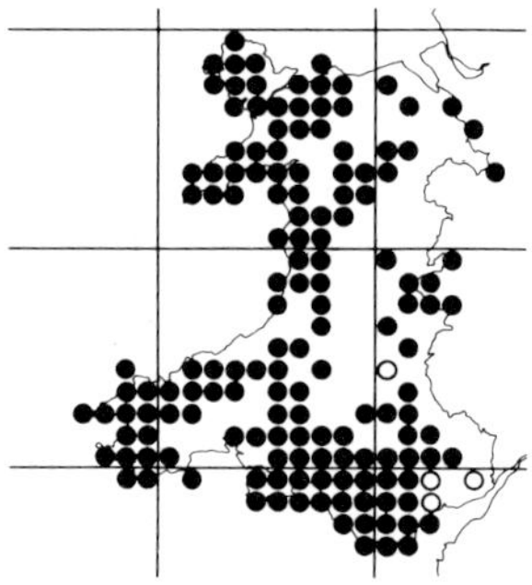

572 Callitriche stagnalis s.s.

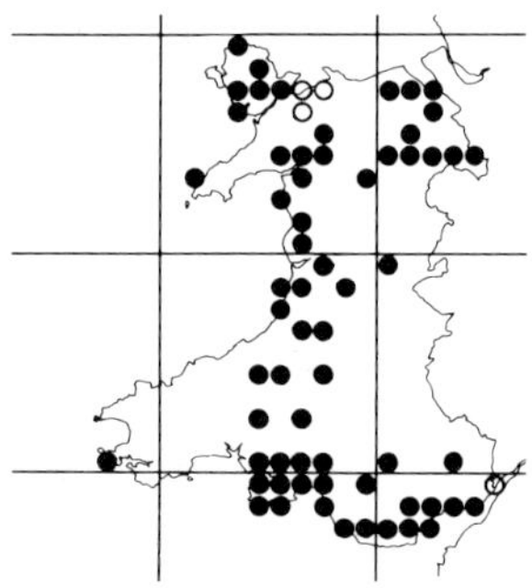

573 Callitriche platycarpa

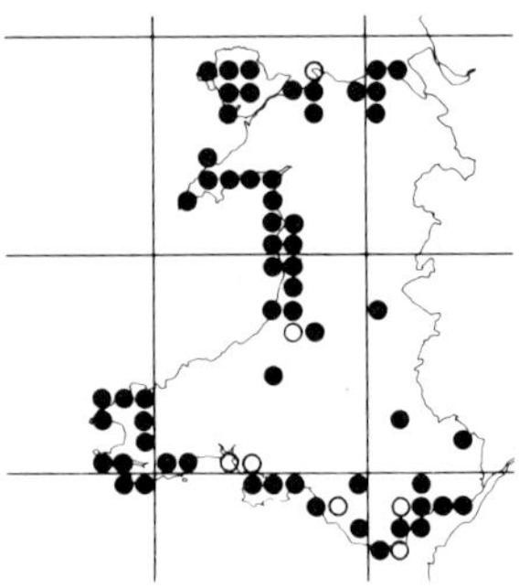

574 Callitriche obtusangula

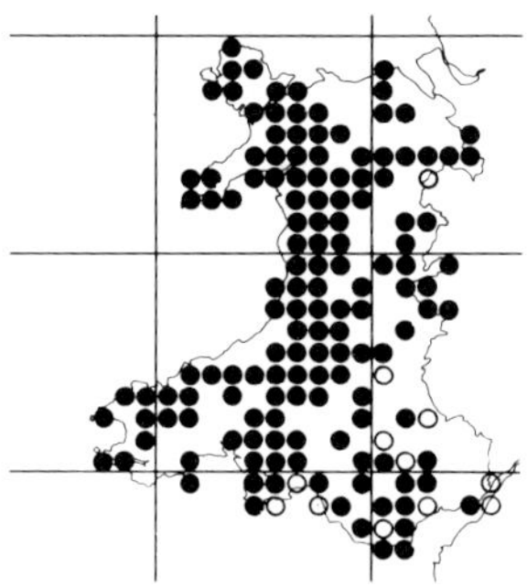

575 Callitriche intermedia agg.

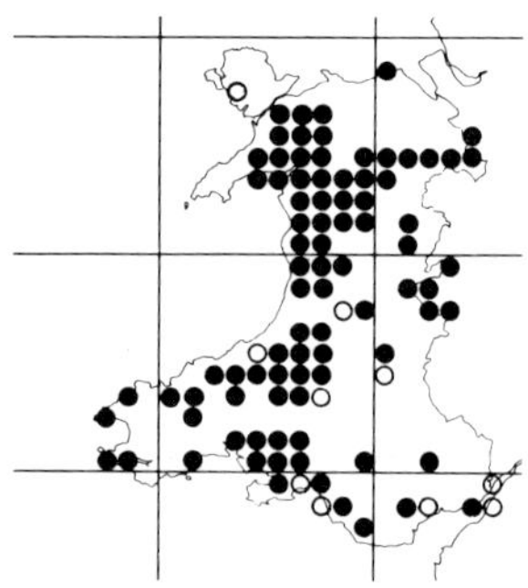

576 Callitriche hamulata

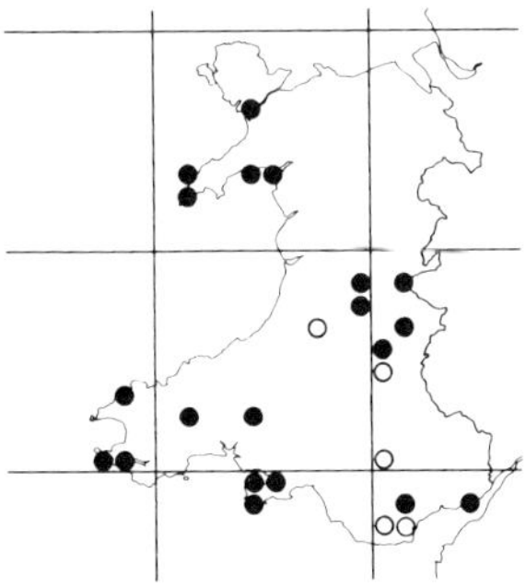

577 Callitriche brutia

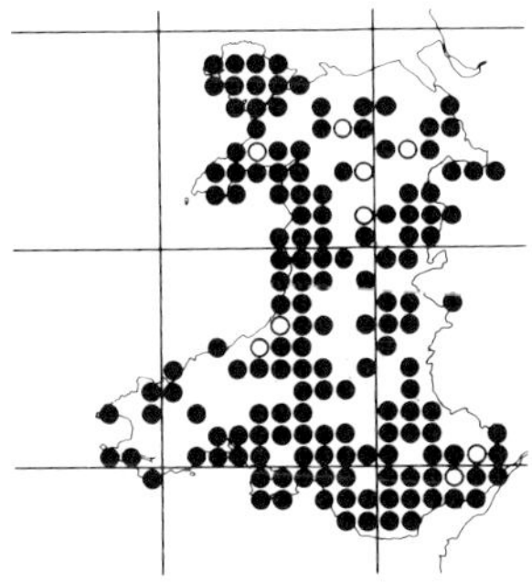

578 Scutellaria galericulata

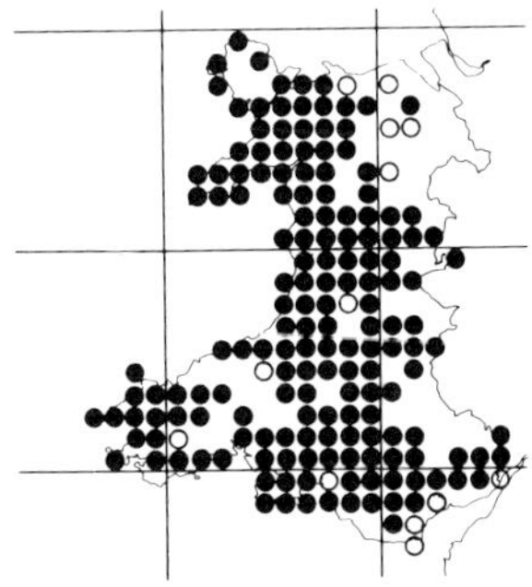

579 Scutellaria minor

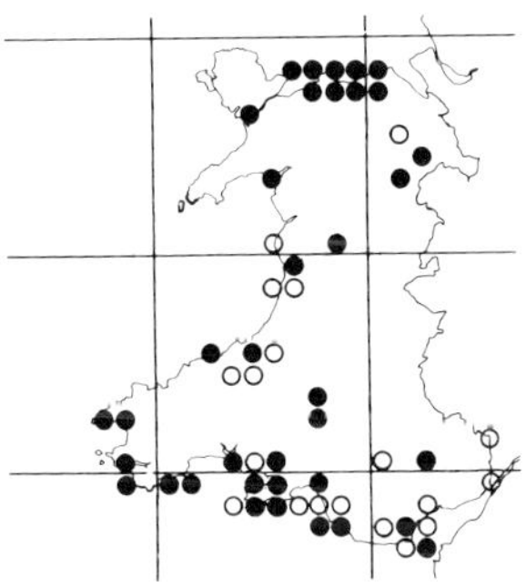

580 Marrubium vulgare

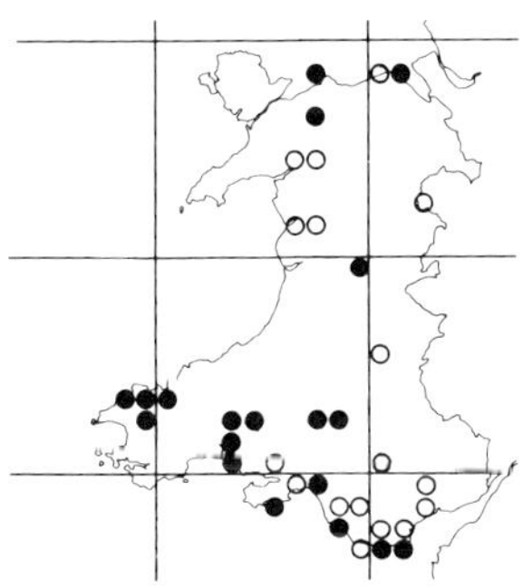

581 Galeopsis angustifolia

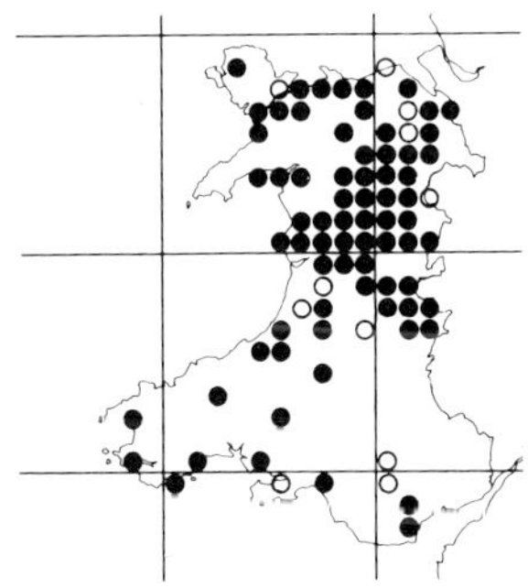

582 Galeopsis speciosa

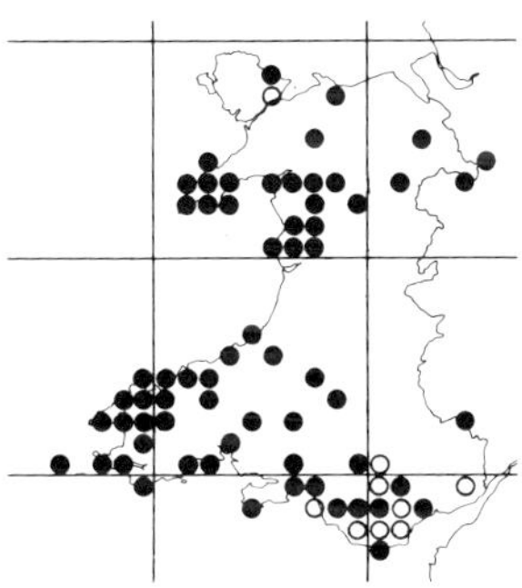

583 Galeopsis bifida

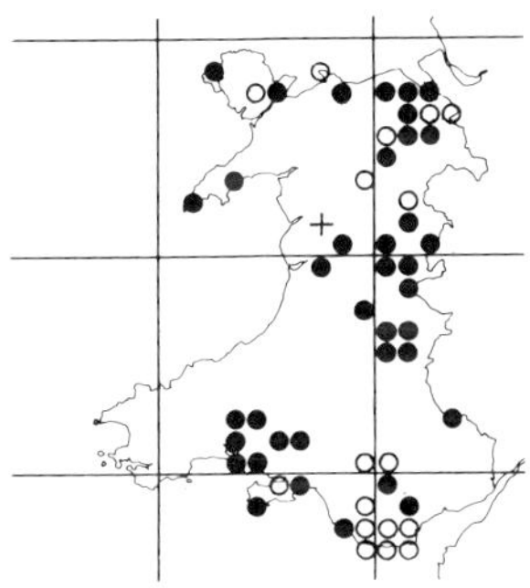

584 Lamium maculatum

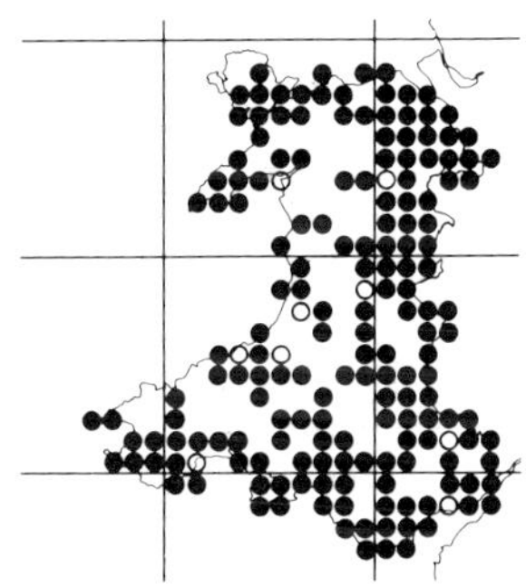

585 Lamium album

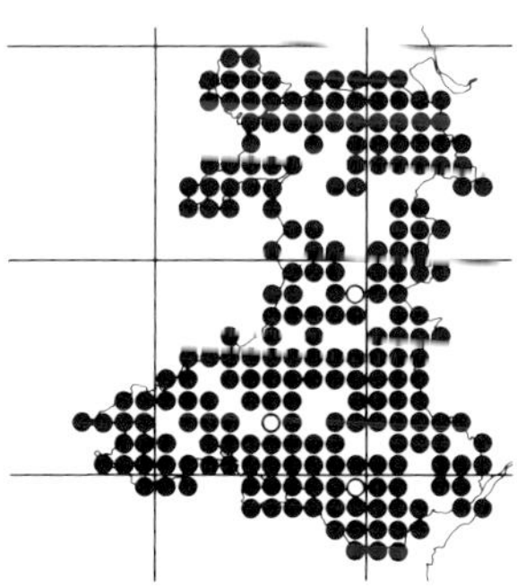

586 Lamium purpureum

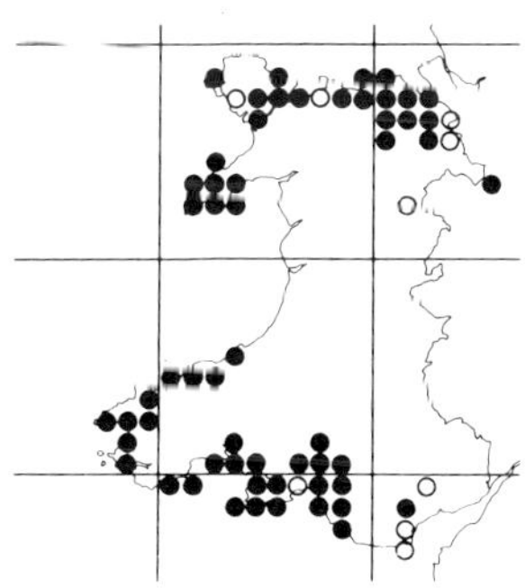

587 Lamium hybridum

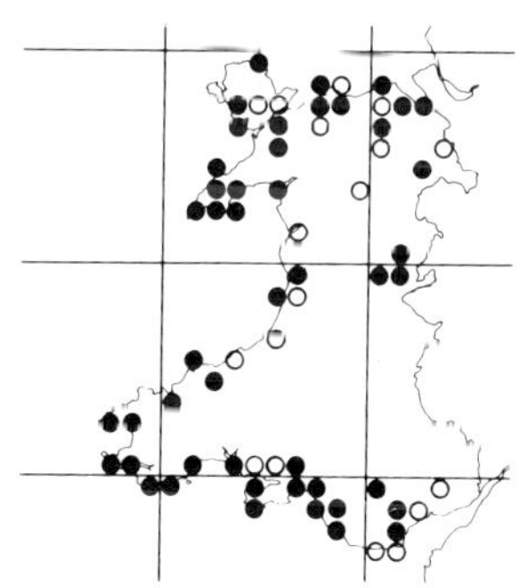

588 Lamium amplexicaule

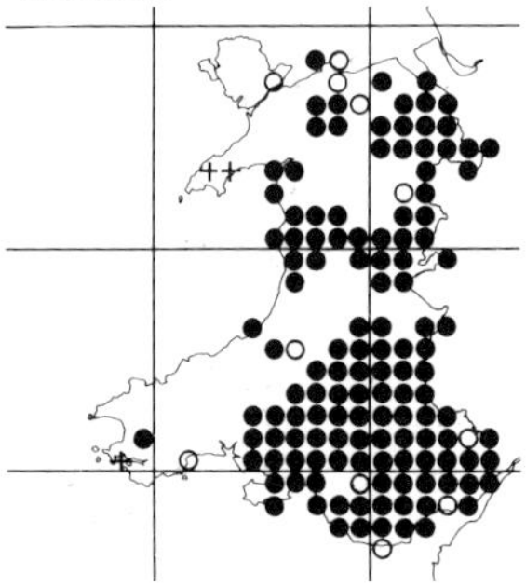

589 Lamiastrum galeobdolon

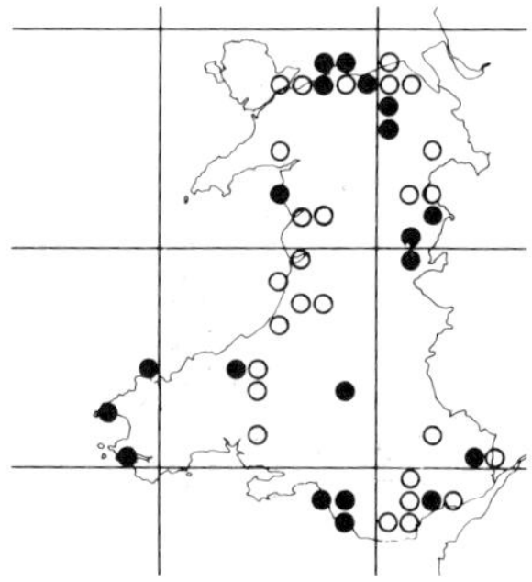

590 Leonurus cardiaca

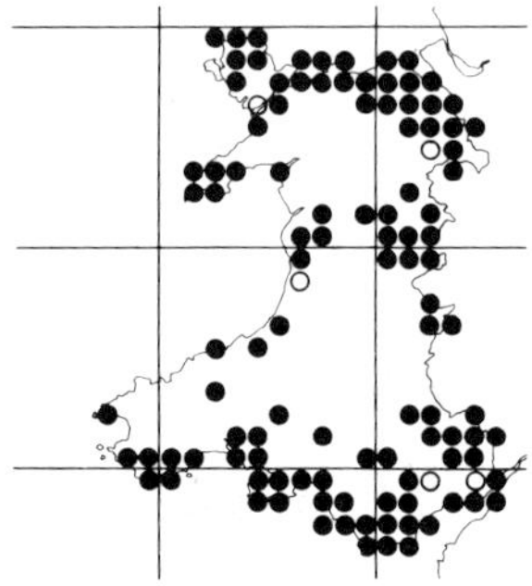

591 Ballota nigra

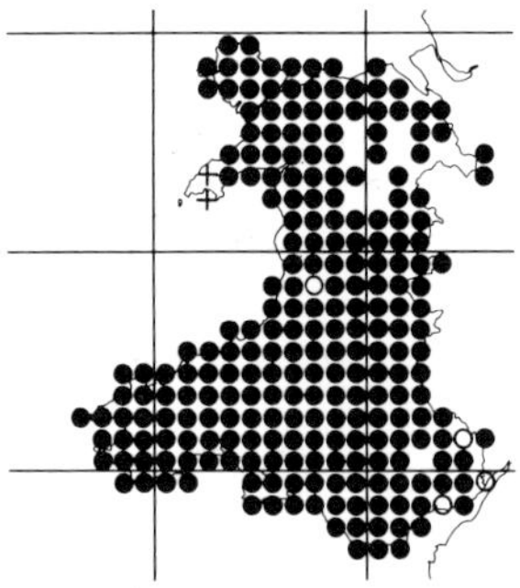

592 Stachys officinalis

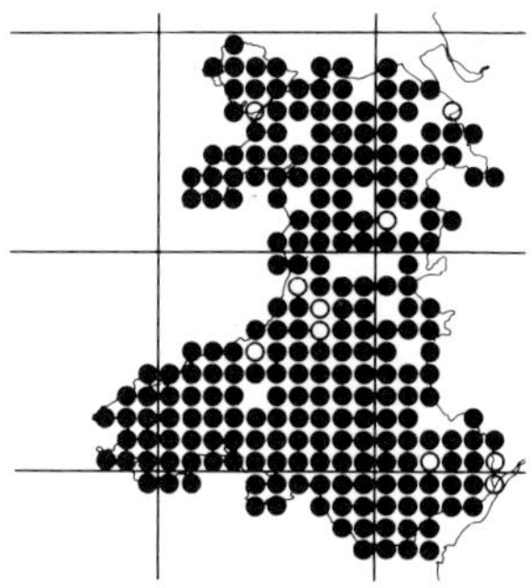

593 Stachys palustris

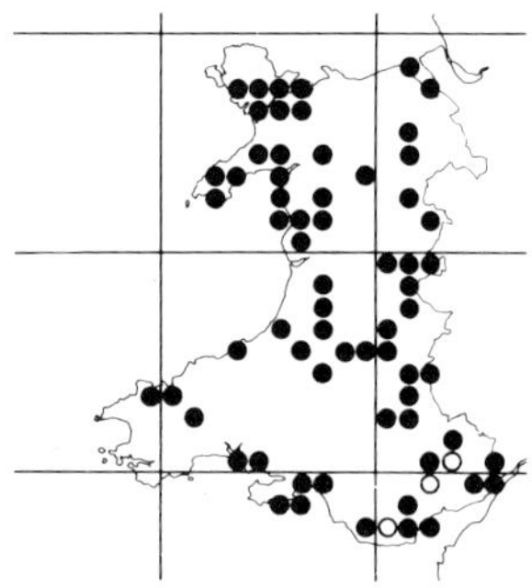

594 Stachys palustris × sylvatica = S. × ambigua

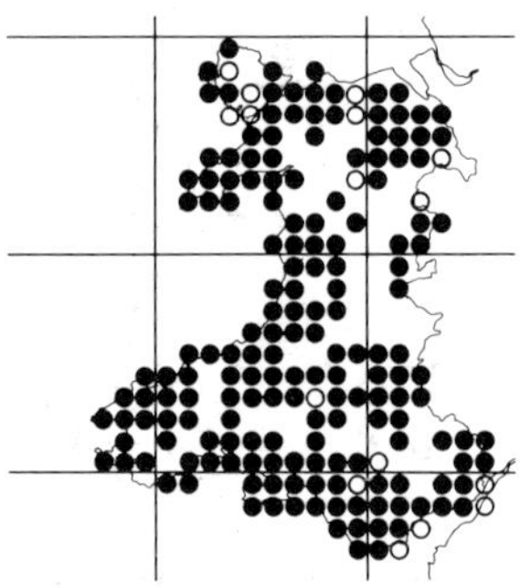

595 Stachys arvensis

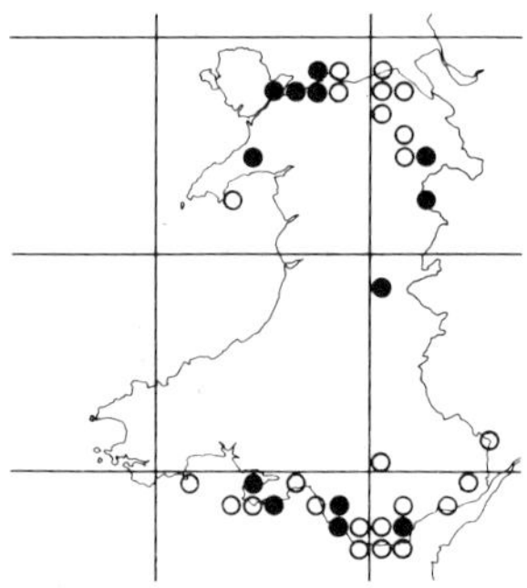

596 Nepeta cataria

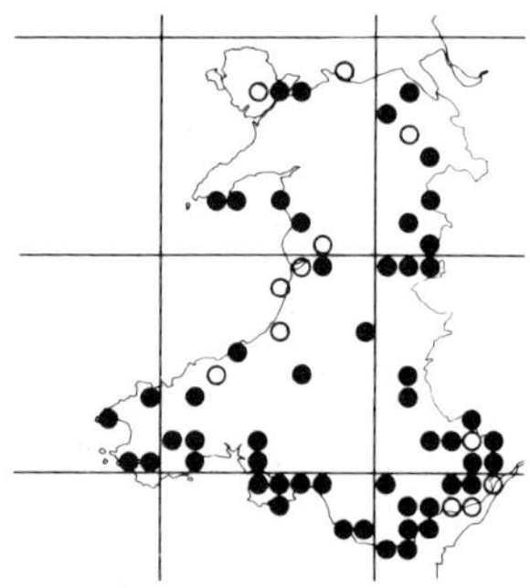

597 Melissa officinalis

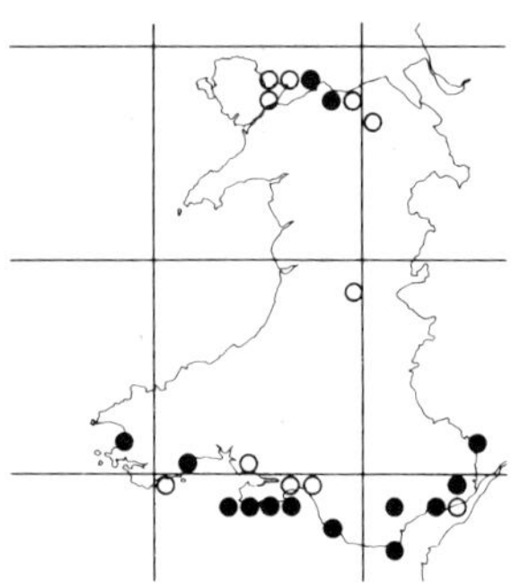

598 Acinos arvensis

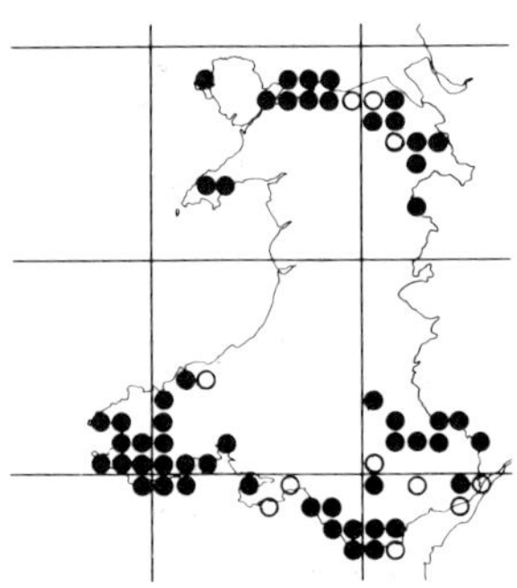

599 Calamintha sylvatica

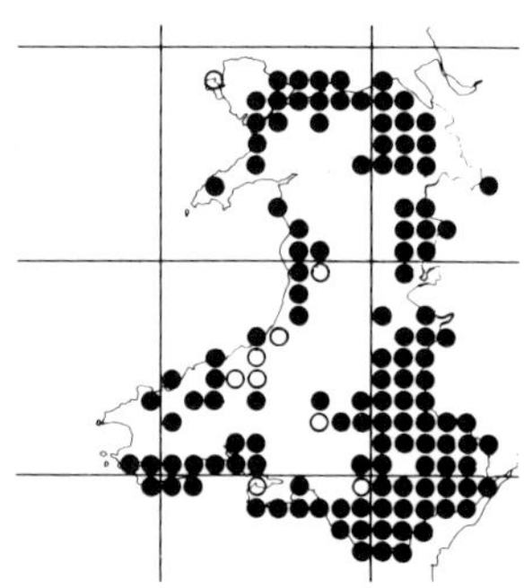

600 Clinopodium vulgare

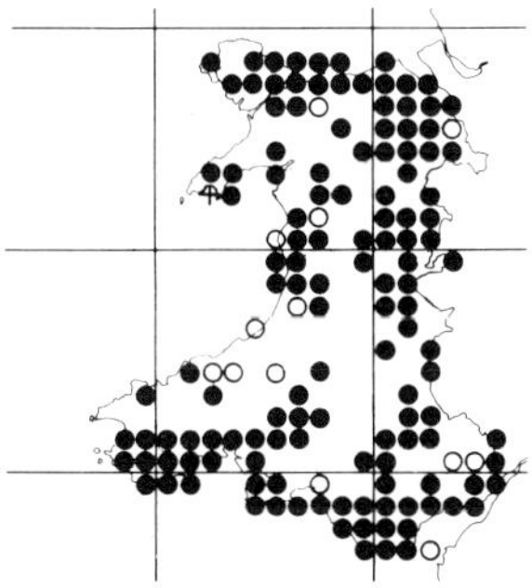

601 Origanum vulgare

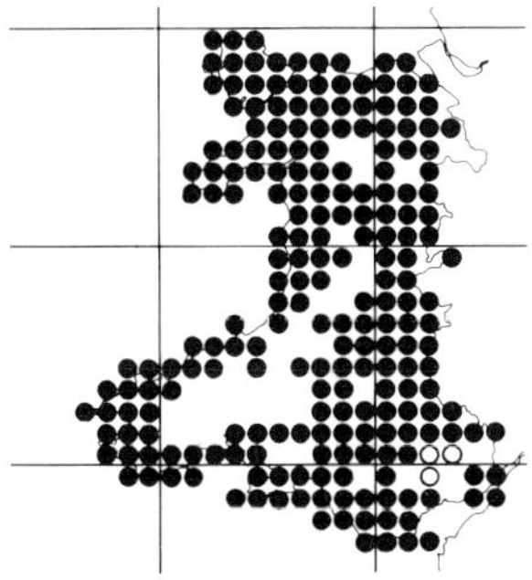

602 Thymus praecox

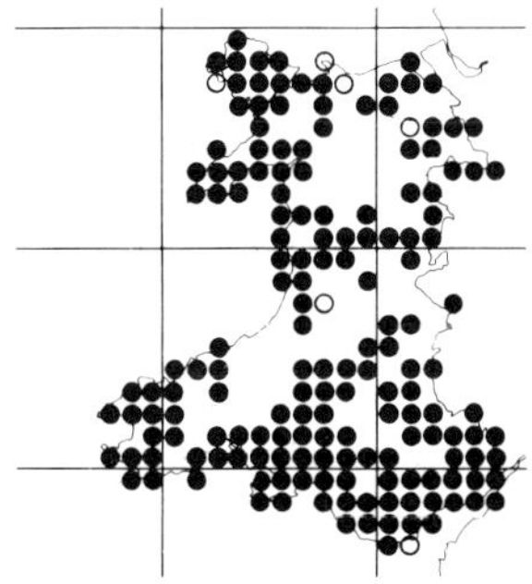

603 Lycopus europaeus

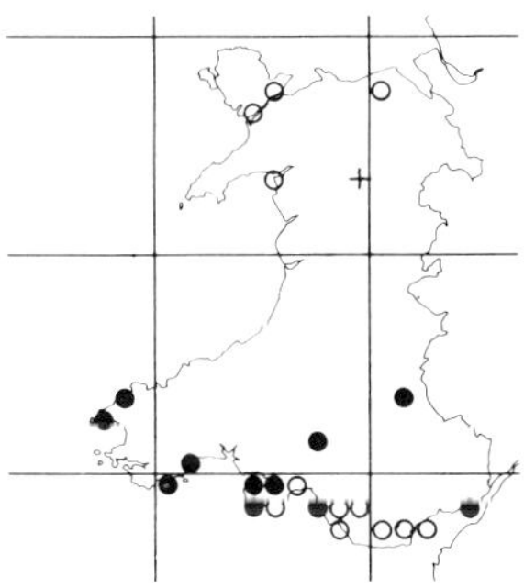

604 Mentha pulegium

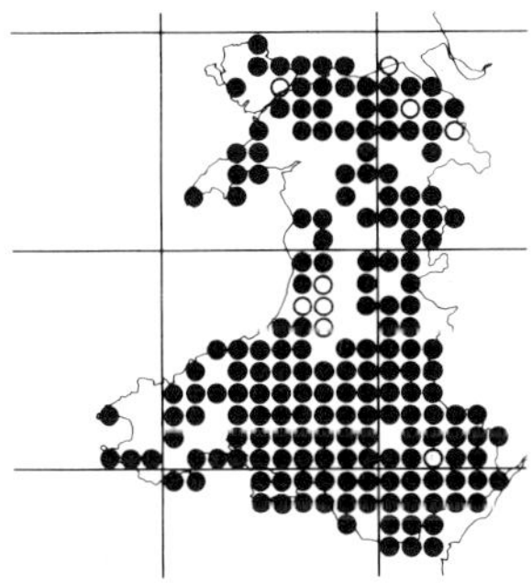

605 Mentha arvensis

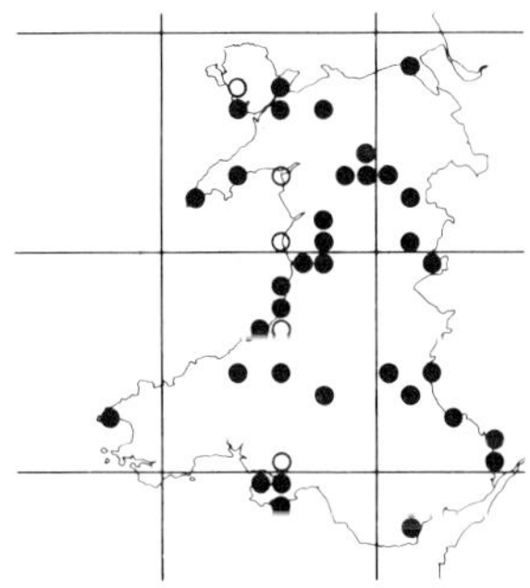

606 Mentha arvensis × spicata
= M. × gentilis

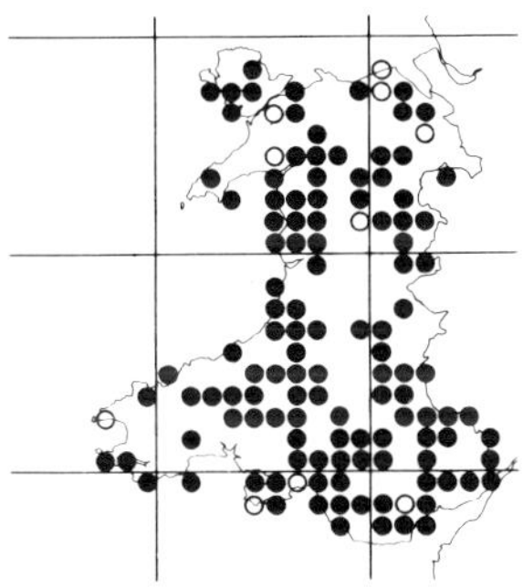

607 Mentha aquatica × arvensis
= M. × verticillata

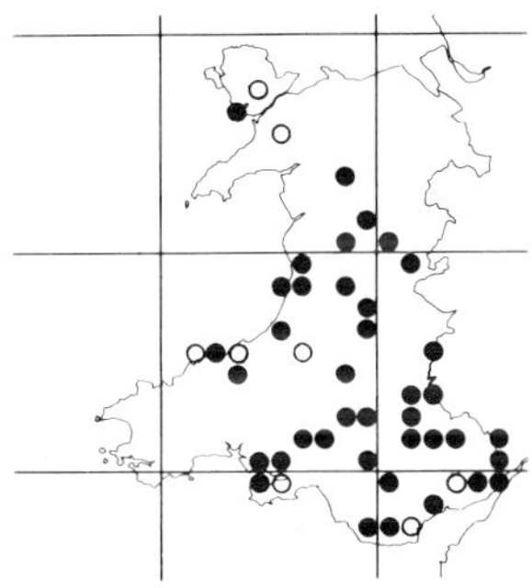

608 Mentha aquatica × arvensis
× spicata = M. × smithiana

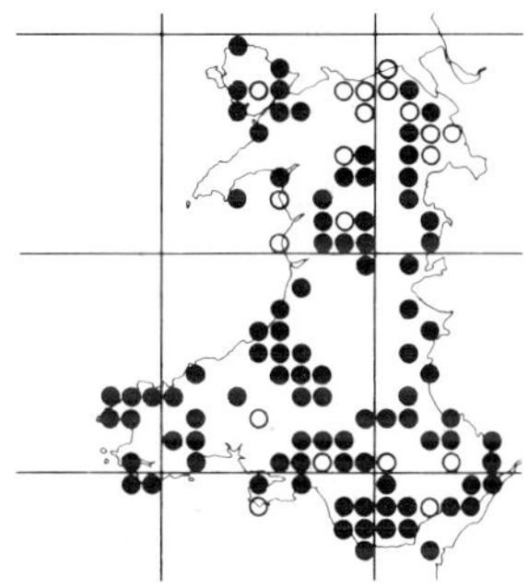

609 Mentha aquatica × spicata
= M. × piperita

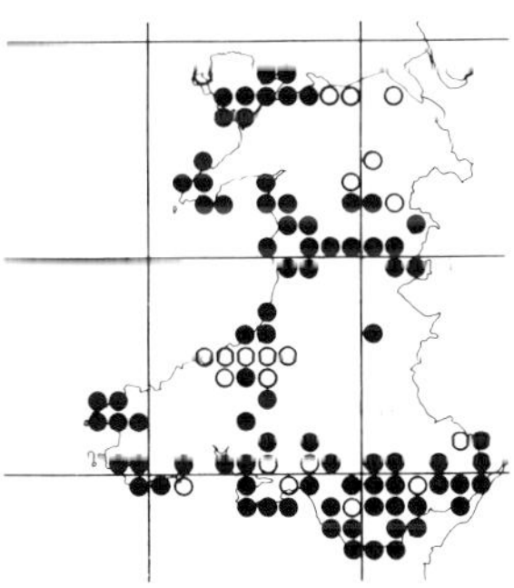

610 Mentha suaveolens

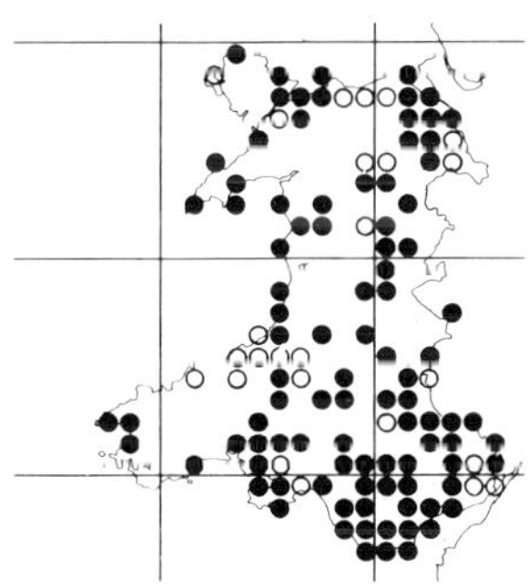

611 Mentha spicata

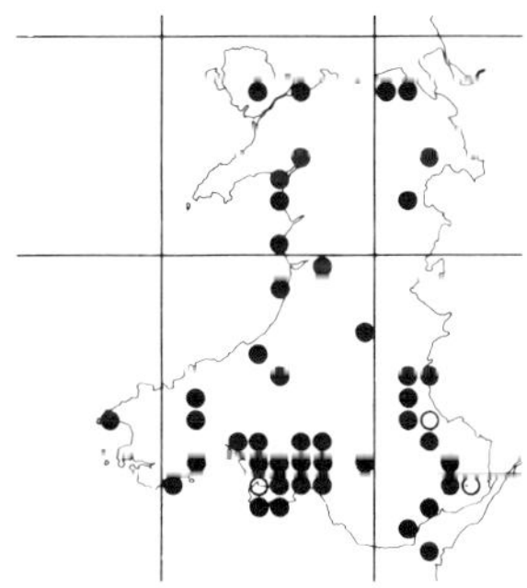

612 Mentha spicata × suaveolens
= M. × villosa

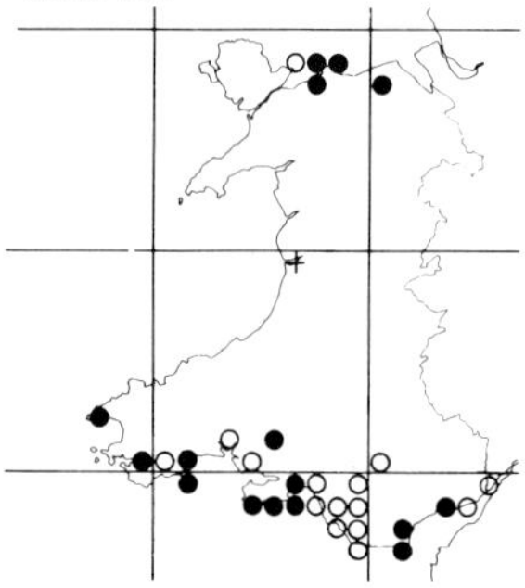

613 Salvia verbenaca

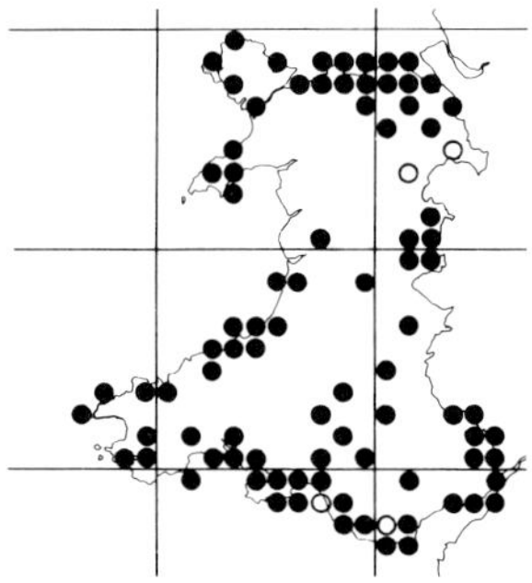

614 Lycium barbarum

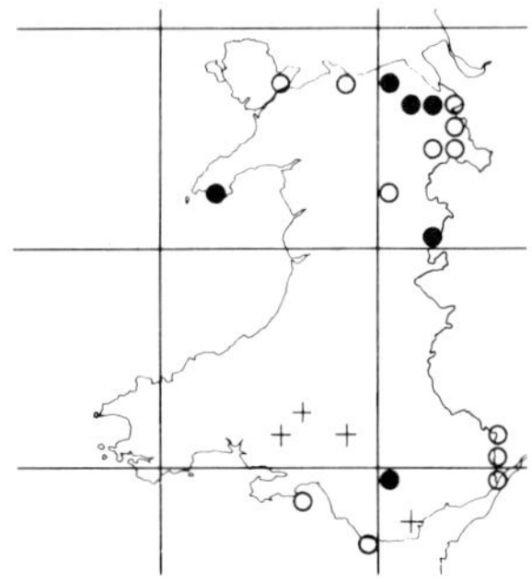

615 Atropa bella-donna

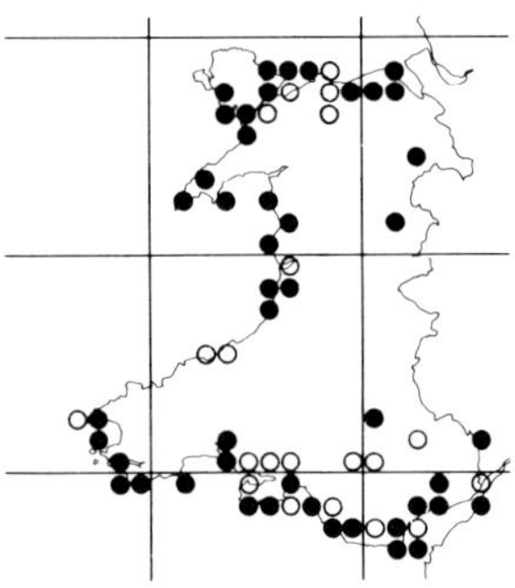

616 Hyoscyamus niger

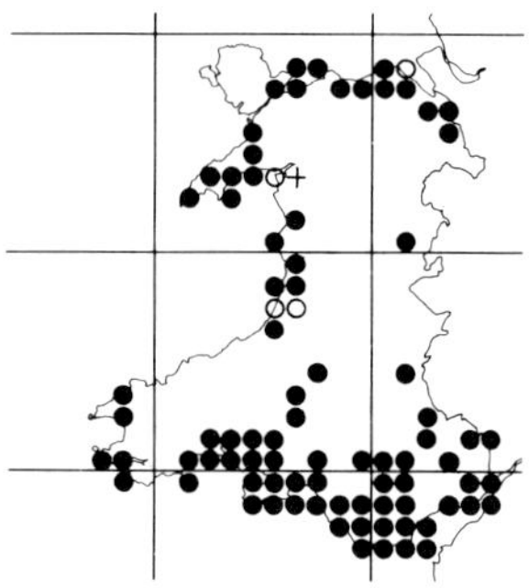

617 Solanum nigrum

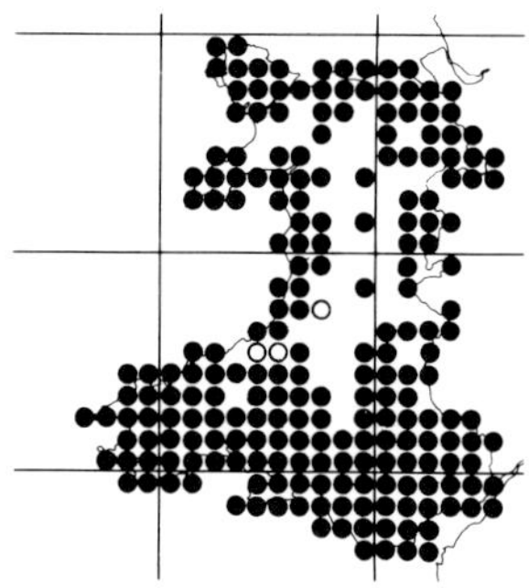

618 Solanum dulcamara

619 Solanum tuberosum

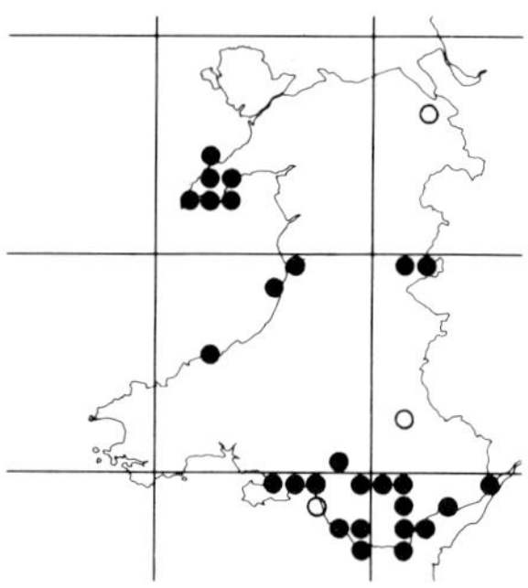

620 Lycopersicon esculentum

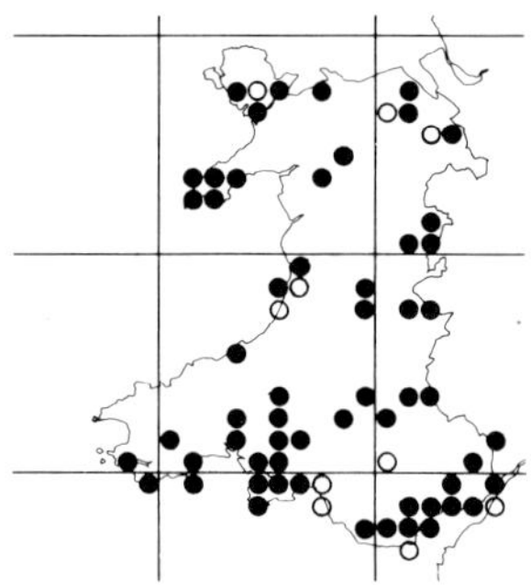

621 Datura stramonium

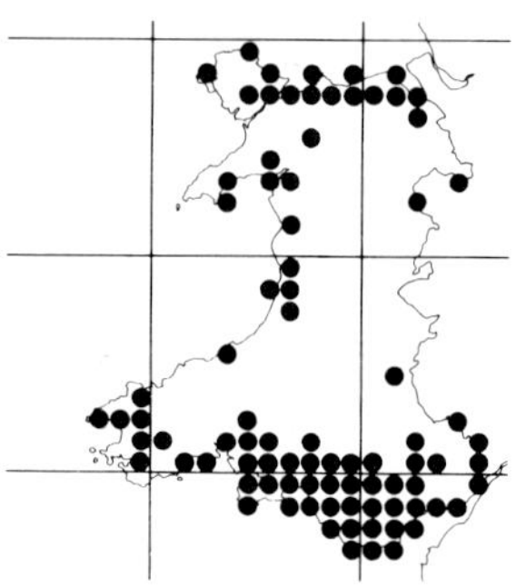

622 Buddleja davidii

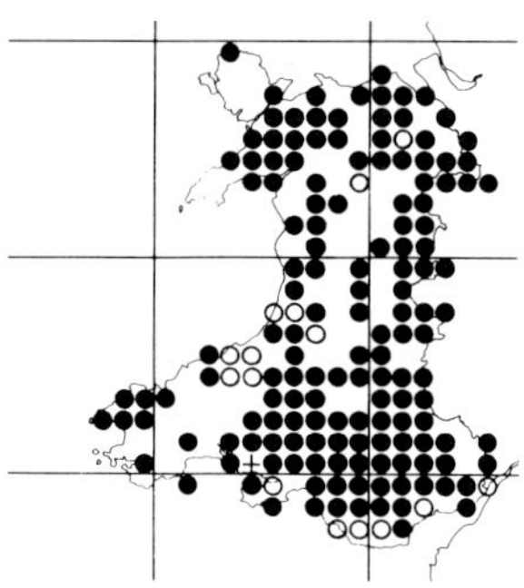

623 Mimulus guttatus

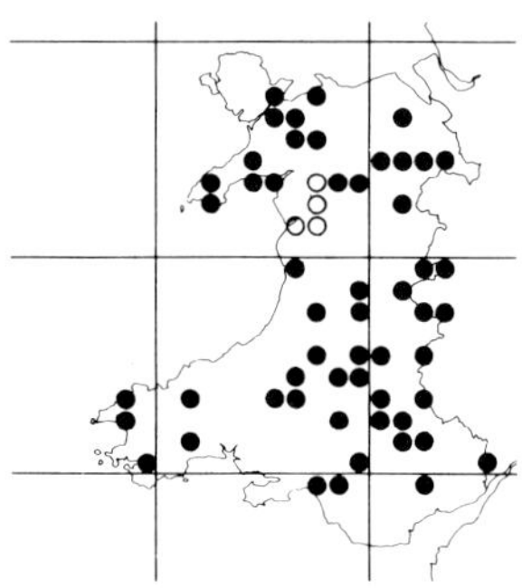

624 Mimulus guttatus × luteus

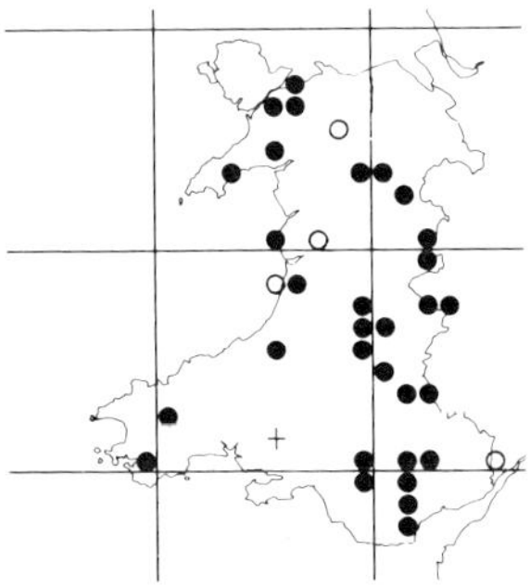
625 Mimulus moschatus

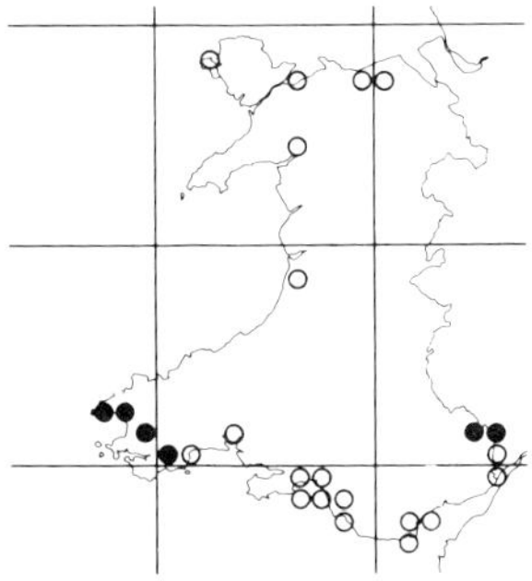
626 Verbascum blattaria

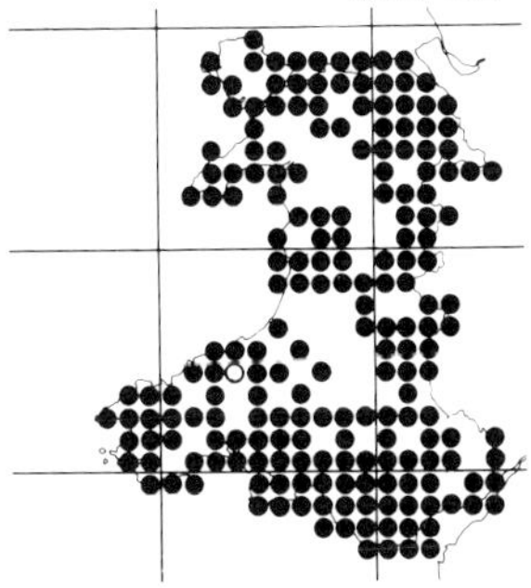
627 Verbascum thapsus

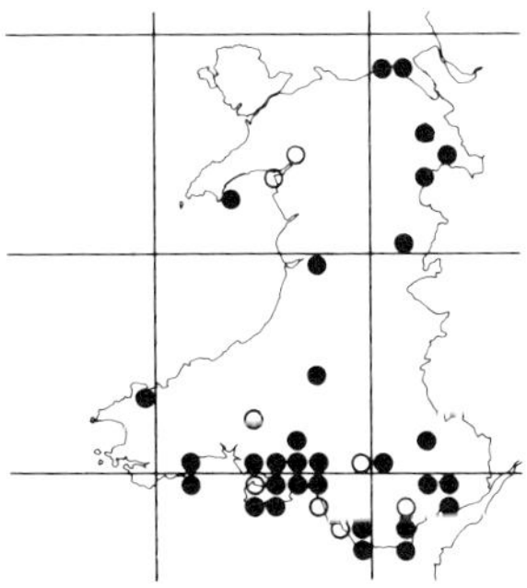
628 Verbascum nigrum

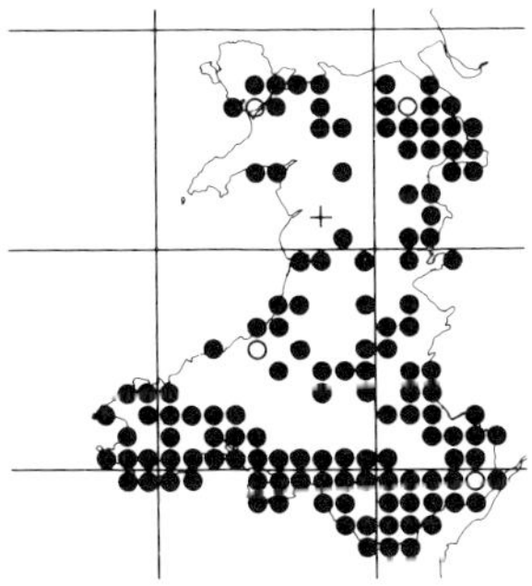
629 Scrophularia auriculata

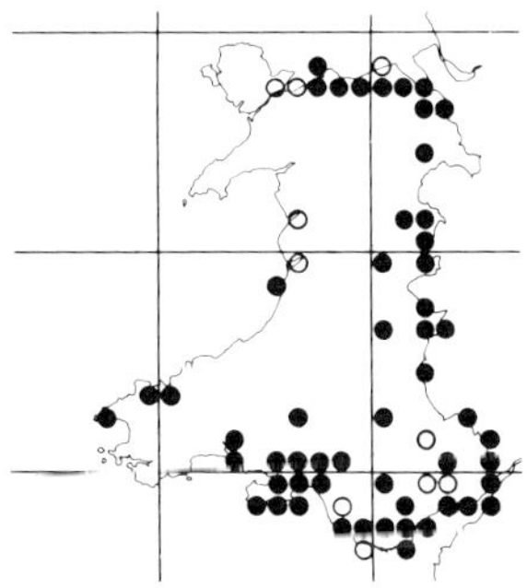
630 Antirrhinum majus

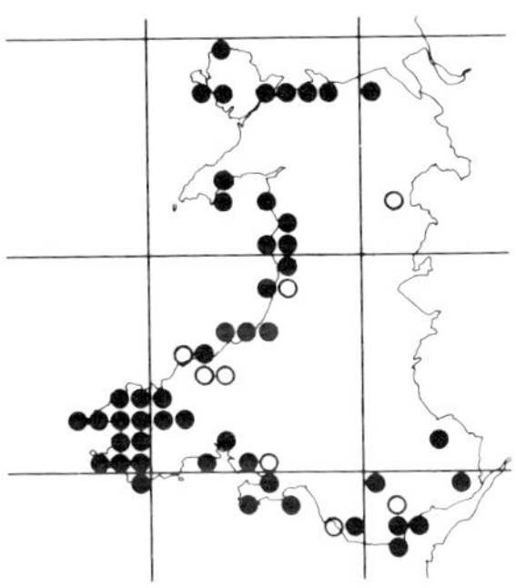
631 Misopates orontium

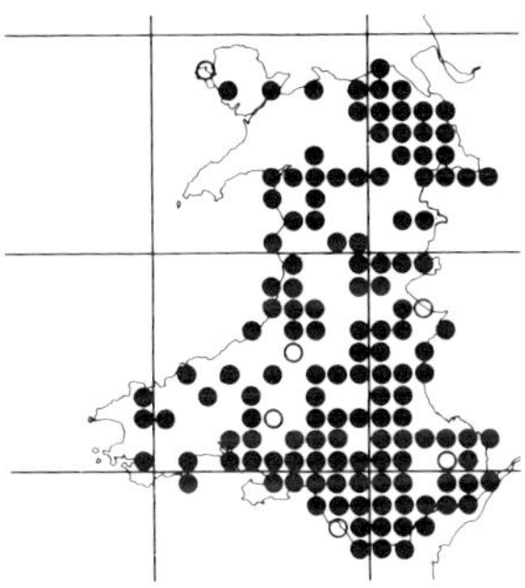
632 Chaenorhinum minus

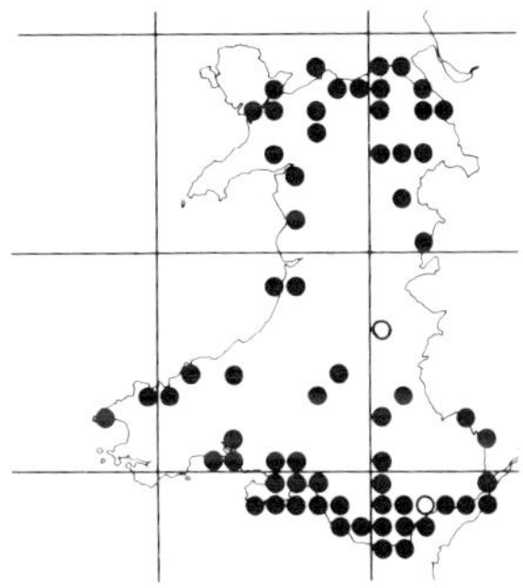
633 Linaria purpurea

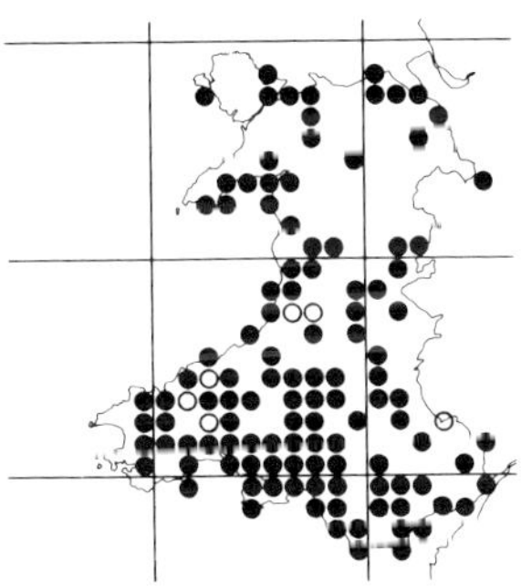
634 Linaria repens

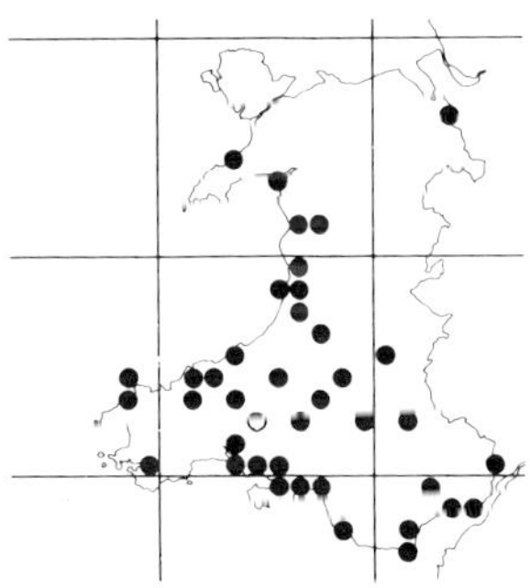
635 Linaria repens × vulgaris
= L. × sepium

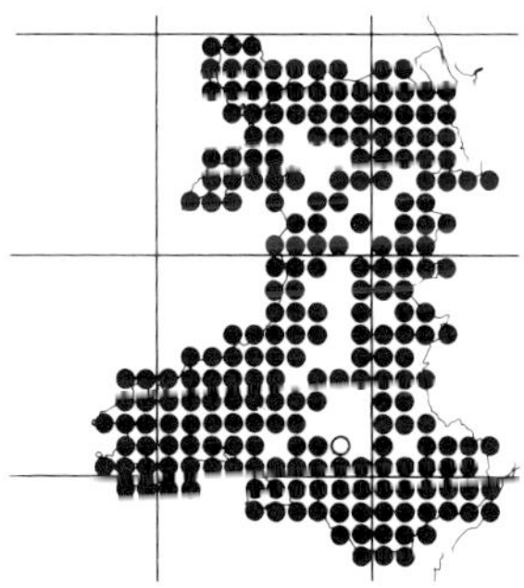
636 Linaria vulgaris

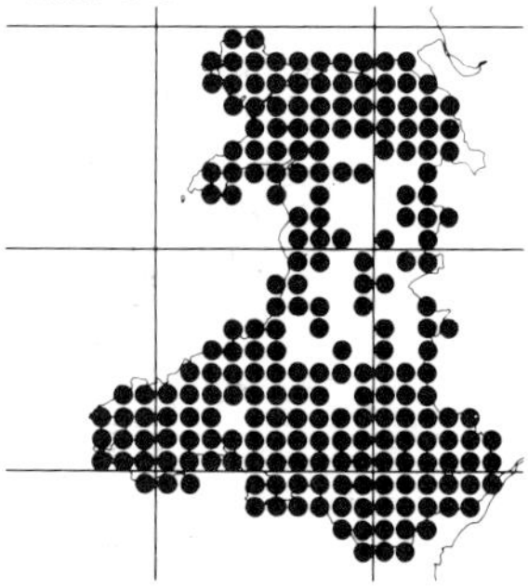

637 Cymbalaria muralis

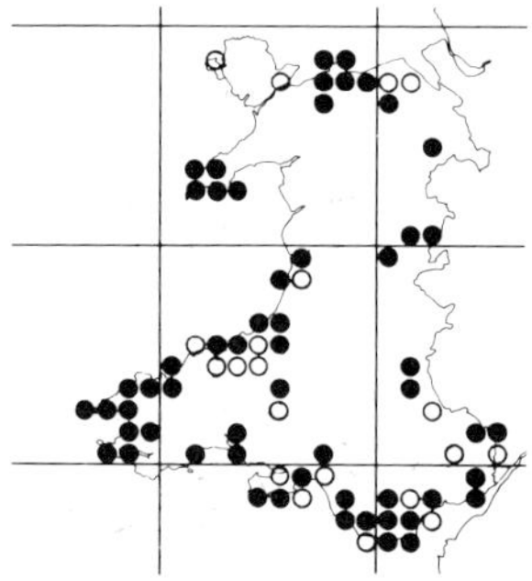

638 Kickxia elatine

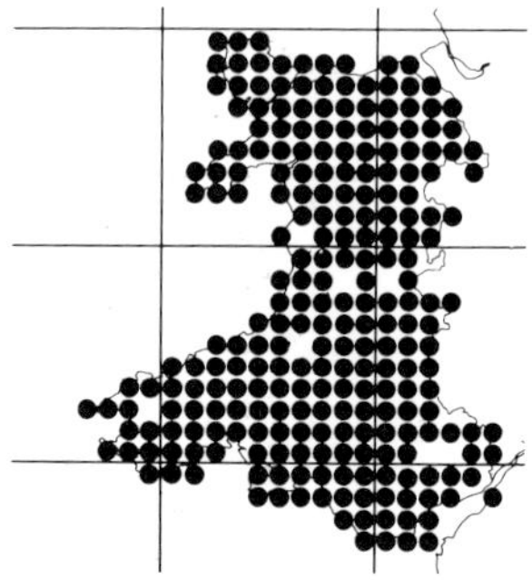

639 Veronica serpyllifolia
subsp. serpyllifolia

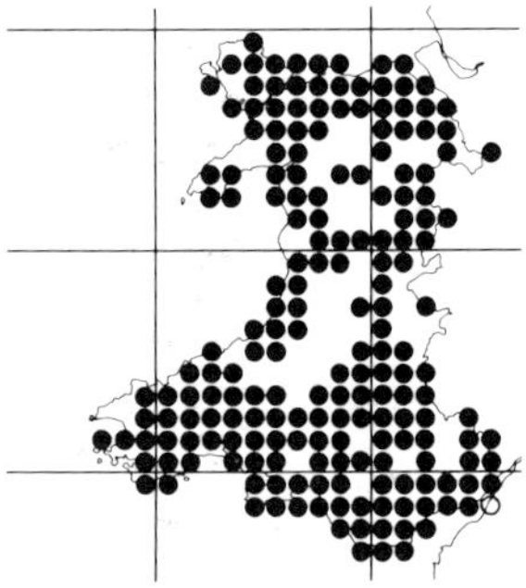

640 Veronica montana

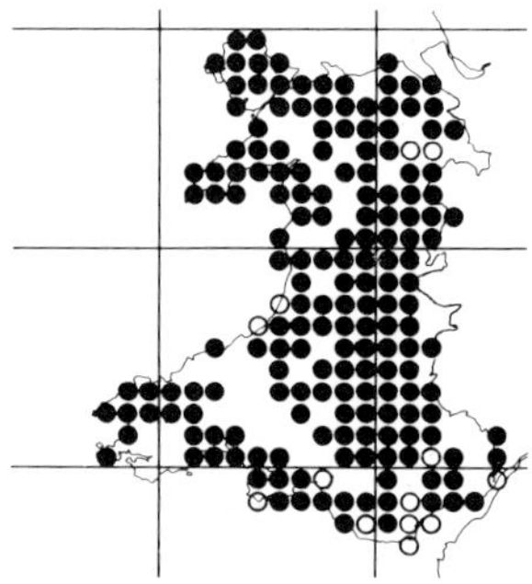

641 Veronica scutellata

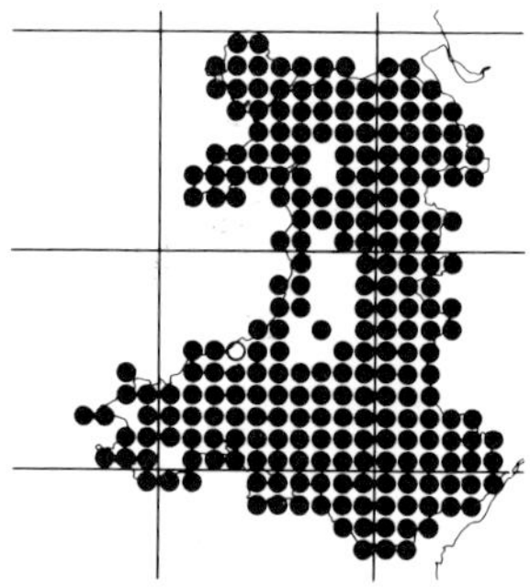

642 Veronica beccabunga

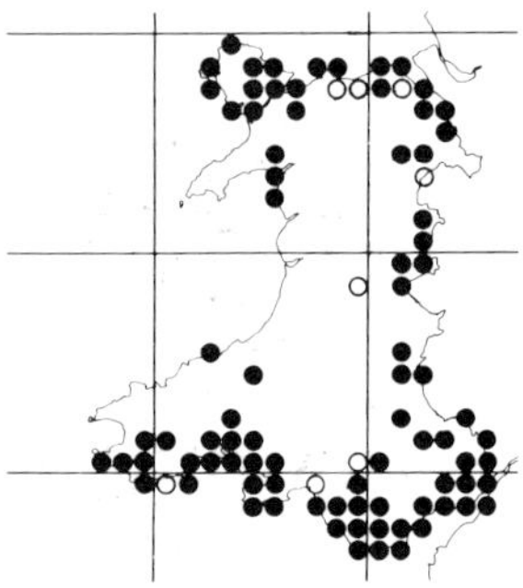

643 Veronica anagallis-aquatica
agg.

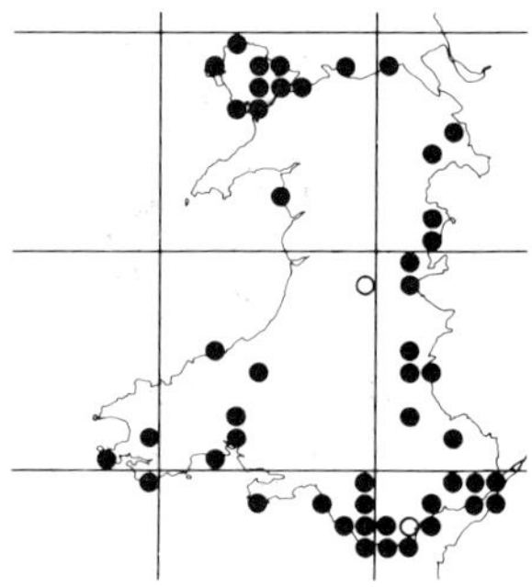

644 Veronica anagallis-aquatica
s.s.

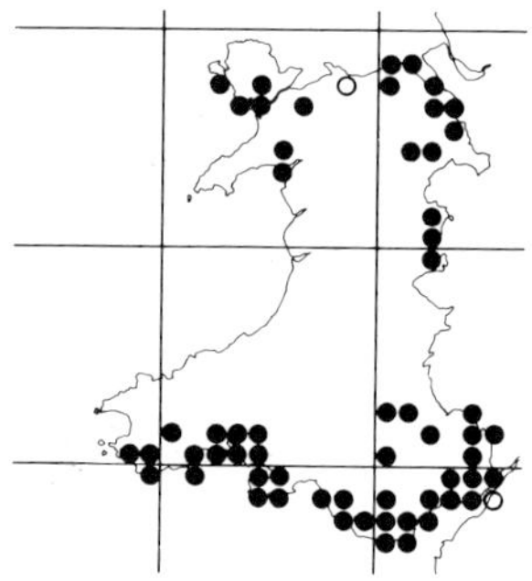

645 Veronica catenata

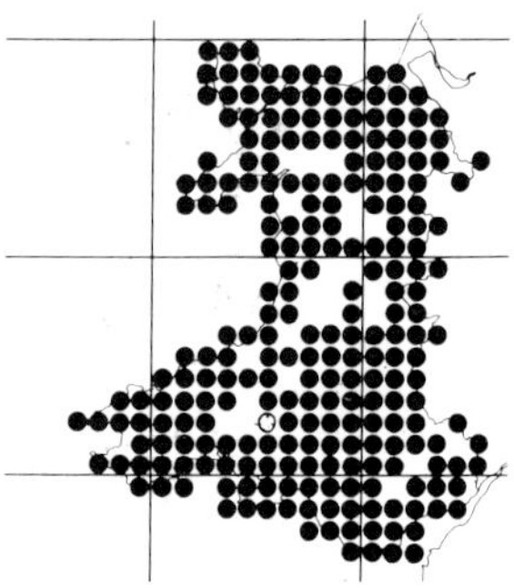

646 Veronica arvensis

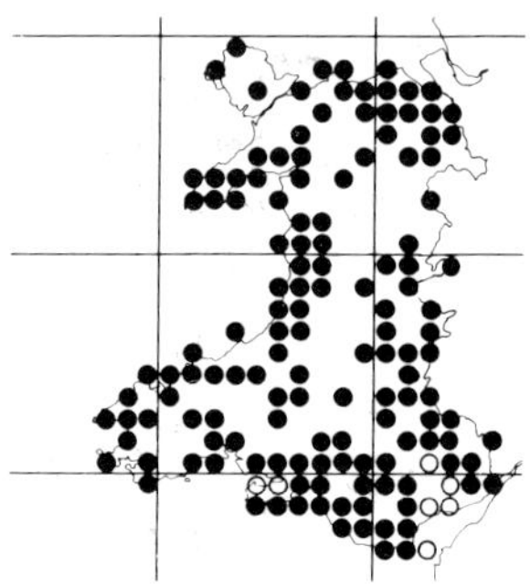

647 Veronica agrestis

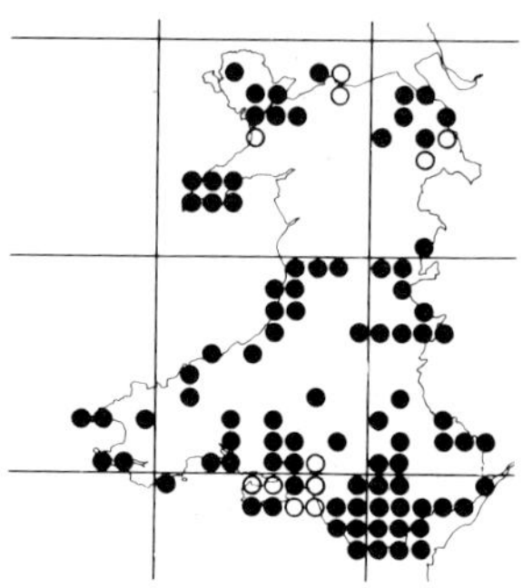

648 Veronica polita

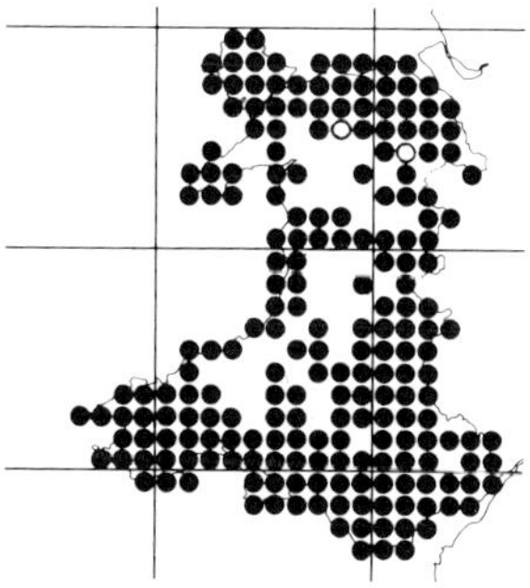

649 Veronica persica

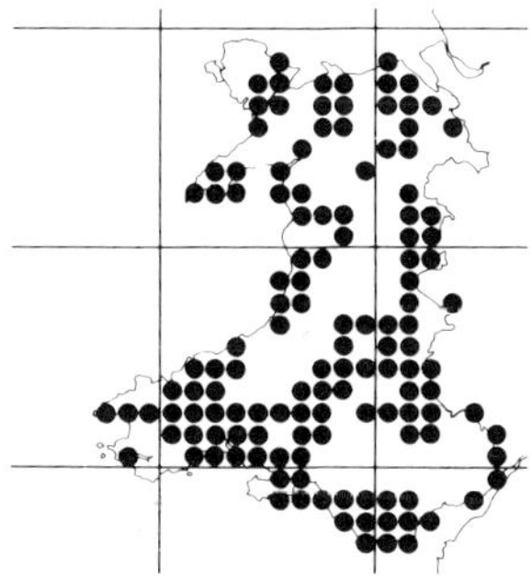

650 Veronica filiformis

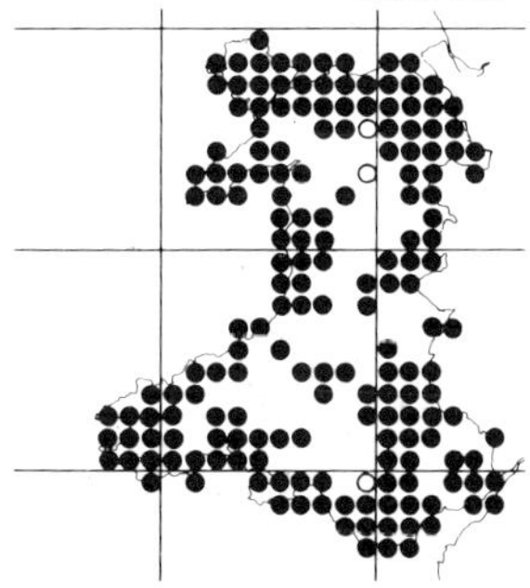

651 Veronica hederifolia

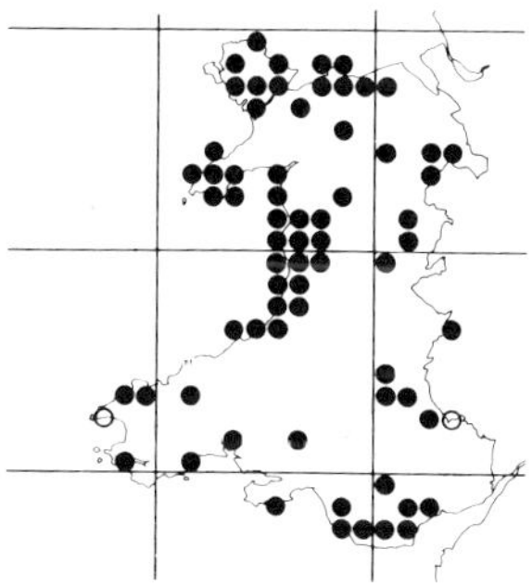

652 Veronica hederifolia subsp. lucorum

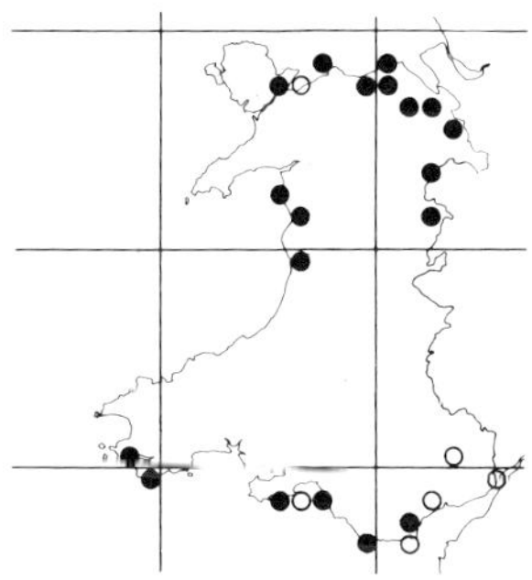

653 Veronica hederifolia subsp. hederifolia

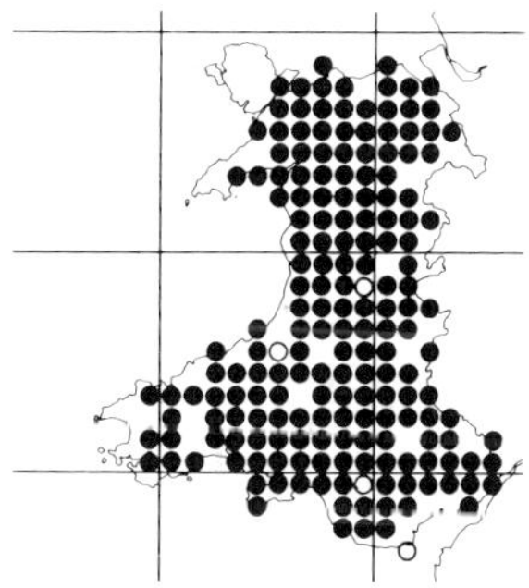

654 Melampyrum pratense

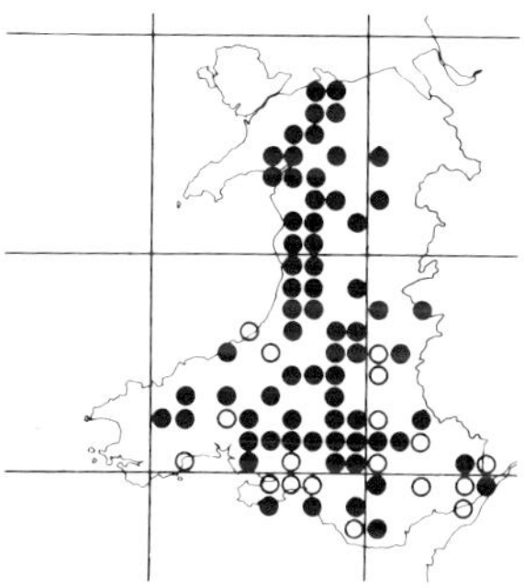

655 Euphrasia rostkoviana

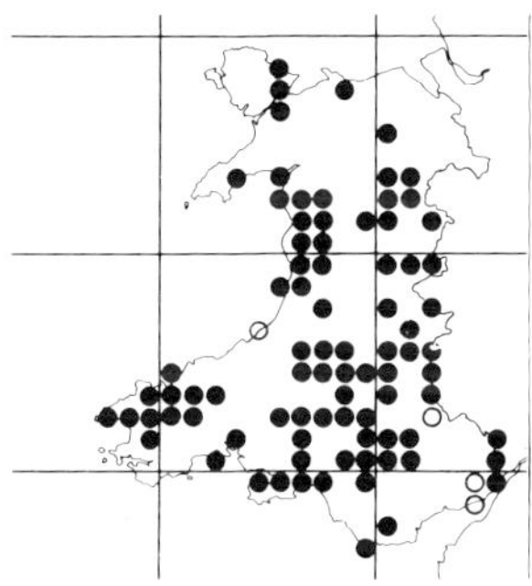

656 Euphrasia anglica

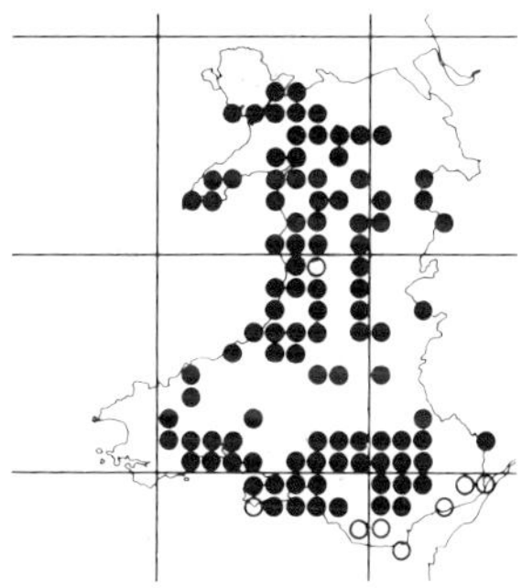

657 Euphrasia arctica

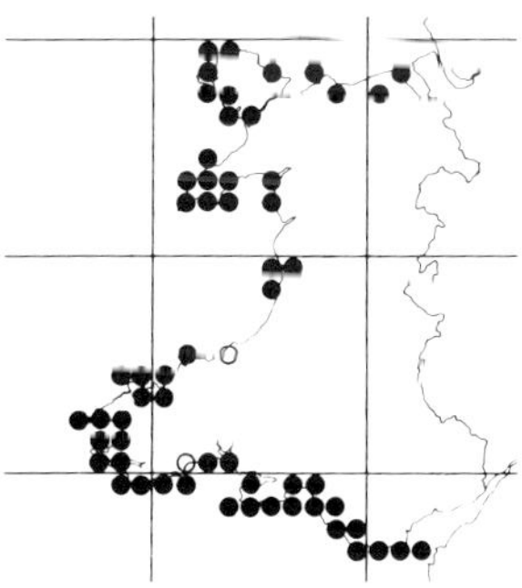

658 Euphrasia tetraquetra

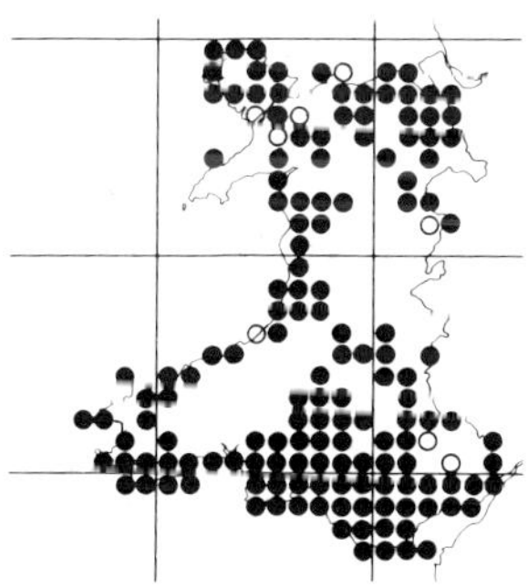

659 Euphrasia nemorosa

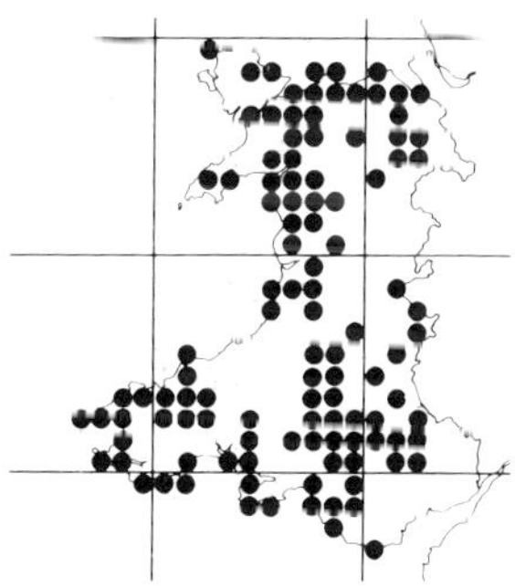

660 Euphrasia confusa

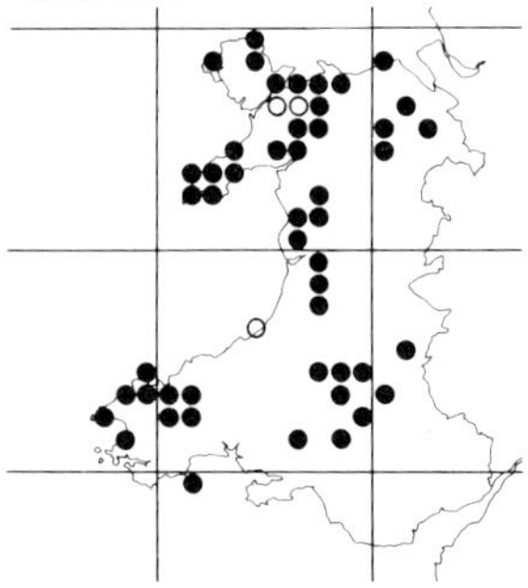

661 Euphrasia micrantha

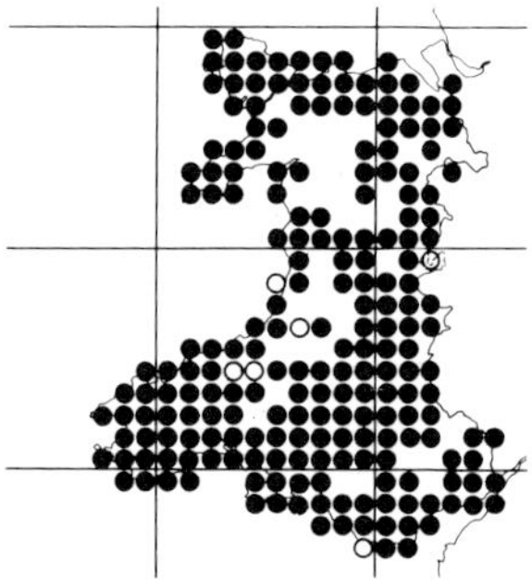

662 Odontites verna

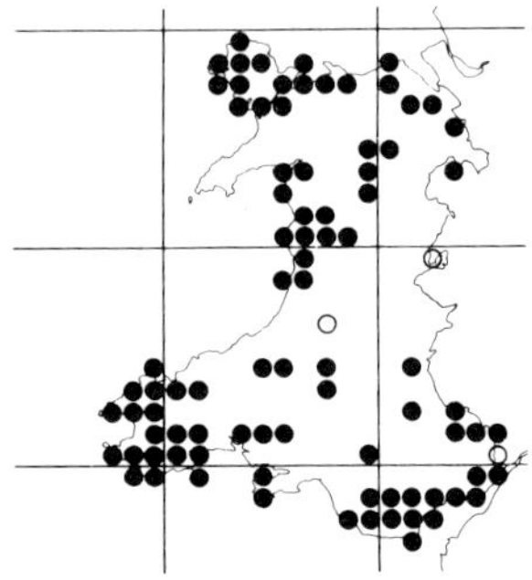

663 Odontites verna subsp. serotina

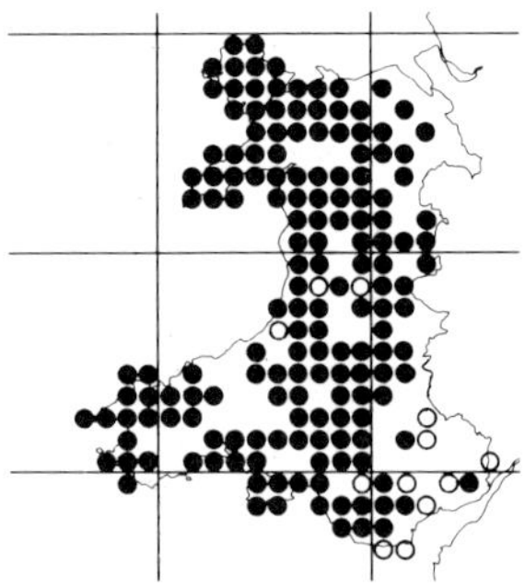

664 Pedicularis palustris

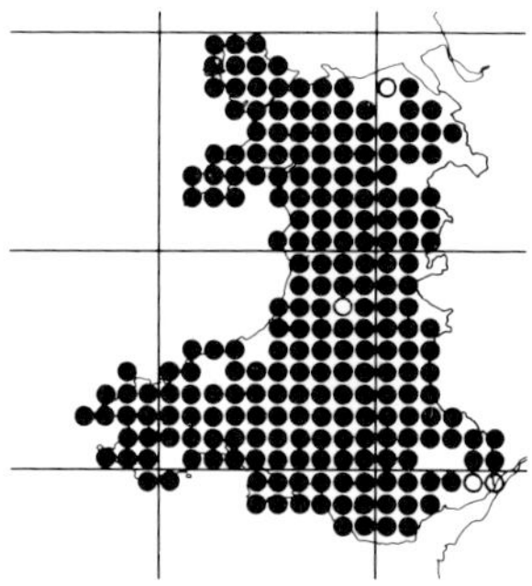

665 Pedicularis sylvatica

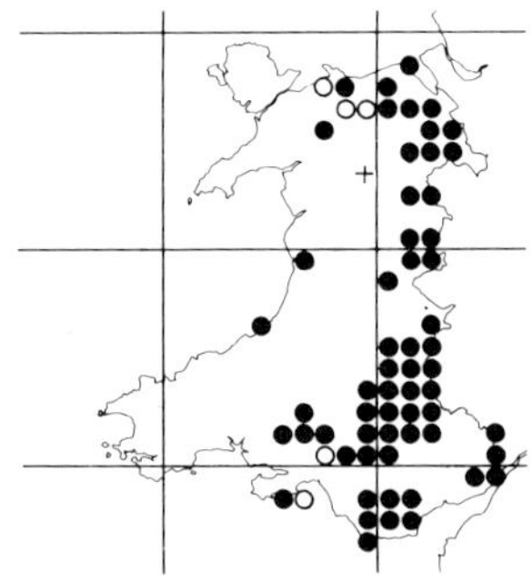

666 Lathraea squamaria

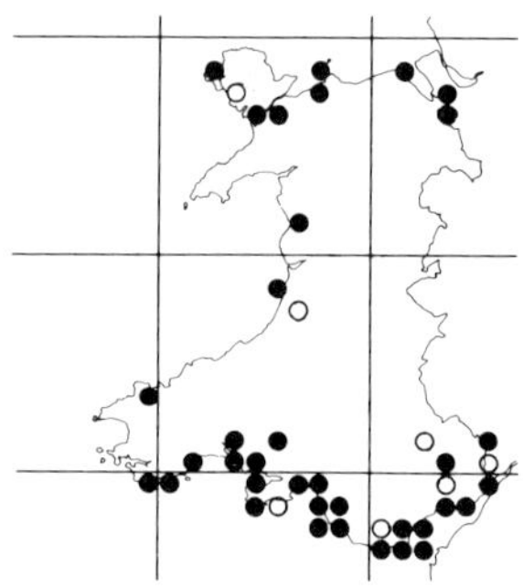

667 Orobanche minor

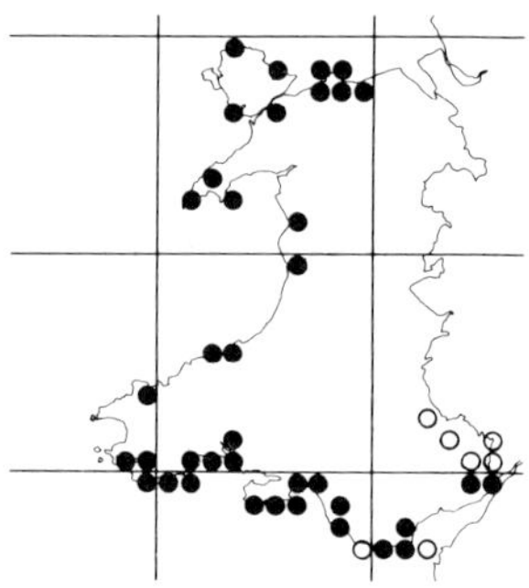

668 Orobanche hederae

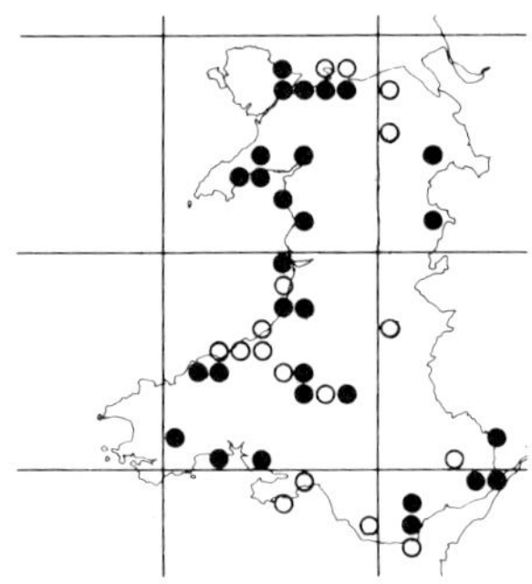

669 Orobanche rapum-genistae

670 Pinguicula vulgaris

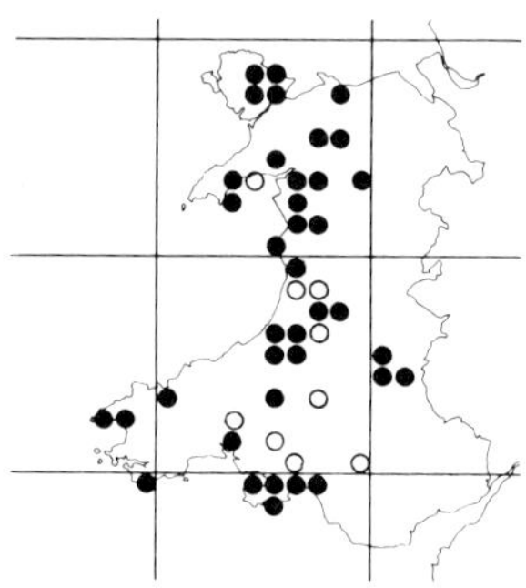

671 Utricularia minor

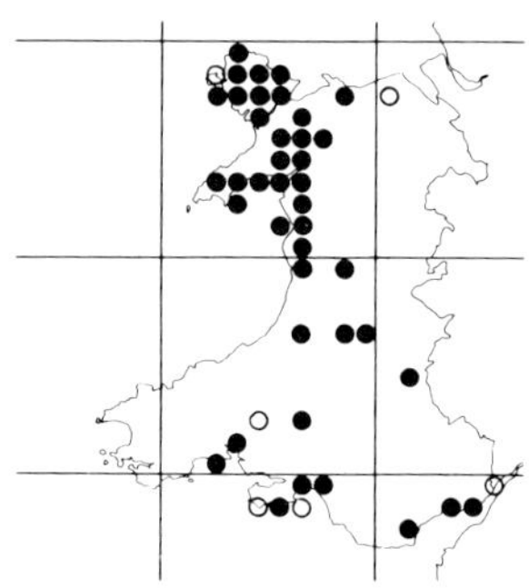

672 Utricularia australis

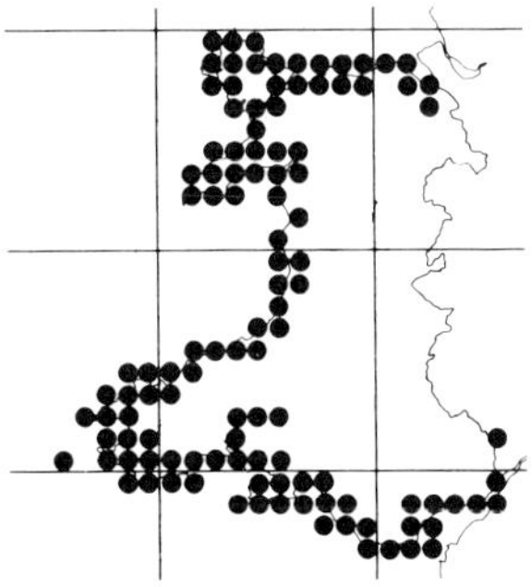

673 Plantago coronopus

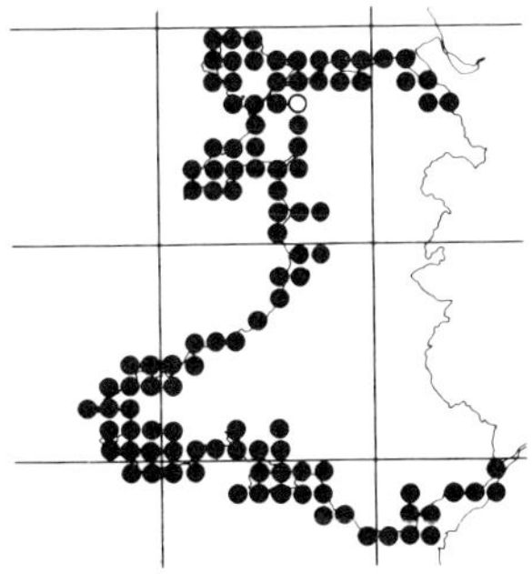

674 Plantago maritima

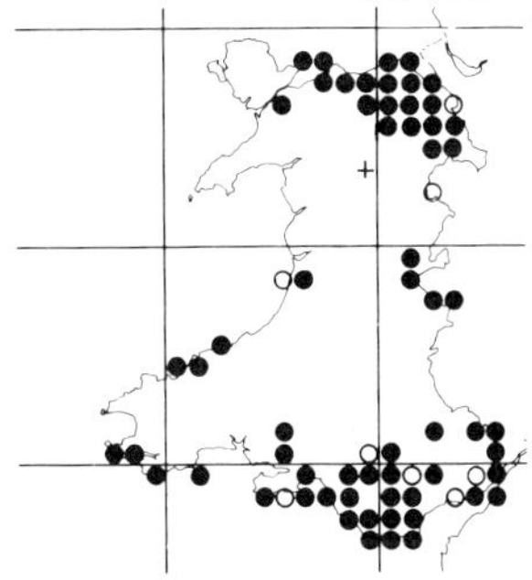

675 Plantago media

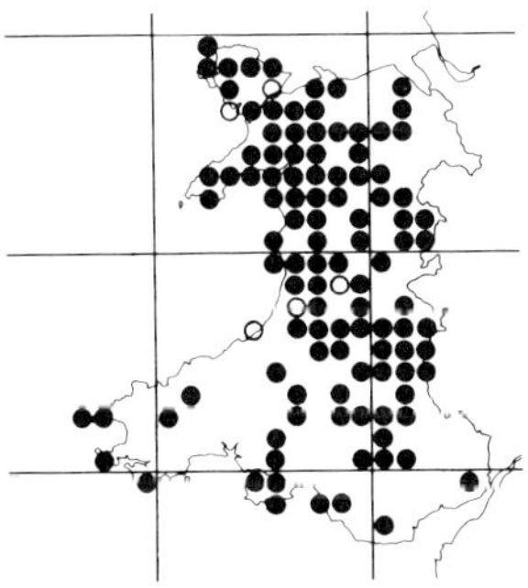

676 Littorella uniflora

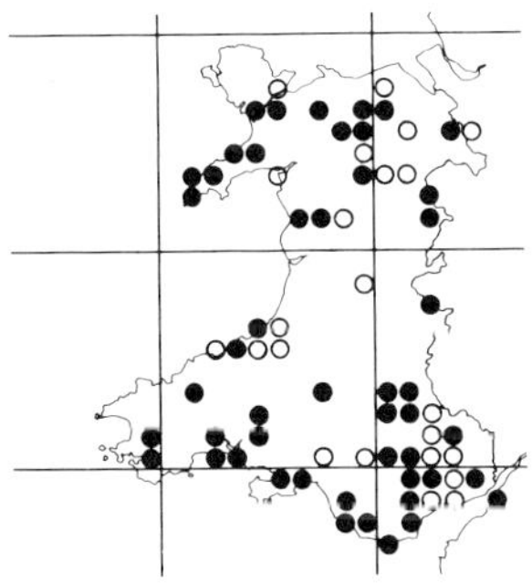

677 Sambucus ebulus

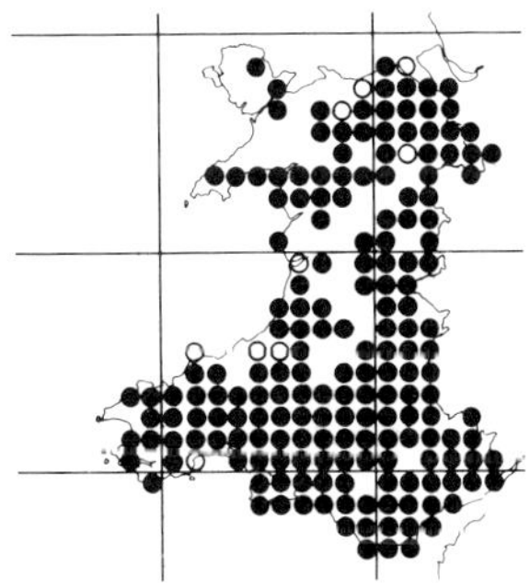

678 Viburnum opulus

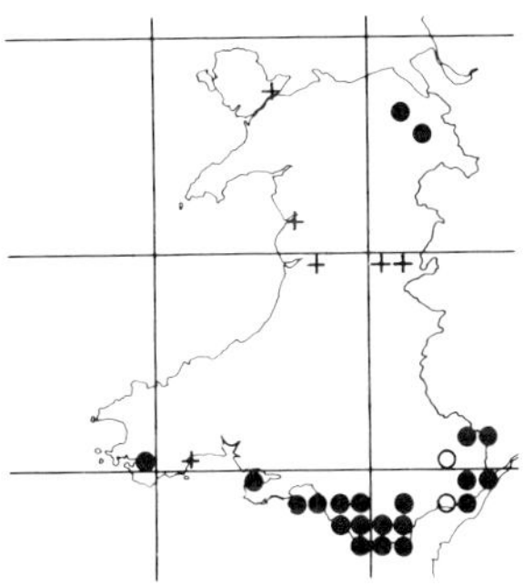

679 Viburnum lantana

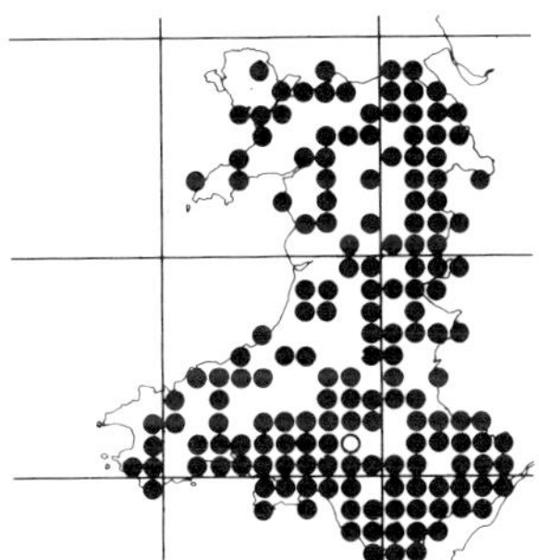

680 Symphoricarpos albus

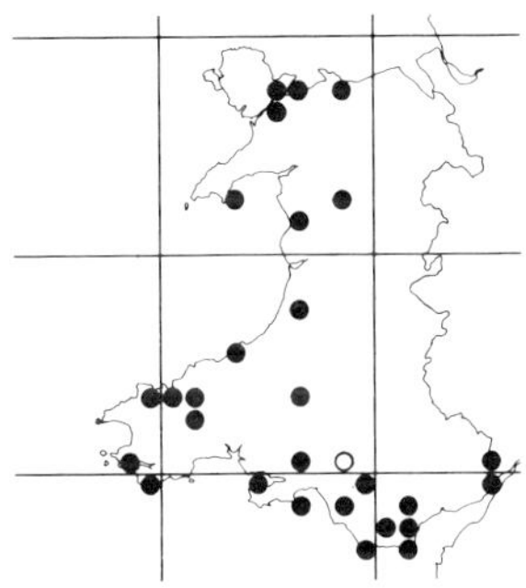

681 Leycesteria formosa

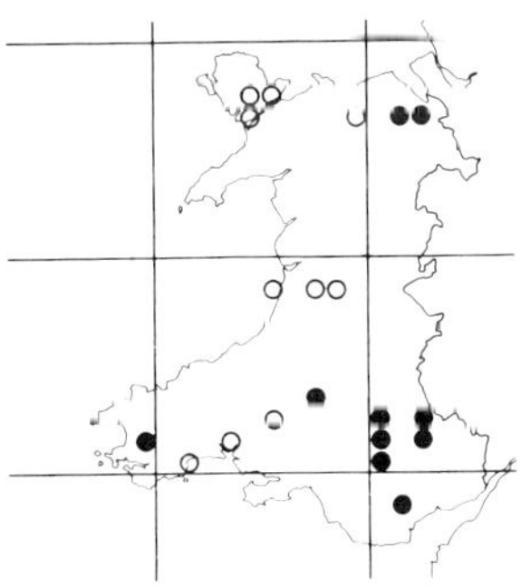

682 Lonicera xylosteum

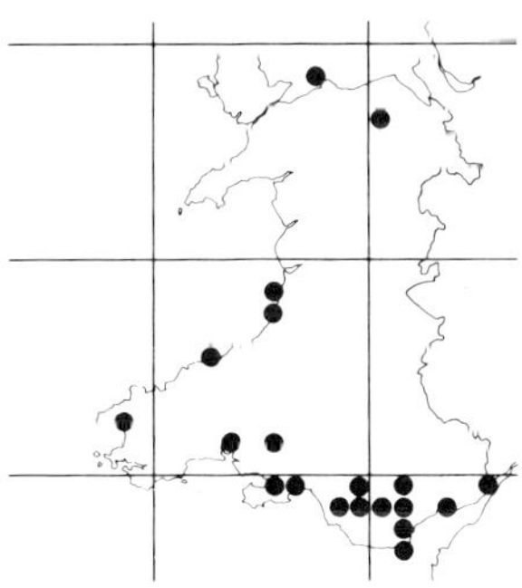

683 Lonicera nitida

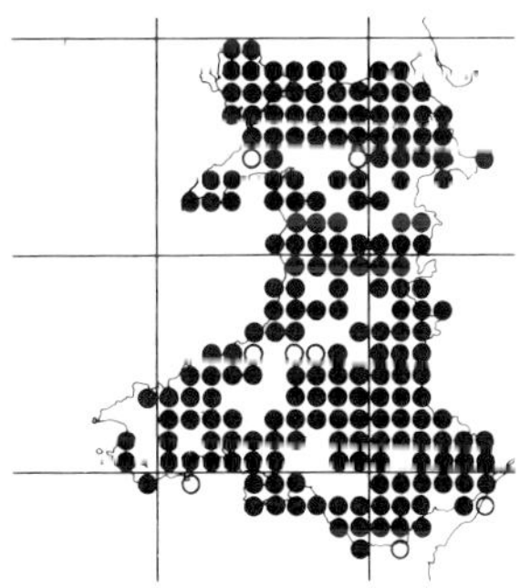

684 Adoxa moschatellina

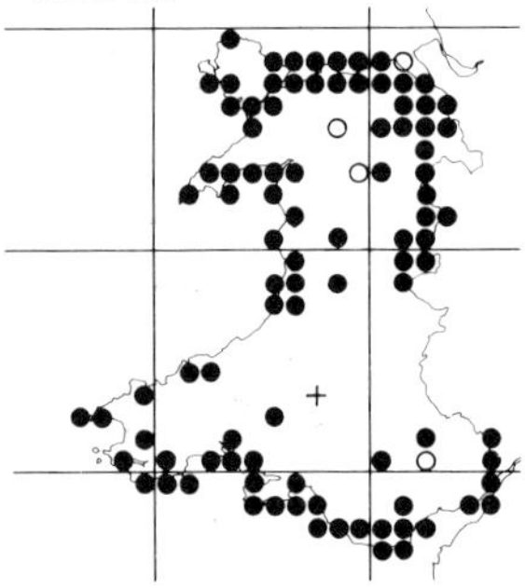

685 Valerianella locusta

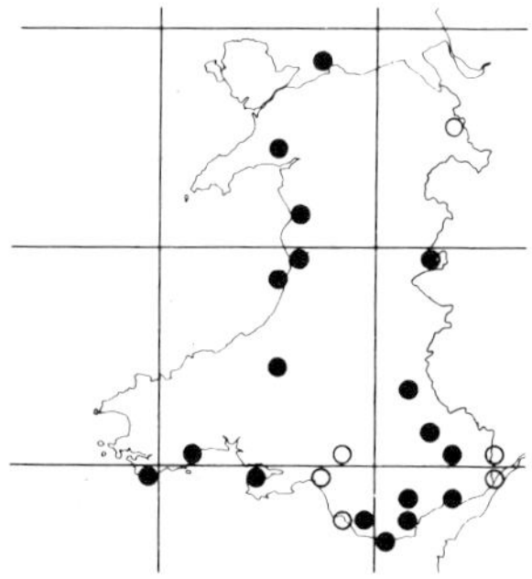

686 Valerianella carinata

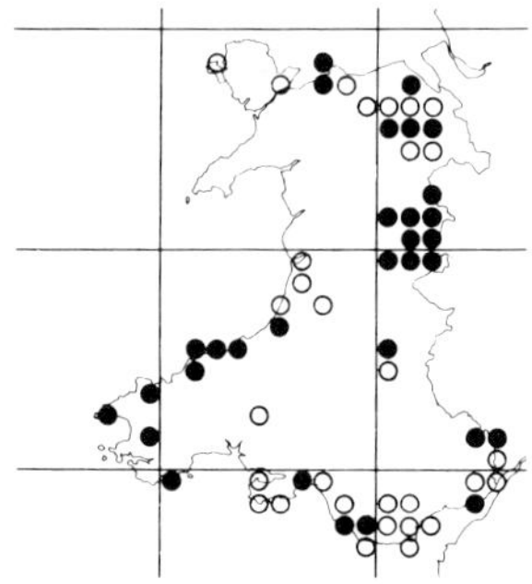

687 Valerianella dentata

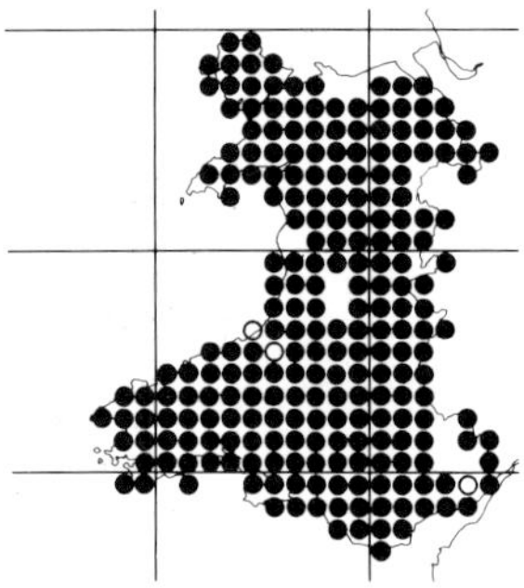

688 Valeriana officinalis

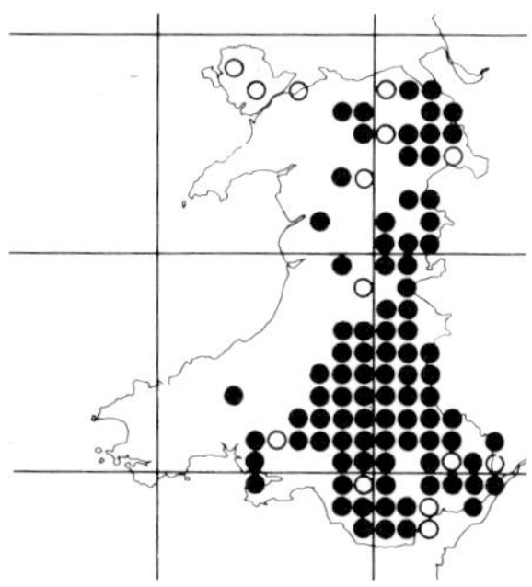

689 Valeriana dioica

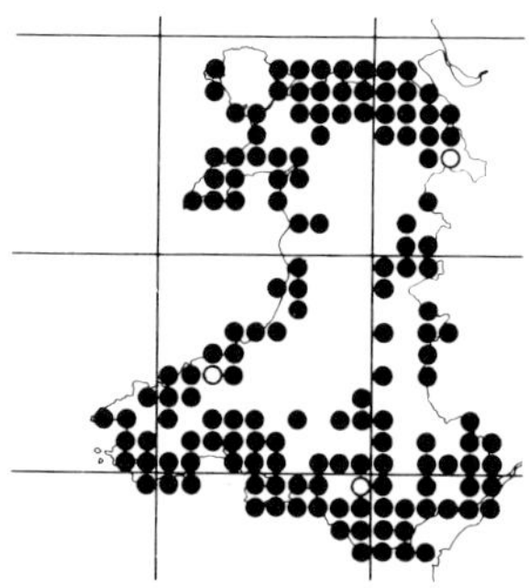

690 Centranthus ruber

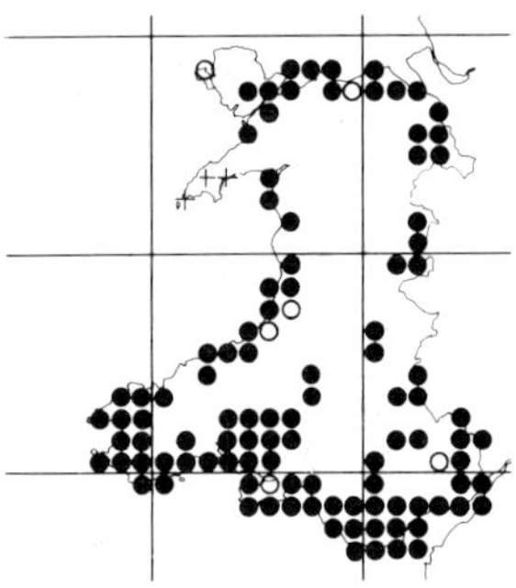

691 Dipsacus fullonum

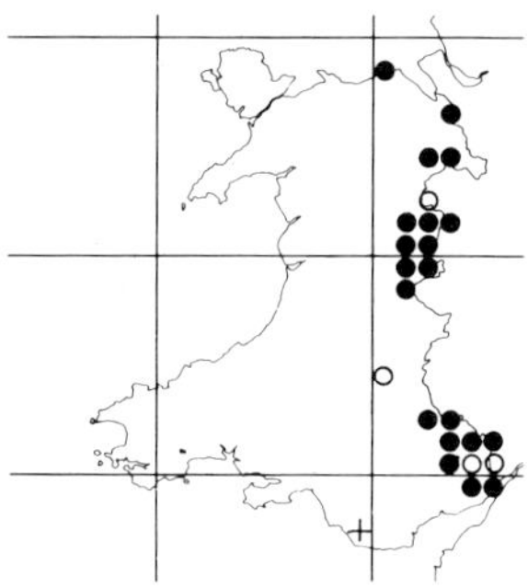

692 Dipsacus pilosus

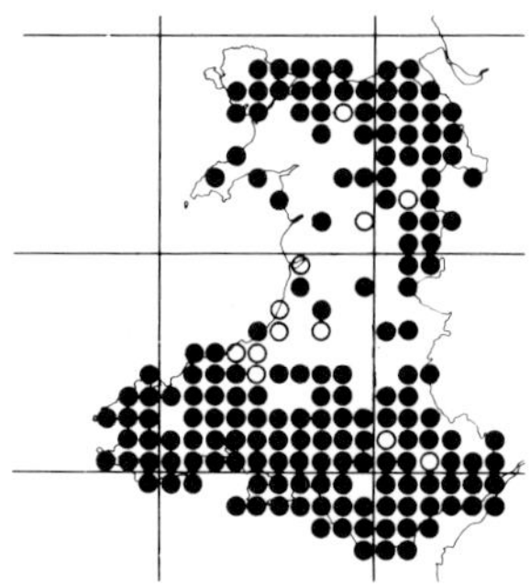

693 Knautia arvensis

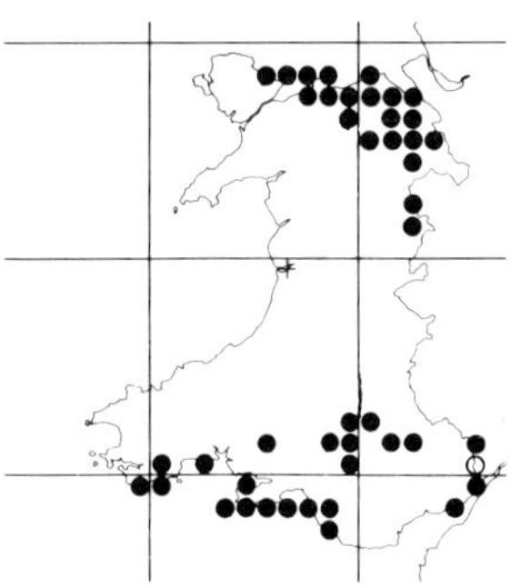

694 Scabiosa columbaria

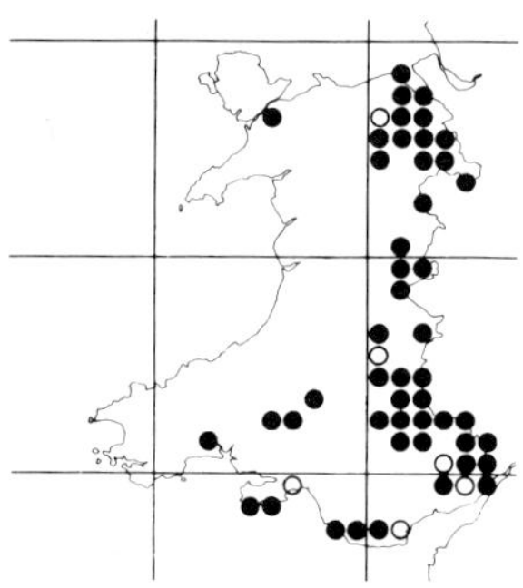

695 Campanula latifolia

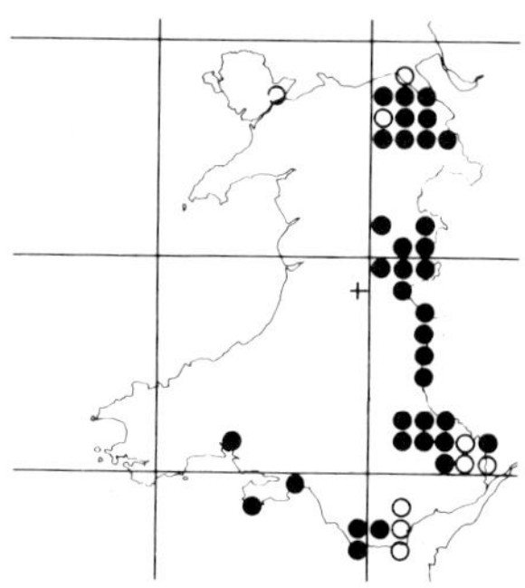

696 Campanula trachelium

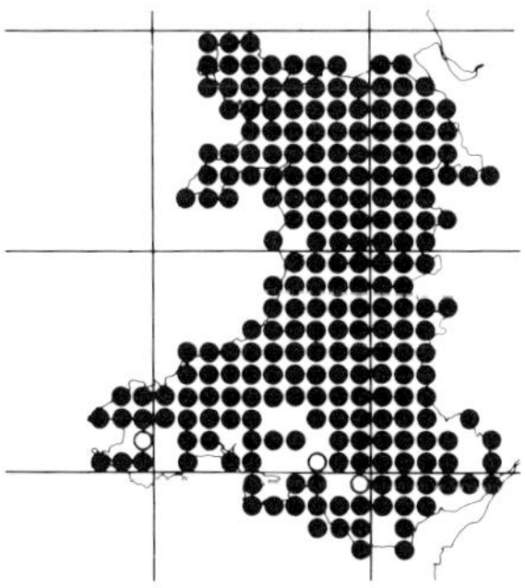

697 Campanula rotundifolia

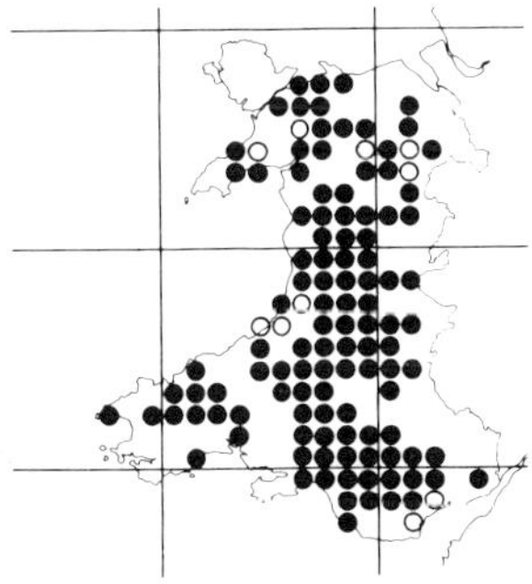

698 Wahlenbergia hederacea

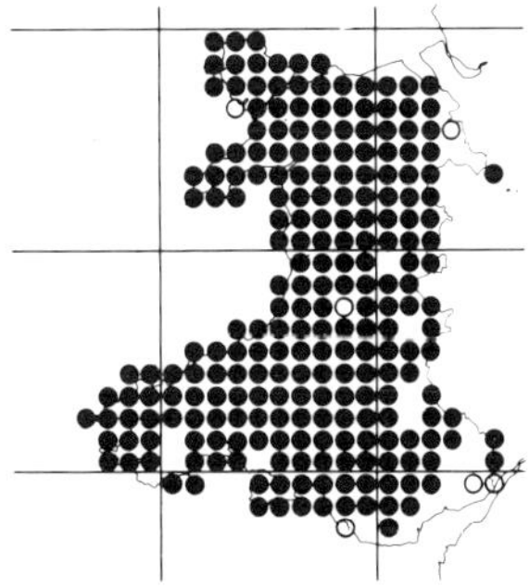

699 Jasione montana

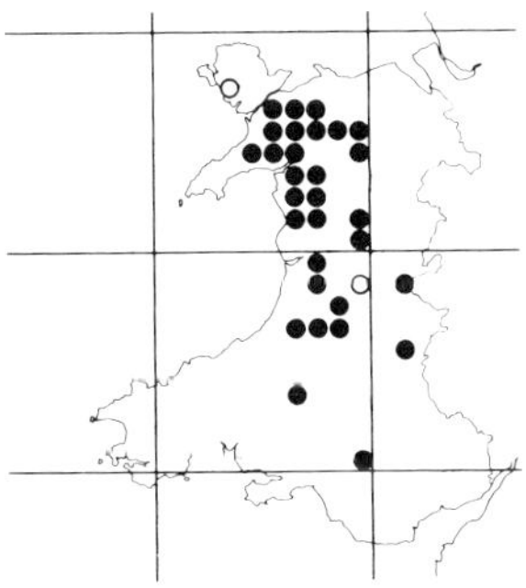

700 Lobelia dortmanna

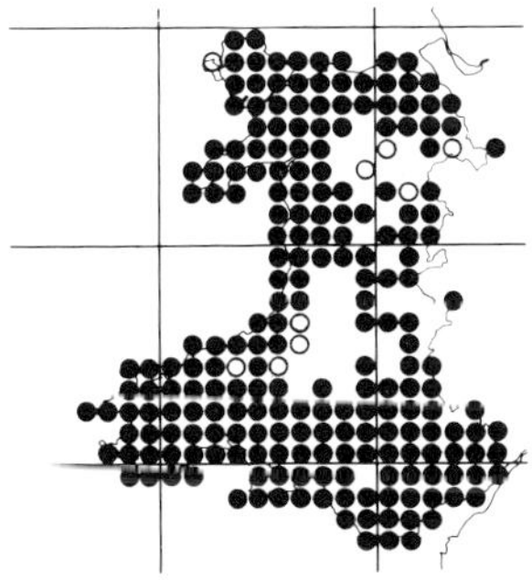

701 Eupatorium cannabinum

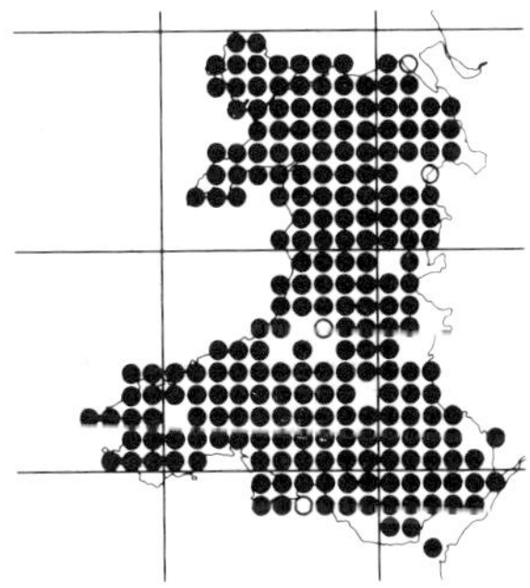

702 Solidago virgaurea

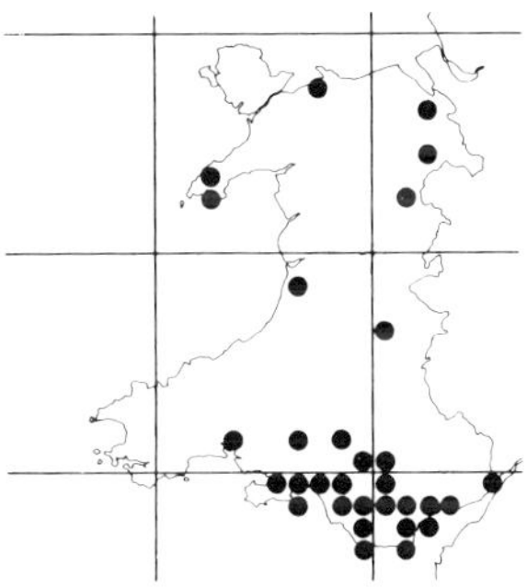

703 Solidago canadensis

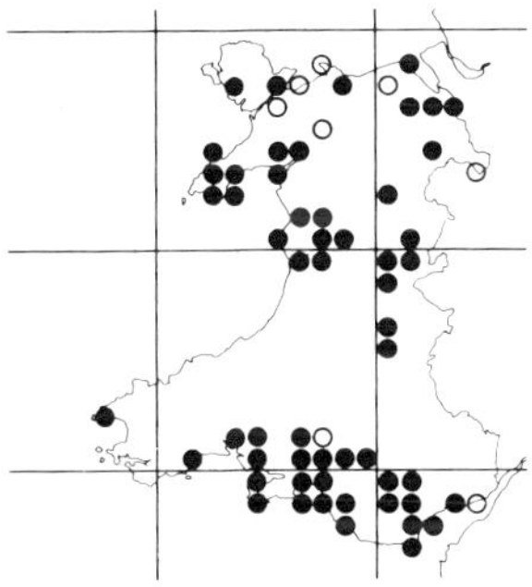

704 Aster novi-belgii

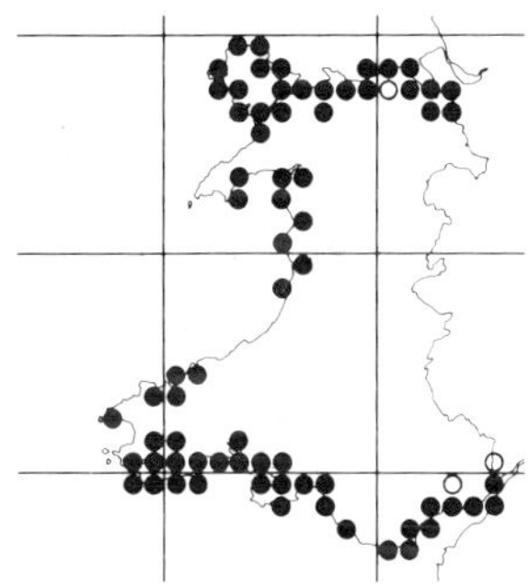

705 Aster tripolium

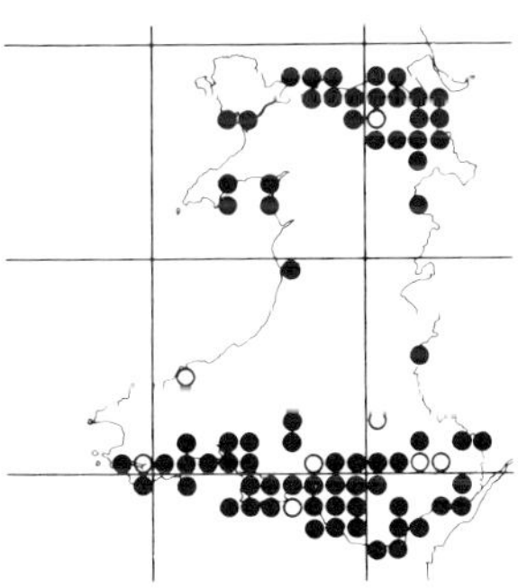

706 Erigeron acer

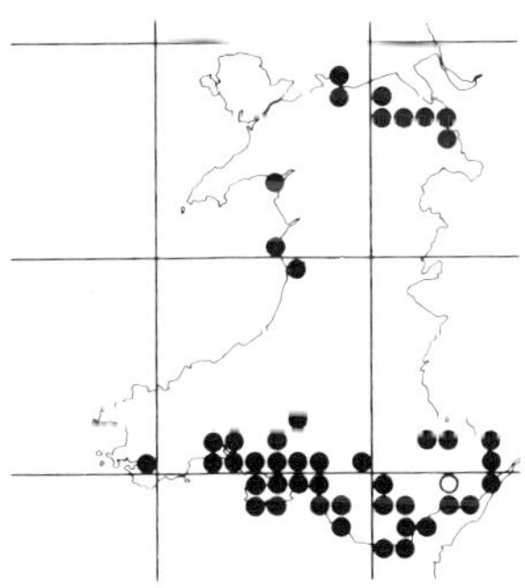

707 Conyza canadensis

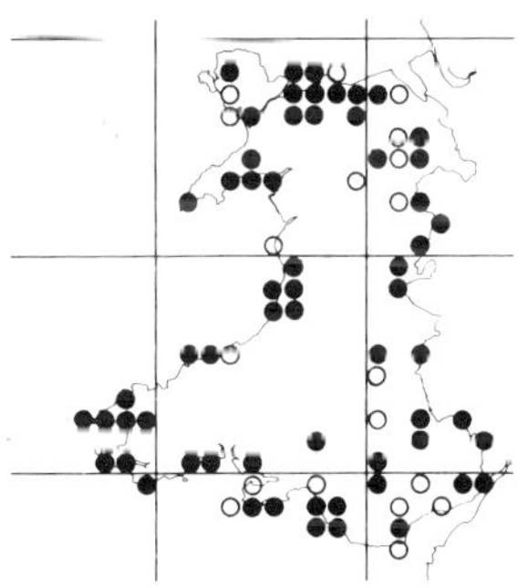

708 Filago vulgaris

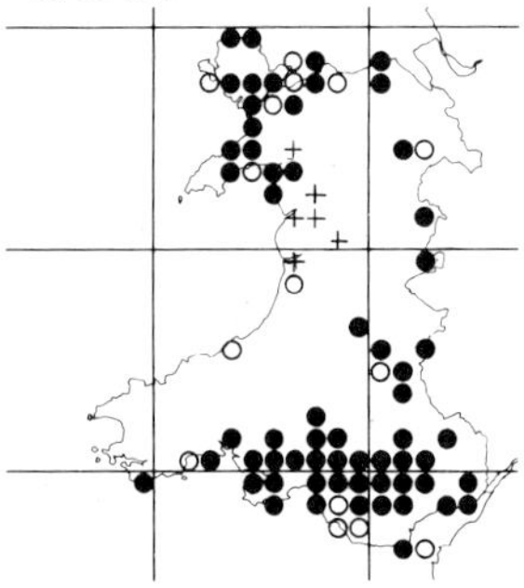

709 Logfia minima

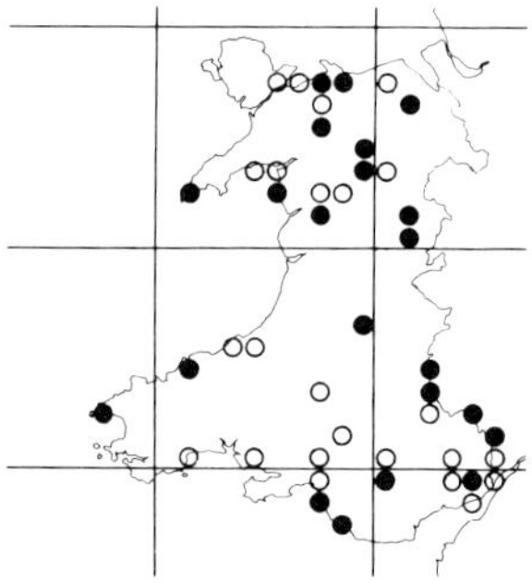

710 Omalotheca sylvatica

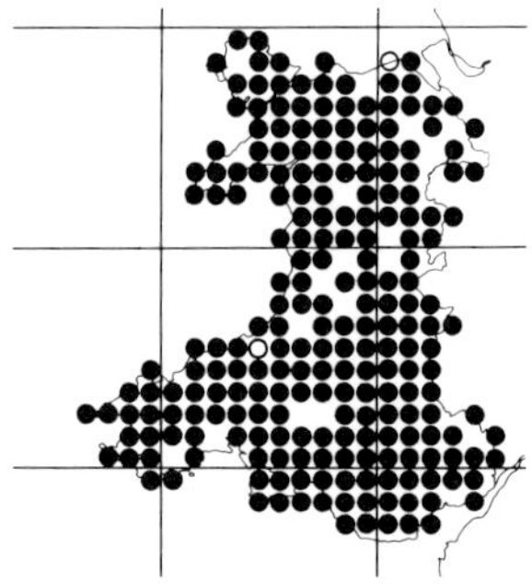

711 Filaginella uliginosa

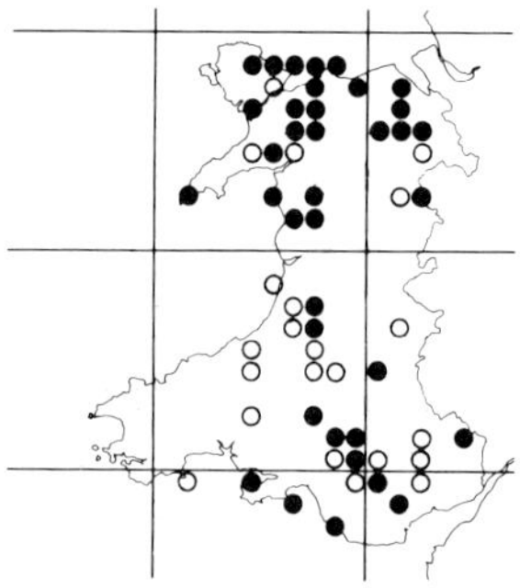

712 Antennaria dioica

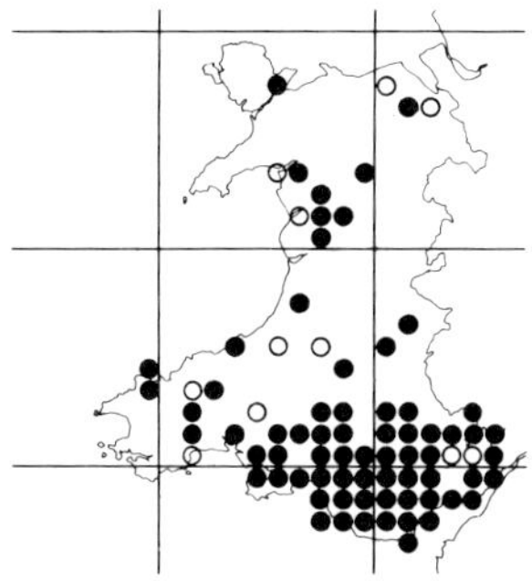

713 Anaphalis margaritacea

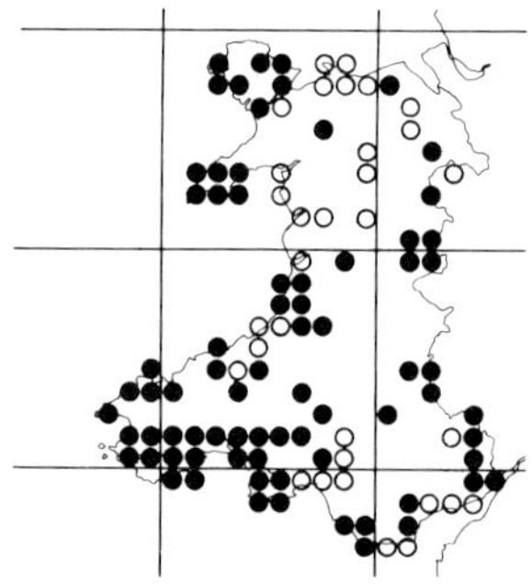

714 Inula helenium

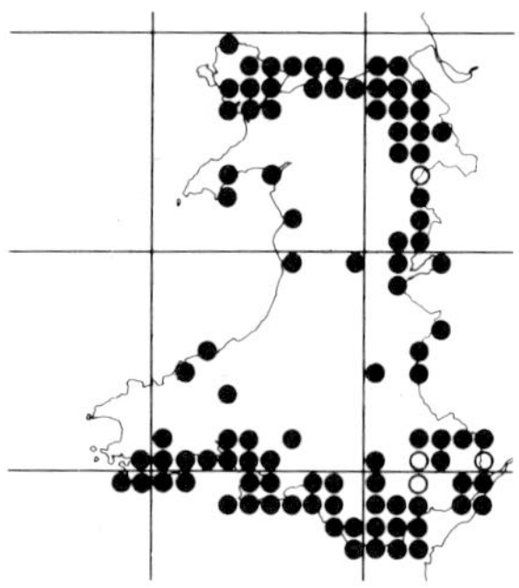

715 Inula conyza

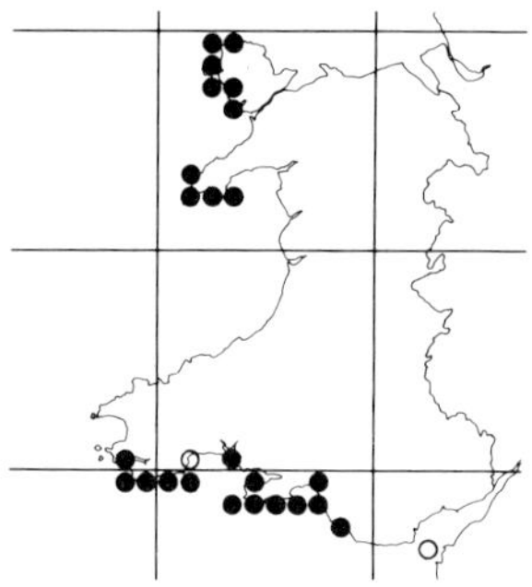

716 Inula crithmoides

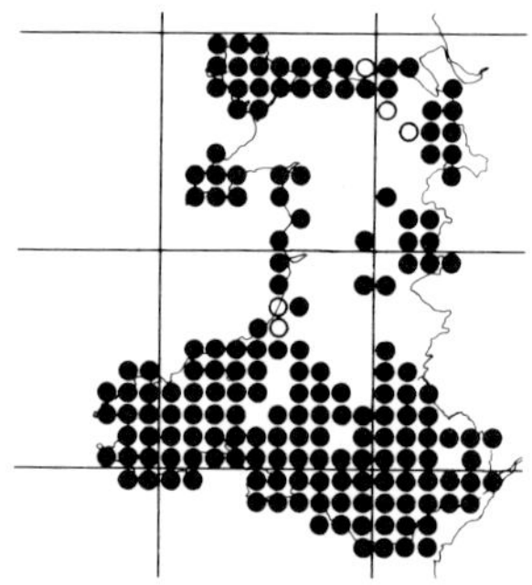

717 Pulicaria dysenterica

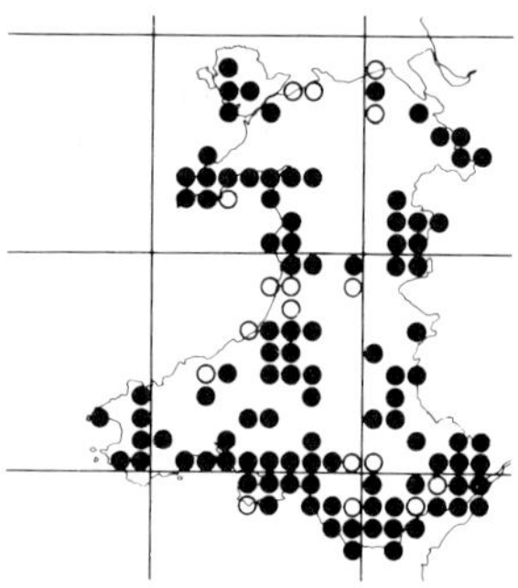

718 Bidens tripartitus

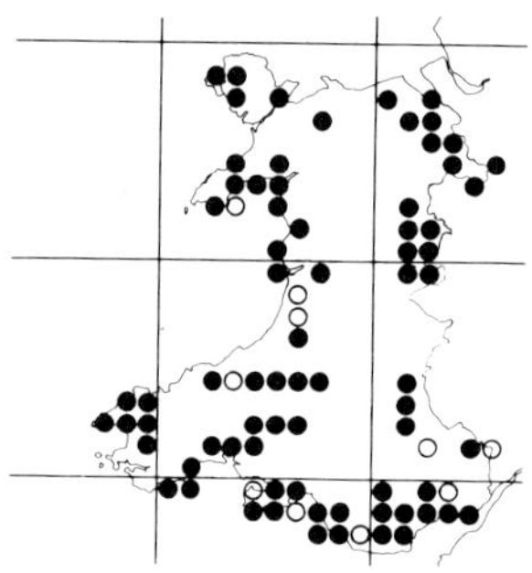

719 Bidens cernua

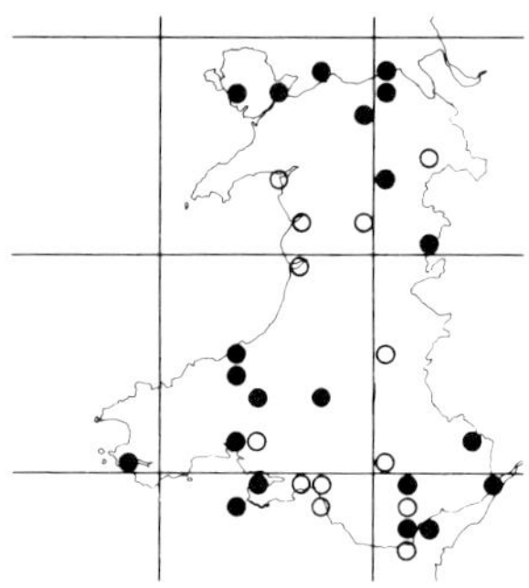

720 Anthemis arvensis

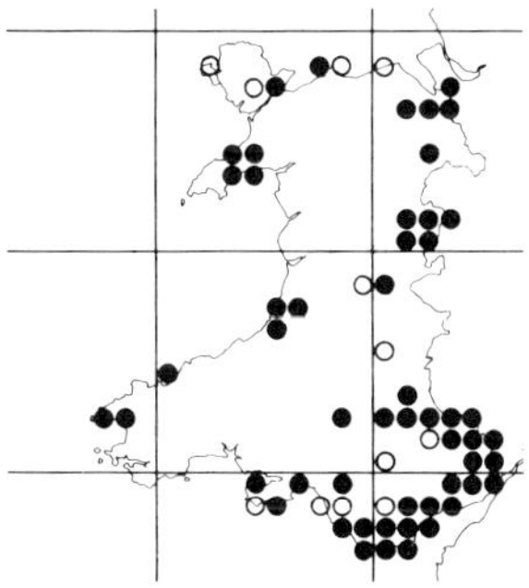
721 Anthemis cotula

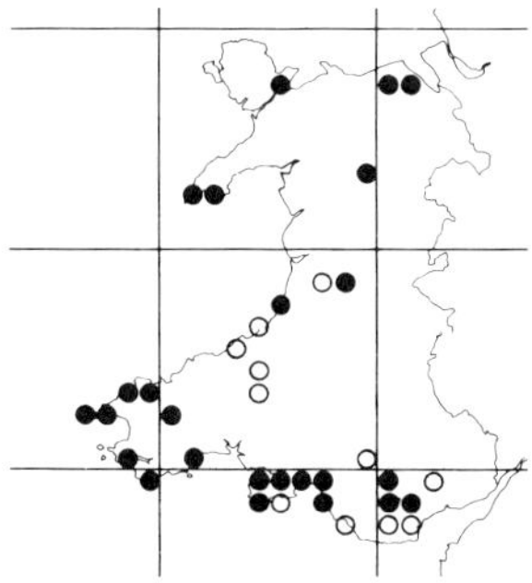
722 Chamaemelum nobile

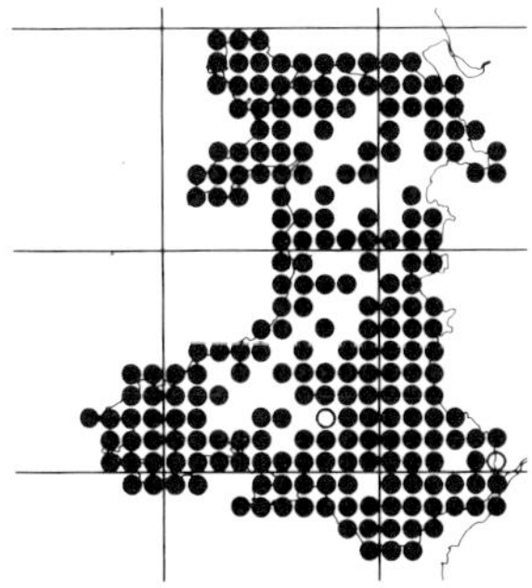
723 Tripleurospermum maritimum agg.

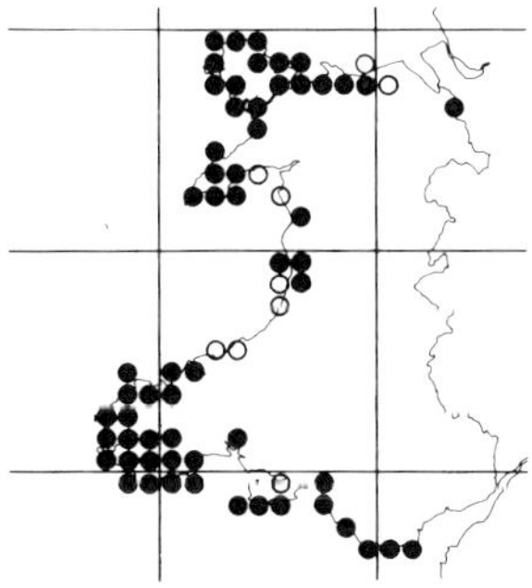
724 Tripleurospermum maritimum s.s.

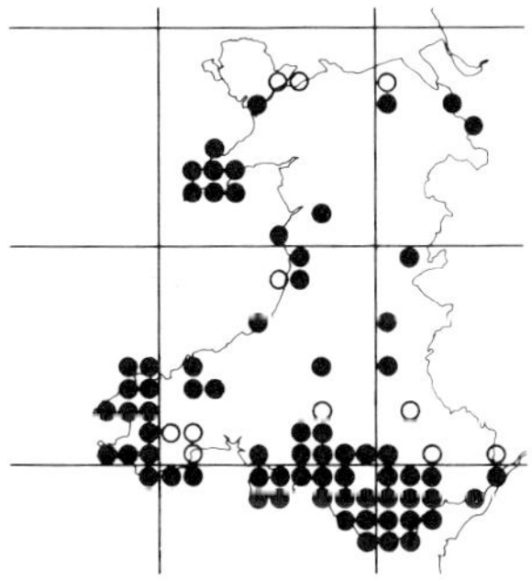
725 Tripleurospermum inodorum

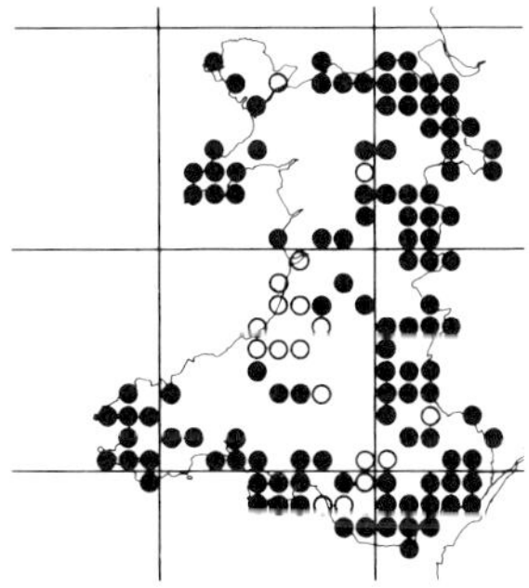
726 Matricaria recutita

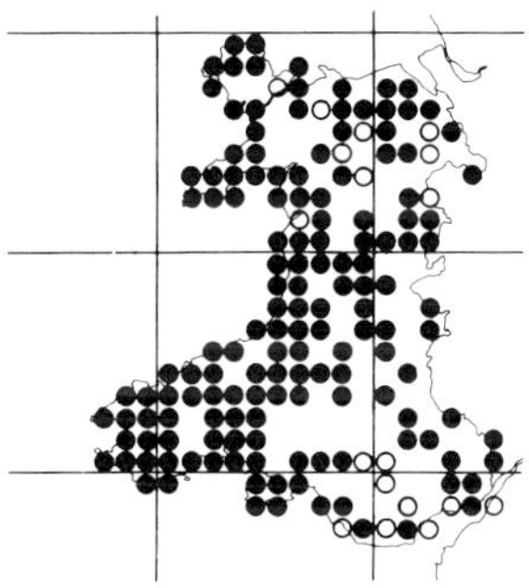
727 Chrysanthemum segetum

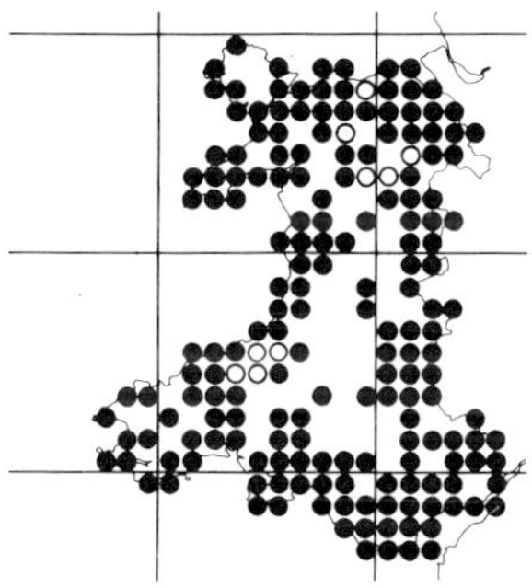
728 Tanacetum vulgare

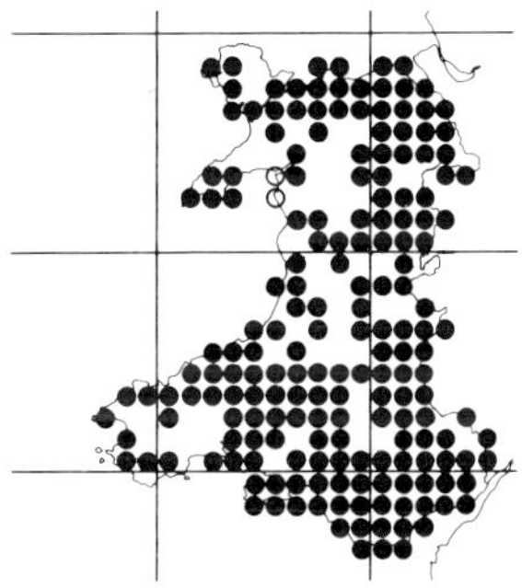
729 Tanacetum parthenium

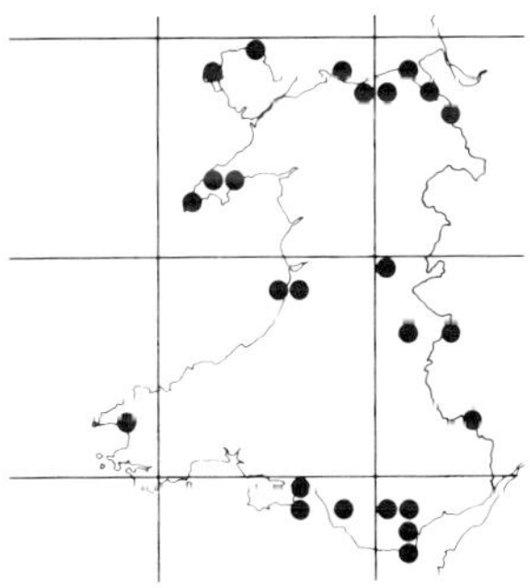
730 Leucanthemum maximum

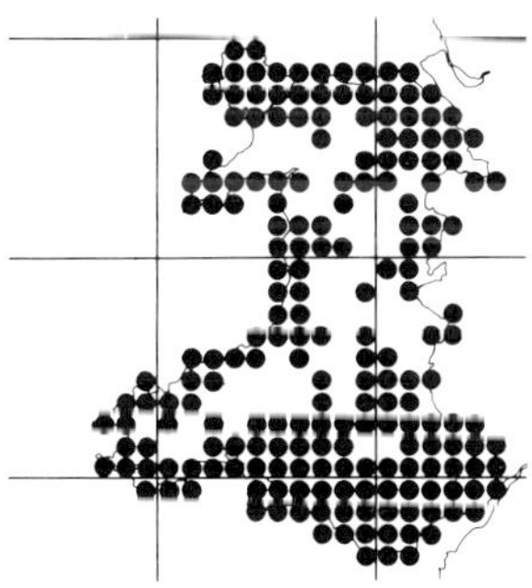
731 Artemisia vulgaris

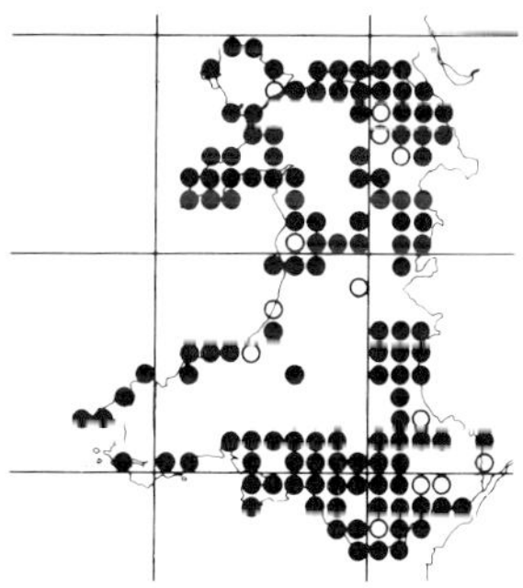
732 Artemisia absinthium

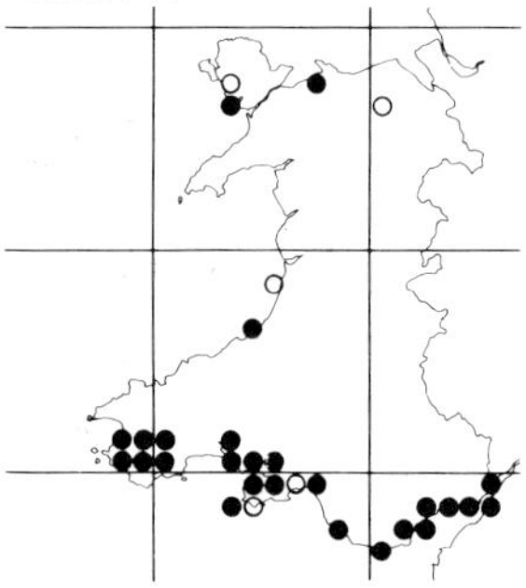

733 Artemisia maritima

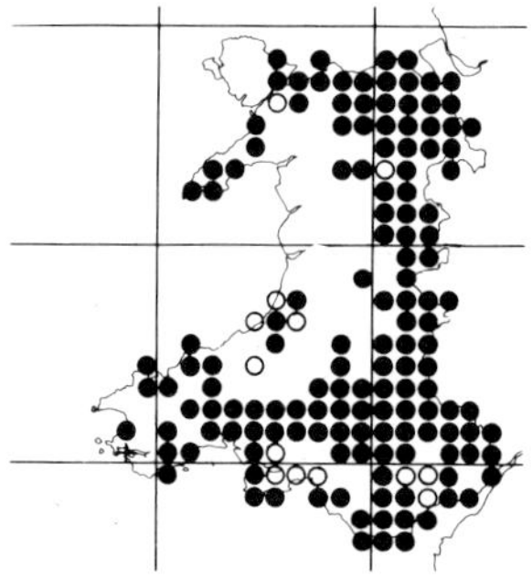

734 Petasites hybridus

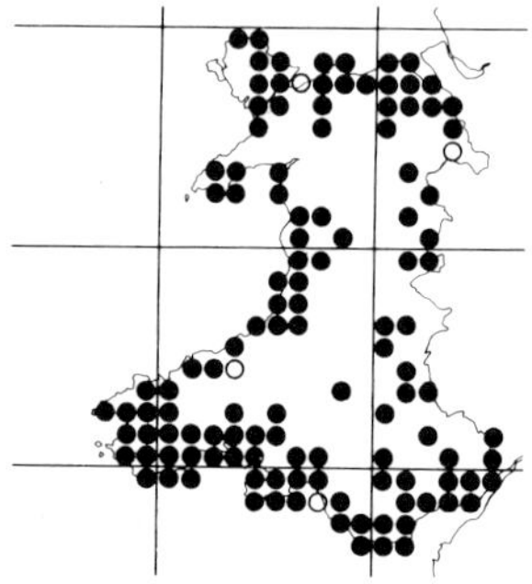

735 Petasites fragrans

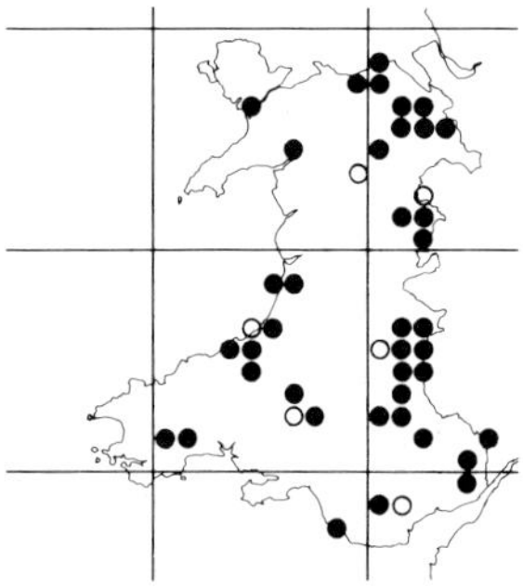

736 Doronicum pardalianches

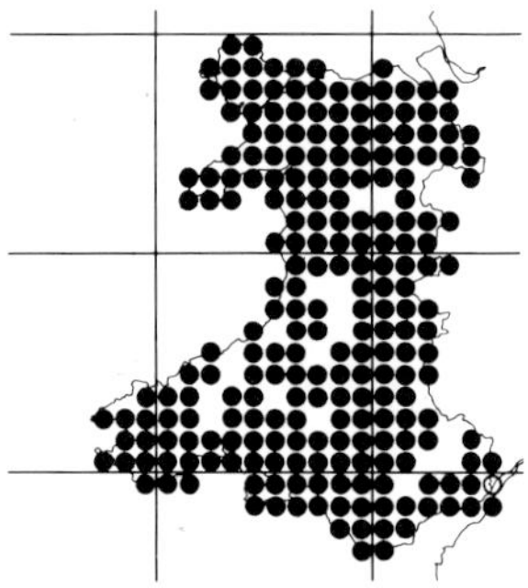

737 Senecio aquaticus

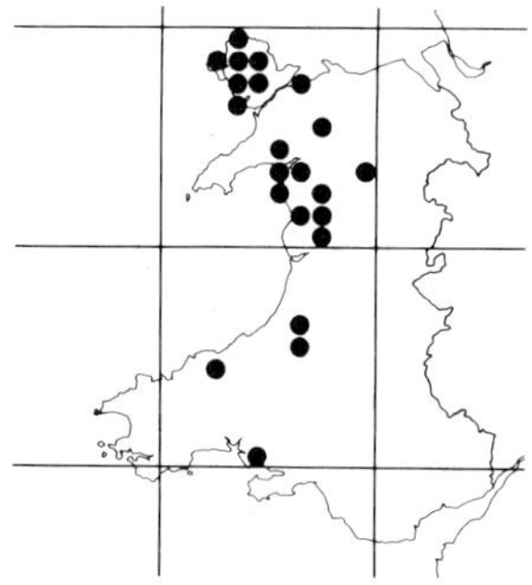

738 Senecio aquaticus
× jacobea
= S. × ostenfeldii

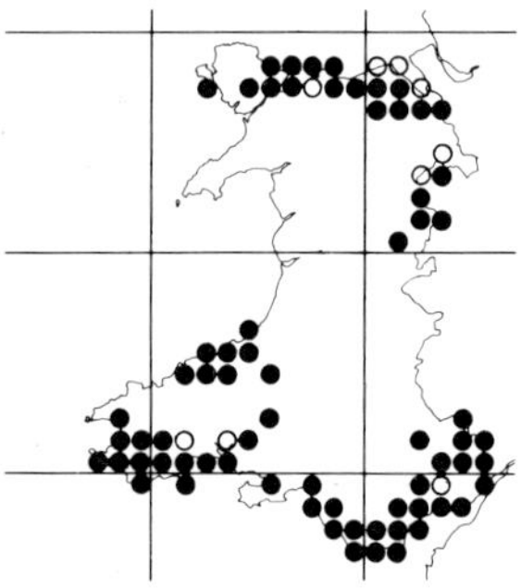

739 Senecio erucifolius

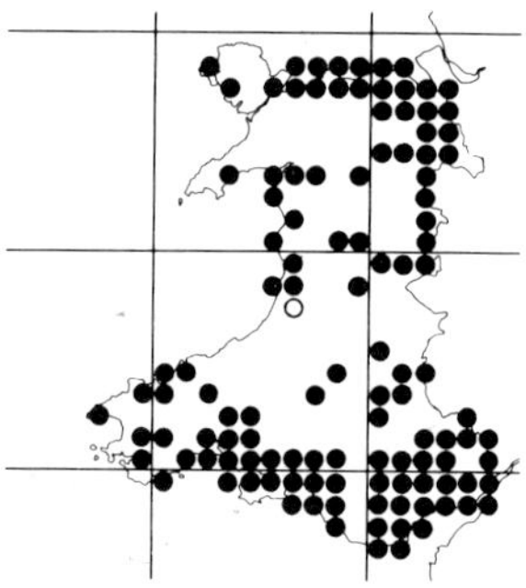

740 Senecio squalidus

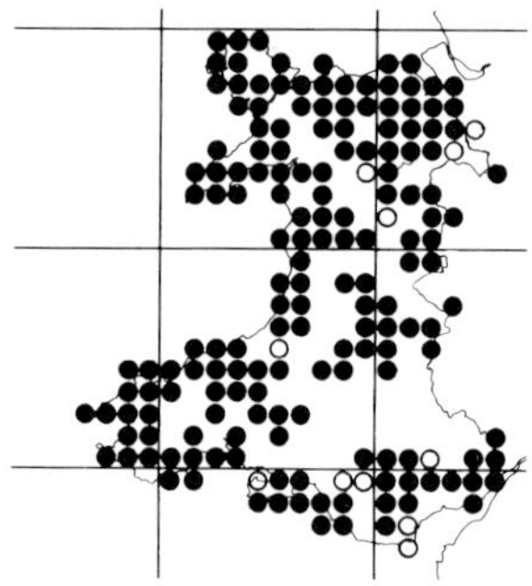

741 Senecio sylvaticus

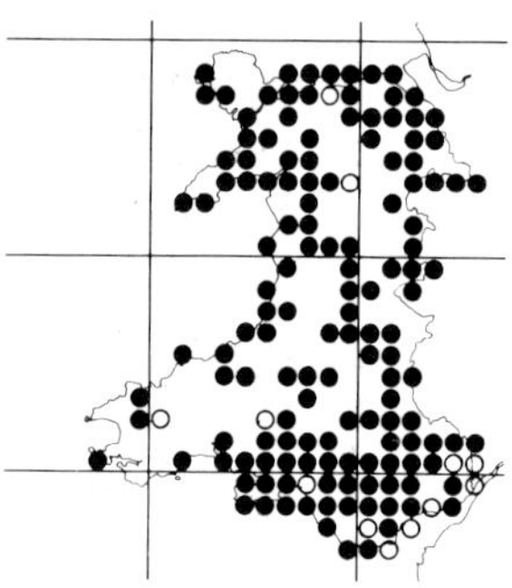

742 Senecio viscosus

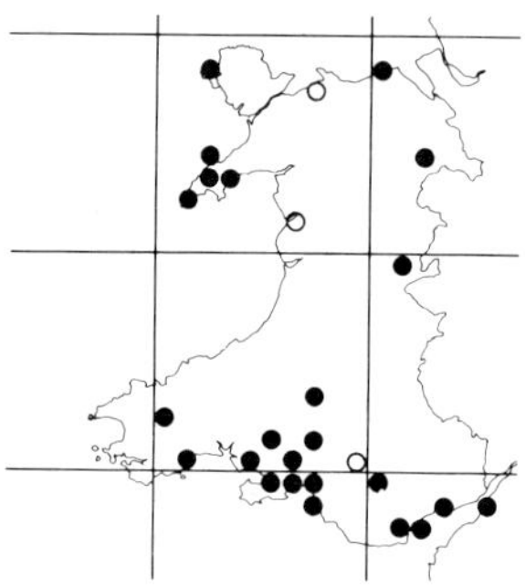

743 Calendula officinalis

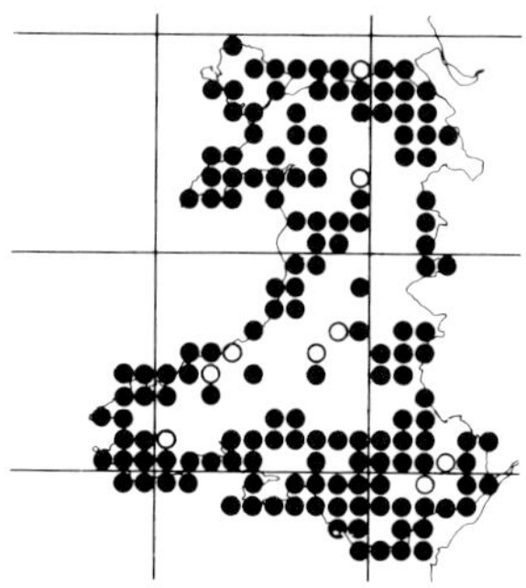

744 Carlina vulgaris

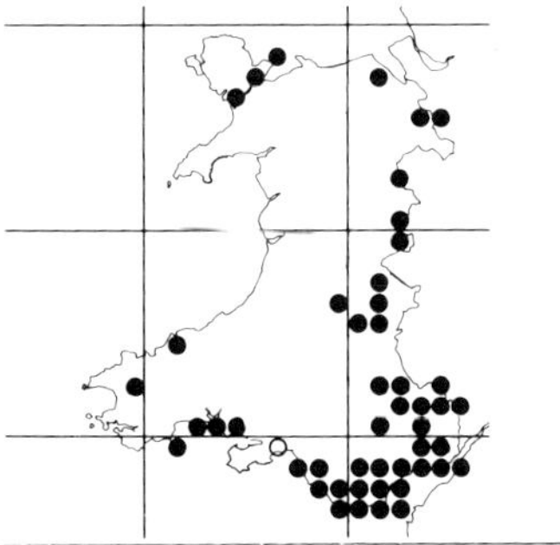

745 Arctium lappa

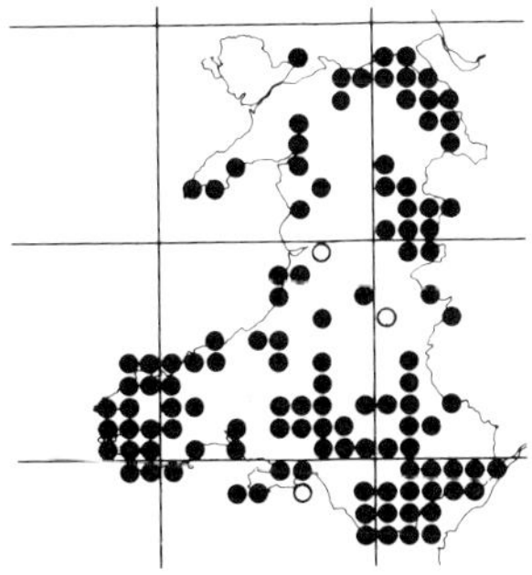

746 Arctium minus s.s.

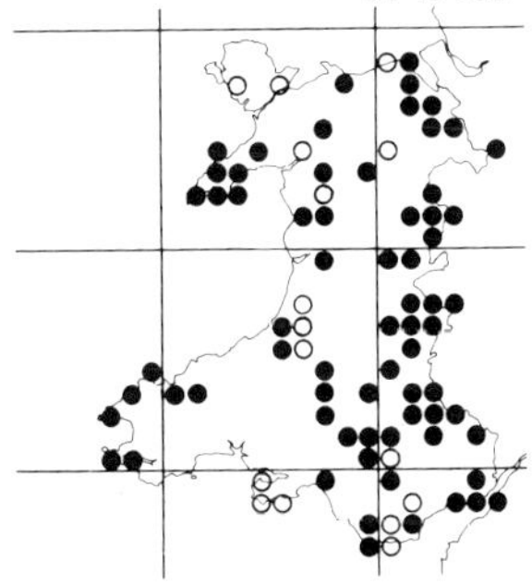

747 Arctium nemorosum

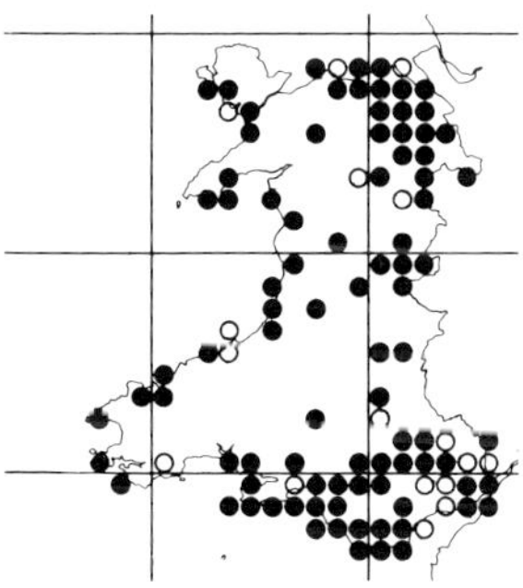

748 Carduus nutans

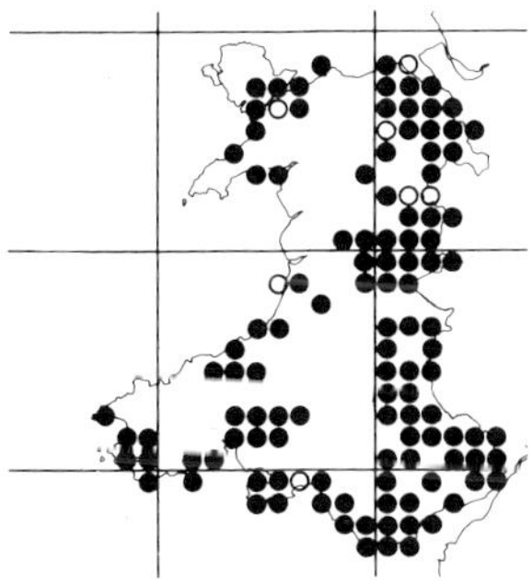

749 Carduus acanthoides

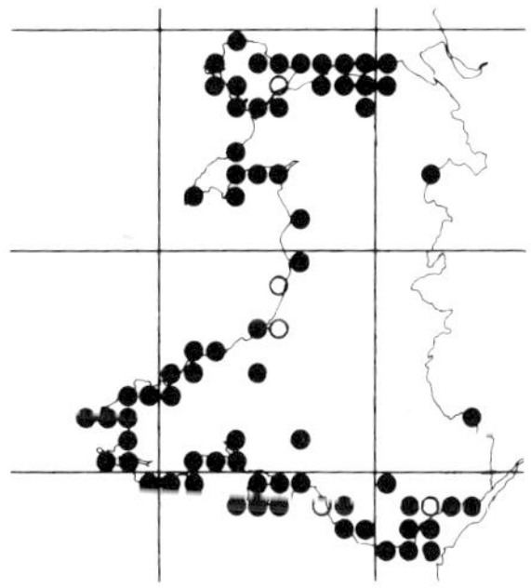

750 Carduus tenuiflorus

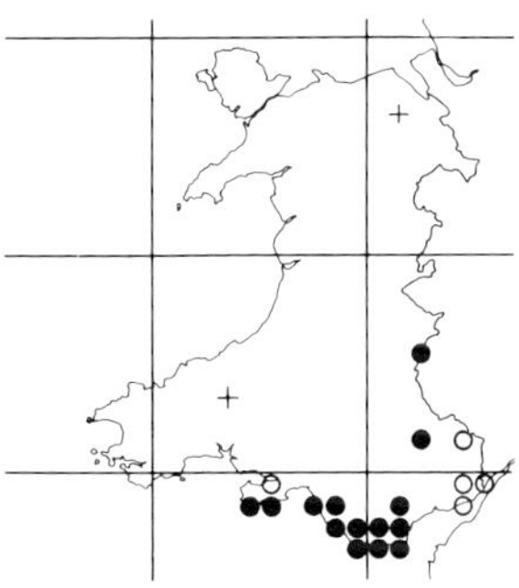

751 Cirsium eriophorum

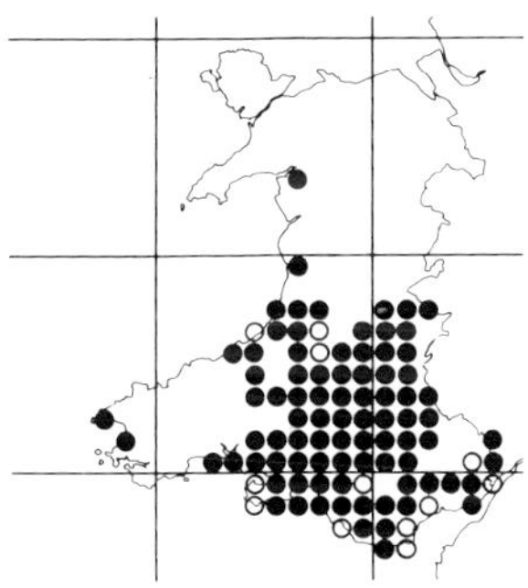

752 Cirsium dissectum

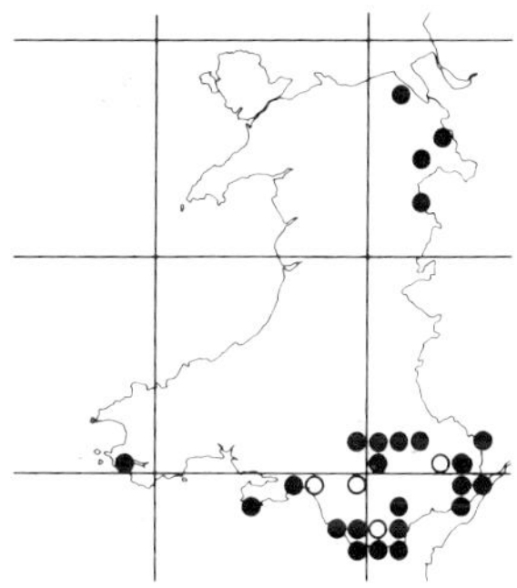

753 Cirsium acaule

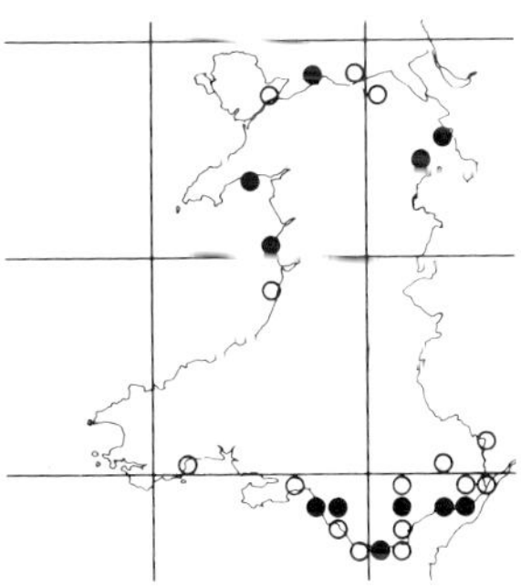

754 Onopordum acanthium

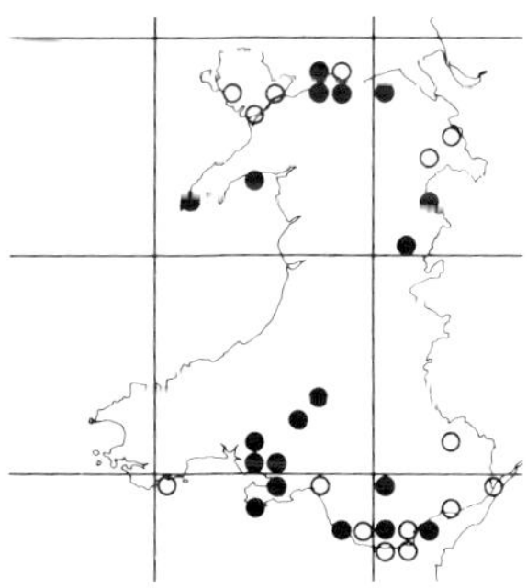

755 Silybum marianum

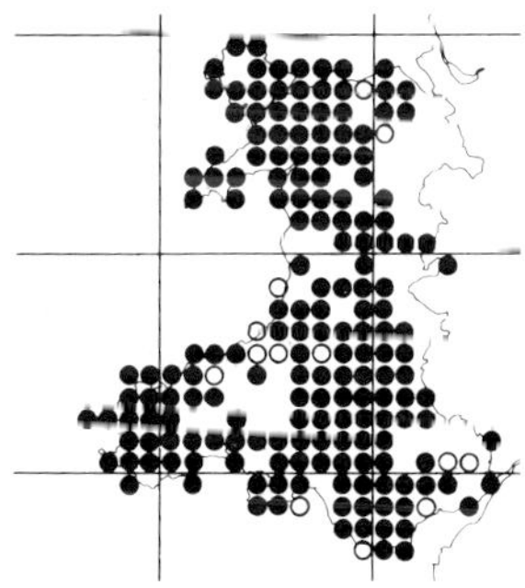

756 Serratula tinctoria

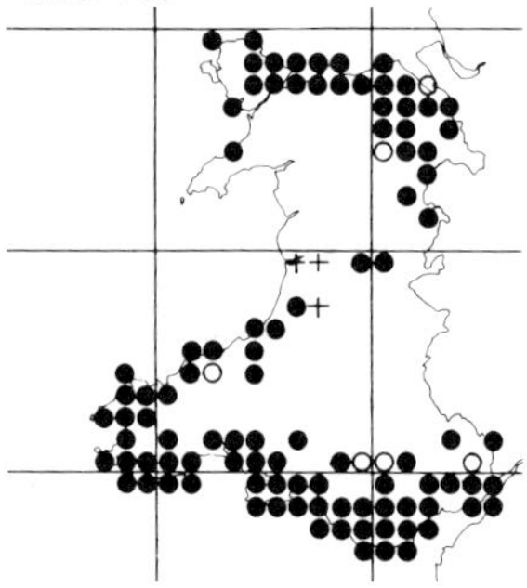

757 Centaurea scabiosa

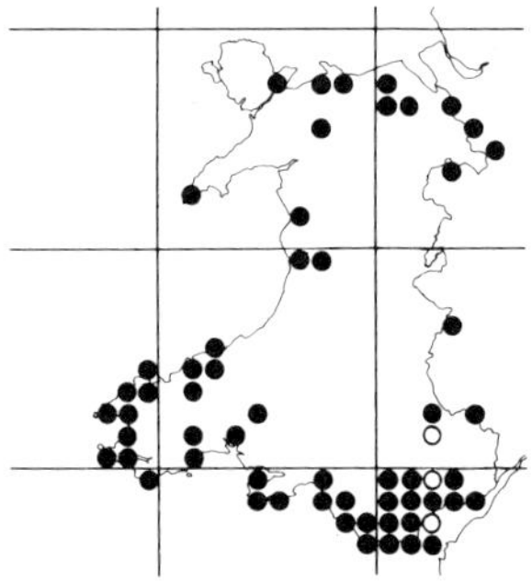

758 Centaurea debeauxii

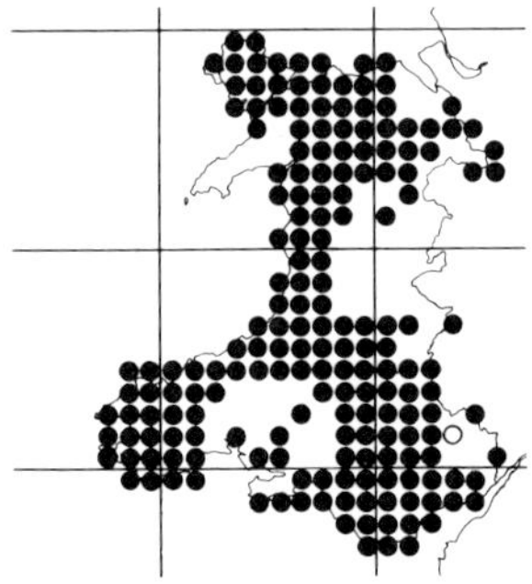

759 Centaurea nigra s.s.

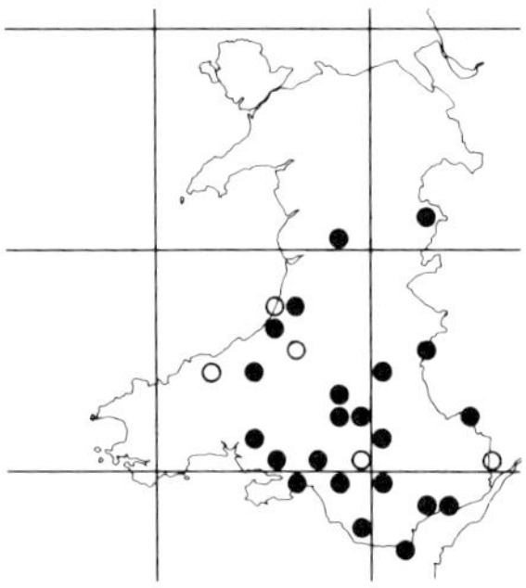

760 Centaurea montana

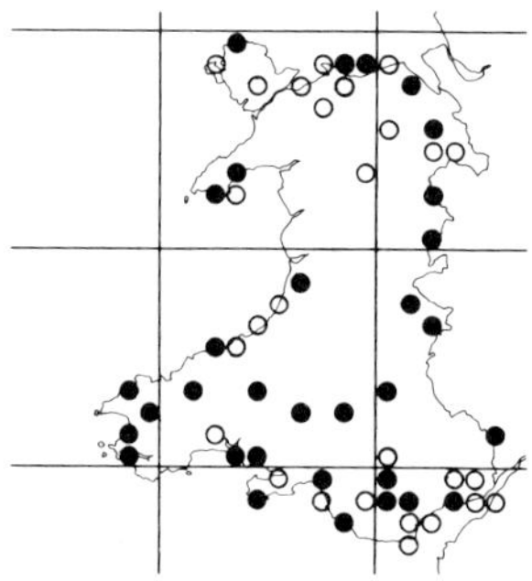

761 Centaurea cyanus

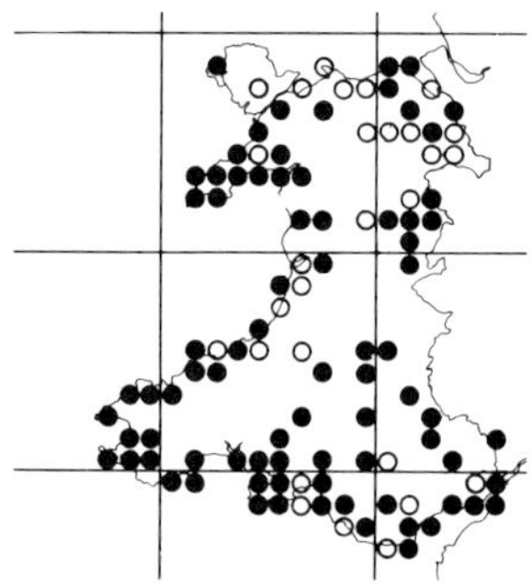

762 Cichorium intybus

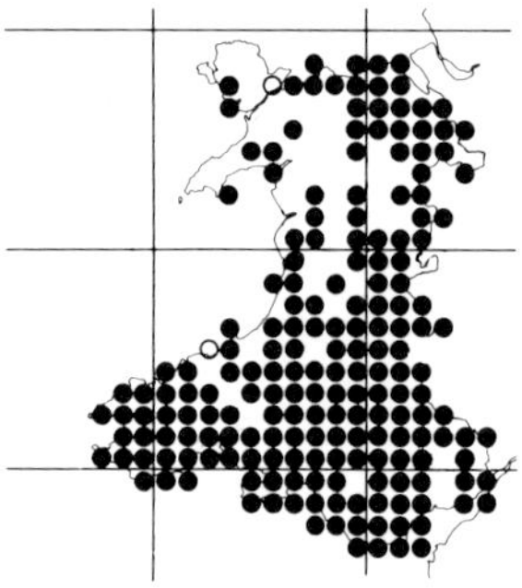

763 Leontodon hispidus

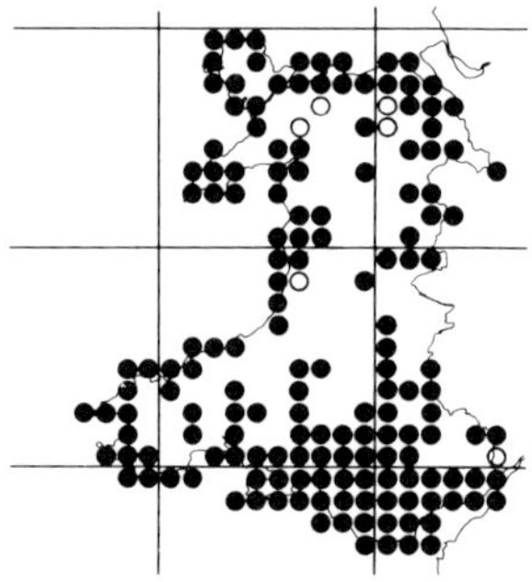

764 Leontodon taraxacoides

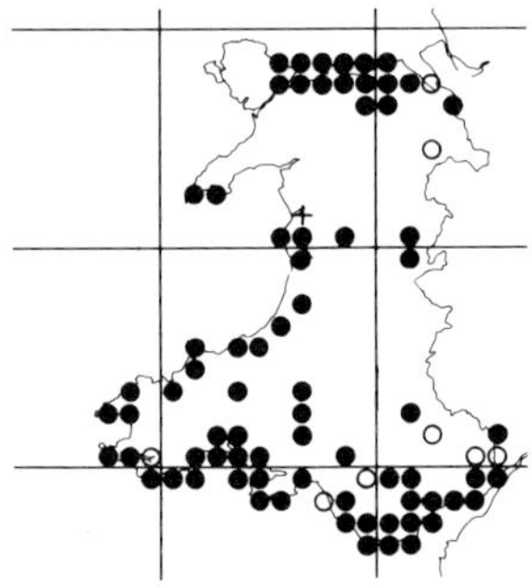

765 Picris echioides

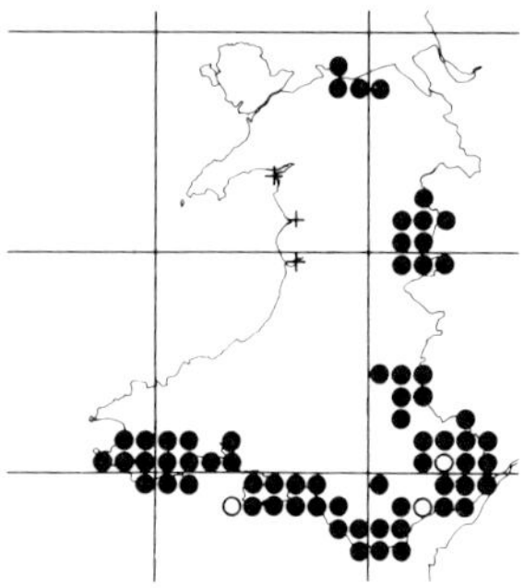

766 Picris hieracioides

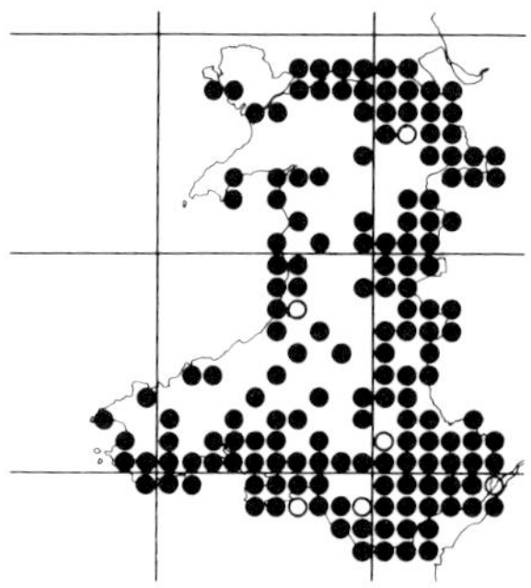

767 Tragopogon pratensis subsp. minor

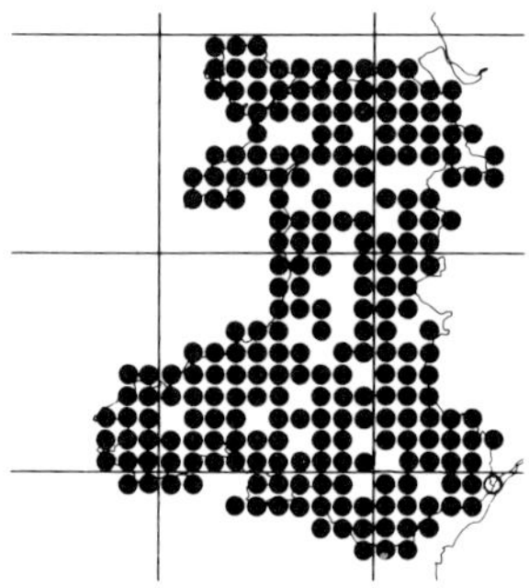

768 Sonchus arvensis

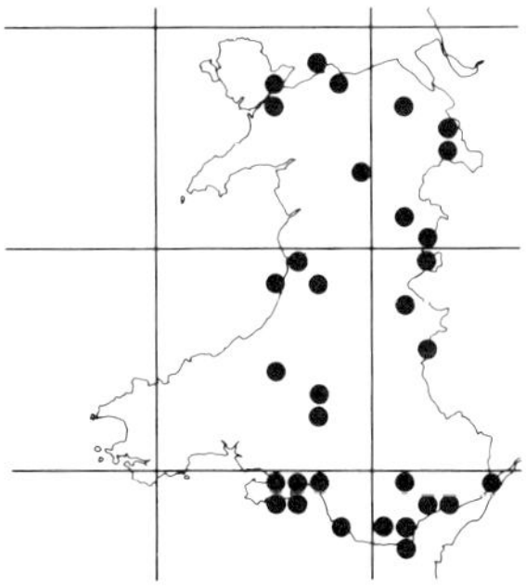

769 Cicerbita macrophylla

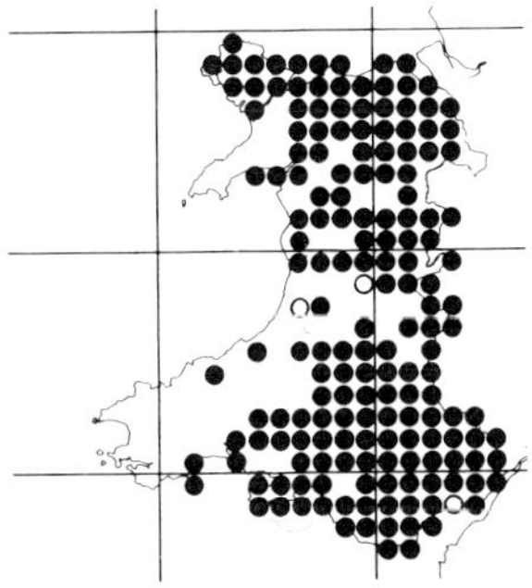

770 Mycelis muralis

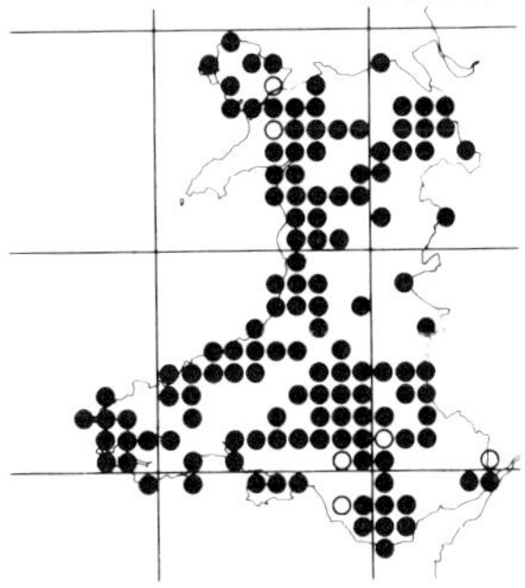

771 Taraxacum Sect. Spectabilia

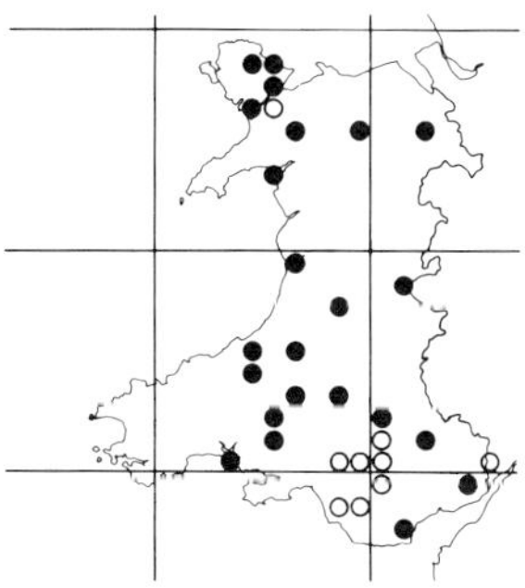

772 Taraxacum faeroense

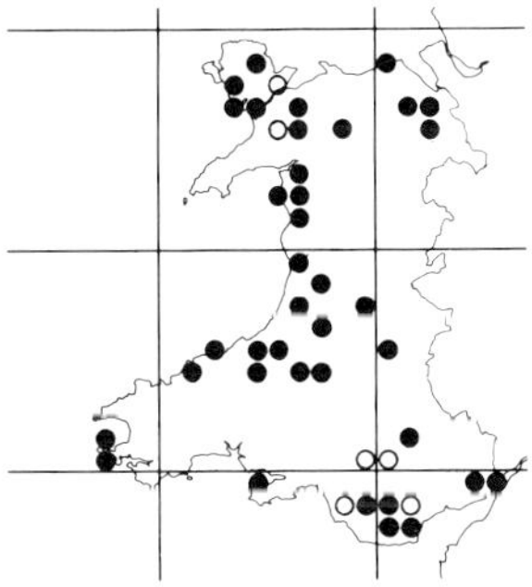

773 Taraxacum nordstedtii

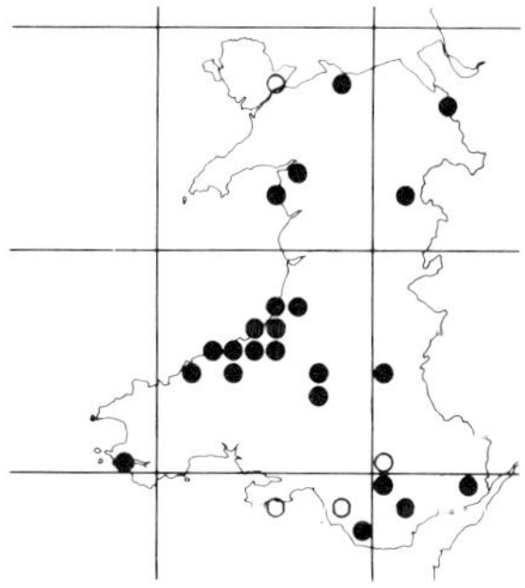

774 Taraxacum adamii

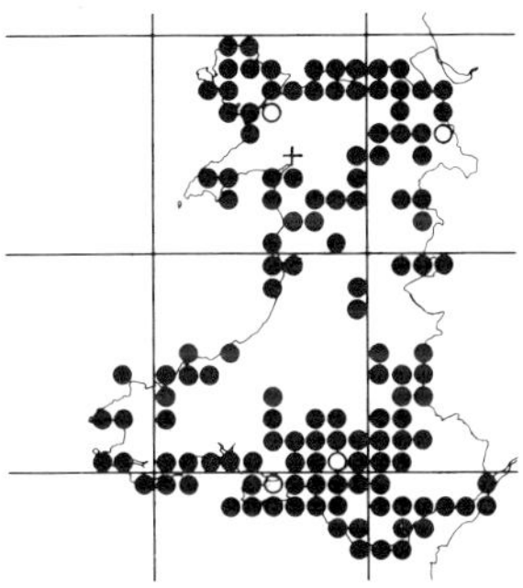

775 Taraxacum Sect. Erythrosperma

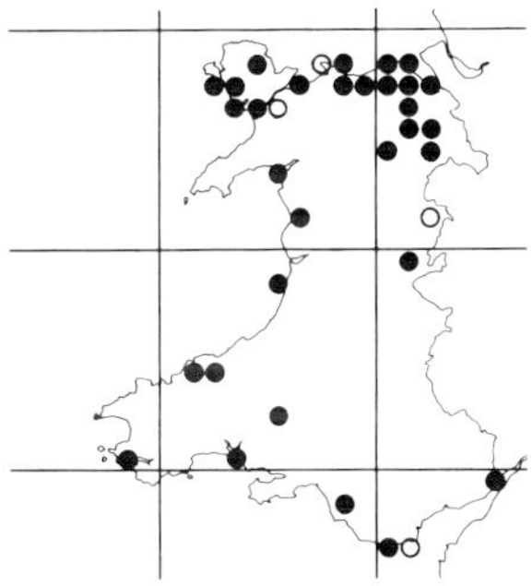

776 Taraxacum brachyglossum

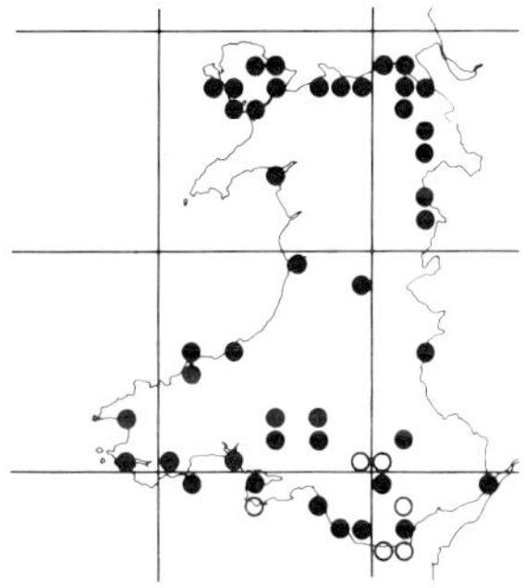

777 Taraxacum oxoniense

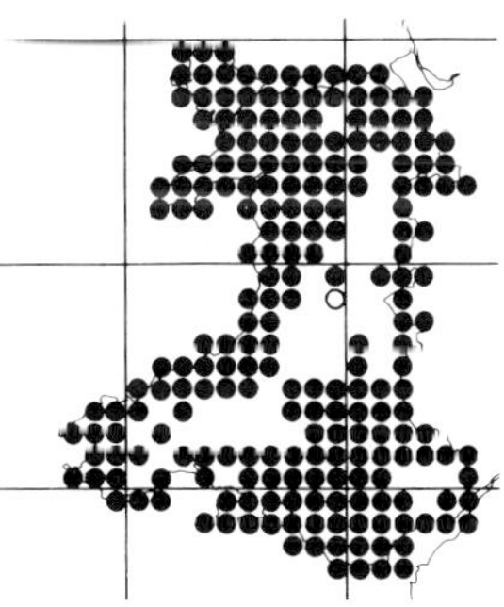

778 Taraxacum Sect. Taraxacum

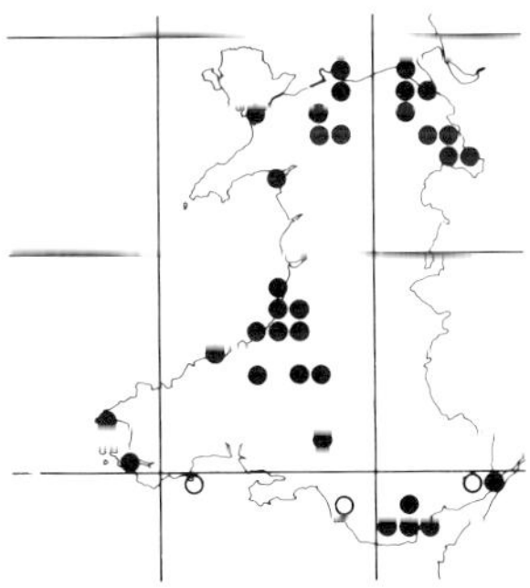

779 Taraxacum hamatum

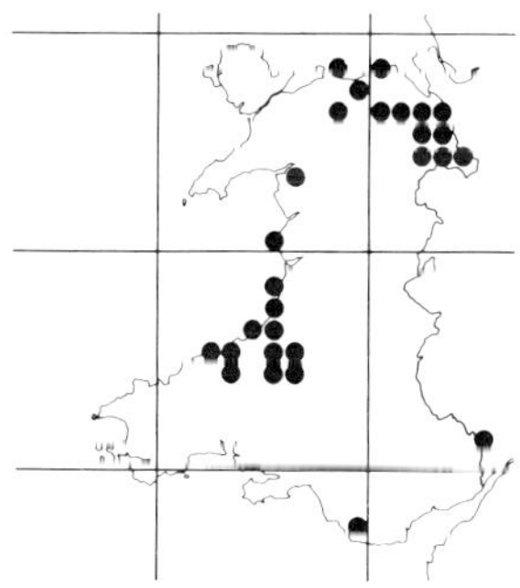

780 Taraxacum pseudohamatum

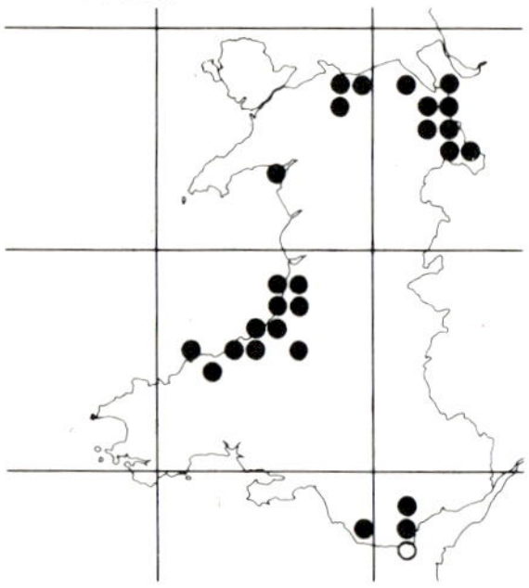

781 Taraxacum hamatiforme

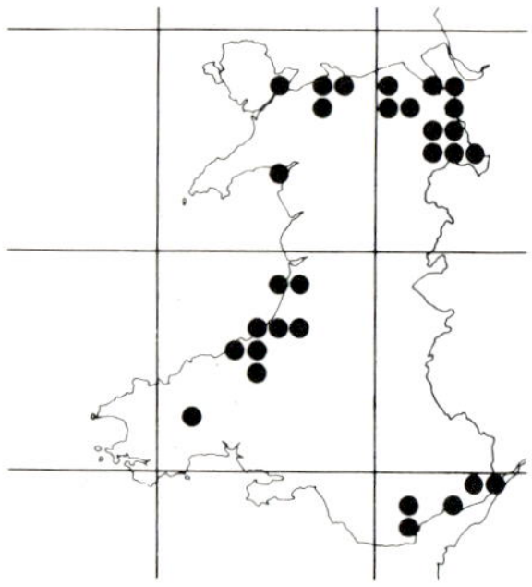

782 Taraxacum subhamatum

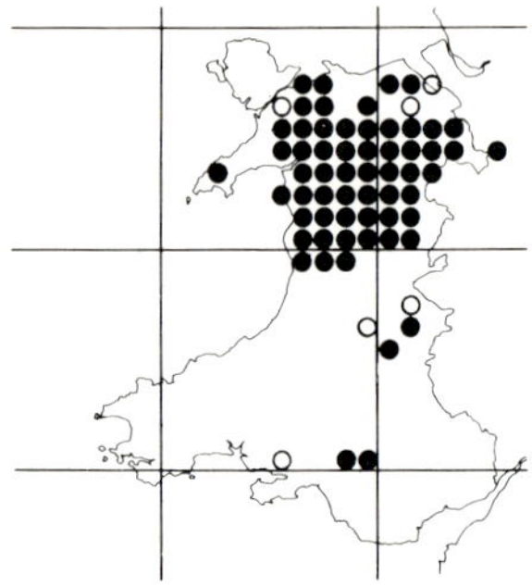

783 Crepis paludosa

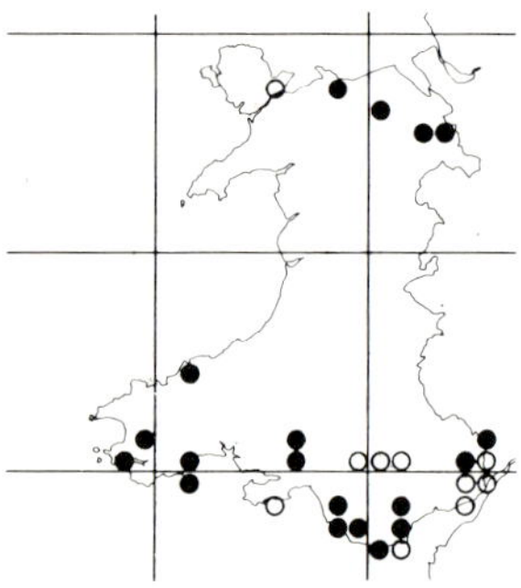

784 Crepis biennis

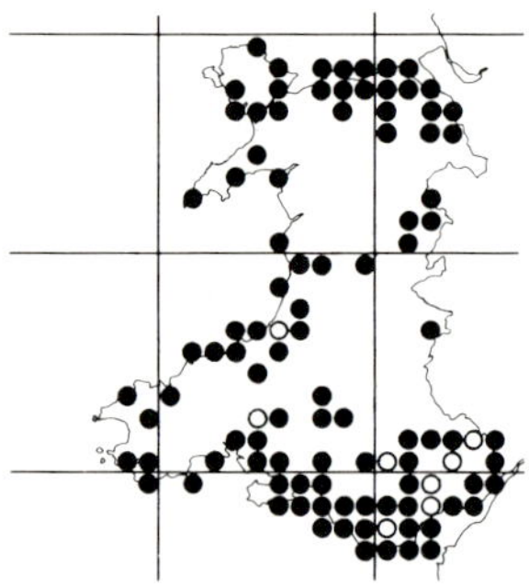

785 Crepis vesicaria

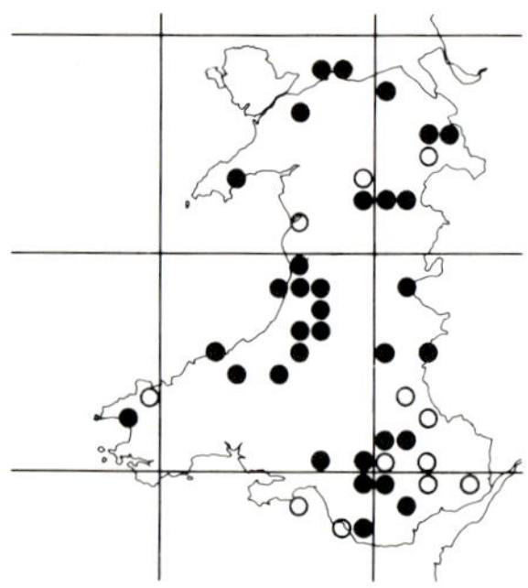

786 Pilosella officinarum subsp. micradenia

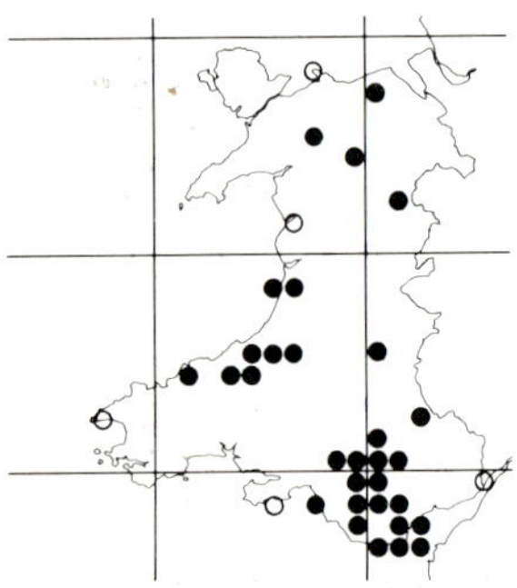

787 Pilosella officinarum subsp. officinarum

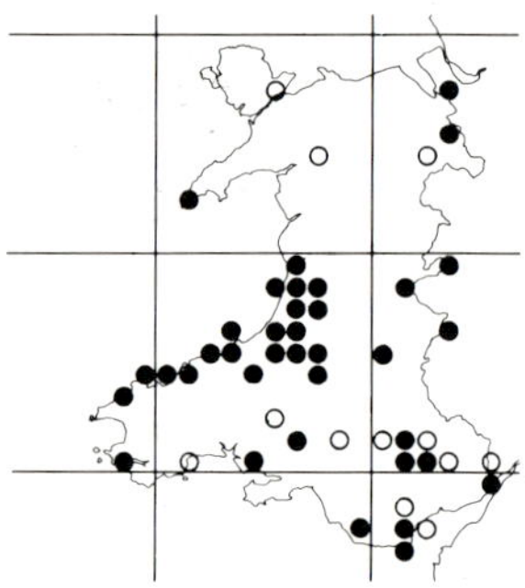

788 Pilosella officinarum subsp. trichosoma

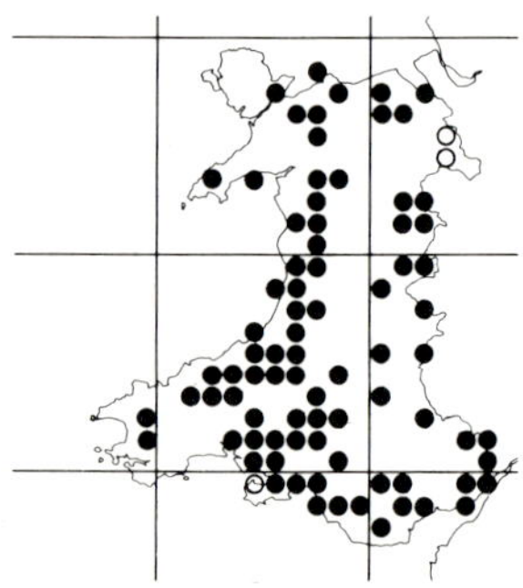

789 Pilosella aurantiaca subsp. carpathicola

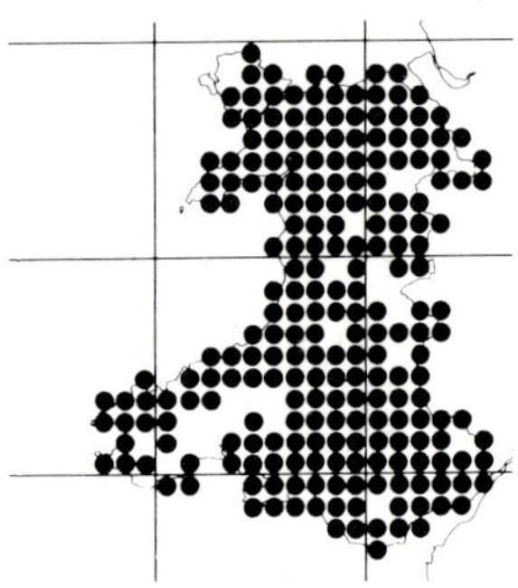

790 Hieracium murorum agg.

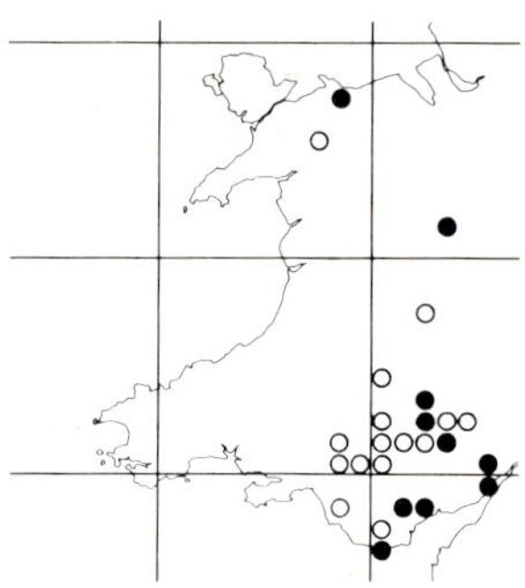

791 Hieracium exotericum agg.

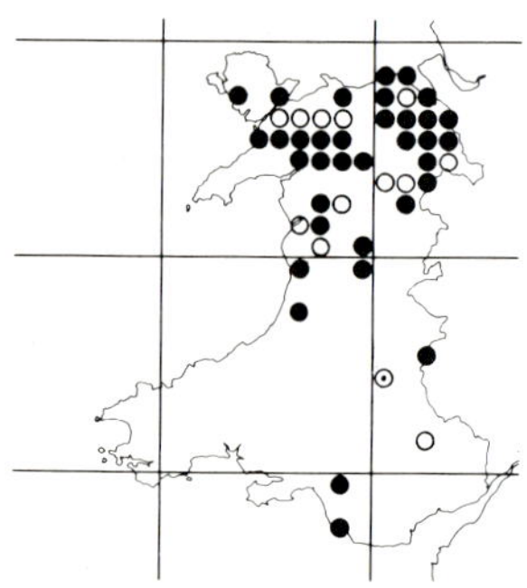

792 Hieracium vulgatum

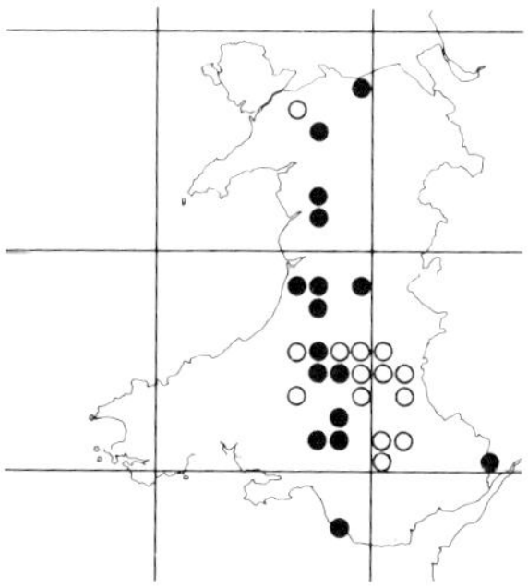

793 Hieracium submutabile

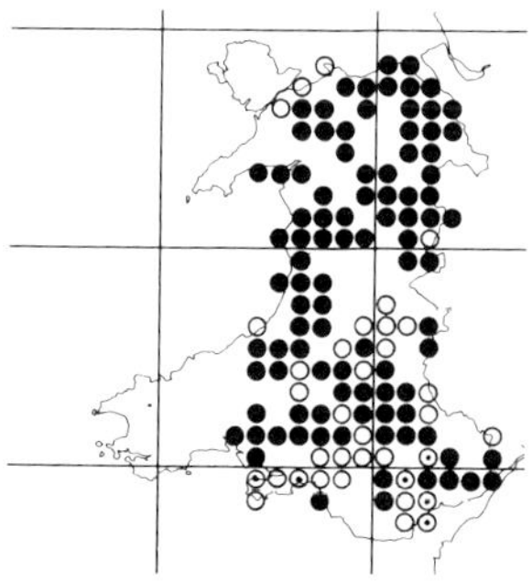

794 Hieracium acuminatum

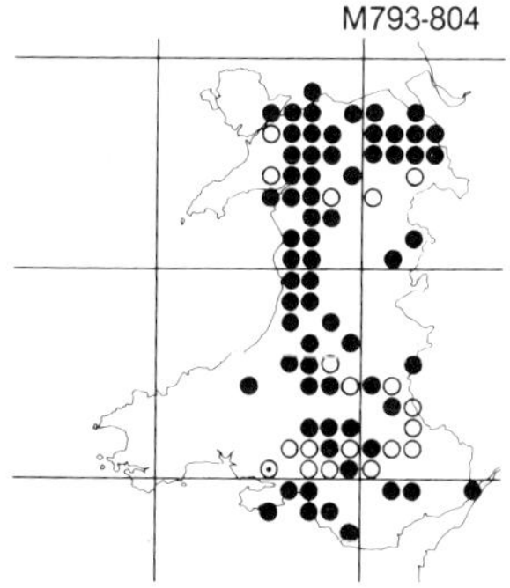

795 Hieracium diaphanum

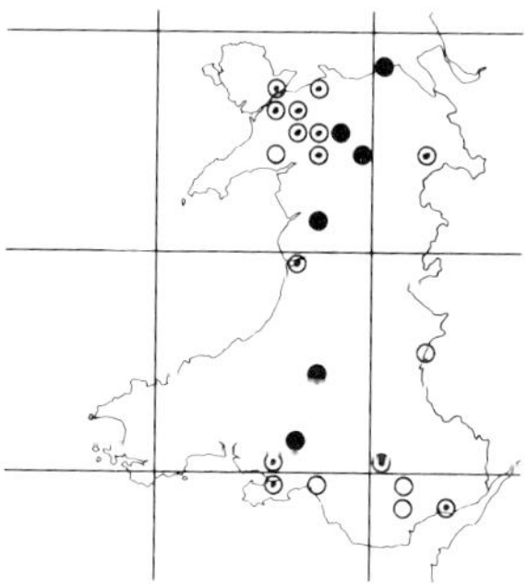

796 Hieracium diaphanoides

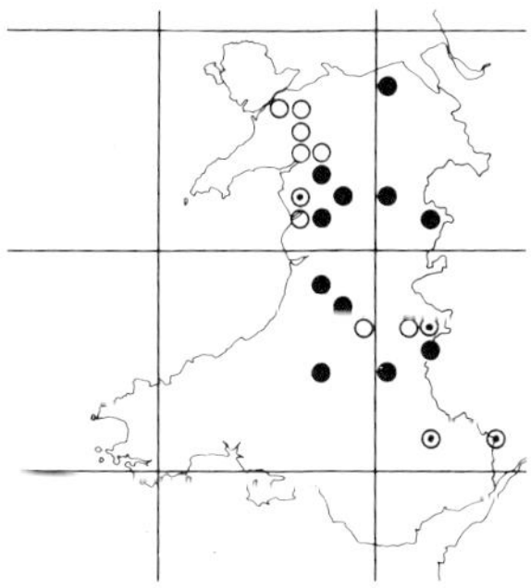

797 Hieracium lasiophyllum

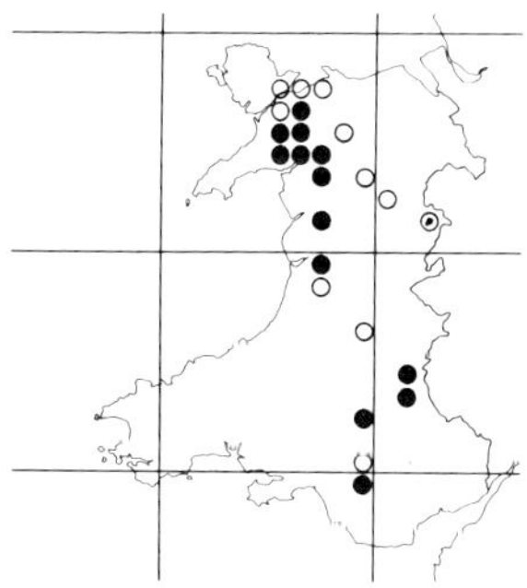

798 Hieracium argenteum

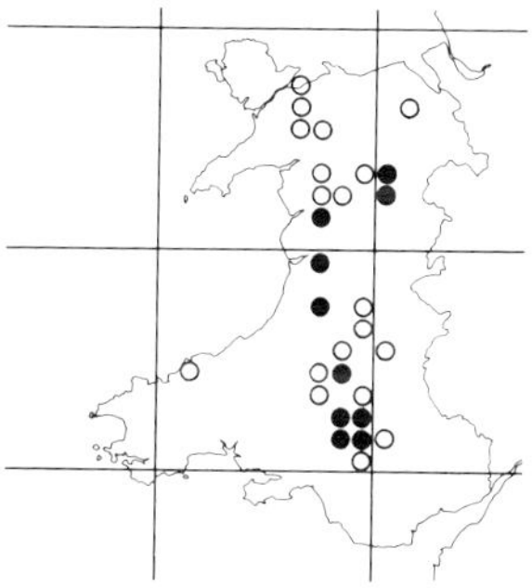

799 Hieracium subcrocatum

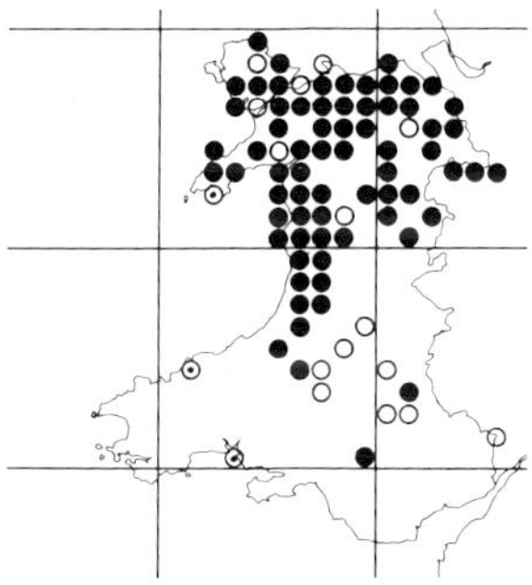

800 Hieracium vagum

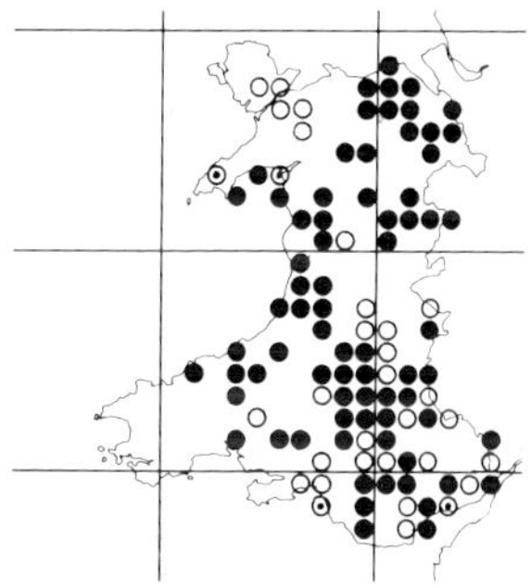

801 Hieracium perpropinquum

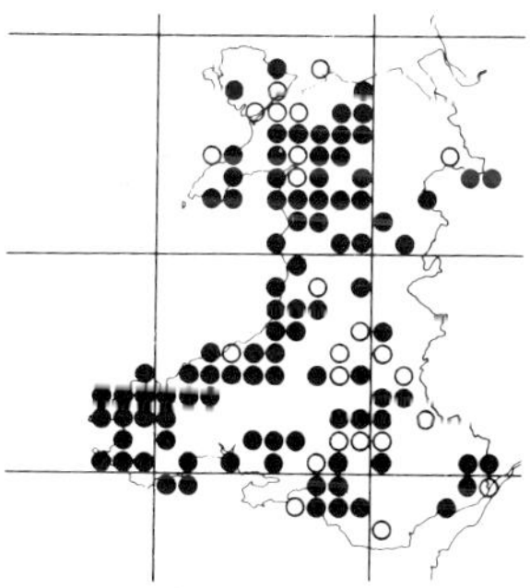

802 Hieracium umbellatum

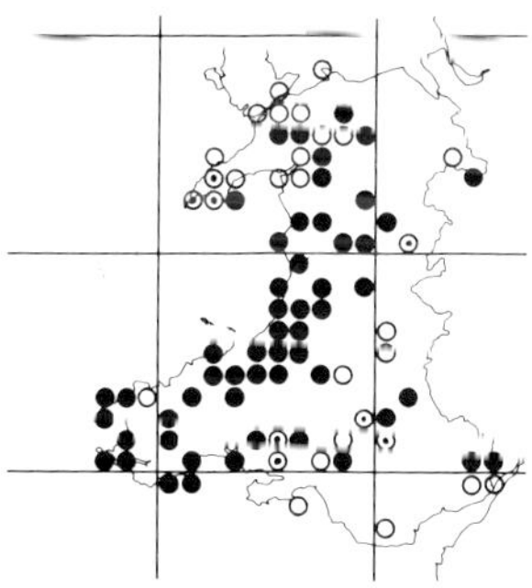

803 Hieracium umbellatum subsp. umbellatum

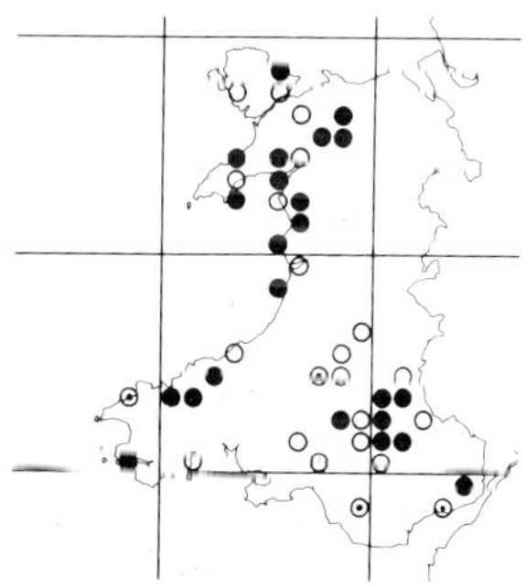

804 Hieracium umbellatum subsp. bichlorophyllum

M805-816

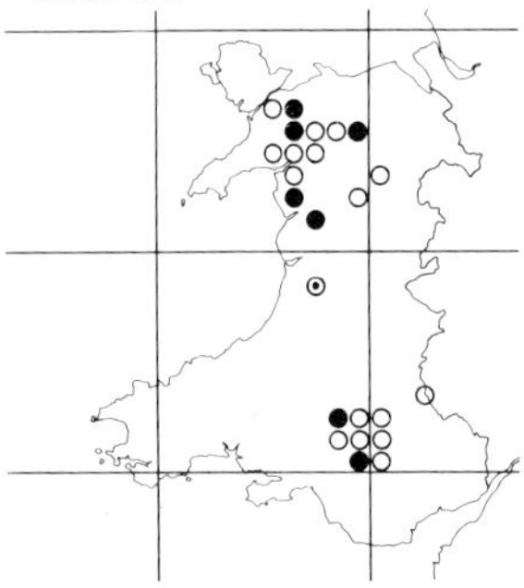

805 Hieracium sparsifolium

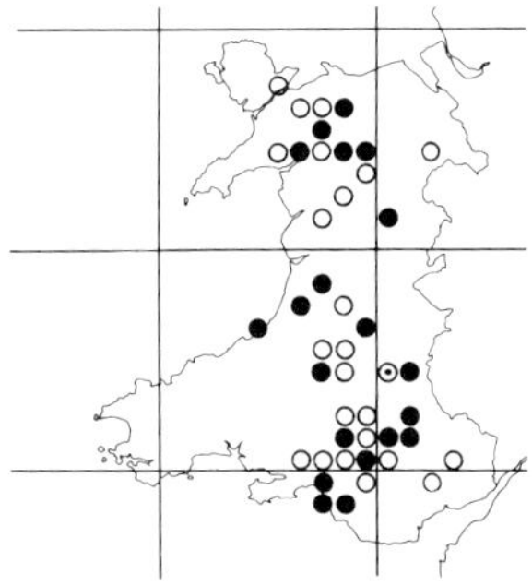

806 Hieracium scabrisetum

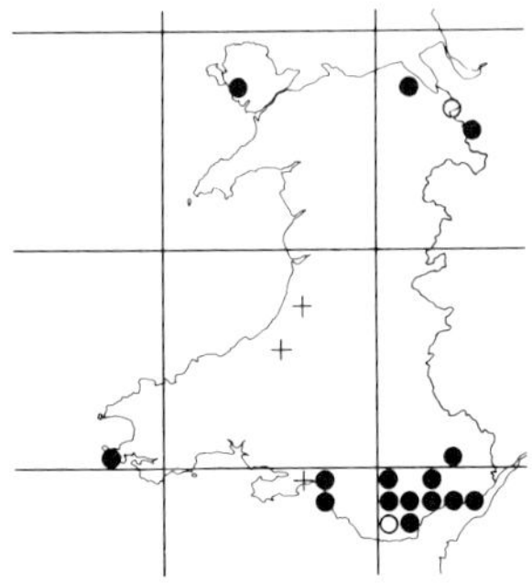

807 Sagittaria sagittifolia

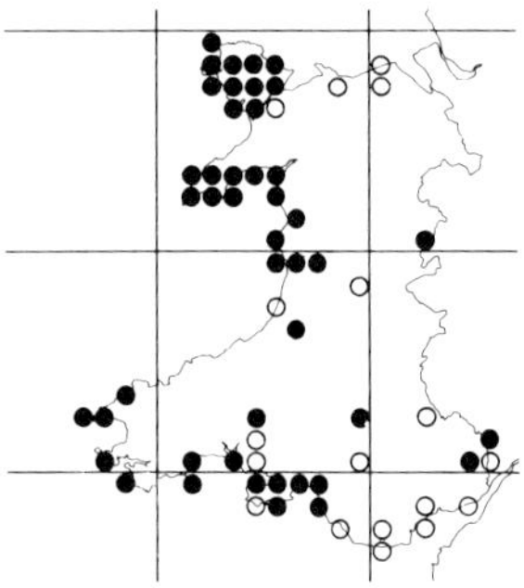

808 Baldellia ranunculoides

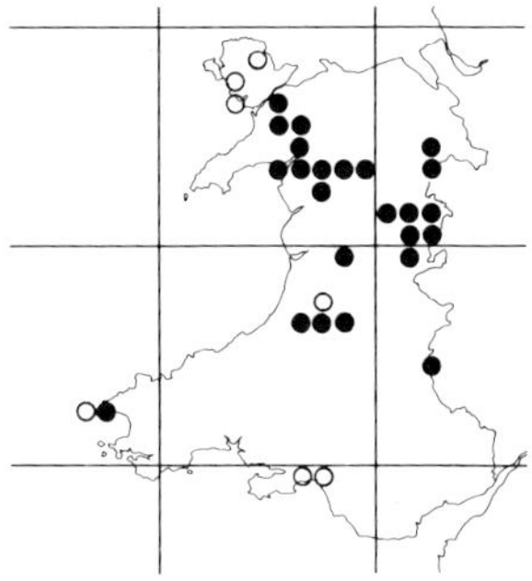

809 Luronium natans

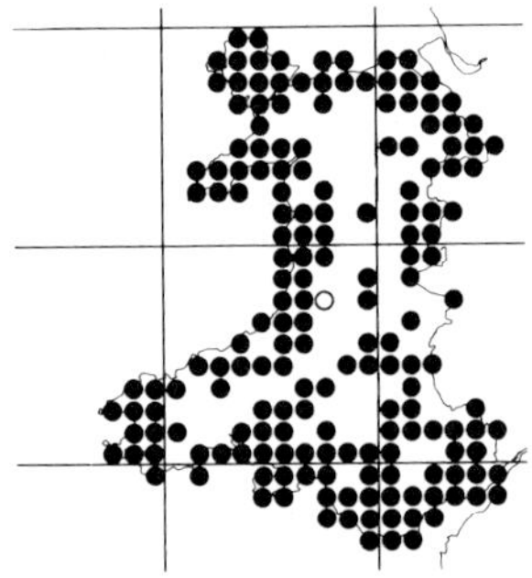

810 Alisma plantago-aquatica

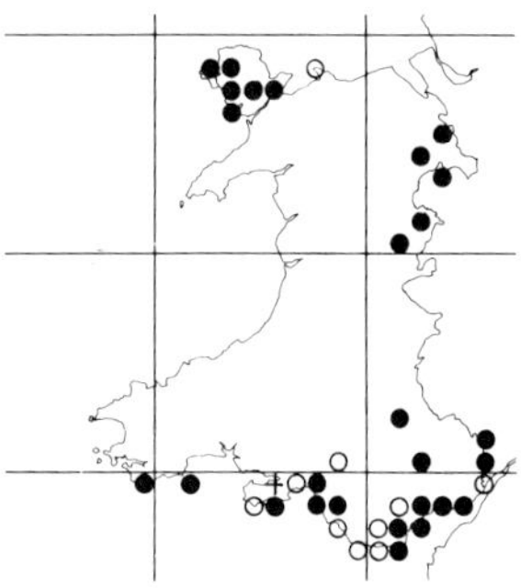

811 Butomus umbellatus

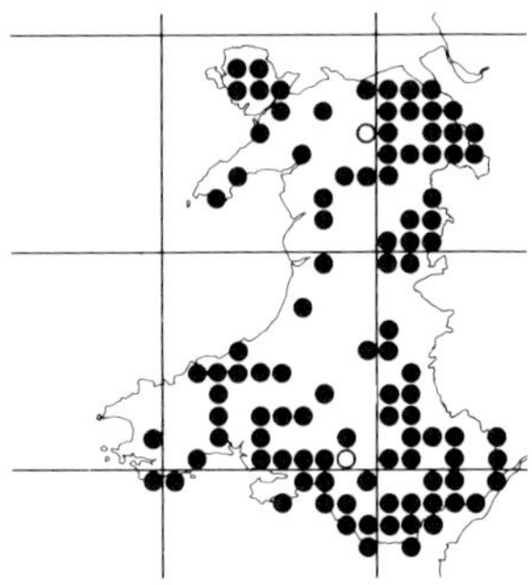

812 Elodea canadensis

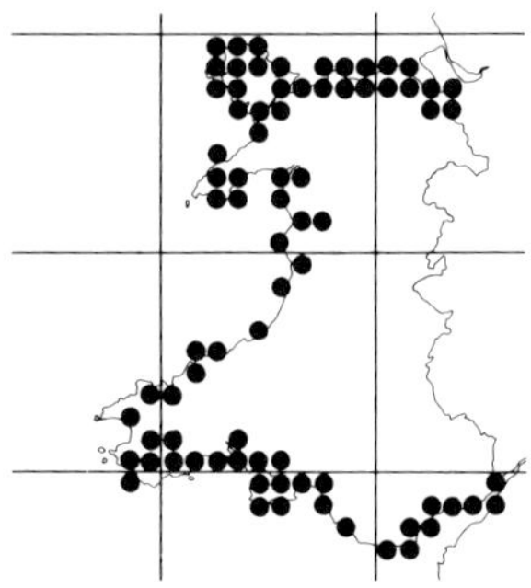

813 Triglochin maritima

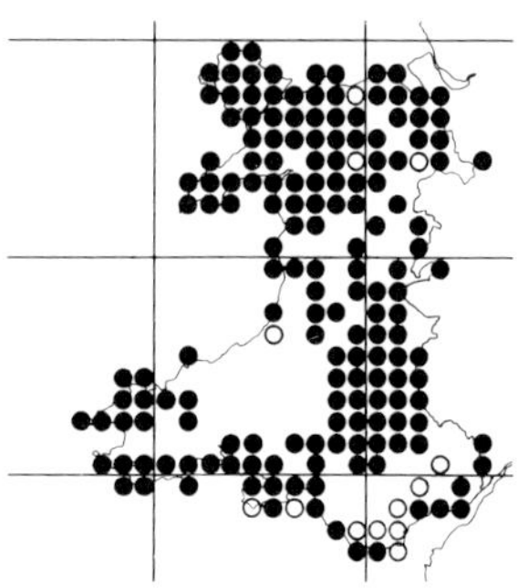

814 Triglochin palustris

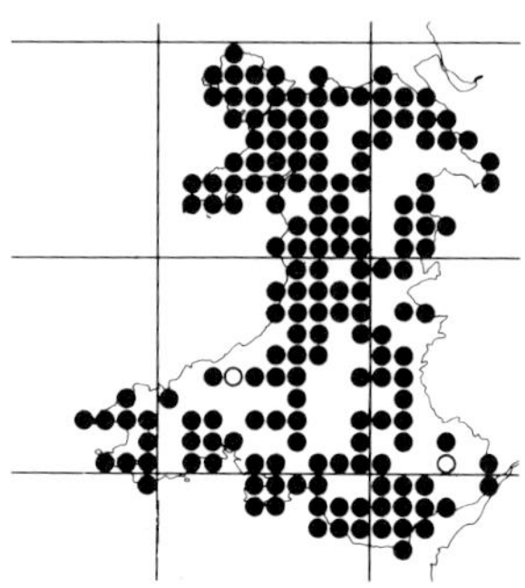

815 Potamogeton natans

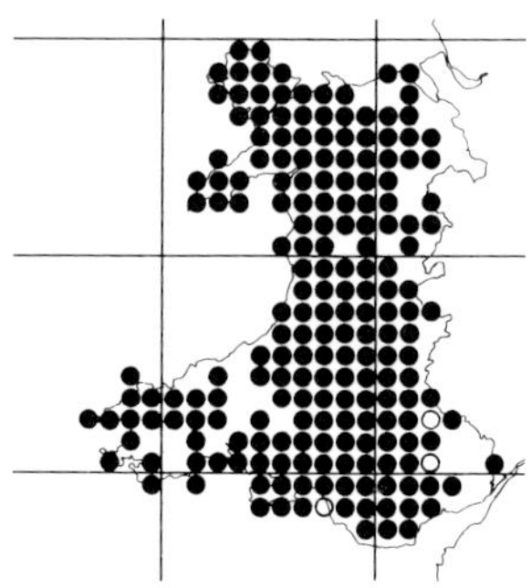

816 Potamogeton polygonifolius

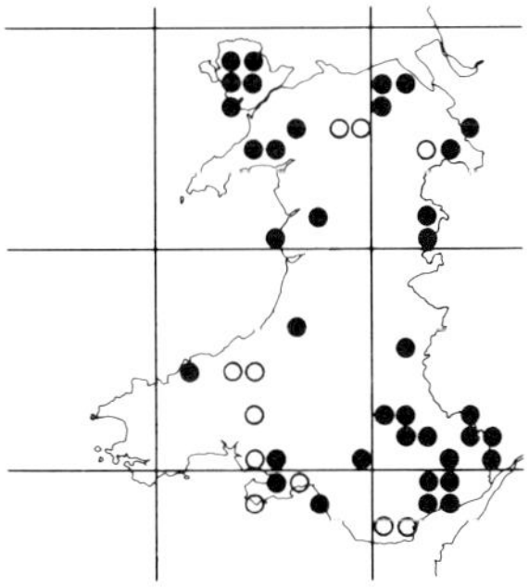

817 Potamogeton perfoliatus

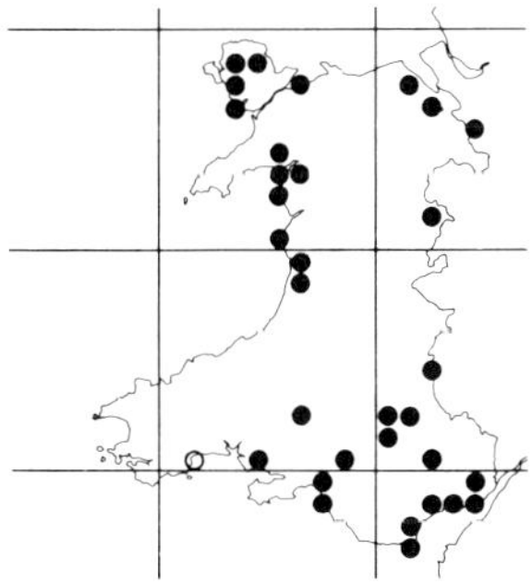

818 Potamogeton pusillus

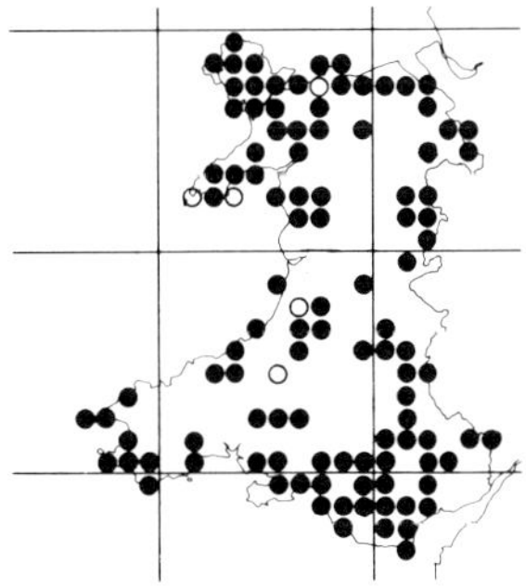

819 Potamogeton berchtoldii

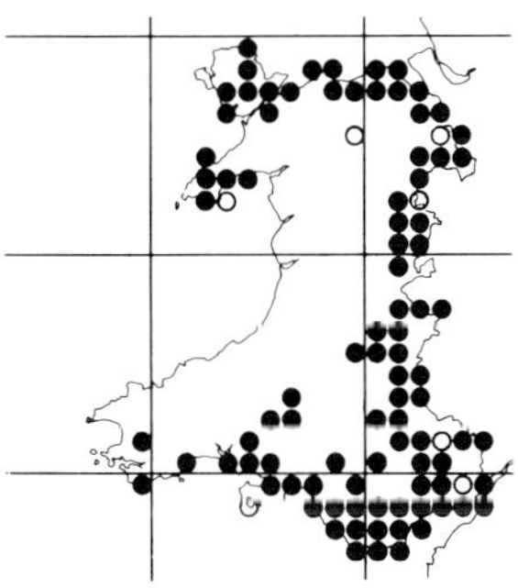

820 Potamogeton crispus

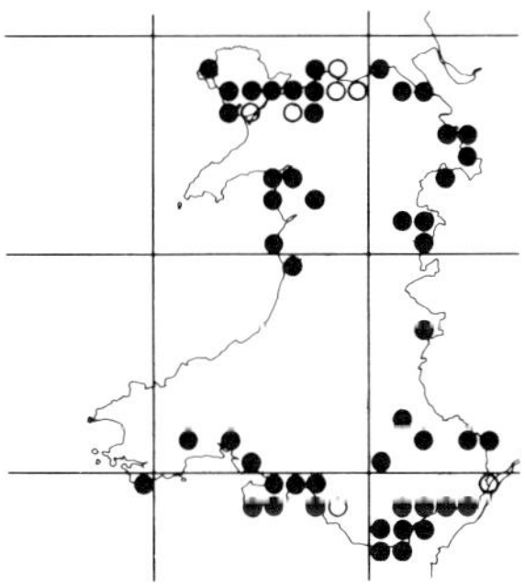

821 Potamogeton pectinatus

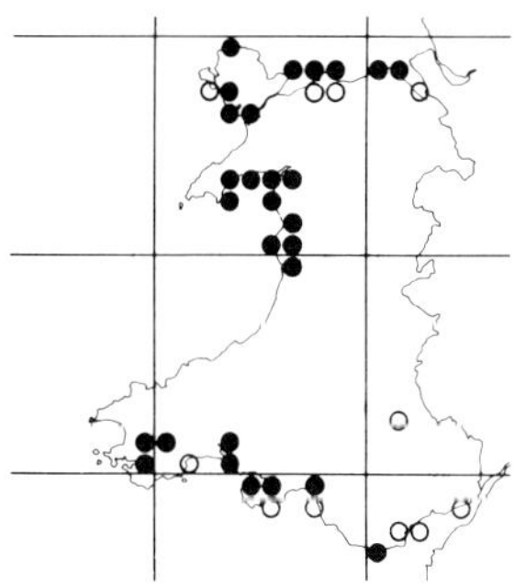

822 Ruppia maritima

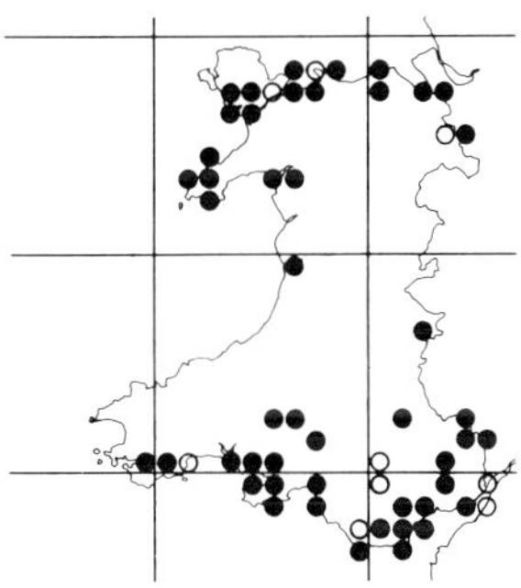

823 Zanichellia palustris

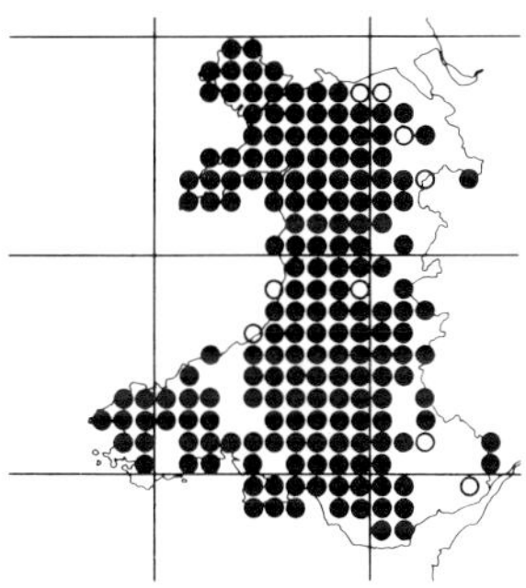

824 Narthecium ossifragum

825 Colchicum autumnale

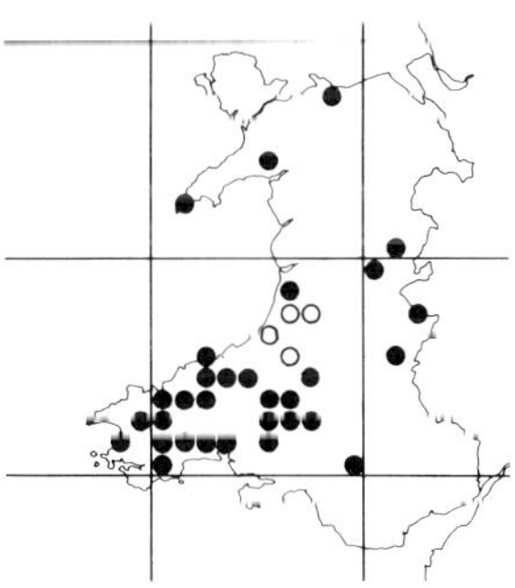

826 Lilium pyrenaicum

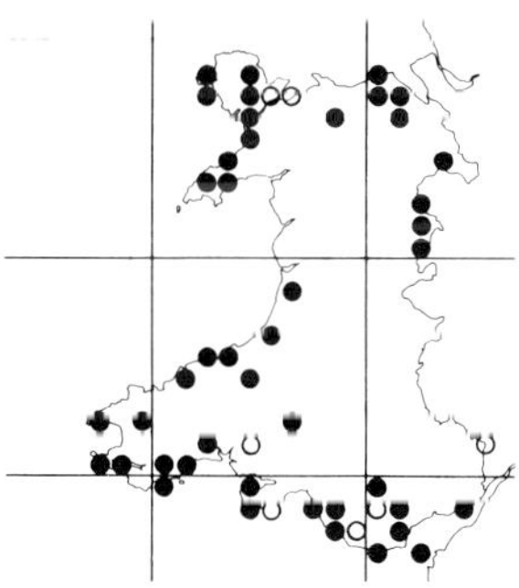

827 Ornithogalum umbellatum

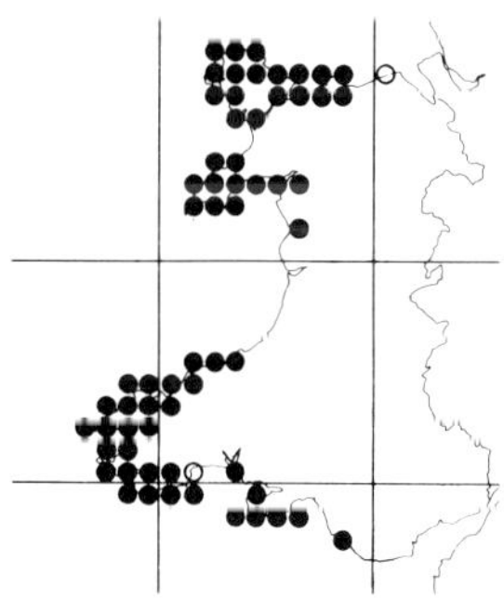

828 Scilla verna

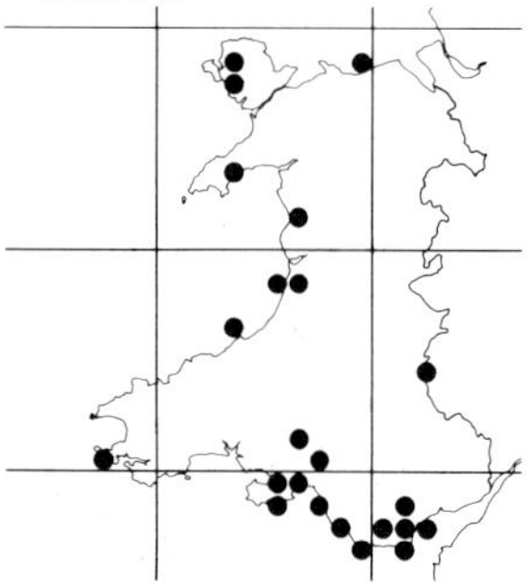

829 Hyacinthoides hispanica

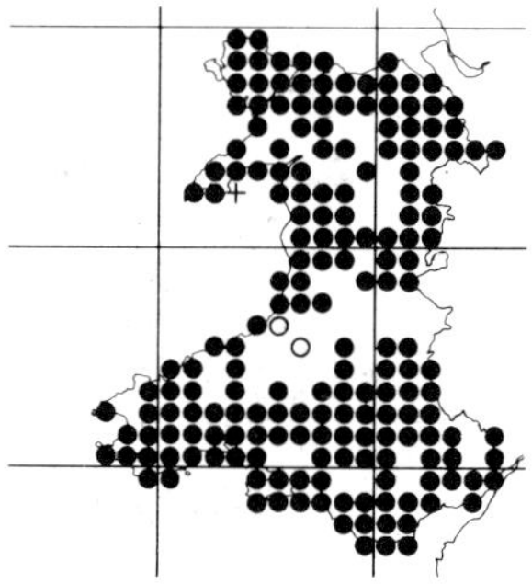

830 Allium ursinum

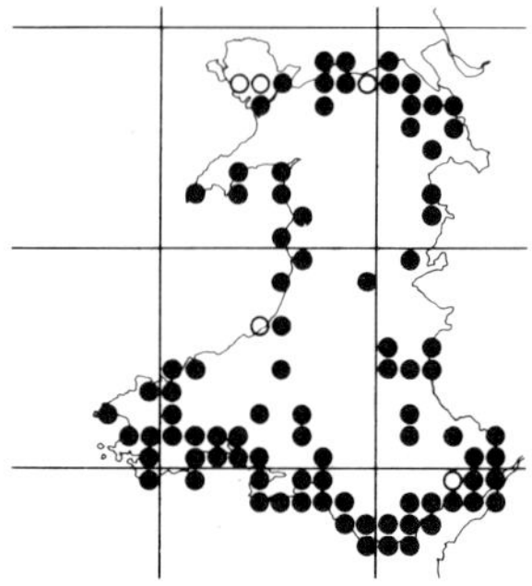

831 Allium vineale

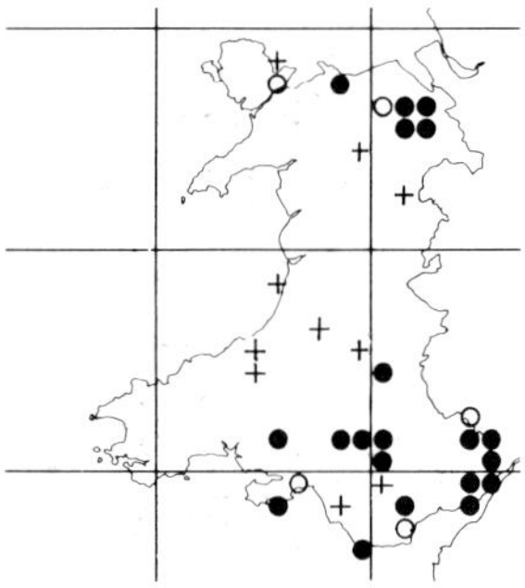

832 Convallaria majalis

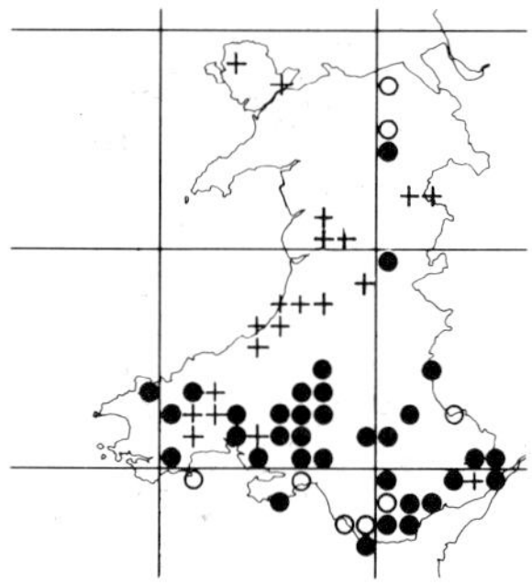

833 Polygonatum multiflorum

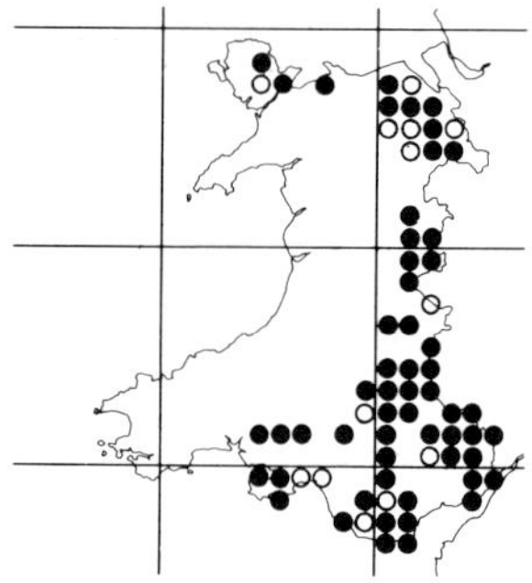

834 Paris quadrifolia

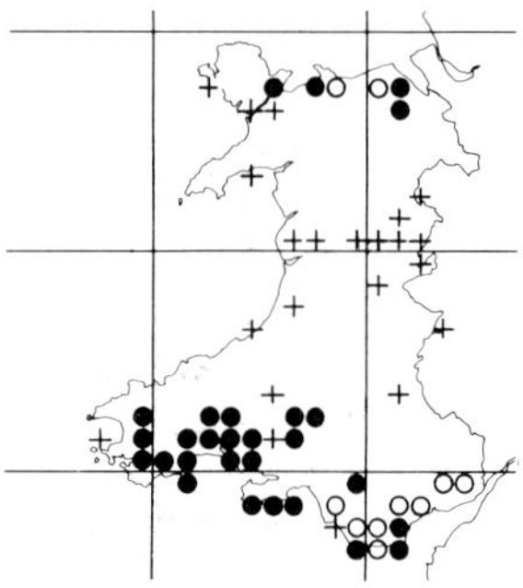

835 Ruscus aculeatus

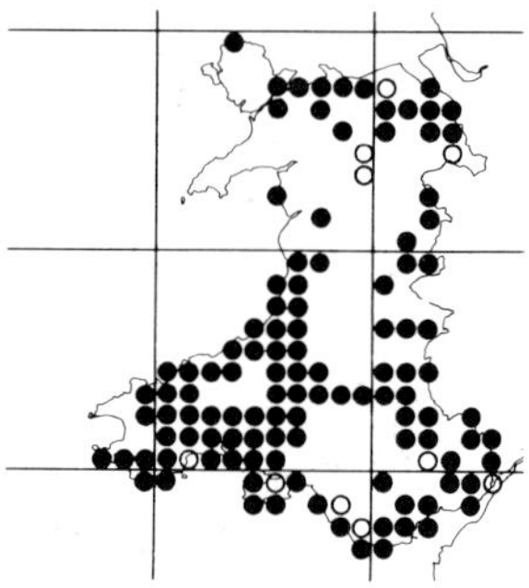

836 Galanthus nivalis

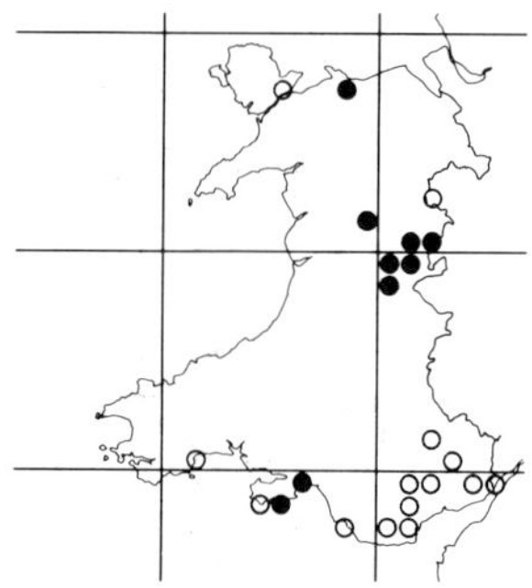

837 Narcissus poeticus

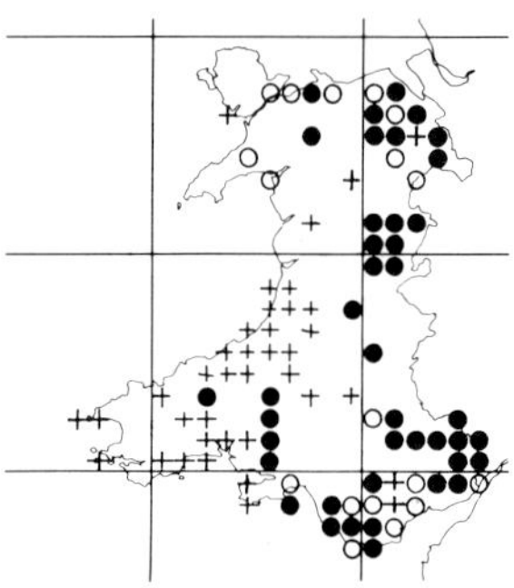

838 Narcissus pseudonarcissus subsp. pseudonarcissus

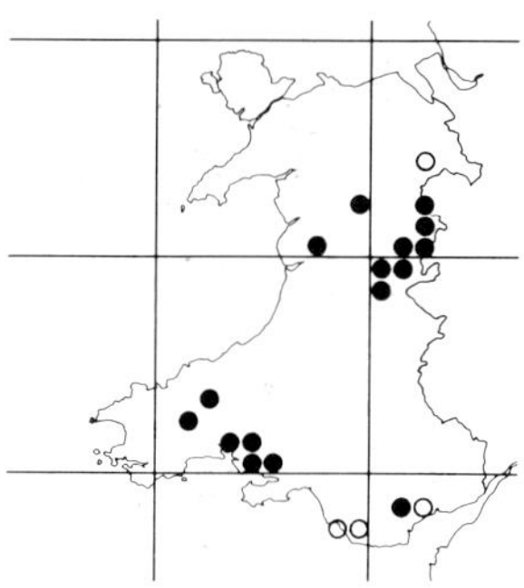

839 Narcissus pseudonarcissus subsp. major

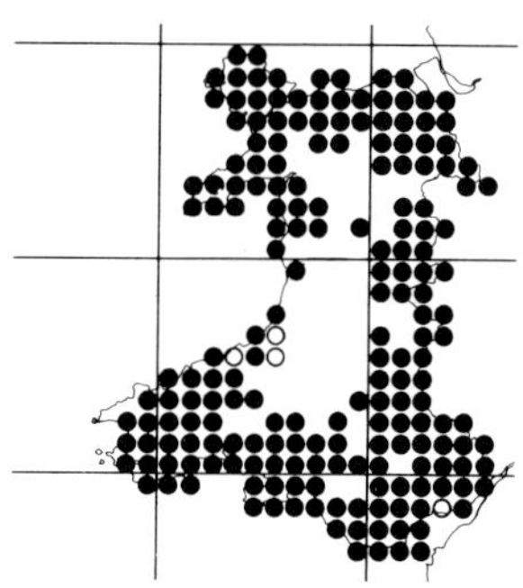

840 Tamus communis

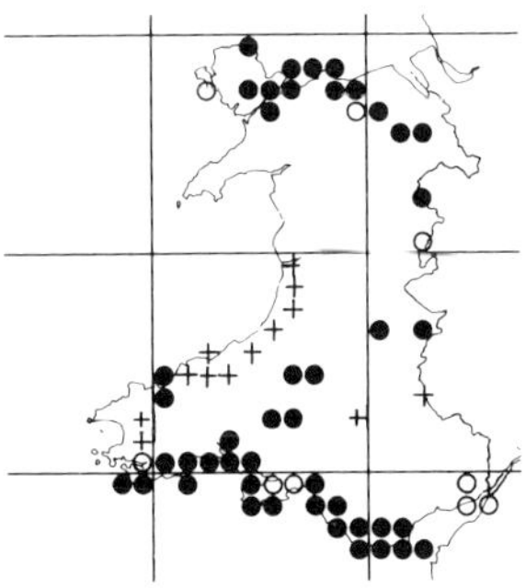

841 Iris foetidissima

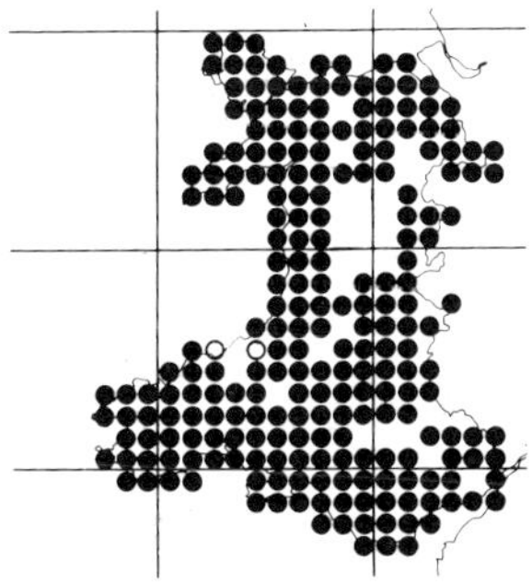

842 Iris pseudacorus

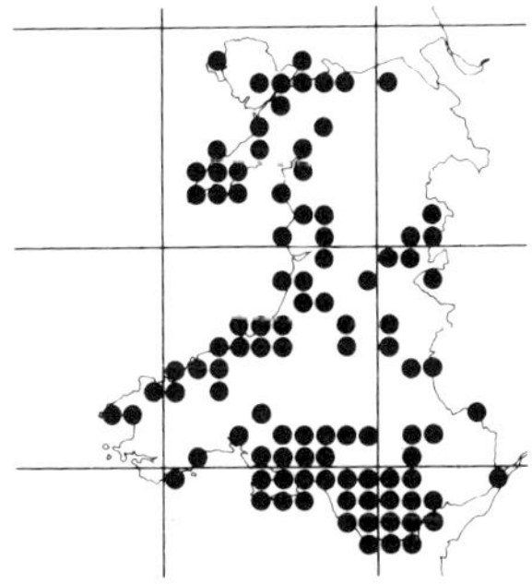

843 Tritonia aurea × pottsii
= T. × crocosmiflora

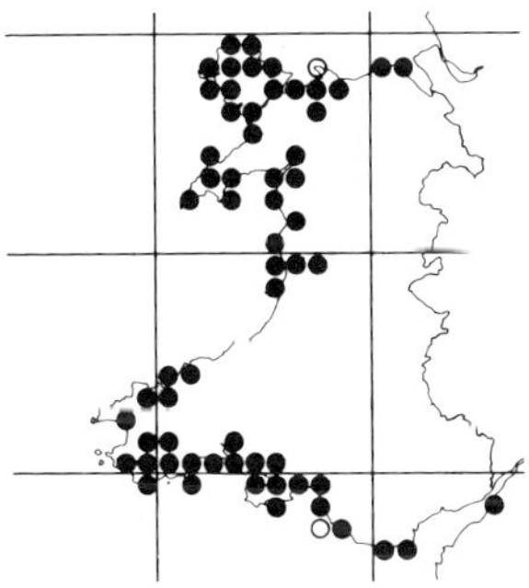

844 Juncus maritimus

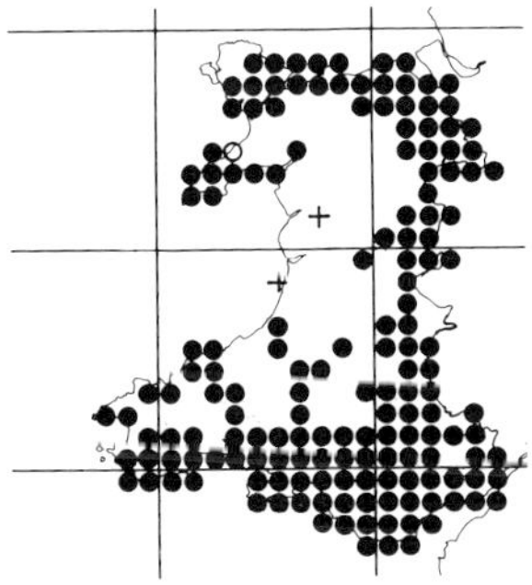

845 Juncus inflexus

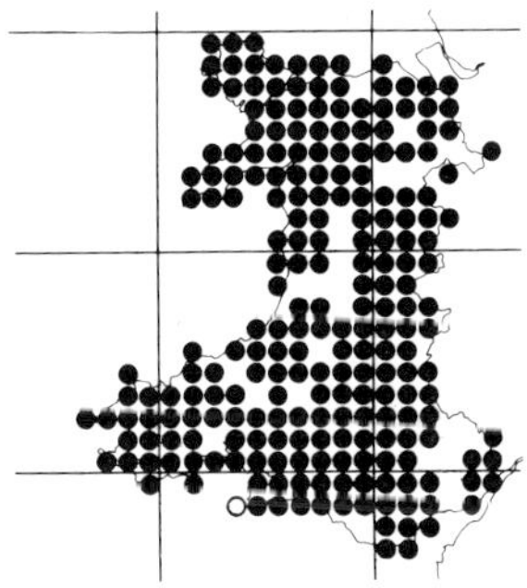

846 Juncus conglomeratus

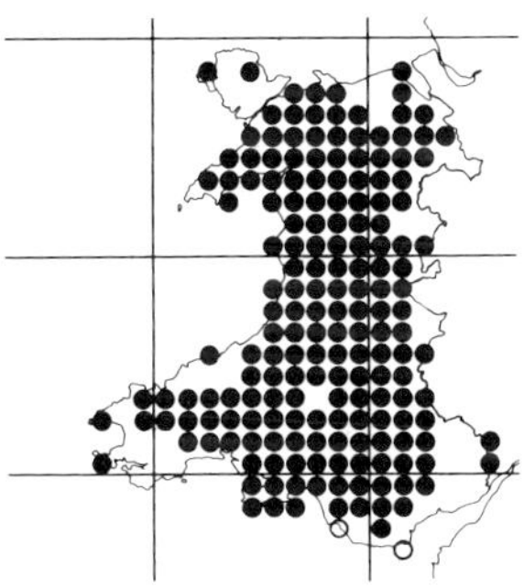

847 Juncus squarrosus

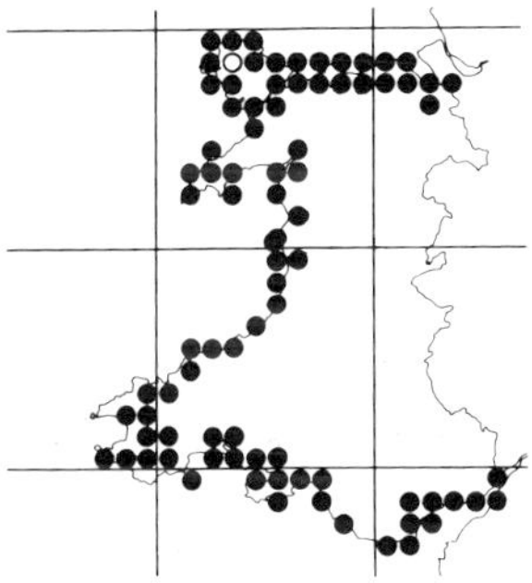

848 Juncus gerardii

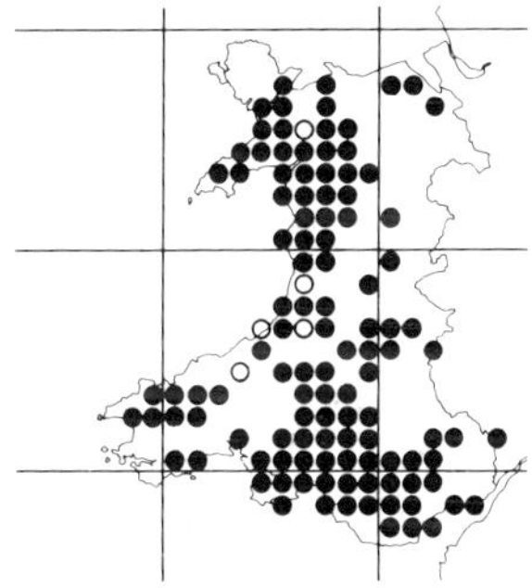

849 Juncus tenuis

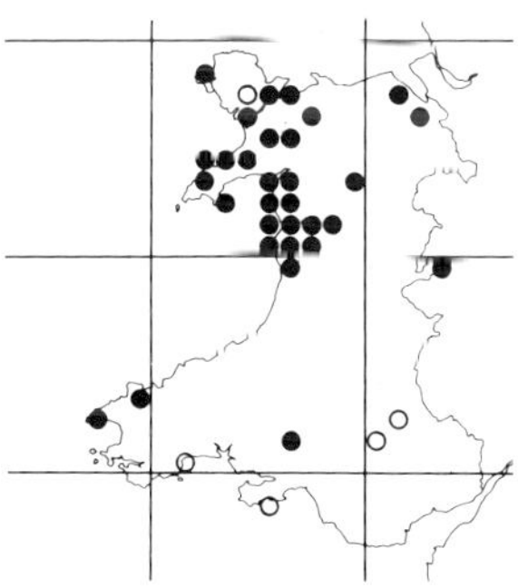

850 Juncus foliosus

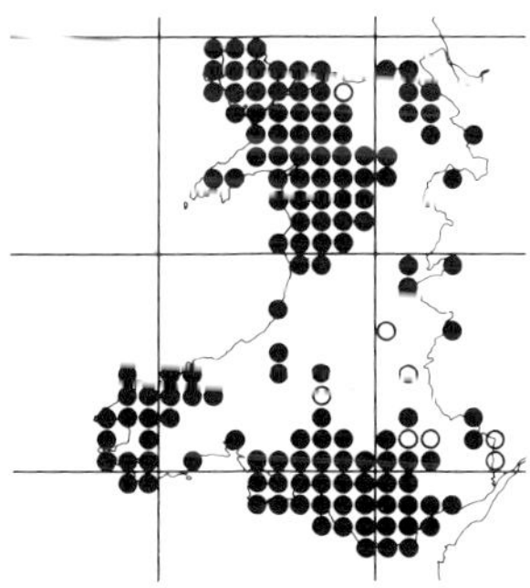

851 Juncus bufonius s.s.

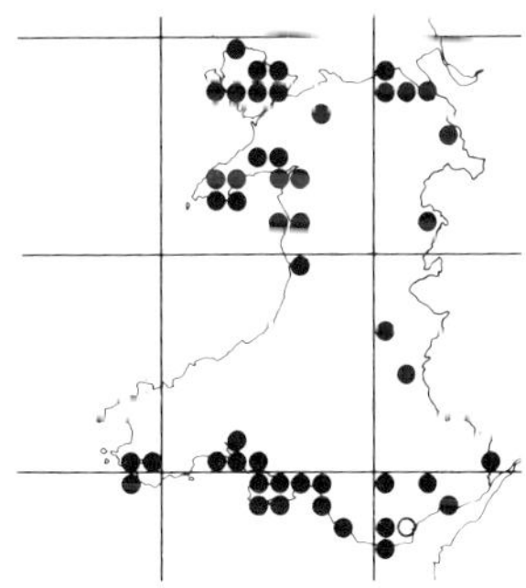

852 Juncus subnodulosus

M853-864

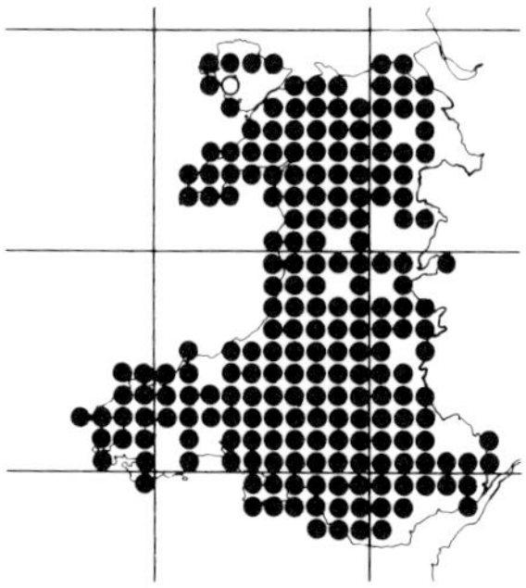
853 Juncus bulbosus agg.

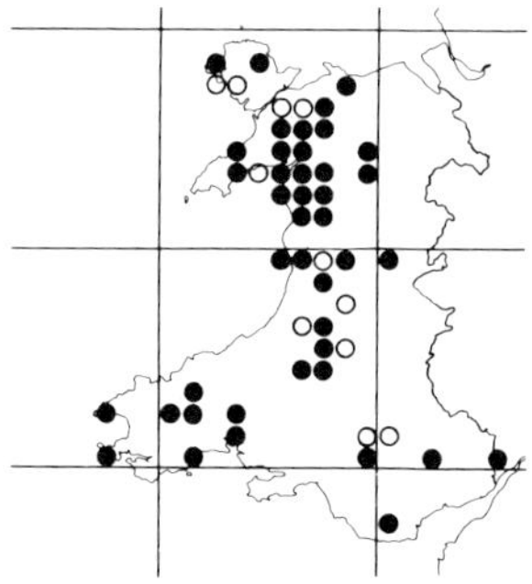
854 Juncus bulbosus s.s.

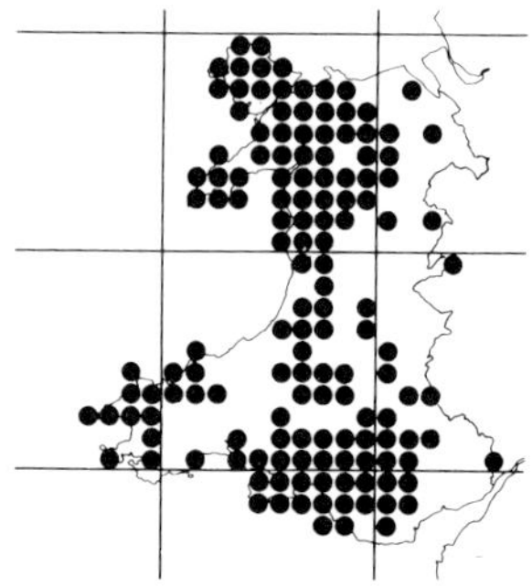
855 Juncus kochii

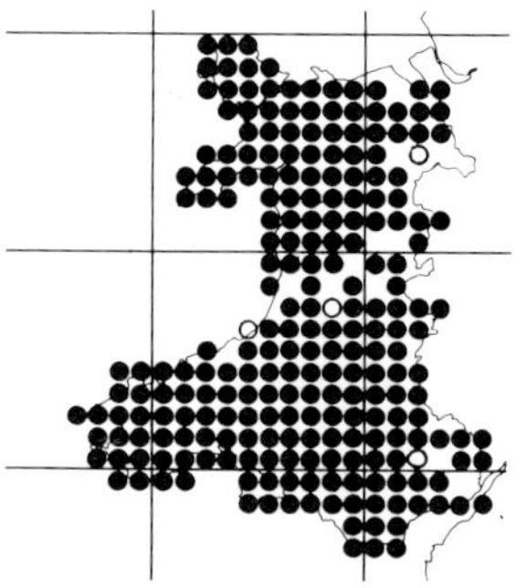
856 Juncus acutiflorus

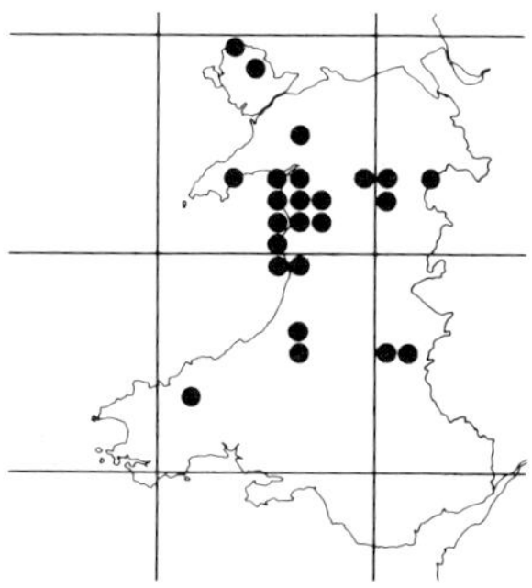
857 Juncus acutiflorus × articulatus = J. × surrejanus

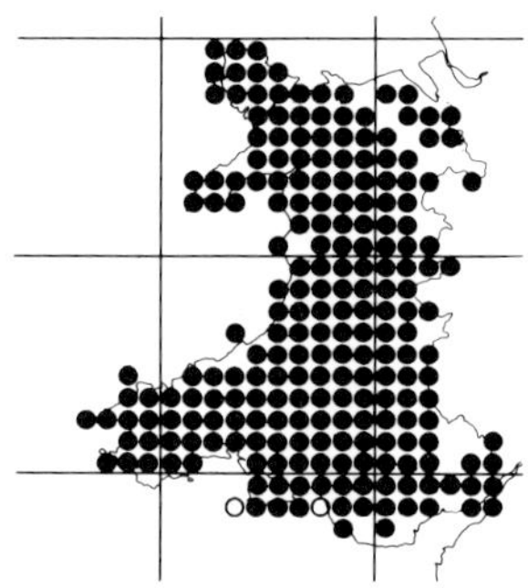
858 Luzula multiflora

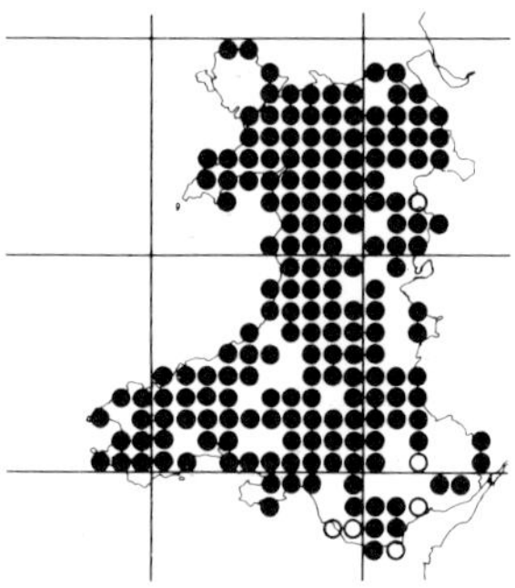
859 Luzula sylvatica

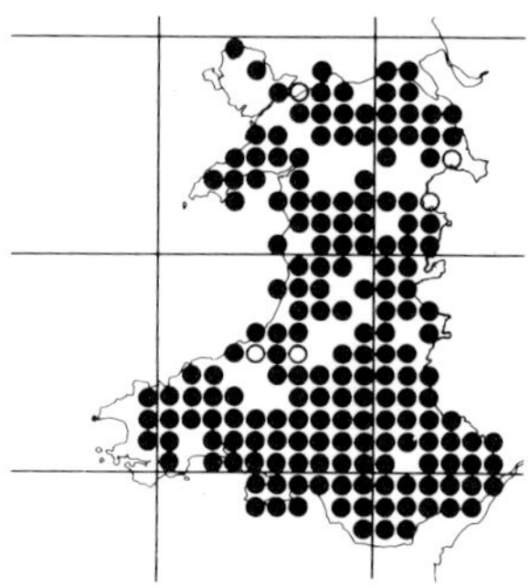
860 Luzula pilosa

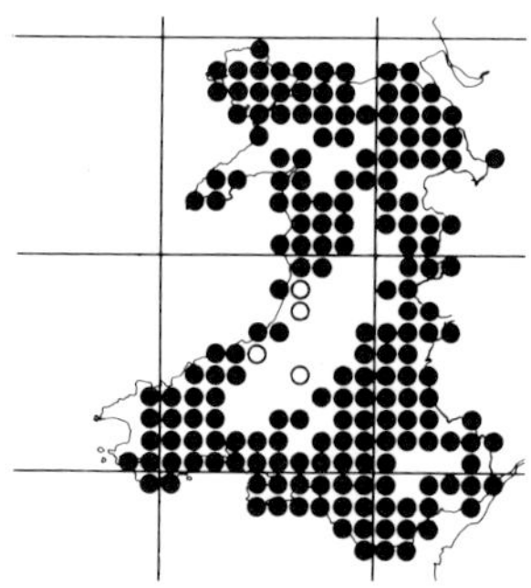
861 Festuca gigantea

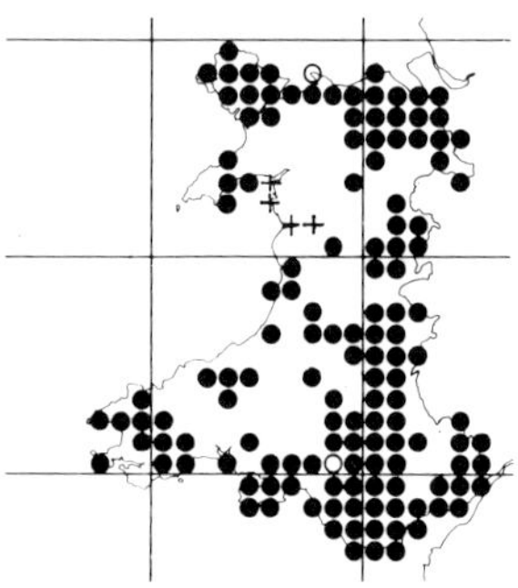
862 Festuca pratensis

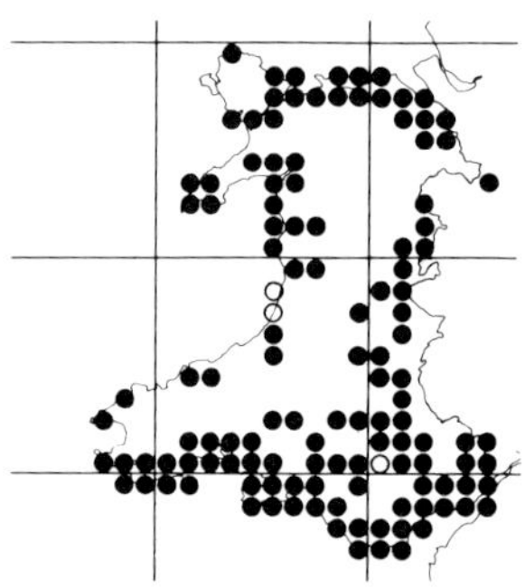
863 Festuca arundinacea

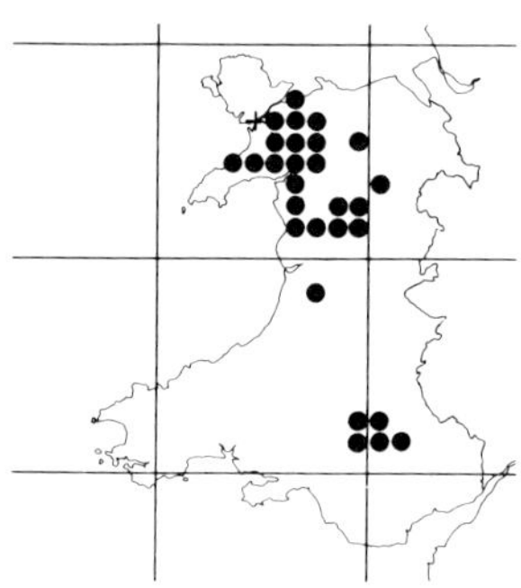
864 Festuca vivipara

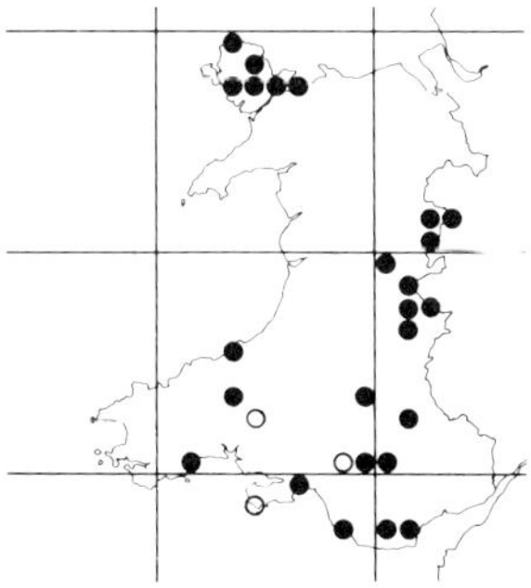

865 Festuca pratensis
× Lolium perenne
= × Festulolium loliaceum

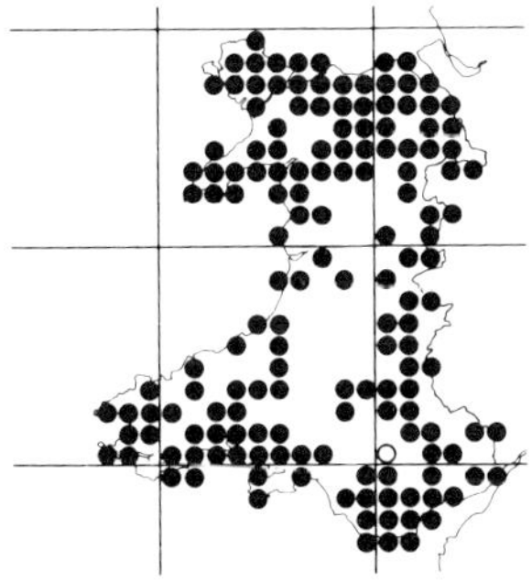

866 Lolium multiflorum

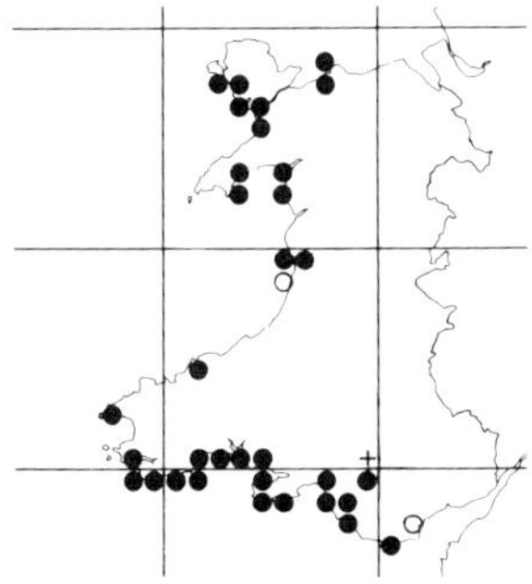

867 Vulpia fasciculata

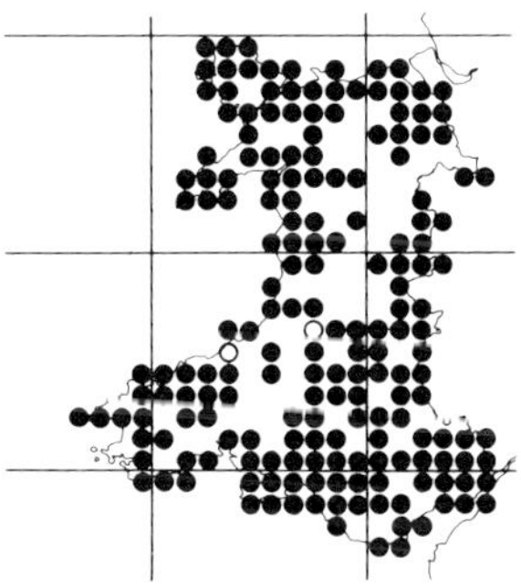

868 Vulpia bromoides

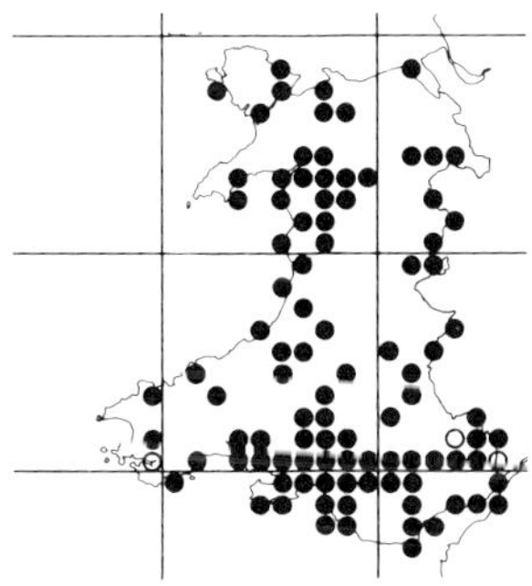

869 Vulpia myuros

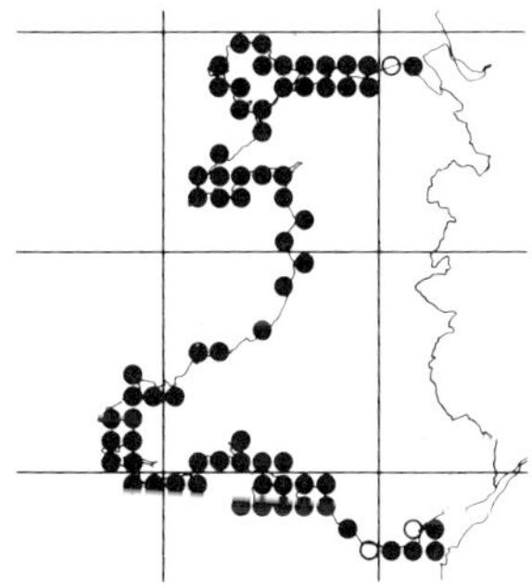

870 Desmazeria marina

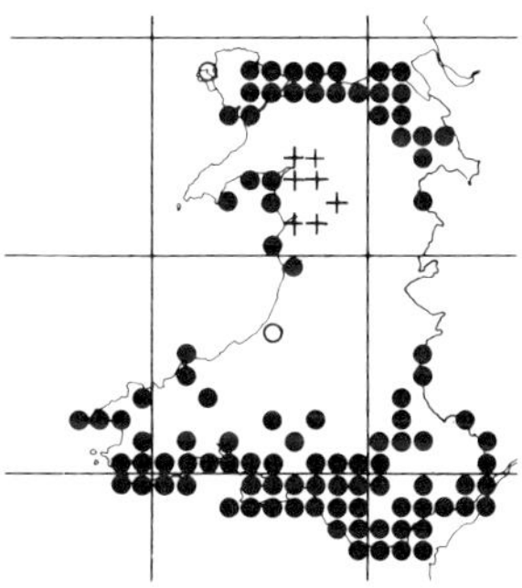

871 Desmazeria rigida

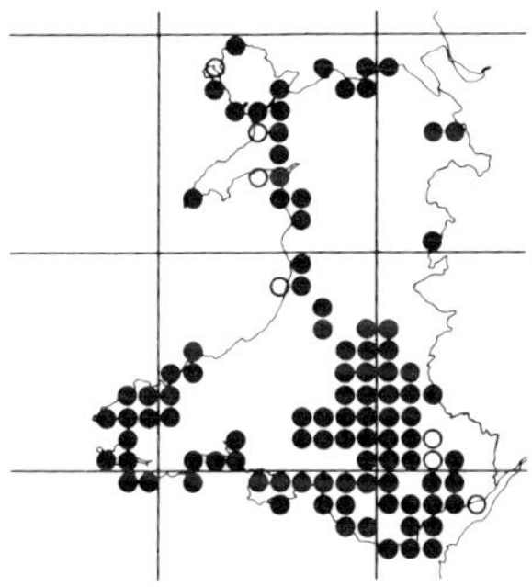

872 Poa subcaerulea

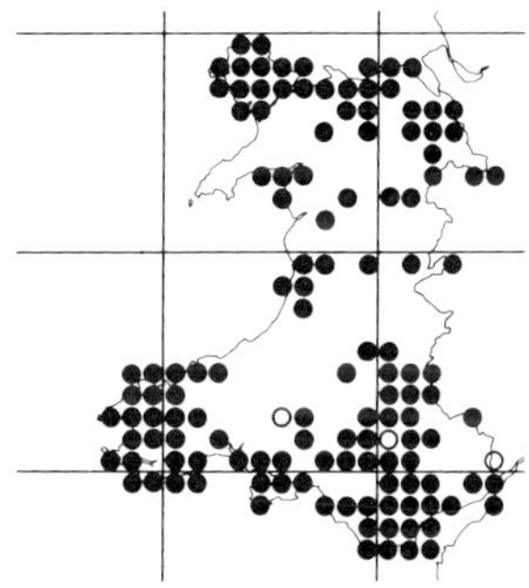

873 Poa pratensis s.s.

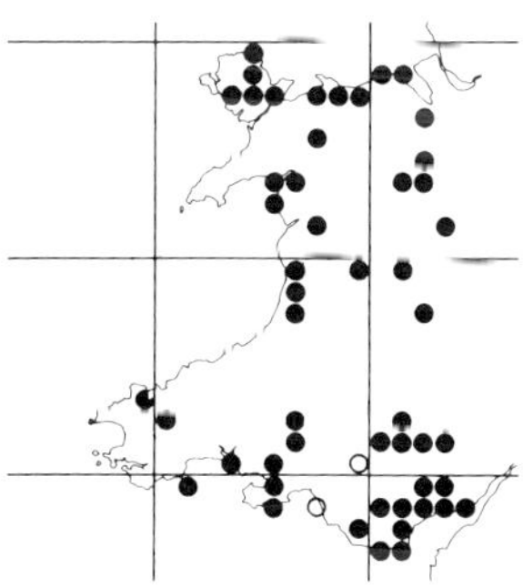

874 Poa angustifolia

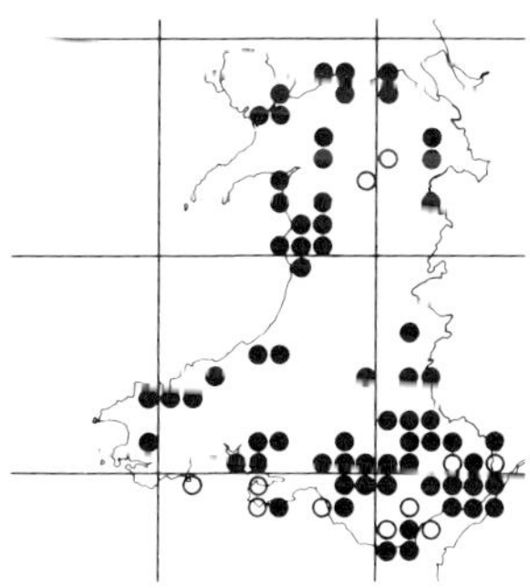

875 Poa compressa

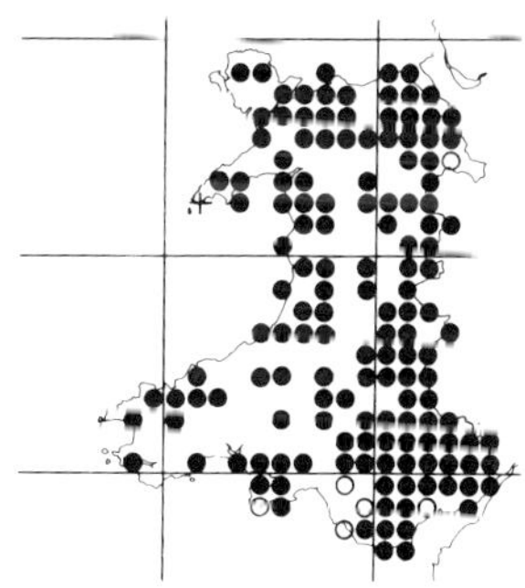

876 Poa nemoralis

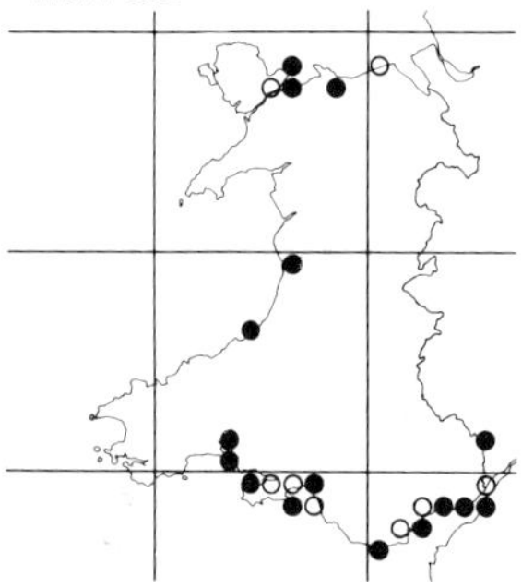

877 Puccinellia distans

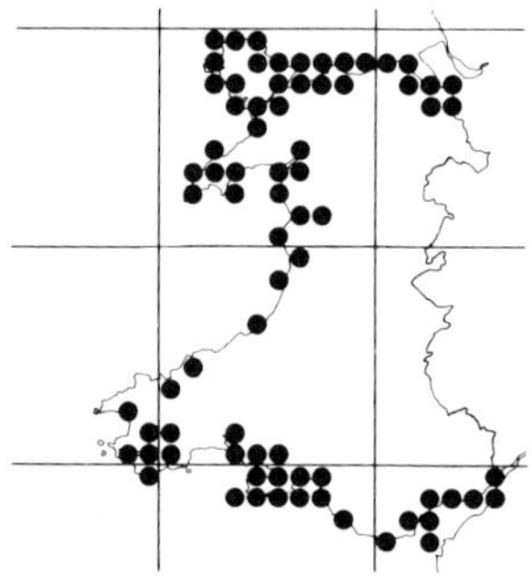

878 Puccinellia maritima

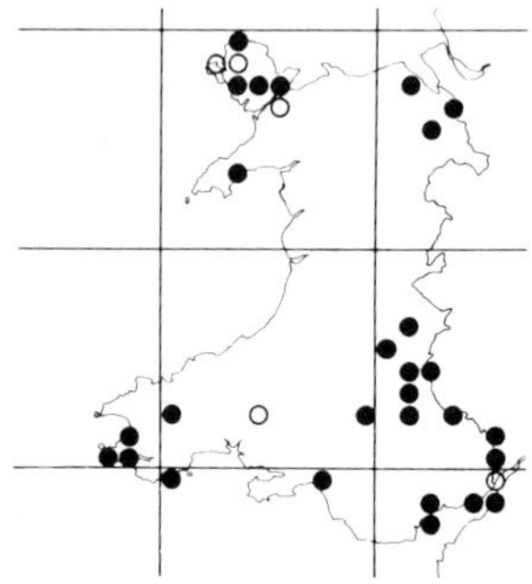

879 Catabrosa aquatica

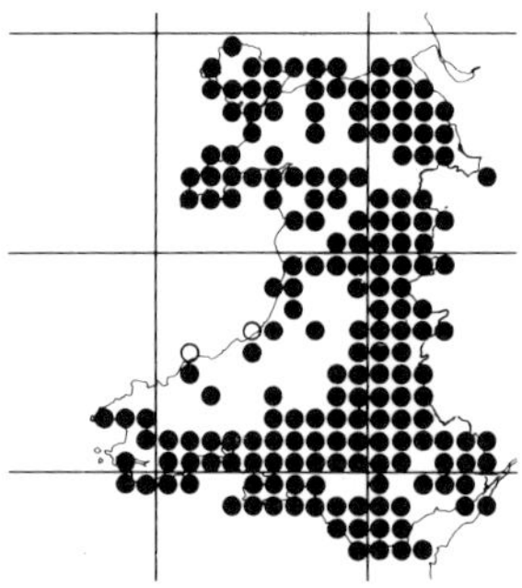

880 Briza media

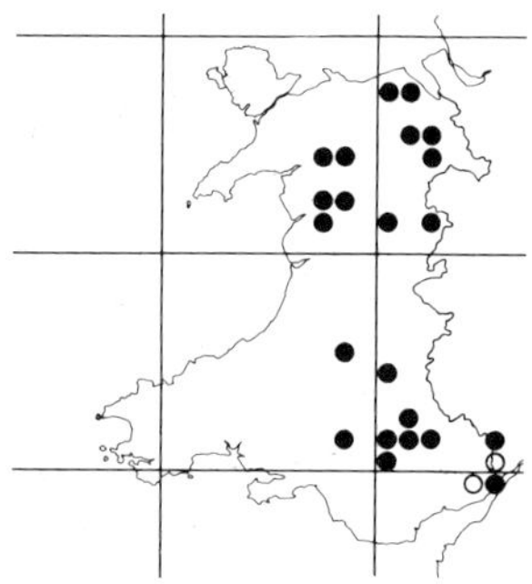

881 Melica nutans

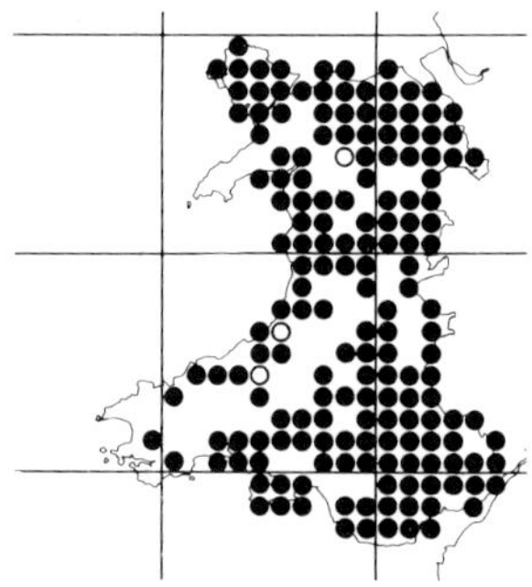

882 Melica uniflora

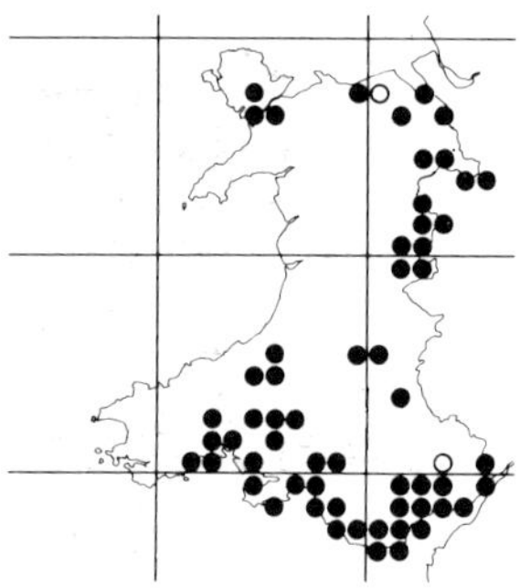

883 Glyceria maxima

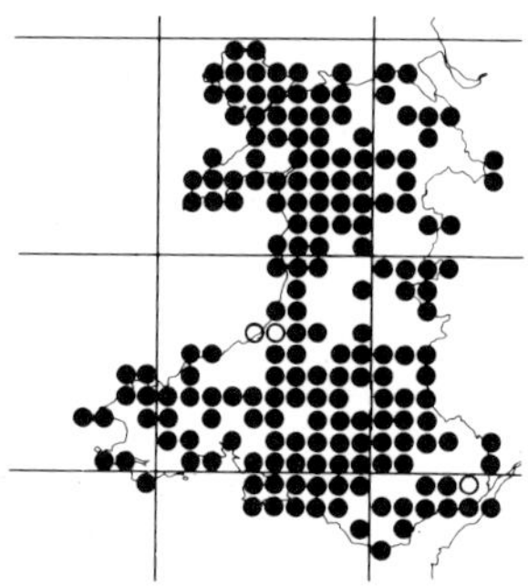

884 Glyceria declinata

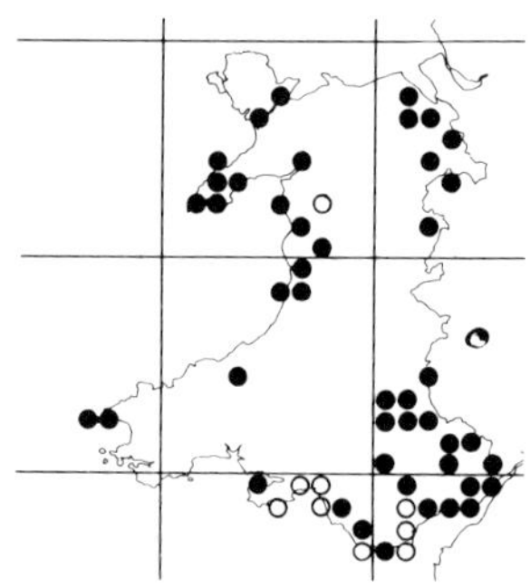

885 Glyceria fluitans × plicata
= G. × pedicellata

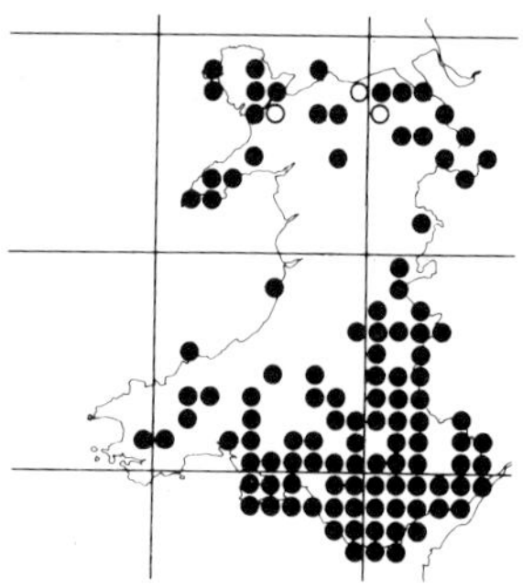

886 Glyceria plicata

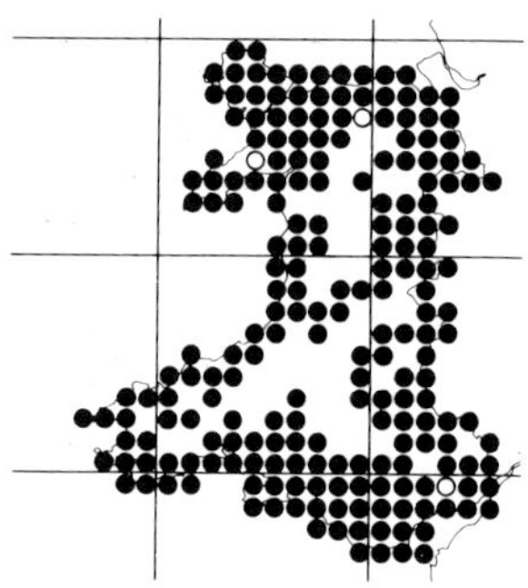

887 Bromus sterilis

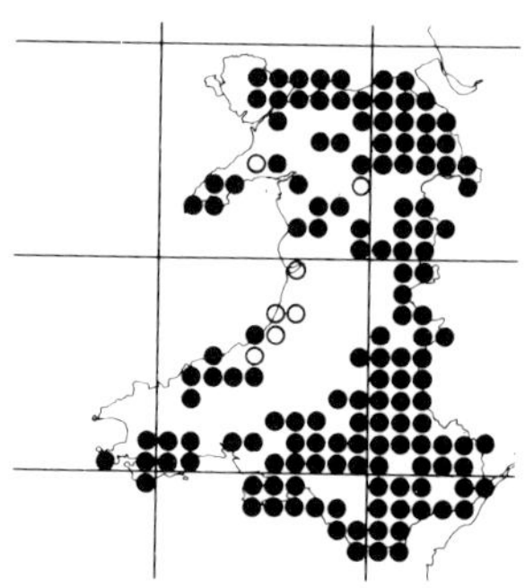

888 Bromus ramosus

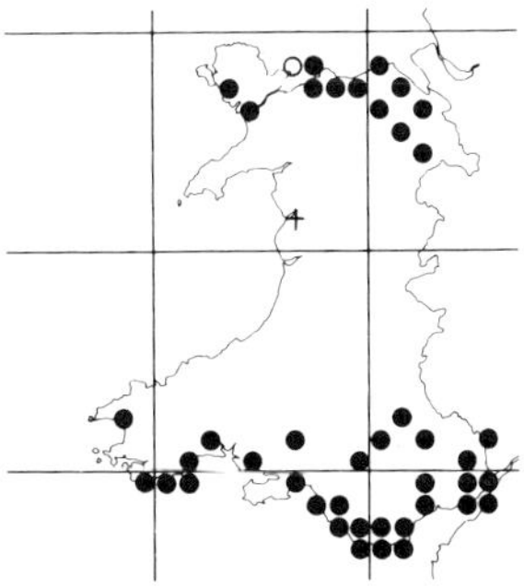

889 Bromus erectus

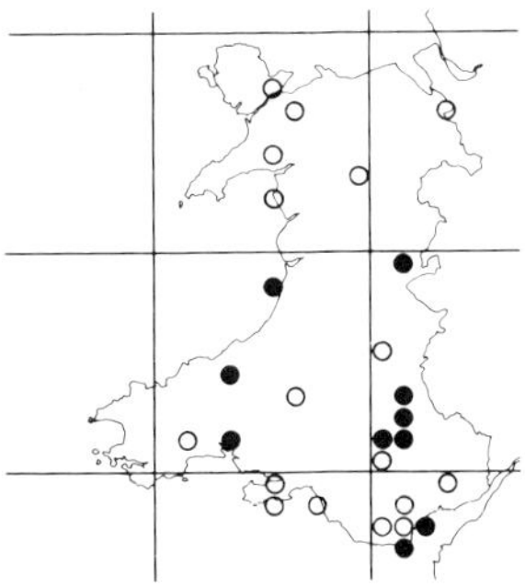

890 Bromus secalinus

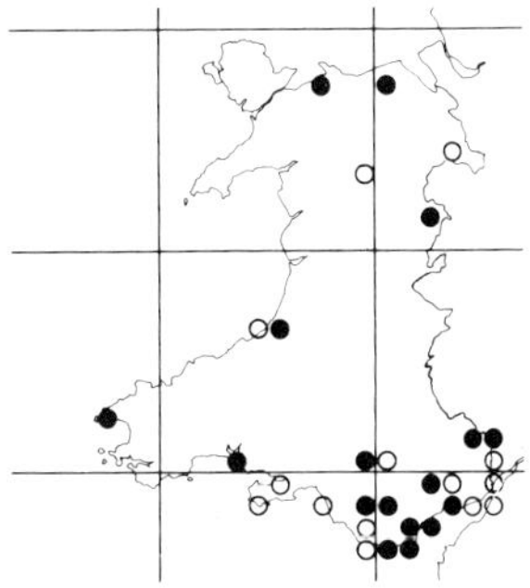

891 Bromus commutatus

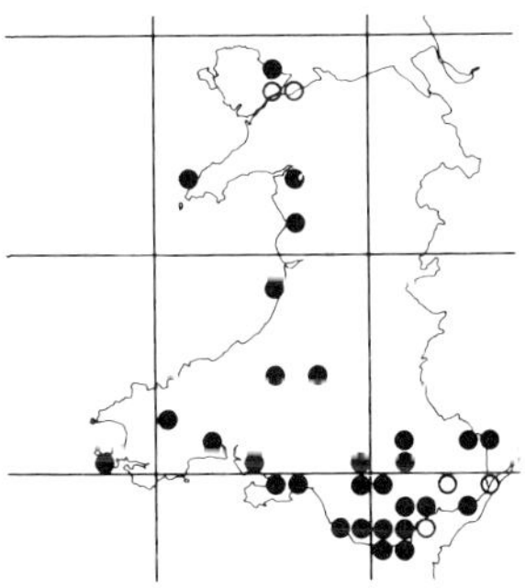

892 Bromus racemosus

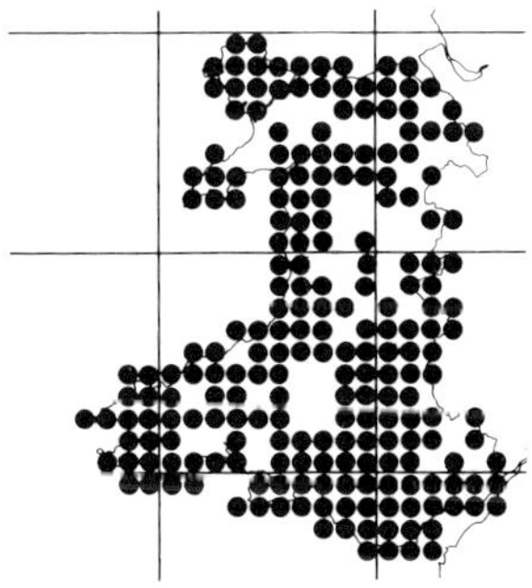

893 Bromus hordeaceus subsp. hordeaceus

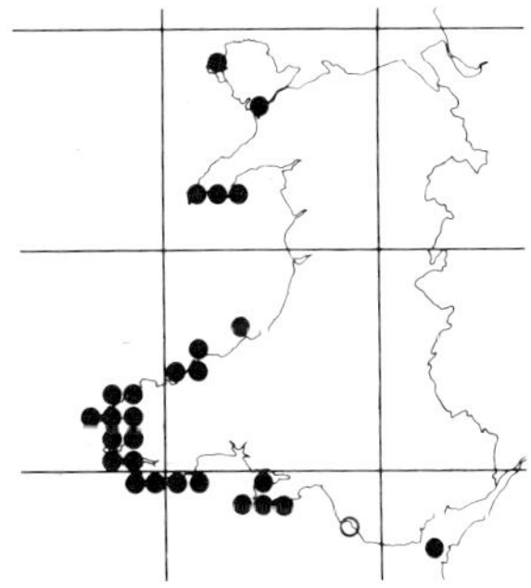

894 Bromus hordeaceus subsp. ferronii

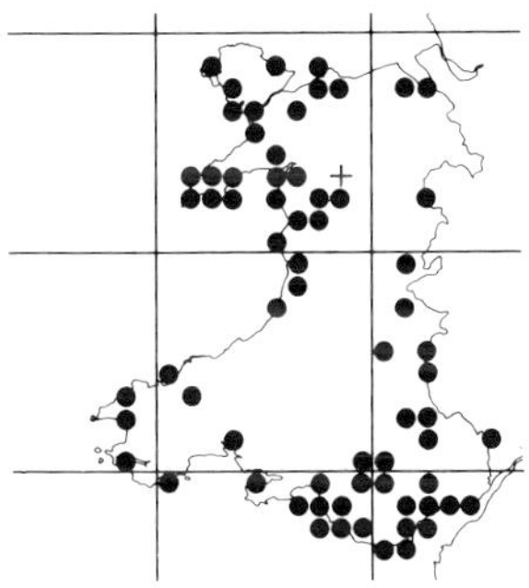

895 Bromus hordeaceus × lepidus = B. × pseudothominii

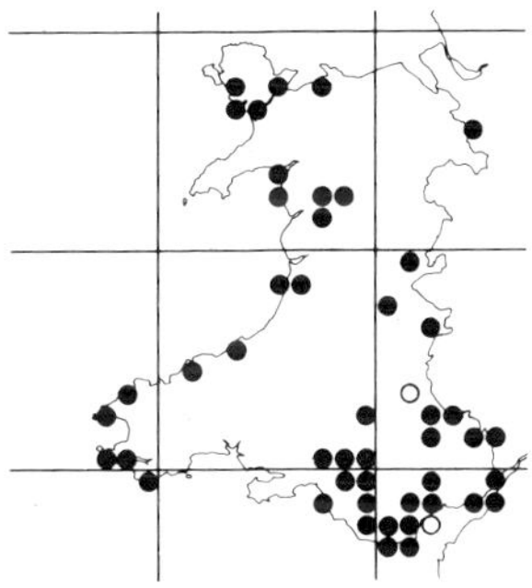

896 Bromus lepidus

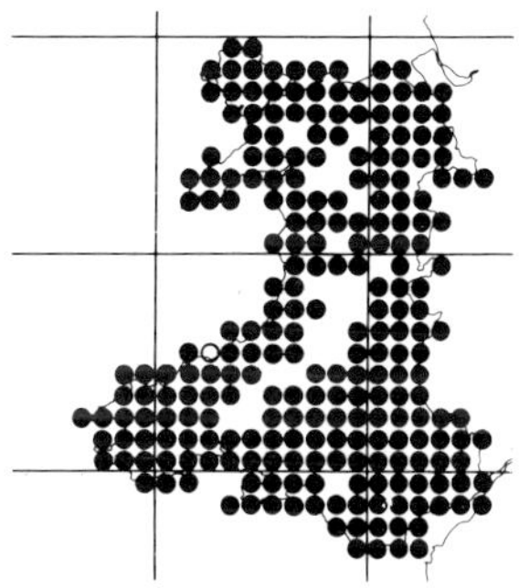

897 Brachypodium sylvaticum

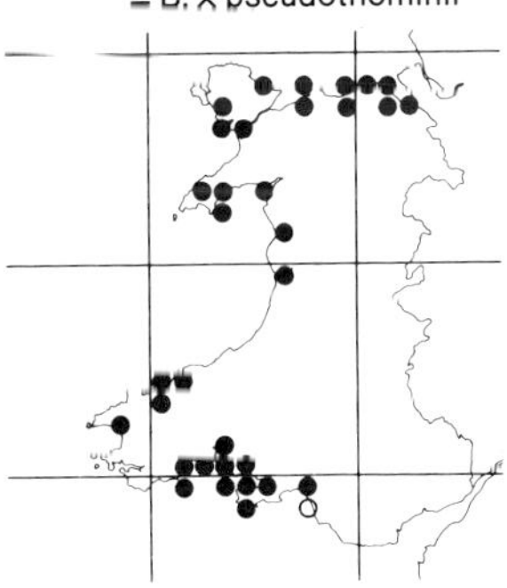

898 Leymus arenarius

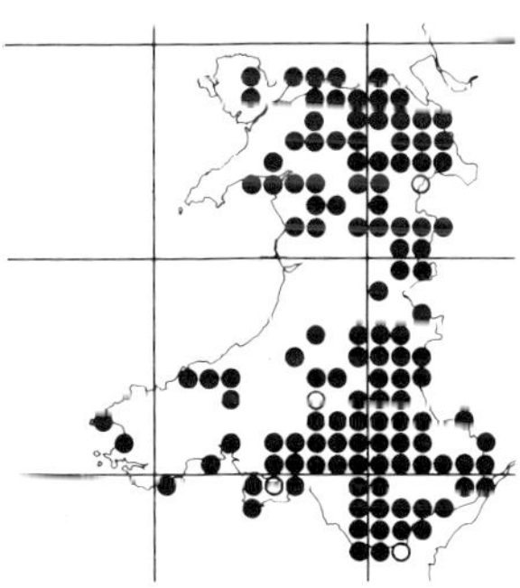

899 Elymus caninus

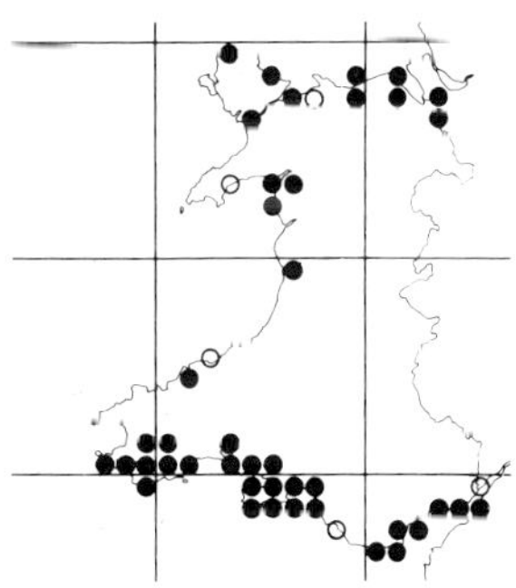

900 Elymus pycnanthus

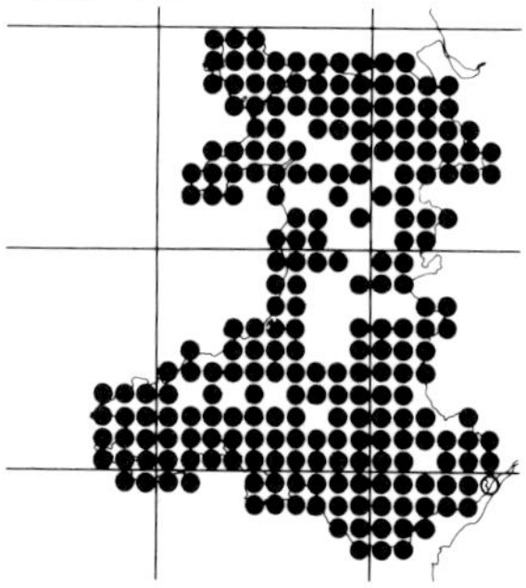

901 Elymus repens

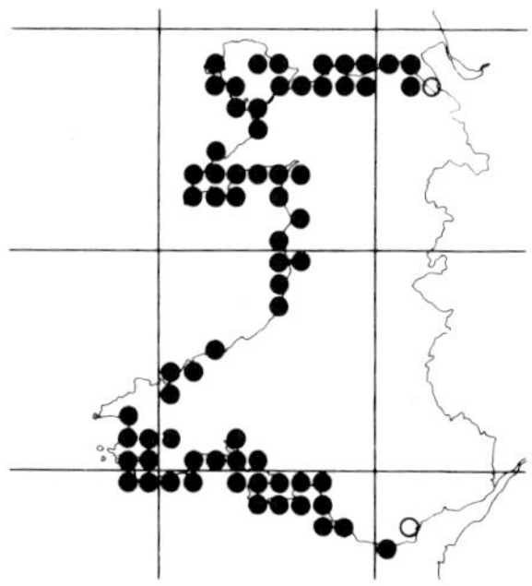

902 Elymus farctus

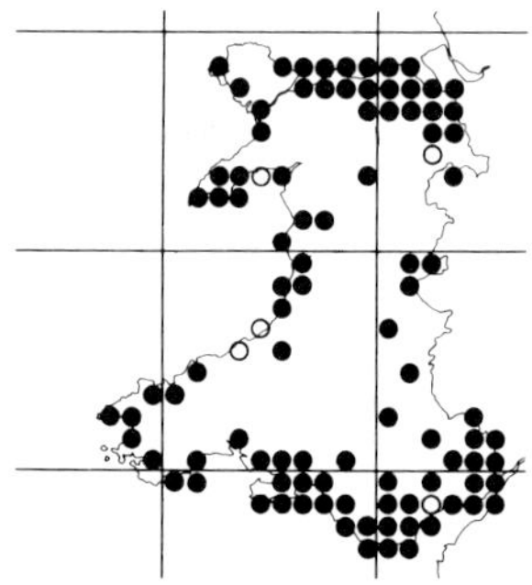

903 Hordeum murinum

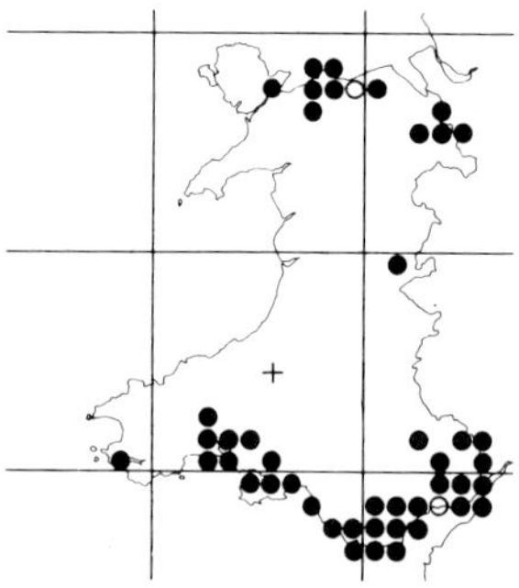

904 Hordeum secalinum

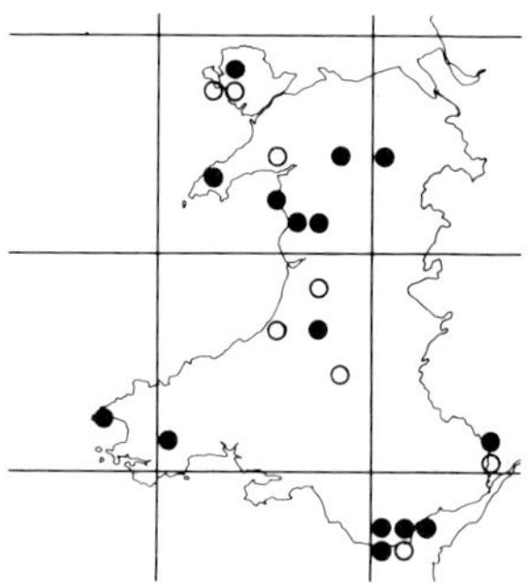

905 Avena strigosa

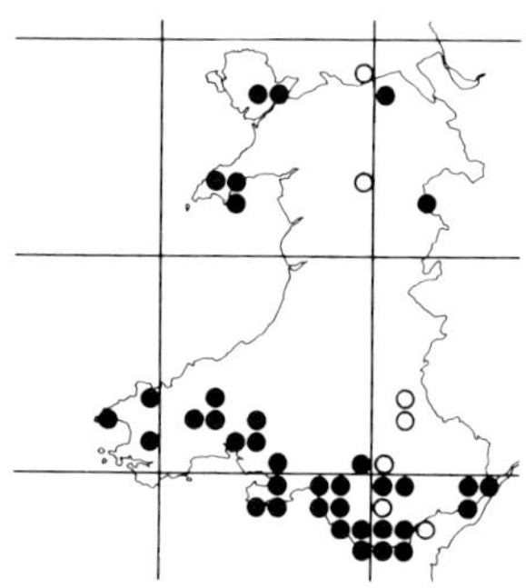

906 Avena fatua

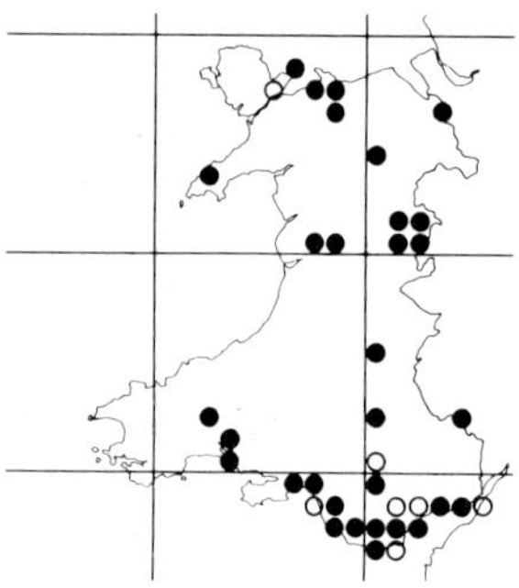

907 Avena sativa

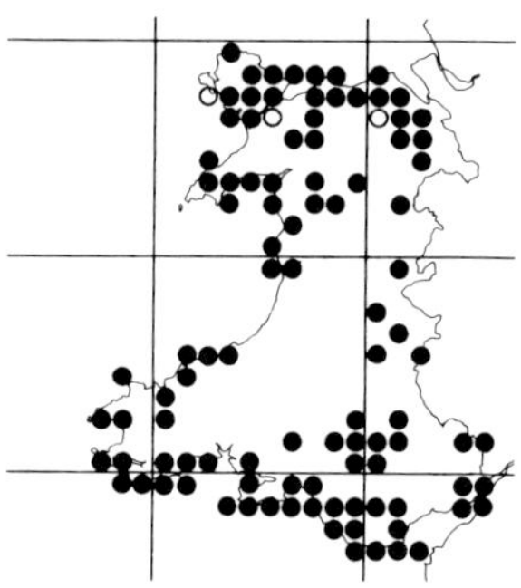

908 Avenula pubescens

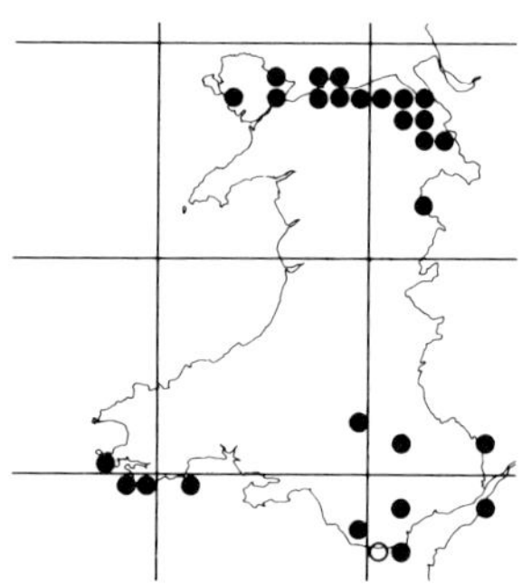

909 Avenula pratensis

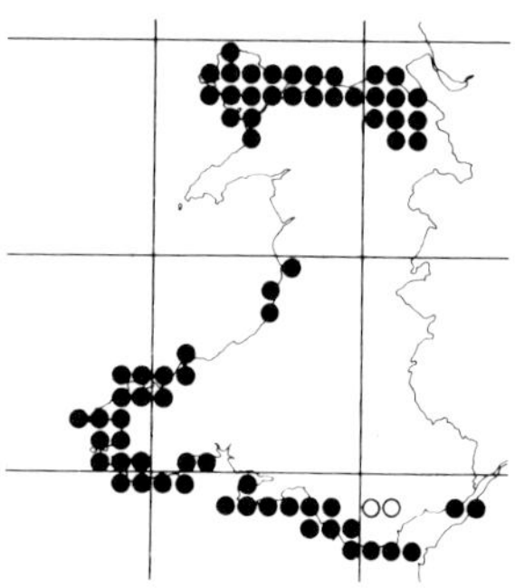

910 Koeleria macrantha

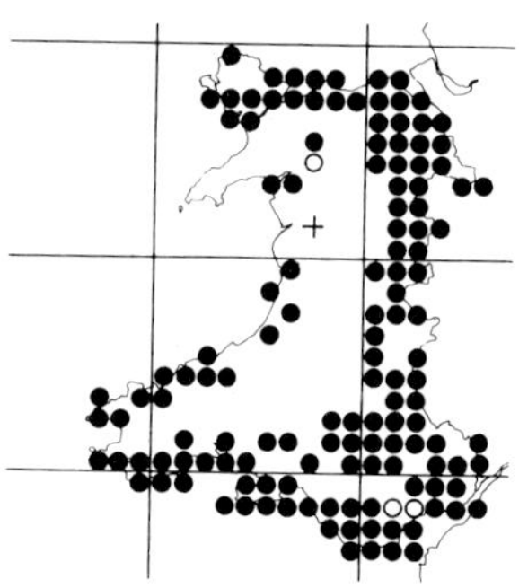

911 Trisetum flavescens

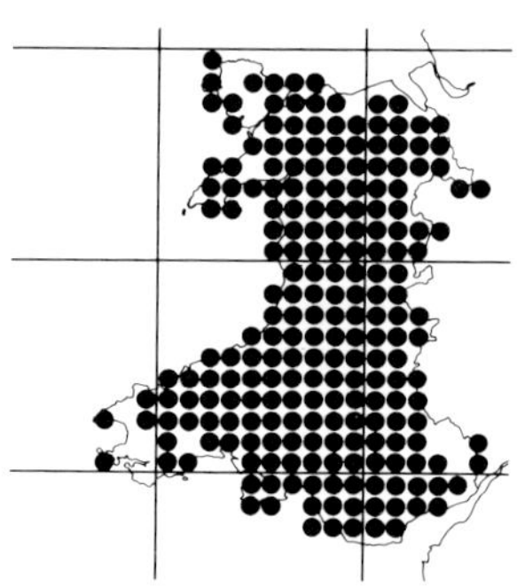

912 Deschampsia flexuosa

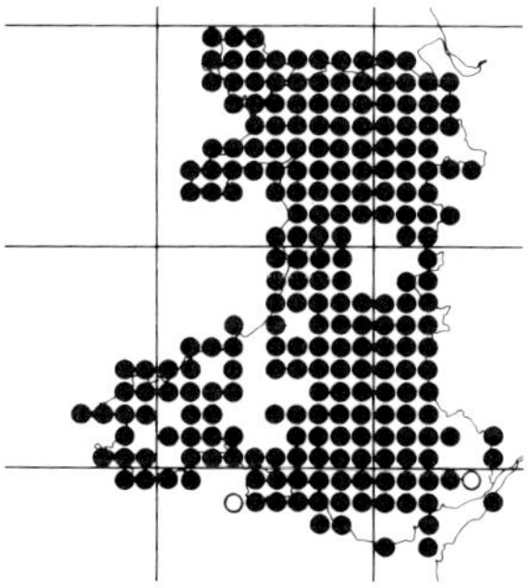

913 Aira praecox

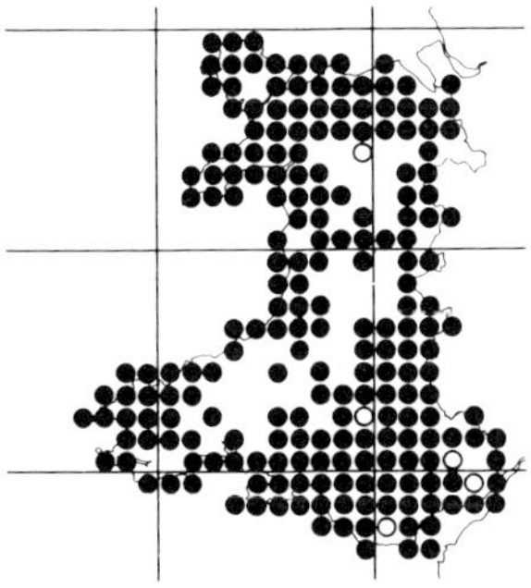

914 Aira caryophyllea

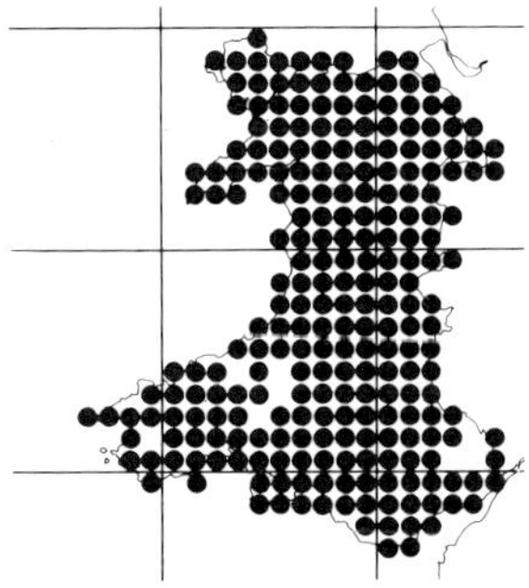

915 Holcus mollis

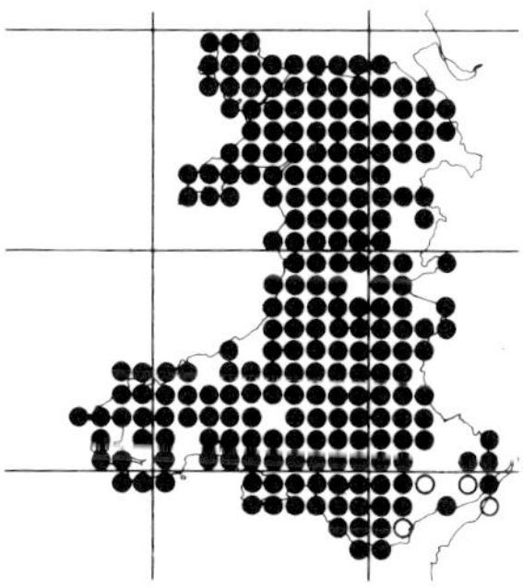

916 Agrostis canina

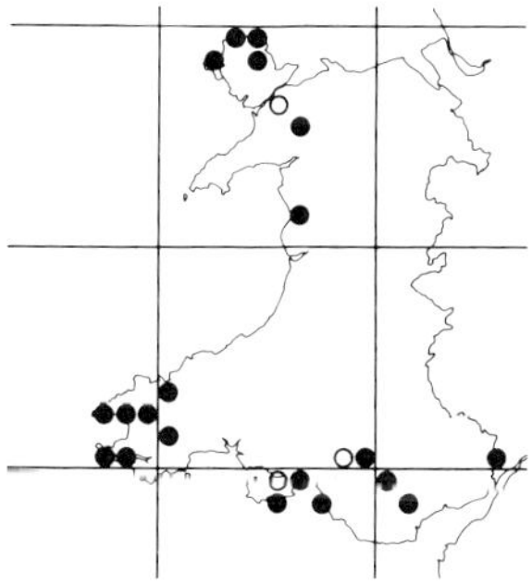

917 Agrostis vinealis

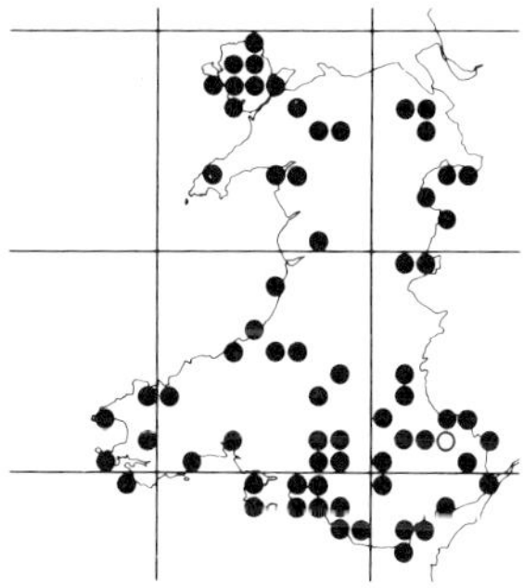

918 Agrostis gigantea

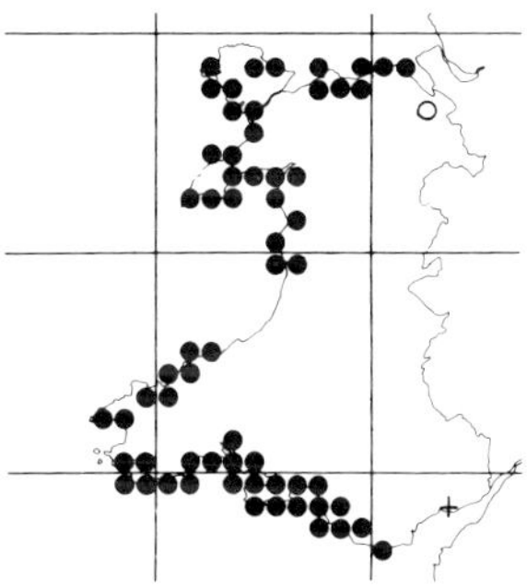

919 Ammophila arenaria
subsp. arenaria

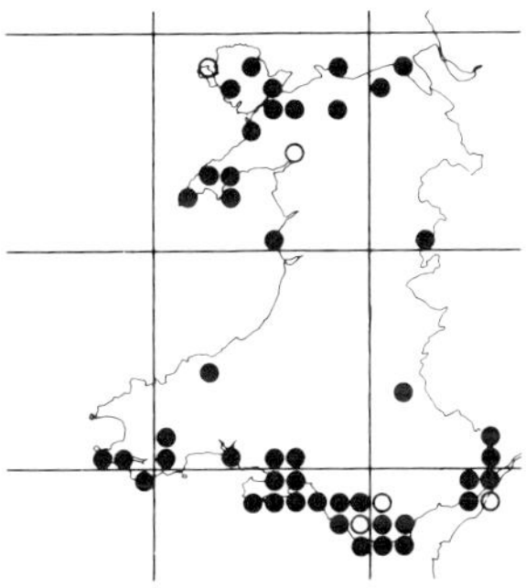

920 Calamagrostis epigejos

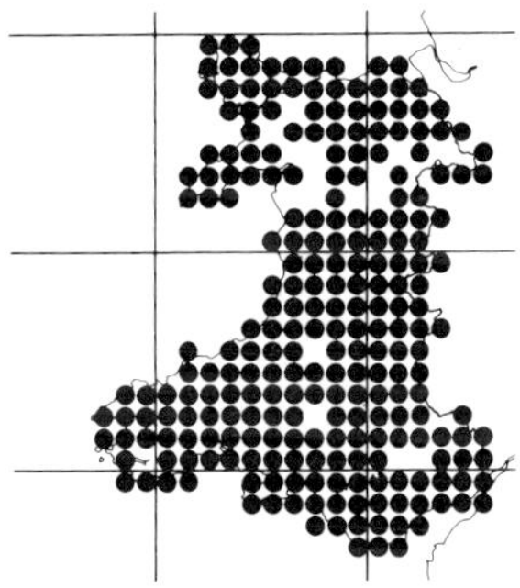

921 Phleum pratense

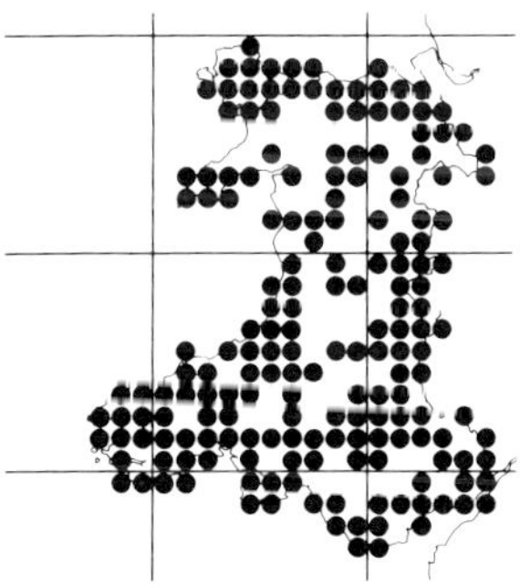

922 Phleum pratense
subsp. pratense

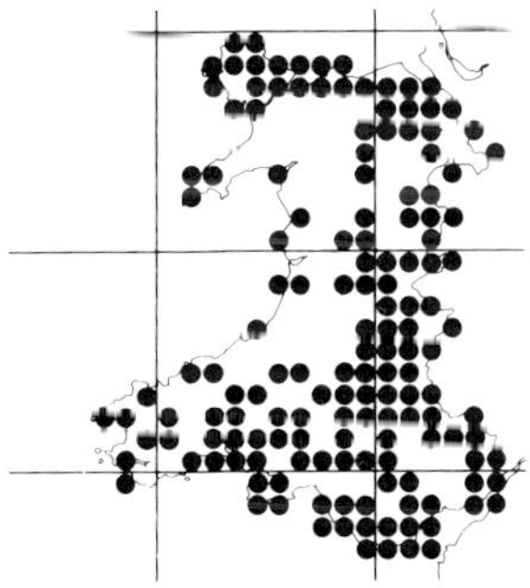

923 Phleum pratense
subsp. bertolonii

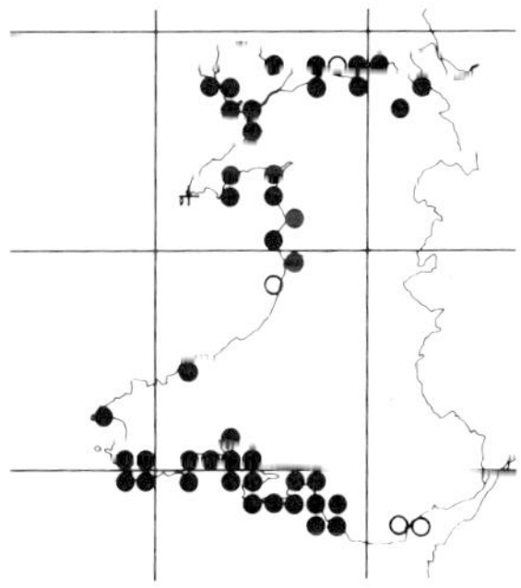

924 Phleum arenarium

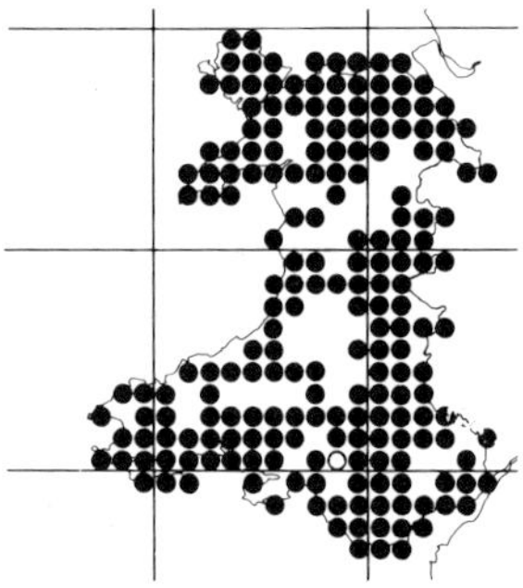

925 Alopecurus pratensis

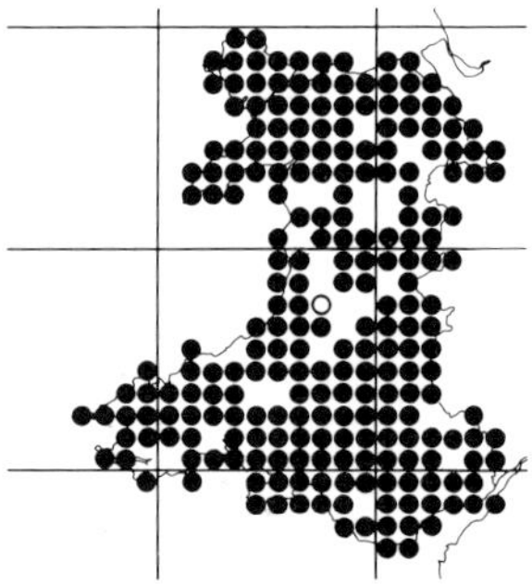

926 Alopecurus geniculatus

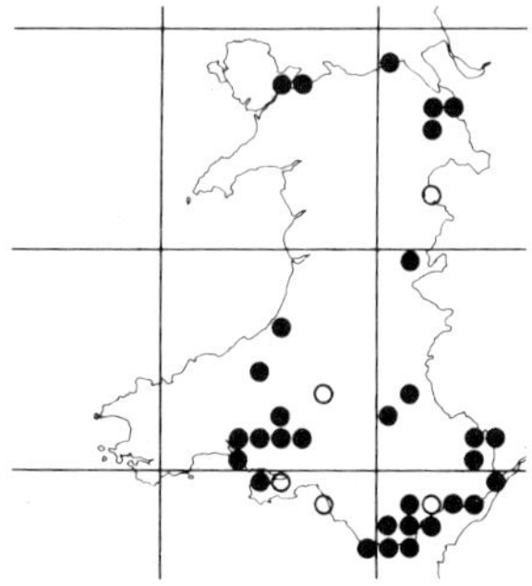

927 Alopecurus myosuroides

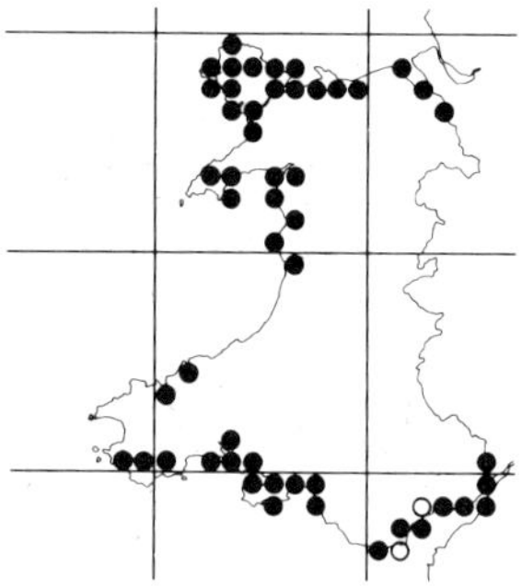

928 Parapholis strigosa

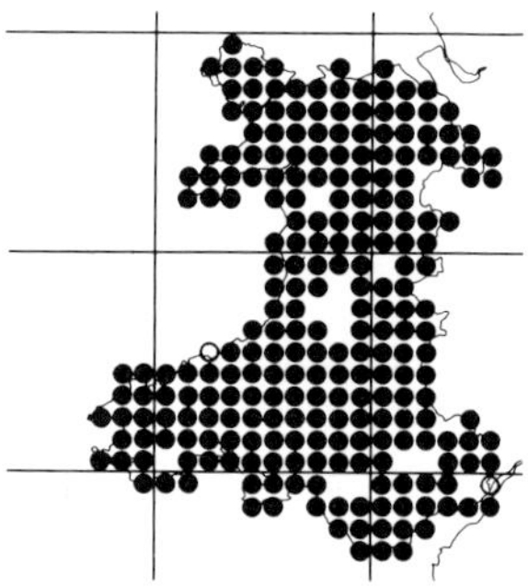

929 Phalaris arundinacea

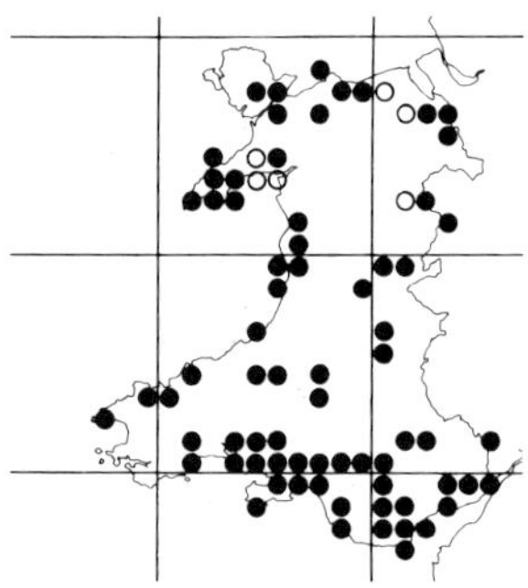

930 Phalaris canariensis

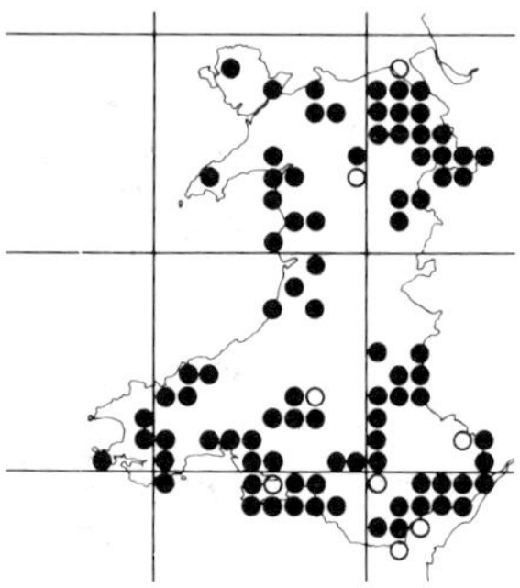

931 Milium effusum

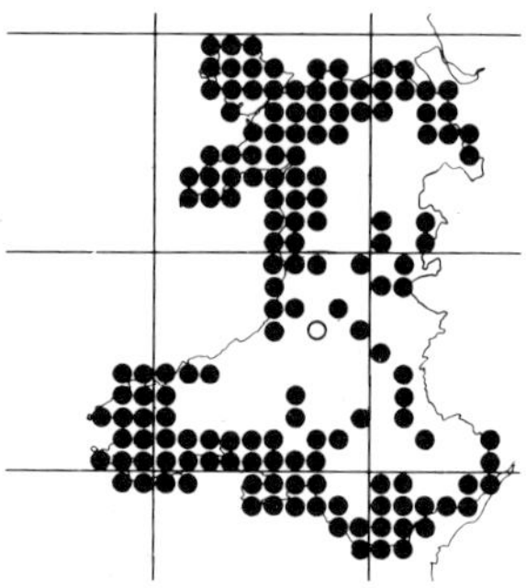

932 Phragmites australis

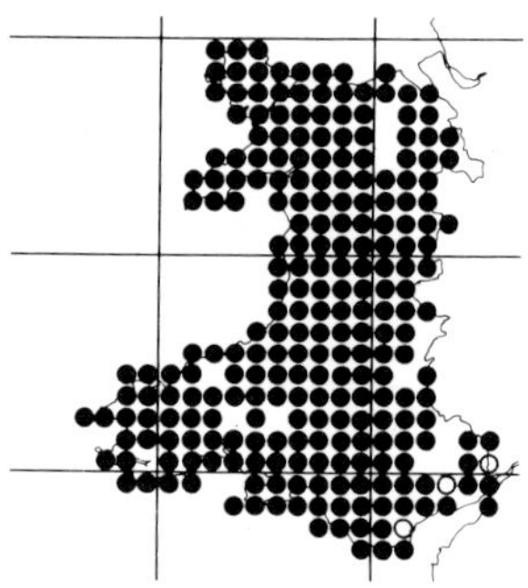

933 Danthonia decumbens

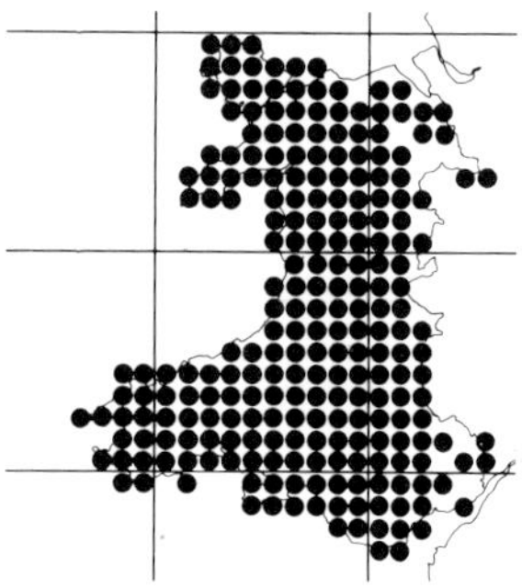

934 Molinia caerulea subsp. caerulea

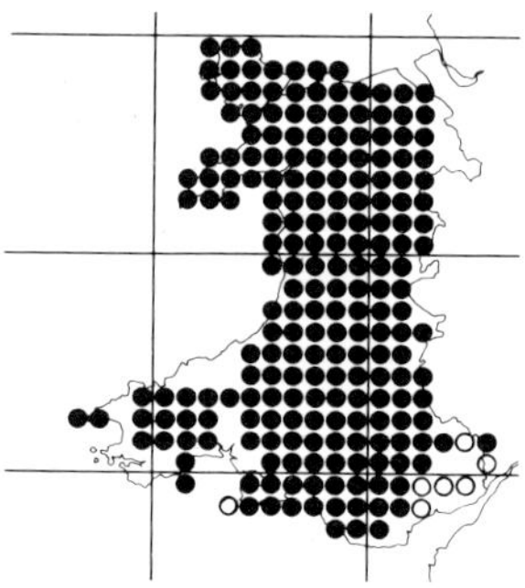

935 Nardus stricta

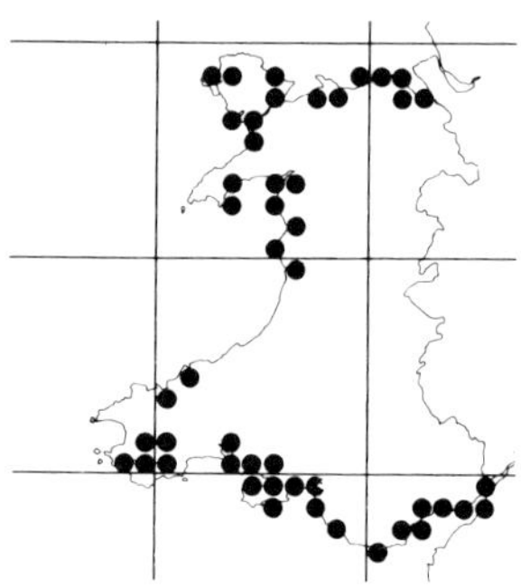

936 Spartina anglica

937 Arum maculatum

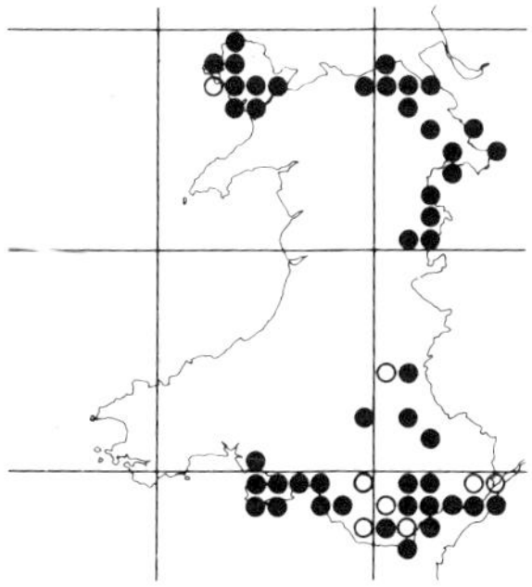

938 Lemna trisulca

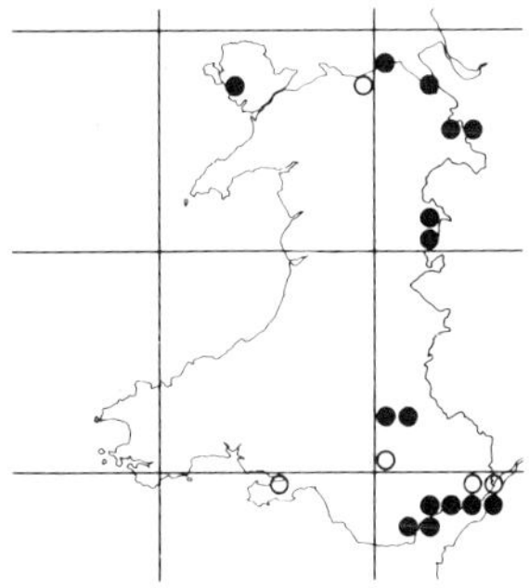

939 Lemna gibba

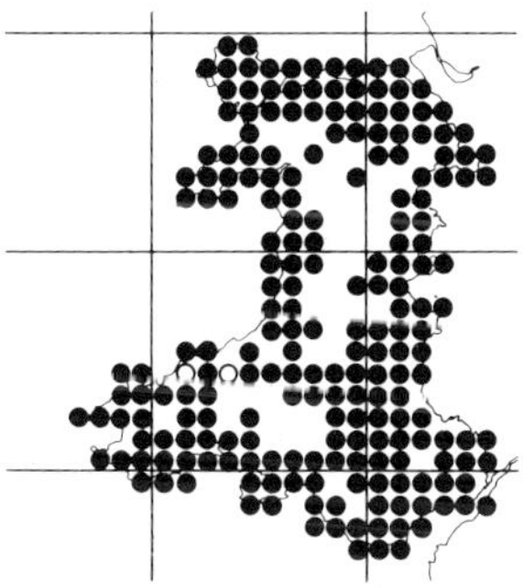

940 Lemna minor

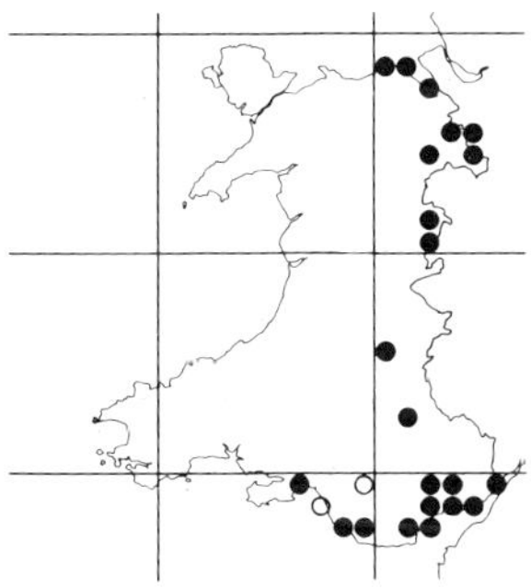

941 Spirodela polyrhiza

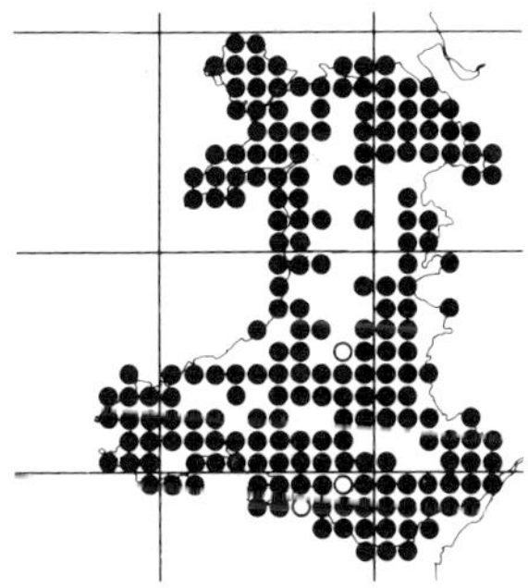

942 Sparganium erectum

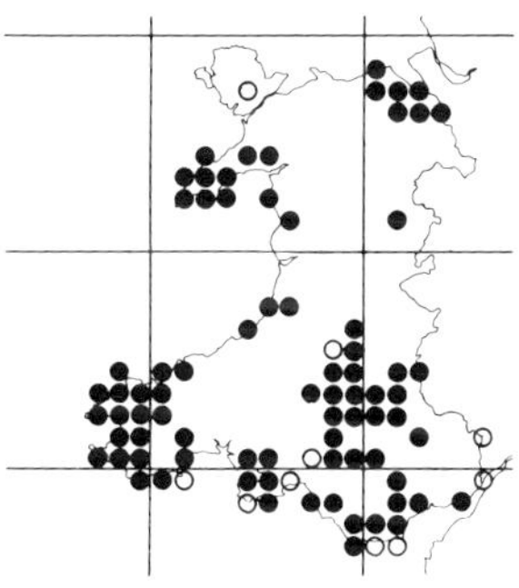

943 Sparganium erectum subsp. erectum

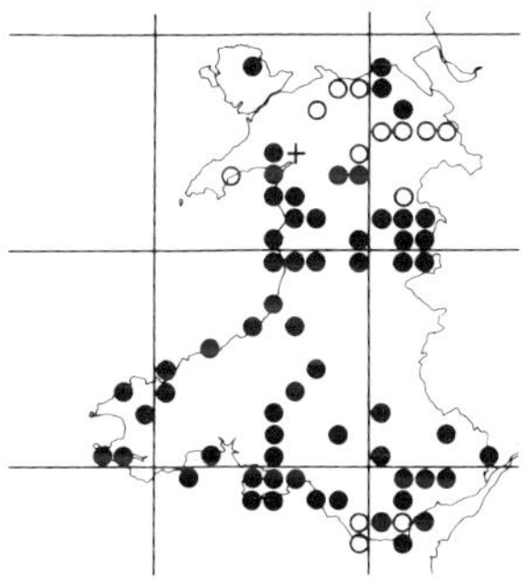

944 Sparganium erectum subsp. neglectum

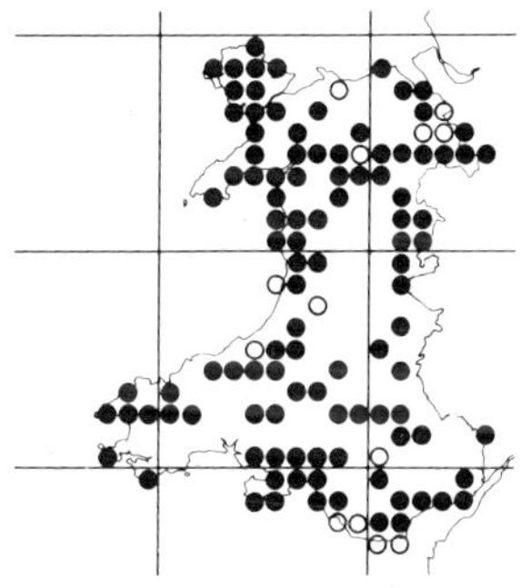

945 Sparganium emersum

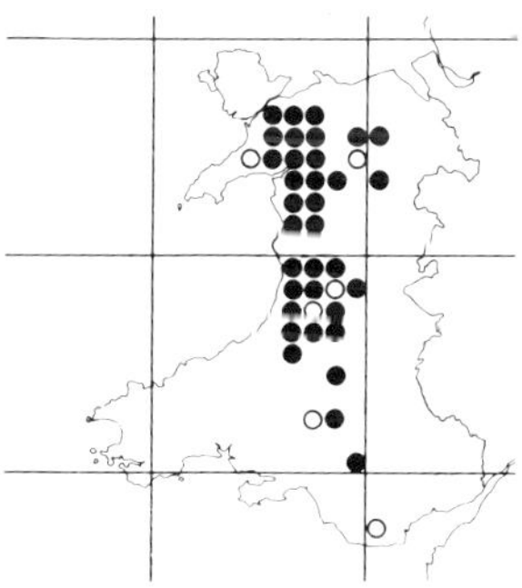

946 Sparganium angustifolium

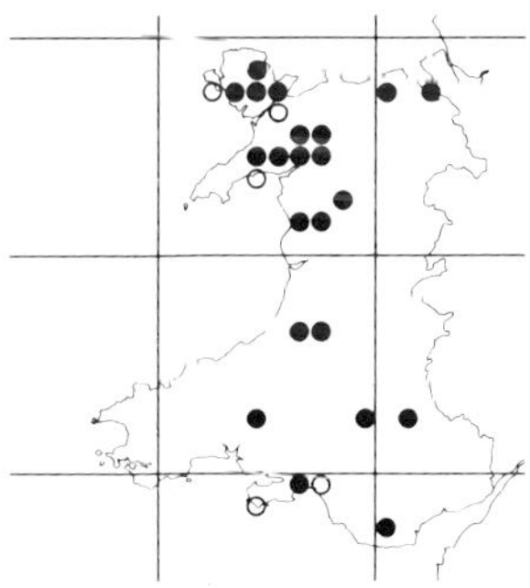

947 Sparganium minimum

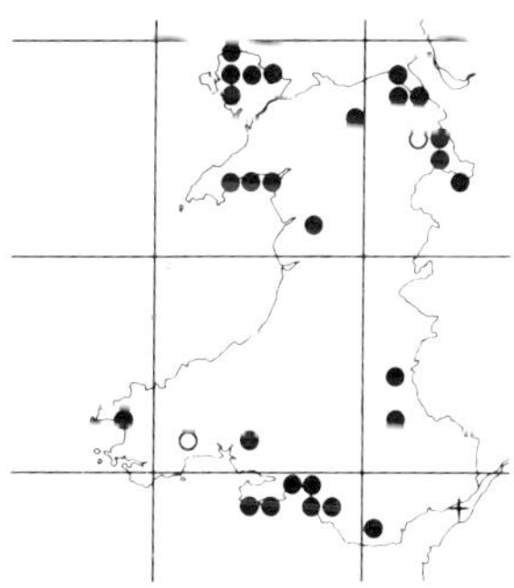

948 Typha angustifolia

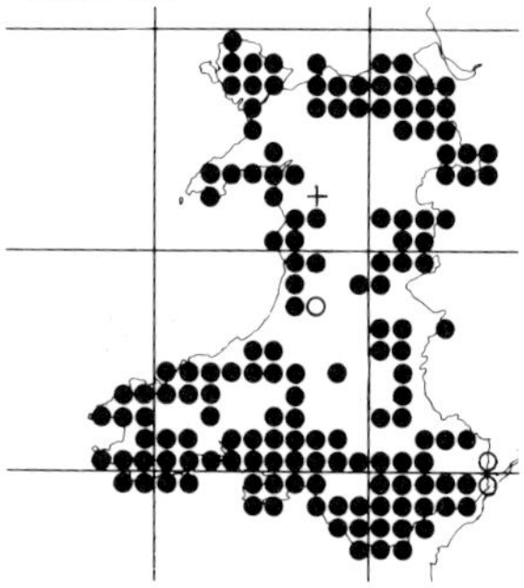

949 Typha latifolia

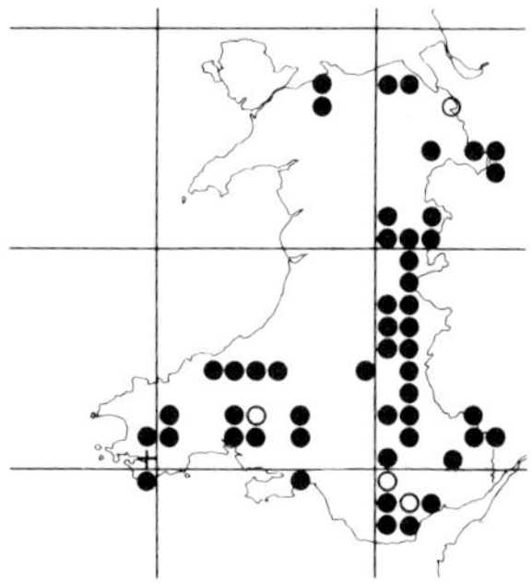

950 Scirpus sylvaticus

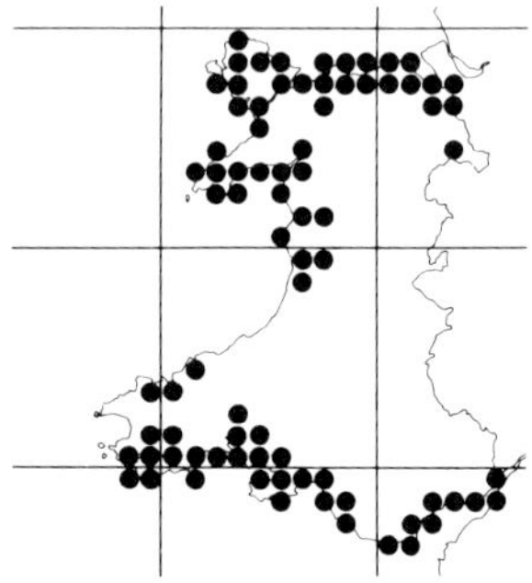

951 Scirpus maritimus

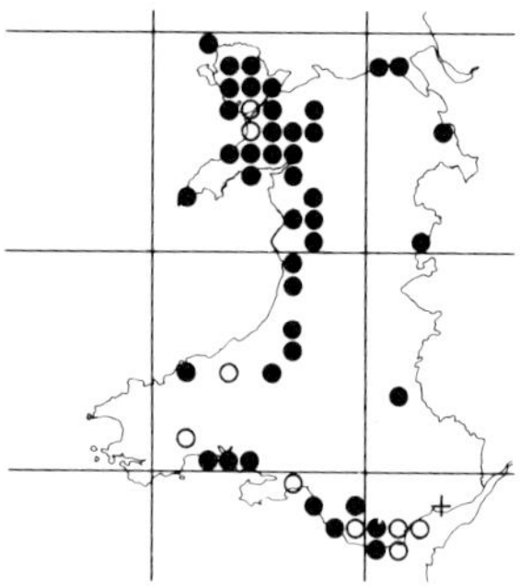

952 Scirpus lacustris subsp. lacustris

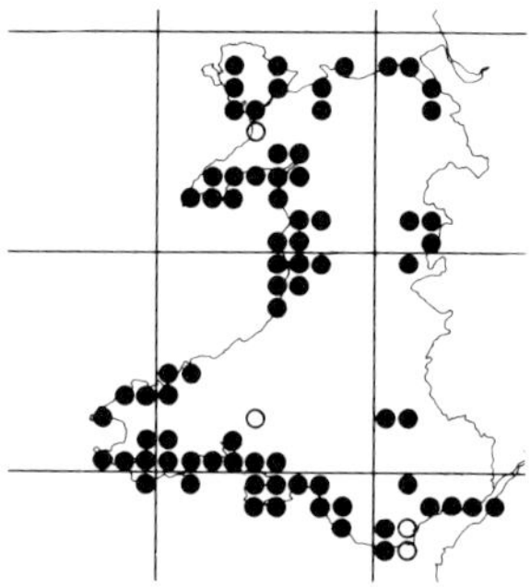

953 Scirpus lacustris subsp. tabernaemontani

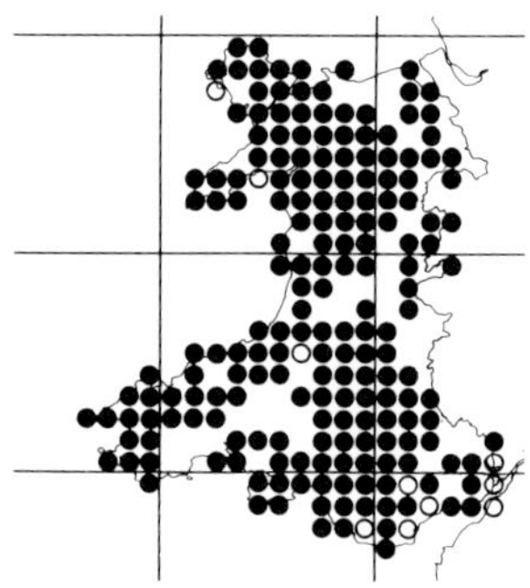

954 Scirpus setaceus

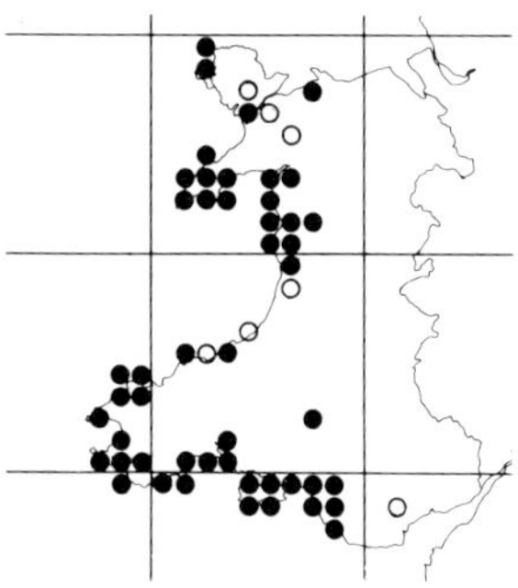

955 Scirpus cernuus

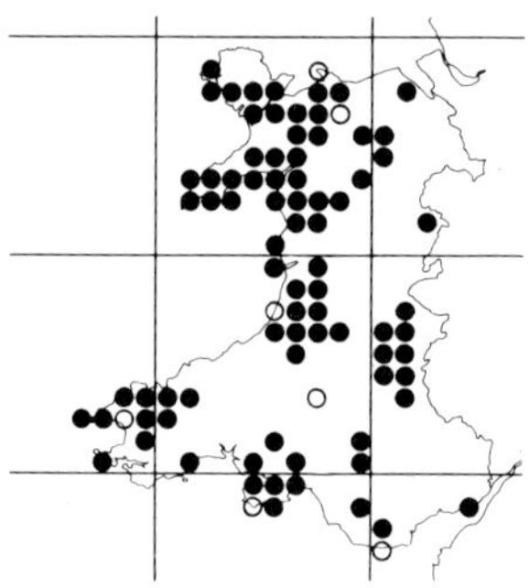

956 Scirpus fluitans

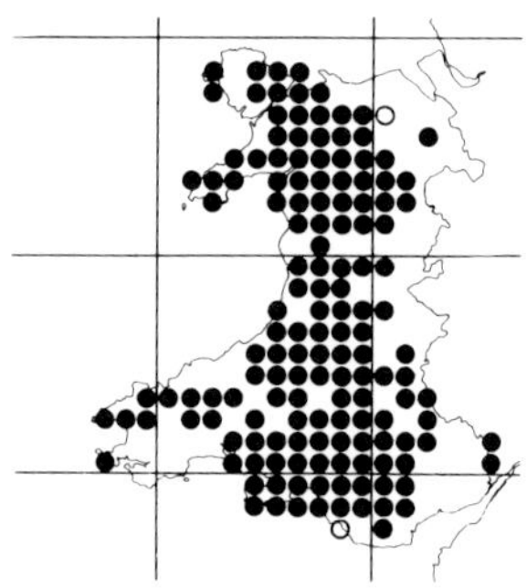

957 Scirpus cespitosus

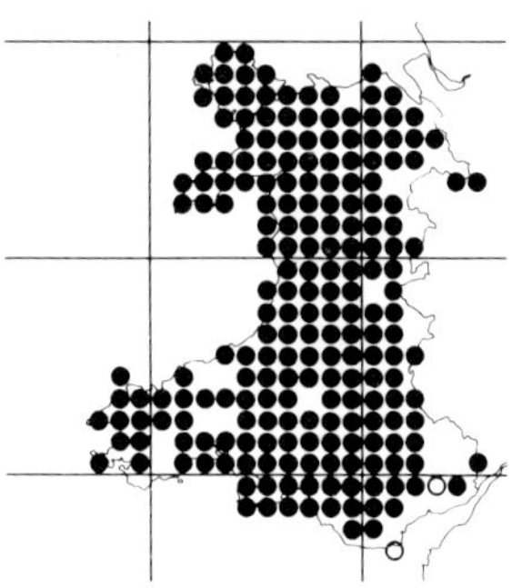

958 Eriophorum angustifolium

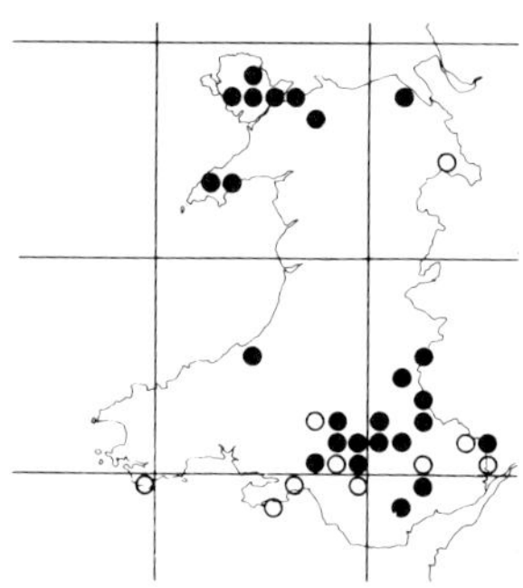

959 Eriophorum latifolium

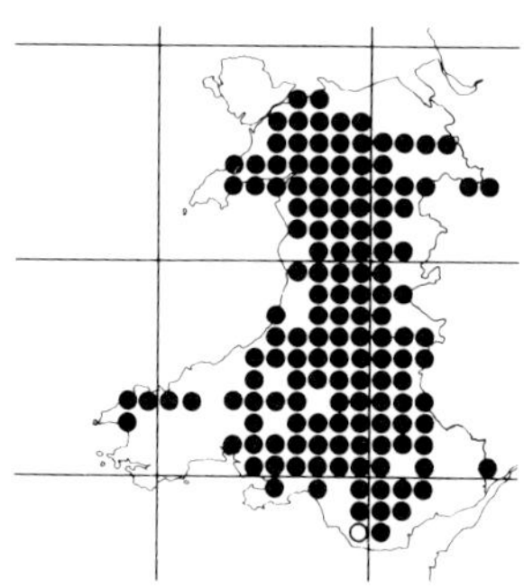

960 Eriophorum vaginatum

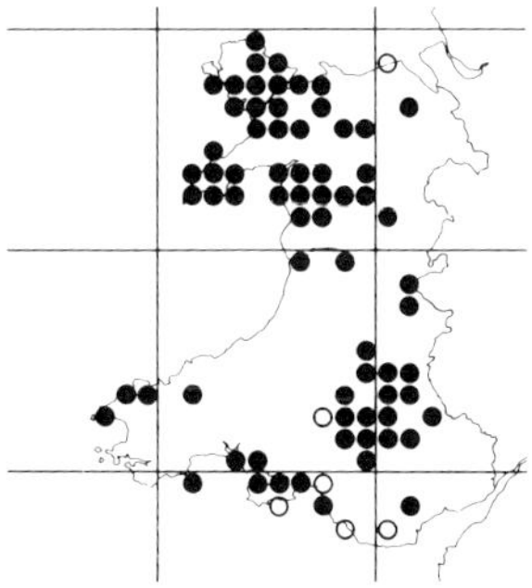

961 Eleocharis quinqueflora

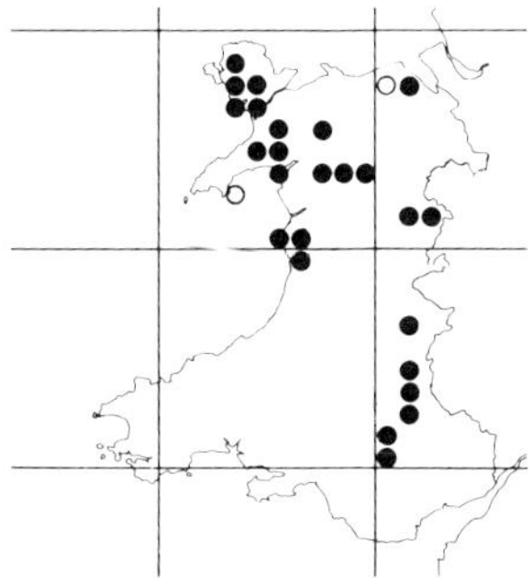

962 Eleocharis acicularis

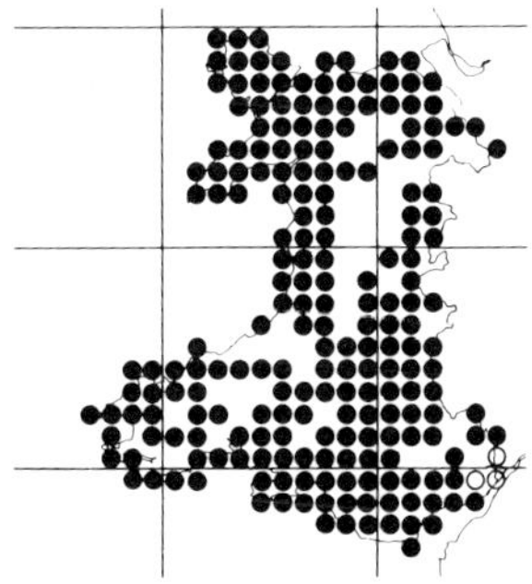

963 Eleocharis palustris

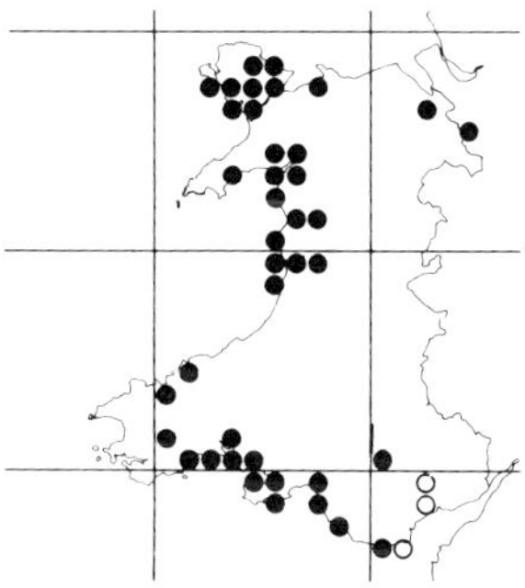

964 Eleocharis uniglumis

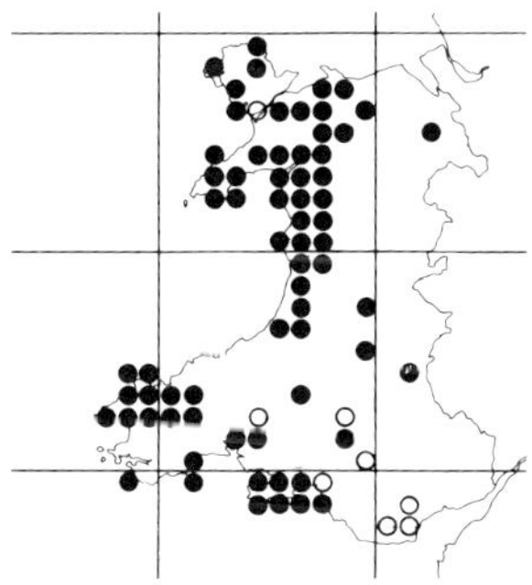

965 Eleocharis multicaulis

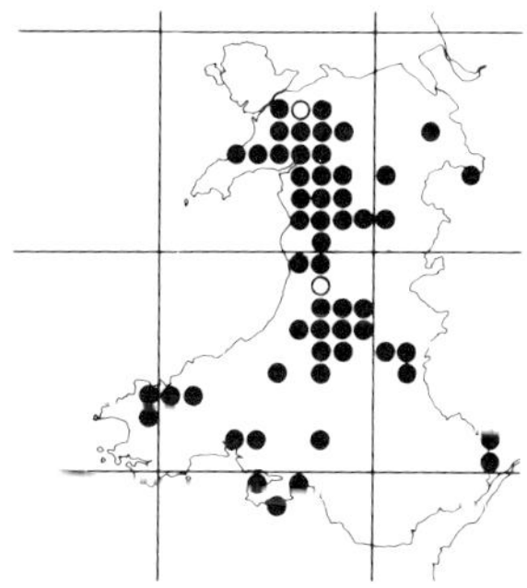

966 Rhynchospora alba

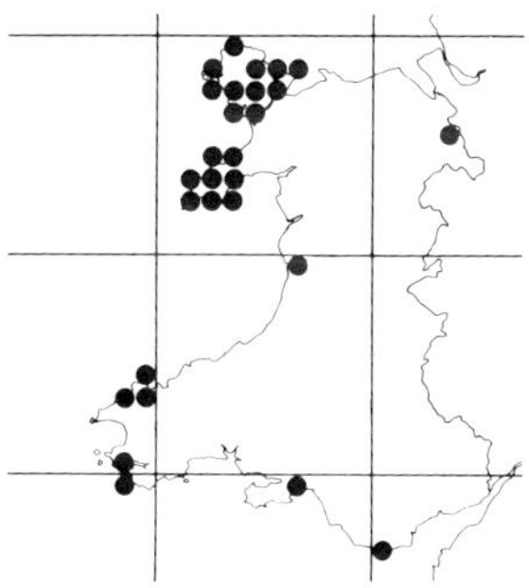

967 Schoenus nigricans

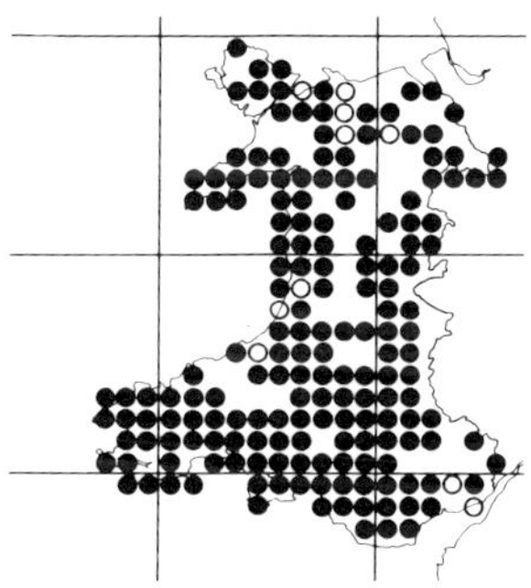

968 Carex paniculata

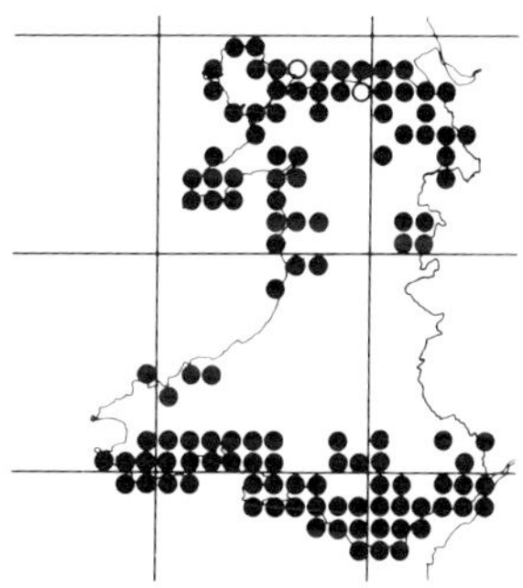

969 Carex otrubae

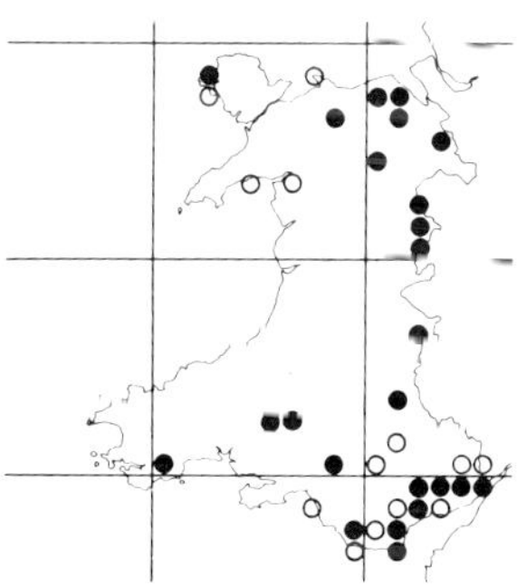

970 Carex spicata

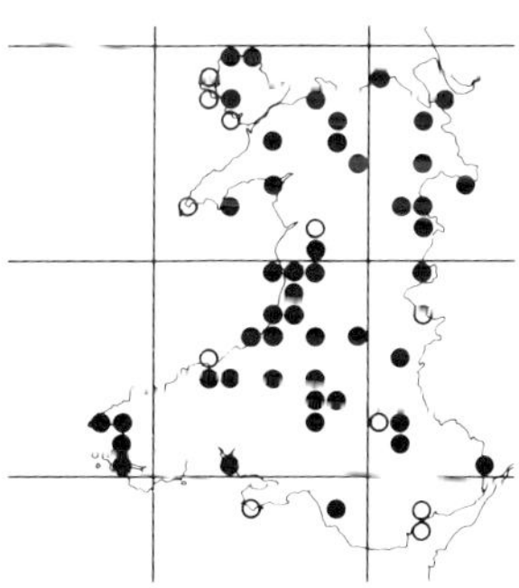

971 Carex muricata subsp. lamprocarpa

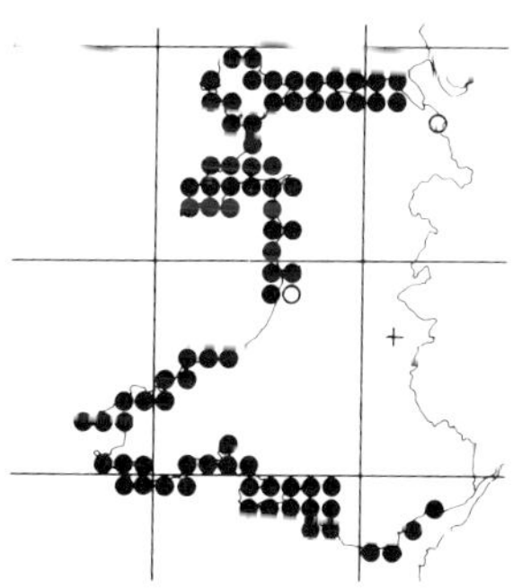

972 Carex arenaria

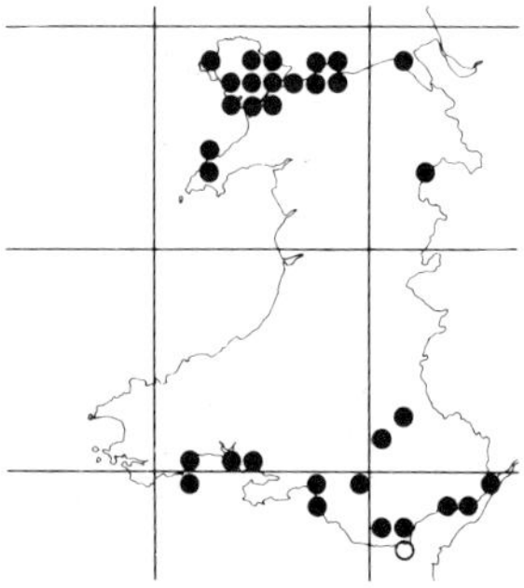

973 Carex disticha

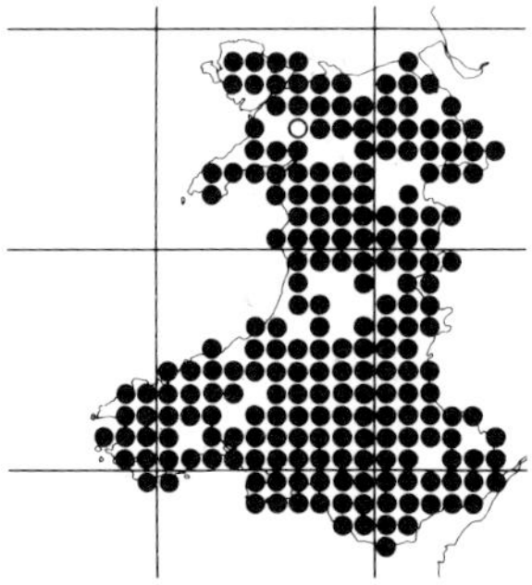

974 Carex remota

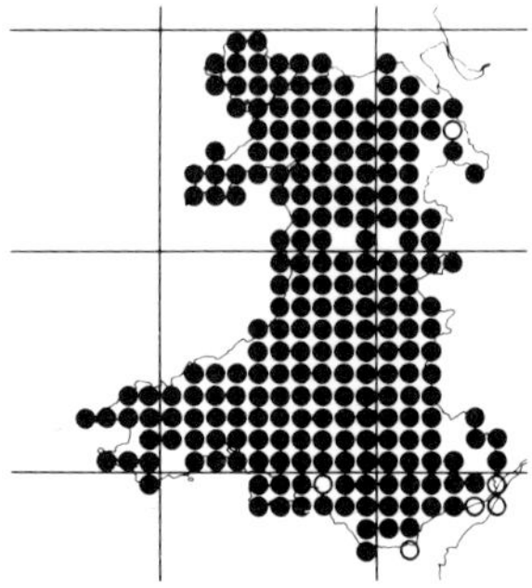

975 Carex ovalis

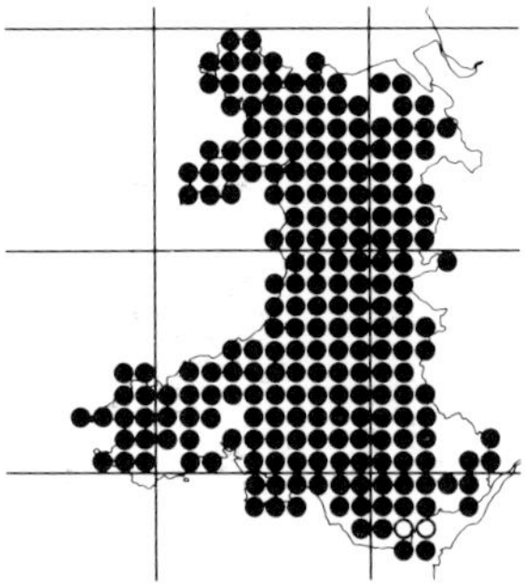

976 Carex echinata

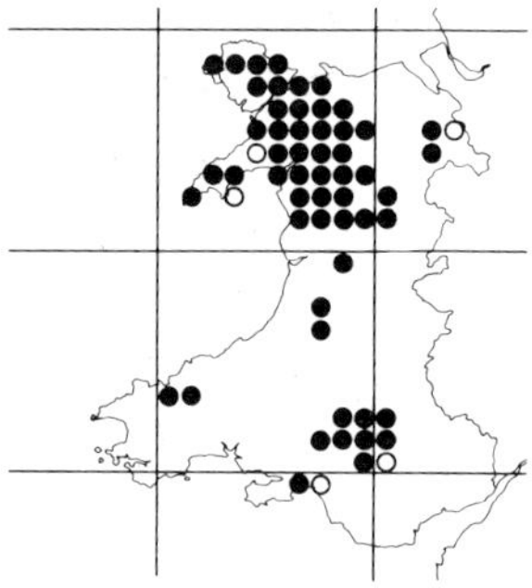

977 Carex dioica

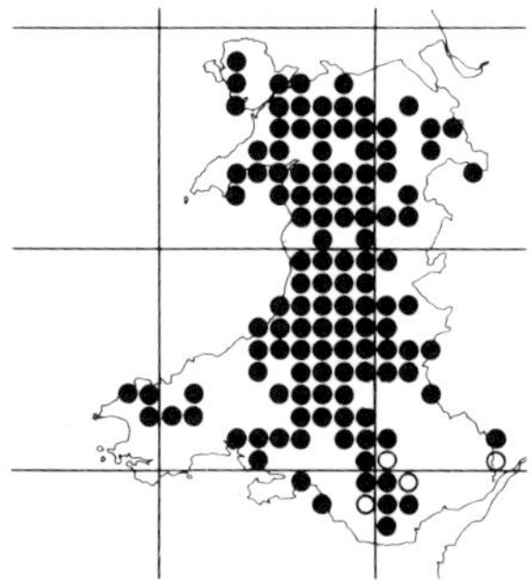

978 Carex curta

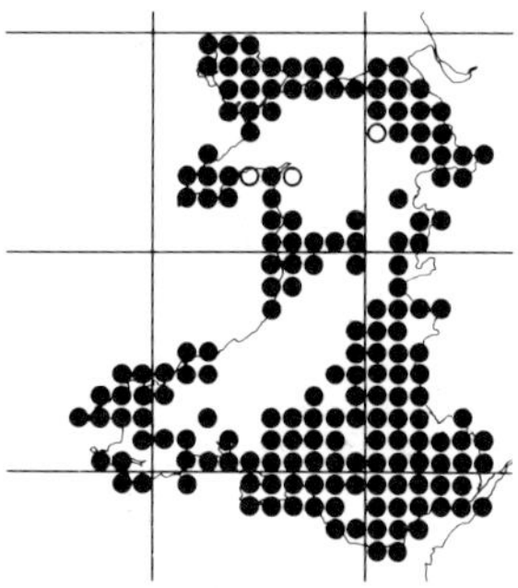

979 Carex hirta

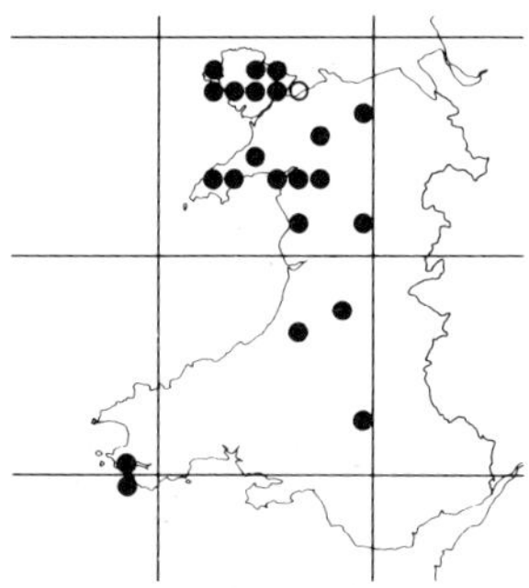

980 Carex lasiocarpa

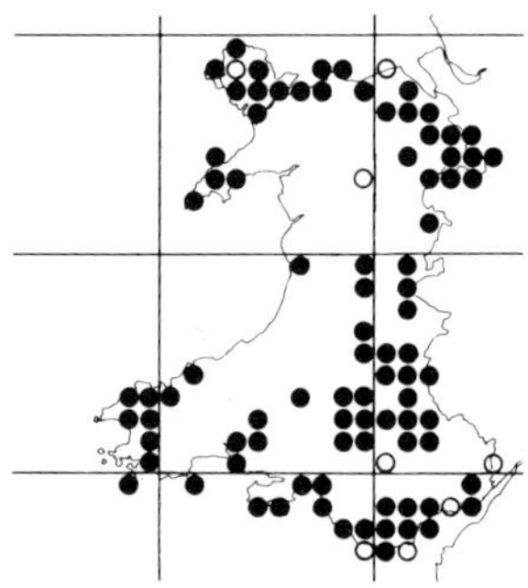

981 Carex acutiformis

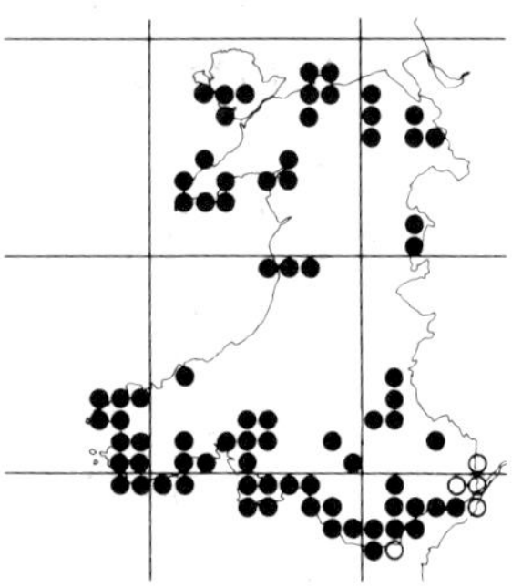

982 Carex riparia

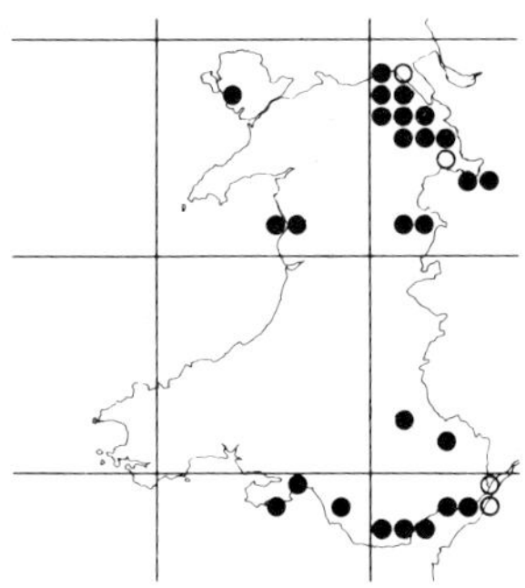

983 Carex pseudocyperus

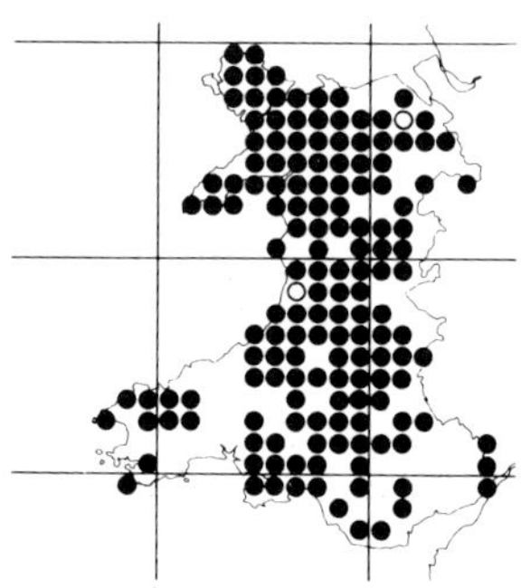

984 Carex rostrata

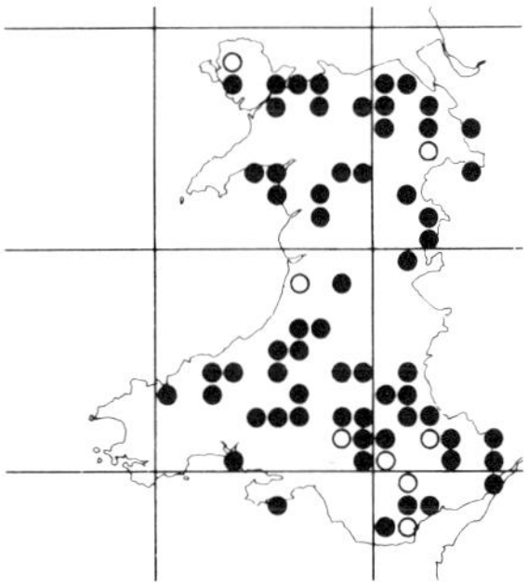

985 Carex vesicaria

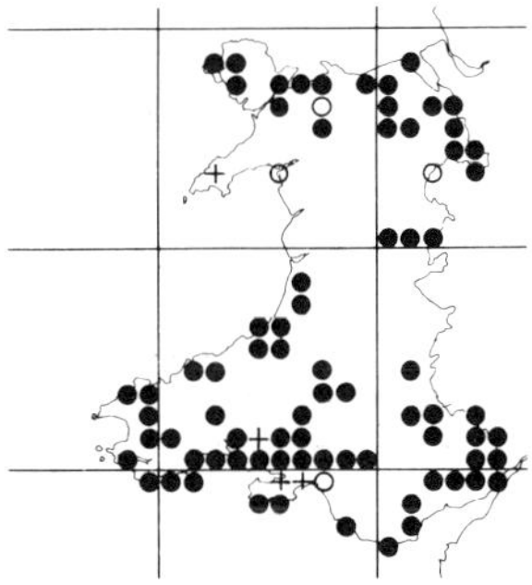

986 Carex pendula

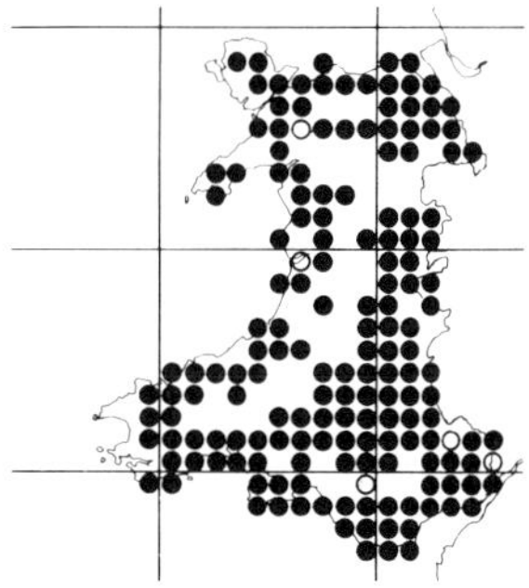

987 Carex sylvatica

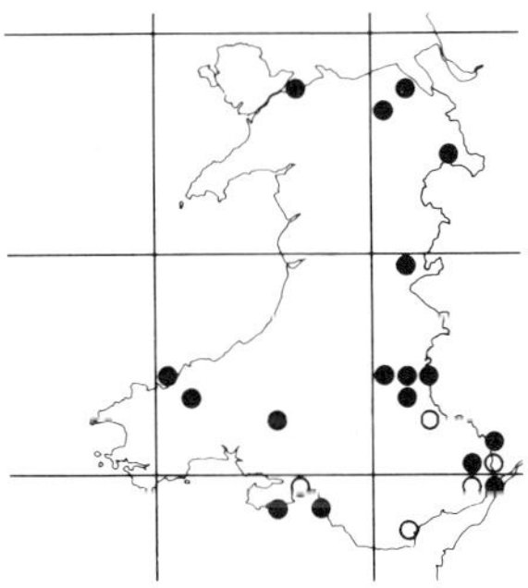

988 Carex strigosa

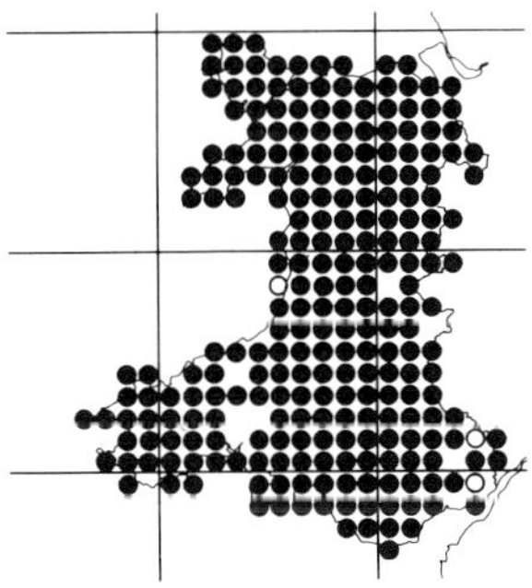

989 Carex panicea

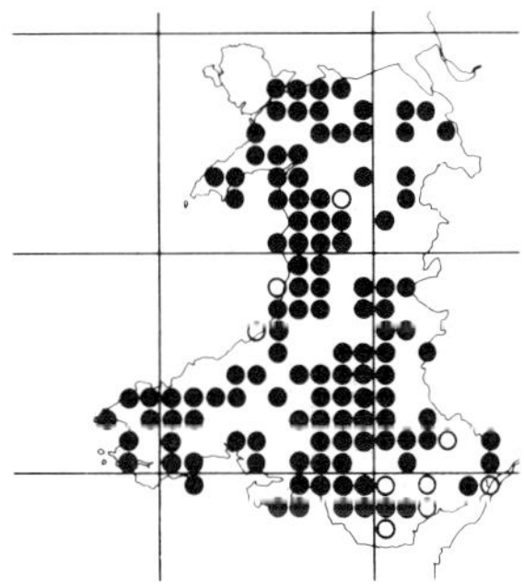

990 Carex laevigata

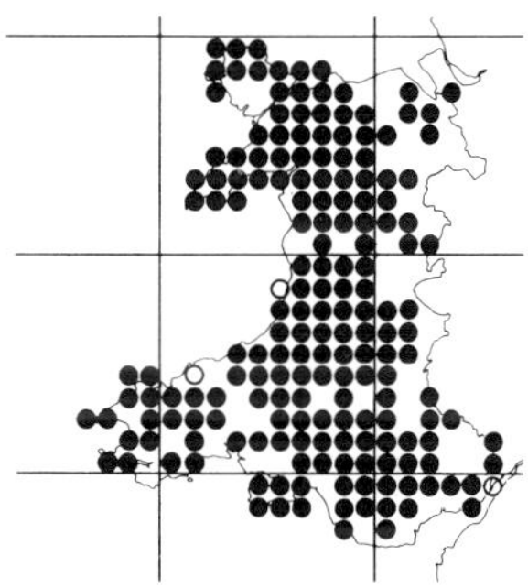

991 Carex binervis

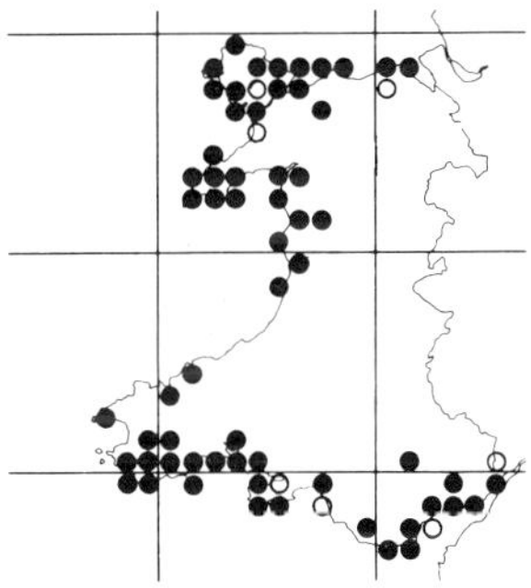

992 Carex distans

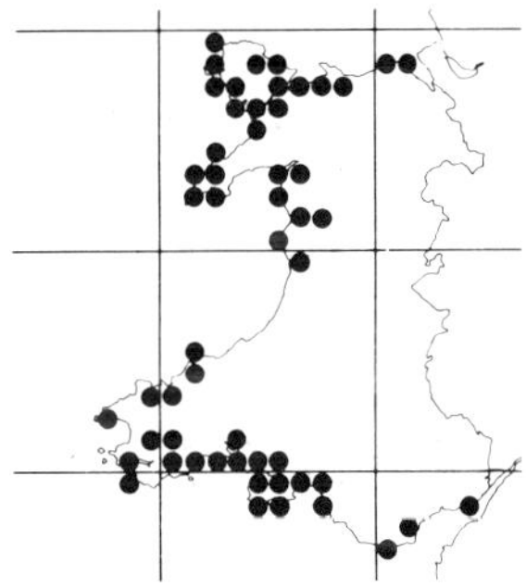

993 Carex extensa

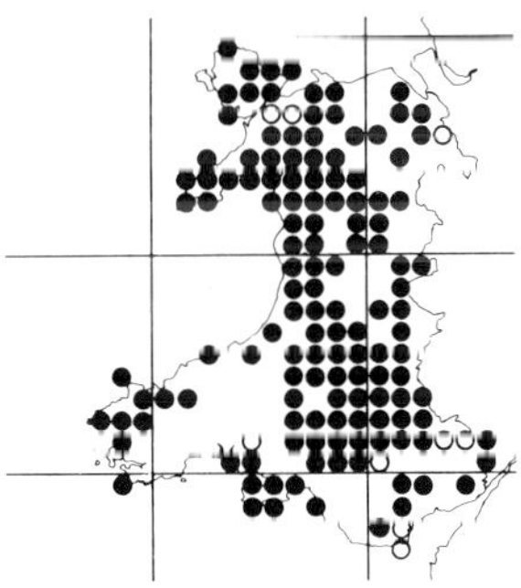

994 Carex hostiana

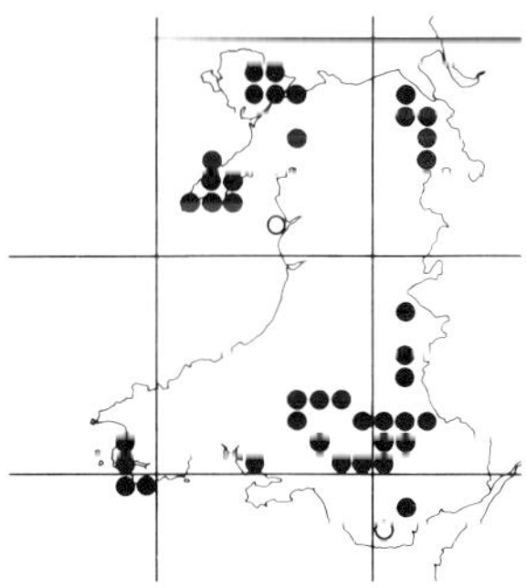

995 Carex lepidocarpa
subsp. lepidocarpa

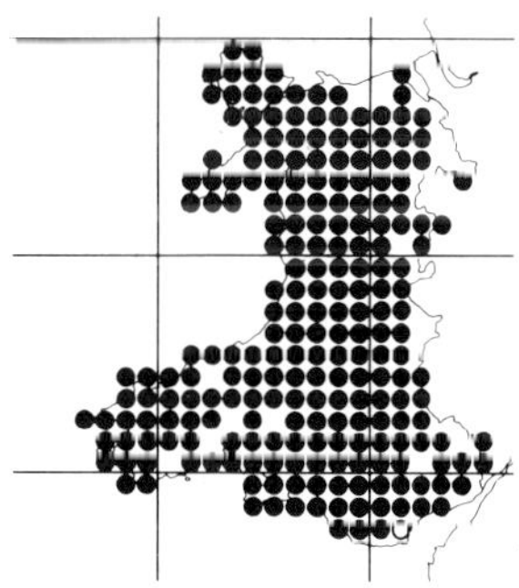

996 Carex demissa

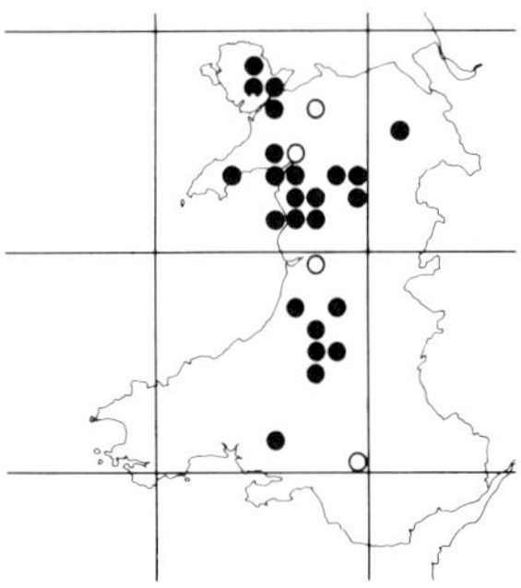

997 Carex demissa × hostiana

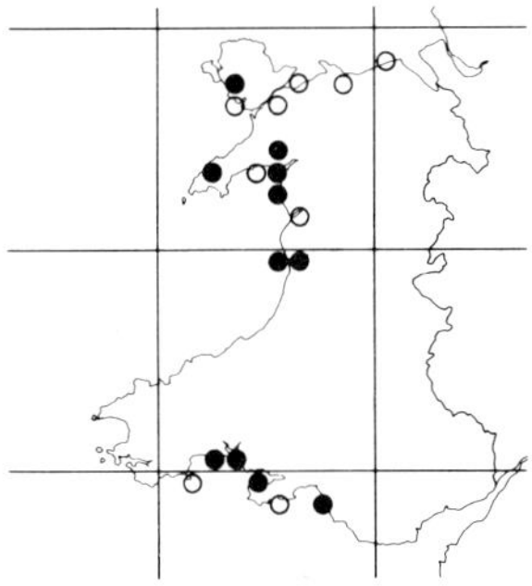

998 Carex serotina subsp. serotina

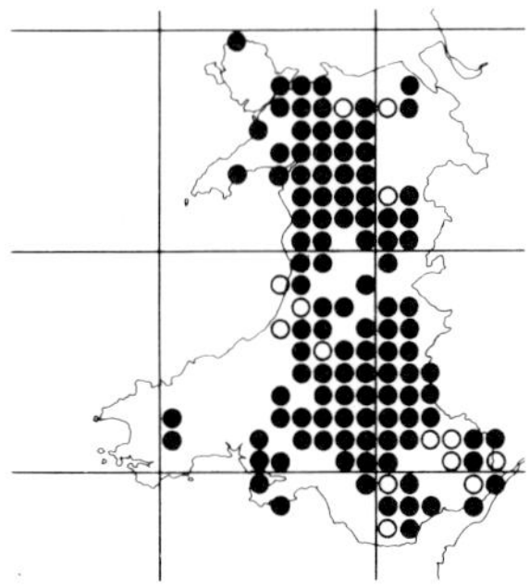

999 Carex pallescens

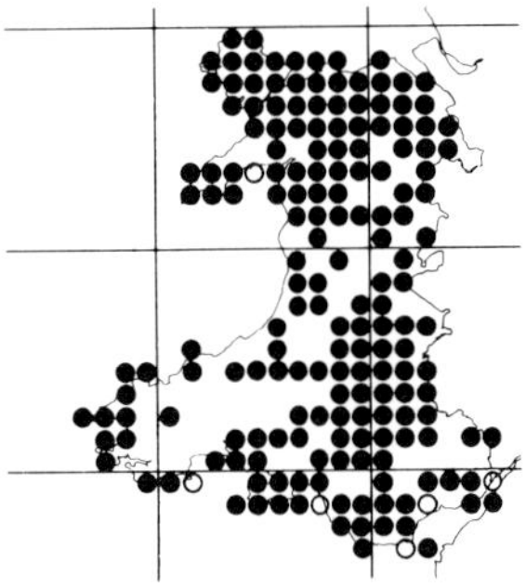

1000 Carex caryophyllea

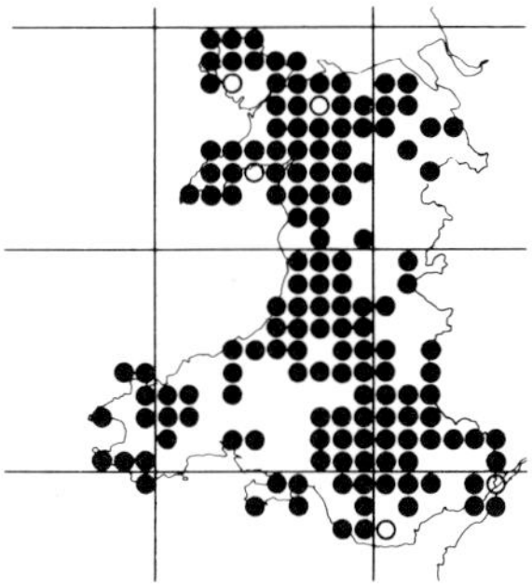

1001 Carex pilulifera

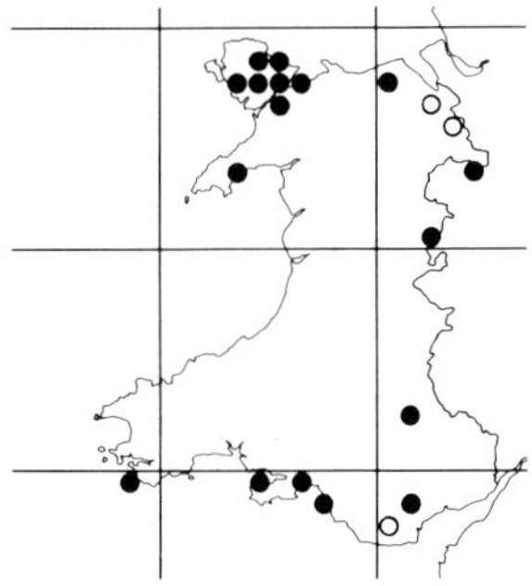

1002 Carex elata

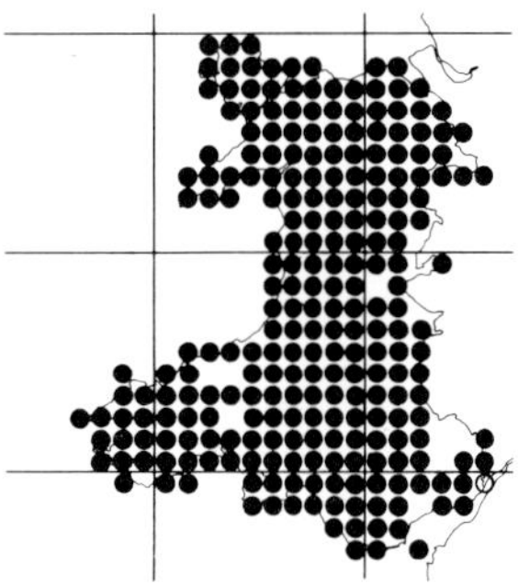

1003 Carex nigra

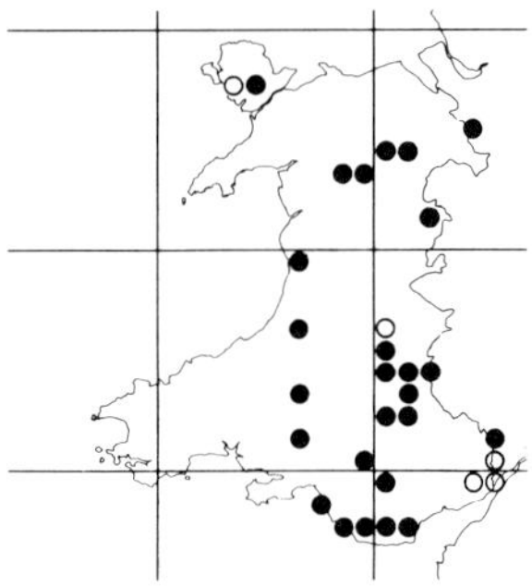

1004 Carex acuta

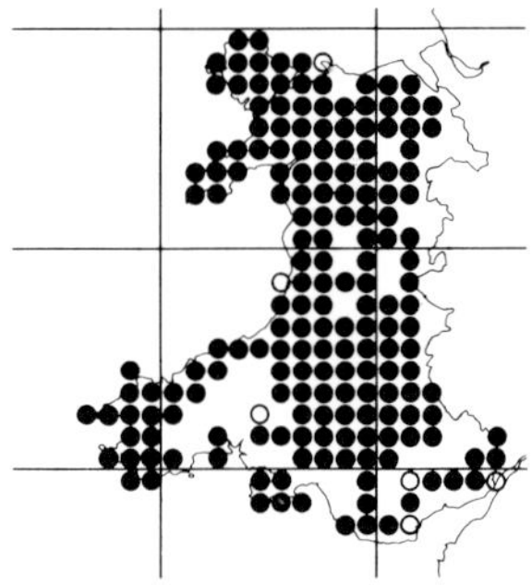

1005 Carex pulicaris

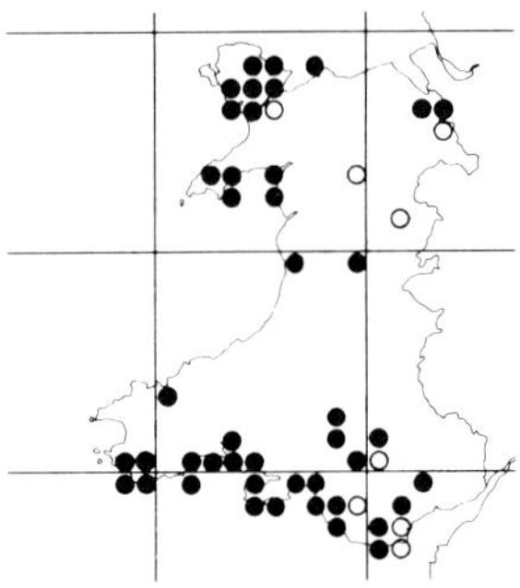

1006 Epipactis palustris

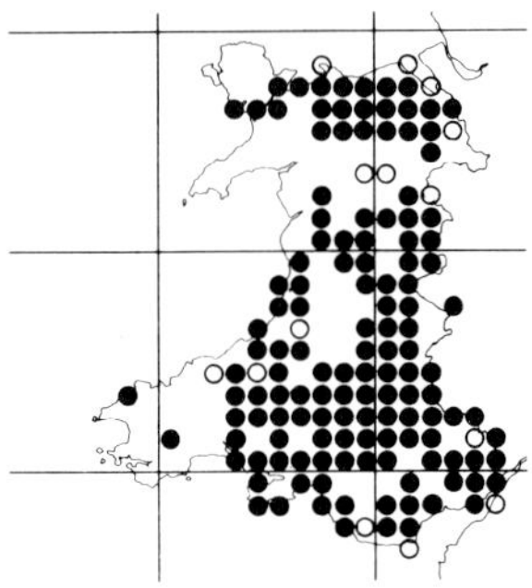

1007 Epipactis helleborine

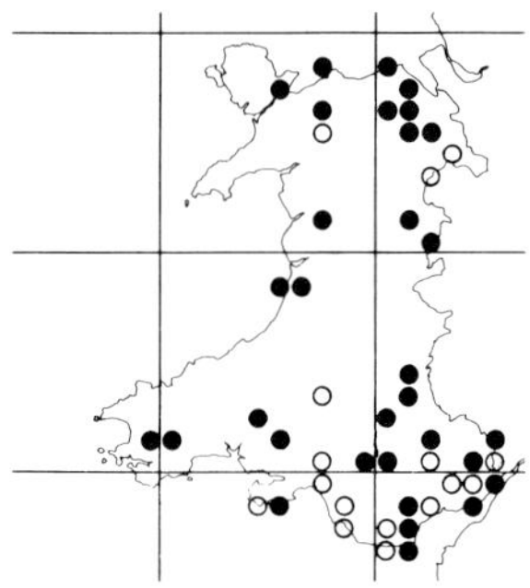

1008 Neottia nidus-avis

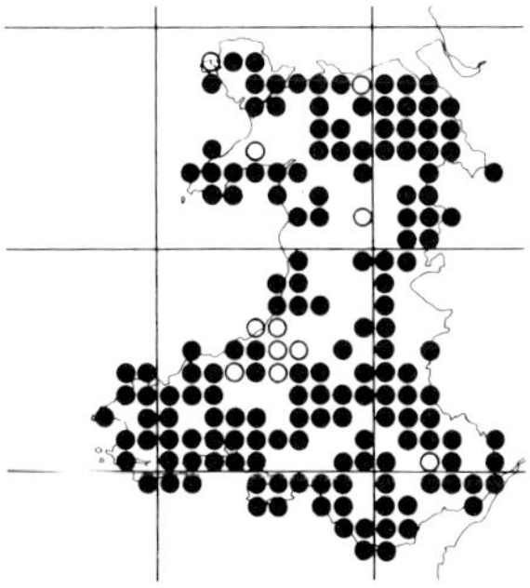

1009 Listera ovata

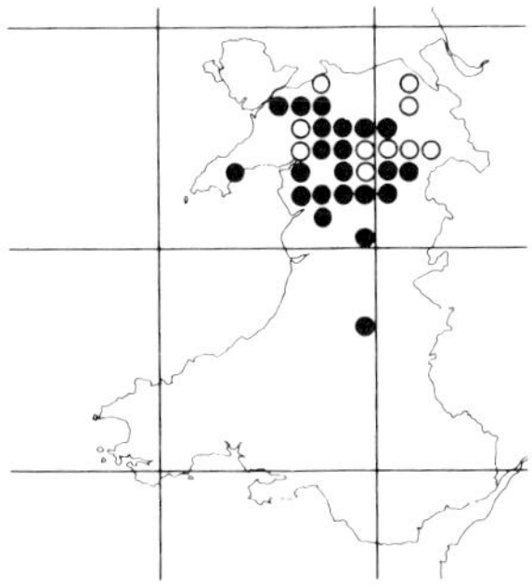

1010 Listera cordata

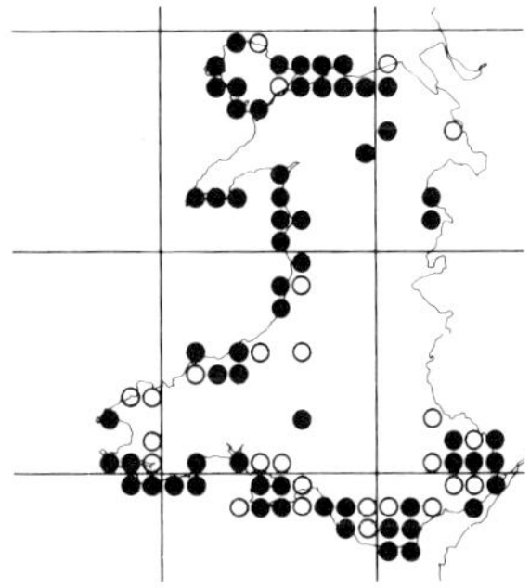

1011 Spiranthes spiralis

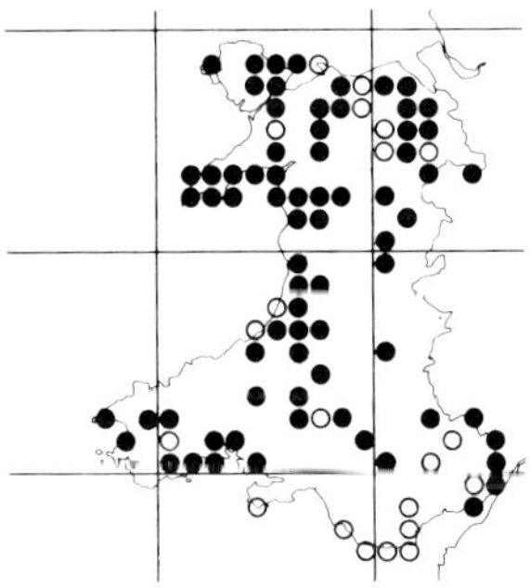

1012 Platanthera bifolia

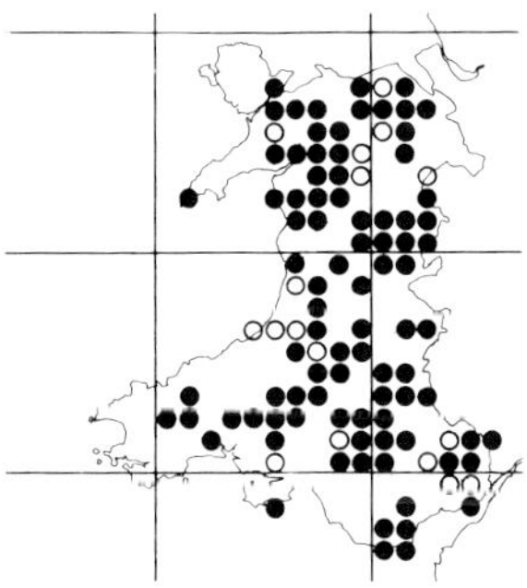

1013 Platanthera chlorantha

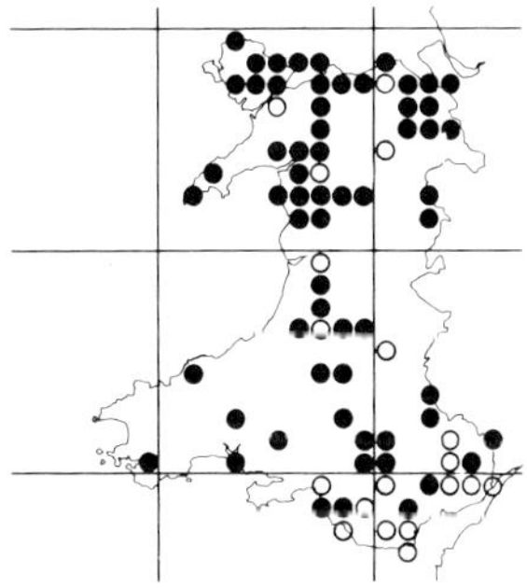

1014 Gymnadenia conopsea subsp. conopsea

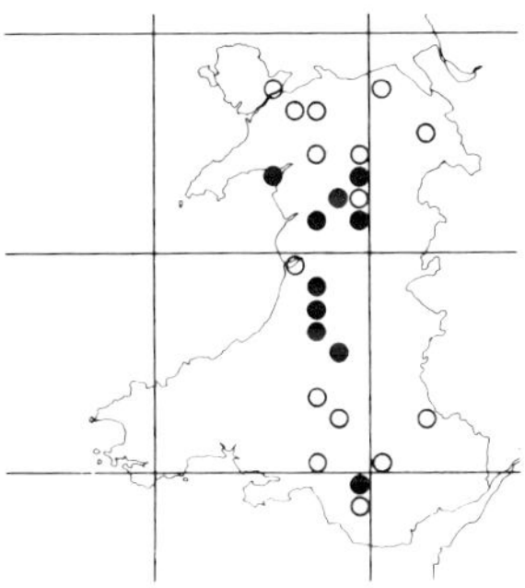

1015 Pseudorchis albida

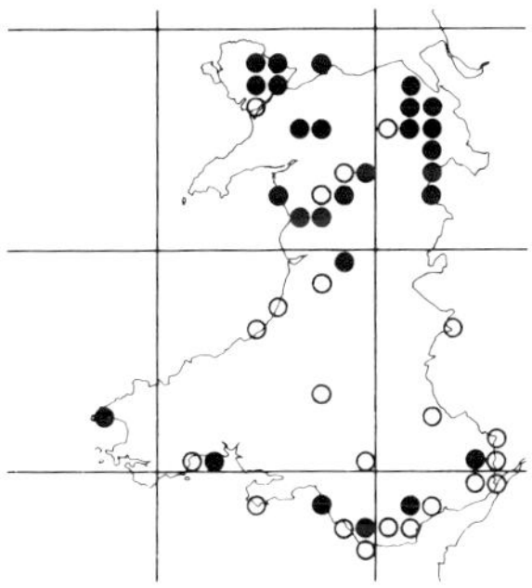

1016 Coeloglossum viride

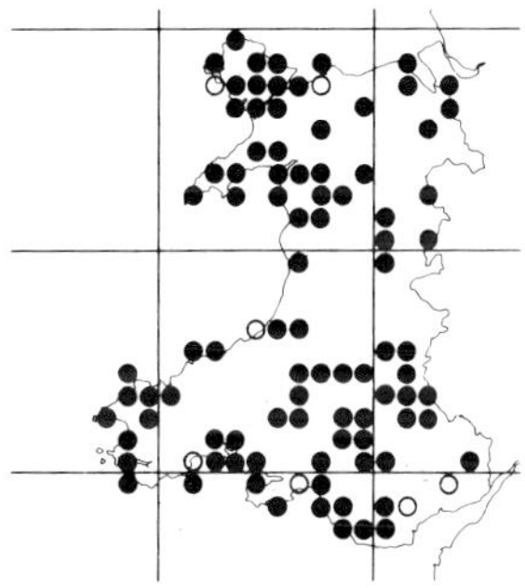

1017 Dactylorhiza incarnata

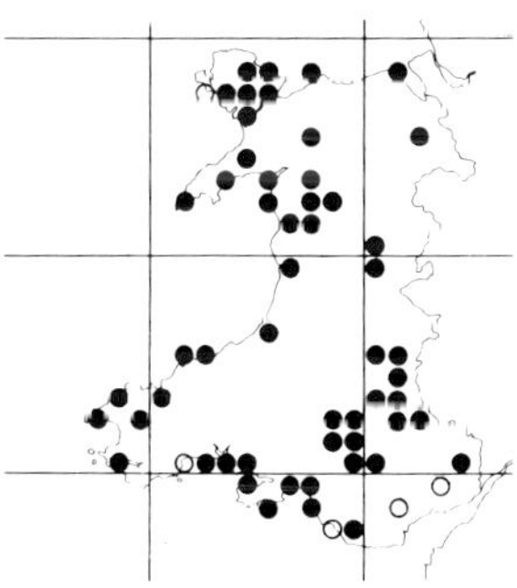

1018 Dactylorhiza incarnata subsp. incarnata

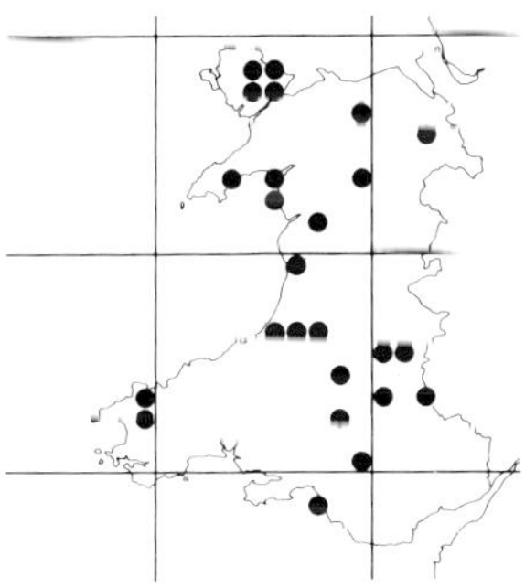

1019 Dactylorhiza incarnata subsp. pulchella

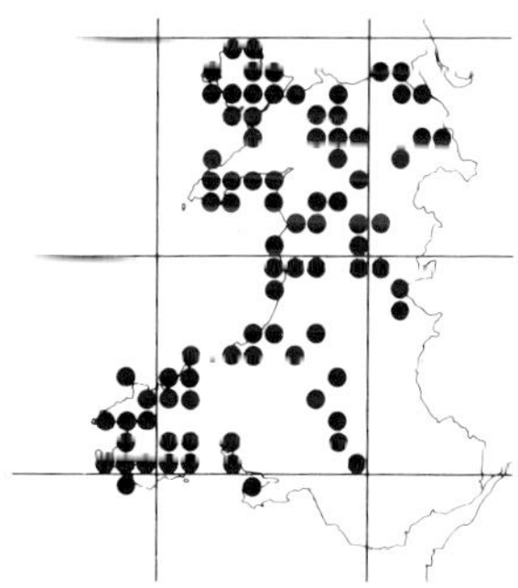

1020 Dactylorhiza majalis subsp. purpurella

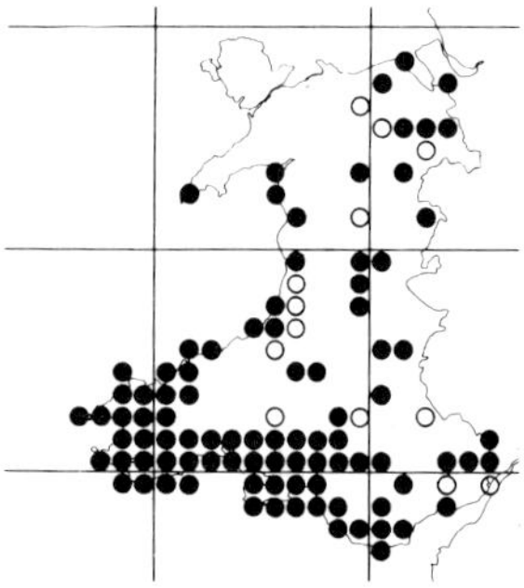

1021 Dactylorhiza majalis subsp. praetermissa

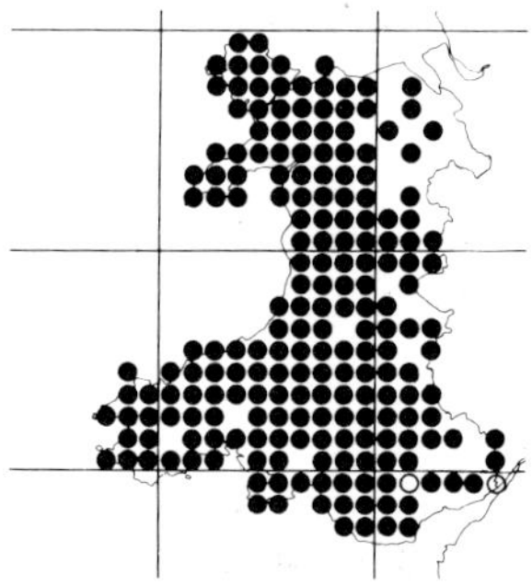

1022 Dactylorhiza maculata

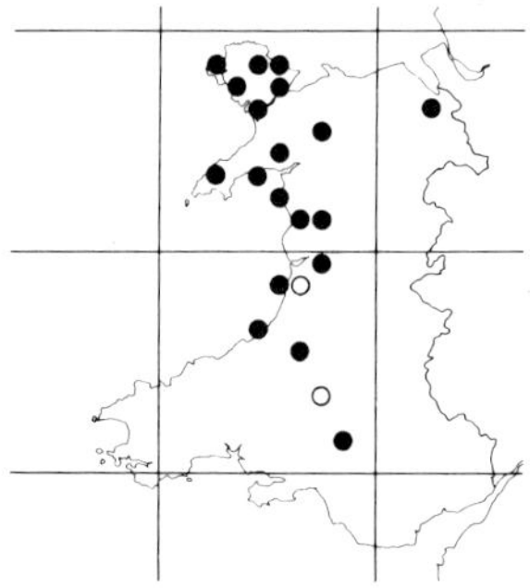

1023 Dactylorhiza maculata × majalis subsp. purpurella

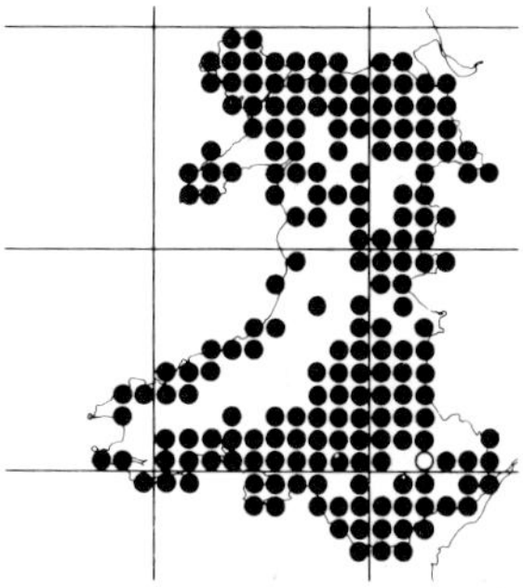

1024 Dactylorhiza fuchsii

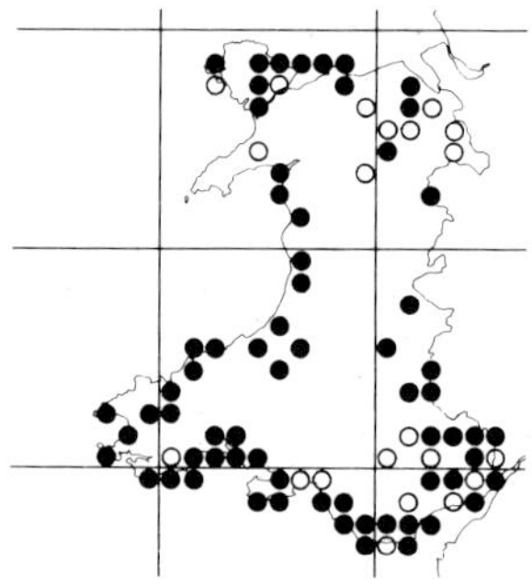

1025 Orchis morio

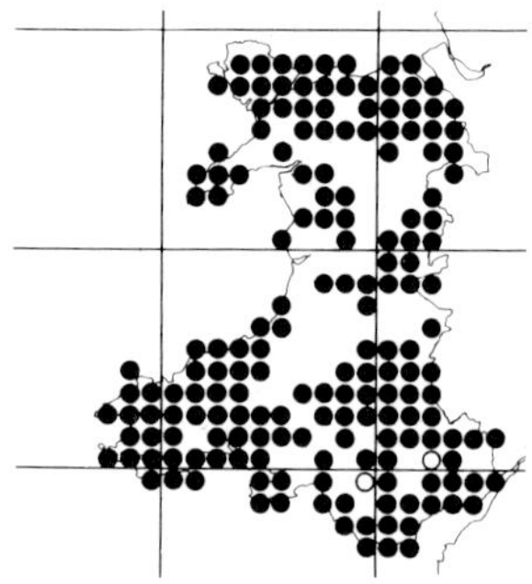

1026 Orchis mascula

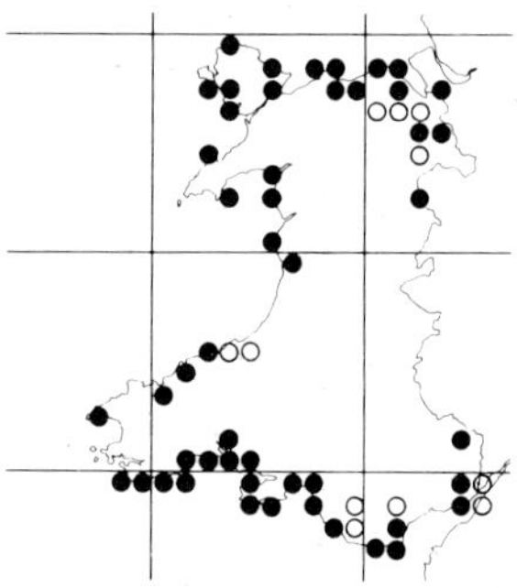

1027 Anacamptis pyramidalis

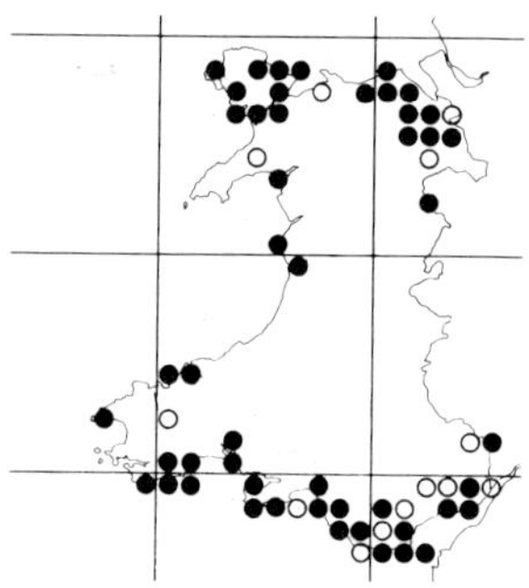

1028 Ophrys apifera

Appendix I: Key to the Abbreviations of Recorders' Initials

This is a complete list of the recorders whose records are referred to in the text; their names are arranged alphabetically in order of their first initials. No account is taken of titles except for ladies and clergymen.

A	Miss Adamson
An	Anonymous
AA	A. Amos
AAi	A. Aiken
AAD	A. A. Dallman
ABC	Miss A. B. Cobbe
ABP	Mrs A. B. Pinkard
ACJ	A. C. Jermy
ACP	Miss A. C. Powell
AD	Miss A. David
ADQA	A. D. Q. Agnew
AEB	A. E. Bush
AEE	A. E. Ellis
AEL	A. E. Lomax
AEW	A. E. Wade
AG	A. L. Grenfell
AGH	A. G. Holder
AGM	A. G. More
AGS	A. G. Spencer
AGSt	A. G. Still
AHGA	A. H. G. Alston
AHLH	A. H. L. Huiskes
AHT	A. H. Trow
AHW	A. H. Wolley-Dod
AJ	A. Jones
AJB	A. J. Bird
AJC	A. J. Crossfield
AJL	A. J. Lees
AJM	A. J. C. Malloch
AJS	A. J. Silverside
AL	Rev. A. Ley
AM	A. Morgan
AMa	Miss A. Mathews
AMc	A. McKenzie
AMP	Mrs A. M. Pell
AMS	A. McG. Stirling
AN	A. Newton
AOC	A. O. Chater
AOH	A. O. Hume
AP	A. Pettet
APC	Miss A. P. Conolly
AR	A. Richards
ARP	A. R. Perry
ARPa	A. R. Parfitt
AS	Ms A. Sumner
AVJ	A. V. Jones
AW	A. Wilson
AWC	A. W. Crump
AWW	A. W. Westrup
B	Rev. W. Bingley
Bry	Mr Bryant
BA	Miss B. Allen
BAM	B. A. Miles
BAW	B. A. Williams
BB	B. Bolton
BDH	B. D. Harding
BF	B. Formstone
BI	B. Ing
BJ	B. Jones
BK	B. King
BL	B. Lloyd
BMCM	Miss B. M. C. Morgan
BMF	Mrs B. M. Frederick
BMH	Mrs B. M. Hope
BMO	Mrs B. M. Oakeshott
BMW	B. M. Watkins
BRF	B. R. Fowler
BS	B. A. Seddon
BW	Mrs B. Welch
Car	Carpenter
Co	Mrs Corrie
CA	C. E. A. Andrews
CAS	C. A. Stace
CASm	C. A. Small
CB	C. Bailey
CBe	Mrs C. C. Begg
CBr	C. E. Britton
CBC	C. B. Clark
CCo	C. Conway
CCB	C. C. Babington
CCJ	C. C. Jones
CCL	C. C. Lacaita
CCT	C. C. Townsend
CDP	C. D. Prime
CDPi	C. D. Piggott
CEB	C. E. Broome
CES	C. E. Salmon
CFH	C. F. Hosking
CGT	C. G. Trapnell
CH	C. Hipkin
CIS	Mrs C. I. Sandwith
CLW	C. L. Walton
CMD	Mrs C. M. Davies
CMG	Miss C. M. Goodman
CML	Mrs C. M. N. LeLacheur
CMT	C. M. Thomas
CPC	C. P. Carroll
CPH	C. P. Hurst
CR	Rev. C. E. Raven
CSc	C. Scotter
CSe	Miss C. Sergeant
CSi	C. A. Sinker
CSN	C. S. Nicholson
CT	C. Titcombe
CTh	C. Thomas
CTV	C. T. Vachell
CW	C. Waterfall
CWe	C. West
CWi	C. Williams
CWW	C. W. Williams
DA	Rev. A. D. Amos
DAJ	D. A. Jones
DAR	D. A. Ratcliffe
DC	Miss D. A. Cadbury
DCo	D. Cottle
DCL	D. C. Lang
DEA	D. E. Allen
DEV	Miss D. E. de Vesian
DGC	D. G. Catcheside
DGJ	D. G. Jones
DH	D. Hamer
DHK	D. H. Kent
DIGV	D. I. Gwynne-Vaughan
DIH	D. I. Hughes
DL	Mrs D. Llewellyn
DM	D. McClintock
DMC	Mrs D. Munro-Cape
DMH	D. M. Howarth
DMHa	D. M. Harrison
DMP	Mrs D. E. M. Paish
DN	D. Neale
DPr	D. Price
DPMG	D. P. M. Guile
DPW	Miss D. P. Williams
DPY	D. P. Young
DR	Dennis Ratcliffe
DRe	Mrs D. Rees
DRH	D. R. Humphries
DRS	D. R. Saunders
DS	Mrs D. Stephenson
DSR	D. S. Ranwell
DTE	D. T. Ettlinger
DTP	D. T. Price
DWB	D. W. Bloodworth
DWC	D. W. Cox

DWH D. W. Horne

E Evans
EA Miss E. Armitage
EAGD E. A. G. Duffy
EB E. Battiscombe
EC E. Cravion
ECh E. Chicken
ECu Mrs E. Curtis
ECH Miss E. C. Howells
ECW E. C. Wallace
ED Miss E. David
EDa Miss E. Davies
EDM E. D. McArthur
EDP Miss E. D. Pugh
EE Mrs E. F. M. Edmonds
EF E. Forster
EFJ Miss E. F. Jones
EFL Rev. E. F. Linton
EFW E. F. Warburg
EGW E. G. Williams
EH E. Huyton
EHC E. H. Chater
EHW E. H. Wolfe
EJ Miss E. Jones
EJo E. Jones
EJL E. J. Lowe
EKH E. K. Horwood
EL E. Lees
ELD E. L. Davies
ELDa Rev. E. L. Dams
EM E. Milne-Redhead
EMJ E. M. Marsden-Jones
EMP Miss E. M. Payne
EMR Miss E. M. Rosser
EMRu E. M. Rutter
EMT Miss E. M. Thomas
EMW E. M. Wakefield
EP E. Phenna
EPo Miss E. Potts
EPAJ E. P. A. Jones
EPB Miss E. P. Beattie
EPE E. P. Evans
ER E. Roberts
ERe Rev. E. M. Reynolds
ESM Rev. E. S. Marshall
EST Miss E. S. Todd
EV Miss E. Vachell
EVW E. V. Watson
EGW E. G. Webster

Fr Miss Fryer
FAC F. A. Currie
FAL F. A. Lees
FAR Rev. F. A. Rogers
FB F. Brent
FBo F. Bos
FCG F. C. G. Gough
FCR F. C. Roper
FD F. Dickenson
FDr F. Drew
FHo Mrs F. Houseman
FHBM Miss F. H. B. Marsh
FHP F. H. Perring
FHW F. H. Webb
FJ F. Jeal
FJH F. J. A. Hort
FLR Mrs F. L. Rees
FLS Rev. F. L. Shaw
FMS F. M. Slater
FNH F. N. Hepper
FP F. Palmer
FRB F. R. Browning
FS F. Stratton
FTJ F. T. Jones
FW Fleming-Williams
FWA F. W. Adams

G Earl of Gainsborough
Gr Mrs Griffith
Gri Griffith
GAH G. A. Holt
GB G. Bentham
GC Mrs G. Crompton
GCD G. C. Druce
GCF Miss G. C. Francis
GCL Mrs G. C. Lambourne
GEv G. Evans
GGG Rev. G. G. Graham
GH G. Hardy
GHB G. H. Bryan
GHT G. H. Tallowin
GK G. Kay
GM G. Milne-Redhead
GMB Miss G. M. Barrett
GMK G. M. Kay
GMR Miss G. M. Reyland
GMSE G. M. S. Easy
GP Miss G. Parry
GR G. Rodway
GRW G. R. Willan
GT G. Taylor
GTr G. Traherne
GTG G. T. Goodman
GW G. Wynne
GWa G. Walters
GWe G. Wesson
GWH G. W. Hollis

H Hodge
HAH H. A. Hyde
HCE H. C. Evans
HCG H. C. Grigg
HCL H. C. Levinge
HD Rev. H. Davies
HDa H. David
HDr H. Drinkwater
HDG H. D. Geldart
HES H. E. Salmon
HFP H. F. Parsons
HGB H. G. Baker
HGH Mrs H. G. Harvey
HH Mrs H. Hipkin
HHa H. Harris
HHan H. Handley
HHH Rev. H. H. Harvey
HHHa H. H. Haines
HHJ H. H. Jones
HHK H. H. Knight
HJF H. J. W. Freer
HJMB H. J. M. Bowen
HJR Rev. H. J. Riddelsdell
HL H. Leyton
HM Miss H. Moseley
HMS H. M. Salmon
HNR H. N. Ridley
HRW H. R. Wakefield
HS H. Spooner
HSo H. Soan
HSM H. S. Marsh
HSR H. S. Redgrove
HT H. Trimen
HTM H. T. Mennell
HTS H. T. Sheldon
HW H. Webster
HWi Rev. H. Williams
HWil H. Williams
HWD H. W. Daltry
HWM H. W. Monington
HWP H. W. Pugsley
HWS Mrs H. W. Sampson

IAW I. A. Williams
IB I. Bullock
IMC I. M. Cross
IMR Miss I. M. Roper
IMV Mrs I. M. Vaughan
IP I. Proctor
IR I. Rees
IRB I. R. Bonner
IRS I. R. Smith
ISN Mrs I. S. Nicholl
IWC I. W. Callan

Ja Miss Jackson
JAG Mrs J. A. Green
JAR J. A. Ratter
JAW J. A. Webb
JAWh J. A. Wheldon
JAWhe J. A. Whellan
JB J. Banks
JBa J. Backhouse

JBe	Mrs J. Berney
JBy	J. Bailey
JBD	J. B. Davy
JBL	Rev. J. B. Lloyd
JBM	Miss J. B. Miller
JC	J. Comont
JCu	J. Cullen
JDB	Miss J. D. Butler
JDG	J. D. Grose
JDM	J. D. Massey
JEA	J. E. Arnett
JEB	Rev. J. E. Beckerlegge
JEBi	J. E. Bicheno
JEBo	J. E. Bowman
JED	J. E. Dandy
JEG	J. E. Griffith
JEH	J. E. Halfhide
JEL	J. E. Lousley
JESJ	J. E. Sibbering-Jones
JEW	J. E. Woodhead
JFGC	J. F. G. Chapple
JFH	J. F. Hall
JFT	J. F. Thomas
JG	J. D. Grey
JGr	J. Griffith
JGH	J. G. Hodgson
JH	J. Holland
JHo	J. D. Hooker
JHC	J. H. Clark
JHL	J. H. Lewis
JHP	J. H. Payne
JHR	J. H. Roberts
JHS	J. H. Salter
JHTa	J. H. Tallis
JIL	Mrs J. I. Littley
JJD	Rev. J. Jones-Davies
JKM	J. K. Morton
JL	J. Loveit
JLB	J. L. Bruce
JLOL	J. L. O'Loughlin
JLW	J. L. Williams
JM	J. MacQueen
JMo	J. Morgan
JMB	J. M. Brummitt
JMG	J. M. Griffith
JMT	J. M. Thomas
JND	J. N. Davies
JNM	J. N. Mills
JO	J. Oates
JOM	J. O. Mountford
JPAJ	J. P. A. Jones
JPC	J. P. Curtis
JPS	J. P. Savidge
JR	J. Ray
JRa	J. E. Raven
JRe	J. Rees
JRg	J. Rogerson
JRC	J. R. Cadman
JRG	J. R. Gates
JRP	J. R. Palmer
JRS	J. R. Shepherd
JS	J. Storrie
JSi	J. Sidebotham
JSy	Mrs J. Sykes
JSH	Rev. J. S. Henslow
JSLG	J. S. L. Gilmour
JSP	J. S. Pritchard
JT	Mrs J. Tann
JTCW	J. T. C. Wilkins
JTCWi	J. T. C. Williams
JW	J. Woods
JWC	J. W. Carr
JWD	J. W. Davies
JWDo	J. W. Donovan
JWG	J. W. G. Gutch
JWL	J. W. Lewis
JWLo	J. W. Long
Kn	Knowlton
KA	Mrs K. Adams
KEB	K. E. Bull
KG	K. M. Goodway
KGA	K. G. Allenby
KJ	K. Jones
KMB	Mrs K. M. Benoit
KMG	Miss K. M. Green
KP	Mrs K. Pickles
KPe	K. Pearce
KR	Mrs K. Richards
KRo	K. Rostański
KS	Mrs K. M. Stevens
LAE	Lady A. Eliot
LB	Miss L. Browning
LE	L. E. Evans
LF	Miss L. Farrell
LHP	L. W. H. Pegler
LJL	L. J. Larsen
LKS	Lady K. Stanley
LL	L. Larsen
LP	L. Pierce
LPe	L. Peddle
LR	Mrs L. Rathbone
LT	Rev. L. Thomas
LW	Miss L. Whitehead
LWD	L. W. Dillwyn
LWF	Miss L. W. Frost
M	J. Motley
McW	Miss M. McC. Webster
MAL	M. A. Lawson
MB	Mrs M. Barnes
MC	M. Cubitt
MCh	Miss M. Chorley
MCo	Miss M. Cobbe
MCr	Mrs M. Craster
MCFP	M. C. F. Proctor
MCK	Miss M. C. Knowles
MD	M. Davies
MDa	M. Davids
MDT	Mrs M. D. Thomas
ME	M. C. R. Edgell
MG	Mary E. Gillham
MGe	M. George
MGBE	Miss M. G. Bentham-Evans
MGC	M. G. Cahn
MH	Mrs M. Humphrey
MHa	Mrs M. Hanford
MHA	Miss M. H. Asterley
MHB	Miss M. H. Bigwood
MHS	Miss M. H. Sykes
MJG	M. J. Godfrey
MJM	M. J. Morgan
MJW	M. J. Wigginton
MM	M. Morris
MP	M. Porter
MPa	Miss M. Patterson
MPe	Mrs M. Perring
MPl	M. Plews
MR	Mrs M. A. E. Richards
MRo	M. Roberts
MRS	Mrs M. R. Shanahan
MSP	Mary S. Percival
MSW	Mrs M. S. Wedgewood
MTH	Miss M. T. Hamilton
MWa	M. Wade
MWe	M. Wenham
N	Mrs New
NB	N. Brown
NC	Mrs N. Cragg
NDS	Rev. N. D. Simpson
NGM	N. G. Marchant
NH	N. T. H. Holmes
NJ	N. Jardine
NMP	N. M. Pritchard
NYS	N. Y. Sandwith
OHB	O. H. Black
OW	Mrs O. Webb
Pa	Mrs Paterson
PC	P. Carpenter
PCH	P. C. Hall
PD	P. Day
PDe	P. Denne
PDS	P. D. Sell
PE	P. Evans
PFY	P. F. Yeo
PH	P. Halligey
PHD	P. H. Dunn
PHH	P. H. Holland
PJ	Miss P. Jones

PLT	P. L. Thomas
PM	Miss P. Morgan
PMB	P. M. Benoit
PMH	P. M. Hall
PMS	P. M. Smith
PN	P. J. M. Nethercott
PS	Ms P. A. Sims
PWa	Mrs P. Walton
PWR	P. W. Richards
QC	Q. Cronk
QONK	Q. O. N. Kay
RA	R. Arthur
RAB	R. A. Boniface
RAC	R. A. Conibear
RAG	R. A. Graham
RB	R. Brown
RBa	R. Baker
RBr	R. Browne
RC	Mrs R. Cotton
RCLH	R. C. L. Howitt
RD	R. Davies
RDr	R Drane
RDP	R. D. Pryce
REH	R. E. Hughes
RFM	R. F. May
RFT	R. F. Towndrow
RGE	R. G. Ellis
RGJ	R. G. Jones
RGW	R. G. Woods
RHD	R. H. Day
RHH	R. H. Hatton
RHR	R. H. Roberts
RIG	R. I. Green
RJHL	R. J. H. Lloyd
RJP	R. J. Pankhurst
RKB	R. K. Brummitt
RL	R. Lewis
RLa	R. Lawton
RLB	R. C. L. Burges
RLS	R. L. Smith
RM	R. Melville
RMd	R. Meade
RMe	R. Meinertzhagen
RMo	R. Moore
RMB	R. M. Burton
RMH	R. M. Harley
RML	R. M. Lockley
RMS	R. M. Speed
RPB	R. P. Bowman
RPT	R. P. Turner
RR	R. Ross
RSF	R. S. R. Fitter
RT	R. Tudor
RU	R. A. E. Ubsdell
RVI	R. V. Innes
RWD	R. W. David
RWR	R. W. Rickards
SAN	S. A. Naylor
SBE	S. B. Evans
SCC	S. C. Clark
SD	S. Daggitt
SDW	S. D. Ward
SF	Mrs S. Furse
SGC	S. G. Charles
SGH	S. G. Harrison
SH	S. Hamilton
SHa	Miss S. Haines
SHB	S. H. Bickham
SIL	Mrs S. I. Leitch
SJ	S. Jones
SJH	S. J. Hughes
SMW	S. M. Walters
SNT	S. N. Tallowin
SP	S. Powell
ST	Mrs S. Titcombe
STy	S. Tyler
SW	S. Waldron
Th	E. J. H. & Miss F. M. Thomas
Tr	Mrs Traherne
TAS	T. A. Stephens
TAWD	T. A. W. Davis
TB	T. Butler
TBF	T. B. Flower
TC	T. Chandler
TCh	T. Chapman
TE	T. Edmondson
TF	T. Fox
TG	T. Glynne
TGa	T. Galway
TGD	T. G. David
TGE	T. G. Evans
TGP	T. Gambier-Parry
TH	Rev. T. Hutchinson
THB	T. H. Blackstock
THE	T. H. Ellis
THT	T. H. Thomas
TJ	T. Johnson
TJW	T. J. Wall
TR	T. Ruddy
TRL	T. R. Lovery
TSt	Rev. T. Stephenson
TT	T. Trought
TW	T. Walford
TWe	T. Westcombe
TWB	T. W. Barker
TWG	T. W. Gissing
TWH	T. W. Hancock
UKD	Miss U. K. Duncan
V	Mrs M. Voyce
VCB	Mrs V. C. Baldwin
VG	Miss V. Gordon
VGE	Mrs V. G. Ellis
VJM	Miss V. J. Macnair
VMP	Mrs V. M. Peel
WAL	Rev. W. A. Leighton
WAS	W. A. Shoolbred
WASl	W. A. Sledge
WAV	W. A. Vice
WBR	W. B. Russell
WBT	W. B. Turrill
WBW	Mrs W. B. Watt
WBY	W B. Yates
WCo	W. M. Condry
WCB	W. C. Barton
WEH	W. E. Hughes
WEJ	Rev. W. E. Jones
WF	Miss W. Frost
WGG	W. G. Griffiths
WGT	W. G. Travis
WH	W. Hodge
WHH	W. H. Hardaker
WHP	Rev. W. H. Painter
WHPu	Rev. W. H. Purchas
WHW	W. H. A. Williams
WI	W. Ingham
WIJ	W. I. Jones
WL	W. Lovegrove
WLB	W. P. J. Le Brocq
WM	W. Morris
WMD	W. M. Davies
WN	W. Nelmes
WP	W. Pamplin
WRL	Rev. W. R. Linton
WSL	W. S. Lacey
WT	W. Turton
WW	W. Wilson
WWh	W. Whitewell
WWB	W. W. Boucher
WWM	Rev. W. W. Mason

Appendix II: List of References

The following list contains only those books, papers and journals that are referred to in the text (by number) or in the Introduction. It is not a complete bibliography of Welsh botany. Some entries are inserted twice, e.g. under author and journal, they have the same reference number which is in italics when out of sequence. The chapter on Climate has its own list of references (p.32).

—. Anonymous (1969). *Enwau Planhigion.* Cardiff.

—. Anonymous (Ed.) (1971). *Reader's Digest Encyclopaedia of Garden Plants and Flowers.* London.

1. Barker, T. W. (n.d., [1905]). *Handbook to the Natural History of Carmarthenshire.* Carmarthen.

2. Barnes, Mrs. M. (1965). Personal communication.

130. Barrett, W. B. (1885). A Contribution towards a Flora of Breconshire. *Journal of Botany,* **23.**

—. Bendelow, V. C. & Hartnup, R. (1980). *Climatic Classification of England and Wales.* Soil Survey Technical Monograph, No. 15. Harpenden.

231. Bennett, A. (1905). Supplement to Topographical Botany. Ed. 2. *Journal of Botany,* **43** (suppl.).

232. Bennett, A., Salmon, C. E. & Matthews, J. R. (1929-30). Second Supplement to Watson's 'Topographical Botany'. *Journal of Botany,* **67** & **68** (suppl.).

3. Benoit, P. M. (1955-81). Personal communications.

4. Benoit, P. M. & Richards, M. (1963). *A Contribution to a Flora of Merioneth.* Ed. 2. Haverfordwest.

5. Benoit, P. M. *et al.,* Unpublished maps prepared for a Flora of Gwynedd project in the 1970's, now abandoned. Copies in Library, National Museum of Wales.

6. Best, F. C. (1972). Personal communication.

7. Biological Records Centre, Monks Wood, Huntingdon. (1955-81).

Botanical Exchange Club. Reports.

8. — 1867.
9. — 1898.
10. — 1906.
11. — 1909.
12. — 1912.

Botanical Society and Exchange Club. Reports.

13. — **4,** 1914.
14. — 1915.
15. — 1916.
16. — **5,** 1917.
17. — 1918.
18. — 1919.
19. — **6,** 1920.
20. — 1921.
21. — 1922.
22. — **7,** 1923.
23. — 1924.
24. — 1925.
25. — **8,** 1926.
26. — 1927.
27. — 1928.

28. — **9,** 1929.
29. — 1930.
30. — 1931.
31. — **10,** 1932.
32. — 1933.
33. — 1934.
34. — **11,** 1935.
35. — 1936.
36. — 1937.
37. — **12,** 1938.
38. — 1939.
39. — 1940.
40. — 1941.
41. — 1942.
42. — 1943.
43. — 1944.
44. — **13,** 1945.
45. — 1946
46. — 1947.
47. *Botanical Record Club. Reports*, 1873-86.

Botanical Society of the British Isles.

— *Abstracts.*

48. — Nos. **1-11,** 1971-81.

— *News.*

49. — Nos. **1-16,** 1972-77.
50. — **17-25,** 1977-80.
51. — **26-28,** 1980-81.

— *Proceedings.*

52. — **1,** 1954-55.
53. — **2,** 1956-57.
54. — **3,** 1958-60.
55. — **4,** 1960-62.
56. — **5,** 1963-64.
57. — **6,** 1965-67.
58. — **7,** 1967-69.

— *Welsh Bulletin.*

59. — Nos. **1-31,** 1964-79.
60. — **32-33,** 1980.
61. — **34-35,** 1981.

— *Year Book.*

62. — 1949.
63. — 1950.
64. — 1951.

65. — 1952.
66. — 1953.
67. Brummitt, J. M. (1964-81). Personal communications.
68. *Caernarfon and Denbigh Herald*, 7/10/1977.

Cardiff Naturalists' Society. Transactions.

69. — **8,** 1876.
70. — **22,** 1890.
—. Carter, P. W. (1946). Botanical exploration in Montgomeryshire. *Montgomeryshire Collections*, **49.**
—. — (1950). A history of botanical exploration in Radnorshire. *The Radnorshire Society. Transactions*, **20.**
—. — (1950). Botanical exploration in Cardiganshire. *Ceredigion*, **1.**
—. — (1951 & 1957). Botanical exploration in Carmarthenshire. *The Carmarthen Antiquary*, **2.**
—. — (1952). Some account of the botanical exploration of Anglesey. *Anglesey Antiquarian Society and Field Club. Transactions.*
—. — (1955). Some account of the History of Botanical Exploration in Glamorgan. *Cardiff Naturalists' Society. Transactions*, **82.**
—. — (c.1955). *Some account of the history of botanical exploration in Pembrokeshire.* Unpublished. Typescript in Library, National Museum of Wales.
—. — (1955). Botanical exploration in Merionethshire. *The Merioneth Miscellany*, **1.**
—. — (1955-56). Some account of the botanical exploration of Caernarvonshire. *Caernarvonshire Historical Society. Transactions*, **16** & **17.**
—. — (1956). Notes on the botanical exploration of Flintshire. *Publications of the Flintshire Historical Society*, **16,** *Flintshire Miscellany*, No. 1.
—. — (1957). A history of botanical exploration in Brecknock. *Brycheiniog*, **3.**
—. — (1960). Some account of the Botanical Exploration of Denbighshire. *Denbighshire Historical Society. Transactions*, **9.**
—. — (c.1960) *Monmouthshire.* Unpublished. Typescript in Library, National Museum of Wales.
71. Chater, A. O. (1964-81). Personal communications.
72. Clapham, A. R., Tutin, T. G. & Warburg, E. F., (1962). *Flora of the British Isles.* Ed. 2. Cambridge.
—. — (1981). *Excursion Flora of the British Isles.* Ed. 3. Cambridge.
73. Clement, E. J. (1975-81). Personal communications.
74. Conolly, Miss A. P. (1955-81). Personal communications.
75. Curtis, J. P. (1970-81). Personal communications.
140. Dallman, A. A. (1907). Notes on the Flora of Flintshire. *Journal of Botany*, **45.**
141. — (1908). Notes on the Flora of Flintshire. *Journal of Botany*, **46.**
143. — (1910). Notes on the Flora of Flintshire. *Journal of Botany*, **48,** (suppl.).
144. — (1911). Further Notes on the Flora of Flintshire. *Journal of Botany*, **49.**
144. — (1911). Notes of the Flora of Denbighshire. *Journal of Botany*, **49.**
146. — (1913). Further Notes on the Flora of Denbighshire. *Journal of Botany*, **51,** (suppl.).
76. Dandy, J. E. *Potamogeton* card index in the British Museum (Natural History).
77. — (1958). *List of British Vascular Plants.* London.
78. Davies, H. (1813). *Welsh Botanology.* London.
79. Davis, T. A. W. (1970). *Plants of Pembrokeshire.* Haverfordwest.
80. — (1957-80). Personal communications.
81. Dillwyn, L. W. (1848). *Materials for a Fauna and Flora of Swansea.* London.
82. Turner, D. & Dillwyn, L. W. (1805). *The Botanists Guide Through England and Wales.* London.

—. Dony, J. G., Perring, F. H. & Rob, C. M. (1980). *English Names of Wild Flowers.* Cambridge.

83. Drinkwater, H. (c.1920). *Catalogue of Drawings.* Unpublished. Manuscript in Library, National Museum of Wales.

84. Druce, G. C. (1928). *British Plant List.* Ed. 2. Arbroath.

85. — (1932). *The Comital Flora of the British Isles.* Arbroath.

86. Edlin, H. L. (Ed.) (1961). *Glamorgan Forests.* London.

87. — (1969). *Snowdonia Forest Park Guide.* London.

88. — (1974). *Dean Forest and Wye Valley.* London.

89. — (1976). *Cambrian Forests.* London.

90. Edmondson, T. (1970-81). Personal communications.

—. Ellis, R. G. (1972). Plant Hunting in Wales — I. *Amgueddfa,* **10.**

—. — (1973). Plant Hunting in Wales — II. *Amgueddfa,* **13.**

—. — (1974). Plant Hunting in Wales — III. *Amgueddfa,* **16.**

91. Evans, T. G. (1970-81). Personal communications.

92. Evans, S. B. (1976-81). Personal communications.

93. Field records, various recorders. (1950-81).

94. *Field Studies,* **1,** 1961.

95. Flora of Glamorgan, field records. 1970-81.

96. Fox, T. (1978). Personal communications.

97. *Freshwater Biology,* **2,** 1972.

98. *Garden, The. Journal of the Royal Horticultural Society,* **106,** 1981.

94. George M. (1961). The Flowering Plants and Ferns of Dale, Pembrokeshire. *Field Studies,* **1.**

—. Gerarde, J. (1597). *The Herball or General History of Plants.* Ed. 1. London.

124. — (1633). *The Herball or General History of Plants.* Ed. 2. Very much enlarged and amended by T. Johnson. London.

99. Gillham, M. E. (1955-81). Personal communications.

—. — (1977). *The Natural History of Gower.* Cowbridge.

—. — (1982). *Swansea Bay's Green Mantle.* Cowbridge.

100. Godfrey, M. J. (1933). *Monograph and Iconograph of Native British Orchidaceae.* Cambridge.

101. Griffith, J. E. (n.d., [1895]). *The Flora of Anglesey and Caernarvonshire.* Bangor.

102. Guile, D. P. M. (1952). *Ecological Survey of the Brecon Beacons National Park.* Ph.D. Thesis, in Library, National Museum of Wales.

—. Hamilton, S. (1909). *The Flora of Monmouthshire.* Newport.

103. Harris, H. (1905). *The Flora of The Rhondda.* Bristol.

The Herbarium of:—

104. — University College of Wales, Aberystwyth, **ABS.**

105. — Ulster Museum, Belfast, **BEL.**

106. — University, Birmingham, **BIRM.**

107. — British Museum (Natural History), London, **BM.**

108. — University, Cambridge, **CGE.**

109. — Royal Botanic Gardens, Kew, **K.**

110. — Merseyside Museum, Liverpool, **LIV.**

111. — Motely, J. formerly in The Royal Institution of South Wales, Swansea, **SWA.** ? lost or destroyed 1940-45.

112. — National Museum of Wales, Cardiff, **NMW.**
113. — University, Oxford, **OXF.**
114. — University College of North Wales, Bangor, **UCNW.**
115. — University College of Swansea and the Royal Institution of South Wales, Swansea, **SWA.**
116. Hignett, M. & Lacey, W. S. (1977). *Plants of Montgomeryshire.* Welshpool.
117. Hipkin, C. & H. (1970-81). Personal communications.
118. Holmes, N. T. H. (1979-81). Records ex River Survey of England and Wales.
119. Hughes, W. E. (1965-70). Personal communications.
120. Hyde, H. A. & Harrison, S. G. (1977). *Welsh Timber Trees.* Ed. 4. Cardiff.
121. Hyde, H. A. & Wade, A. E. (1934). *Welsh Flowering Plants.* Ed. 1. Cardiff.
122. — (1957). *Welsh Flowering Plants.* Ed. 2. Cardiff.
123. Jodrell Botanic Garden, University of Manchester.
124. Johnson, T. (Ed.) (1633). *Gerarde's Herball or General Historie of Plants.* Ed. 2. London.
125. — (1641). *Mercurii Botanici Pars Altera.* London.
126. Jones, D. A. (1898). *A Handbook to the Botany of Merionethshire, North Wales.* Unpublished. Manuscript Library, National Museum of Wales.

Journal of Botany.

127. — **1,** 1863.
128. — **19,** 1881.
129. — **22,** 1884.
130. — **23,** 1885.
131. — **30,** 1892.
132. — **33,** 1895.
133. — **37,** 1899.
134. — **39,** 1901.
135. — **40,** 1902.
136. — **41,** 1903.
137. — **42,** 1904.
138. — **43,** 1905.
139. — **44,** 1906.
140. — **45,** 1907.
141. — **46,** 1908.
142. — **47,** 1909.
143. — **48,** 1910.
144. — **49,** 1911.
145. — **50,** 1912.
146. **51,** 1913.
147. — **55,** 1917.
148. — **58,** 1920.
149. — **69,** 1931.
150. — **70,** 1937.
151. — **80,** 1942.

Journal of Ecology.

152. — **20,** 1932.
153. — **42,** 1954.
154. — **46,** 1958.
155. — **57,** 1969.
156. — **67,** 1979.
157. Kay, Q. O. N. (1969-81). Personal communications.
158. Kent, D. H. (1978). Personal communication.
159. Knight, Mrs. C. (1977). Personal communication.

—. Lightfoot, J. (1777). *Flora Scotica.* London.

—. Lousley, J. E. (1953). The Recent influx of Aliens into the British Flora. In *The Changing Flora of Britain.* Lousley, J. E. (ed.). Oxford.

160. Marsden-Jones, E. M. & Turrill, W. B. (1954). *British Knapweeds.* London.
161. May, R. F. (1967). *A List of the Flowering Plants and Ferns of Carmarthenshire.* Haverfordwest.
162. McClintock, D. (1957). *Supplement to Collins Pocket Guide to Wild Flowers.* Platt.

—. McClintock, D. & Fitter, R. S. R. (1956). *The Pocket Guide to Wild Flowers.* London.

Montgomeryshire Collections.

163. — **5,** 1872.
164. — **15,** 1882.
165. Morgan, J. (1892). *The Flora of Llanbrynmair.* Oswestry.
166. Morris, M. (1955-81). Personal communications.
167. *The Naturalist,* **5,** 1879.
168. *Nature,* **264**, 1976.

Nature in Wales.

169. — **1,** 1955.
170. — **2,** 1956.
171. — **3,** 1957.
172. — **4,** 1958.
173. — **5,** 1959.
174. — **6,** 1960.
175. — **7,** 1961.
176. — **8,** 1962-63.
177. — **9,** 1964-65.
178. — **10,** 1966-67.
179. — **11,** 1968-69.
180. — **12,** 1970-71.
181. — **13,** 1972-73.
182. — **14,** 1974-75.
183. — **15,** 1976-77.
184. — **16,** 1978-79.
185. — **17,** 1980-81.

New Phytologist.

186. — **57,** 1958.

187. — **64,** 1965.

North Western Naturalist.

188. — **17,** 1942.

189. — **18,** 1943.

190. — **20,** 1945.

191. — **21,** 1946.

192. — **22,** 1947.

193. — *New Series,* **1,** 1953.

—. Parry, M. (1969). *Casgliad o Enwau Blodau, Llysiau a Choed.* Cardiff.

194. Perring, F. H. & Sell, P. D. (1968). *Critical Supplement to the Atlas of the British Flora.* London.

195. Perring, F. H. & Walters, S. M. (1976). *Atlas of the British Flora.* Ed. 2. London.

196. — (1962). *Atlas of the British Flora.* Ed. 1 London. Corrections and additions.

197. Personal communications

—. Phillips, R. W. (1926). The First Flora of Breconshire. *The Breconshire Border.* T. R. Phillips (ed.). Talgarth.

Phytologist.

198. — **1,** 1844.

199. — **3,** 1848-50.

200. Porter, M. (1968-81). Personal communications.

201. Pryce, R. D. (1977-81). Personal communications.

202. Pugh, Miss D. (1973-81). Personal communications.

263. Pugsley, H. W. (1948). A Prodromus of the British Hieracia. *Journal of the Linnean Society of London (Botany),* **54.**

203. Ray, J. (1670). *Catalogus Plantarum Angliae.* London.

—. — (1688). *Fasciculus Stirpium Britannicarum.* London.

—. — (1690). *Synopsis Methodica Stirpium Britannicarum.* Ed. 1. London.

—. — (1696). *Synopsis Methodica Stirpium Britannicarum.* Ed. 2. London.

204. Rees, F. L. (1950). *List of Pembrokeshire Plants.* Tenby.

205. Richards, A. J. (1972). The Taraxacum Flora of the British Isles. *Watsonia,* **9,** (suppl.).

206. — (1972-81). Personal communications.

138. Riddelsdell, H. J. (1905). Lightfoot's visit to Wales in 1773. *Journal of Botany,* **43.**

140. — (1907). A Flora of Glamorgan. *Journal of Botany,* **45,** (suppl.).

207. Roberts, E. (c.1945-65). *Plant Lists of Snowdonia,* Unpublished. Copies of manuscripts in Library, National Museum of Wales.

—. Roberts, E. S. (Ed.) (1916). *Llysieulyfr Meddyginiaethol A Briodolir i William Salesbury.* Liverpool.

208. Roberts, R. H. (1955-81). Personal communications.

209. — (1982). *The Flowering Plants and Ferns of Anglesey.* Cardiff.

210. Salter, J. H. (1935). *The Flowering Plants and Ferns of Cardiganshire.* Cardiff.

211. Wade, A. E. (Ed.) (1952). *A Supplement to Dr. J. H. Salter's The Flowering Plants and Ferns of Cardiganshire.* Cardiff.

212. Salway, T. (1863). A List of Plants of Barmouth and its neighbourhood. In *The Tourists and Visitors Handbook and Guide to Harlech, Barmouth, Dolgelly and Towyn.*

213. Sargent, C. (1978-79). Railway plant lists. (ITE/NCC).

214. Savidge, J. P. (1963-81). Personal communications.

97. Seddon, B. A. (1972). Aquatic macrophytes as Limnological indicators. *Freshwater Biology,* **2.**

215. Sell, P. D. (1981). Personal communication.

216. Shoolbred, W. A. (1920). *The Flora of Chepstow.* London.

217. Stace, C. A. (Ed.) (1975). *Hybridization and the Flora of the British Isles.* London.

218. Storrie, J. (1886). *The Flora of Cardiff.* Cardiff.

219. — (c.1887). Manuscript Notes in a copy of *The Flora of Cardiff* in Library, National Museum of Wales.

Swansea Field Naturalists' Society. Proceedings.

220. — **1,** 1927-36.

221. — **2,** 1938-52.

222. — **3,** 1956.

223. Trow, A. H. (Ed.) (1911). *The Flora of Glamorgan.* Cardiff. Originally published as parts in *Transactions of the Cardiff Naturalists' Society* between 1906-1910.

82. Turner, D. & Dillwyn, L. W. (1805). *The Botanists Guide through England and Wales.* London.

—. Tutin, T. G. *et al.,* (Eds.) (1964-1980). *Flora Europaea.* **1-5.** Cambridge.

224. Vachell, E. (c.1900-47). Manuscript notes in Library, National Museum of Wales.

—. — (1936). Glamorgan Flowering Plants and Ferns. In *Glamorgan County History,* **I.** Tattersall, W. M. (ed.). Cardiff.

225. Vaughan, Mrs. I. M. (1957-81). Personal communications.

211. Wade, A. E. (Ed.) (1952). *A Supplement to Dr. J. H. Salter's The Flowering Plants and Ferns of Cardiganshire.* Cardiff.

226 — (1970). *The Flora of Monmouthshire.* Cardiff.

227. — (1958-81). Personal communications.

24. Wade. A. E. & Smith, R. L. (1925). The Adventive Flora of the Port of Cardiff. *Botanical Society and Exchange Club. Reports.* **7.**

25. — (1926). Additions to the Adventive Flora of the Port of Cardiff. *Botanical Society and Exchange Club. Reports.* **8.**

189. Wade, A. E., & Webb, J. A. (1943). Montgomeryshire Plant Records. *North Western Naturalist.* **18.**

228. — (1941). *Materials for a Flora of Montgomeryshire.* Unpublished. Manuscript in Library, National Museum of Wales.

190. — (1945). Radnorshire Plant Records. *North Western Naturalist,* **20.**

229. — (c.1955) *Materials towards a Flora of Radnorshire.* Unpublished. Manuscript in Library, National Museum of Wales.

—. Watson, H. C. (1835). *The New Botanist's Guide.* London.

—. — (1847-1859). *Cybele Britannica.* **I-IV.** London.

230. — (1883). *Topographical Botany.* Ed. 2. London.

231. Bennett, A. (1905). Supplement to Topographical Botany. Ed. 2. *Journal of Botany.* **43** (suppl.).

232. Bennett, A., Salmon, C. E. & Matthews, J. R. (1929-30). Second Supplement to Watson's Topographical Botany. *Journal of Botany.* **67** & **68** (suppl.).

Watsonia. Journal of the Botanical Society of the British Isles.

233. — **1,** 1949-50.

234. — **2,** 1951-53.

235. — **3,** 1953-56.

236. — **4,** 1957-61.

237. — **5,** 1961-63.

238. — **6,** 1964-68.

239. — **7,** 1969.

240. — **8,** 1970-71.

241. — **9,** 1972-73.

242. — **10,** 1974-75.

243. — **11,** 1976-77.

244. — **12,** 1978-79.

245. — **13,** 1980-81.

246. Webb, J. A. (c.1955). *Materials towards a Flora of Breconshire.* Unpublished. Manuscript in University College, Swansea.

247. — (1956). The Flora of Gower. *Proceedings of the Swansea Field Naturalists' Society.* **3.**

188. Whellan, J. A. (1942). Notes on the Flora of South West Wales. 1941. *North Western Naturalist.* **17.**

Wild Flower Magazine.

248. — 1961.

249. — 1965.

250. — 1968.

251. — 1970.

252. — 1973.

253. — 1974.

254. — 1975.

255. — 1976.

256. — 1977.

257. — 1978.

258. — 1979.

263. — 1980.

264. — 1981.

259. Woods, R. G. (1978-81). Personal communications.

260. *Woolhope Club. Transactions.*

261. Wynne, G. (1970-81). Personal communications.

262. Yeo, P. F. (1973). Personal communication.

Wild Flower Magazine.

263. — 1980.

264. — 1981.

265. Pugsley, H. W. (1948). A Prodromus of the British Hieracia. *Journal of the Linnean Society of London (Botany).* **54.**

Appendix III: World Distribution of Welsh Plants

The following list of elements is based on Matthews, J. R. (1955) *Origin and Distribution of the British Flora,* and Hyde, H. A. & Wade, A. E. (1957) *Welsh Flowering Plants* Ed. 2, but the Welsh totals have been brought up to date.

1. **The General Northern Element.** Species common to temperate Europe, Asia, and North America; the few British plants (chiefly aquatics) which are more or less cosmopolitan have also been included here. About 170 species, of which 145 are recorded from Wales.

2. **The Eurasian Element.** Species which are widely and generally distributed in Europe and which occur also in Asia, especially in the west or south-west, and extend in a few instances eastwards to China and Japan. A large component of about 460 species, of which 403 are recorded for Wales.

3. **The General European Element.** Species widely distributed in Europe (but not Asia); many absent from northern and eastern Europe; some extending to North Africa; about 200 species in all, of which some 160 occur in Wales.

4. **The Continental Southern Element.** Species found chiefly in central and southern Europe, the great majority extending also into North Africa and South-west Asia; 114 species in all, of which 97 are recorded for Wales.

5. **The Oceanic West European Element.** Species found almost exclusively in western Europe and apparently favoured by an 'atlantic' climate; over 100 species, of which 96 occur in Wales.

6. **The Oceanic Southern Element.** An essentially South-west European group consisting of species which appear to prefer oceanic conditions and which are confined chiefly to western and southern Europe including the Mediterranean basin; 72 species in all, including 58 found in Wales.

7. **The Mediterranean Element.** Species whose chief centre of distribution is the Mediterranean basin but most of which extend into the Orient; 33 species of which 19 are recorded from Wales.

8. **The Continental Element.** Species which are characteristic of Central Europe but which 'thin out' westwards, though they extend markedly eastwards through Russia and Asia, apparently favouring a continental climate. The element includes our so-called 'steppe' plants and comprises 70 species, of which 56 occur in Wales.

9. **The Continental Northern Element.** Species whose main area of distribution in Europe is central and northern, and which farther south take to the mountains, decrease, or disappear. Many are also found in North America and northern Asia, so that their range is more or less circumboreal; 92 species in all, of which 88 occur in Wales.

10. **The Oceanic Northern Element.** Species which are characteristic of North-west Europe, though some occur also in eastern North America; about 30 species, of which 27 are found in Wales.

11. **The North American Element.** Species which, apart from western Britain, belong to North America; 6 such species are recorded from Britain of which 1 (**Limosella subulata**) is native to Wales.

12. **The Northern Montane Element.** Related to the Arctic-Alpine Element (No. 15) and comprising 26 species which are found in northern Europe and which reappear in mountain areas in central or southern Europe, though they are generally absent from the low-lying plains in between. Most of these species occur also in North America and Asia; 16 are found in Wales.

13. **The Arctic-Subarctic Element.** Species which in Europe are exclusively northern, most of them extending into the subarctic or even the arctic regions, but none occurring in Central Europe; 33 species, 6 of which are recorded from Wales.

14. **The Alpine Element.** Species which are found in mountain areas of Central Europe, but which are absent from northern Europe and the Arctic. A small group of less than 20 species, 10 of which are recorded from Wales.

15. **The Arctic-Alpine Element.** Species whose chief area of distribution lies in the Arctic (i.e. north of the tree limit) or in subarctic regions, and which reappear farther south at relatively high altitudes in the mountains of Europe and either Asia or America or both; 78 species, of which 26 occur in Wales.

16. **Endemic Species.** Species which occur in the British Isles, but nowhere else, are said to be endemic thereto. Most so-called British endemics are races or microspecies related to species of wider distribution, and belong to the critical genera **Hieracium, Taraxacum** and **Rubus**; apart from these, 27 British endemics are admitted, 18 of which occur in Wales and 6 are confined to it.

In the Index, the geographical element is indicated by its initial letters, in brackets, after the specific name. Natives, Denizens, Colonists and those other Established or Naturalized Aliens which are native somewhere in the British Isles, are allocated to a Geographical Element, but not in the few instances where the range outside the British Isles is too imperfectly known.

A – Alpine
AA – Arctic-Alpine
AS – Arctic-Subarctic
C – Continental
CN – Continental Northern
CS – Continental Southern
E – Eurasian
En,B – Endemic, British Isles
En,W – Endemic, Wales
GE – General European
GN – General Northern
M – Mediterranean
NA – North American
NM – Northern Montane
ON – Oceanic Northern
OS – Oceanic Southern
OWE – Oceanic West European

INDEX

To Latin Welsh and English Plant Names and Geographical Elements

Latin Names
Family and generic names are in bold type, synonyms in italics

Welsh Names
Alphabetical order of the first name, e.g. Afal Daear

English Names
Alphabetical order of the second (substantive) name, e.g. Aconite, Winter not Winter Aconite

Geographical Elements
Letters in brackets before the page number. See page 295 for key to abbreviations

During the preparation of this index a few minor spelling errors were noticed in the text, the correct spelling is given here.